Maritime Aluminum Welding

Trainee Guide

Pearson

Boston Columbus Indianapolis New York San Francisco Amsterdam
Cape Town Dubai London Madrid Milan Munich Paris Montreal Toronto Delhi
Mexico City Sao Paulo Sydney Hong Kong Seoul Singapore Taipei Tokyo

NCCER

President and Chief Executive Officer: Don Whyte
President: Boyd Worsham
Vice President: Steve Greene
Chief Operations Officer: Katrina Kersch
Maritime Welding Project Manager: Chris Wilson
Senior Development Manager: Mark Thomas
Senior Production Manager: Tim Davis
Managing Editor: Natalie Richoux
Desktop Publishing Manager: James McKay
Art Manager: Kelly Saddler
Desktop Publishing Specialists: Gene Page, Eric Caraballoso
Editors: Graham Hack, Jordan Hutchinson

Writing and development services provided by Topaz Publications, Liverpool, NY

Lead Writer/Project Manager: Troy Staton
Desktop Publisher: Joanne Hart
Art Director: Alison Richmond
Permissions Editor: Tonia Burke
Writers: Troy Staton, Carol Herbert, Thomas Burke

Pearson Education, Inc.

Director of Alliance/Partnership Management: Kelly Trakalo
Content Producer: Alexandrina B. Wolf
Assistant Content Producer: Alma Dabral
Digital Content Producer: Jose Carchi
Senior Marketing Manager: Brian Hoehl
Composition: NCCER
Text Fonts: Palatino and Univers

Credits and acknowledgments for content borrowed from other sources and reproduced, with permission, in this textbook appear at the end of each module.

10 9 8 7 6 5 4 3 2 1
ScoutAutomatedPrintCode

Pearson

Perfect Bound ISBN-13: 978-0-13-560384-0
ISBN-10: 0-13-560384-6

ScoutAutomatedPrintCode

Preface

Welding is a high-tech industry that can take you to places all over the world. Welds are everywhere — just look around. From ladders to aircraft carriers; from NASCAR to national defense; from the laboratory to sales and repair; the welding trade impacts virtually every industry.

Few career choices offer as many options for employment and opportunities for growth. According to the Department of Labor, "the basic skills of welding are similar across industries, so welders can easily shift from one industry to another, depending on where they are needed most." Welders will continue to be in high demand since there is a continual need to rebuild aging infrastructure such as highways and buildings. In addition, the creation of new power generation facilities will yield new job prospects for trained and skilled welders.

Of course, one of the places to find a prosperous career as a welder is in the maritime industry. Shipyards and repair facilities are plentiful in coastal areas and on major rivers and lake systems. Maritime welding can be done on land or sea and above or underneath the surface of the water. It includes the work on ships, boats, and off-shore oil rigs. It includes the installation and repair of metal structures and components. .

This new edition of *Maritime Welding* introduces the fundamentals of the welding trade. The three levels of this curriculum, plus the Maritime Aluminum Welding curriculum, present an apprentice approach and will help you be knowledgeable, safe, and effective on the job.

Maritime Welding

This *Maritime Welding* curriculum is a product of collaboration between NCCER and the National Maritime Education Council (NMEC). Several maritime welding subject matter experts determined which of NCCER's *Welding* modules are suitable to maritime industry training. This edition of *Maritime Welding* can be used for welding training and can be supplemented by instructors with information specific to the maritime industry.

We invite you to visit the NCCER website at **www.nccer.org** for the latest releases, training information, newsletter, and much more. You can also reference the Pearson product catalog online at **www.nccer.org**. Your feedback is welcome. You may e-mail your comments to **curriculum@nccer.org** or send general comments and inquiries to **info@nccer.org**.

NMEC

The National Maritime Education Council (NMEC) is a multi-regional, industry-centered organization working to address critical workforce issues facing its member yards and the maritime industry. Established in 2012 to provide a centralized voice for the industry as it relates to workforce development, NMEC's primary mission is to lead the maritime industry in the development, promotion, and implementation of a national workforce development system that includes standardized craft training, portable credentials, and journey-level assessments.

Through its partnership with NCCER, NMEC is working in collaboration with its member yards to:

- Establish a system of career awareness and recruiting that will allow the industry to speak with one voice and deliver a consistent message to students, student influencers, the underemployed, women, and returning military.
- Establish and implement an industry-wide, comprehensive, standardized training and credentialing system for maritime craft professionals.
- Grow a pipeline of skilled craft professionals trained and credentialed through an industry-driven, uniform process.
- Improve industry outcomes by developing a better skilled, safer, more productive workforce.

NCCER Standardized Curricula

NCCER is a not-for-profit 501(c)(3) education foundation established in 1995 by the world's largest and most progressive construction companies and national construction associations. It was founded to address the severe workforce shortage facing the industry and to develop a standardized training process and curricula. Today, NCCER is supported by hundreds of leading construction and maintenance companies, manufacturers, and national associations. The NCCER Standardized Curricula was developed by NCCER in partnership with Pearson Education, Inc., the world's largest educational publisher.

Some features of the NCCER Standardized Curricula are as follows:

- An industry-proven record of success
- Curricula developed by the industry for the industry
- National standardization providing portability of learned job skills and educational credits
- Compliance with Office of Apprenticeship requirements for related classroom training (*CFR 29:29*)
- Well-illustrated, up-to-date, and practical information

NCCER also maintains a Registry that provides transcripts, certificates, and wallet cards to individuals who have successfully completed a level of training within a craft in the NCCER Standardized Curricula. *Training programs must be delivered by an NCCER Accredited Training Sponsor in order to receive these credentials.*

Special Features

In an effort to provide a comprehensive user-friendly training resource, we have incorporated many different features for your use. Whether you are a visual or hands-on learner, this book will provide you with the proper tools to get started in the welding industry.

Introduction Page

This page is found at the beginning of each module and lists the Objectives, Performance Tasks, and Trade Terms, for that module. The Objectives list the skills and knowledge you will need in order to complete the module successfully. The Performance Tasks give you an opportunity to apply your knowledge to real-world tasks. The list of Trade Terms identifies important terms you will need to know by the end of the module.

> **29105-15**
> **BASE METAL PREPARATION**
>
> **Objectives**
>
> When you have completed this module, you will be able to do the following:
>
> 1. Identify safety practices related to preparing base metals and describe basic cleaning procedures.
>   a. Identify safety practices related to preparing base metals.
>   b. Describe the basic properties and types of carbon and stainless steel.
>   c. Describe basic metal cleaning procedures and concerns.
> 2. Identify and describe basic weld joint design and types of welds.
>   a. Identify and describe the loads that are routinely placed on weld joints.
>   b. Identify and describe the various types of weld joints.
>   c. Describe a welding procedure specification (WPS) and the information it provides.
> 3. Describe how to prepare joints for welding.
>   a. Describe how to mechanically prepare joints for welding.
>   b. Describe how to thermally prepare joints for welding.
>
> **Performance Tasks**
>
> Under the supervision of your instructor, you should be able to do the following:
>
> 1. Mechanically or hand grind a bevel on the edge of a ¼"- to ¾"-thick mild steel plate (6 to 20 mm metric plate) at 22½ degrees.
> 2. Thermally bevel the edge of a ¼"- to ¾"-thick mild steel plate (6 to 20 mm metric plate) at 22½ degrees.
>
> **Trade Terms**
>
> Annealed, Austenitic, Backing, Base metal, Castings, Distortion, Ductile, Ferritic, Ferrous, Load, Malleable, Martensitic, Melt-through, Nonferrous metal, Oxide, Piping porosity, Porosity, Quench, Root face, Root opening, Surfacing, Tempering, Tensile strength, Weathering steel, Welding procedure specification (WPS), Weldment, Wrought
>
> **Industry Recognized Credentials**
>
> If you are training through an NCCER-accredited sponsor, you may be eligible for credentials from NCCER's Registry. The ID number for this module is 29105-15. Note that this module may have been used in other NCCER curricula and may apply to other level completions. Contact NCCER's Registry at 888.622.3720 or go to **www.nccer.org** for more information.

Notes, Cautions, and Warnings

Safety features are set off from the main text in highlighted boxes and organized into three categories based on the potential danger of the issue being addressed. Notes simply provide additional information on the topic area. Cautions alert you of a danger that does not present potential injury but may cause damage to equipment. Warnings stress a potentially dangerous situation that may cause injury to you or a co-worker.

> **NOTE**
> Some companies have physical activity requirements that must be met by apprentices. These requirements vary from company to company.

> **CAUTION**
> Loosening regulator adjusting screws closes the regulators and prevents damage to the regulator diaphragms when the cylinder valves are opened.

> **WARNING!**
> Never apply heat directly to a cylinder or regulator. This can cause excessive pressure, resulting in an explosion.

Special Features

Features present technical tips and professional practices from the construction industry. These features often include real-life scenarios similar to those you might encounter on the job site.

> **Commonly Available Carbon Electrodes**
>
> The carbon electrodes that are usually readily available from general welding suppliers are the copper-coated round and rectangular electrodes used with direct current electrode positive (DCEP) current. Other types and styles, including jointed,

Going Green

Going Green looks at ways to preserve the environment, save energy, and make good choices regarding the health of the planet. Through the introduction of new construction practices and products, you will see how the "greening of the world" has already taken root.

> **GOING GREEN**
> **Cooling Water**
>
> Welders involved in metal-cutting activities often allow the water used in the process to simply flow into the ground or down the drain. To better protect the environment, water used for cooling the plasma arc cutting equipment should be contained whenever

Did You Know?

The Did You Know? features introduce historical tidbits or modern information about the construction industry. Interesting and sometimes surprising facts about construction are also presented.

> **Did You Know?**
> **PHMSA**
>
> The US Department of Transportation (DOT) established the Pipeline and Hazardous Materials Safety Administration (PHMSA) in 2004 to develop and enforce regulations for the safe, reliable, and environmentally sound operation of the 2.6 million

Case History

Case History features emphasize the importance of safety by citing examples of the costly (and often devastating) consequences of ignoring safety rules and regulations.

> **Case History**
>
> **Mexico City Refinery**
>
> In Mexico City in 1984, a weld on a liquid propane gas (LPG) tank cracked. The escaping gas cloud drifted toward a nearby residential area until it encountered an ignition source. When the gas ignited, the fireball quickly burned its way back to the leak source. The resulting explosion destroyed the refinery and many residential neighborhoods, leaving 503 people dead and more than 4,000 injured.
>
> **The Bottom Line:** Welds must be made cor-

Color Illustrations and Photographs

Full-color illustrations and photographs are used throughout each module to provide vivid detail. These figures highlight important concepts from the text and provide clarity for complex instructions. Each figure is denoted in the text in *italic type* for easy reference.

29101-15_F15.EPS

Figure 15 Worker in a confined space.

Step-by-Step Instructions

Step-by-step instructions are used throughout to guide you through technical procedures and tasks from start to finish. These steps show you not only how to perform a task but also how to do it safely and efficiently

> Follow these steps to make an overhead fillet weld:
>
> *Step 1* Tack two plates together to form a T-joint for the fillet weld coupon.
>
> *Step 2* Tack-weld the coupon so it is in the overhead position.

Trade Terms

Each module presents a list of Trade Terms that are discussed within the text, defined in the Glossary at the end of the module, and reinforced with a Trade Terms Quiz. These terms are denoted in the text with **blue bold type** upon their first occurrence. To make searches for key information easier, a comprehensive Glossary of Trade Terms from all modules is found at the back of this book.

> Metals are classified into two basic groups: **ferrous** metals, which are composed mainly of iron, and **nonferrous metals**, which contain very little or no iron. Ferrous metals include all steel, cast iron, **wrought** iron, **malleable** iron, and **ductile**

Section Review

The Section Review features helpful additional resources and review questions related to the objectives in each section of the module.

> **Additional Resources**
>
> *2014 Technical Training Guide.* Current Edition. Cleveland, OH. USA: The Lincoln Electric Company. www.lincolnelectric.com
>
> *Welding Handbook.* Current Edition. Miami, FL: The American Welding Society.
>
> **2.0.0 Section Review**
>
> 1. Which of the following can be quenched without concern?
>   a. Weld test coupons
>   b. Functional welds
>   c. Practice coupons
>   d. Welds done in the 3G position
> 2. On an open-root V-groove weld in the 4G position, the electrode work angle should be _____.
>   a. 30 degrees
>   b. 45 degrees
>   c. 90 degrees
>   d. varied as needed to tie in

Review Questions

Review Questions are provided to reinforce the knowledge you have gained. This makes them a useful tool for measuring what you have learned.

> **Review Questions**
>
> 1. The A-CAC process uses a high-current electric arc between the base metal and the end of a(n) _____.
>   a. carbon-graphite electrode to melt the base metal
>   b. high-frequency pilot arc to melt the base metal
>   c. plasma jet to melt the base metal
>   d. idle arc to melt the base metal
> 2. An A-CAC torch requires compressed air to blow the _____.
>   a. carbon dioxide away from the cut
>   b. molten metal away from the cut
>   c. nitrogen away from the cut
>   d. carbon arc away from the cut
> 3. When cutting and gouging with A-CAC equipment, the jet orifices of the torch head(s) should be located _____.
>   a. beside the electrode
>   b. above the electrode
>   c. ahead of the electrode
>   d. between the electrode and the base metal surface
> 4. To prevent overheating, the electrode tip to the V-groove head should be no farther than _____.
>   a. 1" (2 cm)
>   b. 2" (5 cm)
>   c. 4" (10 cm)
>   d. 6" (15 cm)
> 5. Light-duty gouging or cutting is typically done with a _____.
>   a. 10A welding machine
>   b. 200A welding machine
>   c. 600A welding machine
>   d. 1,200A welding machine
> 6. Heavy-duty cutting can require a torch airflow rate up to _____.
>   a. 20 cfm (566 lpm)
>   b. 35 cfm (991 lpm)
>   c. 40 cfm (1,133 lpm)
>   d. 50 cfm (1,416 lpm)
> 7. Compressed air is supplied to the torch through a hose attached to a fitting on the _____.
>   a. welding machine
>   b. end of the torch cable
>   c. air jet orifices
>   d. workpiece
> 8. Electrodes designed for A-CAC contain a rare-earth, _____.
>   a. magnet metal
>   b. undercurrent ingredient
>   c. amplifier ingredient
>   d. arc-stabilizing metal
> 9. The most common style of A-CAC electrode used for manual cutting or gouging is the _____.
>   a. flat electrode
>   b. round electrode
>   c. half-round electrode
>   d. round-jointed electrode
> 10. When performing A-CAC cutting or gouging, the minimum groove width is controlled by the _____.
>   a. size of the torch
>   b. size of the electrode
>   c. speed of travel
>   d. angle of the electrode holder
> 11. Because A-CAC activities tend to blow molten metal considerable distances, it is important to station a(n) _____.
>   a. observer in the A-CAC work area
>   b. slag trap in the A-CAC work area
>   c. oxygen monitor in the A-CAC work area
>   d. fire watch in the A-CAC work area
> 12. To wash a projection from a metal surface using A-CAC equipment, the electrode is typically used at a push angle of about _____.
>   a. 30 degrees to the base metal surface
>   b. 45 degrees to the base metal surface
>   c. 55 degrees to the base metal surface
>   d. 90 degrees to the base metal surface

NCCER Standardized Curricula

NCCER's training programs comprise more than 80 construction, maintenance, pipeline, and utility areas and include skills assessments, safety training, and management education.

Boilermaking
Cabinetmaking
Carpentry
Concrete Finishing
Construction Craft Laborer
Construction Technology
Core Curriculum: Introductory Craft Skills
Drywall
Electrical
Electronic Systems Technician
Heating, Ventilating, and Air Conditioning
Heavy Equipment Operations
Highway/Heavy Construction
Hydroblasting
Industrial Coating and Lining Application Specialist
Industrial Maintenance Electrical and Instrumentation Technician
Industrial Maintenance Mechanic
Instrumentation
Insulating
Ironworking
Masonry
Millwright
Mobile Crane Operations
Painting
Painting, Industrial
Pipefitting
Pipelayer
Plumbing
Reinforcing Ironwork
Rigging
Scaffolding
Sheet Metal
Signal Person
Site Layout
Sprinkler Fitting
Tower Crane Operator
Welding

Maritime

Maritime Industry Fundamentals
Maritime Pipefitting
Maritime Structural Fitter

Green/Sustainable Construction

Building Auditor
Fundamentals of Weatherization
Introduction to Weatherization
Sustainable Construction Supervisor
Weatherization Crew Chief
Weatherization Technician
Your Role in the Green Environment

Energy

Alternative Energy
Introduction to the Power Industry
Introduction to Solar Photovoltaics
Introduction to Wind Energy
Power Industry Fundamentals
Power Generation Maintenance Electrician
Power Generation I&C Maintenance Technician
Power Generation Maintenance Mechanic
Power Line Worker
Power Line Worker: Distribution
Power Line Worker: Substation
Power Line Worker: Transmission
Solar Photovoltaic Systems Installer
Wind Turbine Maintenance Technician

Pipeline

Control Center Operations, Liquid
Corrosion Control
Electrical and Instrumentation
Field Operations, Liquid
Field Operations, Gas
Maintenance
Mechanical

Safety

Field Safety
Safety Orientation
Safety Technology

Supplemental Titles

Applied Construction Math
Tools for Success

Management

Fundamentals of Crew Leadership
Project Management
Project Supervision

Spanish Titles

Acabado de concreto: nivel uno
Aislamiento: nivel uno
Albañilería: nivel uno
Andamios
Carpintería: Formas para carpintería, nivel tres
Currículo básico: habilidades introductorias del oficio
Electricidad: nivel uno
Herrería: nivel uno
Herrería de refuerzo: nivel uno
Instalación de rociadores: nivel uno
Instalación de tuberías: nivel uno
Instrumentación: nivel uno, nivel dos, nivel tres, nivel cuatro
Orientación de seguridad
Paneles de yeso: nivel uno
Seguridad de campo

Acknowledgments

This curriculum was revised as a result of the farsightedness and leadership of the following sponsors:

Austal USA
Bollinger Shipyards
Claddagh Enterprises
Central Louisiana Technical College
Continental Maritime
Exelon Generation
Gulf States, Inc., Michigan Division
Ingalls Shipbuilding
KBR Industrial Services
Lee College
Lincoln Electric
Northeast Community College
Northland Pioneer College
Robins & Morton
Tecnico Corporation
TIC-The Industrial Company
Toledo Refining Co. LLC
Zachry Industrial, Inc.

This curriculum would not exist were it not for the dedication and unselfish energy of those volunteers who served on the Authoring Team. A sincere thanks is extended to the following:

Tom Ashley
Paul Bennett
Gerald Bickerstaff
Curtis Casey
Bill Cherry
Gerald Guthrie
Rod Hellyer
Frank Johnson
John Knapp
Jason Langford
Jerry McDaniel
Tim Riley
Kyle Savoie
Jason Scales
Dan Sterry
Holley Thomas
Jerry Trainor
Adam Webb

NCCER Partners

American Fire Sprinkler Association
Associated Builders and Contractors, Inc.
Associated General Contractors of America
Association for Career and Technical Education
Association for Skilled and Technical Sciences
Construction Industry Institute
Construction Users Roundtable
Construction Workforce Development Center
Design Build Institute of America
GSSC – Gulf States Shipbuilders Consortium
ISN
Manufacturing Institute
Mason Contractors Association of America
Merit Contractors Association of Canada
NACE International
National Association of Minority Contractors
National Association of Women in Construction
National Insulation Association
National Technical Honor Society
NAWIC Education Foundation
North American Crane Bureau
North American Technician Excellence
Pearson
Pearson Qualifications International
Prov
SkillsUSA®
Steel Erectors Association of America
U.S. Army Corps of Engineers
University of Florida, M. E. Rinker School of Building Construction
Women Construction Owners & Executives, USA

NCCER Business Partners

Contents

Module Twelve

GMAW – Aluminum Plate

Covers the setup of GMAW equipment for welding aluminum plate. Explains aluminum metallurgy and the characteristics of aluminum welding; how to clean and prepare aluminum plate coupons for welding; and problems often encountered in aluminum welds. Explains GMAW techniques used in aluminum welding. Provides GMAW procedures on how to build weld pads on aluminum plate; how to make fillet welds on aluminum plate in the 1F, 2F, 3F, and 4F positions; and how to make V-groove welds on aluminum plate with backing in the 1G, 2G, 3G, and 4G positions. (Module ID 29401; 30 Hours)

Module Thirteen

GTAW – Aluminum Plate

Covers the setup of GTAW equipment for welding aluminum plate. Explains how to clean and prepare aluminum plate coupons for welding, and how to select the aluminum filler metals and shielding gases used in the GTAW process. Explains GTAW techniques used in aluminum welding. Provides GTAW procedures on how to build weld pads on aluminum plate; how to make fillet welds on aluminum plate in the 1F, 2F, 3F, and 4F positions; and how to make V-groove welds on aluminum plate with backing in the 1G, 2G, 3G, and 4G positions. (Module ID 29402; 30 Hours)

Module Fourteen

GTAW – Aluminum Pipe

Covers the setup of GTAW equipment for welding aluminum pipe. Explains how to clean and prepare aluminum pipe coupons for welding. Addresses GTAW techniques used to make V-groove and modified U-groove welds on aluminum pipe with and without backing. Provides GTAW procedures on how to make V-groove or modified U-groove welds on aluminum pipe in the 2G, 5G, and 6G positions. (Module ID 29403; 50 Hours)

Module Fifteen

GMAW – Aluminum Pipe

Covers the setup of GMAW equipment for welding aluminum pipe. Addresses GMAW techniques used to make V-groove welds on aluminum pipe with and without backing. Explains how to clean and prepare aluminum pipe coupons for welding. Provides GMAW procedures on how to make V-groove welds on aluminum pipe in the 2G, 5G, and 6G positions. (Module ID 29404; 50 Hours)

Glossary

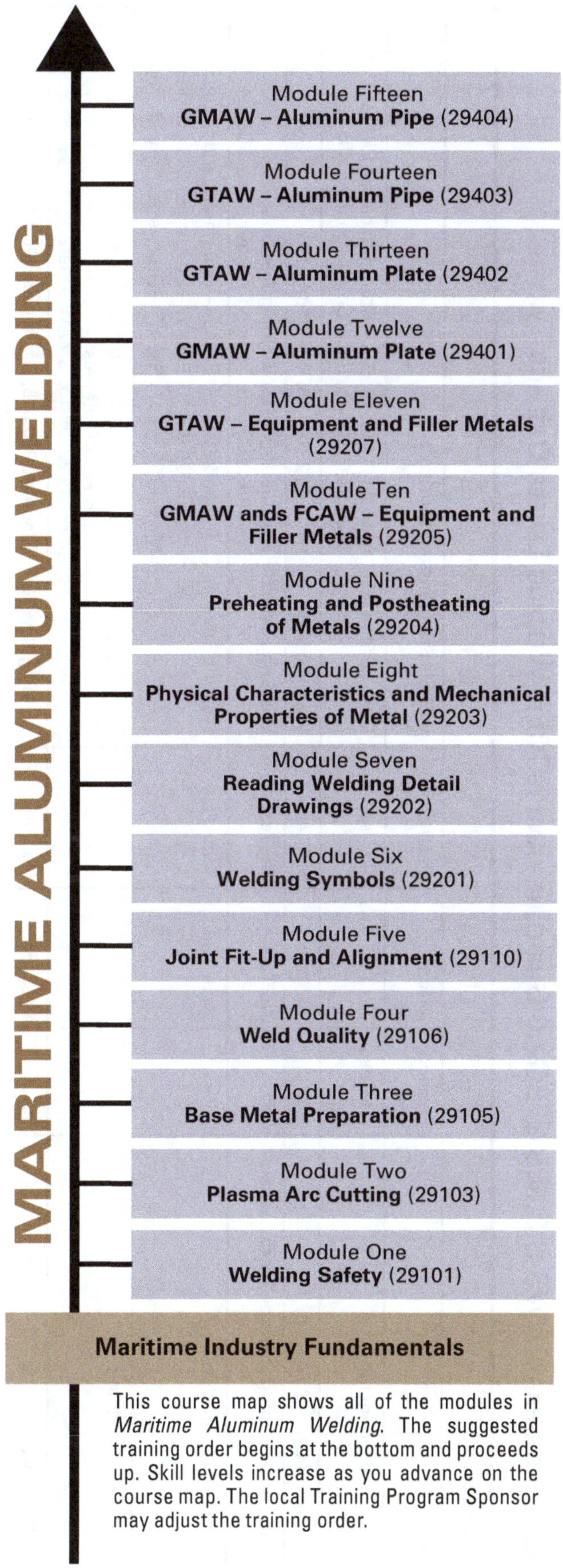

This course map shows all of the modules in *Maritime Aluminum Welding*. The suggested training order begins at the bottom and proceeds up. Skill levels increase as you advance on the course map. The local Training Program Sponsor may adjust the training order.

Correlations Chart: AWS S.E.N.S.E. EG2.0 Guidelines and NCCER Welding Levels One and Two

SENSE Guidelines				NCCER	
Modules	Passing Score	Visual	Destructive	NCCER Standardized Curricula	
1 – Occupational Orientation *	No Test	No Test	No Test	Core Curriculum:	Module 00107-15, Basic Communication Skills; Module 00108-15, Basic Employability Skills
2 – Safety and Health of Welders *	100%	No Test	No Test	Welding Level 1:	Module 29101-15, Welding Safety
3 – Drawing and Welding Symbol Interpretation *	75%	No Test	No Test	Welding Level 2:	Module 29201-15, Welding Symbols; Module 29202-15, Reading Welding Detail Drawings
4 – Shielded Metal Arc Welding	75%	Pass /Fail	Pass /Fail	Welding Level 1:	Module 29107-15, SMAW—Equipment and Setup; Module 29109-15, SMAW—Beads and Fillet Welds; Module 29111-15, SMAW—Groove Welds with Backing
5 – Gas Metal Arc Welding	75%	Pass /Fail	No Test	Welding Level 2:	Module 29205-15, GMAW and FCAW – Equipment and Filler Materials; Module 29209-15, GMAW – Plate
6 – Flux Cored Arc Welding	75%	Pass /Fail	No Test	Welding Level 2:	Module 29205-15, GMAW and FCAW – Equipment and Filler Materials; Module 29210-15, FCAW – Plate
7 – Gas Tungsten Arc Welding	75%	Pass /Fail	No Test	Welding Level 2:	Module 29207-15, GTAW – Equipment and Filler Materials; Module 29208-15, GTAW – Plate
8 – Thermal Cutting Process * †	75%	No Test	No Test		
Unit 1 Manual Oxyfuel Gas Cutting (OFC) †		Pass /Fail	No Test	Welding Level 1:	Module 29102-15, Oxyfuel Cutting
Unit 2 Mechanized Oxyfuel Gas Cutting (OFC) †		Optional	No Test	Welding Level 1:	Module 29105-15, Base Metal Preparation
Unit 3 Manual Plasma Arc Cutting—PAC †		Pass /Fail	No Test	Welding Level 1:	Module 29103-15, Plasma Arc Cutting
Unit 4 Manual Air Carbon Arc Cutting †		Optional	No Test	Welding Level 1:	Module 29104-15, Air-Carbon Arc Cutting and Gouging
9 – Welding Inspection and Testing *	75%	Pass /Fail	No Test	Welding Level 1:	Module 29106-15, Weld Quality

* Required module for Level I Entry Welder Completion (plus one welding process module)

† Completion of Units 1 and 3 minimum

Welding Safety

Overview

Qualified welders are in high demand all over the world. Typical welding tasks may include working in heavy construction; joining pipe sections for oil and natural gas pipelines; building ships; and working in a variety of industrial environments such as power plants, refineries, chemical plants, and manufacturing facilities. This module introduces trainees to the all-important topic of safety in the welding trade.

Module 29101

Trainees with successful module completions may be eligible for credentialing through the NCCER Registry. To learn more, go to **www.nccer.org** or contact us at **1.888.622.3720**. Our website has information on the latest product releases and training, as well as online versions of our *Cornerstone* magazine and Pearson's product catalog.

Your feedback is welcome. You may email your comments to **curriculum@nccer.org**, send general comments and inquiries to **info@nccer.org**, or fill in the User Update form at the back of this module.

This information is general in nature and intended for training purposes only. Actual performance of activities described in this manual requires compliance with all applicable operating, service, maintenance, and safety procedures under the direction of qualified personnel. References in this manual to patented or proprietary devices do not constitute a recommendation of their use.

29101 V5

29101
Welding Safety

Objectives

When you have completed this module, you will be able to do the following:

1. Describe basic welding processes, the welding trade, and training/apprenticeship programs.
   a. Describe basic welding processes and the welding trade.
   b. Describe NCCER standardized training and explain apprenticeship programs.
2. Identify and describe personal protective equipment (PPE) related to the welding trade.
   a. Identify and describe body, foot, and hand protective gear.
   b. Identify and describe ear, eye, face, and head protective gear.
3. Identify and describe welding safety practices related to specific hazards or environments.
   a. Describe the importance of welding safety and identify factors related to accidents.
   b. Describe basic welding safety practices related to the general work area.
   c. Describe hot work permits and fire watch requirements.
   d. Describe confined spaces and their related safety practices.
   e. Identify safety practices related to welding equipment.
   f. Identify and describe respiratory hazards, respiratory safety equipment, and ways to ventilate welding work areas.
   g. Explain the purpose of the SDS/MSDS and how it is used.

Performance Tasks

This is a knowledge-based module; there are no Performance Tasks.

Trade Terms

Arc burn
Bonded
Carbon steel
Electric arc
Electrically grounded
Electrode
Flash burn
Flux
Fume plume
Galvanized steel
Immediate danger to life and health (IDLH)
Purge gas
Shielding gas
Stainless steel
Ultraviolet (UV) radiation
Ventricular fibrillation

Industry Recognized Credentials

If you are training through an NCCER-accredited sponsor, you may be eligible for credentials from NCCER's Registry. The ID number for this module is 29101. Note that this module may have been used in other NCCER curricula and may apply to other level completions. Contact NCCER's Registry at 888.622.3720 or go to **www.nccer.org** for more information.

Contents

Topics to be presented in this module include:

Figures and Tables

SECTION ONE

1.0.0 THE WELDING TRADE

Objective

Describe basic welding processes, the welding trade, and training/apprenticeship programs.

a. Describe basic welding processes and the welding trade.
b. Describe NCCER standardized training and explain apprenticeship programs.

Trade Terms

Carbon steel: An alloy of iron combining iron and a small amount of carbon (usually less than 1 percent) to provide hardness.

Electric arc: The flow of an electrical current across an air gap or gaseous space.

Electrode: The point from which a welding arc is produced. In most welding processes, the electrode is a filler metal which is consumed in the process and becomes part of the weld.

Flux: A material used to dissolve or prevent the formation of oxides and other undesirable substances on a weld joint.

Shielding gas: A gas such as argon, helium, or carbon dioxide used to protect the welding electrode wire from contamination in gas metal arc welding (GMAW) and flux core arc welding (FCAW).

Stainless steel: An iron-based alloy usually containing at least 11 percent chromium.

This module is an extension of the Core Curriculum *Basic Safety* module. Its purpose is to identify general safety considerations that apply to various aspects of welding, along with the steps needed to avoid job-related deaths and injuries. Specific equipment and process hazards are covered in more detail in modules that deal with the use of the equipment or process in a particular application.

1.1.0 Welding Work

Welding is the process of joining similar metals using a high-intensity electric arc at temperatures from 6,000°F (3,316°C) to 10,000°F (5,538°C). The arc melts and fuses the base metals. In most cases, a filler metal is melted along with the base metal in order to strengthen the joint. The finished joint is as strong as, or even stronger than, the base metals that have been joined. Welding is commonly done on carbon steel, stainless steel, and aluminum.

Welders are key contributors to a number of construction and industrial crafts, including building and bridge construction; ship and oil rig construction and repair; heavy industry; and pipeline work. Welding is a craft that requires extensive knowledge and skill as well as a commitment to performing high-quality work and, most importantly, a strong commitment to safety.

There is a lot more to welding than simply striking an electric arc on base metals. Welders must know the characteristics of the metals being welded and the special materials and techniques that apply to the different metals. They must also know how to properly handle and store base metals and filler metals; how to prepare metals for welding; and how to interpret welding symbols and specifications.

The most common type of welding is shielded metal arc welding (SMAW). In the SMAW method, a flux-coated, consumable electrode is used to place the weld (*Figure 1*). SMAW is used primarily for welding iron and steel, including stainless steel.

Gas metal arc welding (GMAW), sometimes called MIG, or metal inert gas welding, uses a consumable wire that is fed through the welding gun to make the weld. It also uses a shielding gas. GMAW is used on steel and other metals. It is widely used in industrial applications, especially in the manufacture of automobiles where robotic welding is common. Flux core arc welding (FCAW) is very similar to GMAW. The main differences are that the FCAW filler metal is a tubular flux-cored wire and shielding gas may or may not be used.

Gas tungsten arc welding (GTAW), sometimes called TIG (tungsten inert gas) welding, uses a non-consumable tungsten electrode to produce the weld. GTAW is commonly used in welding stainless steel and light metals such as aluminum and magnesium.

In addition to welding, welders are often required to cut steel using oxyfuel, air carbon-arc, or plasma-arc equipment (*Figure 2*). Oxyfuel cutting is the most common method. In the oxyfuel method, oxygen and a fuel gas are mixed in a cutting torch to produce a high-intensity flame that can cut through thick steel. Acetylene is the most common fuel gas because it produces the most intense heat. Propane, MAPP® gas, and butane are also used for cutting work. Plasma-arc cutting relies on a jet of superheated gas that is passed through an electric arc. It is used to cut steel as well as other metals. Carbon-arc cutting uses

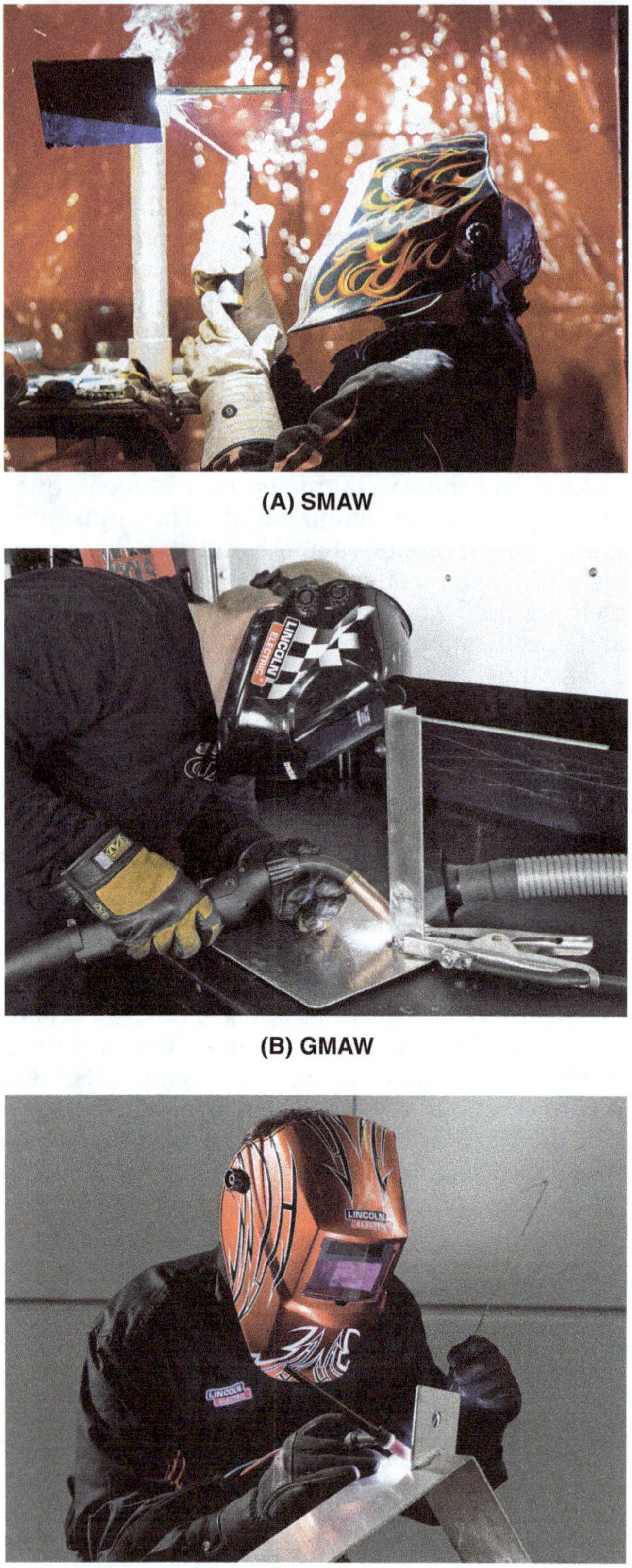

(A) SMAW

(B) GMAW

(C) GTAW

29101-15_F01.EPS

Figure 1 Welding methods.

an electric arc formed between a carbon electrode and the base metal to melt the base metal. Carbon-arc cutting is often used in preparing a base metal for welding.

Welders work on critical structures such as bridges, high-rise buildings, ships, pipelines, and automobiles. As you can imagine, defective welds in any of these applications can lead to a disaster. For that reason, welding work is subjected to intense quality control inspections. The more critical the weld, the more intense the inspection is. Inspectors use a variety of methods to examine welds, including X-ray inspection.

In order to qualify as a welder, a person must undergo extensive training that requires many hours of practice. The training provided by this program will prepare you to weld plate and pipe in a variety of positions and to cut carbon steel plate and pipe using common cutting equipment and techniques.

There are numerous opportunities for welders in heavy construction, manufacturing, ship building, and auto racing. Once you have completed your training and have become a journey-level welder, you can advance in the trade in a variety of ways (see *Figure 3*). The more training and education you complete, the greater the opportunities.

1.2.0 NCCER Standardized Training

NCCER is a not-for-profit education foundation established by the nation's leading construction companies. NCCER was created to provide the industry with standardized construction education materials, the NCCER Standardized Curricula (the welding modules are part of this series), and a system for tracking and recognizing students' training accomplishments—the NCCER Registry.

NCCER also offers accreditation, instructor certification, and craft assessments. NCCER is committed to developing and maintaining a training process that is internationally recognized, standardized, portable, and competency-based.

Working in partnership with industry and academia, NCCER has developed a system for program accreditation that is similar to those found in institutions of higher learning. NCCER's accreditation process ensures that students receive quality training based on uniform standards and criteria. These standards are outlined in NCCER's *Accreditation Guidelines* and must be adhered to by NCCER Accredited Training Sponsors.

(A) OXYFUEL CUTTING

(B) MANUAL PLASMA ARC CUTTING

(C) CARBON ARC GOUGING

29101-15_F02.EPS

Figure 2 Cutting methods.

More than 603 training centers across the United States and eight other countries are proud to be NCCER Accredited Training Sponsors/ Accredited Assessment Centers. Millions of craft professionals and construction managers have received quality construction education through NCCER's network of Accredited Training Sponsors. Every year the number of NCCER Accredited Training Sponsors increases significantly.

A craft instructor is a journey-level craft professional or career and technical educator trained and certified to teach the NCCER Standardized Curricula. This network of certified instructors ensures that NCCER training programs meet the standards of instruction set by the industry. There are more than 5,919 master trainers and 61,698 craft instructors within the NCCER instructor network. More information is available at **www.nccer.org**.

The Welding curriculum developed by NCCER provides participants with portable, industry-recognized credentials. These credentials include transcripts, certificates, and wallet cards that are tracked through NCCER's Registry.

This curriculum provides trainees with industry-driven training and education using a competency-based learning approach. This means that trainees must show the instructor that they possess the knowledge and skills needed to safely perform the hands-on tasks that are covered in each module.

When the instructor is satisfied that a trainee has demonstrated through written exams and hands-on performance testing that he or she has the required knowledge and skills for a given module, that information is sent to NCCER and kept in the Registry. The NCCER Registry can then confirm training and skills for workers as they move from state-to-state, company-to-company, or position-to-position.

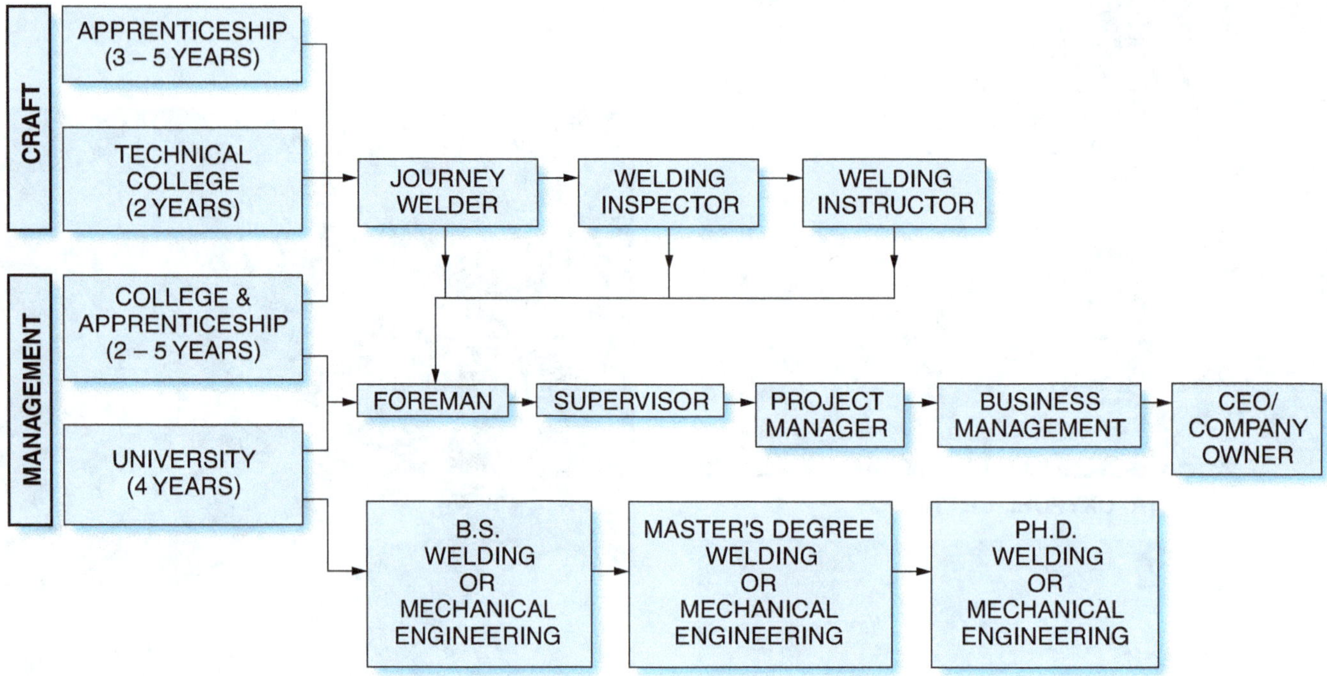

Figure 3 Opportunities in the welding trade.

1.2.1 Apprenticeship Training

Apprentice training goes back thousands of years. Its basic principles have not changed over time. First, it is a means for a person entering the craft to learn from those who have mastered the craft. Second, it focuses on learning by doing; real skills versus theory. Some theory is presented in the classroom. However, it is always presented in a way that helps the apprentice understand the purpose behind the skill that is to be learned.

The US Department of Labor's Office of Apprenticeship sets the minimum standards for training programs across the country. These programs rely on mandatory classroom instruction and on-the-job learning (OJL). They require at least 144 hours of classroom instruction per year and 2,000 hours of OJL per year. In a typical four-year apprenticeship program, trainees spend 576 hours in classroom instruction and 8,000 hours in OJL before receiving certificates issued by registered apprenticeship programs.

NCCER uses the Department of Labor apprenticeship standards as a foundation for comprehensive training that provide trainees with in-depth classroom and OJL experience.

All apprenticeship standards prescribe certain work-related or on-the-job learning. This OJL is broken down into specific tasks in which the apprentice receives hands-on training. In addition, a specified number of hours is required in each task. In a competency-based program, it may be possible to shorten the time required by testing out of specific tasks through a series of written and performance tests.

In a traditional program, the required OJL may be acquired in increments of 2,000 hours per year. Layoffs or illness may affect the duration. The apprentice must log all work time and turn the log in to the apprenticeship committee so that accurate records are maintained.

The classroom instruction and work-related training will not always run concurrently due to such reasons as layoffs, the type of work needed to be done in the field, and other factors. Apprentices with special job experience or coursework may obtain credit toward their classroom requirements. This reduces the total time required in the classroom while maintaining the total 8,000-hour on-the-job training requirement. These special cases depend on the type of program and the regulations and standards under which it operates.

Informal OJL provided by employers is usually less thorough than that provided through a formal apprenticeship program. The degree of training and supervision in this type of program often depends on the size of the employing firm. A small contractor may provide training in only one area, while a large company may be able to provide training in several areas.

For those entering an apprenticeship program, a high school or technical school education is desirable. Courses in shop, mechanical drawing, and

Did You Know?

An Ancient Apprenticeship Contract

Historical records include a contract between a young Greek named Heracles and a weaver. In exchange for food, a tunic (a long shirt), and 20 holidays a year, Heracles worked for the weaver for five years. After two and a half years, he was paid 12 drachmas. In his fifth year, his pay was scheduled to double.

In addition to learning to become a journeyman weaver, Heracles most likely worked seven long days each week and had to clean the shop, build fires, pick up supplies, make deliveries, and do whatever other jobs the master weaver needed to have done. It's also likely that any damage he caused came out of his promised pay.

So how much pay is 12 drachmas? There is no way of knowing for certain, but the word *drachma* means handful—not much pay for almost three years of hard work. Yet Heracles was willing to put in the time to become a skilled master in his chosen trade. As a master weaver, Heracles could look forward to a better income and a better life.

general mathematics are helpful. Manual dexterity, good physical condition, and quick reflexes are important. The ability to solve problems quickly and accurately and to work closely with others is essential. You must also have high respect for safety and its place in the construction environment.

> **NOTE**
> Some companies have physical activity requirements that must be met by apprentices. These requirements vary from company to company.

1.2.2 Youth Training and Apprenticeship Programs

Youth apprenticeship programs are available that allow students to begin their apprenticeship or craft training while still in high school. A student entering the program in the 11th grade may complete as much as one year of the training program by high school graduation. In cooperation with local construction industry employers, students may be able to work in the craft and earn money while still in school. Upon graduation, students can enter the industry at a higher level and with more pay than someone just starting in a training program.

Students participating in apprenticeship training may enter the second level or year of the program wherever it is offered. They may also have the option of applying credits at two-year or four-year colleges that offer degree or certificate programs in their selected field of study.

Did You Know?

Apprenticeships in the United States

In 2012, the US Department of Labor reported that there were more than 440,000 apprentices receiving registered apprenticeship training in 37,000 programs in the United States.

Additional Resources

Arc Welding Safety E205. Latest Edition. Cleveland, OH: Lincoln Electric.

Safety Topics in Welding, Cutting, and Brazing. **www.osha.gov**.

AWS F32M/F32, Ventilation Guide for Weld Fumes. Latest Edition. Miami, FL: American Welding Society.

1.0.0 Section Review

1. Flux core arc welding (FCAW) is similar to _____.
   a. SMAW
   b. GMAW
   c. GTAW
   d. carbon-arc

2. Before graduation, a high school student entering an apprenticeship program in the 11th grade may complete _____.
   a. as much as one year of the program
   b. nearly two years of the program
   c. more than two years of the program
   d. three years of the program

Section Two

2.0.0 Personal Protective Equipment

Objective

Identify and describe personal protective equipment (PPE) related to the welding trade.

a. Identify and describe body, foot, and hand protective gear.
b. Identify and describe ear, eye, face, and head protective gear.

Trade Terms

Arc burn: Burn to the skin produced by brief exposure to intense radiant heat and ultraviolet light.

Flash burn: Burns to the eyes sometimes called welder's flash; caused by exposure to intense radiant heat and ultraviolet light.

Ultraviolet (UV) radiation: Invisible rays capable of causing burns. UV rays from the sun are the cause of sunburn.

29101-15_F04.EPS

Figure 4 Welder with proper protective clothing.

Each person in the shop environment wears general work clothing. Extra protection is needed for each person that is in direct contact with hot materials. Depending on the specific job and conditions, protective equipment can include the following:

- Body protection
- Foot protection
- Hand protection
- Ear protection
- Eye, face, and head protection

2.1.0 Body, Foot, and Hand Protection

Welding produces hazards from molten metal and sparks. For that reason, welders and anyone working in the welding environment must have proper protection.

2.1.1 Body Protection

Basic clothing for welders, as shown in *Figure 4*, should offer protection from flying sparks, heat, and ultraviolet (UV) radiation from electric arcs. Shirts should be made of a tight-weave fabric, have long sleeves and pocket flaps, and be worn with the collar buttoned. Pants must not have cuffs and should fit so they hang straight down the leg, touching the shoe tops without creases. Cuffs and creases can catch sparks, which can cause fires. Never wear polyester or other synthetic fibers; sparks will melt these materials, causing serious burns. Wear wool and cotton, which are more resistant to sparks. Cotton clothing, if worn, should be treated with a fire-retardant.

It is important to follow the manufacturer's recommendations for laundering fire-resistant clothing. Fabric softeners may reduce the effectiveness of the treatment, increasing the flammability. Some fire-resistant clothing may be ruined by repeated washings or incorrect washing. When purchasing fire-resistant clothing, look for fire-resistant labeling on the garment itself, rather than on the packaging alone.

Additional protective coverings are required for the following:

- Cutting operations
- Out-of-position or overhead welding/cutting
- Welding operations that produce spattering molten metal and sparks or large amounts of heat

Figure 5 shows a welder wearing protective clothing and equipment. Protective coverings, usually made of leather because of its durability, are often referred to as leathers. However, because leather is heavy and hot to wear, lightweight fire- and heat-resistant clothing is also available. *Figure 6* shows a variety of leather and fire-resistant items that can be worn in combinations to provide various degrees of covering.

Full jackets offer the most protection. A cape is cooler, but offers only shoulder, chest, and arm protection. Leather pants, in combination with a jacket, will protect a welder's lap and legs. If the weather is hot and full leather pants are uncomfortable, leggings, sometimes called chaps, are available. They are strapped to the legs, leaving the back open. However, an apron alone is cooler and will protect a welder's lap and most of the leg area when squatting, sitting, or bending over a table. In some cases, if the welding is done at or below waist level, only full- or half-sleeve arm covers are required in combination with an apron. The welding position and type of welding operation typically dictate the appropriate personal protective equipment (PPE), in conjunction with company or site policies.

29101-15_F05.EPS

Figure 5 Personal protective equipment (PPE).

Terms

Don't confuse the terms *fire-retardant* and *fireproof*. A product impregnated with a fire-retardant delays ignition. A fireproof product is not supposed to burn. Clothing treated with a fire-retardant may lose some of its protective ability if washed, so it may need retreating after washing.

WARNING!

Sparks of molten metal from welding and metal cutting can reach temperatures greater than 9,900°F (5,500°C) and can travel a significant distance. They can land on the clothing or exposed skin of the welder or others in the immediate area, potentially resulting in burns. Sparks can also land on flammable material a significant distance away, resulting in a fire. For this reason, welders and nearby workers must wear the proper protective clothing. Flammable materials nearby must be covered with non-flammable blankets.

2.1.2 Foot Protection

The Occupational Safety and Health Administration (OSHA) requires that protective footwear be worn when working where falling, rolling, or sharp objects pose a danger of foot injuries and where feet are exposed to electrical hazards. For welders, sparks and droplets of molten metal provide additional hazards. Footwear with leather-reinforced soles or innersoles of flexible metal are recommended. High-top safety shoes or boots should be worn. Ensure that the pant leg covers the tongue and lace area of the footwear. If the tongue and lace area is exposed, wear leather spats (as shown in *Figure 6*) under the pants or leggings and over the front top of the footwear to protect them from sparks or falling molten metal. Spats will prevent sparks from burning through the front of lace-up boots. Protective footwear must comply with *ANSI Z49.1*. Sneakers, tennis shoes, and similar types of footwear must never be worn on the job site.

29101-15_F06.EPS

Figure 6 Leather and fire-resistant protective coverings.

2.1.3 Hand Protection

Gloves are the primary type of hand protection. Gloves must be selected on the basis of the hazards involved in doing the work. Gauntlet-type welding gloves must be worn when welding or cutting to protect against UV rays from an electric arc and heat from any thermal welding/cutting process. The most common glove is the standard leather welding glove shown in *Figure 7*. For heavy-duty welding or cutting, special heat-resistant or heat-reflective gloves are also available.

> **WARNING!**
> Don't weld or cut metal with gloves that have been used to handle petroleum products, especially when oxygen is being used. Also consider that the transfer of petroleum residue to the welding rod can affect the weld quality.

2.2.0 Protection Above the Neck

Special attention must be given to eye and ear safety when welding or cutting. Severe burns to the eyes can result from viewing the welding arc without the proper protection. Ears can be damaged by noise and sparks. The face, head, and neck must also be protected from sparks and heat.

2.2.1 Hearing Protection

Welding areas can be very noisy. If overhead work is being performed, hot sparks can cause burns to the ears and ear canals unless a leather hood is used. Such burns are extremely painful, so it is important to wear ear plugs when welding (*Figure 8*). Disposable earplugs are the most common form of hearing protection used in welding. These devices usually have an outer layer of pliable foam and a core layer of acoustical fiber that filters out harmful noise, yet allows you to hear normal conversation. To use disposable earplugs, simply roll each plug into a cone and insert the tapered end into the ear canal while pulling up on the upper portion of your ear. The earplugs will expand, filling the ear canal and creating a proper fit.

> **WARNING!**
> Plain cotton placed in the ear is not acceptable protection. Only approved ear protection is to be used.

2.2.2 Eye, Face, and Head Protection

The heat and light produced by cutting or welding operations can damage the skin and eyes. Injury to the eyes may result in permanent loss of vision. Oxyfuel cutting and welding can cause eye fatigue and mild burns to the skin because of the infrared heat radiated by the process. Welding or cutting operations involving an electric arc of any kind produce UV light. The UV light can cause a severe **arc burn** to exposed skin and **flash burns** to unprotected eyes. (The sun also produces UV light, some of which is not filtered out by the Earth's atmosphere and can also dam-

Figure 7 Welding gloves.

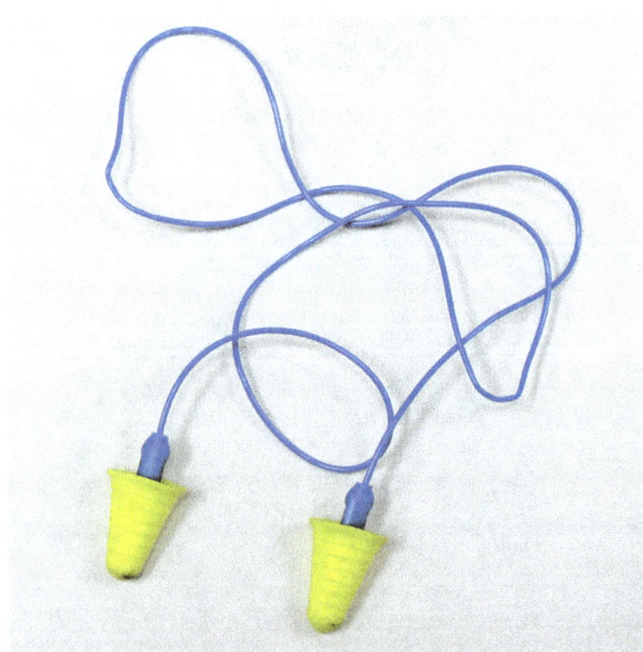

Figure 8 Ear plugs.

age the eyes.) A flash burn can very easily cause permanent damage to the retina of the eye. Welders should never view an electric arc directly or indirectly without wearing a properly tinted lens designed for electric arc use.

> **NOTE**
> Goggles are normally used with a leather soft cap to protect the face and head from UV rays.

As specified in *ANSI/AWS Z49.1*, protective spectacles with side shields, arc goggles, or other approved eye protection must be worn by welders in addition to a helmet or face shield with a filtered lens (*Figure 9*). The spectacles or goggles may have either clear or filtered lenses, depending on the amount of exposure to welding or cutting radiation (see *Table 1*). For overhead oxyfuel cutting operations, a leather hood may be used in place of the face shield to provide protection from sparks and molten metal. Goggles can be paired with a leather soft-cap for cutting and burning. The combination face shield with flip-up lenses can be used at some sites with a lens that is dark enough for the application.

For electric arc operations, a welding helmet with a properly tinted lens (shades 9 to 14) must be worn over safety glasses to provide proper protection. Many varieties of helmets are available; typical styles are shown in *Figure 10*. Some of the helmets are available with additional side-view lenses in a lighter tint so that welders can sense, by peripheral vision, any activities occurring beside them. A tinted lens that flips up to reveal a clear safety lens is a good feature. The clear lens enables the welder to continue using the helmet for protection during chipping and grinding.

Most welding and cutting tasks require the use of safety glasses, chemical-resistant goggles, dust goggles, or face shields. Always check your company's policy for the welding or cutting process being used to find out what type of eye protection is needed.

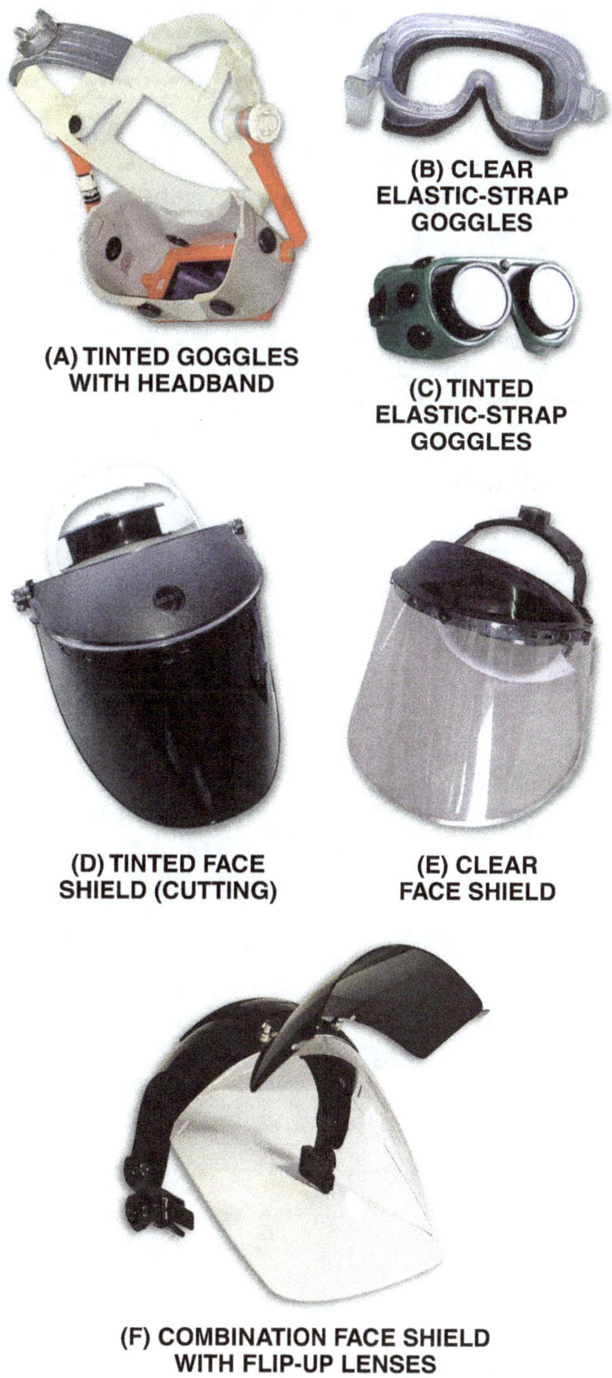

Figure 9 Oxyfuel welding/cutting goggles and face shield combinations.

Table 1 Guide for Shade Numbers

From *AWS F2.2 Lens Shade Selector.* Shade numbers are given as a guide only and may be varied to suit individual needs.

Process	Electrode Size in (mm)	Arc Current (Amperes)	Minimum Protective Shade	Suggested* Shade No. (Comfort)
Shielded Metal Arc Welding (SMAW)	Less than 3/32 (2.4)	Less than 60	7	–
	3/32–5/32 (2.4–4.0)	60–160	8	10
	5/32–1/4 (4.0–6.4)	160–250	10	12
	More than 1/4 (6.4)	250–550	11	14
Gas Metal Arc Welding (GMAW) and Flux Cored Arc Welding (FCAW)		Less than 60	7	–
		60–160	10	11
		160–250	10	12
		250–500	10	14
Gas Tungsten Arc Welding (GTAW)		Less than 50	8	10
		50–150	8	12
		150–500	10	14
Air Carbon Arc Cutting (CAC-A)	(Light)	Less than 500	10	12
	(Heavy)	500–1000	11	14
Plasma Arc Welding (PAW)		Less than 20	6	6 to 8
		20–100	8	10
		100–400	10	12
		400–800	11	14
Plasma Arc Cutting (PAC)		Less than 20	4	4
		20–40	5	5
		40–60	6	6
		60–80	8	8
		80–300	8	9
		300–400	9	12
		400–800	10	14
Torch Brazing (TB)		–	–	3 or 4
Torch Soldering (TS)		–	–	2
Carbon Arc Welding (CAW)		–	–	14

	Plate Thickness		Suggested* Shade No. (Comfort)
	in	mm	
Oxyfuel Gas Welding (OFW)			
Light	Under 1/8	Under 3	4 or 5
Medium	1/8 to 1/2	3 to 13	5 or 6
Heavy	Over 1/2	Over 13	6 or 8
Oxygen Cutting (OC)			
Light	Under 1	Under 25	3 or 4
Medium	1 to 6	25 to 150	4 or 5
Heavy	Over 6	Over 150	5 or 6

*As a rule of thumb, start with a shade that is too dark to see the weld zone. Then go to a lighter shade which gives sufficient view of the weld zone without going below the minimum. In oxyfuel gas welding, cutting, or brazing where the torch and/or the flux produces a high yellow light, it is desirable to use a filter lens that absorbs the yellow or sodium line of the visible light spectrum.

29105-15_T01.EPS

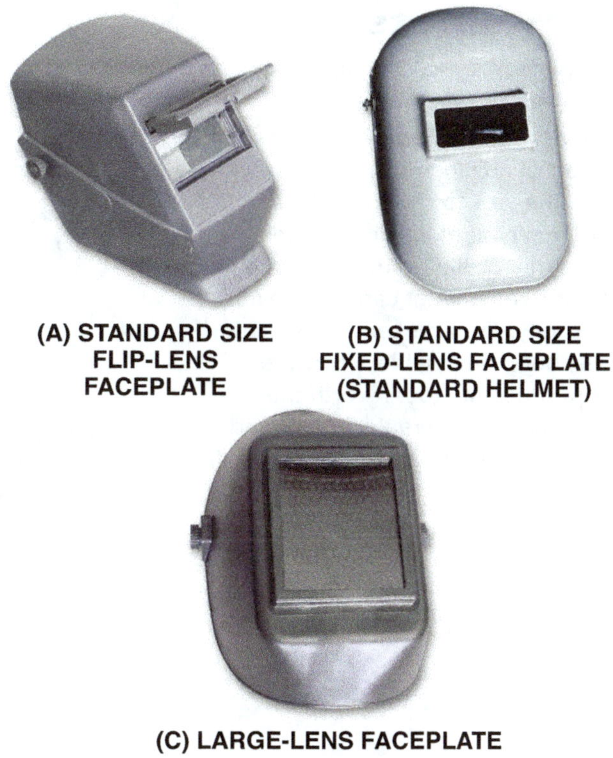

Figure 10 Arc-welding helmets.

Lightweight Welding Gloves

Soft, flexible, leather gloves are used for light-duty work and for operations involving GMAW, GTAW, brazing, soldering, and oxyfuel welding where free hand movement is required. These types of welding produce less heat and spatter than others. Lightweight gloves provide more flexibility for the required motions and pulling the trigger of the welding torch.

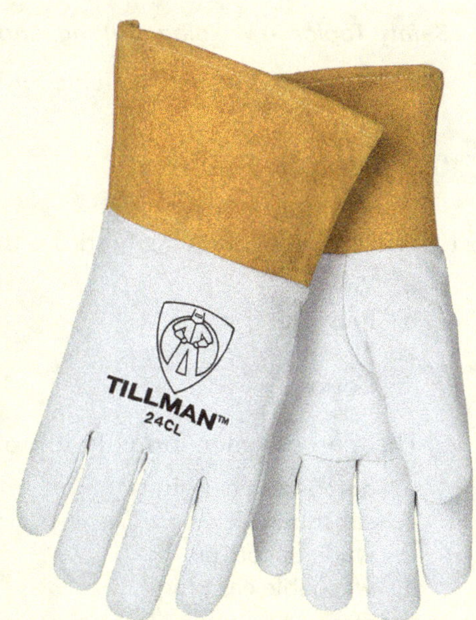

29101-15_SA01.EPS

Auto-Darkening Helmets

Helmets are available with an adjustable auto-darkening lens that automatically changes shades the instant an electric arc starts to occur. Typical shade ranges are 5-14 and 9-13. A sensor just above the lens detects the start of an electric arc and electronically causes the shade of the lens to change in less than $\frac{1}{25,000}$ of a second. The use of this type of lens eliminates the need for the welder to raise the helmet or flip up the faceplate to view the work before striking an arc. The helmet can also be left down for chipping and grinding safety. Auto-darkening helmets cost more than standard helmets, but many welders consider the improvement in quality and productivity to be worth the additional cost. Some require periodic battery replacement and testing to ensure proper operation, but others are solar-powered.

29101-15_SA02.EPS

Additional Resources

ANSI/AWS Z49.1, Safety in Welding, Cutting, and Allied Processes. Miami, FL: American Welding Society.

Arc Welding Safety E205, Latest Edition. Cleveland, OH: Lincoln Electric.

Personal Protective Equipment (PPE) for Welding and Cutting. Miami, FL: American Welding Society.

Safety Topics in Welding, Cutting, and Brazing. **www.osha.gov**.

AWS F32M/F32, Ventilation Guide for Weld Fumes. Latest Edition. Miami, FL: American Welding Society.

2.0.0 Section Review

1. Which of the following fabrics is the best choice for welding clothing?
   a. Rayon
   b. Polyester
   c. Nylon
   d. Cotton

2. The most common form of ear protection is ____ .
   a. disposable ear plugs
   b. ear muffs
   c. wads of cotton
   d. washable ear plugs

SECTION THREE

3.0.0 WELDING SAFETY PRACTICES

Objective

Identify and describe welding safety practices related to specific hazards or environments.

a. Describe the importance of welding safety and identify factors related to accidents.
b. Describe basic welding safety practices related to the general work area.
c. Describe hot work permits and fire watch requirements.
d. Describe confined spaces and their related safety practices.
e. Identify safety practices related to welding equipment.
f. Identify and describe respiratory hazards, respiratory safety equipment, and ways to ventilate welding work areas.
g. Explain the purpose of the SDS/MSDS and how it is used.

Trade Terms

Bonded: The permanent joining of metallic parts to form an electrically conductive path that will assure electrical continuity and the capacity to safely conduct any current likely to be imposed on it.

Electrically grounded: Connected to Earth or to some conducting body that serves in place of the earth.

Fume plume: The fumes, gases, and particles from the consumables, base metal, and base metal coating during the welding process.

Galvanized steel: Carbon steel dipped in zinc to inhibit corrosion.

Immediate danger to life and health (IDLH): As defined by OSHA, this term refers to an atmosphere that poses an immediate threat to life, would cause irreversible adverse health effects, or would impair an individual's ability to escape from a dangerous atmosphere.

Purge gas: An inert gas, such as nitrogen, used to drive oxygen away from a weld site.

Ventricular fibrillation: Irregular heart rhythm characterized by a rapid fluttering heartbeat.

There are a number of special hazards associated with welding, including intense heat, flame, and dangers of exposure to a high-intensity electric arc. Metal cutting brings additional hazards to the workplace, including explosive and flammable gases and the intense flame used to cut metal. These conditions create the need for specialized safety practices that must be followed by those performing the work, as well as those in the vicinity of the work.

3.1.0 Welding Safety Hazards

The welding industry is fortunate to have a well-developed standard for safety: *American National Standards Institute (ANSI) Z49.1, Safety in Welding, Cutting, and Allied Processes*. Everyone involved with welding, cutting, and related processes should be familiar with this document. The document discusses personal protective equipment, ventilation, and fire prevention, as well as welding in confined spaces. It contains precautionary information that is important to welders, supervisors, and managers. The document also details specific welding process safety procedures.

There are many causes of accidents. They can usually be divided into two broad categories: personal and physical. Workers must be aware of the factors that can cause accidents so they can understand the consequences of their actions on the job site.

3.1.1 Personal Factors That Cause Accidents

Accidents can often be traced to personal factors such as poor health, lack of experience, and the improper use of alcohol and medications. Something as simple as excessive use of colognes, after-shaves, or perfumes that contain alcohol can represent a burn hazard. A good general rule is: whenever you feel burning or excessive heat, immediately stop work and check it out.

People who are ill or injured may not be able to concentrate on their work. In addition, they may be physically weakened and unable to handle strenuous work. A person doing work that is inherently dangerous is more likely to have an accident when sick or injured. Mental stress can also play a role in accidents; concentration is the issue. An employee who is worried about a serious personal problem is likely to be distracted and unable to focus on work. Workers need to realize that they endanger themselves and those around them when they are not 100 percent effective. As a worker, you need to be able to recognize when others may not be up to par and take appropriate precautions.

Age and inexperience often play a role in accidents. Insurance company studies show that a person who lacks experience is more likely to take risks that cause accidents. Sometimes accidents occur because someone hasn't had enough experience to learn how to avoid them. In fact, insurance company statistics show that more accidents occur in the under-18 age group than in any other. An inexperienced person often is not able to foresee the outcome of an action, or lacks the knowledge to know what works and what doesn't.

People who consume alcohol or use illegal drugs while working risk their lives and the lives of their co-workers. A drug or alcohol hangover is nearly as bad. Legal prescription and over-the-counter drugs can also cause problems. Some cold remedies and cough medicines contain alcohol or other substances that will make people drowsy. Alcohol and drugs affect a person's coordination, alertness, and decision-making ability. Never use these substances while you are working, and don't work with people who use them on the job.

3.1.2 Physical Factors That Cause Accidents

Accidents are often caused by conditions at the job site. For example, foot and vehicle traffic congestion increases at starting and quitting times. As quitting time approaches, workers are more likely to hurry, possibly taking risks they might not otherwise take. People often slow down right after a meal, so workers may be less attentive after lunch than at other times.

People who learned to put their toys away when they were young probably keep a neat, safe work site as adults. The others are likely to have tools, equipment, scrap material, and other items lying around their work site that could trip people. If you keep a messy work site, eventually you or someone else will be injured because of it. Put away your tools, dispose of scrap materials, and secure your equipment and cables when not in use. If you have flammable or hazardous material at your work site, keep it properly contained and covered. When you're done with it, return it to the designated location for disposal or storage.

Damaged or defective welding equipment is dangerous. For example, a frayed or damaged welding cable could give someone a lethal shock or start a fire. Any damaged or defective equipment should be so tagged and either repaired or replaced. All equipment should be periodically inspected for damage. Make sure that periodic maintenance procedures recommended by the manufacturer are performed on schedule.

3.2.0 Work Area Safety

An important factor in area safety is good housekeeping. The work area should be picked up and swept clean. The floors and workbenches should be free of dirt, scrap metal, grease, oil, and anything that is not essential to accomplishing the given task. Collections of steel, welding electrode stubs, wire, hoses, and cables are difficult to work around and easy to trip over. An electrode caddy can be used to hold the electrodes and stubs. Hooks can be made to hold hoses and cables, and scrap steel should be thrown into scrap bins.

The ideal welding shop should have bare concrete floors and electrically grounded and bonded bare metal walls and ceilings to reduce the possibility of fire. Never weld or cut over wood floors, as this increases the possibility of fire.

It is important to keep flammable liquids as well as rags, wood scraps, piles of paper, and other combustibles out of the welding area.

If you must weld in an enclosed building, make every effort to eliminate anything that could trap a spark. Sparks can smolder for hours and then burst into flames. Regardless of where you are welding, be sure to have a fire extinguisher nearby. Also, keep a bucket of water handy to cool off hot metal (other than a test coupon or functional weld) and quickly douse small fires. If a piece of hot metal must be left unattended, use soapstone to write the word HOT on it before leaving. This procedure can also be used to warn people of hot tables, vises, firebricks, and tools.

Perform cutting activities in the workshop in a properly prepared cutting area. Be aware that welding and cutting sparks can ignite any type of flammable vapors. Prevent hot droplets from landing directly on the torch hoses, as they could melt a hole in the hose.

Whenever welding must be done outside a welding booth, use portable screens to protect other personnel from the bright glare of the welding arc (*Figure 11*). The portable screen also prevents drafts of air from interfering with the stability of the arc. Welding blankets (*Figure 12*) should be used to cover material and equipment that could be burned or otherwise damaged.

The most common welding accident is burned hands and arms. Keep first-aid equipment nearby to treat burns in the work area. Eye injuries can also occur if you are careless. Post emergency phone numbers in a prominent location.

29101-15_F11.EPS

Figure 11 A typical welding screen.

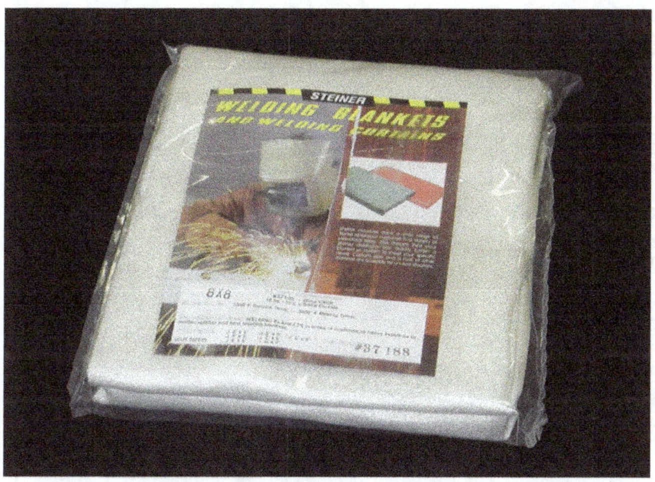

3,000°F (1,620°C) INTERMITTENT, 1,500°F (820°C) CONTINUOUS SILICON DIOXIDE CLOTH

29101-15_F12.EPS

Figure 12 A welding blanket.

The following are some work-area reminders:

- Eliminate tripping hazards by coiling cables and keeping clamps and other tools off the floor.
- Don't get entangled in cables, loose wires, or clothing while you work. You must be able to move freely, especially should your clothing ignite or some other accident occur.
- Clean up oil, grease, or other agents that may ignite and splatter off surfaces while welding.
- Shut off the welding machine and disconnect the power before performing any service or maintenance. A circuit breaker or power disconnect switch should be provided for each machine when hard-wired. Others may be equipped with only a cord and plug.
- Keep the floor free of electrodes and electrode stubs. They could easily cause a slip or fall. All used electrodes should be disposed of in a proper metal container. Throwing electrodes on the floor can result in disciplinary action or even dismissal, due to the safety hazard and poor housekeeping attitude it represents.
- Always remove a SMAW electrode from the electrode holder before laying it down. Leaving a stub in the electrode holder allows it to easily arc to nearby surfaces.
- Work in a dry area, booth, or other shielded area whenever possible.
- Make sure there are no open doors or windows through which sparks may travel to flammable materials.
- Don't coil electrical leads over your body while working on scaffolding.

Conditions on a job site can change quickly. When performing welding or cutting work, it is a good idea to stop occasionally and check your surroundings.

3.3.0 Hot Work Permits and Fire Watches

A hot work permit (*Figure 13*) is an official authorization from the site manager to perform work that may pose a fire hazard. The purpose of the permit is to ensure that a survey process takes place to confirm that the area is safe for cutting or welding, while also establishing accountability for the safety of the job. The permit includes information such as the time, location, and type of work being done. The hot work permit system promotes the development of standard fire safety guidelines. Hot work permits and fire watches are required by OSHA. Permits also help managers keep records of who is working where and at what time. This information is essential in the event of an emergency or at other times when personnel need to be evacuated.

During a fire watch, a person other than the welder or cutting operator must constantly scan the work area for fires. Fire watch personnel must have ready access to fire extinguishers and alarms and know how to use them. When a person is standing fire watch, it should be their only duty.

Cutting operations must never be performed without a fire watch. Whenever oxyfuel cutting equipment is used, there is a great danger of fire. Hot work permits and fire watches are used to minimize this danger. Most sites require the use of hot work permits and fire watches. When they are violated, severe penalties are imposed.

HOT WORK PERMIT

For Cutting, Welding, or Soldering with Portable Gas or ARC Equipment

(References: 1997 Uniform Fire Code Article 49 & National Fire Protection Association Standard NFPA 51B.)

Job Date__________ Start Time__________ Expiration__________ WO #__________

Name of Applicant____________________ Company__________ Phone__________

Supervisor____________________ Phone__________

Location / Description of work ______________________________

IS FIRE WATCH REQUIRED?

1. ______ (yes or no) Are combustible materials in building construction closer than 35 feet to the point of operation?
2. ______ (yes or no) Are combustibles more than 35 feet away but would be easily ignited by sparks?
3. ______ (yes or no) Are wall or floor openings within a 35 foot radius exposing combustible material in adjacent areas, including concealed spaces in floors or walls?
4. ______ (yes or no) Are combustible materials adjacent to the other side of metal partitions, walls, ceilings, or roofs which could be ignited by conduction or radiation?
5. ______ (yes or no) Does the work necessitate disabling a fire detection, suppression, or alarm system component?

YES to any of the above indicates that a qualified fire watch is required.

Fire Watcher Name(s) ____________________________ Phone__________

NOTIFICATIONS

Notify the following groups at least 72 hours prior to work and 30 minutes after work is completed.
Write in names of persons contacted.

Notify in person OR by phone ONLY if question #5 above is answered "yes":

- Facilities Management Fire Alarm Supervisor

Notify by phone or in person: (If by phone, write down name of person and send them a completed copy of this permit.)

- Facilities Management Fire Protection Group
- Environmental Health & Safety Industrial Hygiene Group

SIGNATURES REQUIRED

University Project Manager________________________ Date __________ Phone__________

I understand and will abide by the conditions described in this permit. I will implement the necessary precautions which are outlined on both sides of this permit form. Thirty minutes after each hot work session, I will reinspect work areas and adjacent areas to which spark and heat might have spread to verify that they are fire safe, and contact Facilities Management Alarm Technicians to have any disabled fire protection systems reactivated.

____________________ ____________________ Date __________ Phone__________
Permit Applicant Company or Department

1/17/15

29101-15_F13.EPS

Figure 13 Hot work permit.

WARNING!

Never perform any type of heating, cutting, or welding until you have obtained a hot work permit and established a fire watch. If you are unsure of the procedure, check with your supervisor. Violation of hot work permit and fire watch procedures can result in serious injury or death. OSHA requires that the fire watch be maintained for one-half hour after welding or cutting is finished.

3.4.0 Confined Spaces

A confined space refers to a relatively small or restricted space, such as a storage tank, boiler, or pressure vessel, or small compartments, such as underground utility vaults, small rooms, or the unventilated corners of a room. *Figure 14* shows an example of a confined space entry permit typically found on a job site.

OSHA 29 *CFR* 1910.146 defines a confined space as a space that:

- Is large enough and so configured that an employee can bodily enter and perform assigned work
- Has a limited or restricted means of entry or exit; for example, tanks, vessels, silos, storage bins, hoppers, vaults, and pits
- Is not designed for continuous employee occupancy

OSHA 29 *CFR* 1910.146 further defines a permit-required confined space as a space that:

- Contains or has the potential to contain a hazardous atmosphere
- Contains a material that has the potential for engulfing an entrant
- Has an internal configuration such that an entrant could be trapped or asphyxiated by inwardly converging walls or by a floor that slopes downward and tapers to a smaller cross-section
- Contains any other recognized serious safety or health hazard

For safe working conditions, the oxygen level in a confined space atmosphere must range between 19.5 percent and 23.5 percent by volume, with 21.5 percent being considered the normal level (*Table 2*). Oxygen concentrations below 19.5 percent by volume are considered deficient.

WARNING!

If too much oxygen is ventilated into a confined space, it can be absorbed by the welder's clothing and ignite. If too little oxygen is present, it can lead to the welder's death in minutes. For this reason, make sure confined spaces are ventilated properly for cutting or welding purposes and *never* use oxygen in confined spaces for ventilation purposes.

Before operations are started, the wheels of heavy portable equipment must be securely blocked to prevent accidental movement. Where a welder must enter a confined space through a manhole or other opening, all means must be provided for quickly removing the worker in case of emergency (*Figure 15*). When safety belts and lifelines are used for this purpose, they must be attached to the welder's body so that he or she

Oxyfuel Gas Welding Safety Precautions

- Always light the oxyfuel gas torch flame using an approved torch lighter to avoid burning your fingers.
- Never point the torch tip at anyone when lighting it or using it.
- Never point the torch at the cylinders, regulators, hoses, or anything else that may be damaged and cause a fire or explosion.
- Never lay a lighted torch down on the bench or workpiece, and do not hang it up while it is lighted. If the torch is not in the welder's hands, it must be off.
- To prevent possible fires and explosions, check valves and flashback arrestors must be installed in all oxyfuel gas welding and cutting outfits.
- When cutting with oxyfuel gas equipment, clear the area of all combustible materials.
- Skin contact with liquid oxygen can cause frostbite. Be careful when handling liquid oxygen.
- Never use oxygen as a substitute for compressed air.
- All oxygen cylinders with leaky valves or safety fuse plugs and discs should be set aside and marked for the attention of the supplier. Do not tamper with or attempt to repair oxygen cylinder valves. Do not use a hammer or wrench to open the valves.

Master Card No.________________

1. Work Description

Area_______________________ Equipment Location______________________

Work to be done:

2. Gas Test		Results	Recheck	Recheck
Required	☐ Instrument Check			
☐ Yes	☐ Oxygen % 20.8 Min.			
☐ No	☐ Combustible % LFL			
		Date/Time/Sig.	Date/Time/Sig.	Date/Time/Sig.

3. Special Instructions: ☐ Check issuer before beginning work ☐ None

4. Hazardous Materials: ☐ None What did the line / equipment last contain?

5. Special Protection Required: ☐ None ☐ Forced Air Ventilation

☐ Avoid Skin Contact ☐ Gloves ________________ ☐ Suit ______________________

☐ Goggles or Face Shield ☐ Respirator ___________________ ☐ Safety Harness

☐ Self-Contained Breathing Equipment ☐ Hoseline Breathing Equipment

☐ Other, Specify _________ ☐ Standby - Name: ____________________________________

6. Fire Protection Required: ☐ None ☐ Portable Fire Extinguisher ☐ Fire Hose and Nozzle

☐ Fire Watch ☐ Other, specify:

7. Condition of Area and Equipment

Required Yes	No	THESE KEY POINTS MUST BE CHECKED
		a. Lines disconnected & blinded or where disconnecting is not possible, blinds installed? (Includes drains, vents and instrument leads) and appropriate valves locked out?
		b. Equipment cleaned, washed, purged, ventilated?
		c. Low voltage or GFCI-protected electrical equipment provided?
		d. Explosion-proof electrical equipment provided?
		e. Life lines required to be attached to safety harnesses?

Comments

rev.6/10/15

29101-15_F14A.EPS

Figure 14A Confined space entry permit (sheet 1 of 2).

8. Approval	Date	Time	Permit Authorization Area Supervisor	Permit Acceptance Maint./Contractor Supervisor/Engineer	Date	Time
Issued by						
Endorsed by						
Endorsed by						
Endorsed by						

9. Individual Review

I have been instructed on proper Safety Procedures and proper Confined Space Entry Procedures. I have signed in on the appropriate Master Card and have affixed personal locks on energy isolation devices as appropriate.

Signature of all personnel covered by this permit.

Forward to Production Superintendent 7 days after completion of work.

rev.6/10/15

29101-15_F14B.EPS

Figure 14B Confined space entry permit (sheet 2 of 2).

Table 2 Effects of an Increase or Decrease in Oxygen Levels

Oxygen Level	Effects
> 23.5%	Easy ignition of flammable material such as clothes
19.5% – 21.5%*	Normal
17%	Deterioration of night vision, increased breathing volume, accelerated heartbeat
14% – 16%	Very poor muscular coordination, rapid fatigue, intermittent respiration
6% – 10%	Nausea, vomiting, inability to perform, unconsciousness
< 6%	Spasmodic breathing, convulsive movements, and death in minutes

*There is a safety zone of 2%. Above 23.5% is considered enriched.

cannot be jammed in a small exit opening. An attendant with a pre-planned rescue procedure must be stationed outside to observe the welder at all times and must be capable of putting the rescue operations into effect.

When welding or cutting operations are suspended for any substantial period of time, such as during lunch or overnight, all electrodes must be removed from the holders, and the holders must be carefully located so that accidental contact cannot occur. The welding machines must also be disconnected from the power source.

In order to eliminate the possibility of gas escaping through leaks or improperly closed valves when gas welding or cutting, the gas and oxygen supply valves must be closed, the regulators

29101-15_F15.EPS

Figure 15 Worker in a confined space.

released, the gas and oxygen lines bled, and the valves on the torch shut off when the equipment will not be used for a substantial period of time. Where practical, the torch and hose must also be removed from the confined space. After welding operations are completed, the welder must mark the hot metal or provide some other means of warning other workers.

> **WARNING!**
>
> Nitrogen and argon are used as backing and purge gases for some welding applications. These gases will displace all the oxygen in the space where they are used.

3.5.0 Welding Equipment Safety

Anyone working with welding equipment must be aware of the equipment-related hazards. Of particular concern are the hazards associated with the voltages and electrical currents produced by welding machines. There are additional hazards related to falling equipment and equipment powered by gasoline engines. *Figure 16* shows some examples of safety notices typically found in operator manuals for welding units.

3.5.1 Electrical Hazards

Electric shock from welding and cutting equipment can result in death, injury, and severe burns. Injury can also result if a fall occurs because of the shock. The amount of current that passes through the human body determines the outcome of an electric shock. The higher the voltage, the greater the chance for a fatal shock. Electric current flows along the path of least resistance to return to its source. If you come in contact with a live conductor, you become a load in the circuit and the current will flow though you. *Figure 17* shows how much resistance the human body presents under various circumstances and how this converts into amps or milliamps when the voltage is 110V. Note

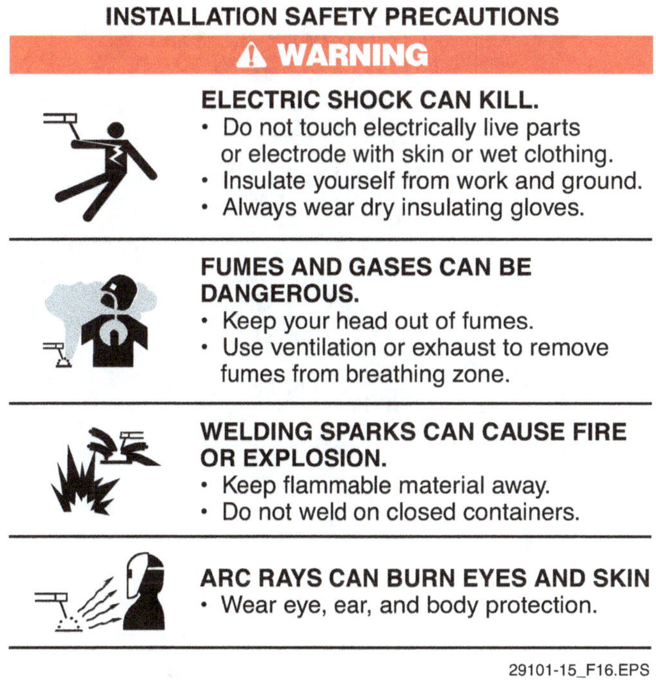

Figure 16 Safety hazard notices.

that the potential for shock increases dramatically if the skin is damp. A cut will also reduce resistance and increase shock potential. Currents of less than 1 amp can severely injure and even kill a person. Contact with a live electrical circuit can also cause severe burns. Under the right conditions, even a small current can cause serious injury or death. The Construction Employers Association has issued the following guidelines regarding the seriousness of electric shock:

- *Greater than 3 milliamps (mA)* – Indirect accident
- *Greater than 10 mA* – Muscle contraction
- *Greater than 30 mA* – Lung paralysis, usually temporary
- *Greater than 50 mA* – Possible ventricular fibrillation
- *100 mA to 4A* – Certain ventricular fibrillation
- *Greater than 4A* – Heart paralysis, severe burns

> **Did You Know?**
>
> **Oxygen-Deficient Air**
>
> People don't realize that entering an oxygen-deficient area can result in immediate death. Even a single breath of oxygen-deficient air can prevent muscles from responding and prevent you from escaping. Persons with physical problems such as heart ailments are at the greatest risk. At a refinery in Corpus Christi, Texas, while the welder was at lunch, a helper entered a pipe in which argon had been used as a backing gas. When the welder returned, he found the helper dead.
>
> Nitrogen is lighter than air and, if allowed, will float upward. Argon, on the other hand, is heavier than air and will settle in low-lying areas. When removing nitrogen from a confined space, open the top to allow it to float out. When removing argon from a confined space, provide some way for it to escape out the bottom or push/pull it out using air and mechanical equipment.

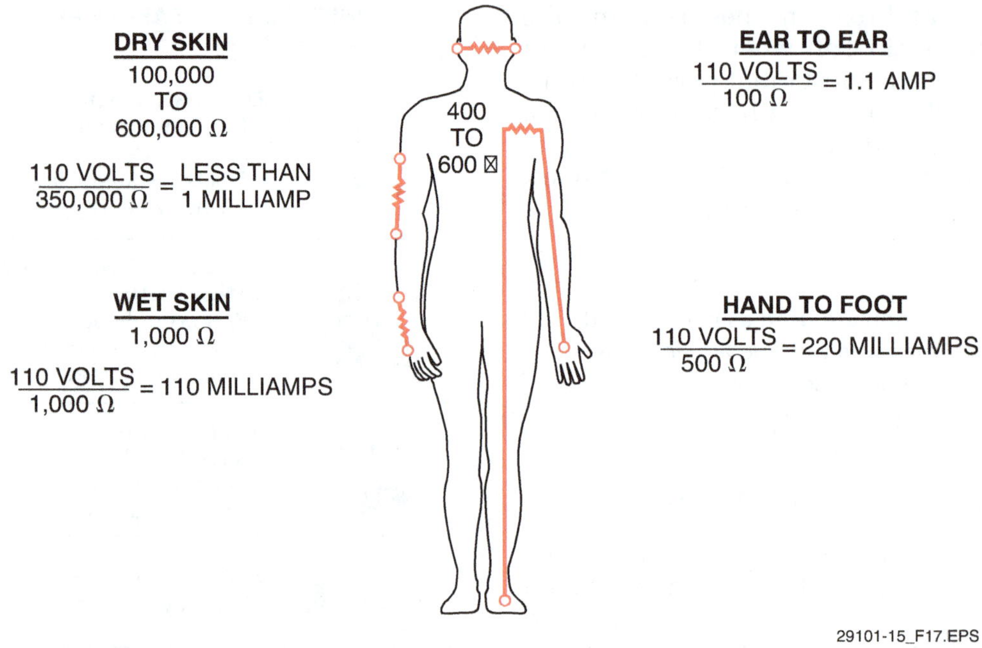

Figure 17 Typical body resistances and current.

Each welder and operator of the equipment must be trained and able to recognize the dangers associated with each particular type of equipment to avoid injuries, fatalities, and other electrical accidents.

It is a good idea for all involved to know first-aid procedures, such as cardiopulmonary resuscitation (CPR) for treating electric shock. Treat electrical burns like any other type of burn by applying ice or a cold compress over the burned area. Prevent infection and cover with a clean, dry dressing. Seek medical attention if necessary.

Welding-related electrical hazards come from two sources. The first, which we will call primary hazards, come from the source voltage that powers the welding machine. This voltage is usually 230 or 460 volts. These are potentials that create the possibility of serious, even fatal, shocks. It is important to know that this voltage exists inside the welding machine any time the machine is plugged in. The only way to disable it is to unplug the machine or turn off the power at the disconnect switch. The welding machine must also be properly grounded. This is done by connecting the ground terminal on the machine to a known ground.

WARNING!

Be sure to follow applicable lockout/tagout procedures when disabling the power to the unit. Also, have a qualified electrician install the welding machine so that it will be correctly wired and grounded. If the welding power supply does not work properly, have a qualified technician repair it. Never open the unit yourself.

Secondary hazards exist in the welding process itself. This type of shock can occur if a person touches part of the electrode circuit, such as a bare spot on the electrode cable, and touches the metal work material with another part of the body at the same time. Such a condition can occur if the welder is leaning against the work while welding. Here are some things to remember that will help keep you safe:

- The voltage is highest at the electrode when you are not welding. Due to the high current flow through the leads, the voltage drops somewhat during welding.
- The jaws and screws on the electrode holder, as well as the stick electrode itself, are electrically energized. Make sure they do not touch your skin or clothing, especially if the clothing is damp or wet.

- Never touch an electrode to a grounded conductive surface because these surfaces will become electrically energized. The electrode and work circuit is electrically energized when the welding machine switch is on.
- Never operate arc-welding equipment on a wet or damp floor.
- Wear dry gloves that are in good condition when welding. Keep an extra pair on hand.
- Keep dry insulation between your body, including your arms and legs, and the metal being welded or a ground such as a metal floor or wet earth. Use plywood or a rubber mat between you and the ground or floor.
- Keep the welding cables and electrode holder in good condition.

If you must work in conditions where you might come into contact with the work while welding, use one of the following types of machines:

- Semiautomatic DC constant-voltage welder
- DC manual (stick) welder
- AC welder with reduced voltage control
- Use a ground fault circuit interrupter (GFCI) for power tools that plug in

3.5.2 Gasoline-Powered Units

Gasoline- or diesel-powered welding machines are often used on sites where there is no direct access to 230- or 460-volt electrical outlets. In fact, this type of unit is very common. They range from small, portable units that look like the portable generators you would see at hardware stores, to trailer-mounted units like the one shown in *Figure 18.* In addition to the electrical hazards previously covered, gasoline- and diesel-powered units have their own particular hazards:

Shock Prevention

- Floors must be dry at all times. Use wooden platforms or rubberized carpet/floor coverings or other insulated material.
- Only qualified electricians may work on electric arc welding machine power connections.
- Workbenches must be grounded.
- With the power off, cable connections must be checked for cracked insulators, loose contacts, and other signs of damage.

- Gasoline-powered units produce deadly carbon monoxide just like automobile engines. They must never be operated indoors or in an enclosed space unless the exhaust is vented to the outdoors.
- Gasoline is highly flammable, so it must be kept in approved containers and stored away from hot work.
- When refueling a generator, turn off any welding or cutting flames. Allow the engine to cool to prevent the gasoline from contacting hot engine parts.

3.6.0 Fumes and Gases

The gases, fumes, and dust produced by welding and cutting processes can be hazardous. Adequate ventilation must be provided to prevent workers from breathing these products. There are three methods available to protect personnel against fumes and gases:

- *Natural ventilation* – The movement of air through the workplace caused by natural forces is often enough to remove fumes and gases.
- *Mechanical ventilation* – If natural ventilation is not adequate, portable or fixed fans can be used to provide the necessary ventilation.
- *Source extraction* – This method uses a mechanical device to capture welding fumes at or near the arc.

4-WHEEL TRAILER WITH MULTI-PROCESS WELDER

29101-15_F18.EPS

Figure 18 Gasoline-powered welding machine.

Regardless of the ventilation method used, it is important to avoid breathing in the welding fumes. This means keeping your head up and away from the **fume plume**, which is the term used to describe the fumes, gases, and particles created by the welding process. The heated fumes and smoke generally rise straight up or move towards the ventilation system opening, where a negative pressure exists. Observe the column of smoke and position yourself to avoid it. A small fan may also be used to divert the smoke, but take care to keep the fan from blowing directly on the work area. The gases from flux and the shielding gases controlled by the welding machine must be present at the welding arc in order to protect the molten metal from the surrounding air. A fan directed at the weld will disturb the process and create flaws.

In addition to the fume plume itself, some consumables and base metals contain toxic materials that require special ventilation. These materials include the following:

- Barium
- Cadmium (found in some steel and fasteners)
- Chromium (found in stainless steel)
- Base metal coatings such as paint
- Cobalt
- Copper
- Manganese
- Nickel
- Silica
- Zinc (found in galvanized metals)

Always check to determine what material you are welding or cutting. Some of the materials listed here produce fumes that are known to cause cancer or can lead to serious respiratory illness. Long-term exposure can even result in death. Some consumables are designated as special ventilation products as a result of the hazards they represent. When welding or cutting these materials, wear approved respiratory equipment and/or ensure that source extraction is used.

Chromium 6, also known as hexavalent chrome, is used in stainless steel. It gives off toxic fumes when heated. Manganese is another hazardous metal used in stainless steel. In addition, it is used in some welding rods, so welders have a greater risk of exposure. Exposure to manganese fumes and dust have been linked to Parkinson's Disease.

Galvanized steel, which is very common, is dipped in zinc to inhibit rust. When flame-cut or welded, it can produce a condition known as zinc chills, which causes flu-like symptoms. For occasional short-term exposure to fumes from zinc- or copper-coated materials, an approved respirator must be used.

In many cases, natural ventilation provides the required level of protection. However, it may be necessary to have a specialist determine if the ventilation is adequate. Some guidelines used in making this determination are as follows:

- The welding area must contain at least 10,000 cubic feet. This represents an area of 22' × 22' × 22' for each welder.
- The ceiling height must be no less than 16 feet.
- Cross-ventilation must not be blocked by partitions, equipment, or other barriers.
- The area must not be designated a confined space.

3.6.1 Source Extraction Equipment

Source extraction, also called local exhaust, can be provided by portable, mobile, or stationary systems such as those shown in *Figure 19*. As a general rule, the farther the extraction hose is from the arc, the greater the volume of air required. There are two classes of source extraction equipment: low vacuum/high volume and high vacuum/low volume.

Low vacuum/high volume equipment can be used in situations where the extraction hose is 6 to 15 inches (≈15 to 38 cm) from the arc (*Figure 20*). It uses extraction hoses, or hose and tubing combinations, with a diameter of 6 to 8 inches (≈15 to 20 cm) to accommodate the higher air volume of 560 to 860 cubic feet per minute (≈16 to 24 cubic meters per minute). The mobile unit shown in *Figure 19* is an example of a low vacuum/high volume system.

High vacuum/low volume systems are designed for close-proximity work and are effective in difficult to reach or confined work spaces. In this case, close proximity means the inlet opening needs to be 2" to 4" (≈5 to 10 cm) from the fume source. There are two types of high vacuum/low volume systems. The fume extraction gun type has a fume extraction nozzle built into the welding gun. Suction nozzles—the second type—are positioned near the weld, usually within 4 inches (10 cm) of the work. Adequate ventilation can be a problem in tight or cramped working quarters. To ensure adequate room ventilation in such cases, local exhaust ventilation should be used to capture fumes. Positioning is very important with these systems, to ensure the immediate area around the weld is not disturbed.

The exhaust hood should be kept close enough to the source of the fumes to ensure their capture.

Figure 19 Source extraction equipment.

29101-15_F20.EPS

Figure 20 Low vacuum/high volume fume extraction.

Welders should recognize that fumes of any type, regardless of their source, should not be inhaled. The best way to avoid problems is to provide adequate ventilation and keep your face out of the fume plume. Respirators must be used. *Figure 21* shows a welder using both a fume extraction system and a powered air-purifying respirator unit for increased safety.

29101-15_F21.EPS

Figure 21 Personal filtration and an extraction system combined.

WARNING!

Fumes and gases can be dangerous. Overexposure can cause nausea, headaches, dizziness, metal fume fever, and severe toxic effects that can be fatal. Studies have demonstrated irritation to eyes, skin, and the respiratory system and even more severe complications. In confined spaces, fumes and gases may cause asphyxiation.

3.6.2 Respirators

Special metals require the use of respirators to protect welders from harmful fumes. Respirator selection is based on the type and concentration of the contaminants. Respirators are grouped into three main types based on how they work to protect the wearer from contaminants. The types are the following:

- Air-purifying respirators
- Supplied-air respirators (SARs)
- Self-contained breathing apparatus (SCBA)

If respirators are used, your employer must offer worker training on respirator fitting and use. Medical screenings should also be provided.

Air-Purifying Respirators – Air-purifying respirators provide the lowest level of protection. They are made for use only in atmospheres that have enough oxygen to sustain life (at least 19.5 percent). Air-purifying respirators use special filters and cartridges to remove specific gases, vapors, and particles from the air. The respirator cartridges contain charcoal, which absorbs certain toxic vapors and gases. When the wearer detects any taste or smell, the charcoal's absorption capacity has been reached and the cartridge can no longer remove the contaminant. The respirator filters remove particles such as dust, mists, and metal fumes by trapping them in the filter material. Filters should be changed when it becomes difficult to breathe. Depending on the contaminants, cartridges can be used alone or in combination with a filter/pre-filter and filter cover. Air-purifying respirators should be used for protection only against the types of contaminants listed on the filters and cartridges and on the National Institute for Occupational Safety and Health (NIOSH) approval label affixed to each respirator carton and replacement filter/cartridge carton. Respirator manufacturers typically classify air-purifying respirators into four groups:

- No-maintenance
- Low-maintenance
- Reusable
- Powered air-purifying respirators (PAPRs)

No-maintenance and low-maintenance respirators are typically used for residential or light commercial work that does not call for constant and heavy respirator use. No-maintenance respirators are typically half-mask respirators with permanently attached cartridges or filters. The entire respirator is discarded when the cartridges or filters are spent. Low-maintenance respirators generally are also half-mask respirators that use replaceable cartridges and filters. However, they are not designed for constant use.

Reusable respirators (*Figure 22*) are made in half-mask and full facepiece styles. These respirators require the replacement of cartridges, filters, and respirator parts. Their use also requires a complete respirator maintenance program.

Powered air-purifying respirators (PAPRs) are made in half-mask, full facepiece, and hood styles. They use battery-operated blowers to pull outside air through the cartridges and filters attached to the respirator. The blower motors can be either mask- or belt-mounted. Depending on the cartridges used, they can filter particulates, dusts, fumes, and mists, along with certain gases and vapors. PAPRs like the one shown in *Figure 23* have a belt-mounted, powered air-purifier unit connected to the mask by a breathing tube. The hood style of PAPR can be seen in *Figure 21*. Many models also have an audible and visual alarm that is activated when airflow falls below the required minimum level. This feature gives an immediate indication of a loaded filter or low battery-charge condition. Units with the blower mounted in the mask do not use the breathing tube and belt-mounted package. Since they are more compact, they typically have less filtration capacity.

Supplied-Air Respirators – Supplied-air respirators (*Figure 24*) provide a supply of breathable air for extended periods of time via a high-pressure hose that is connected to an external source, such as a compressor, compressed-air cylinder, or pump. They provide a higher level of protection in atmospheres where air-purifying respirators are

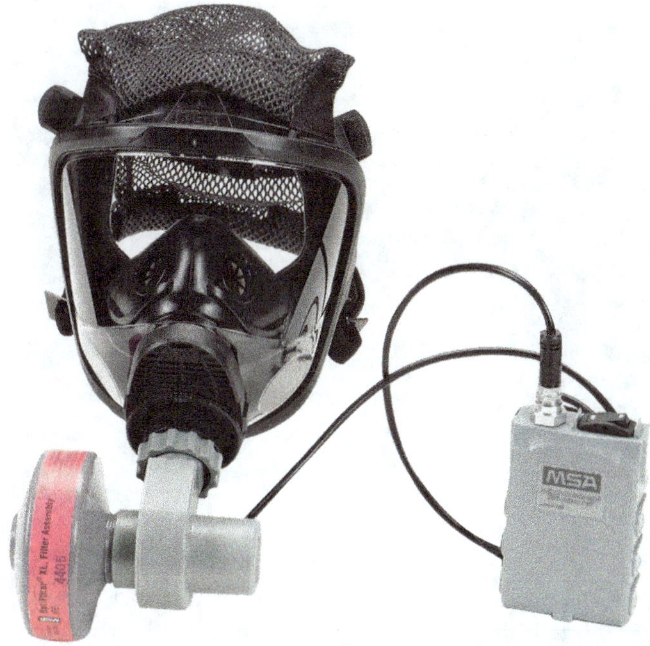

29101-15_F23.EPS

Figure 23 Full facepiece powered air-purifying respirator (PAPR).

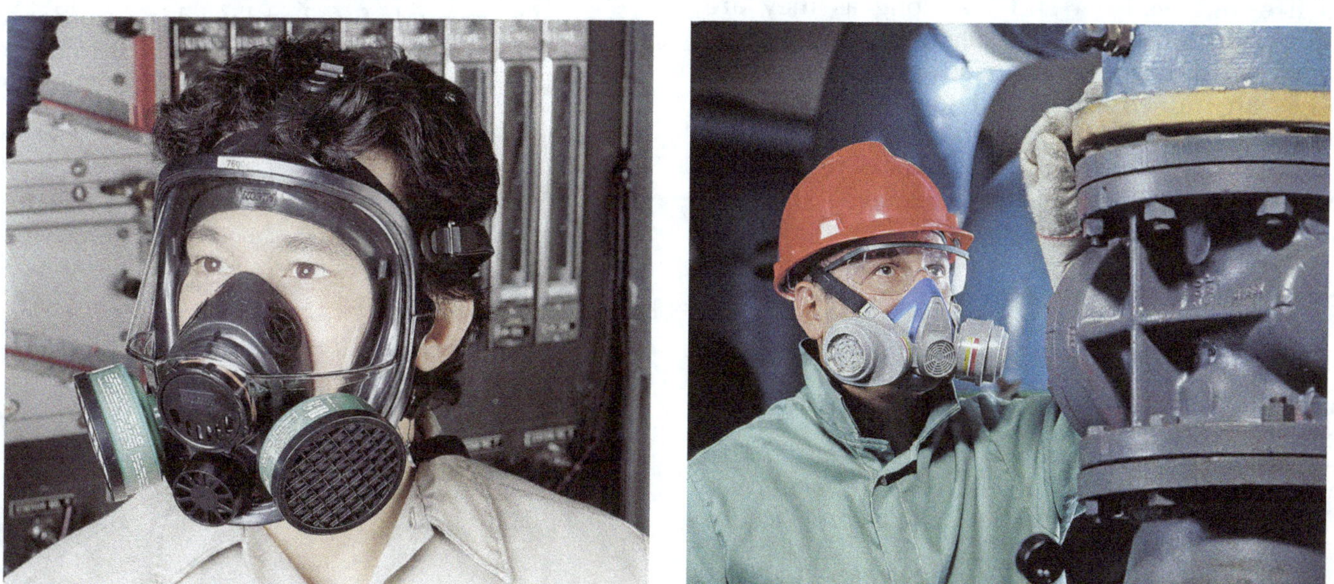

(A) FULL FACEPIECE MASK

(B) HALF MASK

29101-15_F22.EPS

Figure 22 Full-face and half-face air-purifying respirators.

29101-15_F24.EPS

Figure 24 Supplied-air respirator.

not capable of capturing the hazardous fumes. Supplied-air respirators are typically used in toxic atmospheres. Some can be used in atmospheres that are classified as immediately dangerous to life and health (IDLH) as long as they are equipped with an air cylinder for emergency escape. An atmosphere is considered IDLH if it poses an immediate hazard to life or produces immediate, irreversible, adverse effects on health. There are two types of supplied-air respirators: continuous-flow and pressure-demand. The unit shown in *Figure 24* is a continuous-flow model.

The continuous-flow supplied-air respirator provides air to the user in a constant stream. One or two hoses may be used to deliver the air from the air source to the facepiece. Unless the compressor or pump is especially designed to filter the air or a portable air-filtering system is used, the unit must be located where there is breathable air (Grade D or better as described in *Compressed Gas Association [CGA] Commodity Specification G-7.1*). Continuous-flow respirators are made with tight-fitting half-masks or full facepieces. They are also made with hoods. The flow of air to the user may be adjusted either at the air source (fixed flow) or on the unit's regulator (adjustable flow). Pressure-demand supplied-air respirators are similar to the continuous-flow type except that they supply air to the user's facepiece via a pressure-demand valve as the user inhales and fresh air is required. They typically have a two-position exhalation valve that allows the worker to switch between pressure-demand and negative-pressure modes to permit entry into, movement within, and exit from a work area.

Self-Contained Breathing Apparatus (SCBA) – SCBAs (*Figure 25*) can be used in oxygen-deficient atmospheres (below 19.5 percent oxygen); in poorly ventilated or confined spaces; and in IDLH atmospheres. These respirators provide a supply of air for 30 to 60 minutes from a compressed-air cylinder worn on the user's back.

An emergency escape breathing apparatus (EEBA) is a smaller version of an SCBA cylinder. EEBA units are used for escape from hazardous environments and generally provide only a five- to ten-minute supply of air. This type is usually kept in strategic locations where they can be used in the event of an emergency. In facilities that require that they be available, all workers must become familiar with their exact location.

3.6.3 Respiratory Program

In the United States, OSHA regulations, along with your company's procedures, must be followed when selecting the proper type of respirator for a particular job (*Figure 26*). Respirator selection is based on the contaminant present and its concentration level. This determination should be made by a trained safety specialist. The respirator must be properly fitted and used in accordance with the manufacturer's instructions. It must be worn during all times of exposure. Regardless of

29101-15_F25.EPS

Figure 25 Self-contained breathing apparatus (SCBA).

Figure 26 Use the right respirator for the job.

the kind of respirator needed, OSHA regulations require employers to have a respirator protection program consisting of the following:

- Standard operating procedures for selection and use
- Employee training
- Regular cleaning and disinfecting
- Sanitary storage
- Regular inspection
- Annual fit testing
- Pulmonary function testing

As an employee, you are responsible for wearing respiratory protection when needed. Both vapors and fumes can be eliminated in certain concentrations by the use of air-purifying devices, as long as oxygen levels are acceptable. Examples of fumes are smoke billowing from a fire, or the fumes generated when welding. Always check the cartridge on your respirator to make sure it is the correct type to use for the air conditions and contaminants found on the job site.

When selecting a respirator to wear while working with specific materials, first identify the hazardous ingredients contained in the material and their exposure levels, then choose the proper respirator to protect yourself at these levels. Always read the product's safety data sheet (SDS) or material safety data sheet (MSDS). It identifies the hazardous ingredients and should list the type of respirator and cartridge recommended for use with the product.

Limitations that apply to all half-mask (air-purifying) respirators are as follows:

- These respirators do not completely eliminate exposure to contaminants, but they will reduce the level of exposure to below-hazardous levels.
- These respirators do not supply oxygen and must not be used in areas where the oxygen level is below 19.5 percent.
- These respirators must not be used in areas where chemicals provide weak warning signs of their presence, such as a lack of taste or a unique odor.

If your breathing becomes difficult; if you become dizzy or nauseated; if you smell or taste the chemical; or if you have other noticeable effects, leave the area immediately, return to a fresh air area, and seek assistance.

All respirators are useless unless properly fit-tested to each individual. To obtain the best protection from your respirator, you must perform positive and negative fit checks each time you wear it. These fit checks must be done until you have obtained a good face seal.

To perform the positive fit check, do the following:

Step 1 Adjust the facepiece for the best fit, then adjust the head and neck straps to ensure good fit and comfort.

WARNING!

Do not overtighten the head and neck straps. Tighten them only enough to stop leakage. Overtightening can cause facepiece distortion and dangerous leaks.

Step 2 Block the exhalation valve with your hand or other material.

Step 3 Breathe out into the mask to increase the pressure inside.

Step 4 Check for air leakage around the edges of the facepiece.

Step 5 If the facepiece puffs out slightly for a few seconds, a good face seal has been obtained.

To perform a negative fit check, do the following:

Step 1 Block the inhalation valve with your hand or other material.

Step 2 Attempt to inhale.

Step 3 Check for air leakage around the edges of the facepiece.

Step 4 If the facepiece caves in slightly for a few seconds, a good face seal has been obtained.

3.7.0 Safety Data Sheets

Safety data sheets (SDSs) and material safety data sheets (MSDSs) are fact sheets prepared by the chemical manufacturer or importer. All manufacturers of potentially hazardous materials must provide detailed information regarding possible hazards resulting from the use of their product. An SDS/MSDS must be made available to anyone using the product or anyone working in the area where the product is in use. Each product used on a construction site must have an SDS/MSDS readily available for examination by workers. *Figure 27* shows a sample MSDS storage box where the information is available on site. An SDS/MSDS describes the substance, its hazards, safe handling, first aid, and emergency spill procedures.

The SDS replaces the MSDS that has been used for a number of years. The primary difference in the two is the arrangement of the entries; both provide the same basic information and have the same function. The Hazard Communication (HAZCOM) standard requires a new, internationally standardized format for the SDS. The sections of the SDS form include the following:

- *Section 1, Product identification, manufacturer information, recommended uses and restrictions*
- *Section 2, Hazard identification*
- *Section 3, Composition/information on ingredients*
- *Section 4, First aid measures*
- *Section 5, Firefighting information*
- *Section 6, Accidental release measures*
- *Section 7, Handling and storage*
- *Section 8, Exposure controls/personal protection*
- *Section 9, Physical and chemical properties*
- *Section 10, Stability and reactivity*
- *Section 11, Toxicological properties*
- *Section 12, Ecological properties*
- *Section 13, Disposal considerations*
- *Section 14, Transport information*
- *Section 15, Regulatory information*
- *Section 16, Other information*

29101-15_F27.EPS

Figure 27 SDS/MSDS storage box.

Respirator Inspection, Care, and Maintenance

A respirator must be clean and in good condition, and all of its parts must be in place for it to give you proper protection. Respirators must be cleaned every day. Failure to do so will limit their effectiveness and offer little or no protection. For example, suppose you wore the respirator yesterday and did not clean it. The bacteria from breathing into the respirator, plus the airborne contaminants that managed to enter the facepiece, will have made the inside of your respirator unsanitary. Continued use may cause you more harm than good. Remember, only a clean and complete respirator will provide you with the necessary protection. Follow these guidelines:

- Inspect the condition of your respirator before and after each use.
- Do not wear a respirator if the facepiece is distorted or if it is worn and cracked. You will not be able to get a proper face seal.
- Do not wear a respirator if any part of it is missing. Replace worn straps or missing parts before using.
- Do not expose respirators to excessive heat or cold, chemicals, or sunlight.
- Clean and wash your respirator after every time you use it. Remove the cartridge and filter, hand wash the respirator using mild soap and a soft brush, and let it air dry overnight.
- Sanitize your respirator each week. Remove the cartridge and filter, then soak the respirator in a sanitizing solution for at least two minutes. Thoroughly rinse with warm water and let it air dry overnight.
- Store the clean and sanitized respirator in its resealable plastic bag. Do not store the respirator face down. This will cause distortion of the facepiece.

The SDS format is gradually replacing the MSDS and is required for all new products. During the transition, you may find that the data sheets for many existing products are still in the MSDS format.

Figure 28 shows a sample SDS for solvent cement used with PVC pipe products. The most important things to look for on an SDS are the specific hazards, personal protection, handling procedures, and first aid information. Most SDSs have a 24-hour emergency-response number.

Look for the information you would need to use the cement described on the SDS form in *Figure 28*. First locate the hazards. Section 2 shows that the adhesive is flammable and an eye and skin irritant that can cause respiratory irritation, dizziness, and drowsiness.

Next, find out how to minimize these hazards. Section 7 gives general handling and storage information. It indicates that ventilation is needed to reduce hazardous vapors. This can be as simple as a fan in an open window. If ventilation is not enough, respiratory protection is needed. Section 8 tells you how to protect your eyes and skin.

Section 4 lists the first-aid measures for eye contact, skin contact, or inhalation. Section 5 explains fire hazards and firefighting measures. You now have the information you need in case of an emergency.

WARNING!

The SDSs must be kept in the work area and be readily accessible to all workers. Hazardous chemical producers are required to provide an SDS/MSDS in the official language(s) of each country or market where it is being produced and sold. Employers should ensure that workers are capable of reading and understanding the contents of an SDS/MSDS, and take steps to provide the information in a language they can understand. The company's safety officer or competent person should always review the SDS/MSDS before the hazardous material is used.

All SDSs/MSDSs must be kept on site. Ask your supervisor to tell you where they are located and to point out the sections that relate to your job. The health and safety of you and your co-workers depends on it. Companies often post these sheets on a bulletin board or put them in a convenient place near the work area.

At this writing, the MSDS is still in use for many products. The layout of the MSDS may vary somewhat between products and/or manufacturers. This is one of the reasons for the change to the SDS format. *Figure 29* shows a portion of an MSDS for Excalibur 7018-1 welding electrodes. As a welder, it is important to be familiar with the safety precautions related to the consumable products in use. Remember that some consumables contain potentially hazardous substances.

WELD-ON	**GHS SAFETY DATA SHEET**	Date Revised: **DEC 2011**
	WELD-ON® 705™ Low VOC Cements for PVC Plastic Pipe	Supersedes: **FEB 2010**

SECTION I - PRODUCT AND COMPANY IDENTIFICATION

PRODUCT NAME: **WELD-ON® 705™ Low VOC Cements for PVC Plastic Pipe**

PRODUCT USE: Low VOC Solvent Cement for PVC Plastic Pipe

SUPPLIER:

MANUFACTURER: IPS Corporation
17109 South Main Street, Carson, CA 90248-3127
P.O. Box 379, Gardena, CA 90247-0379
Tel. 1-310-898-3300

EMERGENCY: Transportation: CHEMTEL Tel. 800.255-3924, 813-248-0585 (International) **Medical:** Tel. 800.451.8346, 760.602.8703 3E Company (International)

SECTION 2 - HAZARDS IDENTIFICATION

GHS CLASSIFICATION:

Health		Environmental		Physical	
Acute Toxicity:	Category 4	Acute Toxicity:	None Known	Flammable Liquid	Category 2
Skin Irritation:	Category 3	Chronic Toxicity:	None Known		
Skin Sensitization:	NO				
Eye:	Category 2B				

GHS LABEL:  OR  **Signal Word: Danger** **WHMIS CLASSIFICATION:** CLASS B, DIVISION 2

Hazard Statements	Precautionary Statements
H225: Highly flammable liquid and vapor	P210: Keep away from heat/sparks/open flames/hot surfaces – No smoking
H319: Causes serious eye irritation	P261: Avoid breathing dust/fume/gas/mist/vapors/spray
H332: Harmful if inhaled	P280: Wear protective gloves/protective clothing/eye protection/face protection
H335: May cause respiratory irritation	P304+P340: IF INHALED: Remove victim to fresh air and keep at rest in a position comfortable for breathing
H336: May cause drowsiness or dizziness	P403+P233: Store in a well ventilated place. Keep container tightly closed
EUH019: May form explosive peroxides	P501: Dispose of contents/container in accordance with local regulation

SECTION 3 - COMPOSITION/INFORMATION ON INGREDIENTS

	CAS#	EINECS #	REACH Pre-registration Number	CONCENTRATION % by Weight
Tetrahydrofuran (THF)	109-99-9	203-726-8	05-2116297729-22-0000	25 - 50
Methyl Ethyl Ketone (MEK)	78-93-3	201-159-0	05-2116297728-24-0000	5 - 36
Cyclohexanone	108-94-1	203-631-1	05-2116297718-25-0000	15 - 30

All of the constituents of this adhesive product are listed on the TSCA inventory of chemical substances maintained by the US EPA, or are exempt from that listing.
* Indicates this chemical is subject to the reporting requirements of Section 313 of the Emergency Planning and Community Right-to-Know Act of 1986 (40CFR372).
# indicates that this chemical is found on Proposition 65's List of chemicals known to the State of California to cause cancer or reproductive toxicity.

SECTION 4 - FIRST AID MEASURES

Contact with eyes: Flush eyes immediately with plenty of water for 15 minutes and seek medical advice immediately.
Skin contact: Remove contaminated clothing and shoes. Wash skin thoroughly with soap and water. If irritation develops, seek medical advice.
Inhalation: Remove to fresh air. If breathing is stopped, give artificial respiration. If breathing is difficult, give oxygen. Seek medical advice.
Ingestion: Rinse mouth with water. Give 1 or 2 glasses of water or milk to dilute. Do not induce vomiting. Seek medical advice immediately.

SECTION 5 - FIREFIGHTING MEASURES

Suitable Extinguishing Media: Dry chemical powder, carbon dioxide gas, foam, Halon, water fog.
Unsuitable Extinguishing Media: Water spray or stream.
Exposure Hazards: Inhalation and dermal contact
Combustion Products: Oxides of carbon, hydrogen chloride and smoke

	HMIS	NFPA	
			0-Minimal
Health	2	2	1-Slight
Flammability	3	3	2-Moderate
Reactivity	0	0	3-Serious
PPE	B		4-Severe

Protection for Firefighters: Self-contained breathing apparatus or full-face positive pressure airline masks.

SECTION 6 - ACCIDENTAL RELEASE MEASURES

Personal precautions: Keep away from heat, sparks and open flame.
Provide sufficient ventilation, use explosion-proof exhaust ventilation equipment or wear suitable respiratory protective equipment.
Prevent contact with skin or eyes (see section 8).
Environmental Precautions: Prevent product or liquids contaminated with product from entering sewers, drains, soil or open water course.
Methods for Cleaning up: Clean up with sand or other inert absorbent material. Transfer to a closable steel vessel.
Materials not to be used for clean up: Aluminum or plastic containers

SECTION 7 - HANDLING AND STORAGE

Handling: Avoid breathing of vapor, avoid contact with eyes, skin and clothing.
Keep away from ignition sources, use only electrically grounded handling equipment and ensure adequate ventilation/fume exhaust hoods.
Do not eat, drink or smoke while handling.
Storage: Store in ventilated room or shade below 44°C (110°F) and away from direct sunlight.
Keep away from ignition sources and incompatible materials: caustics, ammonia, inorganic acids, chlorinated compounds, strong oxidizers and isocyanates.
Follow all precautionary information on container label, product bulletins and solvent cementing literature.

SECTION 8 - PRECAUTIONS TO CONTROL EXPOSURE / PERSONAL PROTECTION

EXPOSURE LIMITS:

Component	ACGIH TLV	ACGIH STEL	OSHA PEL	OSHA STEL:
Tetrahydrofuran (THF)	50 ppm	100 ppm	200 ppm	
Methyl Ethyl Ketone (MEK)	200 ppm	300 ppm	200 ppm	
Cyclohexanone	20 ppm	50 ppm	50 ppm	

Engineering Controls: Use local exhaust as needed.
Monitoring: Maintain breathing zone airborne concentrations below exposure limits.
Personal Protective Equipment (PPE):
Eye Protection: Avoid contact with eyes, wear splash-proof chemical goggles, face shield, safety glasses (spectacles) with brow guards and side shields, etc. as may be appropriate for the exposure.
Skin Protection: Prevent contact with the skin as much as possible. Butyl rubber gloves should be used for frequent immersion.
Use of solvent-resistant gloves or solvent-resistant barrier cream should provide adequate protection when normal adhesive application practices and procedures are used for making structural bonds.
Respiratory Protection: Prevent inhalation of the solvents. Use in a well-ventilated room. Open doors and/or windows to ensure airflow and air changes. Use local exhaust ventilation to remove airborne contaminants from employee breathing zone and to keep contaminants below levels listed above.
With normal use, the Exposure Limit Value will not usually be reached. When limits approached, use respiratory protection equipment.

29101-15_F28A.EPS

Figure 28A Solvent Cement SDS (sheet 1 of 2).

WELD-ON® **GHS SAFETY DATA SHEET**

WELD-ON® 705™ Low VOC Cements for PVC Plastic Pipe

Date Revised: **DEC 2011**
Supersedes: **FEB 2010**

SECTION 9 - PHYSICAL AND CHEMICAL PROPERTIES

Appearance:	Clear or gray, medium syrupy liquid		
Odor:	Ketone	**Odor Threshold:**	0.88 ppm (Cyclohexanone)
pH:	Not Applicable		
Melting/Freezing Point:	-108.5°C (-163.3°F) Based on first melting component: THF	**Boiling Range:**	66°C (151°F) to 156°C (313°F)
Boiling Point:	66°C (151°F) Based on first boiling component: THF	**Evaporation Rate:**	> 1.0 (BUAC = 1)
Flash Point:	-20°C (-4°F) TCC based on THF	**Flammability:**	Category 2
Specific Gravity:	0.9611 @23°C (73°F)	**Flammability Limits:**	**LEL:** 1.1% based on Cyclohexanone
Solubility:	Solvent portion soluble in water. Resin portion separates out.		**UEL:** 11.8% based on THF
Partition Coefficient n-octanol/water:	Not Available	**Vapor Pressure:**	129 mm Hg @ 20°C (68°F)based on THF
Auto-ignition Temperature:	321°C (610°F) based on THF	**Vapor Density:**	>2 (Air = 1)
Decomposition Temperature:	Not Applicable	**Other Data: Viscosity:**	Medium bodied
VOC Content:	When applied as directed, per SCAQMD Rule 1168, Test Method 316A,VOC content is: ≤ 510 g/l.		

SECTION 10 - STABILITY AND REACTIVITY

Stability: Stable
Hazardous decomposition products: None in normal use. When forced to burn, this product gives off oxides of carbon, hydrogen chloride and smoke.
Conditions to avoid: Keep away from heat, sparks, open flame and other ignition sources.
Incompatible Materials: Oxidizers, strong acids and bases, amines, ammonia

SECTION 11 - TOXICOLOGICAL INFORMATION

Likely Routes of Exposure: Inhalation, Eye and Skin Contact
Acute symptoms and effects:
Inhalation: Severe overexposure may result in nausea, dizziness, headache. Can cause drowsiness, irritation of eyes and nasal passages.
Eye Contact: Vapors slightly uncomfortable. Overexposure may result in severe eye injury with corneal or conjunctival inflammation on contact with the liquid.
Skin Contact: Liquid contact may remove natural skin oils resulting in skin irritation. Dermatitis may occur with prolonged contact.
Ingestion: May cause nausea, vomiting, diarrhea and mental sluggishness.
Chronic (long-term) effects: None known to humans
Toxicity:

	LD_{50}	LC_{50}
Tetrahydrofuran (THF)	Oral: 2842 mg/kg (rat)	Inhalation 3 hrs. 21,000 mg/m^3 (rat)
Methyl Ethyl Ketone (MEK)	Oral: 2737 mg/kg (rat), Dermal: 6480 mg/kg (rabbit)	Inhalation 8 hrs. 23,500 mg/m^3 (rat)
Cyclohexanone	Oral: 1535 mg/kg (rat), Dermal: 948 mg/kg (rabbit)	Inhalation 4 hrs. 8,000 PPM (rat)

Reproductive Effects	Teratogenicity	Mutagenicity	Embryotoxicity	Sensitization to Product	Synergistic Products
Not Established	Not Established	Not Established	Not Established	Not Established	Not Established

SECTION 12 - ECOLOGICAL INFORMATION

Ecotoxicity: None Known
Mobility: In normal use, emission of volatile organic compounds (VOC's) to the air takes place, typically at a rate of ≤ 510 g/l.
Degradability: Biodegradable
Bioaccumulation: Minimal to none.

SECTION 13 - WASTE DISPOSAL CONSIDERATIONS

Follow local and national regulations. Consult disposal expert.

SECTION 14 - TRANSPORT INFORMATION

Proper Shipping Name: Adhesives
Hazard Class: 3
Secondary Risk: None
Identification Number: UN 1133
Packing Group: PG II
Label Required: Class 3 Flammable Liquid
Marine Pollutant: NO

EXCEPTION for Ground Shipping
DOT Limited Quantity: Up to 5L per inner packaging, 30 kg gross weight per package.
Consumer Commodity: Depending on packaging, these quantities may qualify under DOT as "ORM-D".

TDG INFORMATION	
TDG CLASS:	FLAMMABLE LIQUID 3
SHIPPING NAME:	ADHESIVES
UN NUMBER/PACKING GROUP:	UN 1133, PG II

SECTION 15 - REGULATORY INFORMATION

Precautionary Label Information: Highly Flammable, Irritant
Symbols: F, Xi

Ingredient Listings: USA TSCA, Europe EINECS, Canada DSL, Australia AICS, Korea ECL/TCCL, Japan MITI (ENCS)

Risk Phrases:
R11: Highly flammable.
R20: Harmful by inhalation.
R36/37: Irritating to eyes and respiratory system.
R66: Repeated exposure may cause skin dryness or cracking
R67: Vapors may cause drowsiness and dizziness

Safety Phrases:
S9: Keep container in a well-ventilated place.
S16: Keep away from sources of ignition - No smoking.
S25: Avoid contact with eyes.
S26: In case of contact with eyes, rinse immediately with plenty of water and seek medical advice.
S33: Take precautionary measures against static discharges.
S46: If swallowed, seek medical advise immediately and show this container or label.

SECTION 16 - OTHER INFORMATION

Specification Information:
Department issuing data sheet: IPS, Safety Health & Environmental Affairs
E-mail address: <EHSinfo@ipscorp.com>

All ingredients are compliant with the requirements of the European Directive on RoHS (Restriction of Hazardous Substances).

Training necessary: Yes, training in practices and procedures contained in product literature.
Reissue date / reason for reissue: 12/14/2011 / Updated GHS Standard Format
Intended Use of Product: Solvent Cement for PVC Plastic Pipe

This product is intended for use by skilled individuals at their own risk. The information contained herein is based on data considered accurate based on current state of knowledge and experience. However, no warranty is expressed or implied regarding the accuracy of this data or the results to be obtained from the use thereof.

29101-15_F28B.EPS

Figure 28B Solvent Cement SDS (sheet 2 of 2).

LINCOLN ELECTRIC

Date:	1/21/2013	MSDS No.:	US-M292
Trade Name:	Excalibur 7018-1 MR		
Sizes:	All		
Supersedes:	8/29/2011		

MATERIAL SAFETY DATA SHEET

For Welding Consumables and Related Products

Conforms to Hazard Communication Standard 29CFR 1910.1200 Rev. October 1988

SECTION I - IDENTIFICATION

Manufacturer/ Supplier:	The Lincoln Electric Company 22801 St. Clair Avenue Cleveland, OH 44117-1199 (216) 481-8100	**Product Type:**	Covered Electrode
		Classification:	AWS E7018-1H4R

SECTION II - HAZARDOUS MATERIAL (1)

IMPORTANT!

This section covers the materials from which this product is manufactured. The fumes and gases produced during welding with the normal use of this product are covered by Section V; see it for industrial hygiene information.
CAS Number shown is representative for the ingredients listed. All ingredients listed may not be present in all sizes.
(1) The term "hazardous" in "Hazardous Materials" should be interpreted as a term required and defined in the Hazards Communication Standard and does not necessarily imply the existence of any hazard. All materials are listed on the TSCA inventory.

Ingredients:	CAS No.	Wt.%	TLV mg/m^3	PEL mg/m^3
Iron	7439-89-6	15	10*	15*
Limestone and/or calcium carbonate	1317-65-3	10	10*	15
Titanium dioxides	13463-67-7	< 5	10	15
Fluorides (as F)	7789-75-5	< 5	2.5	2.5
Silicates and other binders	1344-09-8	< 5	10*	15*
Manganese and/or manganese alloys and compounds (as Mn)*****	7439-96-5	< 5	0.2	5 (c)
Mineral silicates	1332-58-7	< 5	5**	5**
Silicon and/or silicon alloys and compounds (as Si)	7440-21-3	1	10*	15*
Zirconium alloys and compounds (as Zr)	12004-83-0	1	5	5
Cellulose and other carbohydrates	65996-61-4	0.5	10*	15*
Quartz	14808-60-7	< 0.5	#0.025**	#0.1**
Molybdenum alloys (as Mo)	7439-98-7	< 0.5	10	10
Lithium compounds (as Li)	554-13-2	< 0.5	10*	15*
Carbon steel core wire	7439-89-6	55	10*	15*

Supplemental Information:

(*) Not listed. The OSHA PEL for nuisance particles is 15 milligrams per cubic meter. The ACGIH guideline for total particulate is 10 milligrams per cubic meter. PEL value for iron oxide is 10 milligrams per cubic meter. TLV value for iron oxides is 5 milligrams per cubic meter.

(**) As respirable dust.

(*****) Subject to the reporting requirements of Sections 311, 312, and 313 of the Emergency Planning and Community Right-to-Know Act of 1986 and of 40CFR 370 and 372.

(c) Value is for manganese fume. Present PEL is 5 milligrams per cubic meter (ceiling value). Values proposed by OSHA in 1989 were 1.0 milligrams per cubic meter TWA and 3.0 milligrams per cubic meter STEL (Short Term Exposure Limit).

(#) Crystalline silica (quartz) is on the IARC (International Agency for Research on Cancer) and NTP (National Toxicology Program) lists as posing a carcinogenic risk to humans.

SECTION III - HAZARD DATA

Non Flammable; Welding arc and sparks can ignite combustibles and flammable products. See Z49.1 referenced in Section VI. Product is inert, no special handling or spill procedures required. Not regulated by DOT.

Rev 1/13

(CONTINUED ON SIDE TWO)

29101-15_F29A.EPS

Figure 29A Welding electrode MSDS. (sheet 1 of 2)

Product: Excalibur 7018-1 MR

Date: 1/21/2013

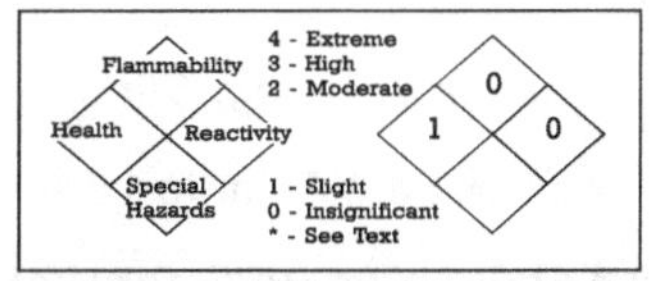

SECTION IV - HEALTH HAZARD DATA

Threshold Limit Value: The ACGIH recommended general limit for Welding Fume NOS - (Not Otherwise Specified) is 5 mg/m^3. ACGIH-1999 preface states that the TLV-TWA should be used as guides in the control of health hazards and should not be used as fine lines between safe and dangerous concentrations. See Section V for specific fume constituents which may modify this TLV. Threshold Limit Values are figures published by the American Conference of Government Industrial Hygienists. Units are milligrams per cubic meter of air.

Effects of Overexposure: Electric arc welding may create one or more of the following health hazards:
Fumes and Gases can be dangerous to your health. Common entry is by inhalation. Other possible routes are skin contact and ingestion.

Short-term (acute) overexposure to welding fumes may result in discomfort such as metal fume fever, dizziness, nausea, or dryness or irritation of nose, throat, or eyes. May aggravate pre-existing respiratory problems (e.g. asthma, emphysema).

Long-term (chronic) overexposure to welding fumes can lead to siderosis (iron deposits in lung) and may affect pulmonary function. Manganese overexposure can affect the central nervous system, resulting in impaired speech and movement. Bronchitis and some lung fibrosis have been reported. Repeated exposure to fluorides may cause excessive calcification of the bone and calcification of ligaments of the ribs, pelvis and spinal column. May cause skin rash. Titanium dioxide is listed by the IARC (International Agency for Research on Cancer) as a Group 2B carcinogen (possibly carcinogenic to humans based on animal studies). Respiratory exposure to the crystalline silica present in this welding electrode is not anticipated during normal use. Respiratory overexposure to airborne crystalline silica is known to cause silicosis, a form of disabling pulmonary fibrosis which can be progressive and may lead to death. Crystalline silica is on the IARC (International Agency for Research on Cancer) and NTP (National Toxicology Program) lists as posing a cancer risk to humans. WARNING: This product, when used for welding or cutting, produces fumes or gases which contain chemicals known to the State of California to cause birth defects and, in some cases, cancer. (California Health & Safety Code Section 25249.5 et seq.)

Arc Rays can injure eyes and burn skin. *Skin cancer has been reported.*
Electric Shock can kill. If welding must be performed in damp locations or with wet clothing, on metal structures or when in cramped positions such as sitting, kneeling or lying, or if there is a high risk of unavoidable or accidental contact with workpiece, use the following equipment: Semiautomatic DC Welder, DC Manual (Stick) Welder, or AC Welder with Reduced Voltage Control.

Emergency and First Aid Procedures: Call for medical aid. Employ first aid techniques recommended by the American Red Cross.
IF BREATHING IS DIFFICULT give oxygen. IF NOT BREATHING employ CPR (Cardiopulmonary Resuscitation) techniques.
IN CASE OF ELECTRICAL SHOCK, turn off power and follow recommended treatment. In all cases call a physician.

SECTION V - REACTIVITY DATA

Hazardous Decomposition Products: Welding fumes and gases cannot be classified simply. The composition and quantity of both are dependent upon the metal being welded, the process, procedure and electrodes used.

Other conditions which also influence the composition and quantity of the fumes and gases to which workers may be exposed include: coatings on the metal being welded (such as paint, plating, or galvanizing), the number of welders and the volume of the worker area, the quality and amount of ventilation, the position of the welder's head with respect to the fume plume, as well as the presence of contaminants in the atmosphere (such as chlorinated hydrocarbon vapors from cleaning and degreasing activities.)

When the electrode is consumed, the fume and gas decomposition products generated are different in percent and form from the ingredients listed in Section II. Decomposition products of normal operation include those originating from the volatilization, reaction, or oxidation of the materials shown in Section II, plus those from the base metal and coating, etc., as noted above.

Reasonably expected fume constituents of this product would include: Primarily iron oxide and fluorides; secondarily complex oxides of manganese, potassium, silicon, sodium, and titanium.

Maximum fume exposure guideline for this product (based on manganese content) is 4.0 milligrams per cubic meter.

Gaseous reaction products may include carbon monoxide and carbon dioxide. Ozone and nitrogen oxides may be formed by the radiation from the arc.

Determine the composition and quantity of fumes and gases to which workers are exposed by taking an air sample from inside the welder's helmet if worn or in the worker's breathing zone. Improve ventilation if exposures are not below limits. See ANSI/AWS F1.1, F1.2, F1.3 and F1.5, available from the American Welding Society, 550 N.W. LeJeune Road, Miami, FL 33126.

SECTION VI AND VII
CONTROL MEASURES AND PRECAUTIONS FOR SAFE HANDLING AND USE

Read and understand the manufacturer's instruction and the precautionary label on the product. Request Lincoln Safety Publication E205. See American National Standard Z49.1, "Safety In Welding, Cutting and Allied Processes" published by the American Welding Society, 550 N.W. LeJeune Road, Miami, FL, 33126 (both available for free download at http://www.lincolnelectric.com/community/safety/) and OSHA Publication 2206 (29CFR1910), U.S. Government Printing Office, Superintendent of Documents, P.O. Box 371954, Pittsburgh, PA 15250-7954 for more details on many of the following:

Ventilation: Use enough ventilation, local exhaust at the arc, or both to keep the fumes and gases from the worker's breathing zone and the general area. Train the welder to keep his head out of the fumes. *Keep exposure as low as possible.*

Respiratory Protection: Use respirable fume respirator or air supplied respirator when welding in confined space or general work area when local exhaust or ventilation does not keep exposure below TLV.

Eye Protection: Wear helmet or use face shield with filter lens shade number 12 or darker. Shield others by providing screens and flash gog gles.

Protective Clothing: Wear hand, head, and body protection which help to prevent injury from radiation, sparks and electrical shock. See Z49.1. At a minimum this includes welder's gloves and a protective face shield, and may include arm protectors, aprons, hats, shoulder protection, as well as dark substantial clothing. Train the welder not to permit electrically live parts or electrodes to contact skin . . . or clothing or gloves if they are wet. Insulate from work and ground.

Disposal Information: Discard any product, residue, disposable container, or liner as ordinary waste in an environmentally acceptable manner according to Federal, State and Local Regulations unless otherwise noted. No applicable ecological information available.

29101-15_F29B.EPS

Figure 29B Welding electrode MSDS. (sheet 2 of 2)

Additional Resources

ANSI/AWS Z49.1, Safety in Welding, Cutting, and Allied Processes. Miami, FL: American Welding Society.

Arc Welding Safety E205. Latest Edition. Cleveland, OH: Lincoln Electric.

Personal Protective Equipment (PPE) for Welding and Cutting. Miami, FL. American Welding Society.

Safety Topics in Welding, Cutting, and Brazing. **www.osha.gov**.

AWS F32M/F32, Ventilation Guide for Weld Fumes. Latest Edition. Miami, FL: American Welding Society.

3.0.0 Section Review

1. The AWS standard for welding and cutting safety is ____ .
   a. *Z49.1*
   b. *CFR* 1929
   c. *CFR* 1910
   d. *Z45.5*

2. The primary purpose of a welding screen is to ____ .
   a. keep non-welders from watching welders at work
   b. protect others from the glare of the welding arc
   c. prevent insects from affecting the welding arc
   d. keep nearby material from catching fire

3. When a worker is assigned the responsibility of providing a fire watch, the worker ____ .
   a. cannot do another task
   b. can retrieve additional consumables for the welder from the shop
   c. can also weld or cut to remain productive
   d. may conduct weld testing

4. Argon is used as a ____ .
   a. breathing gas
   b. cutting fuel gas
   c. solvent
   d. purge gas

5. Which of the following is a *correct* statement about electrical hazards associated with welding?
   a. The jaws of the electrode holder do not carry electrical current.
   b. GFCIs should not be used when welding.
   c. Electrical shocks can cause severe burns.
   d. Welding machines typically operate on 120-volt house current.

6. Which of these metals is *least* likely to produce toxic fumes when heated?
   a. Manganese
   b. Hexavalent chrome
   c. Carbon steel
   d. Galvanized steel

7. In which section of an SDS are hazards identified?
   a. Section 1
   b. Section 2
   c. Section 5
   d. Section 12

SUMMARY

Welders join metal using a high-intensity electric arc to melt and fuse the base metals. Welders may work on critical structures such as pipelines, high-rise buildings, ships, and bridges, so the trade requires a high level of skill and concentration, which can only be achieved through extensive training and practice. Welders also are called upon to cut and gouge metal using equipment that produces a high-intensity flame. The most common cutting method is oxyfuel cutting, in which oxygen is mixed with a fuel gas such as acetylene to produce the cutting flame.

Safety is everyone's responsibility. Proper clothing, footwear, and eye protection are required for safe welding and cutting. Workers who fail to comply with safety rules are subject to dismissal. All welding shops must have established plans for dealing with accidents. Take the time to learn the proper procedures for accident response and reporting before you need to respond to an emergency.

Accidents are very harmful for both employees and employers, and they are often caused by poor behavior and unsafe conditions. Most accidents can be prevented. By knowing and avoiding the behaviors that cause accidents and by keeping working conditions safe, it may be possible to avoid injuries and reduce hazards.

Trying to bluff your way through a job you do not understand is asking for trouble. Even if you think you know the correct procedures, a review may bring out an important part of the job that you may have forgotten. Do not be afraid to ask questions. The responses you receive may help new or less experienced co-workers get answers to questions they may be embarassed to ask.

Materials, fumes, welding radiation exposure, and storing and handling cylinders present particular hazards for the welder. Developing an attitude of safety is an excellent way for every worker to avoid or reduce all of these hazards. Practicing good safety attitudes means that you do the following:

- Report all unsafe conditions and behaviors immediately.
- Keep work areas clean and orderly at all times.
- Immediately report all accidents and injuries, no matter how minor.
- Be certain you completely understand the instructions given before starting work.
- Know how and where medical help may be obtained.
- Wear the required protective devices when working in a hazardous operation area.
- Do not use alcohol or drugs. If you are ill and must take prescribed medication, notify your supervisor immediately.

Review Questions

1. The most common type of welding is ____ .
   a. GTAW
   b. GMAW
   c. FCAW
   d. SMAW

2. A competency-based curriculum is one in which ____ .
   a. the focus is on classroom training
   b. students must demonstrate knowledge and skills
   c. instructors must have four-year degrees
   d. all the training is done on the job

3. Youth apprenticeship programs are available that allow students to begin their apprenticeship or craft training ____ .
   a. in middle school
   b. in high school
   c. instead of their traditional subjects
   d. instead of English classes

4. Body protection while welding is best provided by wearing ____ .
   a. pants with cuffs
   b. polyester clothing
   c. cotton clothing
   d. loose-fitting shirts

5. Which of the following is a *correct* statement?
   a. It is not necessary to wear other eye protection if you are wearing a welding helmet.
   b. The same shade of lens can be used for almost all welding and cutting tasks.
   c. The recommended shade for electric arc operations is 9–14.
   d. Safety glasses with side shields should not be used because they block peripheral vision.

6. Accidents are more likely for employees in the age group ____ .
   a. 18 and under
   b. 19 to 25
   c. 26 to 45
   d. 60 and over

7. A good work area safety practice is ____ .
   a. welding near an open window
   b. welding over wooden floors
   c. using cardboard to control welding/grinding sparks
   d. writing HOT on hot metal before leaving it unattended

8. A hot work permit ____ .
   a. authorizes the performance of work potentially posing a fire hazard
   b. promotes development of standard fire safety guidelines
   c. records unsafe conditions at a job site
   d. helps the manager keep records of hazardous spaces

9. Cutting operations should never be performed without which of these items in the area?
   a. A bucket of sand
   b. A bucket of water
   c. A fire watch
   d. A fire hose

10. A confined space is one that ____ .
    a. has a flammable atmosphere
    b. has unrestricted means of entry or exit
    c. is designed for continuous employee occupancy
    d. is large enough that a worker can enter and work

11. An atmosphere is considered oxygen deficient when the oxygen level is below ____ .
    a. 19.5 percent
    b. 20 percent
    c. 21.5 percent
    d. 23.5 percent

12. The potential for electric shock ____ .
    a. decreases when the skin is damp
    b. remains the same when the skin is damp
    c. increases when the skin is damp
    d. decreases when the skin is cut

13. A potentially toxic coating that is used in galvanized metal is _____ .
    a. zinc
    b. chromium
    c. barium
    d. manganese

14. To make sure a respirator provides proper protection, what must be done each time it is worn?
    a. Only a positive fit check must be performed.
    b. Positive and negative fit checks must be performed.
    c. Only a negative fit check must be performed.
    d. The head and neck straps must be tightened as much as possible

15. An SDS is a form used to _____ .
    a. file a worker's compensation claim with an insurance company
    b. list the contents, hazards, and precautions that pertain to a chemical or material
    c. record unsafe conditions that exist at a job site
    d. recommend changes to an employer safety program

Trade Terms Quiz

Fill in the blank with the correct term that you learned from your study of this module.

1. An electric circuit that is connected to the earth is said to be ________ .
2. Helium or argon is used as a(n) ________ in some welding and cutting operations.
3. The cloud of gases and other materials that rises during the welding process is called the ________ .
4. An irregular heart rhythm characterized by a fluttering heartbeat is called ________.
5. A condition that poses an immediate or delayed threat to life or health is referred to as ________ .
6. When metal parts are joined together to form an electrically conductive path, they are said to be ________ .
7. When skin is exposed to intense ultraviolet light, it can cause a(n) ________ .
8. When carbon is added to steel to increase hardness, the result is ________ .
9. Steel dipped in zinc is referred to as ________ .
10. If your eyes are exposed to intense ultraviolet light, the injury is referred to as a(n) ________ .
11. When nitrogen is used to drive oxygen away from a weld site, it is called a(n) ________ .
12. Chromium, manganese, and other materials are added to steel to produce ________ .
13. The flow of an electrical current across an air gap is called a(n) ________ .
14. The point from which a welding arc is produced is called a(n) ________ .
15. A(n) ________ is a material used to dissolve or prevent the formation of oxides and other undesirable substances on a weld joint.
16. The invisible rays produced by arc welding that can cause burns to the eye are ________ .

Trade Terms

Arc burn
Bonded
Carbon steel
Electric arc
Electrically grounded
Electrode
Flash burn
Flux
Fume plume
Galvanized steel
Immediate danger to life and health (IDLH)
Purge gas
Shielding gas
Stainless steel
Ultraviolet (UV) radiation
Ventricular fibrillation

Cornerstone of Craftsmanship

Bill D. Cherry

Zachry Industrial, Inc.
Manager—Weld Testing

How did you choose a career in the construction industry?
I liked looking at a finished product that resulted from my efforts and knowing that it made an impact on people's lives. Most, if not all, of the structures I welded are still in use today. This gives me a strong sense of accomplishment.

Who inspired you to enter the industry?
I was inspired by the father of a school friend who owned a private construction company.

How important are education and training in construction?
They help me maintain my current position and also to teach the trades I have experience with.

What kinds of work have you done in your career?
After graduating from high school, I started out in the road-boring business. This is where I learned how to weld. My next job was at a coal-fired power plant in San Antonio, Texas where I was exposed to many different welding processes and procedures. After that, I was given the opportunity to learn welding inspection and training. I was able to achieve Certified Welding Instructor (CWI), Certified Welding Educator (CWE), and NCCER Master Welding Instructor certifications. I have also trained students to become welders since that time.

Tell us about your present job.
I primarily test welders for our construction and maintenance jobsites. I also have been training welders for 25 years.

What factors have contributed most to your success?
I attribute my success to one person, who gave me the opportunity to advance my career. Without that one individual, I might still be in an unfulfilling career. One person can make a big difference in your life. I started as a welding helper at Zachary Industrial. I eventually became a certified welder, and then a welding inspector, and now manager of weld testing.

What types of training have you been through?
The majority of my welding skills were self-taught. I did receive formal training to become a Certified Welding Instructor (CWI) and have been through the NCCER Instructor Certification Training Program (ICTP). Of course, on-the-job training also played a major role.

Would you suggest construction as a career to others? Why?
Absolutely. Obviously it is not for everyone, but it can be a very rewarding and lucrative career.

What advice would you give to those new to the field?
My advice to those new to the field is never stop learning and always strive to better yourself. Never just sit back and be content with what you are currently doing. The opportunities are always there. You have to actively pursue them.

Tell us an interesting career-related fact or accomplishment.
I hold the only known certification from NCCER as a Master Welding Instructor. Also, I have done many different types of construction. I have administered more than 10,000 weld tests to date. Another thing that means a lot to me is having former students stop by my shop just to say hello and let me know how well they are doing in the welding field.

How do you define craftsmanship?
I define it as the ability to do quality work the first time around.

Trade Terms Introduced in This Module

Arc burn: Burn to the skin produced by brief exposure to intense radiant heat and ultraviolet light.

Bonded: The permanent joining of metallic parts to form an electrically conductive path that will assure electrical continuity and the capacity to safely conduct any current likely to be imposed on it.

Carbon steel: An alloy of iron combining iron and usually less than 1 percent carbon to provide hardness.

Electric arc: The flow of an electrical current across an air gap or gaseous space.

Electrically grounded: Connected to Earth or to some conducting body that serves in place of the earth.

Electrode: The point from which a welding arc is produced.

Flash burn: Burns to the eyes sometimes called welder's flash; caused by exposure to intense radiant heat and ultraviolet light.

Flux: A material used to dissolve or prevent the formation of oxides and other undesirable substances on a weld joint.

Fume plume: The fumes, gases, and particles from the consumables, base metal, and base metal coating during the welding process.

Galvanized steel: Carbon steel dipped in zinc to inhibit corrosion.

Immediate danger to life and health (IDLH): As defined by OSHA, this term refers to an atmosphere that poses an immediate threat to life, would cause irreversible adverse health effects, or would impair an individual's ability to escape from a dangerous atmosphere.

Purge gas: An inert gas, such as nitrogen, used to drive oxygen away from a weld site.

Shielding gas: A gas such as argon, helium, or carbon dioxide used to protect the welding electrode wire from contamination in GMAW and FCAW welding.

Stainless steel: An iron-based alloy usually containing at least 11 percent chromium.

Ultraviolet (UV) radiation: Invisible rays capable of causing burns. UV rays from the sun are the causes of sunburn.

Ventricular fibrillation: Irregular heart rhythm characterized by a rapid fluttering heartbeat.

Additional Resources

This module presents thorough resources for task training. The following resource material is suggested for further study.

ANSI/AWS Z49.1, Safety in Welding, Cutting, and Allied Processes. Miami, FL: American Welding Society.

Arc Welding Safety E205. Latest Edition. Cleveland, OH: Lincoln Electric.

Personal Protective Equipment (PPE) for Welding and Cutting. Miami, FL: American Welding Society.

Safety Topics in Welding, Cutting, and Brazing. **www.osha.gov**.

AWS F32M/F32, Ventilation Guide for Weld Fumes. Latest Edition. Miami, FL: American Welding Society.

Figure Credits

Justin Poland, Robins and Morton, Module Opener, Figure 1A

The Lincoln Electric Company, Cleveland, OH, USA, Figures 1B, 1C, 2A, 2B, 4, 5, 6A–6C, 18–21, 29, SA02

Zachry Industrial, Inc., Figure 2C

Topaz Publications, Inc., Figures 6D–6F, 7, 8, 9A–9E, 10, 12

John Tillman Company, SA01

AWS F2.2:2001, Lens Shade Selector, reproduced with permission from the American Welding Society (AWS), Miami, FL, USA, Table 1

Sellstrom Manfacturing, Figures 9F, 11

North Safety Products USA, Figure 22A

MSA The Safety Company, Figures 22B, 23–25

Photo courtesy of Justrite Mfg. Co. LLC, Figure 27

Weld-On Adhesives, Inc., a division of IPS Corporation, Figure 28

Section Review Answer Key

Answer	Section Reference	Objective
Section One		
1. b	1.1.0	1a
2. a	1.2.2	1b
Section Two		
1. d	2.1.1	2a
2. a	2.2.1	2b
Section Three		
1. a	3.1.0	3a
2. b	3.2.0	3b
3. a	3.3.0	3c
4. d	3.4.0	3d
5. c	3.5.1	3e
6. c	3.6.0	3f
7. b	3.7.0	3g

NCCER CURRICULA — USER UPDATE

NCCER makes every effort to keep its textbooks up-to-date and free of technical errors. We appreciate your help in this process. If you find an error, a typographical mistake, or an inaccuracy in NCCER's curricula, please fill out this form (or a photocopy), or complete the online form at **www.nccer.org/olf**. Be sure to include the exact module ID number, page number, a detailed description, and your recommended correction. Your input will be brought to the attention of the Authoring Team. Thank you for your assistance.

Instructors – If you have an idea for improving this textbook, or have found that additional materials were necessary to teach this module effectively, please let us know so that we may present your suggestions to the Authoring Team.

NCCER Product Development and Revision

13614 Progress Blvd., Alachua, FL 32615

Email: curriculum@nccer.org
Online: www.nccer.org/olf

❑ Trainee Guide ❑ Lesson Plans ❑ Exam ❑ PowerPoints Other ________________

Craft / Level: Copyright Date:

Module ID Number / Title:

Section Number(s):

Description:

Recommended Correction:

Your Name:

Address:

Email: Phone:

Plasma Arc Cutting

Overview

Welders must be familiar with the task of plasma arc cutting. This module presents information related to the plasma arc cutting process, including safety procedures, setup, gas types, flow rates, and equipment.

Module 29103

Trainees with successful module completions may be eligible for credentialing through the NCCER Registry. To learn more, go to **www.nccer.org** or contact us at **1.888.622.3720**. Our website has information on the latest product releases and training, as well as online versions of our *Cornerstone* magazine and Pearson's product catalog.

Your feedback is welcome. You may email your comments to **curriculum@nccer.org**, send general comments and inquiries to **info@nccer.org**, or fill in the User Update form at the back of this module.

This information is general in nature and intended for training purposes only. Actual performance of activities described in this manual requires compliance with all applicable operating, service, maintenance, and safety procedures under the direction of qualified personnel. References in this manual to patented or proprietary devices do not constitute a recommendation of their use.

29103 V5

29103
PLASMA ARC CUTTING

Objectives

When you have completed this module, you will be able to do the following:

1. Explain plasma arc cutting processes and identify related safety precautions.
   a. Describe the plasma arc cutting processes.
   b. Identify safety practices related to plasma arc cutting.
2. Identify and describe plasma arc cutting equipment.
   a. Identify and describe plasma arc power units.
   b. Identify and describe plasma arc torches and accessories.
   c. Identify and describe plasma arc cutting gases and gas control devices.
3. Describe how to set up, safely operate, and care for plasma arc cutting equipment.
   a. Describe how to set up plasma arc cutting equipment and the adjacent work area.
   b. Describe how to safely operate plasma arc cutting equipment.
   c. Describe how to care for plasma arc cutting equipment.

Performance Tasks

Under the supervision of your instructor, you should be able to do the following:

1. Set up plasma arc cutting equipment.
2. Set the amperage and gas pressures or flow rates for the type and thickness of metal to be cut using plasma arc equipment.
3. Square-cut metal using plasma arc equipment.
4. Bevel-cut metal using plasma arc equipment.
5. Pierce and cut slots in metal using plasma arc equipment.
6. Dismantle and store the equipment.

Trade Terms

Amperage
Dross
Duty cycle
Phase
Plasma
Potential
Solenoid valve

Industry Recognized Credentials

If you are training through an NCCER-accredited sponsor, you may be eligible for credentials from NCCER's Registry. The ID number for this module is 29103. Note that this module may have been used in other NCCER curricula and may apply to other level completions. Contact NCCER's Registry at 888.622.3720 or go to **www.nccer.org** for more information.

Contents

Topics to be presented in this module include:

Figures and Tables

Section One

1.0.0 Plasma Arc Cutting Processes

Objective

Explain plasma arc cutting processes and identify related safety precautions.

a. Describe plasma arc cutting processes.
b. Identify safety practices related to plasma arc cutting.

Trade Terms

Dross: A waste byproduct of molten metal.

Plasma: A fourth state of matter (not solid, liquid, or gas) created by heating a gas to such a high temperature that it boils electrons off the gas molecules (ionization).

Plasma arc cutting (PAC) uses a jet of plasma to pierce, cut, and gouge metal. The plasma is created by superheating gas in an electric arc. The temperature of the plasma jet is hot enough to melt almost any metal. The process can be set up for manual or mechanized operation (*Figure 1*) and, in most cases, is faster and more efficient than any other cutting method. PAC offers the following advantages:

- Cuts both ferrous and nonferrous metals
- Produces minimal dross
- Creates a minimal heat-affected zone
- Causes little or no distortion
- Possesses high cutting speeds
- Produces a very narrow kerf

The plasma arc cutting setup (*Figure 2*) requires an electrical power supply and a cutting gas, which can be compressed air. Depending on the equipment, a separate shielding or cooling gas may also be required. PAC torches convert the cutting gas into plasma by passing the gas through an electric arc inside the torch. The torch has a relatively small opening (orifice), which constricts the arc and high-pressure gas flow. This results in the plasma stream exiting the torch as a supersonic jet hotter than any flame. Depending on the current flow, the plasma cutting jet may reach temperatures of 30,000°F (16,650°C) or higher. When the jet contacts metal, it transfers its tremendous heat, causing the metal to melt instantly. The high-velocity jet blasts the molten metal away, forming a hole, groove, or gouge. Be advised that this high-velocity jet blast can carry molten material 40 feet (12 m) away from the workpiece.

(A) MANUAL PLASMA ARC CUTTING

(B) MECHANIZED PLASMA ARC CUTTING

29103-15_F01.EPS

Figure 1 Plasma arc cutting (PAC).

1.1.0 Plasma Arc Cutting Process

There are two types of plasma arc cutting processes: transferred arc and non-transferred arc.

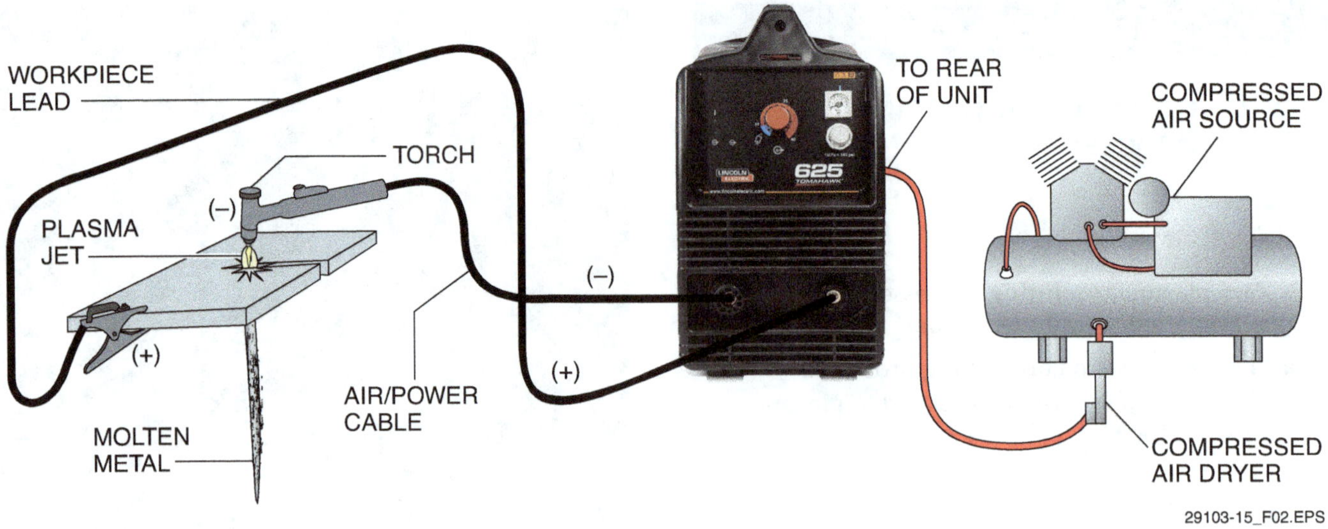

Figure 2 Plasma arc cutting setup diagram.

The transferred arc process is the most common and is used to cut materials that are electrically conductive, such as metals. The non-transferred arc process is less common and is used to cut materials that are not electrically conductive, such as ceramics and concrete.

1.1.1 Transferred Arc Process

In the transferred-arc process, the workpiece is part of the electrical circuit. The initial arc is established between the electrode and the workpiece via an arc transfer process. A number of different arc-initiation methods are used for the transfer process. These include contact starting, discharge-starting to the workpiece, and discharge-starting using a pilot arc. Of these methods, discharge-starting using a pilot arc is the most common. In this method, a low-voltage, low-current, high-frequency (HF) arc is established between the electrode and the torch nozzle inside the torch. This small arc between the torch electrode and torch nozzle is called a pilot arc. The pilot arc is used to ionize the air jet, resulting in plasma generation. When the torch is near the workpiece, a higher-voltage, current-cutting arc is transferred and established between the torch electrode and the work via the plasma. The higher-voltage, current-cutting arc maintains the plasma jet and the pilot arc is then shut off. For the transferred-arc process to work, one lead of the power supply (the positive lead) must be connected to the metal to be cut. *Figure 3* shows a schematic of the transferred-arc process.

1.1.2 Non-Transferred Arc Process

In the non-transferred arc process, the entire arc is established within the torch, between the torch electrode and the torch nozzle. The material being cut is not electrically connected to the arc circuit, since it is typically non-conductive material. The non-transferred arc process is not normally found in facilities that cut only metal. *Figure 4* shows a schematic of the non-transferred arc process.

1.2.0 Safety Practices

The following is a summary of safety procedures and practices that must be observed while cutting. Keep in mind that this is a summary only. Complete safety coverage is provided in the *Welding Safety* module. If you have not completed that module, do so before continuing. Above all, be sure to wear appropriate protective clothing and equipment when welding or cutting.

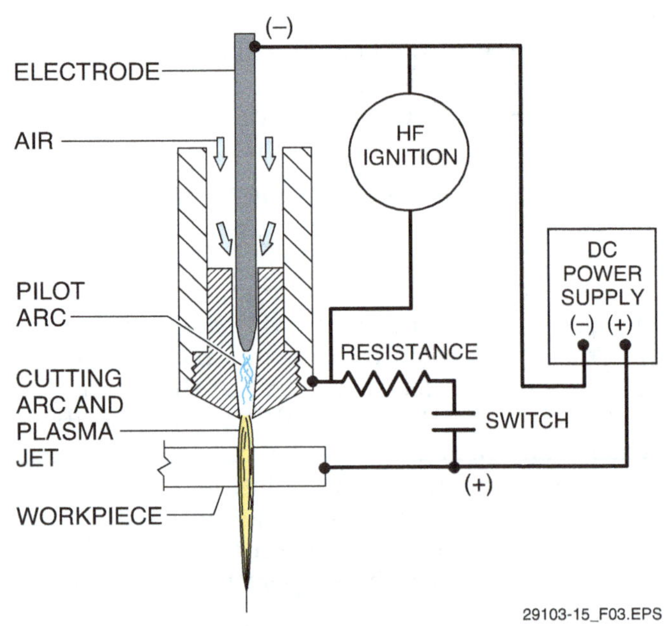

Figure 3 Schematic of the transferred-arc process.

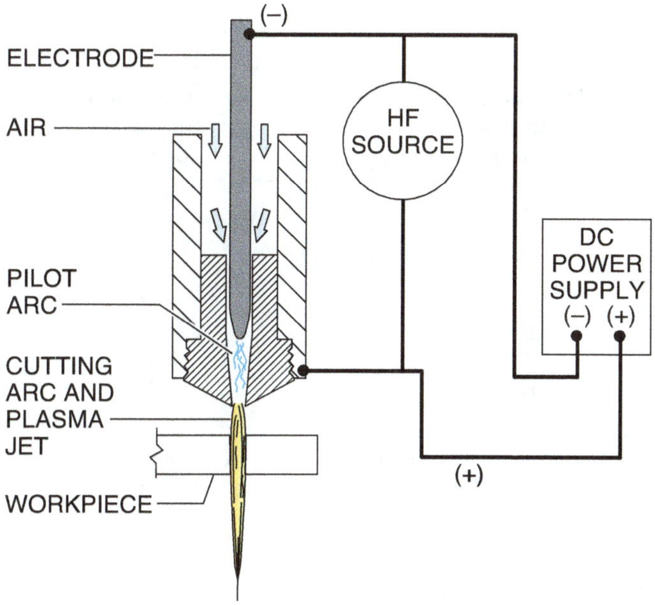

Figure 4 Schematic of the non-transferred arc process.

A plasma arc cutting process that uses air or nitrogen and is set up correctly is generally safer than oxyfuel processes. There is no chance of flashbacks and other dangers caused by flammable gases such as those used in oxyfuel processes. However, the voltages used, the extremely high temperatures of the cutting jet, and the intense ultraviolet radiation of the arc are hazardous. In addition to the safety practices provided in the following section, the following practices should also be observed:

- Make sure that all sources of energy to the PAC equipment are turned off before disassembling a plasma torch to replace any consumable components. Open-circuit DC voltages as high as 400 volts may be present between the torch components and the workpiece when they are connected.
- The extremely high temperatures of the plasma cutting jet can quickly and easily cut through gloves and flesh, resulting in severe injuries.

WARNING!

Working with energized PAC equipment is extremely hazardous. Always disconnect the power supply and all other sources of energy before servicing or otherwise working with plasma cutting equipment. Never place your hands under the workpiece near the cutting jet. Even with protective clothing, you can be severely injured.

1.2.1 Protective Clothing and Equipment

Welding activities can cause injuries unless you wear all of the protective clothing and equipment that is designed specifically for the welding industry. The following information includes important safety guidelines about protective clothing and equipment:

- Always use safety glasses with a full face shield or a helmet. The glasses, face shield, or helmet lens must have the proper light-reducing tint for the type of welding or cutting being performed. Use corrective lenses if needed; some companies will not allow contact lenses to be worn during cutting or welding activities. Never view an electric arc directly or indirectly without using a properly tinted lens.
- Wear proper protective leather and/or flame-retardant clothing along with welding gloves that will protect you from flying sparks and molten metal.
- Wear high-top safety shoes or boots. Make sure that the tongue and lace area of the footwear will be covered by a pant leg or other protective material. Boots with no laces at all are generally preferred for both welding and cutting activities. If the tongue and lace area is exposed, or if the footwear must be protected from burn marks, wear leather spats under your pants or chaps and over the top front area of the footwear.
- Wear a 100-percent cotton cap with no mesh material included in its construction. The bill of the cap points to the rear. If a hard hat is required for the environment, use one that allows the attachment of rear deflector material and a face shield. A hard hat with a rear deflector is generally preferred when working overhead, and may be required by some employers and job sites.

WARNING!

Do not wear a cap with a button in the middle. The conductive metal button beneath the fabric represents a safety hazard.

- Wear a face shield over snug-fitting cutting goggles or safety glasses for gas welding or cutting. Either the face shield or the lenses of the welding goggles must be an approved shade for the application. A welding hood equipped with a properly tinted lens is also acceptable. See *Table 1* for a list of lens shades based on arc current.

Table 1 Plasma Arc Cutting – Guide to Lens Shade Numbers

From *AWS F2.2 Lens Shade Selector.* Shade numbers are given as a guide only and may be varied to suit individual needs.

Process	Arc Current (Amperes)	Minimum Protective Shade	Suggested* Shade No. (Comfort)
Plasma Arc Cutting (PAC)	Less than 20	4	4
	20–40	5	5
	40–60	6	6
	60–80	8	8
	80–300	8	9
	300–400	9	12
	400–800	10	14

*As a rule of thumb, start with a shade that is too dark to see the weld zone. Then go to a lighter shade which gives sufficient view of the weld zone without going below the minimum. In oxyfuel gas welding, cutting, or brazing where the torch and/or the flux produces a high yellow light, it is desirable to use a filter lens that absorbs the yellow or sodium line of the visible light spectrum.

29103-15_T01.EPS

- Wear earplugs to protect ear canals from sparks. Wear hearing protection to protect against the consistent sound of the torch.

WARNING!

Ear protection is essential to protect ears from the noise of the torch. Other personal protective equipment (PPE) must be worn to protect the operator from hot metal and slag.

1.2.2 Fire/Explosion Prevention

Welding activities usually involve the use of fire or extreme heat to melt metal. Whenever fire is used, it must be controlled and contained. Welding or cutting activities are often performed on vessels that may once have contained flammable or explosive materials. Residues from those materials can catch fire or explode when a welder begins work on such a vessel. The following are fire and explosion prevention guidelines associated with welding:

- Never carry matches or gas-filled lighters in your pockets. Sparks can cause the matches to ignite or the lighter to explode, causing serious injury.
- Always comply with all site and/or employer requirements for a hot-work permit and a fire watch.
- Never use oxygen to blow dirt or dust off clothing. The oxygen can remain trapped in the fabric for a time. If a spark hits the oxygen in the fabric, the clothing can burn rapidly and violently.
- Make sure that any flammable material in the work area is moved or shielded by a fire-resistant covering.
- Approved fire extinguishers must be available before attempting any heating, welding, or cutting operations. Make sure the extinguisher is charged, the inspection tag is valid, and any individual that may be required to operate it knows how to do so.
- Never release a large amount of oxygen or use oxygen in place of compressed air. The presence of oxygen around flammable materials or sparks can cause rapid and uncontrolled combustion. Keep oxygen away from oil, grease, and other petroleum products.
- Never release a large amount of fuel gas, especially acetylene. Methane and propane are heavier than air and tend to migrate to and concentrate in low areas. As a result, they can ignite at a considerable distance from the release point. Acetylene is lighter than air but is even more dangerous than methane; when mixed with air or oxygen, it will explode at much lower concentrations than any other common fuel gas.
- To prevent fires, maintain a neat and clean work area, and make sure that any metal scrap or slag is cold before disposal.

Before cutting or welding containers such as tanks or barrels, find out if they contained any explosive, hazardous, or flammable materials, including petroleum products, citrus products, or chemicals that decompose into toxic fumes when heated. Proper procedures for cutting or welding hazardous containers are described in the *American Welding Society (AWS) F4.1, Safe Practices for the Preparation of Containers and Piping for Welding and Cutting,* and *ANSI Z49.1.* As a standard practice, always clean and then fill any tanks or barrels with water, or purge them with a flow of inert gas such as nitrogen to displace any oxygen.

> **WARNING!**
> Welding or cutting must never be performed on drums, barrels, tanks, vessels, or other containers until they have been emptied and cleaned thoroughly, eliminating all flammable materials and all substances (such as detergents, solvents, greases, tars, or acids) that might produce flammable, toxic, or explosive vapors when heated. Do not assume that a container that has held combustibles is clean and safe until proven so by proper tests. Do not weld in places where dust or other combustible particles are suspended in air or where explosive vapors are present.

Containers must be cleaned by steam cleaning, flushing with water, or washing with detergent until all traces of the material have been removed.

> **WARNING!**
> Clean containers only in well-ventilated areas. Vapors can accumulate during cleaning, causing explosions or injury.

After cleaning the container, fill it with water or a purging gas, such as carbon dioxide, argon, or nitrogen to displace the explosive fumes. Air, which contains oxygen, is displaced from inside the container by the water or inert gas. Without oxygen, combustion cannot take place.

A water-filled vessel is the best alternative. When using water, position the container to minimize the air space. When using an inert gas, provide a vent hole so the inert gas can push the air and other vapors out to the atmosphere. Keep in mind, though, that even these precautions do not guarantee the absence of flammable materials inside. For that reason, these types of activities should not be done without proper supervision and the use of proper testing methods.

1.2.3 Work Area Ventilation

Vapors and fumes tend to rise in the air from their sources. Welders often have to work above the welding area where the fumes are being created. Welding fumes can cause personal injuries. Good work area ventilation helps to remove the vapors and to protect the welder. The following is a list of work area ventilation guidelines to consider before and during welding activities:

- Make sure confined-space procedures are followed before conducting any welding or cutting in the confined space.
- Always perform cutting or welding operations in a well-ventilated area. Cutting or welding operations involving materials such as coatings that contain cadmium, lead, zinc, or chromium will result in toxic fumes. For cutting or welding of such materials, always wear an approved respirator with an appropriate filter as directed by your employer.
- Make sure confined spaces are ventilated properly for cutting or welding purposes.
- Never use pure oxygen from cylinders for the purpose of ventilation or breathing air.

Career Opportunities

The growth of the automobile racing industry has created career opportunities for workers who are skilled at welding and cutting metal. In addition to fabrication of race cars at the shop, there is a demand for at-the-track repair work. Lincoln Electric offers a Motorsports Welding School to address the special needs of the racing community.

29103-15_SA01.EPS

Additional Resources

AWS F3.2M/F3.2, Ventilation Guide for Weld Fume. Latest Edition. Miami, FL: American Welding Society.

Plasma Cutters Handbook: Choosing Plasma Cutters, Shop Safety, Basic Operation, Cutting Procedures, Advanced Cutting Tips, CNC Plasma Cutters, Troubleshooting, and Sample Projects. Eddie Paul. New York, NY: Penguin Group.

1.0.0 Section Review

1. The plasma arc cutting process type most commonly used is the ____ .
   a. electrical process
   b. non-transferred arc process
   c. oxyfuel pilot process
   d. transferred arc process

2. When operating plasma arc cutting equipment, it is acceptable to wear a hat made of ____ .
   a. polyester
   b. nylon
   c. 100-percent cotton
   d. wool

SECTION TWO

2.0.0 PLASMA ARC CUTTING EQUIPMENT

Objective

Identify and describe plasma arc cutting equipment.

a. Identify and describe plasma arc power units.
b. Identify and describe plasma arc torches and accessories.
c. Identify and describe plasma arc cutting gases and gas control devices.

Trade Terms

Amperage: The unit of measure used for the intensity of an electric current; often abbreviated to *amp* or *amps*.

Duty cycle: The percentage of time a plasma arc cutting machine can cut without overheating within a ten-minute period.

Phase: In a three-phase power supply system, the sine waves of the voltage on each of the three separate conductors is displaced from each of the others by 120 degrees, although all conductors are carrying alternating current at the same frequency and are synchronized. Each of the three sine waves is referred to as a phase of the power source.

Potential: The relative electrical voltage difference between two points of reference; the electromotive force or difference in potential between two points, expressed in volts.

Solenoid valve:: A valve used to control the flow of gases or liquids that is opened or closed by the action of an energized electromagnet.

There are many different manufacturers of PAC equipment. The following sections provide general information common to most equipment types. For information on the operation of specific equipment at your site, always refer to the manufacturer's documentation.

Basic plasma arc cutting equipment includes the following items:

- A power unit, where the process is controlled
- Plasma arc cutting torch with torch cable
- Workpiece lead assembly
- Plasma, shielding and cooling gases, and gas control components

2.1.0 PAC Power Units

Plasma-arc cutting is performed using direct-current, electrode-negative (DCEN) power. This means that the electrode of the torch has a negative potential compared to the workpiece lead. The power units are designed specifically for plasma arc cutting, and they are available in many sizes and configurations. The smallest and simplest PAC units are designed for light-duty manual cutting of sheet metal and light gauge plate. They typically plug into a single-phase 115VAC or 230VAC outlet. Typical maximum cutting amperages for these light-duty units range from 14 to 30 amperes. Their duty cycle is usually in the 35 percent range. The simplest units use filtered, compressed air for plasma and cooling gas. They have a high-frequency generator for easy arc initiation. Some of these small air units contain their own air compressor, to supply air for both cooling and plasma generation.

Commercial-grade portable units are available that can operate on various single-phase or three-phase power supplies. These are rated for cutting material up to 1¼" (3.2 cm) thick at duty cycles of 50 to 60 percent. At maximum current and operating at a reduced duty cycles, they are capable of severing material up to 1½" (3.8 cm) thick. These units normally require an external air or gas supply for the torch. The size of the unit needed is determined by the type and thickness of the metal to be cut.

Console controls are relatively simple. They usually include most of the following controls and indicators:

- Power On/Off switch
- Power On and/or unit-ready indicator
- Air or gas On/Off switch or setting (for monitoring and adjusting torch air or gas flow, or for purging lines)
- Output current control
- Power unit trouble indicator(s)

Figure 5 shows a typical portable commercial-grade plasma arc cutting unit.

The Hazards of PAC

Ultraviolet radiation, hot particle matter, and noise are all hazards associated with PAC, but they are manageable with the proper equipment. Devices such as water tables and water mufflers use water to control such hazards in mechanized plasma cutting applications.

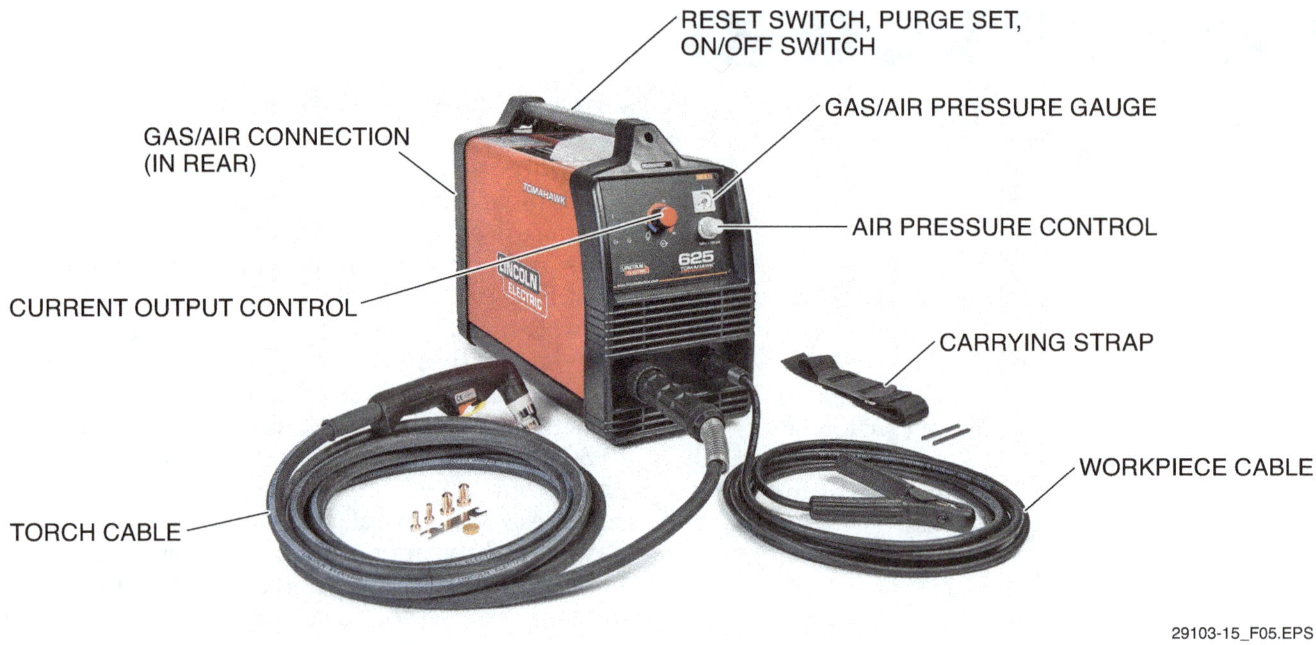

Figure 5 Commercial-grade plasma arc cutting unit.

Larger industrial units are used for heavy-duty manual cutting and general-duty mechanized cutting. These units usually operate on a 230/460VAC three-phase power supply, producing DC output currents up to 750 amperes. Their duty cycle is usually 100 percent, meaning that they are capable of constant operation. They can cut stainless steel and aluminum up to 2½" (6.4 cm) thick and carbon steel up to 3" (7.6 cm) thick. Some units are water-cooled, often containing their own closed-loop torch cooling system.

The larger PAC units are often dual-flow, mechanized units that operate with several types of gases. One gas is used for cutting, and the other is used as a shield around the cutting area, to improve the quality of the cut. Some units may use only one of the gases for the pilot arc.

Although gouging is not the primary use for plasma arc torches, some commercial and industrial units may be used for gouging when fitted with a gouging torch, or a torch with gouging components. Plasma arc gouging is sometimes referred to using the nonstandard acronym PAG.

Controls on the larger units typically include a power On/Off switch, output current control and ammeter, several gas flow controls and gauges, gas selector controls, local/remote switch, and open-circuit voltmeter. A leather cable cover accessory is often used to protect the torch cable from damage from the resulting hot or molten metal. The covers are typically designed with snaps or Velcro® to secure it around the torch cable.

2.1.1 Workpiece Clamp

The workpiece clamp (*Figure 6*) provides the connection between the end of the workpiece lead and the workpiece. The area on the workpiece where the clamp is being attached must be cleaned well enough for good metal-to-metal contact. Workpiece clamps are mechanically connected to the welding lead, and they come in a variety of shapes and sizes. Workpiece clamps are rated by the current they can carry without overheating. If a workpiece clamp needs to be replaced, be sure to select one that is rated at or above the output capacity of the PAC power unit.

When PAC will be used to cut equipment containing electrical or electronic components, batteries, bearings, or seals, the workpiece clamp must be properly located to prevent damage to associated components. If the cutting current passes through any type of bearing, seal, valve, contacting surface, or electrical or electronic component, it could cause severe damage to the item from arcing and overheating.

Cutting Ratings

Manufacturers may rate the cutting capability of their equipment in different ways. Some may rate it at the listed duty cycle, and others may rate it at the maximum cutting (severing) capability at a much reduced duty cycle. When selecting a power supply, make sure that you understand the manufacturer's rating system.

Control Interlocks

Interlocks may be incorporated into the controls of PAC systems. An interlock common in PAC systems monitors the air supply pressure. If the plasma torch is operated without an adequate supply of air, internal arcing may damage the torch. For this reason, a gas pressure safety switch is included in the circuit. Adequate gas pressure must be present before the torch will operate. The safety switch also shuts down the torch if there is a gas supply failure during cutting. Another interlock is an over-temperature switch that shuts down the unit if the duty cycle is exceeded and the unit overheats. Some units are also equipped with a retaining cup switch that shuts down the unit if the retaining cup holding the tip, nozzle, and swirl ring in the torch is loose.

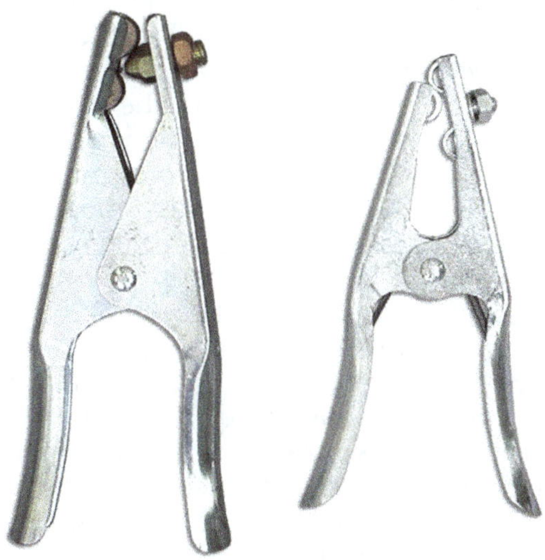

29103-15_F06.EPS

Figure 6 Typical workpiece clamps.

Carefully evaluate the cutting task and position the workpiece clamp so that the cutting current will not pass through any sensitive components or across gaps. If in doubt, ask your supervisor for assistance before proceeding.

CAUTION

To prevent damage from arcing across gapped or contacting surfaces, make sure the path from the workpiece clamp to the torch arc does not pass through these surfaces. Cutting current can severely damage bearings, seals, valves, or contacting surfaces. Position the workpiece clamp to prevent the cutting current from passing through them.

Disconnect the grounded battery lead on any mobile equipment that is being repaired using the plasma arc cutting process. This prevents the cutting current from causing battery explosion or battery damage. Never operate PAC equipment near a battery.

Do not cut near batteries. A spark could cause a battery to explode, showering the area with battery acid. Always remove batteries from the work area. Also, workpiece leads must never be connected to pipes carrying flammable, volatile, or corrosive materials. The cutting current could cause overheating or sparks, resulting in an explosion or fire.

CAUTION

The cutting current can destroy electronic or electrical equipment. Have an electrician check the equipment and, if necessary, isolate the system before cutting.

2.2.0 PAC Torches

The most common plasma arc cutting torch uses a pilot arc and the transferred arc process. A high-frequency pilot arc is first started between the electrode and the nozzle. A small amount of gas is injected into the arc chamber where it is heated to a plasma and escapes through the nozzle as a fine jet. However, the torch current circuit is designed so that the full voltage, or potential, and current are not available in the torch between the internal electrode and nozzle. The base metal to be cut is connected to the power unit by the workpiece clamp to make a complete circuit. This gives the workpiece a higher potential difference than the nozzle. When the torch is moved near the workpiece, the higher-current arc extends (transfers) to the workpiece because of its higher potential. The control unit automatically increases the amperage and gas flow(s) to produce the longer cutting arc, and disables the pilot arc. *Figure 7* shows the result of the plasma transfer process that takes place between the torch and the workpiece.

GOING GREEN

Cooling Water

Welders involved in metal-cutting activities often allow the water used in the process to simply flow into the ground or down the drain. To better protect the environment, water used for cooling the plasma arc cutting equipment should be contained whenever possible and sent to a waste treatment facility.

Figure 7 A transferred plasma cutting arc.

Plasma arc torches may be handheld for manual cutting or machine mounted for mechanized cutting. The heaviest duty torches are usually mechanized and water-cooled. Handheld torches are usually light-duty to heavy-duty, with heavy-duty models typically being water-cooled. Some torches are equipped with separate passages and orifices for cooling and shielding gases. Torches have a button or lever on the torch handle to start the pilot arc. A foot pedal can also be used to start the arc. When a foot pedal is used, the switch on the control panel is set for remote operation.

The typical PAC torch usually contains a replaceable retaining cup, a copper electrode with a hafnium or thoriated tungsten insert, a gas swirl ring, and a nozzle. Nozzles have various-sized holes for cutting metals of different thicknesses. The torch also contains coolant passages for gas or water cooling. *Figure 8* shows a typical handheld PAC torch and its consumable components.

Mechanized PAC torches are designed to be operated on automated carriers. They can usually be used in the same carriers designed for oxyfuel cutting heads. However, any carrier used must be capable of moving at the much greater speeds required for effective plasma arc cutting.

Water-injection PAC torches inject pressurized water to create a whirlpool around the jet, below the point where the arc narrows in the torch's arc cavity. The water swirls around the plasma jet at high velocity and compresses the jet even more. The narrower jet cuts with a more uniform melt rate and leaves the kerfs of stainless steel and titanium free of dross. This type of torch is normally mechanized.

Several styles of shrouds, heat shields, nozzles, and electrodes are usually available for a given model of PAC torch. These components vary in style and size to adapt to different metal types and thicknesses. Nozzles and other components for gouging are offered to fit some of the heavier duty torches. Electrode types and styles vary with torch manufacturers.

The most common parts that are changed in a torch are the electrode and the nozzle. The electrode is changed when it becomes contaminated. The nozzle is changed when it becomes worn and no longer true to its original shape or size, or when metal of a different type or thickness needs to be cut. When performing low-amperage cutting, it is possible to drag the torch on the workpiece using a drag shield or standoff post guide without causing damage or sacrificing the cut. Cutting at current levels above 60 amps or for an extended time usually requires a standoff guide, which attaches to the cup and holds the tip at the required distance from the workpiece. *Figure 9* shows two common styles. The standoff distance, or the distance between the torch tip and the workpiece, should be kept between 1⁄16" and 1⁄4" (1.6 to 6.4 mm) depending on the thickness of the workpiece.

Auto-Voltage Adjustments

Some newer PAC units sense the input voltage and frequency when they are connected to a power source. The units continuously and automatically reconfigure themselves to operate properly with the connected power. This eliminates improper initial power connections and provides continuous adjustment to changing input voltage so that the PAC arc current remains stable during cutting operations. This feature also increases the flexibility of the unit to operate in various locations.

2.3.0 PAC Gases and Controls

All PAC units require one or more types of gases. The specific type of gases required and the controls necessary to set and adjust them vary with the type, size, and manufacturer of the PAC equipment.

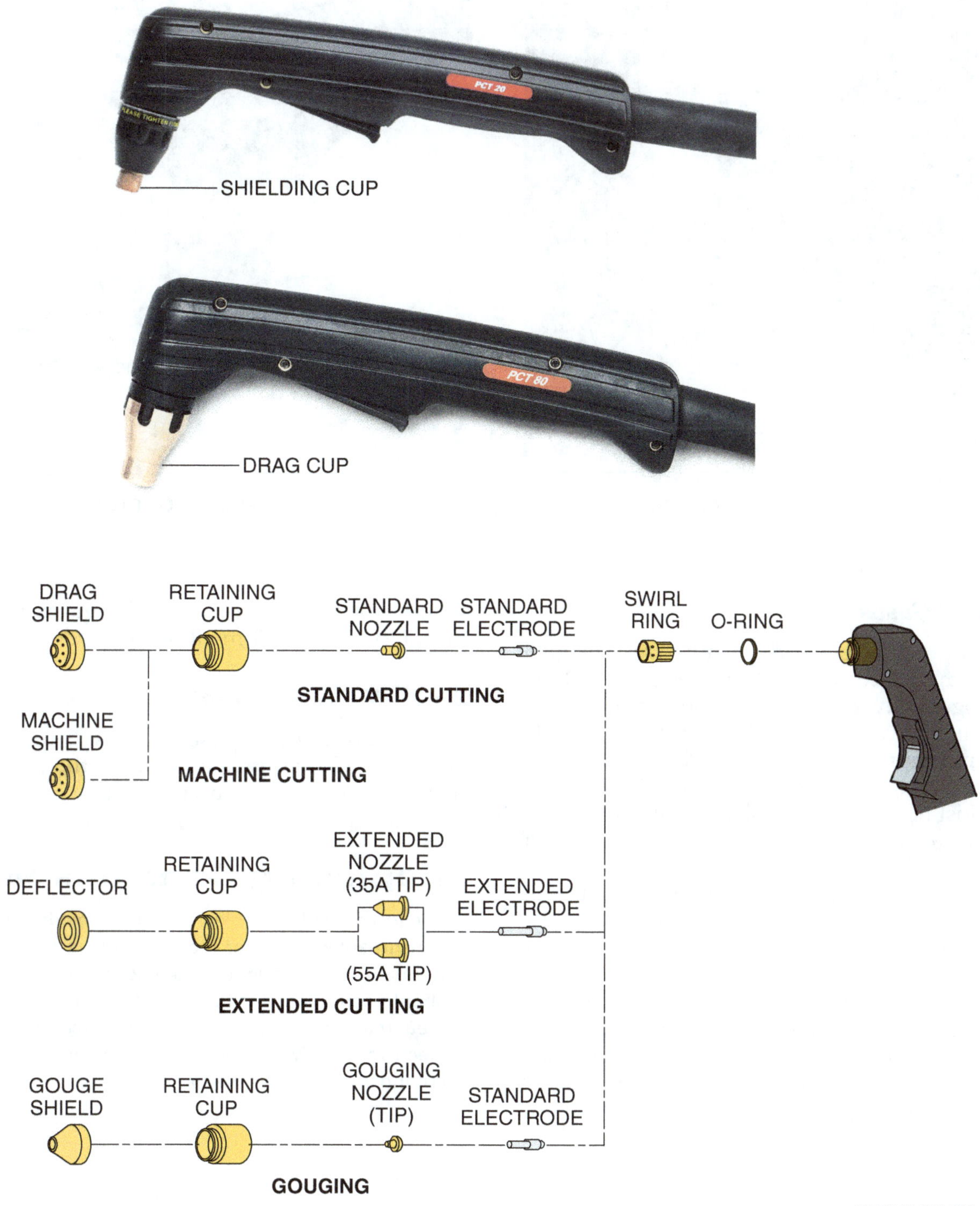

Figure 8 Handheld plasma torches.

(A) STANDOFF ROLLER GUIDE

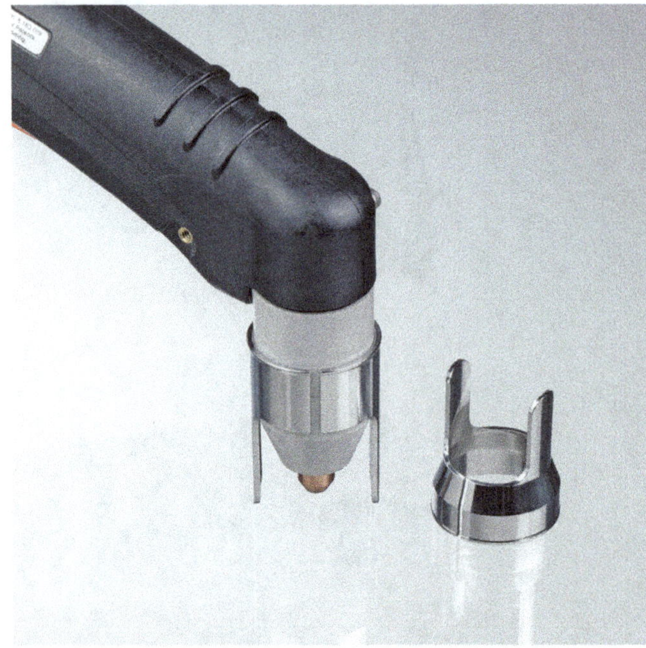

(B) STANDOFF POST GUIDE

29103-15_F09.EPS

Figure 9 Standoff guides.

2.3.1 PAC Gases

Several different gases and gas mixtures are used with plasma arc cutting units. Gases used include air, nitrogen, oxygen, argon and argon mixtures, hydrogen, and carbon dioxide. The simplest units use clean compressed air to cut carbon steel, stainless steel, and aluminum up to about 3⁄16" (4.8 mm) thick. Most heavy-duty air plasma units are rated to cut carbon steel and stainless steel up to about 1¼" (32 mm) thick and possibly sever material up to 1¾" (44.5 mm) thick. Dual-flow PAC units can operate with two different gases. One gas may be used for the pilot arc and for cutting, and another for shielding or cooling. *Table 2* provides general recommendations for plasma and shield gases.

Specific gas requirements for a particular PAC unit or system are specified in the manufacturer's operating instructions manual and/or the information tags on the unit. Refer to these specifications, and use only the gases specified for the type of equipment being used.

2.3.2 PAC Gas Controls

Plasma arc cutting gases are used to generate plasma, cool the torch, and shield the cut against corrosion. Most gas sources must have their pres-

Replacing Consumables

The torch nozzle and electrode are considered consumables, which are components requiring frequent replacement due to deterioration, and both should be replaced at the same time. It is a good idea to always have a replacement torch nozzle and electrode handy. This will reduce downtime when cutting with the plasma unit.

Contact-Start or Discharge-Start Torches

High-frequency pilot arc starting of a PAC torch can interfere with nearby electronic circuits, including computers and microprocessor controllers. To eliminate the use of a high-frequency start, some newer PAC torches use an internal contact or capacitor-discharge pilot arc starting system. One such contact-start system has a spring-loaded electrode that is in contact with the nozzle when the torch is off. When a trigger is pressed, the PAC unit senses the shorted electrode via high-current flow and opens the gas solenoid. Gas pressure then forces the electrode away from the nozzle. This action draws an arc that establishes the plasma flow. The arc instantly transfers to the workpiece because of the higher voltage difference between the electrode and the workpiece.

Table 2 Recommended Plasma/Shield Gas Combinations

Material	Air/Air	O_2/Air	N_2/CO_2	N_2/Air	H35/N_2
Carbon Steel	Most Economical Good Cut Quality Good Speed Good Gouging Good Weldability	Best Cut Quality Maximum Cut Speed Best Weldability	Some Dross Long Electrode Life	Not Recommended	Best Gouging Long Electrode Life Some Dross
Stainless Steel	Most Economical Good Speed Some Dross	Not Recommended	Good Cut Quality Good Gouging Minimal Dross Long Electrode Life	Long Electrode Life Lowest Shield Gas Cost	Best Cut Quality Best Gouging Minimal Dross Long Electrode Life Cuts Thicker Material
Aluminum	Most Economical Good Speed Some Dross	Not Recommended	Good Cut Quality Good Gouging Minimal Dross Long Electrode Life	Not Recommended	Best Cut Quality Best Gouging Minimal Dross Long Electrode Life Cuts Thicker Material

O_2 = Oxygen
N_2 = Nitrogen
CO_2 = Carbon Dioxide
H35 = Mixture of 35% Hydrogen and 65% Oxygen

29103-15_T02.EPS

sures reduced before they can be used. The type of regulator used to control the gas pressure depends on the type of PAC equipment being used and the base metal being cut.

When a compressed-air supply is being used, a heavy-duty air filter and dryer is required to eliminate oil and moisture from the pressurized air. The high-pressure shop air must be reduced by an external regulator, or internally by the PAC, to the required working pressure. Also, depending on the equipment being used, the gas must be provided at a given flow rate for the task at hand, generally measured in cubic feet of gas per hour (cfh). Always check to be sure that the air supply will provide the flow rate required for the PAC equipment being used. If the airflow rate is too low, it will cause poor cuts and overheating of the torch. Refer to the manufacturer's specifications for the required flow rate for the equipment being used. To conserve consumables, make sure to have air dryers in the compressed air system as well. Moisture in the compressed air can damage the equipment.

Gases other than a local compressed air source are normally supplied in pressurized cylinders at 1,500 to 2,000 psi (103 to 138 bars). The high cylinder pressures are reduced by pressure regulators (*Figure 10*) to a lower working pressure. A gas hose connected to the outlet of the regulator delivers the gas to the PAC unit. Most heavy-duty systems require two high-pressure cylinders and two pressure regulators. One cylinder contains the cutting gas and the other a shielding or cooling gas.

GOING GREEN

Fumes and Gases

The various gases used in PAC activities, as well as the fumes generated from these activities, are often released into the atmosphere without much thought about their effect on the environment. To help protect the environment, efforts are needed to better contain the release of PAC gases and fumes. Whenever possible, PAC work should be performed using some type of capture and recovery system so that the fumes and gases are not released into the atmosphere.

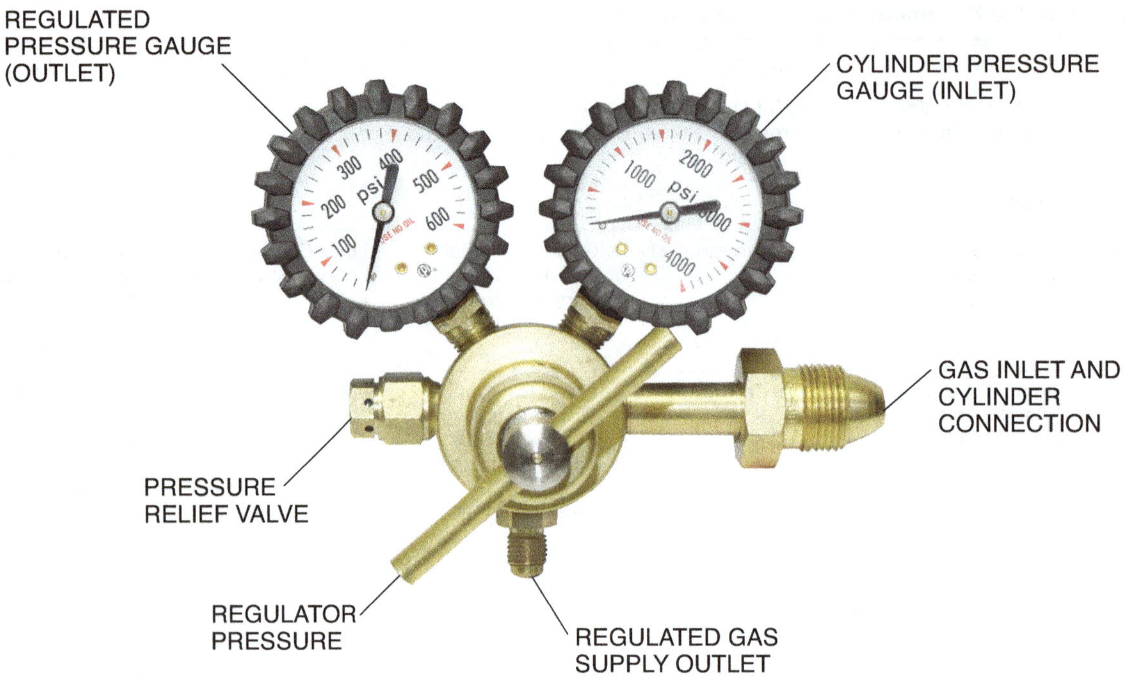

Figure 10 Gas cylinder pressure regulator.

The PAC process was originally developed to cut nonferrous metals using inert gases. Advances in technology led to using air and/or oxygen as the plasma gas so steel could be cut using PAC.

Some systems are able to manage two gas sources to enable the user to manage different applications. *Figure 11* shows a unit commonly known as a dual-flow PAC with two internal gas pressure regulators.

Gas flow to the torch is usually controlled by a **solenoid valve**. Solenoid valves are usually mounted inside the unit and are controlled by a pushbutton on the torch. There is one solenoid valve for each gas source.

Many smaller PAC units come equipped with a built-in air compressor. These light-duty machines are portable and run on 110V power, which makes them ideal for use at remote job sites. *Figure 12* shows a light-duty unit.

Figure 11 Industrial dual-flow PAC unit.

Plasma Torch Cutting Guides

Many manufacturers furnish cutting guide accessories that can be used on several models of their torches. A circle-cutting guide kit can also be used as a freehand standoff roller guide.

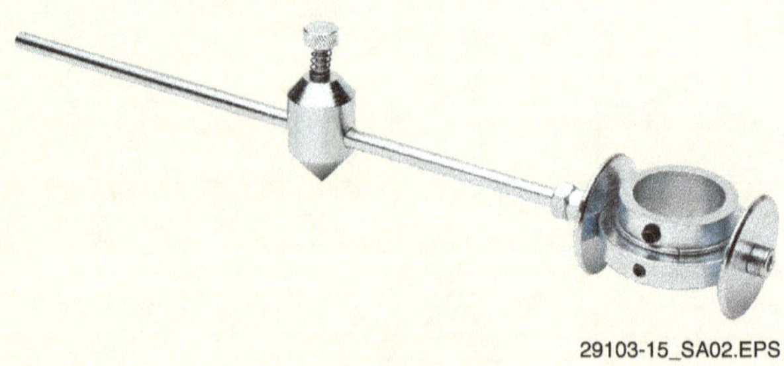

29103-15_F12.EPS

Figure 12 Light-duty PAC unit in use.

Regulator Selection

When selecting a regulator for PAC systems, first determine the required gas pressure from the PAC unit specifications. The desired pressure should be around the center of the gas pressure regulators adjustment range for the best accuracy. For example, if the desired gas pressure to the PAC is 50 psig (345 kPa), a regulator with a 0 to 100 adjustment range would be an ideal choice. However, note that a match between the regulator and the equipment is rarely this perfect. The idea is to avoid operating at the low or high end of the regulator's range, as these areas are where the least amount of accuracy and control occur in a pressure regulator.

Additional Resources

Plasma Cutters Handbook: Choosing Plasma Cutters, Shop Safety, Basic Operation, Cutting Procedures, Advanced Cutting Tips, CNC Plasma Cutters, Troubleshooting, and Sample Projects. Eddie Paul. New York, NY: Penguin Group.

2.0.0 Section Review

1. The duty-cycle range for a light-duty unit is typically ____ .
   a. 35 percent
   b. 50 percent
   c. 75 percent
   d. 90 percent

2. The typical plasma arc cutting torch usually contains a replaceable retaining cup, comprised of a copper electrode with a(n) ____ .
   a. magnesium tungsten insert
   b. aluminum or thoriated tungsten insert
   c. steel tungsten alloy insert
   d. hafnium or thoriated tungsten insert

3. Some types of PAC torches inject water to create a whirlpool of water around the jet to ____ .
   a. cut with a more uniform melt rate
   b. ensure the cut will have a wider kerf
   c. allow the torch to be used on a carrier
   d. minimize the amount of dross from the cut

Section Three

3.0.0 Plasma Arc Cutting Equipment Operation and Care

Objective

Describe how to set up, safely operate, and care for plasma arc cutting equipment.

a. Describe how to set up plasma arc cutting equipment and the adjacent work area.
b. Describe how to safely operate plasma arc cutting equipment.
c. Describe how to care for plasma arc cutting equipment.

Performance Tasks

1. Set up plasma arc cutting equipment.
2. Set the amperage and gas pressures or flow rates for the type and thickness of metal to be cut.
3. Square-cut metal using plasma arc cutting equipment.
4. Bevel-cut metal using plasma arc equipment.
5. Pierce and cut slots in metal using plasma arc cutting equipment.
6. Dismantle and store the equipment.

The following section explains how to correctly set up PAC equipment and use it for cutting. Various types of cutting techniques are also presented.

3.1.0 Setting Up PAC Equipment

PAC equipment is relatively easy to set up because it is supplied as a complete system. However, the following are some things to consider before cutting is attempted:

- The unit must have the rated power and duty cycle to cut the intended material and thickness.
- The required power characteristics of the PAC unit, including the phase, voltage, and amperage must be provided.
- The appropriate nozzle must be installed in the torch for the task at hand.
- The required gas or gases must be available and at the required pressures and flow rates.
- Spare consumable components must be on hand.

3.1.1 Preparing the Work Area for PAC

The PAC control unit and workpiece must be located close enough to each other for the torch cable to comfortably reach the workpiece and still provide maneuverability. If possible, the workpiece should be located at a comfortable height and position for the torch operator. The cutting area should be well ventilated and cleared of all combustible material.

Plasma arc cutting or gouging can spray molten metal for considerable distances, possibly as far as 40 ft (12 m). It is important that everything combustible is removed from the range of the sprayed metal. If necessary, flame-resistant shields or curtains should be erected to protect workers and equipment from ultraviolet arc rays and metal splatter. Some employers and/or job sites may require a hot-work permit to be completed and a fire watch to be posted.

WARNING!

PAC produces fumes and gases that can be harmful to your health. The composition and rate of generation of fumes and gases depend on factors such as arc current, cutting speed, material being cut, and gases used. Various paints and coatings can also create or contribute to the production of harmful fumes. The fume and gas byproducts usually consist of ozone, oxides of the metal being cut, and oxides of nitrogen. These fumes must be removed from the work area through a capture and exhaust or recovery system. Check all applicable local codes; some codes may require that exhausted fumes be filtered before being vented to the atmosphere.

Eye and ear protection are essential when performing PAC as a safeguard against noise and to prevent metal spray from entering the operator's eyes and ears. If the site ventilation is not adequate to keep the smoke and fumes away from the operator, a respirator may also be required.

WARNING!

Ear protection is essential to protect hearing. Other PPE must be worn to protect the operator from hot metal and slag.

3.1.2 Setting the Correct Cutting Amperage

The cutting amperage depends on the type of equipment being used, the type and thickness of the material being cut, and the type of gas being used (*Figure 13*). Light-gauge sheet steel may require as little as 7 amps of current, while 2" (5.1 cm) aluminum plate may require 250 amps. Always refer to the manufacturer's recommendations to identify the correct amperage for the specific equipment and job. Using the incorrect amperage will result in a poor-quality cut. If the amperage is too high, severe damage to the torch may also occur.

3.1.3 Setting Gas Parameters and Installing Gas Cylinders

Several different gases may be used on the same system for cutting the same or different types and thicknesses of metals. In many cases, the same gas or mixture will not be used for both thick and thin metals of the same type. Refer to the manufacturer's instructions or the unit data plate for specific gas recommendations.

Some gases used in cutting can be dangerous if accidentally released. Welders routinely have to remove empty cylinders and replace them with filled cylinders, increasing the potential for a release.

When connecting a gas cylinder to the PAC unit, follow these steps:

Step 1 After identifying the correct gas type, place it adjacent to the unit and secure it in an upright position.

Step 2 Remove the protective valve cap.

(A) (B)

29103-15_F13.EPS

Figure 13 Gas and amperage settings vary according to material type and thickness.

WARNING!

If oxygen is being used, be sure any cloth used on or near the fittings or regulators is free of oil or grease. Oil or grease mixed with oxygen can explode, causing personal injury or death.

Step 3 While standing to one side of the cylinder valve outlet, and with safety glasses in place, quickly crack open the valve a small amount to blow out any dirt or debris. Reclose immediately to prevent unnecessary gas loss.

WARNING!

To avoid personal injury, always stand with the cylinder between you and the attached regulator when opening valves.

Step 4 Connect the regulator to the cylinder valve and tighten with the proper wrench. Do not over-tighten. Connect the hose from the PAC unit to the regulator outlet.

Step 5 Open the cylinder valve very slowly at first, then open fully. Observe the regulator inlet gauge to determine the pressure remaining in the cylinder.

Step 6 Slowly adjust the regulator to the desired output pressure. The pressure usually has to fall into a range to satisfy the requirements of the PAC unit. The PAC unit regulates the final pressure to the torch.

CAUTION

Do not use pliers or pipe wrenches to tighten the regulator's connections. Use an adjustable wrench or correctly sized open-end wrench. Also, be careful not to over-tighten the connections because they are made of soft bronze and can be damaged easily.

3.2.0 Operating PAC Equipment

PAC equipment is relatively simple to operate. The cutting unit's operating instructions provide information for properly setting gas pressure and flow rates, output current, torch standoff distance, and torch travel speed (for mechanized cutting) for various metal types and thicknesses.

To operate the PAC equipment, follow these general steps. Note that this is a general reference only. Refer to the manufacturer's documentation for specific instructions on the PAC equipment being used.

Step 1 Make sure that the capacity of the PAC power unit matches or exceeds the work to be done. Also ensure that the power source for the unit is sufficient, based on the unit data plate.

Step 2 Identify the location of the primary disconnect for the electrical source to be used in case of emergency.

Step 3 Check to make sure that the required gas(es) is connected. If a pressurized cylinder is used, ensure that it has adequate pressure for the work to be done. If compressed air will be used, ensure that it is also connected and at the appropriate pressure.

WARNING!

Some stainless steel contains chromium 6, also known as hexavalent chrome. This metal will give off toxic fumes when heated. Always use proper respiratory protection when cutting this type of stainless steel.

Step 4 Plug the unit in (if required), energize the PAC unit power source, and turn the unit on.

Step 5 In accordance with the PAC manufacturer's instructions, adjust the gas regulator pressure and/or flow rate to the values recommended for the type and thickness of metal being cut.

Step 6 Set the output current to the value recommended in the manufacturer's instructions for the type and thickness of metal being cut.

Step 7 Attach the work lead clamp to the workpiece to be cut.

CAUTION

If cutting on machinery or other assembled equipment, be sure to position the work lead clamp so that the cutting current will not pass through seals, bearings, or other contacting surfaces that could be damaged from heat or arcing. Also isolate any electrical components from the PAC system/workpiece circuit.

Step 9 Prepare the work area as previously described.

WARNING!

Be advised that various gases are extremely flammable, and PAC processes operate at extreme temperatures. The appropriate fire suppression equipment must be stationed in close proximity for emergency use.

Step 10 Put on the appropriate PPE.

Step 11 Hold the torch at the recommended standoff distance above the point where the cut is to begin, lower your hood, and press the torch arc-start button or foot pedal. As soon as cutting begins, move the torch at a 0-degree travel angle with the recommended speed along the cutting line.

3.2.1 Square-Cutting Metal

Square edges can be cut with PAC by holding the torch at 90 degrees (also referred to as a 0-degree work angle) to the metal surface while it is advanced smoothly, with no side-to-side movement. A straightedge or metal angle can be used to guide and steady the torch. The straightedge can be aligned and quickly clamped down at each end. When cutting thicker metals, the cut does taper slightly as it narrows with depth. This can

Safety Caps

Safety caps should be kept with cylinders at all times. Immediately after removing a cylinder's regulator assembly, reinstall the safety cap over the cylinder valve. This protects the cylinder valve if the cylinder should fall over. If the cylinder valve is suddenly sheared off, the cylinder may propel itself like a rocket. This could result in serious injury or death.

29103-15_SA03.EPS

result in cuts where one or both faces are not quite square with the plate face. The angle of the face is referred to as the cut angle. To compensate for the cut angle on thick materials, the torch should be slightly angled, leaning toward the scrap portion of the workpiece, to produce a square edge on the desired piece. All the taper will then be occur on the scrap side (*Figure 14[A]*).

3.2.2 Bevel-Cutting Metal

Bevel cuts are made with the same technique used for square cuts, except that the torch is held at the required bevel angle. A length of angle iron, laid on its open face and shimmed as necessary, can be used to support the torch's side at the required angle. *Figure 14(B)* shows how to use a metal angle to hold a bevel angle.

3.2.3 Piercing and Slot-Cutting in Metal

PAC can be used to pierce and cut slots in metal in any position. Full, fire-resistant body protection, including ear protection, is recommended when piercing or cutting metal in situations where out-of-position orientations could cause the metal splash to fall on the torch operator.

To pierce very thin metal, hold the torch directly over the point to be pierced at a 0-degree work angle, and press the arc button. As soon as the jet passes completely through the workpiece, move the torch head in a smooth circle (or other pattern) to produce the desired diameter or shape for the hole. When piercing thicker materials, rotate the torch to a 10-degree work angle. This will help to prevent metal from splashing directly up into the torch. When the jet passes through the workpiece, rotate the torch back to a 0-degree work angle and complete the cut.

When cutting slots or patterns, a template can be used to precisely guide the torch tip.

Gas Pressure and Flow Rates

PAC unit gas pressures and/or flow rates vary with equipment design, gas type, and cut depth (metal thickness). Typical gas flow rates vary from 15 cfh for nitrogen when cutting 0.1" (2.5 mm) aluminum, carbon steel, or stainless steel, up to 62 cfh of argon and 31 cfh of nitrogen when cutting 3" (7.6 cm) carbon steel, or 2½" (6.4 cm) aluminum or stainless steel. Refer to the manufacturer's instructions for the correct gas pressure and flow rates for each situation.

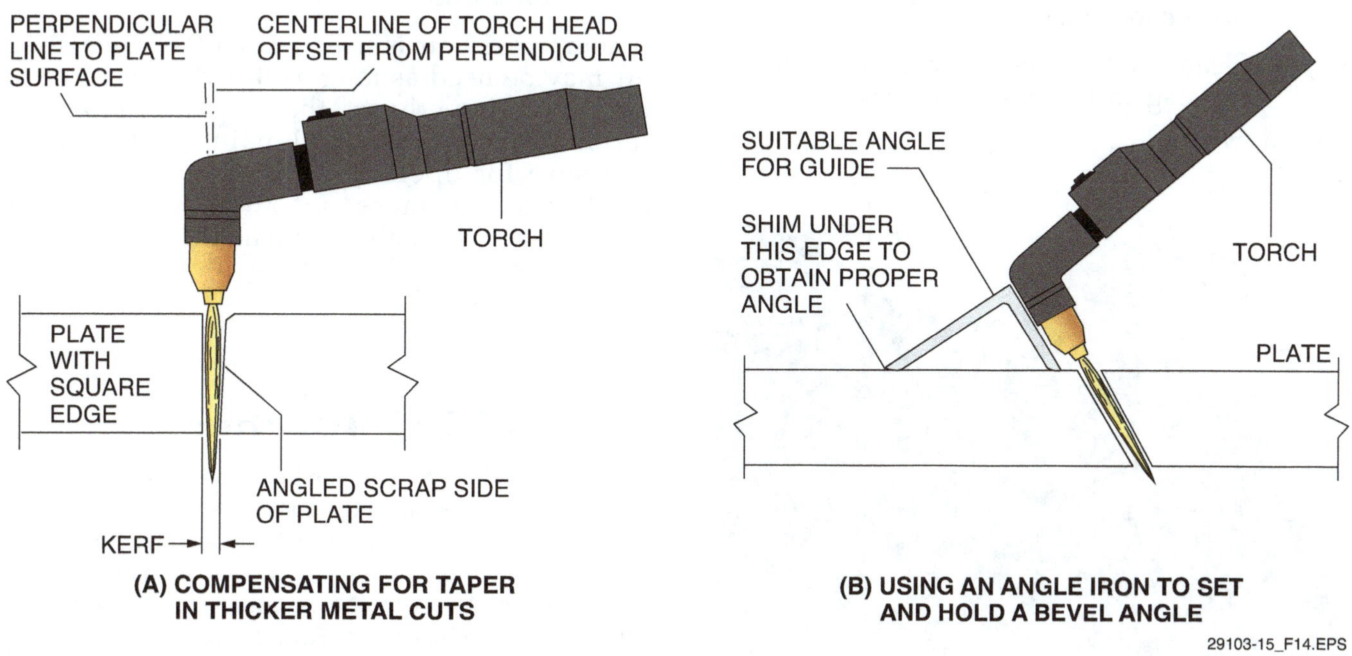

Figure 14 Compensating for taper and holding bevel angle in thicker metal cuts.

3.2.4 Proper Equipment Storage and Housekeeping

Proper equipment storage and housekeeping are essential for work efficiency and safety. When finished using the PAC equipment, follow these steps to store the equipment and maintain good housekeeping:

Step 1 Turn off the power source.

Step 2 Turn off all gas sources, bleed pressure slowly and carefully, and disconnect the lines from the PAC unit. When using cylinders, be sure to close the bottle valve, remove the regulator assembly, and install the protective cap before moving the cylinder.

WARNING!

If cylinders are to be removed, be sure to remove the pressure regulators and replace the cylinder caps before releasing and moving the cylinders.

Step 3 Unplug and coil the power unit cord and all gas lines.

Step 4 Coil the torch cable and the workpiece lead.

Step 5 Return the PAC equipment to its proper storage location.

Step 6 Clean off the work area and sweep up any slag and debris in the area.

Figure 15 shows a PAC unit that is properly stored.

29103-15_F15.EPS

Figure 15 Properly stored PAC unit.

Starting the Torch on a Thick Workpiece

Similar to the oxyfuel cutting of a thick workpiece, it is easier to begin a manual cut at the edge by rotating the PAC torch slightly so that the plasma cutting jet begins cutting at the top edge only. The torch is then rotated vertically so that the cutting jet slices into the remaining thickness before the torch is moved along the cutting line.

3.3.0 Care of Plasma Arc Cutting Equipment

As with all equipment, regular maintenance should be performed according to the manufacturer's recommendations. The cutting equipment should not be altered or modified in any way, except for those changes identified in the manufacturer's literature. The plasma machine requires relatively little maintenance, but its consumables must be inspected regularly and replaced as needed. Minor external repairs usually involve these basic parts only.

3.3.1 Understanding Basic Operating Concerns

If compressed air is used with a PAC unit, shop air may be used as long as it is dry and free of oil. When properly set, the air pressure regulator on the power unit will maintain the correct pressure for operation. The air supply must be dry; only compressed air that has passed through a reliable air dryer should be used. The compressed air supply is typically filtered at the power unit as well.

Swirl Ring Torches

Most swirl rings on PAC torches produce a clockwise tangential plasma swirl as viewed from the torch to the work surface. If the torch is at a 0-degree work angle, the clockwise swirl causes the right side of the kerf to be essentially square and the left side to be angled. For a counterclockwise swirl, the opposite is true.

> **WARNING!**
> Always switch off the machine power, unplug the power cord, and disconnect gas sources before changing consumable parts or performing any repairs on the power unit. If the power unit is directly connected to a line disconnect switch, move the line disconnect switch to the Off position. All sources of energy must be disconnected from the unit before maintenance is performed.

Your ability to change and replace parts will depend on your knowledge of the procedures contained in the manufacturer's literature for the equipment in question.

Perform the following procedures, and always abide by the manufacturer's guidelines for removal and replacement of worn or damaged parts:

- Check the shield for any signs of wear.
- The shield should be clean and clear of metal debris.
- Unscrew the shield and check the inside for wear. If the holes are blocked, attempt to clean them using a torch tip cleaner.
- If the shield cannot be cleaned, replace it.
- Inspect the retaining cup for damage and replace as necessary.
- Examine the nozzle for wear or damage. If the hole in the nozzle is worn or oval-shaped, replace the nozzle.
- Remove the electrode and check the tip for pitting. If pits exist and are more than 1⁄16" (1.6 mm) deep, replace the electrode.
- Inspect the swirl ring, which should be clean, and the holes on the sides free of clogs. If the swirl ring is damaged, replace it.
- Check the O-ring. Apply a small amount of lubricant (provided in a repair kit) to the O-ring before reinstalling it. The O-ring should make a firm seal when all parts are assembled and tight.

3.3.2 Moisture

A significant concern when using PAC equipment is that moisture may enter the unit through the air supply. This can cause the torch to sputter or hiss. If your PAC equipment experiences moisture problems, first drain the air filter/moisture separator bowl. The filter element may also need to be replaced. Refer to the manufacturer's literature for cleaning and replacement of the filter element.

3.3.3 Cooling Filter

Many power units contain an air filter housing to ensure the air supply is clean. Refer to the manufacturer's literature and clean the filter element periodically as required. An air filter that is neglected may cause the PAC unit to overheat and shut down unexpectedly.

Additional Resources

Plasma Cutters Handbook: Choosing Plasma Cutters, Shop Safety, Basic Operation, Cutting Procedures, Advanced Cutting Tips, CNC Plasma Cutters, Troubleshooting, and Sample Projects. Eddie Paul. New York, NY: Penguin Group.

3.0.0 Section Review

1. To remove, attach, or tighten the regulator's connection to a gas cylinder, you would use ____.
   a. no tools; this connection is to be hand tightened only
   b. a correctly sized wrench with an added extension for leverage to apply maximum force
   c. a properly sized pipe wrench adjusted to fit the regulator fitting
   d. an adjustable wrench or correctly sized open-end wrench
2. To pierce very thin metal with a plasma torch, you should ____.
   a. hold the torch directly over the point to be pierced at a 0-degree work angle
   b. start from the bottom, piercing the metal from below
   c. hold the torch directly over the point to be pierced at 10-degree work angle
   d. hold the torch directly over the point to be pierced at a 45-degree work angle
3. If the PAC equipment exhibits a moisture problem, you should first ____.
   a. attach a vacuum pump to the air filter/moisture separator bowl assembly
   b. replace the filter element and air filter/moisture separator bowl assembly
   c. drain the air filter/moisture separator bowl
   d. stop using air and operate the system using argon for at least 12 hours of operation

Summary

Plasma arc cutting is useful for cutting and piercing many types of metals. Unlike oxyfuel cutting, it can be used to cut aluminum, magnesium, copper, nickel, and stainless steel. PAC is usually much faster than oxyfuel cutting, and it produces little or no distortion or alteration zone changes. The cuts are fast and clean, with little or no dross, and there is no chance of carbon inclusions, as can happen with carbon arc cutting. However, because the torch jet produces metal spray, good site fire-prevention practices and the consistent use of appropriate PPE are essential.

With an understanding of metals and the knowledge to correctly set up a PAC unit for the task at hand, workers can quickly become competent in the use of PAC equipment to produce excellent cuts.

Review Questions

1. Plasma arc cutting uses a jet of plasma that can reach temperatures as high as ____ .
   a. 10,000°F (5,537°C)
   b. 20,000°F (11,093°C)
   c. 25,000°F (13,871°C)
   d. 30,000°F (16,650°C)
2. What are the two primary types of plasma arc cutting processes?
   a. Transferred arc and non-transferred arc
   b. Transferred arc and phased arc
   c. Phased arc and non-transferred arc
   d. Oxyfuel pilot arc and non-transferred arc
3. In the transferred arc process, the pilot arc is used to ____ .
   a. cut the metal
   b. ionize the air jet
   c. establish contact with the metal
   d. keep the torch heated
4. In a non-transferred arc process, the material being cut is ____ .
   a. adjacent to the arc circuit
   b. dependent on the arc circuit
   c. electrically connected to the arc circuit
   d. not electrically connected to the arc circuit
5. Which of the following statements regarding PAC safety is true?
   a. If a cap or hat is worn, it should be constructed of 100-percent polyester or nylon.
   b. The electric arc can be safely observed without tinted lenses, but it is not recommended for an extended period.
   c. Boots with no laces at all are generally preferred for both welding and cutting activities.
   d. Flammable materials can remain near the cutting activity, due to the cool nature of the plasma arc.
6. Light-duty PAC equipment typically requires a ____ .
   a. 115VAC, single-phase power source
   b. 220VAC, three-phase power source
   c. 277VAC, single-phase power source
   d. 440VAC, three-phase power source
7. Some heavy-duty PAC systems ____ .
   a. require no torch cooling
   b. are water-cooled
   c. use very little current
   d. weigh less than 5 pounds (2.3 kg)
8. The workpiece clamp is rated by the ____ .
   a. voltage that it can carry without overheating
   b. capacitance that it can carry without overheating
   c. amperage that it can carry without overheating
   d. gas volume that it uses at maximum load
9. Severe damage from arcing and overheating can be caused by the PAC current passing through a ____ .
   a. workpiece
   b. cable
   c. cutting table
   d. bearing
10. Mechanized PAC torches are designed to be ____ .
    a. handheld
    b. mounted in a vise
    c. used at a slow speed
    d. mounted on automated carriers
11. Gases other than compressed air are normally supplied in pressurized cylinders at ____ .
    a. 200 psi (14 bars) to 400 psi (28 bars)
    b. 800 psi (55 bars) to 1,000 psi (69 bars)
    c. 1,000 psi (69 bars) to 1,500 psi (103 bars)
    d. 1,500 psi (103 bars) to 2,000 psi (138 bars)
12. If site ventilation is not adequate to keep fumes away from the operator, a(n) ____ .
    a. fan must be used
    b. respirator may be required
    c. fire watch must be used
    d. airline must be used
13. Light-gauge sheet steel may be cut using a cutting amperage as low as ____ .
    a. 7 amps
    b. 9 amps
    c. 15 amps
    d. 20 amps

14. Square edges can be cut in thin materials by holding the torch at ____ .
    a. 30 degrees to the metal surface
    b. 45 degrees to the metal surface
    c. 75 degrees to the metal surface
    d. 90 degrees to the metal surface

15. When inspecting an electrode before use, it should be replaced if pits exist and are deeper than ____ .
    a. 1⁄32" (0.8 mm)
    b. 1⁄16" (1.6 mm)
    c. 1⁄8" (3.2 mm)
    d. 3⁄16" (4.8 mm)

Trade Terms Quiz

Fill in the blank with the correct term that you learned from your study of this module.

1. A fourth state of matter created by heating a gas to such a high temperature that it boils electrons off the gas molecules is called _______ .
2. The relative electrical voltage difference between two points of reference is called _______ .
3. A(n) _______ can control the flow of gases and liquids, and is operated by an electromagnet.
4. The percentage of time a PAC unit can cut without overheating within a 10-minute period is called its _______ .
5. Each of the three sine waves of a power supply that uses three conductors, all carrying alternating current, is referred to as a(n) _______ .
6. The unit of measure for electrical current is _______ .
7. A waste byproduct of molten metal resulting from plasma cutting processes is known as _______ .

Trade Terms

Amperage
Dross
Duty cycle
Phase
Plasma
Potential
Solenoid valve

Performance Accreditation Tasks

The American Welding Society (AWS) School Excelling through National Skills Standards Education (SENSE) program is a comprehensive set of minimum Standards and Guidelines for Welding Education programs. The following performance accreditation tasks are aligned with and designed around the SENSE program.

The Performance Accreditation Tasks (PATs) correspond to and support the learning objectives in *AWS EG2.0, Guide for the Training and Qualification of Welding Personnel: Entry-Level Welder*.

Note that in order to satisfy all learning objectives in *AWS EG2.0*, the instructor must also use the PATs contained in the second level of the NCCER Welding curriculum.

PATs 1 and 2 correspond to *AWS EG2.0, Module 8 – Thermal Cutting Processes, Unit 3 – Manual Plasma Arc Cutting (PAC)*, Key Indicators 3, 4, and 5.

PATs provide specific acceptable criteria for performance and help to ensure a true competency-based welding program for students.

The following tasks are designed to develop your competency with a plasma arc cutting torch. Practice each task until you are thoroughly familiar with the procedure.

As you complete each task, take it to your instructor for evaluation. Do not proceed to the next task until instructed to do so by your instructor.

PLASMA ARC CUTTING

Using electrically conductive material, lay out and cut the shape shown in the figure in the horizontal position. When it is finished, cut your initials into the shape.

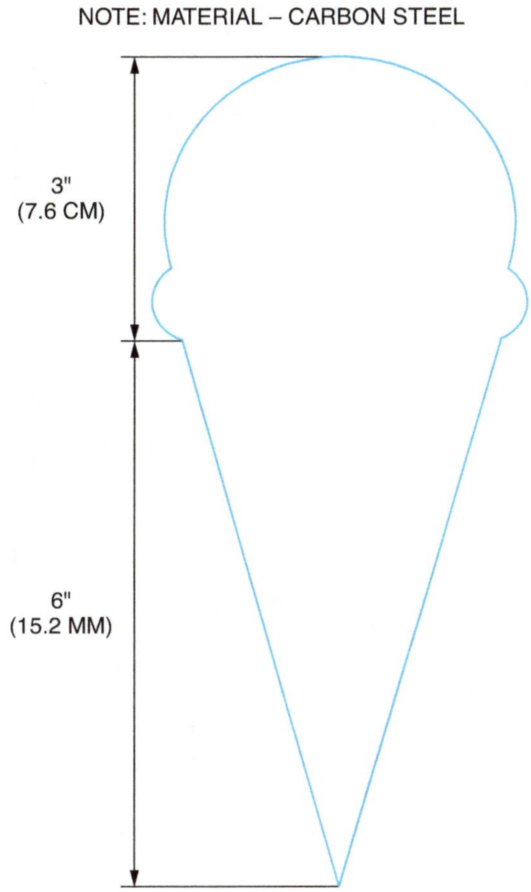

29103-15_A02.EPS

Criteria for Acceptance:

- Dimensions ±¹⁄₁₆" ____________
- Minimal amount of dross sticking to plate which can be easily removed ____________

PLASMA ARC CUTTING

Using electrically conductive material, lay out and cut the shape shown in the figure in the flat position.

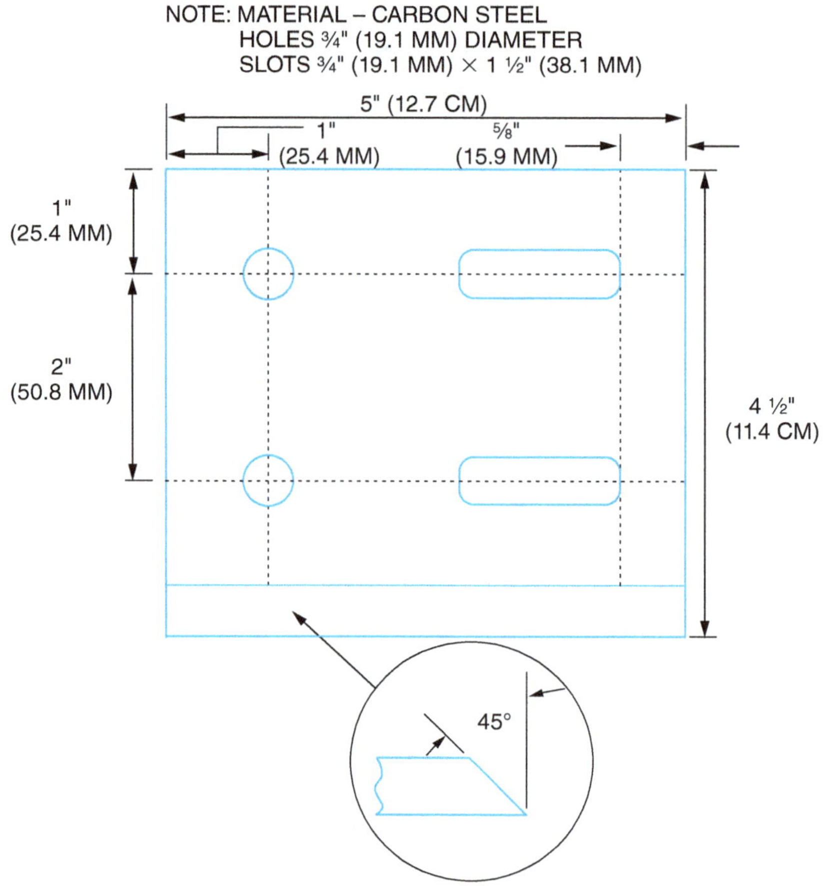

29103-15_A01.EPS

Criteria for Acceptance:

- Outside dimensions ±1⁄16" __________
- Inside (holes and slots) dimensions ±1⁄16" __________
- Square ±2° __________
- Bevel ±2° __________
- Minimal amount of dross sticking to plate which can be easily removed __________
- Square kerf face with minimal notching not exceeding ±1⁄16" deep __________

Cornerstone of Craftsmanship

Curtis Casey

Northland Pioneer College
Emeritus Welding Faculty

Curtis received his welding training in the US Navy, where he was originally slated to be trained as a submarine power plant operator. However, those jobs were filled when he joined, so he chose advanced training as a welder. Curtis welded on SSBN class submarines for six years, working on high-tensile hulls, pressure pipe, and nuclear components. After the Navy, a new opportunity opened up at Bechtel Power Corporation, where he became a Certified Welding Inspector. For the next 15 years, Curtis traveled around the United States helping to build nuclear power plants. He also found the time to earn a Welding Technology degree on the G.I. Bill and to take on supervisory responsibilities. Not long after joining the faculty at Northland Pioneer College, Curtis was appointed chairperson of the Welding Department, which now serves hundreds of students each year.

How did you choose a career in the construction industry?
I took the opportunity to train while in the US Navy's nuclear welding program after graduating from high school. There I got a chance to weld on submarines and other naval vessels.

Who inspired you to enter the industry?
My stepfather inspired me to enlist in the US Navy where I received advanced training in welding. He was a career sailor who always had great stories about seeing the world.

What types of training have you been through?
In the Navy, I received training in the metals trades to be a hull maintenance technician. I finished welding school, learning to weld on both common and exotic materials. After the military I went to college while I worked as a welding inspector in nuclear power plants and got a chance to teach weld inspection as an adjunct professor while attending college. From then on, I was hooked on training and found the opportunity to teach welding full time at Northland Pioneer College. Currently I am a welding consultant wearing many hats within the industry from inspecting to auditing to teaching and consulting.

How important are education and training in construction?
Education is an ongoing process; life-long learning is another way to say it. Education and training are the tools we fit ourselves with to be able to create and mold our careers. Without them our direction in life could easily be aimless and unproductive.

How important are NCCER credentials to your career?
NCCER credentialing gives me an edge over many of my colleagues and peers because of the clout the organization has within the industry. Many companies now are requiring these credentials just to get your foot in their door.

How has training/construction impacted your life?
Simply put, it has given me great sense of satisfaction knowing that I can design and build on ideas that come to my mind. I can take a project from inception to creation because of my training.

What kinds of work have you done in your career?
I started my adult working life as a welder, then worked as a welding inspector, a field welding engineer, a welding business owner, a welding instructor, welding department chair, welding curriculum developer, and welding consultant. I think welding sums it up pretty well. I still maintain my AWS Certified Welding Instructor (CWI) and Certified Welding Educator (CWE) certifications, and I'm actively involved with my students in SkillsUSA, a national student leadership organization.

Tell us about your present job.
I'm currently consulting in the welding arena while occasionally inspecting and auditing other NCCER training and assessment centers.

What do you enjoy most about your job?
My passion has been teaching welding at Northland Pioneer College for the past 15 years, and now in semi-retirement. I love the versatility in what I do. One day I could be doing R & D at a fabrication shop, and the next I could be inspecting welds or auditing programs. It is refreshing to see the newest technology being put to use at certain companies; in this job I see it firsthand.

What factors have contributed most to your success?
Work ethics! I learned at an early age to be responsible for my own successes and failures. I pitched papers on a paper route when I was 11 years old. That taught me to be dependable and gave me exposure to the business world and having to deal with customers, good and bad.

Would you suggest construction as a career to others? Why?
If they were interested in seeing an idea come to fruition from design to completion and if they were willing to get their hands dirty, then I would recommend construction. One can receive great satisfaction and get a sense of worth when they take part in the creation of something that people can actually use in their lives.

What advice would you give to those new to the welding field?
Try it out through a job shadowing program to see if you like it. Then if you dream about it that night, you probably have a passion for it. Then simply follow your passion! Get the right training, surround yourself with like-minded people, keep a good attitude, network a lot, and go to work!

Tell us an interesting career-related fact or accomplishment.
I have had the chance to teach close to 1,000 people the craft that has given me my career, and have seen many former students return to share their success in the industry. That's an emotional paycheck worth the effort.

How do you define craftsmanship?
Developing a good balance between technical and artistic skills.

Trade Terms Introduced in This Module

Amperage: The unit of measure used for the intensity of an electric current; often abbreviated to *amp* or *amps*.

Dross: A waste byproduct of molten metal.

Duty cycle:: The percentage of time a plasma arc cutting machine can cut without overheating within a ten-minute period.

Phase: In a three-phase power supply system, the sine waves of the voltage on each of the three separate conductors is displaced from each of the others by 120 degrees, although all conductors are carrying alternating current at the same frequency and are synchronized. Each of the three sine waves is referred to as a phase of the power source.

Plasma: A fourth state of matter (not solid, liquid, or gas) created by heating a gas to such a high temperature that it boils electrons off the gas molecules (ionization).

Potential: The relative electrical voltage difference between two points of reference; the electromotive force or difference in potential between two points, expressed in volts.

Solenoid valve: A valve used to control the flow of gases or liquids that is opened or closed by the action of an energized electromagnet.

Additional Resources

This module presents thorough resources for task training. The following resource material is suggested for further study.

Plasma Cutters Handbook: Choosing Plasma Cutters, Shop Safety, Basic Operation, Cutting Procedures, Advanced Cutting Tips, CNC Plasma Cutters, Troubleshooting, and Sample Projects. Eddie Paul. New York, NY: Penguin Group.

AWS F3.2M/F3.2, Ventilation Guide for Weld Fume. Latest Edition. Miami, FL: American Welding Society.

Figure Credits

The Lincoln Electric Company, Cleveland, OH, USA, Module Opener, Figures 1A, 2 (photo), 5, 8 (photos), 11, 12, 15, SA01

Courtesy of BUG-O Systems Inc., Figure 1B

AWS F2.2:2001, Lens Shade Selector, reproduced with permission from the American Welding Society (AWS), Miami, FL, USA, Table 1

Topaz Publications, Inc., Figures 6, SA03

ESAB Welding and Cutting Products, Figures 7, 13A

Courtesy of Miller Electric Mfg. Co., Figure 9A

Victor Technologies, Figure 9B, 13B

Courtesy of Uniweld Products, Figure 10

The Eastwood Company, SA02

Section Review Answer Key

Answer	Section Reference	Objective
Section One		
1. d	1.1.0	1a
2. c	1.2.1	1b
Section Two		
1. a	2.1.0	2a
2. d	2.2.0	2b
3. a	2.2.0	2c
Section Three		
1. d	3.1.3	3a
2. a	3.2.3	3b
3. c	3.3.2	3c

NCCER CURRICULA — USER UPDATE

NCCER makes every effort to keep its textbooks up-to-date and free of technical errors. We appreciate your help in this process. If you find an error, a typographical mistake, or an inaccuracy in NCCER's curricula, please fill out this form (or a photocopy), or complete the online form at **www.nccer.org/olf**. Be sure to include the exact module ID number, page number, a detailed description, and your recommended correction. Your input will be brought to the attention of the Authoring Team. Thank you for your assistance.

Instructors – If you have an idea for improving this textbook, or have found that additional materials were necessary to teach this module effectively, please let us know so that we may present your suggestions to the Authoring Team.

NCCER Product Development and Revision

13614 Progress Blvd., Alachua, FL 32615

Email: curriculum@nccer.org

Online: www.nccer.org/olf

❑ Trainee Guide ❑ Lesson Plans ❑ Exam ❑ PowerPoints Other ____________

Craft / Level: Copyright Date:

Module ID Number / Title:

Section Number(s):

Description:

Recommended Correction:

Your Name:

Address:

Email: Phone:

Base Metal Preparation

Overview

This module introduces information related to the preparation of metals for welding, including the identification of common base metals, the physical characteristics and mechanical properties of different metals, and cleaning techniques. Welding joint preparation, codes, and specifications related to metal preparation are also described.

Module 29105

Trainees with successful module completions may be eligible for credentialing through the NCCER Registry. To learn more, go to **www.nccer.org** or contact us at **1.888.622.3720**. Our website has information on the latest product releases and training, as well as online versions of our *Cornerstone* magazine and Pearson's product catalog.

Your feedback is welcome. You may email your comments to **curriculum@nccer.org**, send general comments and inquiries to **info@nccer.org**, or fill in the User Update form at the back of this module.

This information is general in nature and intended for training purposes only. Actual performance of activities described in this manual requires compliance with all applicable operating, service, maintenance, and safety procedures under the direction of qualified personnel. References in this manual to patented or proprietary devices do not constitute a recommendation of their use.

29105 V5

29105

Base Metal Preparation

Objectives

When you have completed this module, you will be able to do the following:

1. Identify safety practices related to preparing base metals and describe basic cleaning procedures.
   a. Identify safety practices related to preparing base metals.
   b. Describe the basic properties and types of carbon and stainless steel.
   c. Describe basic metal cleaning procedures and concerns.
2. Identify and describe basic weld joint design and types of welds.
   a. Identify and describe the loads that are routinely placed on weld joints.
   b. Identify and describe the various types of weld joints.
   c. Describe a welding procedure specification (WPS) and the information it provides.
3. Describe how to prepare joints for welding.
   a. Describe how to mechanically prepare joints for welding.
   b. Describe how to thermally prepare joints for welding.

Performance Tasks

Under the supervision of your instructor, you should be able to do the following:

1. Mechanically or hand grind a bevel on the edge of a ¼"- to ¾"-thick mild steel plate (6 to 20 mm metric plate) at 22½ degrees.
2. Thermally bevel the edge of a ¼"- to ¾"-thick mild steel plate (6 to 20 mm metric plate) at 22½ degrees.

Trade Terms

Annealed
Austenitic
Backing
Base metal
Castings
Distortion
Ductile
Ferritic
Ferrous
Load
Malleable
Martensitic
Melt-through
Nonferrous metal
Oxide
Piping porosity
Porosity
Quench
Root face
Root opening
Surfacing
Tempering
Tensile strength
Weathering steel
Welding procedure specification (WPS)
Weldment
Wrought

Industry Recognized Credentials

If you are training through an NCCER-accredited sponsor, you may be eligible for credentials from NCCER's Registry. The ID number for this module is 29105. Note that this module may have been used in other NCCER curricula and may apply to other level completions. Contact NCCER's Registry at 888.622.3720 or go to **www.nccer.org** for more information.

Contents

Topics to be presented in this module include:

Figures and Tables

Section One

1.0.0 Basic Welding Safety and Cleaning

Objective

Identify safety practices related to preparing base metals and describe basic cleaning procedures.

a. Identify safety practices related to preparing base metals.
b. Describe the basic properties and types of carbon and stainless steel.
c. Describe basic metal cleaning procedures and concerns.

Trade Terms

Annealed: To free from internal stress by heating and gradually cooling.

Austenitic: Containing austenite (a solid solution of carbon, or of carbon and other elements in a ferrous alloy) added through heating.

Base metal: As used in this module, metal to be welded, cut, or brazed.

Castings: Something cast; any article that has been cast in a mold.

Distortion: The expansion and contraction of welded parts caused by the heating and subsequent cooling of the weld joint.

Ductile: Able to undergo change of form without breaking.

Ferritic: Steel containing less than 0.10 percent carbon and is magnetic. This steel can't be hardened via heat treatment.

Ferrous: Containing iron.

Malleable: Capable of being extended or shaped by hammering or by pressure from rollers.

Martensitic: Steel that shares some characteristics with ferritic, but has a higher level of carbon, up to a full 1 percent. It can be tempered and hardened and is used where strength is more important than a resistance to oxidation.

Nonferrous metal: A metal, such as aluminum, copper, or brass, lacking sufficient quantities of iron to have any effect on its properties.

Oxide: The scale that forms on metal surfaces when they are exposed to oxygen or air containing oxygen.

Piping porosity: A form of porosity having a length greater than its width and that is approximately perpendicular to the weld face.

Porosity: Gas pockets, or voids in the weld metal.

Quench: To cool suddenly by plunging into a liquid.

Tempering: To impart strength or toughness to (steel or cast iron) by heating and cooling.

Weathering steel: Steel alloy that, under specific conditions, is designed to form a very dense oxide layer on its outer surfaces, which retards further oxidation.

Weldment: An assembly that is fastened together by welded joints.

Wrought: Produced or shaped by beating with a hammer, as with iron.

To ensure that the safest and highest quality welds are produced, and to comply with welding codes, base metals must be properly prepared prior to welding. Safety must be priority one, and proper safety precautions add to the welder's organizational skills during the preparation process. The type of preparation required depends on the governing code requirements, the base metal type, the condition of the base metal, the welding process to be used, and the equipment available.

1.1.0 Basic Welding Safety

Complete safety coverage is provided in the *Welding Safety* module. If you have not completed that module, do so before continuing.

Welders routinely use manual and powered tools to clean and shape metals. *Figures 1* and *2* are examples of metal being prepared for welding. Everyone entering the welding trade starts by doing basic tasks, which includes preparing the metals for welding.

When using tools to clean metals prior to welding, wires from rotating wire brushes and particles from abrasive discs can become flying objects. Metal particles ground from the workpieces are usually hot and can cause burns, severe eye injuries, or fires. Welding screens or similar protective devices are used to keep the hot particles from igniting combustible materials and harming other workers nearby. Most companies require a fire extinguisher and a fire watch posted near the welding and cutting activities. To prevent injury from the flying particles and debris, always wear the proper personal protective equipment (PPE).

Before working on welding equipment and power tools, ensure that they are disconnected from the power supply. When the welding machine is energized, the open circuit between the electrode holder and the workpiece clamp represents a significant electrical hazard.

While much of the preparation work is performed in the shop where parts can be secured in clamping devices, some of it must be performed in the field on structures or equipment. Welders must sometimes perform at heights where fall prevention and fall arrest equipment is required. Always make sure that fall protection devices are undamaged and the inspections are current before using them.

The following is a summary of safety procedures and practices that must be observed while cutting, cleaning, shaping, and welding. Keep in mind that this is a summary. Complete safety coverage is provided in the *Welding Safety* module. If you have not completed that module, do so before continuing.

1.1.1 Protective Clothing and Equipment

Welding activities can cause injuries unless you wear all of the protective clothing and equipment that is designed specifically for the welding industry. The following information includes important safety guidelines about protective clothing and equipment:

29105-15_F01.EPS

Figure 1 Cutting metal with an oxyfuel torch.

29105-15_F02.EPS

Figure 2 Grinding metal.

- Always use safety glasses with a full face shield or a helmet. The glasses, face shield, or helmet lens must have the proper light-reducing tint for the type of welding or cutting being performed. Use corrective lenses if needed; some companies will not allow contact lenses to be worn during cutting or welding activities. Never view an electric arc directly or indirectly without using a properly tinted lens.
- Wear proper protective leather and or flame-retardant clothing along with welding gloves that will protect you from flying sparks and molten metal (*Figure 3*).

29105-15_F03.EPS

Figure 3 Appropriate personal protective equipment (PPE).

- Wear high-top safety shoes or boots. Make sure that the tongue and lace area of the footwear will be covered by a pant leg or other protective material. Boots with no laces at all are generally preferred for both welding and cutting activities. If the tongue and lace area is exposed, or if the footwear must be protected from burn marks, wear leather spats under your pants or chaps and over the top front area of the footwear.
- Wear a 100-percent cotton cap with no ventilation mesh material included in its construction. The bill of the cap points to the rear. If a hard hat is required for the environment, use one that allows the attachment of rear deflector and a face shield. A hard hat with a rear deflector is generally preferred when working overhead, and may be required by some employers and job sites.

> **WARNING!**
> Do not wear a cap with a button in the middle. The conductive metal button beneath the fabric represents a safety hazard.

- Wear a face shield over snug-fitting cutting goggles or safety glasses for gas welding or cutting. Either the face shield or the lenses of the welding goggles must be an approved shade for the application. A welding hood with a properly tinted lens is also acceptable. See *Table 1* for a list of lens shades based on arc current.
- Wear earplugs to protect ear canals from sparks. Wear hearing protection to protect against the consistent sound of the torch.

> **WARNING!**
> Ear protection is essential to protect ears from the noise of the torch. Other personal protective equipment (PPE) must be worn to protect the operator from hot metal and slag.

1.1.2 Fire and Explosion Prevention

Welding activities usually involve the use of fire or extreme heat to melt metal. Although most work site fires are caused by cutting torches and the hot material they produce, welding sparks and spatter can also start fires easily. Whenever fire is used, it must be controlled and contained. Welding or cutting activities are often performed on vessels that may once have contained flammable or explosive materials. Residues from those materials can catch fire or explode when a welder begins work on such a vessel. The following are fire and explosion prevention guidelines associated with welding:

- Never carry matches or gas-filled lighters in your pockets. Sparks can cause the matches to ignite or the lighter to explode, causing serious injury.
- Always comply with any site requirement for a hot-work permit and/or a fire watch.
- Never use oxygen to blow dirt or dust off clothing. The oxygen can remain trapped in the fabric for a time. If a spark hits the oxygen in the fabric, the clothing can burn rapidly and violently.
- Never release a large amount of oxygen or use oxygen in place of compressed air. The presence of oxygen around flammable materials or sparks can cause rapid and uncontrolled combustion. Keep oxygen away from oil, grease, and other petroleum products.
- Make sure that any flammable material in the work area is moved or shielded by a fire-resistant covering.
- Approved fire extinguishers must be available before attempting any heating, welding, or cutting operations. Make sure the extinguisher is charged, the inspection tag is valid, and any individual that may be required to operate it knows how to do so.
- Never release a large amount of fuel gas, especially acetylene. Propylene and propane are heavier than air and tend to migrate to and concentrate in low areas. As a result, they can ignite at a considerable distance from the release point. Acetylene is lighter than air but is even more dangerous than methane; when mixed with air or oxygen, it will explode at much lower concentrations than any other common fuel gas.
- To prevent fires, maintain a neat and clean work area, and make sure that any metal scrap or slag is cold before disposal.

Before cutting or welding containers such as tanks or barrels, find out if they contained any explosive, hazardous, or flammable materials, including petroleum products, citrus products, or chemicals that decompose into toxic fumes when heated. Proper procedures for cutting or

Table 1 Guide for Shade Numbers

From *AWS F2.2 Lens Shade Selector*. Shade numbers are given as a guide only and may be varied to suit individual needs.

Process	Electrode Size in (mm)	Arc Current (Amperes)	Minimum Protective Shade	Suggested* Shade No. (Comfort)
Shielded Metal Arc Welding (SMAW)	Less than 3/32 (2.4)	Less than 60	7	–
	3/32–5/32 (2.4–4.0)	60–160	8	10
	5/32–1/4 (4.0–6.4)	160–250	10	12
	More than 1/4 (6.4)	250–550	11	14
Gas Metal Arc Welding (GMAW) and Flux Cored Arc Welding (FCAW)		Less than 60	7	–
		60–160	10	11
		160–250	10	12
		250–500	10	14
Gas Tungsten Arc Welding (GTAW)		Less than 50	8	10
		50–150	8	12
		150–500	10	14
Air Carbon Arc Cutting (CAC-A)	(Light)	Less than 500	10	12
	(Heavy)	500–1000	11	14
Plasma Arc Welding (PAW)		Less than 20	6	6 to 8
		20–100	8	10
		100–400	10	12
		400–800	11	14
Plasma Arc Cutting (PAC)		Less than 20	4	4
		20–40	5	5
		40–60	6	6
		60–80	8	8
		80–300	8	9
		300–400	9	12
		400–800	10	14
Torch Brazing (TB)		–	–	3 or 4
Torch Soldering (TS)		–	–	2
Carbon Arc Welding (CAW)		–	–	14

	Plate Thickness		Suggested* Shade No. (Comfort)
	in	mm	
Oxyfuel Gas Welding (OFW)			
Light	Under 1/8	Under 3	4 or 5
Medium	1/8 to 1/2	3 to 13	5 or 6
Heavy	Over 1/2	Over 13	6 or 8
Oxygen Cutting (OC)			
Light	Under 1	Under 25	3 or 4
Medium	1 to 6	25 to 150	4 or 5
Heavy	Over 6	Over 150	5 or 6

*As a rule of thumb, start with a shade that is too dark to see the weld zone. Then go to a lighter shade which gives sufficient view of the weld zone without going below the minimum. In oxyfuel gas welding, cutting, or brazing where the torch and/or the flux produces a high yellow light, it is desirable to use a filter lens that absorbs the yellow or sodium line of the visible light spectrum.

29105-15_T01.EPS

welding hazardous containers are described in the *American Welding Society (AWS) F4.1, Safe Practices for the Preparation of Containers and Piping for Welding and Cutting,* and *ANSI Z49.1.* As a standard practice, always clean and then fill any tanks or barrels with water, or purge them with a flow of inert gas such as nitrogen to displace any oxygen.

WARNING!

Welding or cutting must never be performed on drums, barrels, tanks, vessels, or other containers until they have been emptied and cleaned thoroughly, eliminating all flammable materials and all substances (such as detergents, solvents, greases, tars, or acids) that might produce flammable, toxic, or explosive vapors when heated. Do not assume that a container that has held combustibles is clean and safe until proven so by proper tests. Do not weld in places where dust or other combustible particles are suspended in air or where explosive vapors are present.

Containers must be cleaned by steam cleaning, flushing with water, or washing with detergent until all traces of the material have been removed.

WARNING!

Clean containers only in well-ventilated areas. Vapors can accumulate during cleaning, causing explosions or injury.

After cleaning the container, fill it with water or a purging gas, such as carbon dioxide, argon, or nitrogen to displace the explosive fumes. Air, which contains oxygen, is displaced from inside the container by the water or inert gas. Without oxygen, combustion cannot take place.

A water-filled vessel is the best alternative. When using water, position the container to minimize the air space. When using an inert gas, provide a vent hole so the inert gas can push the air and other vapors out to the atmosphere. Keep in mind, though, that even these precautions do not guarantee the absence of flammable materials inside. For that reason, these types of activities should not be done without proper supervision and the use of proper testing methods.

GOING GREEN

Welding Water

To better protect the environment, send any water used in a welding/cutting process or used to clean tanks to a waste treatment facility instead of simply allowing it to run into a storm drain or onto the ground.

1.1.3 *Work Area Ventilation*

Vapors and fumes tend to rise in the air from their sources. Welders often have to work above the welding area where the fumes are being created. Welding fumes can cause personal injuries. Good work area ventilation helps to protect the welder by removing the vapors. The following is a list of work area ventilation guidelines to consider before and during preparation activities:

- Make sure confined space procedures are followed before conducting any welding or cutting in a confined space.
- Always perform cutting or welding operations in a well-ventilated area. Cutting or welding operations involving materials such as coatings that contain cadmium, lead, zinc, or chromium will result in toxic fumes. For cutting or welding of such materials, always wear an approved respirator with an appropriate filter as directed by your employer.
- Grinding wheels containing aluminum oxide, silicon carbide and other materials have a composition that requires the use of respiratory protection. Consult the SDS/MSDS for warnings before starting any grinding activities.
- Make sure confined spaces are ventilated properly for cutting or welding purposes.
- Never use oxygen in confined spaces for purposes of ventilation.

GOING GREEN

Welding Fumes

To better protect the environment, perform any welding activities under a vapor extraction system that can filter the welding fumes before they reach the atmosphere.

1.2.0 Properties of Steel

Metals are a collection of chemical elements bound together by natural forces or manmade production processes. The chemical elements in some metals make them stronger and give them the ability to withstand stress or force better than other metals. The composition of common welding base metals varies widely, from metals made of essentially one metallic element, to alloys (mixtures) of metallic and nonmetallic components.

Metals are classified into two basic groups: ferrous metals, which are composed mainly of iron, and nonferrous metals, which contain very little or no iron. Ferrous metals include all steel, cast iron, wrought iron, malleable iron, and ductile (nodular) iron. Nonferrous metals and their alloys include the light metals (aluminum, magnesium, titanium); the heavy metals (copper, nickel, lead, tin, zinc); and the precious metals (platinum, gold, silver). Ferrous metals contain mostly iron and some carbon. Special steels may contain as much as 2 percent carbon. Carbon steel, the largest group of the ferrous-based metals, generally contains less than 1.7 percent carbon.

At low levels of carbon content, steel has some of the properties of cast iron, but is more ductile and more easily formed. Above 1.4 percent carbon, steel becomes stronger and more wear-resistant, although it becomes less ductile (difficult to form) and more difficult to weld. The higher carbon content lowers the melting temperature. The carbon content of a particular carbon steel product is the primary factor in determining its use.

Steel castings usually contain a higher percentage of carbon than rolled plate and other rolled shapes. Casting is done instead of rolling or shaping due to the loss of ductility in high-carbon steels. The casting foundry can usually provide the alloy formula, including the carbon content.

Steels, including carbon steels, low-alloy carbon steels, alloy steels, and stainless steels, are classified by various systems, such as by specification

Carbon and Alloy Steels

Steel is an alloy of iron and carbon typically containing less than 1 percent carbon. All steels also contain varying amounts of other elements such as manganese, silicon, phosphorus, sulfur, and oxygen. In addition, some standard alloy steels can also contain elements such as nickel, chromium, and molybdenum. There are currently about 3,500 different grades of steel with many different properties. Seventy percent of these steels have been developed in the past 25 years. Most of the steels produced today are carbon and alloy steels. Ninety-five percent of the construction and fabrication metals used worldwide consist of these materials.

29105-15_SA01.EPS

number and grade, or by manufacturer's trade name and number. Steel classification systems include those developed by the American Iron and Steel Institute (AISI), the American Society for Testing Materials International (ASTM), and the Unified Numbering System (UNS), as well as manufacturer's trade names and identification numbers. Stainless steels have their own classification systems. They are sometimes referred to by the percentages of their chromium and nickel content, such as 18/8, 25/20, and 18/10, but this system has largely been replaced by the AISI stainless steel classification system. Common classification systems for steels are explained in the following sections.

1.2.1 Carbon Steel Classification

The principal classification system for carbon steels is the AISI numerical designation of Standard Carbon and Alloy Steels. It was originally a four-digit system developed by the Society of Automotive Engineers (SAE) for carbon steels commonly used in structural shapes, plate, strip, sheet, and welded tubing. Five digits are now used for some alloys. The classification system may be referred to as the AISI, SAE, or the AISI and SAE system.

The AISI and SAE designations are essentially the same, except that the AISI system sometimes uses a letter prefix to indicate the manufacturing process that produced the steel. The following are examples of letter designations:

- A – Open-hearth steel
- B – Acid-Bessemer carbon steel
- C – Basic open-hearth carbon steel
- D – Acid open-hearth carbon steel
- E – Electric furnace steel

The absence of a letter prefix indicates basic open-hearth or acid-Bessemer carbon steel. The prefix letter (if any) designates the manufacturing process. The first two numerical digits represent the series (type and class) of steel. The third and fourth (and sometimes fifth) numerical digits specify the approximate percentage of carbon content.

The following examples explain the AISI and SAE numbers for carbon steels:

- *AISI number C1020*:
  C = Indicates basic open-hearth carbon steel
  10 = Carbon steel, nonresulfurized
  20 = Contains approximately 0.20 percent carbon
- *AISI number E2512*:
  E = Indicates electric furnace steel
  25 = Designates steel alloyed with approximately 5 percent nickel
  12 = Designates steel containing approximately 0.12 percent carbon
- *AISI number E52100*:
  E = Indicates electric furnace steel
  5 = Contains either approximately 0.50 percent, 1.00 percent, or 1.45 percent chromium, designated by the next digit: 0, 1, or 2, respectively
  2 = Designates approximately 1.45 percent chromium
  100 = Designates approximately 1 percent carbon

The common group classification of carbon steels is primarily based on carbon content. Plain carbon steels (AISI series 10XX through 11XX) are basically iron-carbon alloys. High-strength low alloy (HSLA) carbon steels (AISI series 13XX through 98XX) have small amounts of alloying elements to improve strength, hardness, and toughness or to increase resistance to oxidation, heat, and environmental damage. The plain carbon steels are commonly grouped as follows:

- *Low-carbon* – 0.10 percent to 0.15 percent carbon, 0.25 percent to 1.50 percent manganese
- *Mild-carbon* – 0.15 percent to 0.30 percent carbon, 0.60 percent to 0.70 percent manganese
- *Medium-carbon* – 0.30 percent to 0.50 percent carbon, 0.60 percent to 1.65 percent manganese
- *High-carbon* – 0.50 percent to 1.00 percent carbon, 0.30 percent to 1.00 percent manganese

The HSLA carbon steels are alloyed with one or more of the elements manganese, nickel, chromium, or molybdenum to provide higher strength, better toughness, weldability, and, in some cases, greater resistance to oxidation. As the percentage of these elements increases, changes in the weldability of the low-alloy steels make electrode selection and welding procedures more critical. HSLA steels include the following:

- Low-nickel steels
- Low-nickel chrome steels
- Low-manganese steels
- Low-alloy chromium steels
- **Weathering steels**

Another class of steel in the AISI series is sulfur steel. It is not an alloy steel but is included in the series.

1.2.2 Common Grade Stainless Steel Classifications

Stainless steels are typically classified as being a common grade or a specialty grade. Common-grade stainless steels (oxidation-resistant steels) are iron-based alloys that normally contain at least 11 percent chromium. Other alloying elements, including nickel, carbon, manganese, and silicon, may be present in varying quantities, depending on the specific type of stainless steel, to enhance its physical and mechanical properties. The main physical characteristics of all stainless steels are their resistance to oxidation and heat. Some have good low- and high-temperature mechanical properties. When compared with mild steels, stainless steels have the following characteristics:

- Lower coefficients of thermal conductivity that increase the chances of distortion
- Higher coefficients of thermal expansion that increase the chances of distortion
- Higher electrical resistances that increase the tendency to build up heat from welding current

Common-grade stainless steels are classified by their grain structures. The type of grain structure is determined by the specific alloy content of the stainless steel and its heat treatment during manufacture. Based on their microcrystalline structures, the classification of common-grade stainless steels is divided into these three groups:

- Austenitic
- Ferritic
- Martensitic

Austenitic stainless steels are non-magnetic in the annealed condition and not able to be hardened by heat treatment. However, they can be hardened significantly by cold working. Austenitic stainless steels combine excellent oxidation and heat resistance with good mechanical properties over a broad temperature range. Austenitic stainless steels make up the largest of the three common-grade stainless steel groups. For this reason, austenitic stainless steel is encountered most often by welders. They include the AISI 200 and 300 series stainless steels that all contain significant amounts of both chromium and nickel.

Austenitic steels are sometimes further subdivided in two classifications based on their compositions: chromium-nickel-magnesium (AISI 200 series) and chromium-nickel (AISI 300 series). The range of applications for austenitic stainless steels includes housewares, containers, industrial piping and vessels, and architectural facades. Austenitic stainless steels are also available in low-carbon (L-grade) and high-carbon (H-grade) types. The letter L after a stainless steel type, such as 304L, indicates a low-carbon content of 0.03 percent or under. Similarly, the letter H after a stainless steel type indicates a high-carbon content, ranging between 0.04 and 0.10 percent. L-grades are typically used when annealing after welding is impractical, such as in the field where pipe and fittings are welded. H-grades are used when the steel will be subjected to extreme temperatures; the higher carbon content helps it retain strength.

Ferritic stainless steels are always magnetic. They are hardened to some extent by cold working, not by heat treatment. Ferritic stainless steels combine oxidation and heat resistance with fair mechanical properties over a narrower temperature range than austenitic stainless steels. Ferritic stainless steels are straight chromium stainless steels containing 11.5 to 27 percent chromium, about 1 percent manganese, and little or no nickel. The carbon content is 0.20 percent or less. Examples of ferritic stainless steels include AISI types 405, 409, 430, 442, 444, and 446. They typically are used for decorative trim, sinks, and automotive applications, particularly exhaust systems.

Martensitic stainless steels are also magnetic and can be hardened by quenching and tempering. They are excellent for use in mild environments such as the atmosphere, freshwater, steam, and with weak acids. However, they are not resistant to severely corrosive solutions. Martensitic stainless steels comprise two groups: martensitic and chromium-martensitic stainless steels.

Martensitic stainless steels contain only 4 to 6 percent chromium and no nickel. Chromium-martensitic stainless steels contain from 11.5 to 18 percent chromium, about 1 percent manganese, and in some cases 0 to 2.5 percent nickel. Because of the low chromium and no nickel, martensitic stainless steels are not considered true stainless steels, although their oxidation resistance is much greater than mild carbon steels, even at elevated temperatures.

Examples of chromium-martensitic stainless steels are AISI types 403, 410, 414, 420, 422, 431, and 440. Examples of martensitic stainless steels are AISI types 501 and 502.

1.3.0 Base Metal Cleaning

All base metals must be cleaned before welding. Even new materials that may look clean often are not. They pick up contaminants and surface oxidation during shipping and handling. When performing maintenance welding, a welder often comes across components that have been exposed to surface contamination such as oxidation, paint, oil, and grease. The heat generated by the welding process could cause paints and coatings to give off toxic fumes, endangering the worker performing the welding.

Do not assume a metal's surface is oxidation-free just because it is shiny. Oxidation on stainless steel and aluminum is difficult to see; it may have a white color or it may be transparent. To ensure quality welds and to conform to code requirements, surface contaminants and oxides must be removed prior to welding. In general, the codes state that the surface to be thermally cut or welded must be clean and free from paint, oil, rust, scale, and other material that would be harmful either to the weld or to the base metal when heat is applied. Cleaning is typically performed by mechanical and or chemical means.

Most metal has some surface oxidation. Oxidation occurs when metal is exposed to oxygen in the air or from other sources. Some oxidation exists as a thin, hard film that merely stains the metal and, in fact, acts as a retardant to further oxidation. Examples of this type of oxidation are found on stainless steel, copper, and aluminum. The thin, hard layer of oxidation bonds tightly to the surface and protects the metal.

A more recognizable type of oxidation found on mild steel is called rust (*Figure 4*). Rust is a very coarse type of oxidation that tends to flake easily, exposing more of the base metal to oxidation. Alloys such as chromium and copper are often added to mild steel to protect it from this coarse type of oxidation. These types of steels are called alloy steels. The alloys cause the rust to be very fine, bonding to the surface and protecting the base metal beneath. Weathering steel is an example of a copper-alloyed steel that protects itself from oxidation by forming a hard, tough layer of brown oxidation (rust) on its surface. Weathering steels are used outdoors and require no painting.

29105-15_F04.EPS

Figure 4 Examples of oxidation.

Regardless of the type of contamination, it must be removed prior to welding. Any type of contamination can cause weld defects. The most common defect caused by surface contamination is porosity. Examples of porosity are shown in *Figure 5*. Porosity occurs when gas pockets or voids appear in the weld metal. When the porosity has a length greater than its width and is approximately perpendicular to the weld face, it is called piping porosity. Piping porosity is formed as the gas pocket or bubble floats toward the surface of the weld, leaving a void. The gas pocket is trapped as the weld metal solidifies, but as the next layer of weld is deposited, the bubble often continues to float up. Porosity may not always be visible on the surface of the weld.

Surface contamination can easily result from any of several causes, including the following:

- Use of gloves contaminated with oil or other substances
- Allowing alloys to come into contact with carbon steel objects, including chain falls, jack stands, and forklifts
- The use of chloride-based markers to write on the base metal

Because of the risk of contamination, metals such as carbon steel, stainless steel, and aluminum must be stored separately. Aluminum must be stored indoors.

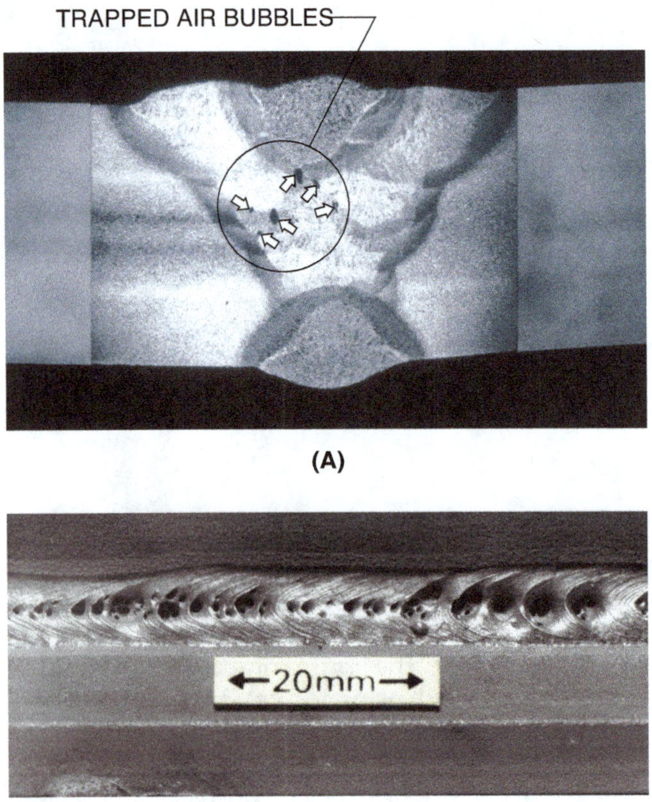

Figure 5 Porosity.

1.3.1 *Mechanical Cleaning*

Mechanical cleaning is the most common method of removing surface contamination. Tools used for mechanical cleaning include hand tools, power tools, and special sandblasting equipment. When performing mechanical cleaning, be sure to wear safety glasses and a face shield for protection from the flying particles produced during the cleaning operation. In addition, special clothing is required for sandblasting.

Flexible scrapers are used to remove dirt and grease from weldments or weldment components. Rigid scrapers can be used to remove hardened paint or dirt.

Wire brushes are used to remove paint and light-to-medium surface oxidation. Wire brushes are available with bristles made from a variety of metals, including carbon steel, stainless steel, and brass. Use only stainless steel wire brushes on stainless steel and aluminum, and make sure that the stainless steel wire brush used on these metals has never been used with another metal. Once brushes, grinding wheels, and files have been used on carbon steel and other metals, they should never be used on stainless steel or aluminum. The remaining particles of other metals on the tools will contaminate stainless steel and aluminum, as will a wire brush made of carbon steel. The contamination leads to weld defects.

Files are another way to remove surface oxidation. When filing aluminum or stainless steel, do not use files that have been used on other metals. Before using a file, always be sure it has a handle installed on the file tang. Use a fine file and keep it clean to prevent scratching the base metal surface. Take care not to damage or scratch the base metal outside the weld zone. Clean files by using stainless steel brushes or file cards that either are new or have only been used to brush the same type of metal (*Figure 6*). Even particles from a file card can contaminate a file, which in turn can contaminate metals on which the file is used.

Tight oxidation must be removed by filing, grinding, or sandblasting. For large jobs or jobs where speed is important, power tools are most efficient. Power tools may be electrically or pneumatically powered.

> **NOTE**
> Sandblasting is not used on aluminum.

Figure 6 File and cleaning tools.

> **WARNING!**
> To contain sparks when cleaning metal with power tools, use a welding screen. Always wear the appropriate PPE when using power tools.

Angle grinders (*Figure 7*) are very effective in removing large areas of surface contamination. Die grinders and small angle grinders work very well for small areas such as weld grooves and bevel edges. Grinders have various attachments that can be used for special applications. These attachments include grinding disks and wheels, wire brushes, rotary files, and flapper wheels (*Figure 8*).

> **WARNING!**
> If a grinding wheel or disk shatters while in use, the pieces of the wheel or disk become deadly projectiles as the weights of the pieces are multiplied by the centrifugal force and the speed of the grinding motor. Never grind without full face protection. Avoid standing in the orbital path of the wheel or disk.

Grinding disks and wheels are made for specific types of metals. Be sure you have the correct disk for the type of metal being ground. Always use aluminum oxide disks for grinding aluminum or stainless steel. For stainless steel or aluminum, do not use wheels that have been used on other metals. They will contaminate the surface of aluminum and stainless steel.

When using wire brush attachments on stainless steel or aluminum, brushes with stainless steel bristles must be used. Only use stainless steel brushes that have not been used on other metals. Remember that carbon steel bristles will contaminate stainless steel and aluminum. Over time, the contamination will reveal itself by iron oxide (rust) developing at the contamination sites of the joint.

When grinding, care must be taken to prevent grinding the base metal below the minimum allowable base metal thickness. If this should happen, the base metal will have to be discarded or, if allowable, built up with weld metal. Both of these alternatives are expensive.

Cutoff wheels are used to cut through metals and other materials. They are made for specific types of metals, or for other materials such as stone. Be sure you have the correct cutoff wheel for the type of material being cut.

Weld flux chippers and needle scalers are usually pneumatic rather than electric. These tools are used by welders to clean surfaces and remove dross from cuts and welds. Weld flux chippers, and needle scalers are also excellent for removing paint, heavy scale, or hardened dirt, but they are not very effective for removing surface oxidation. Weld flux chippers have a single chisel; needle scalers have about 18 to 20 blunt steel needles about 10 inches (≈25 cm) long. Many weld flux chippers can be converted to needle scalers with an attachment. A weld flux chipper and a needle scaler are shown in *Figure 9*.

Figure 7 Handheld grinders.

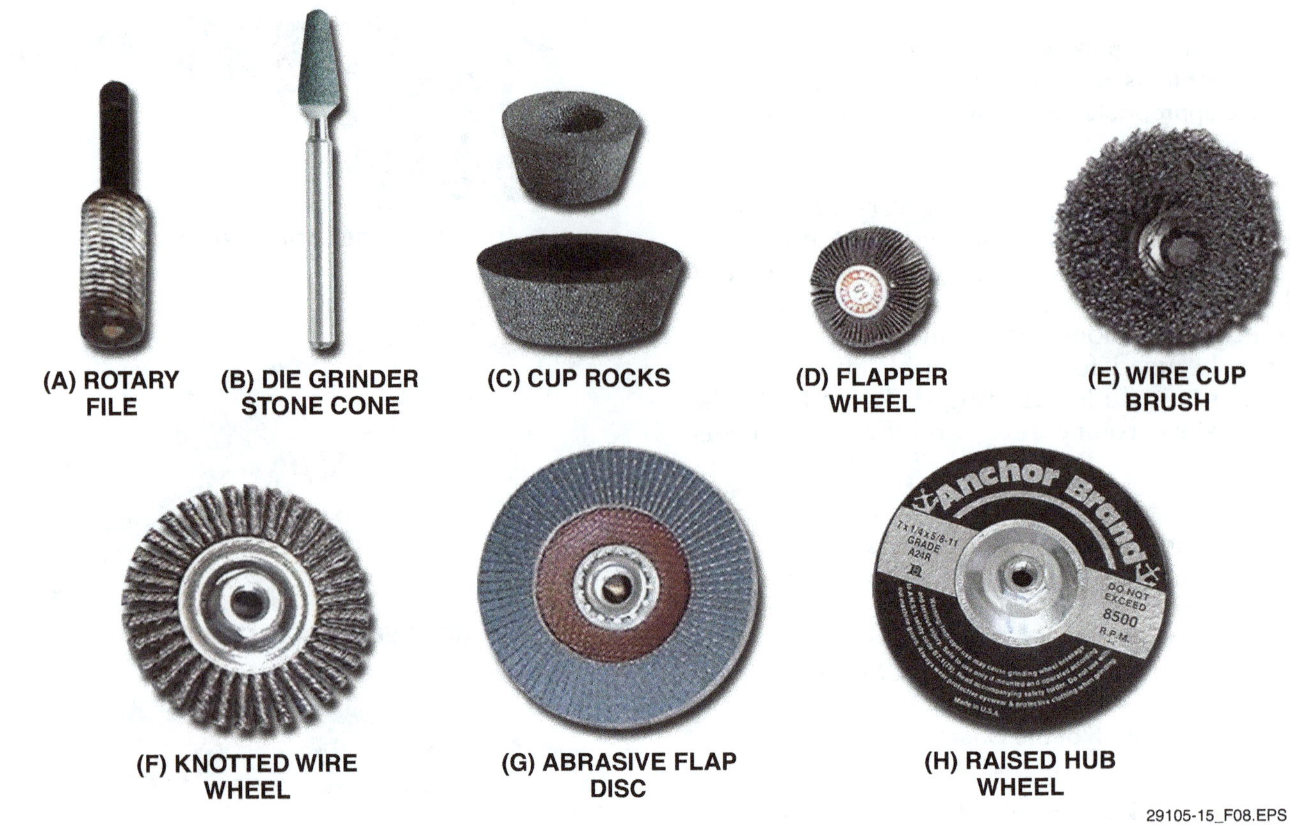

Figure 8 Grinder attachments.

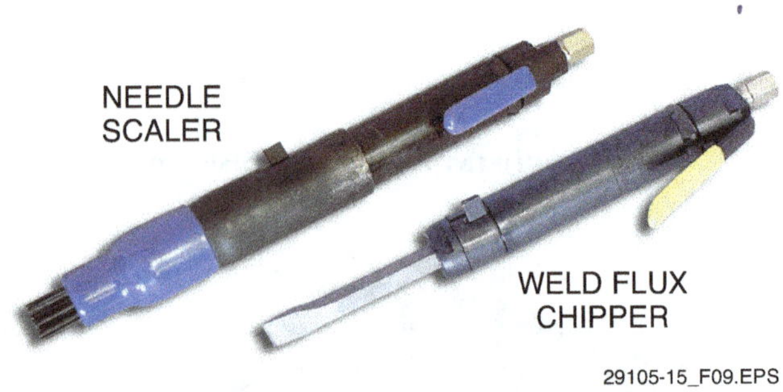

Figure 9 Weld flux chipper and needle scaler.

1.3.2 Chemical Cleaning

There will be times when mechanical cleaning is not enough. Residues left on a metal can contaminate the weld. Even the oil from the skin of the welder handling the metal can contaminate a weld. Welders may have to use some form of chemical cleaner to ensure that the metals being prepared are very clean.

It is essential that only approved cleaners are used. Some chemicals can create hazardous fumes during the welding process. Always follow the prescribed specifications of the approved cleaners and the related SDS/MSDS. The two chemical cleaners most often used by welders are acetone and butanone. Butanone is better known as methyl ethyl ketone, or MEK.

WARNING!

Do not use brake part cleaning fluid to clean surfaces prior to welding. Some brake part cleaning fluids contain chlorinated hydrocarbons. Phosgene gas is formed by the decomposition of chlorinated hydrocarbon solvents by ultraviolet radiation. Phosgene gas reacts with moisture in the lungs to produce hydrogen chloride, which in turn destroys lung tissue. For this reason, any use of chlorinated solvents for any purpose should be kept away from welding operations.

Acetone is a flammable, colorless liquid. Although a manufactured product, it can also be found in some natural environments. It is a primary ingredient in nail polish remover. Anyone who has worked in an auto repair shop has probably used acetone to remove over-sprayed paint from car windows. It will dissolve resins, epoxies, and many plastics. It is a great degreaser that is often used to prepare metals before painting.

From a safety standpoint, acetone is an irritant. Because it can cause permanent eye damage (corneal clouding), workers handling it or working near it must wear safety goggles. Acetone vapors should be avoided, and a suitable respirator is recommended. Prolonged exposure may eventually cause liver damage. Check the product SDS or MSDS closely to ensure that you understand how to safely work with acetone.

MEK is a manufactured organic chemical. It is a colorless liquid that dissolves substances such as gums, resins, vinyl film coatings, and other similar materials that may still be on the metals. MEK is highly flammable, but it presents much less of a health risk than acetone. However, anyone working with MEK must review the SDS or MSDS to make sure all safety precautions are followed.

WARNING!

Acetone and MEK are both extremely flammable. Keep these products away from all potential sources of ignition.

After the metal parts have been mechanically cleaned, the chemical cleaner is applied, usually with a brush or spray. The chemical is allowed to work for a short time before being removed. A clean cloth is used to remove the cleaning agent along with any contamination loosened from the metal. When cleaning metals with chemicals, be aware that the contaminants being removed may give off vapors that could be dangerous. Be sure to use the proper PPE, including respiratory protection in some conditions.

Additional Resources

Steel Metallurgy for the Non-Metallurgist. John D. Verhoeven; Prepared under the direction of the ASM International Technical Book Committee. ASM International® Materials Park, OH 44073-0002, USA. **www.asminternational.org**

American Iron and Steel Institute, 25 Massachusetts Avenue, NW Suite 800, Washington, DC 20001, USA. **www.steel.org**

Association for Iron & Steel Technology, 186 Thorn Hill Road Warrendale, PA, USA. **www.aist.org**

1.0.0 Section Review

1. Grinding wheels containing aluminum oxide, silicon carbide and other materials have a composition that require _____.
   a. a simple dust mask that is comfortable for the wearer
   b. a mask made of 100-percent cotton, such as a handkerchief, or similar type of cover
   c. consultation of the SDS/MSDS for recommended respiratory protection
   d. a safety shield only; a respiratory mask is not necessary

2. Which letter prefix, when used in the AISI numbering system, indicates acid-Bessemer carbon steel?
   a. The letter A
   b. The letter B
   c. The letter C
   d. The letter D

3. When brushing aluminum, use stainless steel brushes that have only been used on _____.
   a. stainless steel
   b. aluminum
   c. bronze
   d. carbon steel

SECTION TWO

2.0.0 JOINT DESIGN AND WELD TYPES

Objective

Identify and describe basic weld joint design and types of welds.

a. Identify and describe the loads that are routinely placed on weld joints.
b. Identify and describe the various types of weld joints.
c. Describe a welding procedure specification (WPS) and the information it provides.

Trade Terms

Backing: A weldable or non-weldable material used behind a root opening to allow defect-free welding at the open root of a joint.

Load: The amount of force applied to a material or a structure.

Melt-through: Complete joint penetration.

Root face: A small flattened area on the end of a bevel for a groove weld.

Root opening: The space between the base metal pieces at the bottom or root of the joint.

Surfacing: The application by welding, brazing, or thermal spraying of a layer of material to a surface to obtain desired properties or dimensions.

Tensile strength: The measure of the ability of a material to withstand a longitudinal stress without breaking.

Welding procedure specification (WPS): The document containing all the detailed methods and practices required to produce a sound weld.

Welded joints are selected primarily for the safety and strength required for the conditions that will be encountered. When selecting the joint for a particular application, many factors must be taken into account, including load considerations, the environment in which it will be used, materials, processes, and cost.

2.1.0 Load Considerations

The loads in a welded steel structure or component are transferred through the welded joint. The welded joint must be designed to withstand the stresses caused by the loads.

Any or all of these forces may act on welded parts. When two parts are welded together, in whatever manner, they are intended to become one part. If the surface areas of the two parts being welded are not properly prepared, the welds may fail when one or more of these forces is applied. The following are brief explanations of load terms as they relate to welds. Refer to *Figure 10* for a visual representation of these forces:

- *Tensile* – When metals are pulled in opposite directions by forces (F_1 vs. F_2), they will eventually break. Their strength to remain one is referred to as their tensile strength. When metals are heated, as when they are welded, the molecules in the metals change. When the metals are cooled, their molecules change again. These heating and cooling cycles can weaken the tensile qualities of metals. When two pieces of metal are welded end-to-end, they become one piece. When they are pulled at each end, the tensile strength of the two pieces, and the weld, are tested. If the weld is a good weld, one of the base metal pieces will usually break before the weld breaks.
- *Compression* – The reduction of volume or mass. Metals can be compressed. If you strike a steel nail or bolt with a hammer, you can eventually compress the nail or bolt into a different shape. When two pieces of metal are welded end-to-end and forces (F_1 and F_2) are applied inwardly on each end, the pieces and their weld are compressed.
- *Bending* – To bend is to make curved. For example, you may be able to manually bend a section of straight lightweight flat bar around a 6-inch (DN150) pipe. The pipe in this case is a force (F_2) pushing against the center of the bar as you pull or apply force (F_1) in the opposite direction at each end of the bar. The result of such action is that the flat bar becomes curved. When two pieces of metal are welded end-to-end, they may be tested in what is known as a bend test.
- *Torsion* – Twisting. Torsion is the amount of stress the welded bars can withstand as a counterclockwise force (F_1) is applied at one end and a clockwise force (F_2) is applied at the other end. For example, the driveshaft of a truck is twisted as it transmits the rotational movements of the drive motor and transmission to the drive axle.

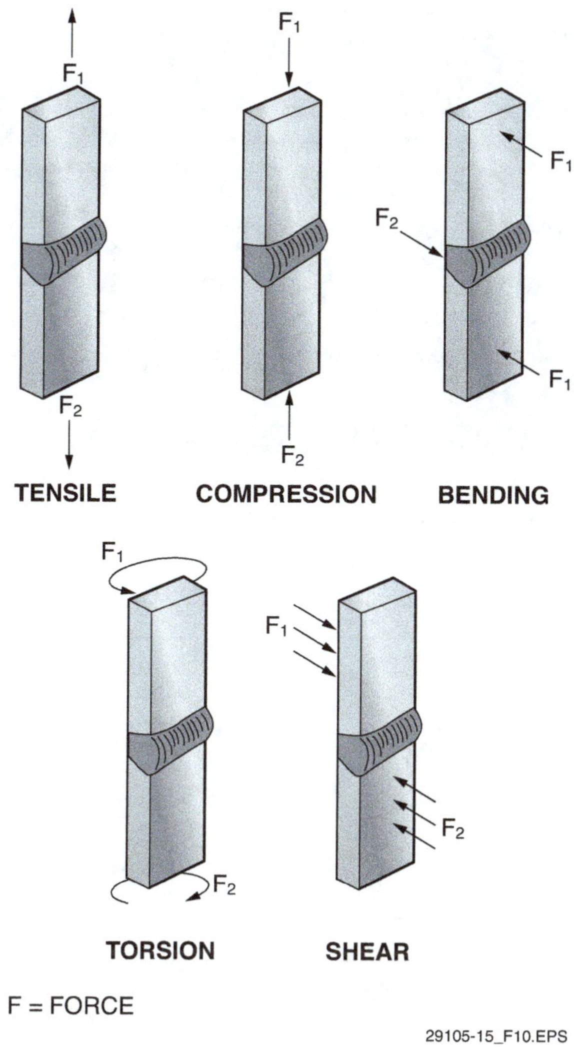

Figure 10 Forces that act on welded joints.

- *Shear* – To cut off or trim. When a piece of sheet metal is placed into a press brake machine, one side of the sheet metal is resting against a stationary surface of the machine. That stationary surface is applying some force to that side of the sheet metal. That stationary force is shown as F_1. When the brake is activated, a moveable part of the machine applies force to the opposite side of the sheet metal. F_2 represents the forced created by that moveable part. When the moveable force (F_2) is increased, it eventually shears or cuts the metal because the stationary surface (F_1) is applying force to the opposite side of the sheet metal.

2.2.0 Types of Welds

Most types of welds require some degree of base metal edge preparation depending on the type of joint being welded. Common types of welds include the following: surfacing, plug or slot, fillet, square groove, bevel groove, V-groove, J-groove, and U-groove. If a particular type of weld requires a specified root opening, it is typically shown on the welding drawing or welding procedure specification (WPS). Base metals may require that some type of bevel be cut or ground on one or both edges, which form a V-groove when they are mated. Edge preparations will be discussed for each type of weld as it is covered in this section.

There are five basic types of joints (*Figure 11*). They are the butt joint, lap joint, T-joint, edge joint, and corner joint (both inside and outside corners). Each of these joints may be accomplished using various types of welds.

2.2.1 Surfacing Welds

Surfacing is the only type of weld that is applied to just one base metal surface, since it is not used to join two or more components. It is used either to build up a base surface that has become worn below the desired thickness or dimension, or to add material to a base to increase its strength and stability. An example of surfacing is shown in *Figure 12*. When you begin making your first welds, you will build what is referred to as a pad—a section of base metal with weld beads run parallel to each other along its length. This is an example of surface welding.

Did You Know?

Surfacing

Surfacing, as defined by the American Welding Society, is the application by welding, brazing, or thermal spraying of a layer of material to a surface to obtain desired properties or dimensions, as opposed to creating a joint between two or more pieces.

Before applying the first layer of multi-layered welds, the base metal surface must be cleaned to remove contaminants, such as oxidation and dirt. When applying the surfacing weld using an oxyfuel process, the surface should be evenly preheated before surfacing to eliminate warping of the base material.

For proper penetration and layering to take place, each new layer of weld surfacing must be properly chipped, scraped, and thoroughly cleaned before applying additional layers of surfacing.

Surfacing is often applied in the tool and die industry to rebuild relatively expensive tools and dies that have worn down. It is also used in the repair of heavy equipment to build up areas that have experienced excessive wear.

29105-15_F12.EPS

Figure 12 Surfacing welding on an inclined vertical surface.

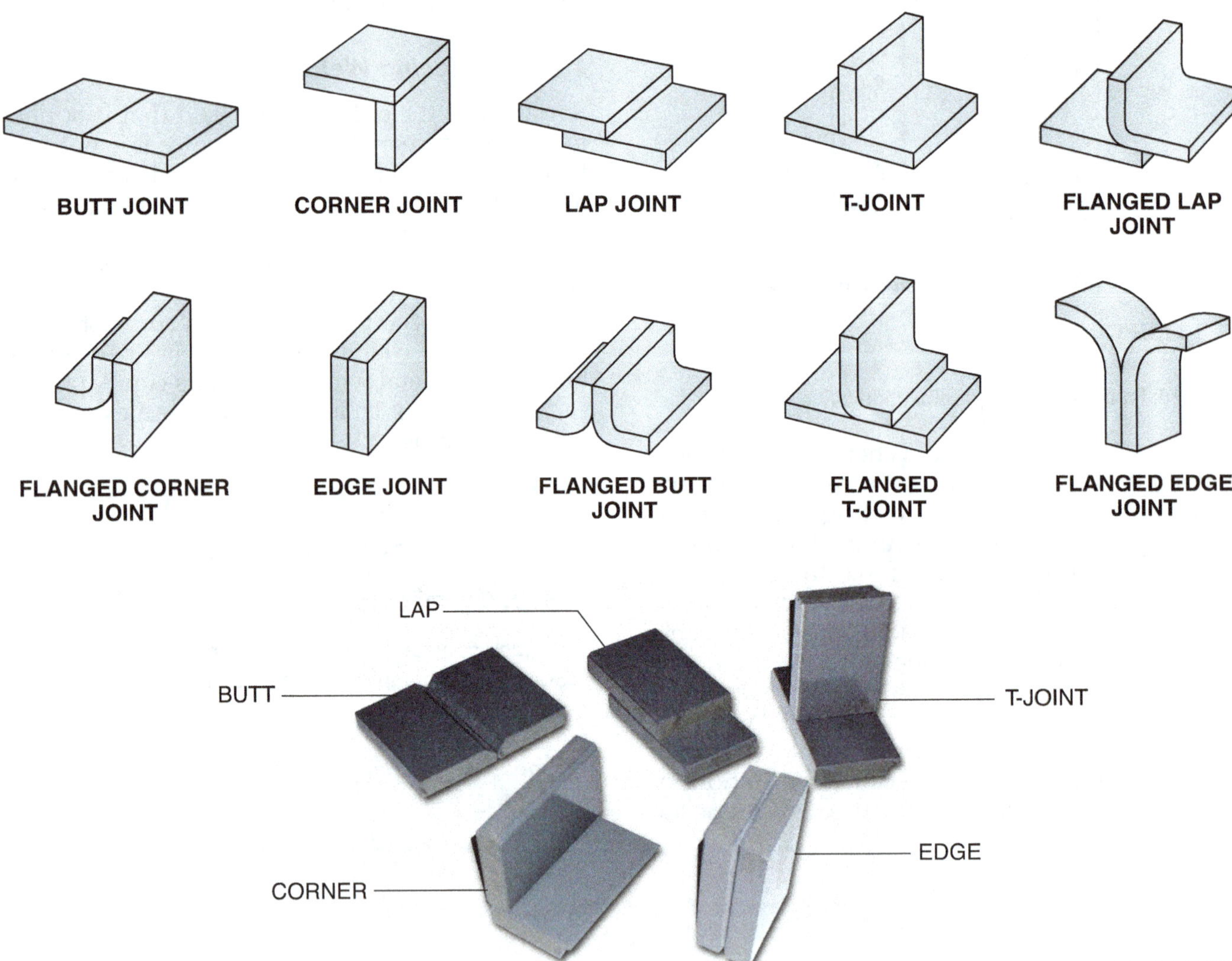

29105-15_F11.EPS

Figure 11 Five basic joint types with variations.

2.2.2 Plug and Slot Welds

Plug and slot welds are used to join metal pieces when the edges cannot be welded. The hole or slot can either be completely or partially filled with weld material when joining it to the other base metal. If the hole is round, the weld is referred to as a plug weld. If the weld is slotted or elongated, it is called a slot weld. *Figure 13* illustrates how a plug or slot weld may be used on a lap joint. A typical application is to overfill the hole or slot, then grind the excess material to a level, finished surface.

Preparation of the base metals for this type of weld requires proper cleaning and removal of any oxides or dirt from both pieces. The hole or slot is then drilled to receive the weld. The walls of the hole or slot must also be cleaned of any oxides, dirt, and oils prior to welding.

Plug and slot welds are commonly used where a finished surface is required, or where overall dimensions are not permitted to exceed those of the plate thickness.

2.2.3 Fillet Welds

Fillet welds may be applied to lap joints, T-joints, or corner joints. The fillet weld does not require any base metal edge preparation other than removing contaminants by cleaning with a wire brush or other appropriate cleaning tools. The two base metals are positioned together with or without a root opening, and one or more passes (welding beads) are applied at the intersection of the two base metal edges (*Figure 14*). Whenever more than one pass is made on a fillet weld—or on any weld—the previous pass must be cleaned of all contaminants such as dross, slag, and oxides before the next pass is applied.

When using a fillet weld on outside corner welds two types of fillet welds may be used. The half-lap joint (*Figure 15*) is easier to assemble, requires less welding material, and is less likely to burn through the corner. The half-lap does require a second weld on the inside of the corner. If a half-lap fit is used, allowances in the plate dimensions must be made for the lap.

The other fillet weld is sometimes called the corner-to-corner joint. The corner-to-corner joint is difficult to assemble because neither plate can support the other, and care must be used when welding to prevent burning through the corner.

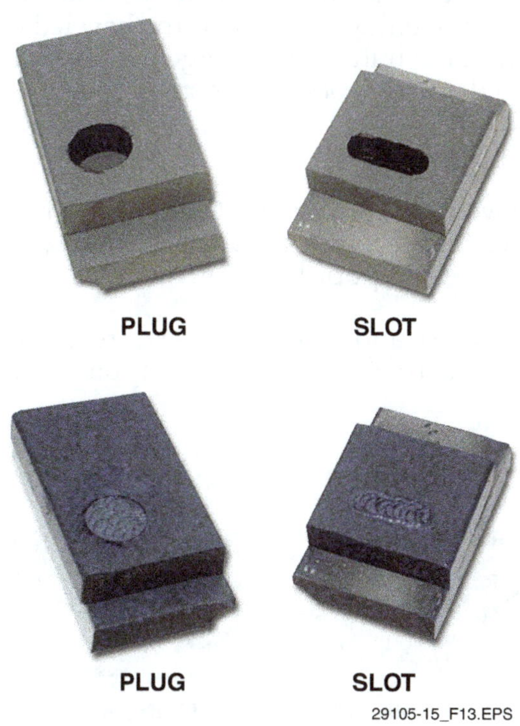

Figure 13 Plug and slot welds.

Surface Welding

Certain shafts located within gearboxes used in the marine industry turn at relatively high rpms (revolutions per minute) and will corrode and wear rapidly if subjected to salt water. It is not always cost effective to discard and replace each shaft with a new one. Surface welding may be used to bring the shafts back to the desired dimensions by building them up and re-machining them. They are then classified as rebuilt shafts. All rotating parts must be balanced before being installed. Otherwise they will vibrate, which in turn will destroy bearings and may cause the part itself to break.

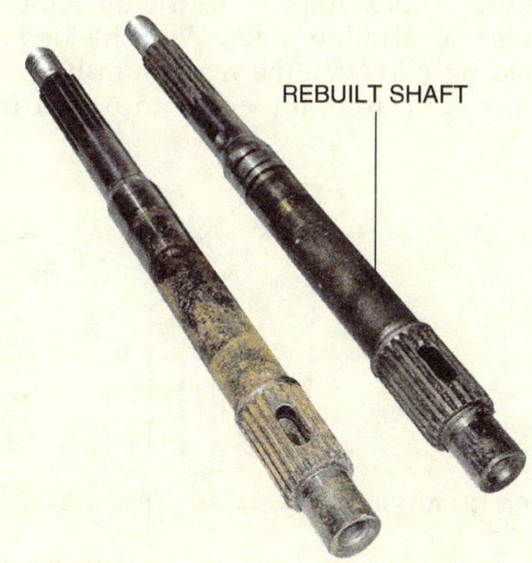

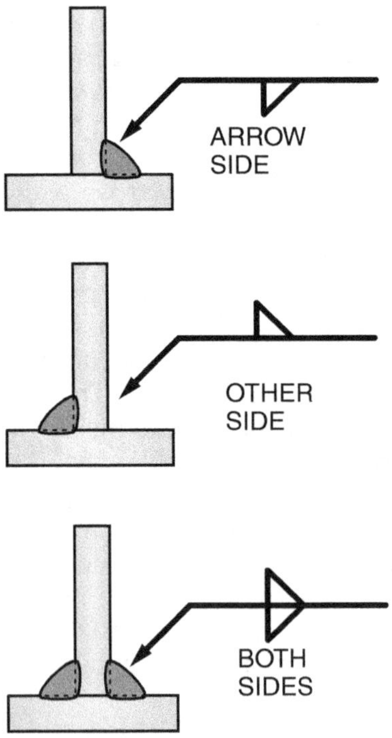

Figure 14 Fillet welds and drawing symbols.

2.2.4 Square-Groove Welds

Square-groove welds can be used with butt joints, corner joints, T-joints, and edge joints. The difference between the fillet and the square groove is that a square groove requires a root opening to be set or prepared between the two base pieces prior to welding, as shown in *Figure 16*. This allows penetration into a greater portion of the surface area of the two base metals. A partial joint-penetration weld will have much less strength than a complete joint-penetration weld.

Welding codes impose restrictions for complete joint-penetration welds. With shielded metal arc welding (SMAW), the maximum base metal thickness is ¼" (roughly equal to 6 mm metric

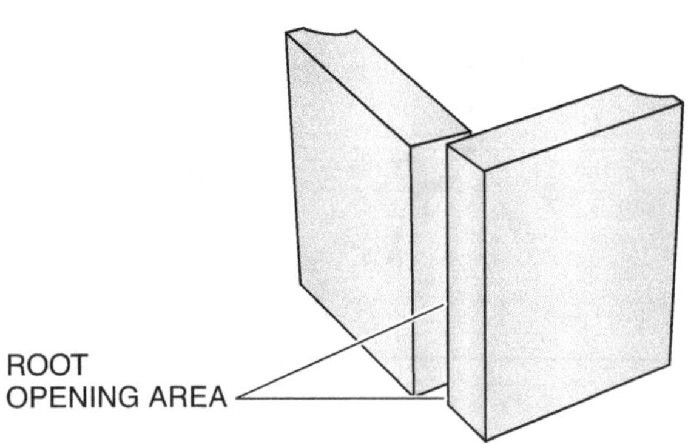

Figure 15 Fit-up of half-lap fillet welds.

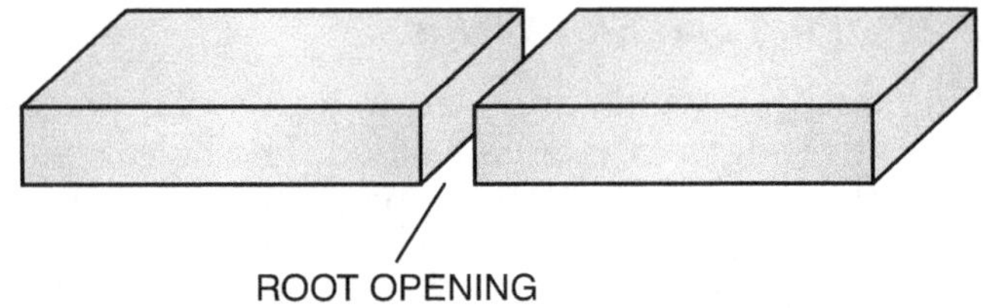

Figure 16 Preparation for a square groove on a butt weld.

steel plate), and welding from both sides is required. In addition, a root opening of half the thickness of the base metal is required, and the root of the first weld must be gouged before the second weld is made. For gas metal arc welding (GMAW) and flux-cored arc welding (FCAW), the maximum base metal thickness is ⅜" (≈10 mm metric plate), with the same requirements for welding from both sides and for back gouging the first pass.

Welding codes and specifications will be discussed in greater detail later in this module.

2.2.5 Single-Bevel Groove Welds

Single-bevel groove welds are commonplace in the welding industry because the base metal preparation is relatively simple and provides greater surface area penetration than the square groove. The single-bevel groove weld can be applied to all five of the basic weld joints. Typically, a single bevel is cut on the edge of one of the base metal pieces. The preparation may include a specified root opening along with the beveled cut, as shown in the butt weld arrangement in *Figure 17*. Although the root opening is small, it can be seen.

2.2.6 V-Groove Welds

V-groove welds are generally applied to inside or outside corner joints, butt joints, or edge joints. This is a very popular joint for both pipe and plate.

V-grooves and bevel grooves are more economical to prepare than other types of welds.

Figure 17 Workpieces for corner and butt bevel-groove welds.

Figure 18 shows a prepared V-groove on a butt joint to be welded. Typically, a 22½- or 30-degree bevel angle must be cut on both workpiece edges to prepare the base metals to receive the weld. In preparing inside or outside corner joints or butt joints for V-groove welds, the angle may not extend from the top of the bevel edge to the bottom of the bevel edge on each piece, but may be cut short of the bottom corner if a root face is required. In this edge preparation, the flat surface from the point where the angle stops to the bottom of the piece is referred to as the root face.

2.2.7 Single Versus Double V-Groove Welds

When possible, a double V-groove weld should be used in place of a single V-groove (*Figure 19*).

Although it may seem unusual, the double V-groove requires half the weld metal compared to the single V-groove. Welding from both sides also reduces distortion, because the forces of distortion work against each other.

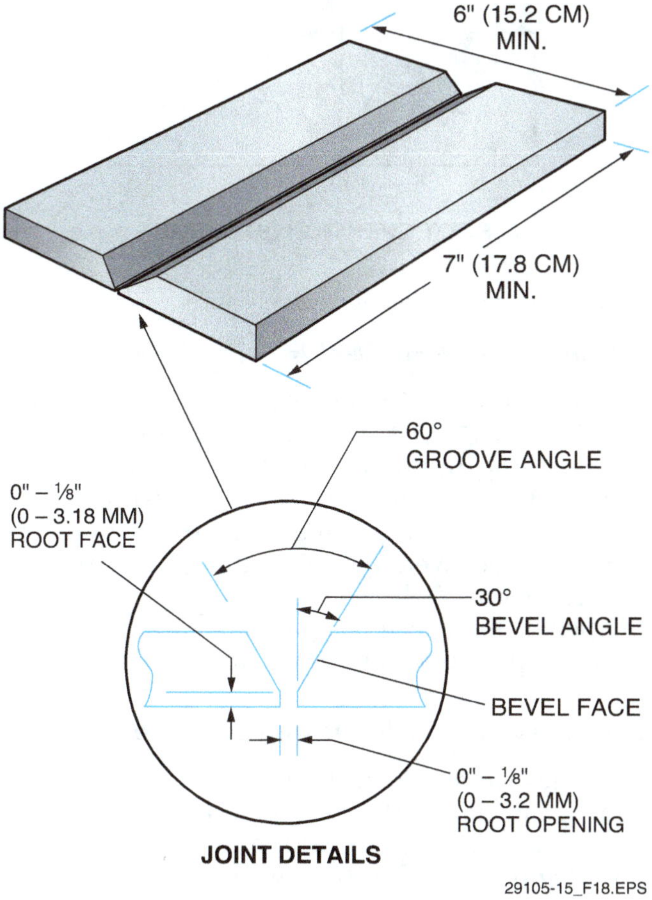

Figure 18 Open root V-groove weld.

2.2.8 J- and U-Groove Welds

J- and U-groove welds (*Figure 20*) are similar. The main difference is that a J-groove weld requires only one base metal to have its edge grooved in the form of a J. A U-groove is formed by preparing two matching J-grooves, one on each base metal edge. Even though both types require much less weld material than bevel or V-groove welds, they require much more preparation time because of the shape of the grooves. A J-groove weld can be applied to all five types of weld joints. The U-groove can only be used on butt, inside and outside corner, and edge joints.

2.2.9 Groove Angles and Root Openings

The purpose of the groove angle is to allow access to the root of the weld. The root preparation is sized to control melt-through.

Increasing or decreasing the root preparation (root opening and root face) will result in excess melt-through or insufficient root penetration. As a general rule, as the groove angle decreases, the root opening increases to compensate (*Figure 21*). Root faces are used with open-root joints, but not when metal backing strips are used.

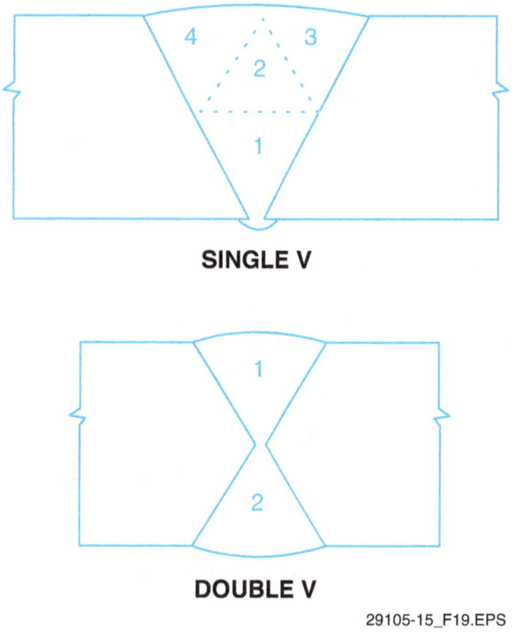

Figure 19 Comparison of double and single V-grooves.

Figure 20 J- and U-groove welds.

Root Face or Land

The root face is also widely known as the land. The American Welding Society (AWS) defines the land or part of the root face as an extension on the root face. The "table" formed on the extension is considered the land.

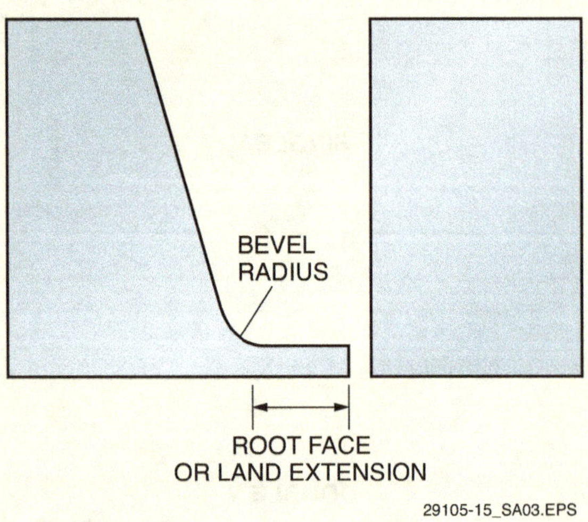

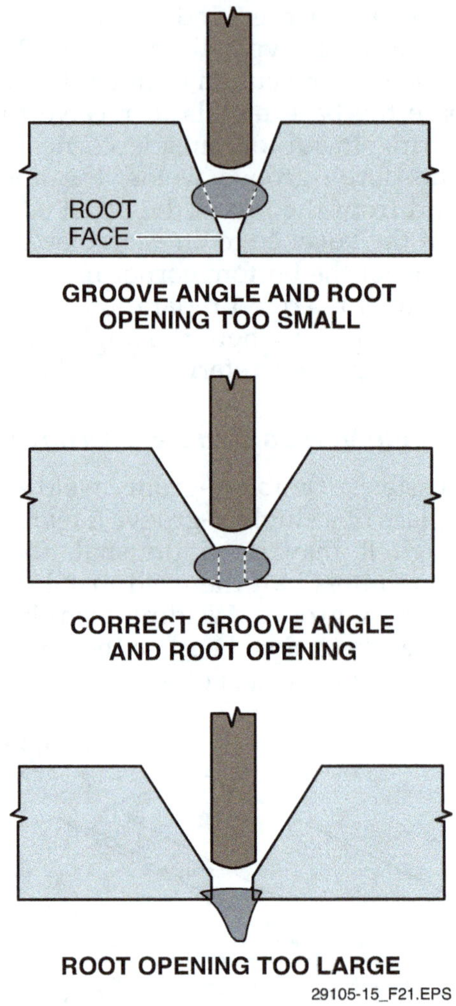

Figure 21 Groove angles and root openings.

2.2.10 Open-Root Welds

For open-root welds on plate, the groove angle should be 60 degrees, and the maximum size of the root opening and root face is ⅛" (3.2 mm). Remember that the 60-degree groove angle results from a 30-degree bevel angle on each mating workpiece. For open-root welds on pipe, the groove angle is 60 degrees or 75 degrees, depending on the code or specifications used. *Figure 22* shows groove angles for open root welds on plate and pipe.

2.2.11 Welds with Backing on Plate

Backing for plate can be strips made from the same material as the base metal, or flux-coated tape, fiberglass-coated tape, ceramic tape, or gas (*Figure 23*). The WPS will specify when to use backing and what type is to be applied.

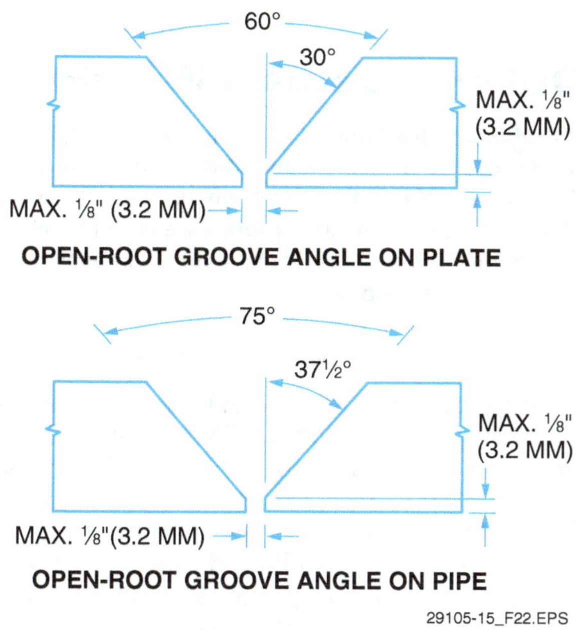

Figure 22 Open root joint preparation.

When gas is used as a backing, the same groove angles and root preparation used for open-root welds are used. When flux-coated tape, fiberglass-coated tape, or ceramic tape is used as a backing, the groove angle should be 60 degrees, the maximum root opening should be 3⁄16" (4.8 mm), and the maximum root face should also be 3⁄16" (4.8 mm). When applying tape, be sure to clean the surface to ensure that the tape adheres tightly, and make sure that the tape is centered on the weld joint.

When strips made from the same material as the base metal are used, a groove angle of 45 degrees, a root opening of ¼" (6.4 mm), and a feather edge (no root face) are used. The backing strip must be thick enough and wide enough to prevent burn-through at the root and to absorb and dissipate the heat of the root pass.

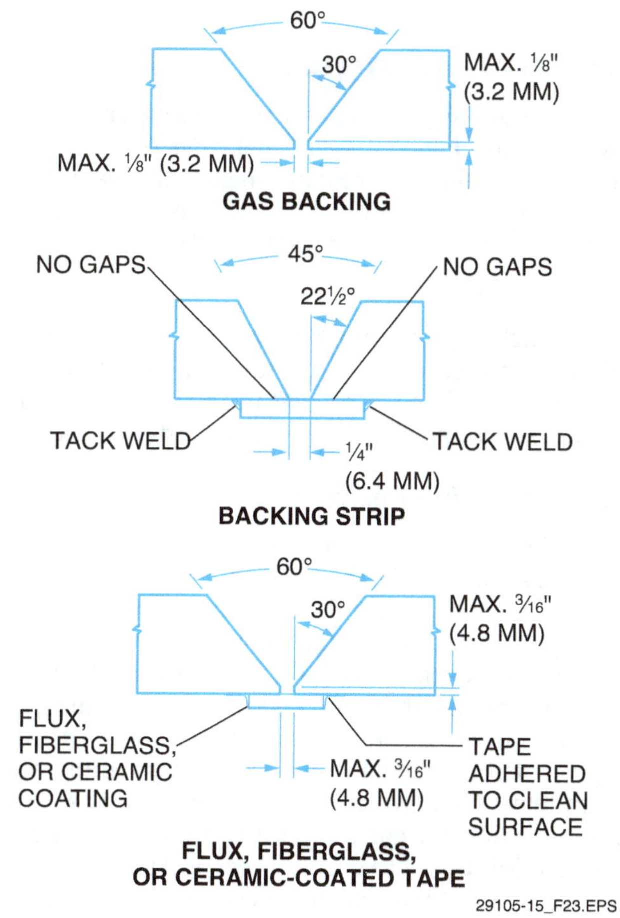

Figure 23 Plate joint preparation for backing.

NOTE

Check the WPS or site procedures for the backing size requirements for your site.

Backing Gas Hazards

Nitrogen and argon are used as backing and purge gases for some welding applications. These gases will displace all the oxygen in the space where they are used. At a refinery in Corpus Christi, Texas, while the welder was at lunch, a helper entered a large section of pipe in which argon had been used as a backing gas. When the welder returned, he found the helper lifeless in the pipe.

Nitrogen is lighter than air and, if allowed, will float upward. Argon, on the other hand, is heavier than air and will settle in low-lying areas. When removing nitrogen from a confined space, open the top to allow it to float out. When removing argon from a confined space, provide some way for it to flow out of the bottom. This is often impractical, however; forced ventilation is by far the most effective method.

Groove Angles

If the groove angle is larger than necessary, it will require additional weld metal to fill. This will increase the time and cost to complete the weld, and also increase distortion. If the groove angle is too small, it may result in weld defects.

The backing strip must be tack welded so that it will be held securely in place during welding. It must also be centered on the groove and be in close contact (no gaps) with the back of the plate being welded.

2.2.12 Welding Position

It is easier and faster to weld groove welds in the flat position and fillets in the flat or horizontal position. Always try to position the weldment so that welding is performed in the flat position. If a weldment is an assembly of parts, look for welds that can be made on the bench before tacking the assembly together. When welding pipe, weld as many fittings as possible before tacking the pipe in position, and try to leave final weld joints that will be the most accessible and the easiest to perform. When preparing weld joints, try to prepare the joint so that welding takes place from the most accessible position.

2.3.0 Codes and Welding Procedure Specifications

A welding code is a detailed listing of the rules and principles that apply to specific welded products. Codes ensure that safe and reliable welded products will be produced and that persons associated with the welding operation will be safe. Clients specify which welding codes should be followed when they place their orders or award project contracts.

Out-of-Position Welding

Out-of-position welding basically includes any welding position other than flat and horizontal. Welding out-of-position generally requires the use of smaller electrodes with more passes to fill the joint. Since it is more difficult, it also increases the likelihood of a weld defect.

If codes are mandated, a WPS must be written for each weld procedure. A WPS (*Figure 24*) is a written set of instructions for producing reliable welds.

The WPS includes the type of joint to be used, as well as the type of weld and any groove preparation that may be required for that particular weld. Each WPS is written and tested in accordance with a particular welding code or specification. All welding requires that acceptable industry standards be followed, but not all welds are governed by codes. The requirement for use of a WPS is often listed on blueprints as a note or in the tail of the welding symbol. If you are unsure whether or not the welding being performed requires a WPS, you should not proceed until you verify this information.

Note that the WPS shown in *Figure 24* uses metric dimensions. Electrodes and filler metals are basically the same size internationally, but they are called out using the metric equivalent of the size in most areas outside of the United States. For example, the 1.6 mm filler metal—wire in this case, for TIG welding—shown on the WPS is actually 1⁄16" wire; the 1.2 mm filler is 3⁄64" filler metal. These are the rounded metric equivalents of their actual manufactured Imperial size.

Weld Procedure Number	30 P1 TIG 01 Issue A
Qualifying Welding Procedure (WPAR)	WP T17/A

Manufacturer:	National Fabs Ltd 25 Lane End Birkenshaw Leeds	Method Of Preparation and Cleaning:	Machine and Degrease
		Parent Metal Specification:	Grade 304L Stainless Steel
		Parent Metal Thickness	3 to 8mm Wall
		Pipe Outside Diameter	25 to 100mm
Location:	Workshop	Welding Position:	All Positions
Welding Process:	Manual TIG	Welding Progression:	Upwards
Joint Type:	Single Sided Butt Weld		

Joint Design

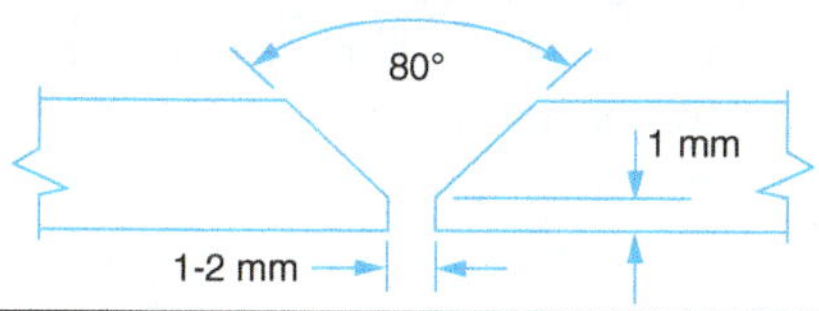

Welding Sequences

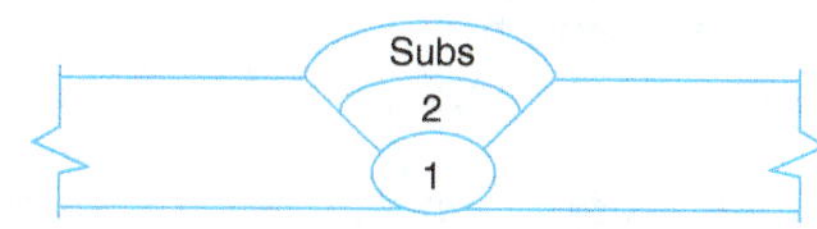

Run	Process	Size of Filler Metal	Current A	Voltage V	Type of Current/Polarity	Wire Feed Speed	Travel Speed	Heat Input
1 2 And Subs	TIG TIG	1.2 mm 1.6 mm	70 - 90 80 - 140	N/A	DC– DC–	N/A	N/A	N/A

Welding Consumables:-	
Type, Designation Trade Name:	BS 2901 Part 2 : 308S92
Any Special Baking or Drying:	No
Gas Flux:	Argon 99.99% Purity
Gas Flow Rate - Shield:	8 - 12 LPM
- Backing:	5 LPM
Tungsten Electrode Type/ Size:	2% Thoriated 2.4mm Dia
Details of Back Gouging/Backing:	Gas Backing
Preheat Temperature:	5°C Min
Interpass temperature:	200°C Max
Post Weld Heat Treatment Time, temperature, method: Heating and Cooling Rates*:	Not Required

Production Sequence

1. Clean weld and 25mm borders to bright metal using approved solvent.
2. Position items to be welded ensuring good fit up and apply purge.
3. Tack weld parts together using TIG, tacks to at least 5mm min length.
4. Deposit root run using 1.2mm dia. wire.
5. Inspect root run internally.
6. Complete weld using 1.6mm dia wire using stringer beads as required.
7. 100% Visual inspection of completed weld

Revision History

Date	Issue	Changes	Authorization
26/11/2000	A	First changes	Jack Straw *Jack Straw*

29105-15_F24.EPS

Figure 24 Example of a WPS.

Additional Resources

2014 Technical Training Guide. Current Edition. The Lincoln Electric Company, Cleveland, OH. USA. **www.lincolnelectric.com**

Welding Handbook. Current Edition. Miami, FL: The American Welding Society.

2.0.0 Section Review

1. When a metal piece is pulled in opposite directions to determine its breaking point, it is a test of its ____.
   a. tensile strength
   b. torsion strength
   c. shearing strength
   d. compression strength

2. To achieve proper penetration for each layer during multi-layer surface welding, ____.
   a. the previous weld must be cleaned before each layer application
   b. no cleaning is required since the heat of welding will remove any impurities
   c. post-heating the base metal is needed before layer application
   d. every layer must first be gouged

3. The requirement for use of a WPS is often listed ____.
   a. within the company's hand book
   b. on blueprints, notes, or on the tail of the welding symbol
   c. in the SDS/MSDS for the electrode
   d. on the base metal

SECTION THREE

3.0.0 WELDING JOINT PREPARATION

Objective

Describe how to prepare joints for welding.

a. Describe how to mechanically prepare joints for welding.
b. Describe how to thermally prepare joints for welding.

Performance Tasks

1. Mechanically or hand grind a bevel on the edge of a ¼"- to ¾"-thick mild steel plate (6 to 20 mm metric plate) at 22½ degrees.
2. Thermally bevel the edge of a ¼"- to ¾"-thick mild steel plate (6 to 20 mm metric plate) at 22½ degrees.

There are many different ways to prepare a joint for welding. The joint can be prepared mechanically using nibblers, grinders, or cutters. It can also be prepared thermally using an oxyfuel, plasma arc, or carbon arc cutting torch. The method used to prepare the joint will generally be determined by the base metal type, ease of use, and code or procedure specifications.

Before starting a joint preparation, determine if the joint is covered by a WPS or site quality standard. If this is the case, the joint type and welding parameters will be specified. These specifications must be followed; the guidance provided by a WPS is mandatory. If you are unsure whether or not a joint is covered by a WPS or site quality standard, check with your supervisor before proceeding.

3.1.0 Mechanical Joint Preparation

Mechanical joint preparation is most often used on alloy steel, stainless steel, and nonferrous metal piping or plate. It is slower than thermal methods such as oxyfuel, carbon arc, and plasma arc, but has the advantages of high precision, low heat input, and the absence of oxides commonly left by thermal methods.

3.1.1 Grinders

Handheld electric or air-operated grinders are used in welding shops, and even more often in the field, to cut and bevel pipe and plate to prepare them for welding. *Figure 25* shows a handheld angle grinder being used to prepare a stainless steel elbow for welding.

WARNING!

Cutoff and grinding wheels must be selected based on the material to be cut or beveled. Using an improper wheel may damage the wheel or workpiece, and can create hazardous conditions for the operator should the wheel shatter during operation. Refer to the manufacturer's recommendations and warnings when selecting grinding wheels. Ensure that the maximum revolutions per minute (rpm) rating of the grinding or cutoff wheel is equal to or greater than the speed of the grinder in use.

29105-15_F25.EPS

Figure 25 Preparing a weld joint.

3.1.2 Pipe Beveling Machines

Welded piping is found extensively in all industries. Nearly every piece of pipe that is welded to a fitting, to another piece of pipe, or to any other connection requires that the end be cut square and beveled according to specifications before welding. Welded pipe fittings are typically beveled from the factory and simply need to be cleaned of paint and other contaminants before welding. If a mechanical cutting and beveling process is used (*Figure 26*), the procedure may be done using an electric, hydraulic, or pneumatic beveling machine. Many of these are portable machines that operate much like a lathe, and they are often referred to as pipe-end prep lathes. They have mandrels that hold various cutting tools that cut metal away, as well as numerous adjusting mechanisms that make it easy to set the bevel angle and depth. Various models are available to cut and bevel both large and small pipe. Boiler tube ends are typically prepared using these smaller machines. When specifications call for boiler tubing to be welded, the machine can face, square, and chamfer the ends of the tubes to prepare them.

29105-15_F26.EPS

Figure 26 Pipe beveling machine.

WARNING!

Always follow all manufacturers' safety procedures when using grinders and cutoff machines. Failure to follow the manufacturer's safety recommendations could result in serious personal injury.

Special mechanical cutoff machines (*Figure 27*) are also made to be mounted on the outside of the pipe. They can be mounted with a ring or a special chain with rollers. A cutoff blade or tool is used to make the cut.

3.1.3 Nibblers

Nibblers prepare the edge of a plate or pipe with a reciprocal punch that cuts or bites off a chip with each stroke. The bevel angle is set by adjusting the nibbler. A nibbler must have access to an edge on which it will be used (*Figure 28*).

WARNING!

Always follow all manufacturers' safety procedures when using nibblers and cutters. Failure to follow the manufacturer's safety recommendations could result in serious personal injury.

3.2.0 Thermal Joint Preparation

Thermal joint preparation involves preparing a joint with the oxyfuel or plasma arc processes. It can be done using the air-carbon arc process, but the result is typically poor. Because of carbon deposits and the inaccuracy of the result, the air-carbon arc cutting process is best used for gouging seams or cracks only.

29105-15_F27.EPS

Figure 27 Pipe cutoff machine.

Thermal Joint Preparation Precautions

When preparing a joint with the oxyfuel, plasma arc, or air-carbon arc cutting processes, all dross must be removed prior to welding. Any dross remaining on the joint during welding will cause porosity in the weld. Joints prepared with a carbon arc cutting torch must be carefully inspected for carbon deposits. It is common for small carbon deposits to be left behind as the carbon electrode is consumed during the cutting or gouging operation. These carbon deposits will cause defects such as hard spots, loss of ductility, and cracking in the weld. Before welding, use a grinder to clean surfaces prepared with most thermal processes.

The torch used for oxyfuel or plasma arc cutting can be handheld or mounted on a motorized or manual carriage. These are often referred to as track burners. The motorized carriage for plate cutting runs on flat tracks positioned on the surface of the plate to be cut (*Figure 29*). The carriage has an On or Off switch, forward or reverse switches, and a speed adjustment. The torch position can be adjusted as necessary for the proper bevel angle. The carriage can carry a special oxyfuel or plasma arc cutting torch designed to fit into the torch holder.

Other equipment designs are used for cutting pipe (*Figure 30*). A steel ring or special chain with rollers is attached to the outside of the pipe.

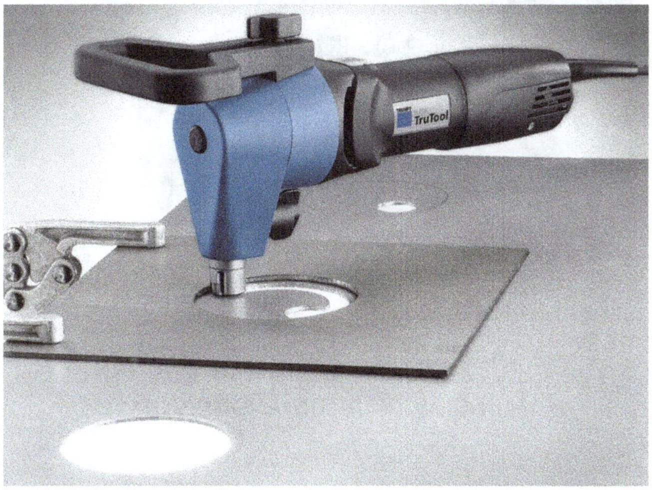

29105-15_F28.EPS

Figure 28 Nibbler.

29105-15_F29.EPS

Figure 29 Track burner thermally cutting and beveling plate.

The torch is carried around the pipe on a torch holder that is powered by electricity, air, or a hand crank. The bevel angle is set by pivoting the torch holder, and the torch is adjusted vertically with a handwheel. An out-of-round attachment can be used to compensate for pipe that is out-of-round.

For very large-diameter pipe, special internal cutting and beveling equipment is available (*Figure 31*). The motorized unit shown here uses rare-earth magnets in the wheels, developing 300 pounds (136 kg) of force. No rings, bands, or chains are required. It can also be used for smaller pipe by traveling on the outside surface, and for long cuts on plate.

Automatic Pipe Bevelers

Automatic versions of pipe bevelers and cutters are an improvement over pattern cutters and other similar equipment. This is because there is no need to reset the preheat flame before each cut. Once an automatic pipe beveler has been initially set up, all subsequent cuts can be made with the same settings. These systems also save time and cutting gases, because when the system is switched off, all gases are immediately shut off as well.

GEAR-DRIVEN UNIT (HANDCRANK)

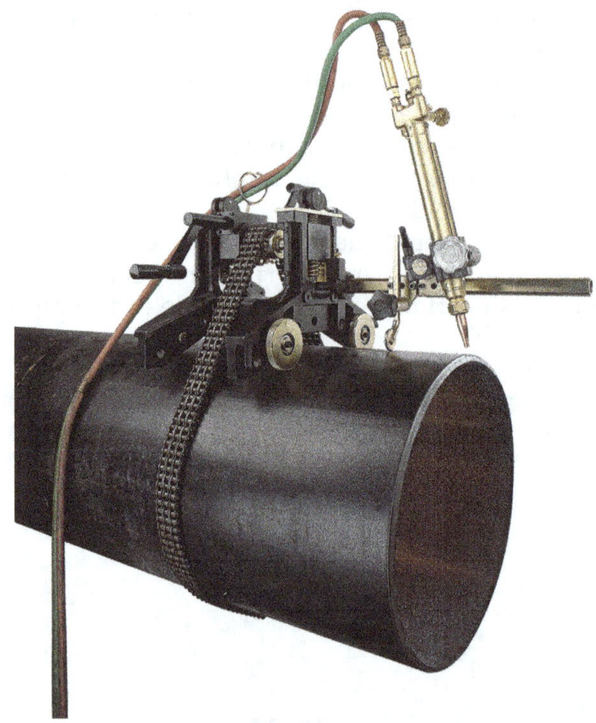

CHAIN-DRIVEN UNIT (HANDCRANK)

29105-15_F30.EPS

Figure 30 Pipe cutting and beveling equipment.

29105-15_F31.EPS

Figure 31 Thermal cutting from inside the pipe.

Additional Resources

2014 Technical Training Guide. Current Edition. The Lincoln Electric Company, Cleveland, OH 44117-1199 USA. **www.lincolnelectric.com**

Welding Handbook. Current Edition. Miami, FL: The American Welding Society.

3.0.0 Section Review

1. Which of the following is an advantage of mechanical joint preparation over the thermal method?
   a. It is more precise with less heat input.
   b. It is faster and less labor intensive.
   c. Mechanically prepared joints require less weld metal.
   d. Mechanically prepared joints require fewer weld passes.

2. Which of the following is a thermal process seldom used due to the deposits left behind which can lead to weld defects?
   a. Air-carbon arc cutting
   b. Oxyfuel cutting
   c. Plasma arc cutting
   d. Pipe-end prep lathe

SUMMARY

The importance of safety cannot be overemphasized within the welding environment. It should be the highest priority at all times.

Performing chemical or mechanical cleaning removes base metal surface contamination. To create reliable weld joints that meet specifications, proper cleaning techniques must always be used.

A great deal of load stress may be placed on a weld joint. Understanding the different types of stress helps engineers to determine what type of welding joint and technique is best for the application. The specifications for a given weld joint are communicated to the welder through the welding procedure specification (WPS).

Weld joints can be prepared mechanically and thermally. Mechanical methods do have advantages, but they can also be labor-intensive. Thermal methods include the use of oxyfuel cutting processes, with the torch mounted on a carriage that allows it to rotate smoothly around the pipe to make a consistent cut and/or bevel.

Review Questions

1. Never directly or indirectly view an electric arc without _____.
   a. having your eyes examined
   b. wearing sunglasses
   c. being at least 10 feet away
   d. using a properly tinted lens
2. Medium-carbon steel contains _____.
   a. 0.10 percent to 0.15 percent carbon
   b. 0.15 percent to 0.30 percent carbon
   c. 0.30 percent to 0.50 percent carbon
   d. 0.50 percent to 1.00 percent carbon
3. A type of copper-alloyed steel that protects itself from damaging oxidation by creating a hard, tough layer of oxidation on its surface is called _____.
   a. spring steel
   b. weathering steel
   c. mild steel
   d. stainless steel
4. A pneumatic tool that uses a number of blunt steel needles to remove surface oxidation is called a _____.
   a. needle scaler
   b. weld flux chipper
   c. needle nose chipper
   d. contamination tattoo scaler
5. Torsion applied to a weld joint refers to a force _____.
   a. pulling in opposite directions
   b. twisting the metal in opposite directions
   c. pushing against a stationary surface
   d. attempting to create a bend
6. The five basic types of joints are butt, lap, corner, T-joint, and _____.
   a. chamfer
   b. square
   c. bevel
   d. edge
7. Lap joints, T-joints, or corner joints can be welded using _____.
   a. surface welds
   b. butt welds
   c. fillet welds
   d. spot welds
8. U-groove welds are formed by preparing two matching J-grooves that require _____.
   a. less preparation time but use more weld material than a V-groove weld
   b. more preparation time but use less weld material than a V-groove weld
   c. more preparation time and use more weld material than a V-groove weld
   d. no preparation time and use the same weld material as a V-groove weld
9. Which of the following materials is *not* a common joint backing material?
   a. The same base metal type
   b. A gas
   c. Ceramic tape
   d. Cast iron
10. When a backing strip of the same material as the base metal is used, the root face should be _____.
    a. a feather edge
    b. 1⁄16" (1.6 mm)
    c. 1⁄8" (3.2 mm)
    d. 1⁄4" (6.4 mm)
11. Grinding wheels selected for use must be rated equal to or greater than the grinder's _____.
    a. current
    b. horsepower
    c. rotational speed
    d. air pressure
12. Before a pipe can be butt-welded to a fitting or other connection, its edge must be _____.
    a. burnished
    b. hardened
    c. ground to fit
    d. cut square and beveled
13. Beveling equipment made for pipe is sometimes called _____.
    a. J-prep cutters
    b. ID-prep cutter
    c. pipe-end prep lathes
    d. compound-bevel prep lathes

14. Which of the following processes produces the least desirable result, and is therefore *not* a popular choice for general metal cutting?
    a. Plasma arc cutting
    b. Band saw
    c. Oxyfuel cutting
    d. Air-carbon arc cutting

15. The best approach to gouging seams or cracks is _____.
    a. using a nibbler
    b. carbide-chisel cutting
    c. grinder cutting
    d. air-carbon arc cutting

Trade Terms Quiz

Fill in the blank with the correct term that you learned from your study of this module.

1. An assembly that is fastened together by weld joints is referred to as a(n) ________.
2. A weldable or non-weldable material used behind a root opening to allow defect-free welding at the open root of a joint is a(n) ________.
3. The document containing all the detailed methods and practices required to produce a sound weld is a(n) ________.
4. The primary metal to be welded, cut, or brazed is referred to as the ________.
5. The steel alloy that, under specific conditions, is designed to form a very dense oxide layer on its outer surfaces to retard further oxidation is called ________.
6. The expansion and contraction of welded parts caused by the heating and subsequent cooling of the weld joint is ________.
7. The space between the base metal pieces at the bottom or root of a joint is called a(n) ________.
8. A steel that contains less than 0.10 percent carbon and is magnetic is called ________.
9. The amount of force applied to a material or structure is called a(n) ________.
10. A small flattened area on the end of a bevel for a groove weld is called a(n) ________.
11. Complete joint penetration results in ________.
12. Gas pockets, or voids in the weld metal are called ________.
13. A metal, such as aluminum or brass, lacking sufficient quantities of iron to have any effect on its properties is called a(n) ________.
14. A form of porosity having a length greater than its width and is approximately perpendicular to the weld face is known as a(n) ________.
15. The scale that forms on metal surfaces when they are exposed to the oxygen or air containing oxygen is called a(n) ________.
16. A steel that shares some characteristics with ferritic, but boasts higher levels of carbon, up to a full 1 percent is called ________.
17. To impart strength or toughness to steel or cast iron by heating and cooling is referred to as ________.
18. When a metal is heated and cooled to free it from internal stress, it is said to be ________.
19. Any article that has been removed from a mold is a(n) ____ .
20. A metal that contains austenite (a solid solution of carbon, or of carbon and other elements, in a ferrous alloy), added through heating is called ________.
21. When a metal is ________, it is able to undergo a change of form without breaking.
22. To ________ a piece of metal, you will need a bucket of water.
23. A metal that has been shaped by beating it with a hammer, as is common with iron, is referred to as being ________.
24. Only ________ metals contain any substantial iron.
25. The measure of the ability of a material to withstand a longitudinal stress, without breaking is considered the material's ________.

26. The application by welding, brazing, or thermal spraying of a layer of material to a surface to obtain desired properties or dimensions is _______.

27. Capable of being extended or shaped by hammering or by pressure from rollers, this material is considered to be _______.

Trade Terms

Annealed
Austenitic
Backing
Base metal
Castings
Distortion
Ductile
Ferritic
Ferrous
Load
Malleable
Martensitic
Melt-through
Nonferrous metal
Oxide
Piping porosity
Porosity
Quench
Root face
Root opening
Surfacing
Tempering
Tensile strength
Weathering steel
Welding procedure specification (WPS)
Weldment
Wrought

Appendix

Performance Accreditation Tasks

The American Welding Society (AWS) School Excelling through National Skills Standards Education (SENSE) program is a comprehensive set of minimum Standards and Guidelines for Welding Education programs. The following performance accreditation is aligned with and designed around the SENSE program.

The Performance Accreditation Tasks (PATs) correspond to and support the learning objectives in *AWS EG2.0, Guide for the Training and Qualification of Welding Personnel: Entry-Level Welder*.

Note that in order to satisfy all learning objectives in *AWS EG2.0*, the instructor must also use the PATs contained in the second level of the NCCER Welding curriculum.

PAT 1 corresponds to no *AWS EG2.0* reference.

PAT 2 corresponds to *AWS EG2.0, Module 8 – Thermal Cutting Processes, Unit 2 – Manual OFC (e.g., Track Burner)*, Key Indicators 4, 5, and 6.

PATs provide specific acceptable criteria for performance and help to ensure a true competency-based welding program for students.

The following tasks are designed to develop your competency in preparing base metal. Practice each task until you are thoroughly familiar with the procedure.

As you complete each task, take it to your instructor for evaluation. Do not proceed to the next task until instructed to do so by your instructor.

PREPARE PLATE JOINTS MECHANICALLY

Using a nibbler, cutter, or grinder, mechanically or manually prepare the edge of a ¼"- to ¾"-thick carbon steel plate (≈6 to 20 mm metric plate) with a bevel angle of 22½ or 30 degrees, at the discretion of the instructor.

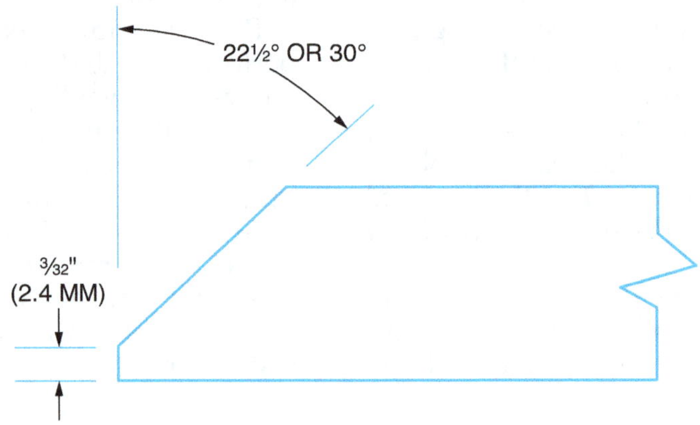

NOTE: BASE METAL = CARBON STEEL PLATE

29105-15_A01.EPS

Criteria for Acceptance:

- Bevel angle ±2½" __________
- Bevel face smooth and uniform to 1⁄16" (1.6 mm) __________
- Root face ±1⁄32" (±0.8 mm) __________

PREPARE PLATE JOINTS THERMALLY

Using oxyfuel or plasma arc cutting equipment, thermally prepare the edge of a ¼"- to ¾"-thick carbon steel plate (≈6 to 20 mm metric plate) with a bevel angle of 22½ or 30 degrees, at the discretion of the instructor.

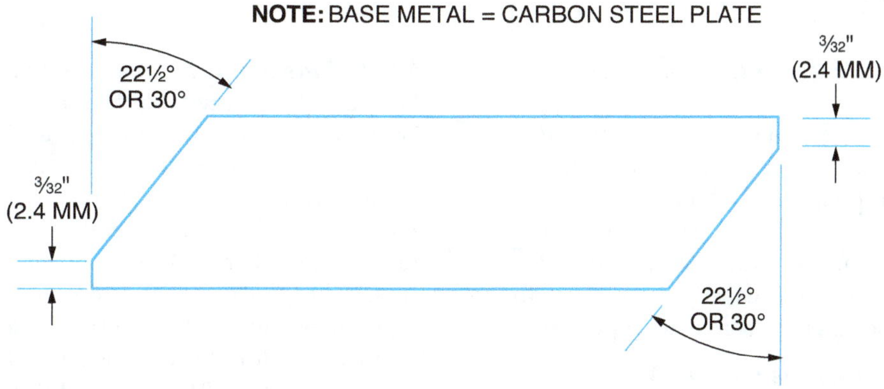

Criteria for Acceptance:

- Bevel angle ±2½" __________
- No dross __________
- Minimal notching not exceeding 1⁄16" (1.6 mm) deep on the kerf face __________
- Root face ±1⁄32" (±0.8 mm) __________

Frank Johnson

PBF Energy Toledo Refinery Inc.
Welding Trainer and CWE Weld Procedure Writer

How did you choose a career in the construction industry?
I took welding courses at Penta County Vocational High School. Upon graduation, I passed the state pipe welding test, an accomplishment which was rare for a man my age. I worked as a welder for a year before joining the military. Though I didn't work as a welder during this time, I received increased pay in the military because of my success at passing the pipe test.

Who inspired you to enter the industry?
The reason for choosing this career is directly related to who inspired me to enter the industry. My grandfather owned a weld shop and I began welding when I was a young boy of just 13. My grandfather pushed me to read and learn as much about welding as possible so that I could solve problems as they arose. I was also inspired by my hobby of steam tractors because I had to weld them whenever they were in need of repair.

How important are education and training in construction?
I think training is absolutely essential in order to be a good welder. It is very important to know the codes, variables, and demands of the industry and to be able to perform any type of welding in any type of situation. I feel that the NCCER curriculum is important because I use it to train welders. The NCCER standardized curriculum, developed by expert welders, ensures that employees are trained properly.

What kinds of work have you done in your career?
After my discharge from the service, I started and completed a boilermaking apprenticeship. Six months into my apprenticeship, I was recognized as a skilled welder, which enabled me to be paid as a journeyman. I worked at various refineries and power plants, including the Davis Bessie Nuclear Plant. At that time, the pipefitters at the company needed welders. As a result, I welded at Davis Bessie for about a year, then worked for 2½ years as a pipefitter. I earned journeyman pipefitter and boilermaker journey level cards. Next, I worked at Enrico Fermi Nuclear Power Plant in Monroe, Michigan. While there, I was asked by Detroit Edison to work with their welding engineer because he qualified 95 percent of their welding procedures. I am now a certified welder, journey-level boilermaker, and pipefitter. I have created and qualified 17 welding procedures.

Tell us about your present job.
In my capacity as a trainer, I test and qualify welders, write welding procedures, qualify welding procedures, and serve as a certified welding instructor. The work is challenging and I really enjoy it.

What factors have contributed most to your success?
Factors that have contributed the most to my success are my natural curiosity, as well as my experience working in a nuclear plant where I admired the excellent craftsmanship of seasoned welders. Their professionalism gave me the desire to achieve their level of expertise. In that environment, quality standards are extremely high, so only top-notch welders will succeed there.

Would you suggest construction as a career to others? Why?
I would suggest construction as a career because if you are a hard worker you can attain a well-paying, secure job. I believe that a career in construction is full of challenges that craftworkers can overcome. They can also experience the satisfaction of seeing a project through, which yields them the satisfaction of putting a project into operation and seeing it work as it should. When recommending a career in construction, I've said that it is "Good work, challenging work, and satisfying work that will support you with a good wage."

What advice would you give to those new to the field?
My advice to those new to the field is: "Don't be afraid to try. Even if you don't succeed the first time, you will make it eventually, and you will be successful."

Tell us an interesting career-related fact or accomplishment.
I serve as an advisor to the Historical Boiler Board for the State of Ohio and on the National Board for Historical Boilers, an honor bestowed upon me due to my successful welding career. I was asked to serve as the subject matter expert on the committee that shapes the Welding curriculum for NCCER. Looking back, I never thought I'd be in a position like this. Welding has taken me a long way.

How do you define craftsmanship?
A craftsman is one who takes pride in his or her work.

Trade Terms Introduced in This Module

Annealed: To free from internal stress by heating and gradually cooling.

Austenitic: Containing austenite (a solid solution of carbon or of carbon and other elements in a ferrous alloy), added through heating.

Backing: A weldable or nonweldable material used behind a root opening to allow defect-free welding at the open root of a joint

Base metal: Metal to be welded, cut, or brazed.

Castings: Something cast; any article that has been cast in a mold.

Distortion: The expansion and contraction of welded parts caused by the heating and subsequent cooling of the weld joint.

Ductile: Able to undergo change of form without breaking.

Ferritic: Steel containing less than 0.10 percent carbon and is magnetic. This steel can't be hardened via heat treatment.

Ferrous: Containing iron.

Load : The amount of force applied to a material or a structure.

Malleable: Capable of being extended or shaped by hammering or by pressure from rollers.

Martensitic: Steel that shares some characteristics with ferritic, but has a higher levels of carbon, up to a full 1 percent. It can be tempered and hardened and is used where strength is more important than a resistance to oxidation.

Melt-through: Complete joint penetration.

Nonferrous metal: A metal, such as aluminum, copper, or brass, lacking sufficient quantities of iron to have any effect on its properties.

Oxide: The scale that forms on metal surfaces when they are exposed to oxygen or air containing oxygen.

Piping porosity: A form of porosity having a length greater than its width and that is approximately perpendicular to the weld face.

Porosity: Gas pockets, or voids in the weld metal.

Quench: To cool suddenly by plunging into a liquid.

Root face: A small flattened area on the end of a bevel for a groove weld.

Root opening: The space between the base metal pieces at the bottom or root of the joint.

Surfacing: The application by welding, brazing, or thermal spraying of a layer of material to a surface to obtain desired properties or dimensions.

Tempering: To impart strength or toughness to (steel or cast iron) by heating and cooling.

Tensile strength: The measure of the ability of a material to withstand a longitudinal stress without breaking.

Weathering steel: Steel alloy that, under specific conditions, is designed to form a very dense oxide layer on its outer surfaces, which retards further oxidation.

Welding procedure specification (WPS): The document containing all the detailed methods and practices required to produce a sound weld.

Weldment: An assembly that is fastened together by welded joints.

Wrought: Produced or shaped by beating with a hammer, as with iron.

Additional Resources

This module presents thorough resources for task training. The following resource material is suggested for further study.

2014 Technical Training Guide. Current Edition. The Lincoln Electric Company, Cleveland, OH. USA. **www.lincolnelectric.com**

American Iron and Steel Institute, 25 Massachusetts Avenue, NW Suite 800, Washington, DC 20001, USA. **www.steel.org**

Association for Iron & Steel Technology, 186 Thorn Hill Road Warrendale, PA, USA. **www.aist.org**

Steel Metallurgy for the Non-Metallurgist. John D. Verhoeven; Prepared under the direction of the ASM International Technical Book Committee. ASM International® Materials Park, OH 44073-0002. **www.asminternational.org**

Welding Handbook. Current Edition. Miami, FL: The American Welding Society.

Figure Credits

Topaz Publications, Inc., Module Opener, Figures 4, 6, 8A, 8C–8H, 9, 11 (photos), 12, 13, 17, 20, 25, 29, SA02

The Lincoln Electric Company, Cleveland, OH, USA, Figures 1–3

Reproduced by permission TWI Ltd., Figure 5

Courtesy of Milwaukee Electric Tool Corporation, Figures 7A, 7B

Courtesy of Chicago Pneumatic Tool Co., Figures 7C, 7D

Courtesy of Dremel, Figure 8B

Courtesy of Tri Tool Inc., Figures 26, 27

TRUMPF Inc., Figure 28

Courtesy of Mathey Dearman, Figures 30, 31

AWS F2.2:2001, Lens Shade Selector, reproduced with permission from the American Welding Society (AWS), Miami, FL, USA, Table 1

Photo Courtesy of U.S. Army, SA01

Section Review Answer Key

Answer	Section Reference	Objective
Section One		
1. c	1.1.3	1a
2. b	1.2.1	1b
3. b	1.3.1	1c
Section Two		
1. a	2.1.0	2a
2. a	2.2.1	2b
3. b	2.3.0	2c
Section Three		
1. a	3.1.0	3a
2. a	3.2.0	3b

NCCER CURRICULA — USER UPDATE

NCCER makes every effort to keep its textbooks up-to-date and free of technical errors. We appreciate your help in this process. If you find an error, a typographical mistake, or an inaccuracy in NCCER's curricula, please fill out this form (or a photocopy), or complete the online form at **www.nccer.org/olf**. Be sure to include the exact module ID number, page number, a detailed description, and your recommended correction. Your input will be brought to the attention of the Authoring Team. Thank you for your assistance.

Instructors – If you have an idea for improving this textbook, or have found that additional materials were necessary to teach this module effectively, please let us know so that we may present your suggestions to the Authoring Team.

NCCER Product Development and Revision

13614 Progress Blvd., Alachua, FL 32615

Email: curriculum@nccer.org
Online: www.nccer.org/olf

❑ Trainee Guide ❑ Lesson Plans ❑ Exam ❑ PowerPoints Other ____________

Craft / Level: Copyright Date:

Module ID Number / Title:

Section Number(s):

Description:

Recommended Correction:

Your Name:

Address:

Email: Phone:

Weld Quality

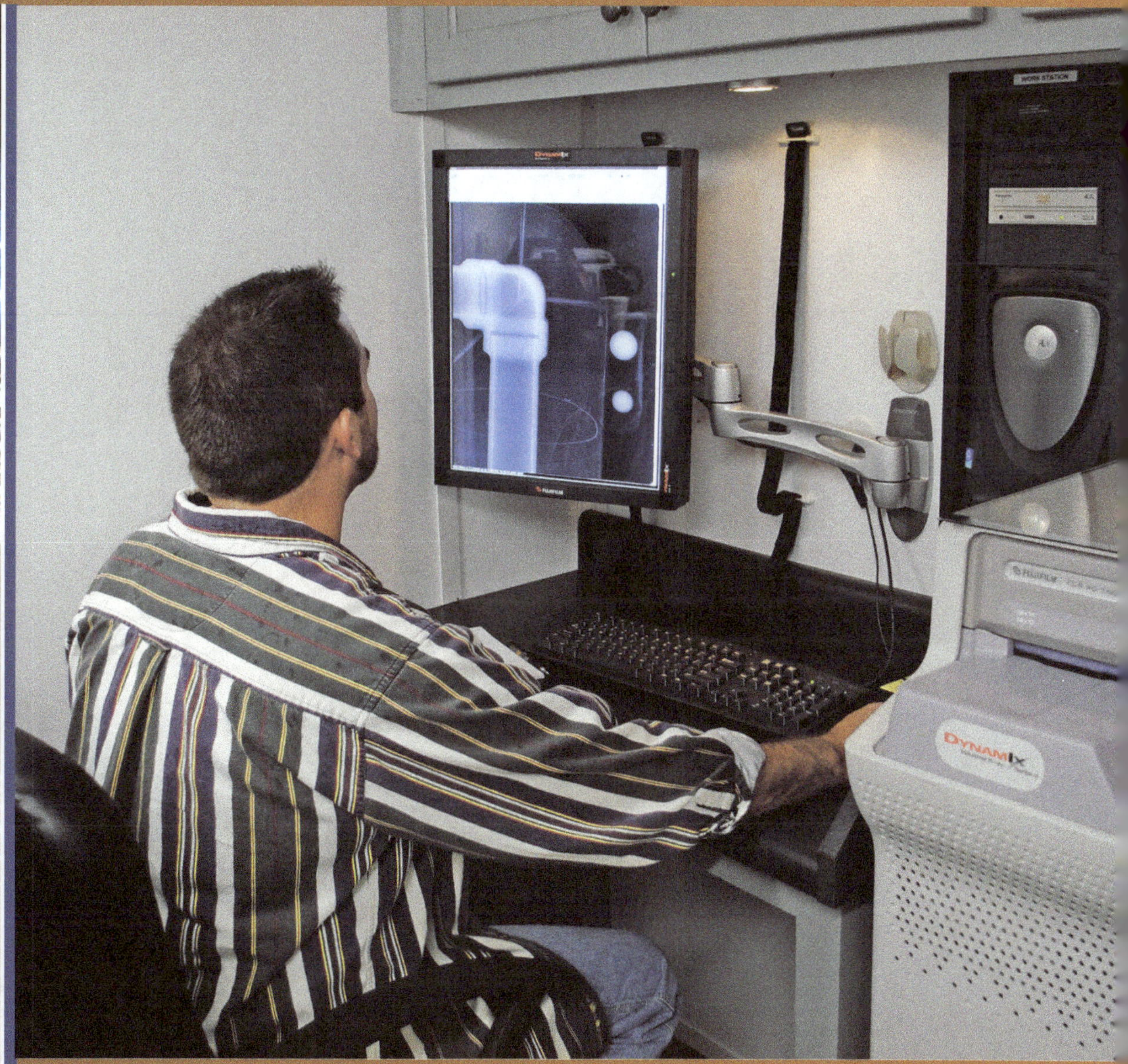

OVERVIEW

The ability to make high-quality welds that will consistently pass inspection is a necessary skill for all welders. An important part of achieving this level of skill is the ability to recognize weld imperfections, as well as their causes.

Module 29106

Trainees with successful module completions may be eligible for credentialing through the NCCER Registry. To learn more, go to **www.nccer.org** or contact us at **1.888.622.3720**. Our website has information on the latest product releases and training, as well as online versions of our *Cornerstone* magazine and Pearson's product catalog.

Your feedback is welcome. You may email your comments to **curriculum@nccer.org**, send general comments and inquiries to **info@nccer.org**, or fill in the User Update form at the back of this module.

This information is general in nature and intended for training purposes only. Actual performance of activities described in this manual requires compliance with all applicable operating, service, maintenance, and safety procedures under the direction of qualified personnel. References in this manual to patented or proprietary devices do not constitute a recommendation of their use.

29106 V5

29106
WELD QUALITY

Objectives

When you have completed this module, you will be able to do the following:

1. Identify and describe the various code organizations that apply to welding and their basic elements.
   a. Identify the various welding code organizations and their sponsoring organizations.
   b. Identify and describe the basic provisions of welding codes.
2. Identify and describe weld discontinuities and their causes.
   a. Identify and describe discontinuities related to porosity and inclusions.
   b. Identify and describe discontinuities that result in cracking.
   c. Identify and describe discontinuities related to joint penetration, fusion, and undercutting.
   d. Identify and describe acceptable and unacceptable weld profiles.
3. Describe various nondestructive and destructive weld examination practices.
   a. Describe basic visual inspection methods including measuring devices and liquid penetrants.
   b. Describe magnetic particle and electromagnetic inspection processes.
   c. Describe the radiographic and ultrasonic inspection processes.
   d. Describe destructive testing processes.
4. Describe the welder performance testing process.
   a. Describe the qualification of welders by position.
   b. Describe welder qualification testing to meet American Welding Society (AWS) and American Society of Mechanical Engineers (ASME) requirements.
   c. Describe the process for completing a weld test.

Performance Task

Under the supervision of your instructor, you should be able to do the following:

1. Perform a visual inspection (VT) on a fillet and/or groove weld and complete an inspection report.

Trade Terms

Arc blow
Brazing
Code
Defect
Discontinuity
Embrittled
Essential variables
Hardenable materials
Homogeneity
Inclusion
Laminations
Non-essential variables
Notch toughness
Procedure qualification record (PQR)
Radiographic
Specification
Standard
Supplemental essential variables
Transducer
Underbead cracking
Welding procedure qualification

Industry Recognized Credentials

If you are training through an NCCER-accredited sponsor, you may be eligible for credentials from NCCER's Registry. The ID number for this module is 29106. Note that this module may have been used in other NCCER curricula and may apply to other level completions. Contact NCCER's Registry at 888.622.3720 or go to **www.nccer.org** for more information.

Contents

Topics to be presented in this module include:

Figures and Tables

Figures and Tables (Continued)

Section One

1.0.0 Welding Codes and Provisions

Objective

Identify and describe the various code organizations that apply to welding and their basic elements.

a. Identify the various welding code organizations and their sponsoring organizations.
b. Identify and describe the basic provisions of welding codes.

Trade Terms

Brazing: A method of joining metal using heat and a filler metal with a melting point above 842°F (450°C). Unlike welding, the base metal is not melted during the brazing process.

Code: A document that establishes the minimum requirements for a product or process. Codes can be, and often are, adopted as laws.

Essential variable: Items in a welding procedure specification (WPS) that cannot be changed without requalifying the WPS.

Non-essential variable: Items in a welding procedure specification (WPS) that can be changed without requalifying the WPS.

Notch toughness: The ability of a material to absorb energy in the presence of a flaw such as a notch.

Procedure qualification record (PQR): The document containing the results of the nondestructive and destructive testing required to qualify a welding procedure specification (WPS).

Specification: A document that defines in detail the work to be performed or the materials to be used in a product or process.

Standard: A document that defines how a code is to be implemented.

Supplemental essential variable:: The variables that must be considered when notch toughness requirements are invoked.

Welding procedure qualification:: A demonstration through testing that welds made following a specific process can meet prescribed standards.

Welding criteria have been established in codes and standards produced by a number of organizations. In addition, each country may have its own welding standards. Welding codes govern welding activities, qualification requirements, and tests that can be performed on weldments to identify imperfections in welds. Standards define how the code requirements are to be achieved. Welders must be familiar with the codes and standards that apply to their work.

1.1.0 Codes and Standards Governing Welding

Welding work is governed by codes, standards, and specifications. A welding code is a set of requirements covering permissible materials, service limitations, fabrication, inspection, testing procedures, and qualification of welders. Welding codes ensure that safe and reliable welded products are produced and that persons associated with the welding operation are safe. Codes are often adopted into law. A standard is a document that defines how a code is to be implemented. Standards are developed by organizations that bring together professionals from all areas of an industry to create standards for that industry. Specifications are detailed instructions for producing a product or performing a particular task. One important difference to note is that standards represent the cooperative effort of people from all parts of an industry, and are published and maintained by organizations established for that purpose. Specifications, on the other hand, can be produced by an individual company or organization without industry consensus and can apply to a single project.

Clients specify in the contract the codes, standards, and specifications that will be used on the project. Since there are a number of codes and each is updated periodically, welders must know which welding codes and code year apply to the project on which they are working. In addition to codes and standards, there are generally specifications that apply to each job. A specification is a document that defines in detail how the work is to be performed, but it is not intended to replace codes and standards.

Codes and standards that apply to welding safety and quality have been developed and are published by a number of nationally recognized agencies. To eliminate the necessity of writing a code for each new job, sections of these existing codes are referenced by the project contract. Agencies and societies that have established widely used welding codes and standards in the United States include the following:

- American Society of Mechanical Engineers (ASME)
- American Welding Society (AWS)
- American Petroleum Institute (API)
- American National Standards Institute (ANSI)

It is beyond the scope of this module to present all the details of the codes and standards organizations mentioned here. However, some of the most important will be discussed along with their impact on the industry. There are also a number of other organizations that have developed codes and standards used internationally. Some of those organizations will also be introduced in this section.

1.1.1 American Society of Mechanical Engineers

The American Society of Mechanical Engineers publishes two codes that welders must be aware of: the *ASME Boiler and Pressure Vessel Code and ASME B31, Code for Pressure Piping.* Both of these codes are endorsed by the American National Standards Institute.

The *ASME Boiler and Pressure Vessel Code (BPVC)* contains eleven sections. The sections most frequently referenced by welders are as follows:

- *Section II, Material Specifications* – This section contains the specifications for acceptable ferrous (Part A) and nonferrous (Part B) base metals and for acceptable welding and brazing filler metals and fluxes (Part C). Many of these specifications are identical to and have the same number designation as AWS specifications for welding consumables. This section is used to match base metals and filler metals.
- *Section V, Nondestructive Examination* – This section covers the methods and standards for nondestructive examination of boilers and pressure vessels.
- *Section IX, Welding and Brazing Qualifications* – This section covers the qualification of welders, welding operators, brazers, and brazing operators. It also covers the welding and brazing procedures that must be used for welding or brazing boilers or pressure vessels. This section of the code is often cited in other codes and standards as the welding qualification standard.

ASME B31, Code for Pressure Piping, consists of eight sections. Each section gives the minimum requirements for the design, materials, fabrication, erection, testing, and inspection of a particular type of piping system. In particular, *B31.1, Power Piping,* covers power and auxiliary service systems for electric generation stations. *B31.3, Process Piping,* covers chemical plant and petroleum refining piping. Other sections include the following:

- *B31.2, Fuel Gas Piping*
- *B31.4, Pipeline Transportation Systems for Liquid Hydrocarbons and Other Liquids*
- *B31.5 Refrigeration Piping and Heat Transfer Components*
- *B31.8 Gas Transmission and Distribution Piping Systems*
- *B31.9 Building Services Piping*
- *B31.11 Slurry Transportation Piping Systems*

All sections of *ASME B31, Code for Pressure Piping,* require qualification of the welding procedures and testing of welders and welding operators. Some sections require these qualifications to be performed in accordance with Section IX of the *ASME Boiler and Pressure Vessel Code,* while in others it is optional.

The Importance of Welding Inspection

It is important that you learn to recognize the various types of weld discontinuities and identify their causes so that you can be the first line of defense against problem welds.

Welding is used to join metals in many critical applications, including buildings, bridges, pipelines, motor vehicles, and heavy machinery. Failure of a weld in any of these applications could mean loss of life or major property damage. For this reason, welding inspectors will frequently check your work to make sure your welds meet the standards established for the project. The greater the potential hazard or cost risk, the more frequent and intense the inspections will be.

Inspection Standards

Among the many documents and standards published by the American Welding Society are documents covering weld inspection. These include the following:

- *AWS B1.10, Guide for the Nondestructive Inspection of Welds*
- *AWS B1.11, Guide for the Visual Inspection of Welds*
- *AWS B5.1, Specification for the Qualification of Welding Inspectors*
- *AWS Welding Inspection Technology*

1.1.2 American Welding Society

The American Welding Society publishes numerous welding-related documents. These documents include codes, standards, specifications, recommended practices, and guides. *AWS D1.1, Structural Welding Code – Steel,* is the code most frequently referenced. It covers welding and qualification requirements for welded structures of carbon and low-alloy steels. It is not intended to apply to pressure vessels, pressure piping, or base metals less than ⅛" (<3.2 mm metric plate) thick. Other commonly used AWS welding codes include the following:

- *AWS D1.2 Structural Welding Code – Aluminum*
- *AWS D1.3 Structural Welding Code – Sheet Steel*
- *AWS D1.6 Structural Welding Code – Stainless Steel*
- *AWS D1.5 Bridge Welding Code*

1.1.3 American Petroleum Institute

The American Petroleum Institute (API) publishes numerous documents in all areas related to petroleum production. *API 1104, Standard for Welding of Pipelines and Related Facilities,* applies to arc and oxyfuel gas welding of piping, pumping, transmission, and distribution systems for petroleum. It presents methods for making acceptable welds by qualified welders using approved welding procedures, materials, and equipment. It also presents methods for ensuring proper analysis of weld quality.

1.1.4 American National Standards Institute

The American National Standards Institute (ANSI) is a private organization that does not actually prepare standards. Instead, it adopts standards that it feels are of value to the public interest. ANSI standards deal with dimensions; ratings; terminology and symbols; test methods; and performance, as well as safety specifications for materials, equipment, components, and products in many fields, including construction. Many codes used today have been adopted as ANSI standards.

1.1.5 Other Standards Organizations

A number of national and international organizations publish standards and specifications governing welding. These organizations include the following:

- *International Standards Organization (ISO)* – This is an independent, non-governmental standards organization with 163 member countries. ISO identifies itself as the world's largest developer of voluntary standards, including nearly 300 welding standards.
- *ASTM International* – Formerly known as the American Society for Testing and Materials, ASTM International develops standards aimed at improving product quality and safety and facilitating market access and trade.
- *American Society for Nondestructive Testing (ASNT)* – ASNT is an organization for professionals involved in nondestructive testing. ASNT provides a forum for the exchange of NDT-related technical information; provides educational materials and programs; and publishes standards related to NDT. The key ASNT standards are as follows:
  - *ANSI/ASNT CP-189, ASNT Standard for Qualification and Certification of Nondestructive Testing Personnel*
  - *ANSI/ASNT CP-105, ASNT Standard Topical Outlines for Qualification of Nondestructive Testing Personnel*

Many countries produce their own standards. Examples include the GOST in Russia; the Canadian Welding Bureau; the European Union (CEN), and many others.

Did You Know?

PHMSA

The US Department of Transportation (DOT) established the Pipeline and Hazardous Materials Safety Administration (PHMSA) in 2004 to develop and enforce regulations for the safe, reliable, and environmentally sound operation of the 2.6 million miles of pipeline operated by 3,000 companies in the United States. Among other PHMSA missions is the inspection of pipelines to ensure compliance with safety regulations.

1.1.6 Maritime Welding Guides and Specifications

The AWS website lists the guides and specifications associated with marine welds and provides the following summary statements for each code:

- *AWS D3.5-93R, Guide for Steel Hull Welding* – This guide provides information to users in the marine construction industry about the best and most practical methods to weld steel hulls for ships, barges, mobile offshore drilling units, and other marine vessels. It provides information on steel plates, shapes, castings, and forging, as well as their selection and weldability. It discusses welding processes and proper design for welding. Hull construction is presented in terms of preparation of materials, erection and fitting, and control of distortion. Qualification of procedures and personnel are outlined, and inspection methods are discussed. A common shipyard problem, stray current protection, is discussed as is the health and safety of the workforce. Supplementary non-mandatory appendices are provided for informational purposes.
- *AWS D3.6M, Specification for Underwater Welding* – This specification covers the requirements for welding structures or components under the surface of the water. It includes welding in both dry and wet environments. Sections 1 through 6 constitute the general requirements for underwater welding, while Sections 7 through 10 contain the special requirements applicable to four individual classes of weld: Class A – Comparable to above-water welding; Class B – For less critical applications; Class C – Where load bearing is not a primary consideration; and Class O – To meet the requirements of another designated code or specification.
- *AWS D3.7, Guide for Aluminum Hull Welding* – This guide provides information on the welding of seagoing aluminum hulls and other structures in marine construction. Included are sections on hull materials, construction preparation, welding equipment and processes, qualification requirements, welding techniques, and safety precautions.

Anyone involved in marine welding must be knowledgeable of these welding codes. When in doubt, find a current copy of the applicable code and review it. In addition, welders working on US Navy shipbuilding projects must recognize that the Navy has their own welding specifications, which can differ significantly from those used in commercial work. The main point to remember is to always ensure that the work is being done to the correct codes, standards, and specifications.

1.1.7 Quality Workmanship

The codes and standards discussed in this module were written to ensure that welders consistently make quality welds. Many weldments will be examined and tested. However, due to time and cost, it is not feasible to examine every one, so inspection is done on a sampling basis. Regardless, quality workmanship is expected in every weld a welder makes.

The welder should be able to work with site representatives to ensure that quality work is achieved. To do this, the site organizational structure needs to be understood. If quality problems arise, the welder needs to follow the appropriate chain of command to eliminate any problems. However, there may be instances when the welder should bypass the chain of command. Examples of such instances include the following:

- You have been directed to perform an unsafe act. If you cannot resolve the matter with your immediate supervisor, it is your responsibility to go to the general foreman, superintendent, project manager, or safety engineer.
- You have been directed to perform a weld that requires a specific certification, and you are not certified to perform it. If you cannot resolve the matter with your immediate supervisor, it is your responsibility to go to the general foreman, superintendent, project manager, or quality engineer.

1.2.0 Basic Provisions of Welding Codes

A major function of welding codes is to establish qualification for operators to perform a particular welding operation. All welding codes provide detailed information about qualification in the following general areas:

- Welding procedure qualification
- Welder performance qualification
- Welding operator qualification

Machine welding is covered in some codes but is not common to all codes. Each type of qualification mentioned is different and is subject to different requirements.

> **NOTE**
> The information in this module is provided as a general guideline only. Check with your supervisor if you are unsure of the codes and specification requirements for your project.

1.2.1 Welding Procedure Qualification

A welding procedure is a written document that contains materials, methods, processes, electrode types, techniques, and all other necessary and relevant information about the weldment. Welding procedures must be qualified before they can be used. Welding procedure qualification has nothing to do with the skills of the individual welder, but deals only with the process itself.

Welding procedure qualifications are limiting instructions written to explain how a welding operation will be done. These limiting instructions are listed in a document known as a welding procedure specification (WPS). *Figure 1* shows a sample WPS. The purpose of the WPS is to define and document in detail the variables related to project-specific welds. The WPS lists the following information in detail:

- Base metals to be joined by welding
- Filler metal to be used
- Range of preheat and postheat treatment
- Thickness and other variables for each welding process

WPS variables are identified either as essential or nonessential variables. Essential variables are items in the welding procedure specification that cannot be changed without requalifying the welding procedure. Supplemental essential variables are those variables that must be considered when notch toughness requirements are invoked.

Essential variables vary by code or welding process. Refer to the specific code or welding process for the relevant essential variables. The following are some of the essential variables in a welding procedure.

- Filler metal classification
- Material thickness
- Joint design
- Type of base metal
- Welding process
- Current type
- Pre- and post-heat treatment

Nonessential variables are items in the WPS that may be changed within a range identified by the code, but that do not affect the qualification status.

Examples of nonessential variables that may be changed without having to requalify the welding procedure include the following:

- Amperage
- Travel speed
- Shielding gas flow (if applicable)
- Electrode and filler wire size
- Rod travel angle

> **CAUTION**
> Do not change any essential or supplemental essential variable without discussing it with your supervisor.

The WPS is qualified for use by welding test coupons as the WPS instructs, and then by testing the coupons in accordance with the applicable code. A test weld is made and test coupons are cut from it. The test coupons are used to make tensile tests, root bends, and face bends as required by the code. The test results are then recorded on a document known as a procedure qualification record (PQR). This documents the tested specimen, testing method, and the results. If the weldment produced by the WPS-guided procedure meets the code requirements, the procedure becomes qualified. Under most codes, each WPS must have a matching PQR to document the quality of the weld produced. *Figure 2* shows a sample PQR.

The methods used to qualify procedures are more detailed and thorough than those used to qualify either welders or welding machine operators. This is because procedures must qualify physical and metallurgical properties.

ANNEX B—SAMPLE FORMS

WELDING PROCEDURE SPECIFICATION (WPS) Yes ☐
PREQUALIFIED ________ QUALIFIED BY TESTING X
or PROCEDURE QUALIFICATION RECORDS (PQR) Yes ☐

Identification # PQR 231
Revision 1 Date 12-1-07 By W. Lye
Authorized by J. Jones Date 1-18-08
Type—Manual ☐ Semi-Automatic ☐
Machine ☐ Automatic ☐

Company Name Red Inc.
Welding Process(es) FCAW
Supporting PQR No.(s) –

JOINT DESIGN USED
Type: Butt
Single ☒ Double Weld ☐
Backing: Yes ☒ No ☐
Backing Material: ASTM A131A
Root Opening 1/4" Root Face Dimension –
Groove Angle: 35° Radius (J–U) –
Back Gouging: Yes ☐ No ☒ Method –

BASE METALS
Material Spec. ASTM A131
Type or (Grade) A
Thickness: Groove 1" Fillet –
Diameter (Pipe) –

FILLER METALS
AWS Specification A5.20
AWS Classification E71T-1

SHIELDING
Flux – Gas CO_2
Composition
Electrode-Flux (Class) Flow Rate 30-40 cfh
Gas Cup Size 5/8" or 3/4"

PREHEAT
Preheat Temp., Min 75° Ambient
Interpass Temp., Min 75°F Max 350° F

POSITION
Position of Groove: 4G Fillet: –
Vertical Progression: Up ☐ Down ☐

ELECTRICAL CHARACTERISTICS
Transfer Mode (GMAW) Short-Circuiting ☐
Globular ☒ Spray ☐
Current: AC ☐ DCEP ☒ DCEN ☐ Pulsed ☐
Other
Tungsten Electrode (GTAW)
Size:
Type:

TECHNIQUE
Stringer or Weave Bead: Stringer
Multi-pass or Single Pass (per side) Multipass
Number of Electrodes 1
Electrode Spacing Longitudinal –
Lateral –
Angle –

Contact Tube to Work Distance 3/4-1"
Peening None
Interpass Cleaning: Wire Brush

POSTWELD HEAT TREATMENT
Temp. N.A.
Time

WELDING PROCEDURE

Pass or Weld Layer(s)	Process	Filler Metals		Current		Volts	Travel Speed	Joint Details
		Class	Diam.	Type & Polarity	Amps or Wire Feed Speed			
1	FCAW	E71T-1	.045"	DC+	180	26	8	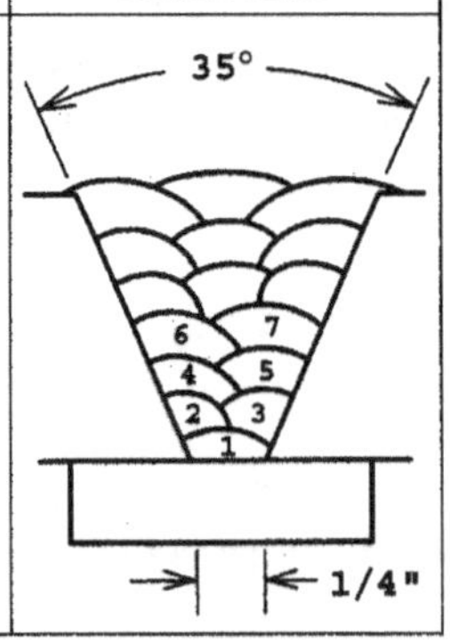
2-8	"	"	"	"	200	27	10	
9-11	"	"	"	"	200	27	11	
12-15	"	"	"	"	200	27	9	
16	"	"	"	"	200	27	11	

Form E-1 (Front)

29106-15_F01.EPS

Figure 1 Example of a welding procedure specification (WPS).

Procedure Qualification Record (PQR) # 231
Test Results

TENSILE TEST

Specimen No.	Width	Thickness	Area	Ultimate Tensile Load, lb	Ultimate Unit Stress, psi	Character of Failure and Location
231-1	.75"	1.00"	.75"	52 500	70 000	Ductile
231-3	.75"	1.00"	.75"	52 275	69 700	Ductile

GUIDED BEND TEST

Specimen No.	Type of Bend	Result	Remarks
231-2	Side	Pass	
231-4	Side	Pass	Small (< 1/16") opening acceptable
231-6	Side	Pass	
231-5	Side	Pass	

VISUAL INSPECTION
Appearance acceptable
Undercut acceptable
Piping porosity none
Convexity none
Test date 12-3-2002
Witnessed by D. Davis

Radiographic-ultrasonic examination
RT report no.: D231 Result passed
UT report no.: ______ Result ______

FILLET WELD TEST RESULTS
Minimum size multiple pass Macroetch
1. ______ 3. ______
2. ______

Maximum size single pass Macroetch
1. ______ 3. ______
2. ______

Other Tests

All-weld-metal tension test

Tensile strength, psi 83,100
Yield point/strength, psi 72,600
Elongation in 2 in., % 28
Laboratory test no. PW 231

Welder's name W. T. Williams Clock no. 261 Stamp no. ______

Tests conducted by RED Inc. & ABC Testing Laboratory

Test number PQR 231

Per D. Miller

We, the undersigned, certify that the statements in this record are correct and that the test welds were prepared, welded, and tested in conformance with the requirements of Section 4 of AWS D1.1/D1.1M, (2002 (year)) Structural Welding Code—Steel.

Signed RED Inc.
Manufacturer or Contractor
By R. M. Boncrack
Title Q.C. Mgr.
Date 12-15-2002

29106-15_F02.EPS

Figure 2 Example of a procedure qualification record (PQR).

GOING GREEN

Test Coupons

When welding coupons are no longer needed or usable, they should be properly recycled, to save the company money and to keep salvageable items out of the landfills.

1.2.2 Welder Performance Qualification

Once a procedure has been qualified, the welder using it must be qualified to use that procedure by passing a welding performance qualification (WPQ) test. Because no single performance test can qualify welders for all the different types of welding that must be done, a welder may be required to pass a number of performance qualification tests. Performance tests used to qualify welders are covered later in this module.

1.2.3 Welding Machine Operator Qualification

When automatic welding equipment is used, the operators of the equipment must demonstrate their ability to set up and monitor the equipment so that it will produce acceptable welds. The codes also contain qualification tests for these operators.

Additional Resources

2014 Technical Training Guide. Current Edition. Cleveland, OH, USA: The Lincoln Electric Company. **www.lincolnelectric.com**

1.0.0 Section Review

1. Which of the following documents is often adopted into law?
   a. Specifications
   b. Codes
   c. Standards
   d. WPSs

2. A WPQ test is used to qualify a welder to perform a specific procedure.
   a. True
   b. False

Section Two

2.0.0 Weld Discontinuities and Their Causes

Objective

Identify and describe weld discontinuities and their causes.

a. Identify and describe discontinuities related to porosity and inclusions.
b. Identify and describe discontinuities that result in cracking.
c. Identify and describe discontinuities related to joint penetration, fusion, and undercutting.
d. Identify and describe acceptable and unacceptable weld profiles.

Trade Terms

Arc blow: The deflection of an arc from its intended path by magnetic forces.

Defect: A discontinuity or imperfection that renders a part of the product or the entire product unable to meet minimum acceptable standards or specifications.

Discontinuity: A change or break in the shape or structure of a part that may or may not be considered a defect, depending on the code.

Embrittled: Metal that has been made brittle and that will tend to crack with little bending.

Hardenable materials: Metals that have the ability to be made harder by heating and then cooling.

Homogeneity: The quality or state of having a uniform structure or composition throughout.

Inclusion: Foreign matter introduced into and remaining in a weld.

Underbead cracking: Cracking in the base metal near the weld, but under the surface.

Codes and standards define the quality requirements necessary to achieve the integrity and reliability of a weldment. These quality requirements help ensure that welded joints are capable of serving their intended function for the expected life of the weldment. Weld discontinuities can prevent a weld from meeting the minimum quality requirements.

AWS defines a discontinuity as an interruption of the typical structure of a weldment, such as a lack of homogeneity in the mechanical, metallurgical, or physical characteristics of the material or weldment. A discontinuity is not necessarily a defect. A defect found during inspection will require the weld to be rejected. A single excessive discontinuity or a combination of discontinuities can make the weldment defective (unable to meet minimum quality requirements). However, a weld can have one or more discontinuities and still be acceptable.

The welder should be able to identify discontinuities and understand the effect they have on weld integrity. Some can be determined from visual inspection. Those that are internal to the weldment can only be detected through other testing methods. The most common weld discontinuities are the following:

- Porosity
- Inclusions
- Cracks
- Incomplete joint penetration
- Incomplete fusion
- Undercuts
- Arc strikes
- Spatter
- Unacceptable weld profiles

Ideally, a weld should not have any discontinuities, but most will have one or more. When evaluating a weld, it is important to note the type, size, and location of the discontinuity. Any one of these factors, or all three, can change a discontinuity into a defect, requiring the weld to be rejected during the inspection process. For example, discontinuities located at stress points tend to expand and thus have higher risk than those in other locations. Surface or near-surface discontinuities may be more harmful than similarly shaped internal discontinuities.

2.1.0 Porosity and Inclusions

Porosity and inclusions are weld defects that generally result from improper welding technique. Welders must recognize the causes of these defects and the techniques required to avoid them.

2.1.1 Porosity

Porosity is the presence of voids or empty spots in the weld metal. It is the result of gas pockets being trapped in the weld as it is being made. As the molten metal hardens, the gas pockets form voids.

Unless the gas pockets work up to the surface of the weld and burst open before the metal hardens, porosity cannot be detected through visual inspection.

Porosity can be grouped into the following major types:

- *Linear porosity* – May be aligned along a weld interface, the root of a weld, or a boundary between weld beads (*Figure 3[A]*).
- *Uniformly scattered porosity* – May be located throughout single-pass welds or throughout several passes in multiple-pass welds (*Figure 3[B]*).
- *Clustered porosity* – A localized grouping of pores that result from improperly starting or stopping the welding.
- *Piping porosity* – Normally extends from the root of the weld toward the face. These elongated gas pores are also called wormholes. They do not often extend to the surface, and the porosity cannot be visually detected (*Figure 3[C]*).

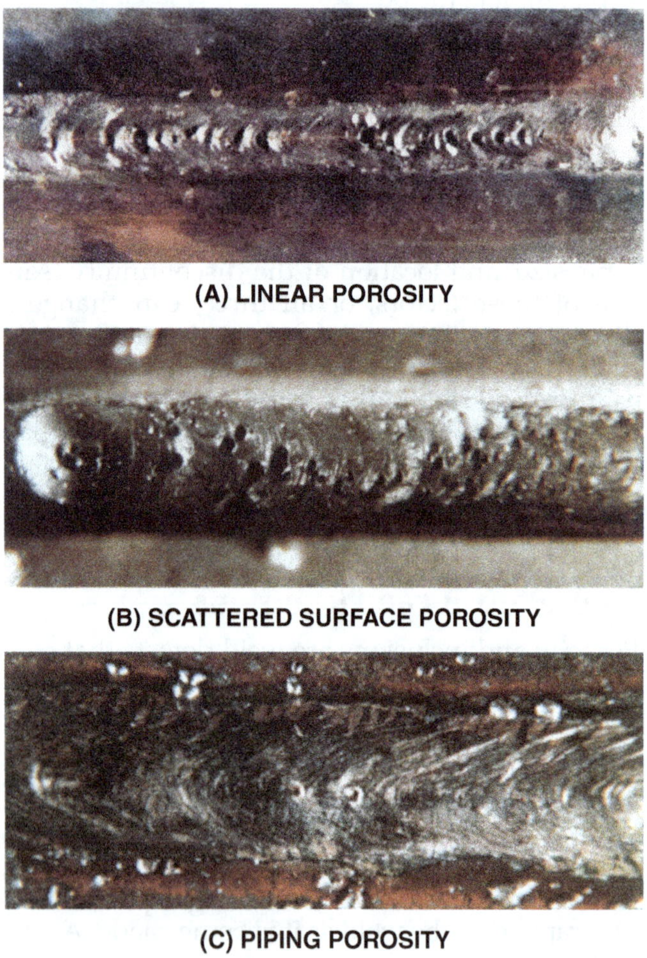

Figure 3 Examples of porosity.

Most porosity is caused by improper welding technique or contamination. Improper welding techniques may cause an inadequate amount of shielding gas to be formed. As a result, parts of the weld site are left unprotected. Oxygen in the air at the weld site, or moisture in the flux or on the base metal that dissolves in the weld pool, can become trapped and produce porosity.

The intense heat of the weld can decompose paint, dirt, oil, or other contaminants, producing hydrogen. This gas can become trapped in the solidifying weld pool and produce porosity.

Excessive porosity has a serious effect on the mechanical properties of the joint. Although some codes permit a certain amount of porosity in welds, it is best to have as little as possible. This can be accomplished by properly cleaning the base metal, avoiding excessive moisture in the electrode covering, and using proper welding techniques. Any porosity that occurs must be ground out until it is removed.

2.1.2 Inclusions

Inclusions are foreign matter trapped in the weld metal (*Figure 4*), between weld beads, or between the weld metal and the base metal. Inclusions are sometimes jagged and irregularly shaped. Sometimes they form in a continuous line. This concentrates stresses in one area and reduces the structural integrity (strength) of the weld.

Inclusions generally result from faulty welding techniques, improper access to the joint for welding, or both. A typical example of an inclusion is slag, which normally forms over a deposited weld. If the electrode is not manipulated correctly, the force of the arc will cause some of the slag particles to be blown into the molten pool. If the pool solidifies before the inclusions can float to the top, they become lodged in the metal, producing a discontinuity. Sharp notches in joint boundaries or between weld passes also can result in slag entrapment.

Inclusions are more likely to occur in out-of-position welding because the tendency is to keep the molten pool small and allow it to solidify rapidly to prevent it from sagging. Tungsten inclusions may occur in GTAW.

With proper welding technique, along with use of the correct electrode and the proper setting, inclusions can be avoided or kept to a minimum. Other preventive measures include the following:

- Positioning the work to maintain slag control
- Changing the electrode to improve control of molten slag

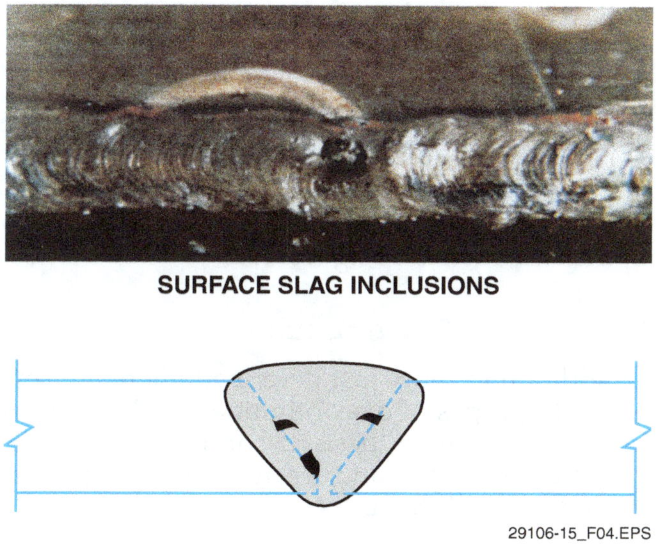

Figure 4 Examples of nonmetallic inclusions.

- Thoroughly removing slag between weld passes
- Grinding or sanding the weld surface if it is rough and likely to entrap slag
- Removing heavy mill scale or rust on weld preparations
- Avoiding the use of electrodes with damaged coverings

2.2.0 Cracks

Cracks are narrow breaks that occur in the weld metal, in the base metal, or in the crater formed at the end of a weld bead (*Figure 5*). Cracks occur when localized stresses exceed the ultimate strength of the metal. Cracks are generally located near other weld or base metal discontinuities.

2.2.1 Weld Metal Cracks

Three basic types of cracks can occur in weld metal: transverse, longitudinal, and crater. As seen in *Figure 5*, weld metal cracks are named to correspond with their location and direction.

Transverse cracks run across the face of the weld and may extend into the base metal. They are more common in joints that have a high degree of restraint.

Longitudinal cracks are usually located in the center of the weld deposit. They may be the continuation of crater cracks or cracks in the first layer of welding. Cracking of the first pass is likely to occur if the bead is thin. If this cracking is not eliminated before the other layers are deposited, the crack will progress through the entire weld deposit.

Case History

Mexico City Refinery

In Mexico City in 1984, a weld on a liquid propane gas (LPG) tank cracked. The escaping gas cloud drifted toward a nearby residential area until it encountered an ignition source. When the gas ignited, the fireball quickly burned its way back to the leak source. The resulting explosion destroyed the refinery and many residential neighborhoods, leaving 503 people dead and more than 4,000 injured.

The Bottom Line: Welds must be made correctly and inspected properly to prevent injury, loss of property, and sometimes death.

Crater cracks have a tendency to form in the crater whenever the welding operation is interrupted. These cracks usually proceed to the edge of the crater and may be the starting point for longitudinal weld cracks. Crater cracks can be minimized or prevented by filling craters to a slightly convex shape prior to breaking the welding arc.

Figure 6 shows various kinds of weld metal cracks. Weld metal cracking can usually be reduced by taking one or more of the following actions:

- Improve the contour or composition of the weld deposit by changing the electrode manipulation or electrical conditions
- Increase the thickness of the deposit and provide more weld metal to resist the stresses by decreasing the travel speed
- Reduce thermal stress by preheating
- Use low-hydrogen electrodes
- Balance shrinkage stress by sequencing welds
- Avoid rapid cooling conditions

2.2.2 Base Metal Cracks

Base metal cracking usually occurs within the heat-affected zone of the metal being welded. The possibility of cracking increases when working with hardenable materials. These cracks usually occur along the edges of the weld and through the heat-affected zone into the base metal. Types of base metal cracking include underbead cracking and toe cracking.

Underbead cracks are limited mainly to steel. They are usually found at regular intervals under the weld metal and usually do not extend to the surface. Because of this, they cannot be detected by visual inspection.

Hot and Cold Cracks

Hot cracks occur while the weld is solidifying. They can be caused by insufficient ductility at high temperature. Cold cracks occur after the weld has solidified. They are often caused by improper welding techniques.

Toe cracks are generally the result of strains caused by thermal shrinkage acting on a heat-affected zone that has been embrittled. They sometimes occur when the base metal cannot handle the shrinkage strains that are imposed by welding.

Base metal cracking can usually be reduced or eliminated by one of the following methods:

- Controlling the cooling rate by preheating
- Controlling heat input
- Using the correct electrode
- Controlling welding materials
- Properly matching the electrode filler metals to the base metals being welded

2.3.0 Other Discontinuities

In addition to those just described, there are several other types of discontinuities. Welders must learn to recognize these discontinuities, understand their causes, and learn the techniques required to avoid them.

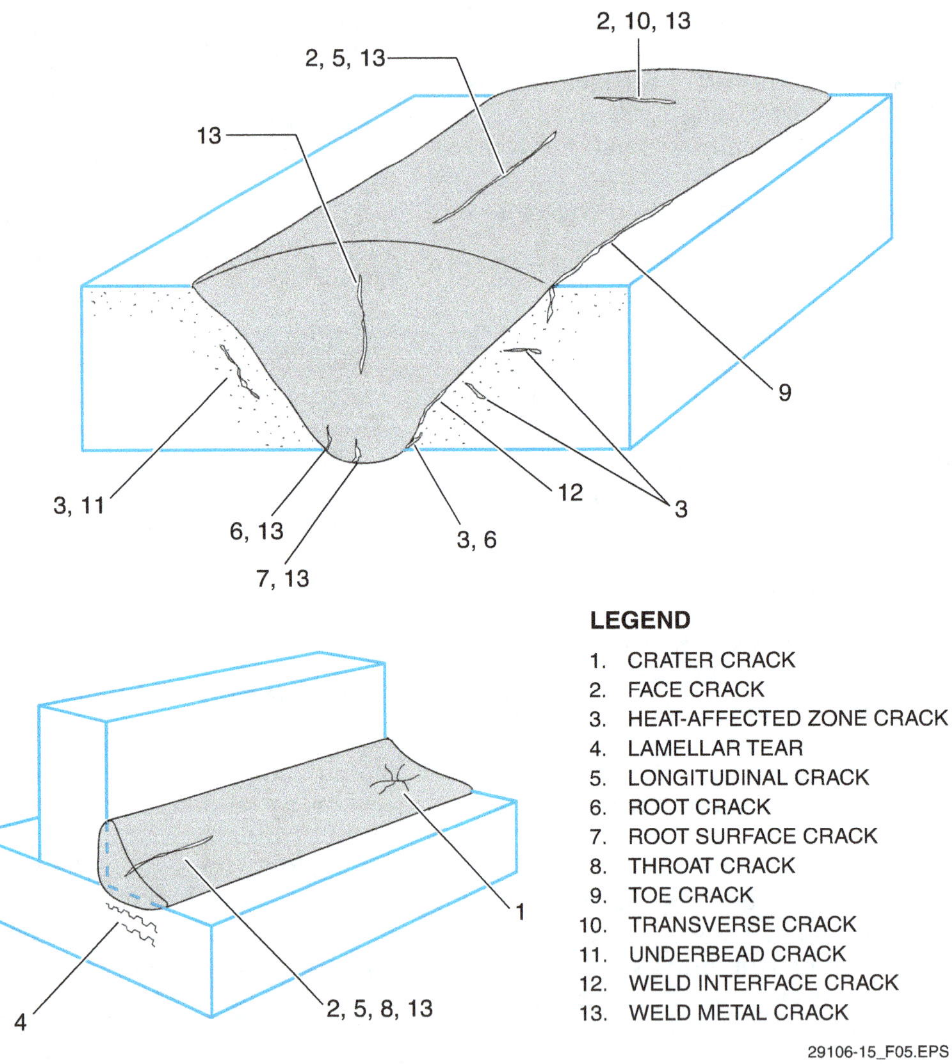

Figure 5 Types of weld metal and base metal cracks.

Figure 6 Examples of weld metal cracks.

2.3.1 Incomplete Joint Penetration

Incomplete joint penetration (*Figure 7*) occurs when the filler metal fails to penetrate and fuse with an area of the weld joint. Incomplete penetration will cause weld failure if the weld is subjected to tension or bending stresses.

Insufficient heat at the root of the joint is a frequent cause of incomplete joint penetration. If the metal being joined first reaches the melting point at the surfaces above the root of the joint, molten metal may bridge the gap between these surfaces and screen off the heat source before the metal at the root melts.

Improper joint design is another leading cause of incomplete joint penetration. If the joint is not prepared or fitted accurately, an excessively thick root face or an insufficient root gap may cause incomplete penetration. Incomplete joint penetration is likely to occur under the following conditions:

- If the root face dimension is too big, even though the root opening is adequate
- If the root opening is too small
- If the included angle of a V-groove is too small

Figure 8 is a labeled diagram showing correct and incorrect joint designs.

Even if the welding heat is correct and the joint design is adequate, incomplete penetration can result from poor control of the welding arc. Examples of poor control include the following:

- Using an electrode that is too large
- Excessive travel speed
- Using a welding current that is too low

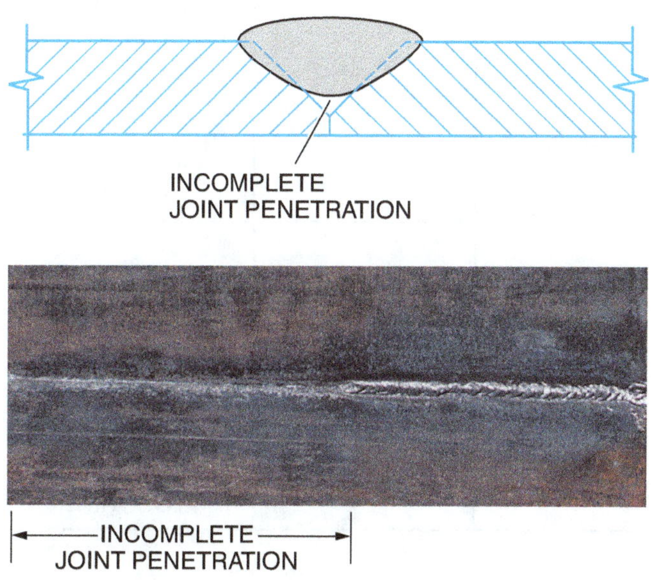

Figure 7 Incomplete joint penetration.

Incomplete penetration is always undesirable in welds, especially in single-groove welds where the root of the weld is subject either to tension or bending stresses. It can lead directly to weld failure or can cause a crack to start at the unfused area.

2.3.2 Incomplete Fusion

Many welders confuse incomplete joint penetration with incomplete fusion. It is possible to have good penetration without complete root fusion. Incomplete fusion is the failure of a welding process to fuse, or join together, layers of weld metal or weld metal and base metal.

Incomplete fusion may occur at any point in a groove or fillet weld, including the root of the weld. Often, the weld metal simply rolls over onto the plate surface. This is generally referred to as overlap. In many cases, the weld has good fusion at the root and at the plate surface, but because of poor technique and insufficient heat, the toe of the weld does not fuse. *Figure 9* shows incomplete fusion and overlap. Causes for incomplete fusion include the following:

- Insufficient heat as a result of low welding current, high travel speeds, or an arc gap that is too close
- Wrong size or type of electrode
- Failure to remove oxides or slag from groove faces or previously deposited beads
- Improper joint design
- Inadequate gas shielding
- Improper electrode angle
- **Arc blow**

Incomplete fusion discontinuities affect weld joint integrity in much the same way as porosity and slag inclusion.

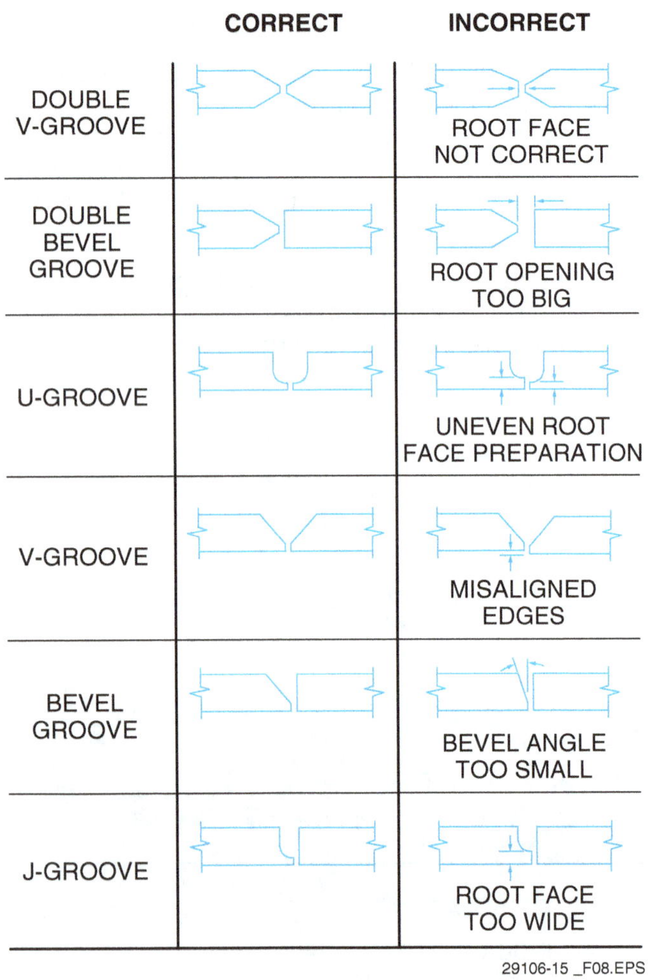

Figure 8 Correct and incorrect joint designs.

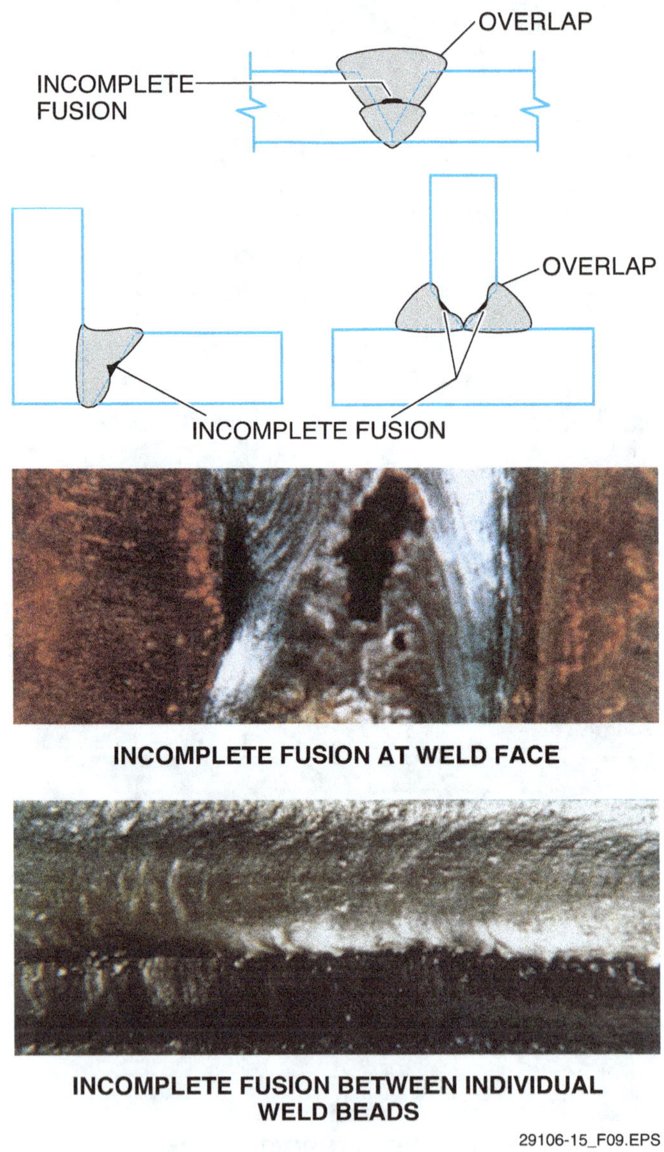

Figure 9 Incomplete fusion and overlap.

2.3.3 Undercut

Undercut is a groove melted into the base metal beside the weld. It is the result of the arc removing more metal from the joint face than is replaced by weld metal. On multilayer welds, it may also occur at the point where a layer meets the wall of a groove (*Figure 10*).

Undercutting is usually caused by improper electrode manipulation. Other causes of undercutting include the following:

- Using a current adjustment that is too high
- Having an arc gap that is too long
- Failing to fill up the crater completely with weld metal
- Improper electrode angle
- Incorrect rod travel speed

Most welds have some undercut that can be found upon careful examination. When it is controlled within the limits of the specifications and does not create a sharp or deep notch, undercut is usually not considered a weld defect. However, when it exceeds the limits, undercutting can be a serious defect because it reduces the strength of the joint.

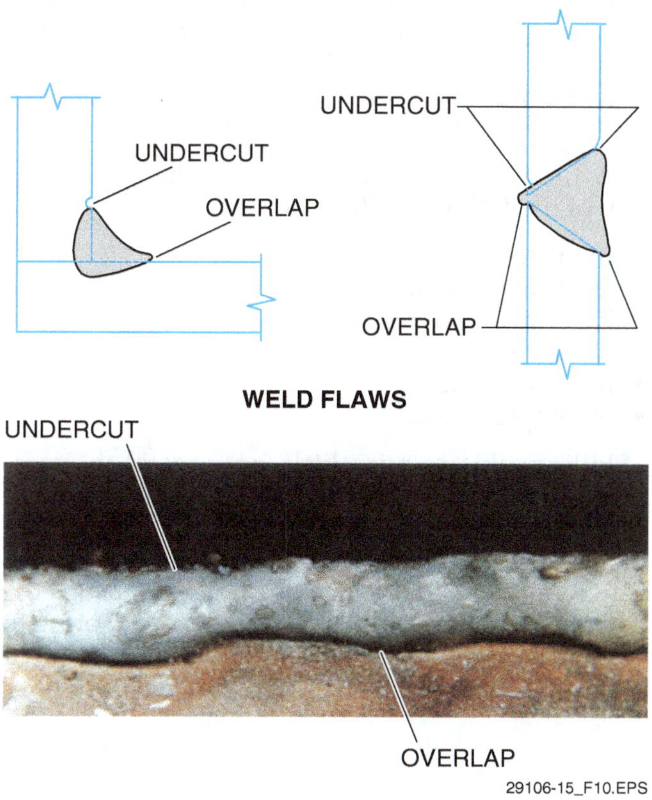

Figure 10 Undercut and overlap.

Causes of Incomplete Joint Penetration

Incomplete joint penetration is generally associated with groove welds. It may result from insufficient welding heat, improper joint design (too much metal for the welding arc to penetrate), or poor control of the welding arc.

2.3.4 Arc Strikes

Arc strikes (*Figure 11*) are small, localized points where surface melting occurs away from the joint. These spots may be caused by accidentally striking the arc in the weld zone or by faulty ground connections. The weld zone is the area where the weld bead should be, and/or where the weld metal is to be deposited.

Striking an arc on base metal that will not be fused into the weld metal should be avoided. Arc strikes can cause hardness zones in the base metal and can become the starting point for cracking. Arc strikes can cause a weld to be rejected.

2.3.5 Spatter

Spatter is made up of very fine particles of metal on the plate surface adjoining the weld area. It is usually caused by high current; a long arc; an irregular and unstable arc; or improper shielding. Spatter makes a poor appearance on the weld and base metal and can make it difficult to inspect the weld. Spatter can also cause coating failure.

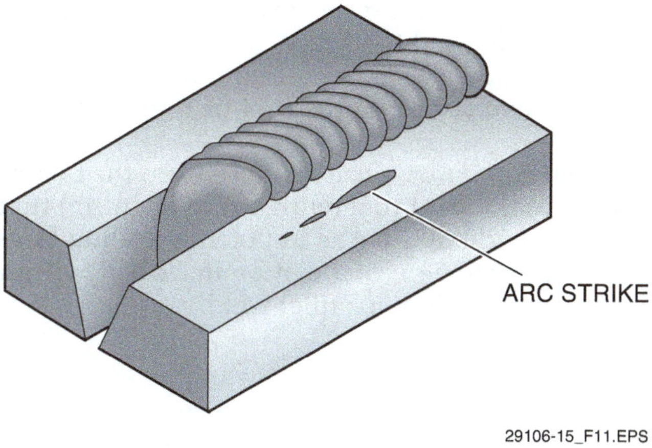

Figure 11 Arc strike.

2.4.0 Acceptable and Unacceptable Weld Profiles

The profile of a finished weld can affect the performance of the joint under load as much as other discontinuities affect it. This applies to the profile of a single-pass weld and to a layer of a multiple-pass weld. An unacceptable profile for a single-pass or multiple-pass weld could lead to the formation of discontinuities such as incomplete fusion or slag inclusions as the other layers are deposited. *Figure 12* shows acceptable and unacceptable weld profiles for both fillet and groove welds.

2.4.1 Fillet Welds

A fillet weld is a weld that is approximately triangular in cross section and is used with T-, lap, and corner joints. The sizes and locations of fillet welds are given as welding symbols. The two types of fillet welds are convex and concave (*Figure 13*). A convex fillet weld has its surface bowed out like the outside surface of a ball. A concave fillet weld has its surface bowed in like the inside surface of a bowl.

The following terms are used to describe a fillet weld:

- *Weld face* – The exposed surface of the weld
- *Leg* – The distance from the root of the joint to the toe of a fillet weld
- *Weld toe* – The junction between the face of a weld and the base metal
- *Weld root* – The point shown in cross section at which the weld metal intersects with the base metal and extends farthest into the weld joint
- *Size* – The leg lengths of the largest right triangle that can be drawn within the cross section of a fillet weld
- *Actual throat* – The shortest distance from the root of the weld to its face
- *Effective throat* – The minimum distance, minus any convexity, from the root of the weld to its face
- *Theoretical throat* – The distance from the beginning of the joint root (with a zero opening) that is perpendicular to the hypotenuse of the largest right triangle that can be inscribed within the cross section of a fillet weld.

Preferred Fillet Weld Contours

In single-pass, weave-bead fillet welds where two workpieces are being joined at an angle (not lap joints), flat or slightly convex faces are usually preferred because weld stresses are more uniformly distributed through the fillet and workpieces.

Fillet welds may be either equal leg or unequal leg (*Figure 14*). The face may be slightly convex, flat, or slightly concave. Welding codes require that fillet welds have a uniform concave or convex face, although a slightly non-uniform face is acceptable. The maximum convexity of a fillet weld or individual surface bead is dependent on the width of the weld face or individual surface bead, as follows:

- If the weld face or individual surface bead is ≤ 5⁄16" (8 mm), the maximum convexity is 1⁄16" (1.6 mm).
- If the weld face or individual surface bead is > 5⁄16" (8 mm) and < 1" (25 mm), the maximum convexity is 1⁄8" (3 mm).
- If the weld face or individual surface bead is ≥ 1", the maximum convexity is 3⁄16" (5 mm).

A fillet weld is unacceptable if the profile has any of the following discontinuities:

- Insufficient throat
- Excessive convexity
- Excessive undercut
- Overlap
- Insufficient leg
- Incomplete fusion

Fillet welds require little base metal preparation except for cleaning the weld area and removing all dross from cut surfaces. Any dross from oxyfuel, plasma arc, or carbon arc cutting will cause porosity in the weld. For this reason, the codes require that all dross be removed prior to welding.

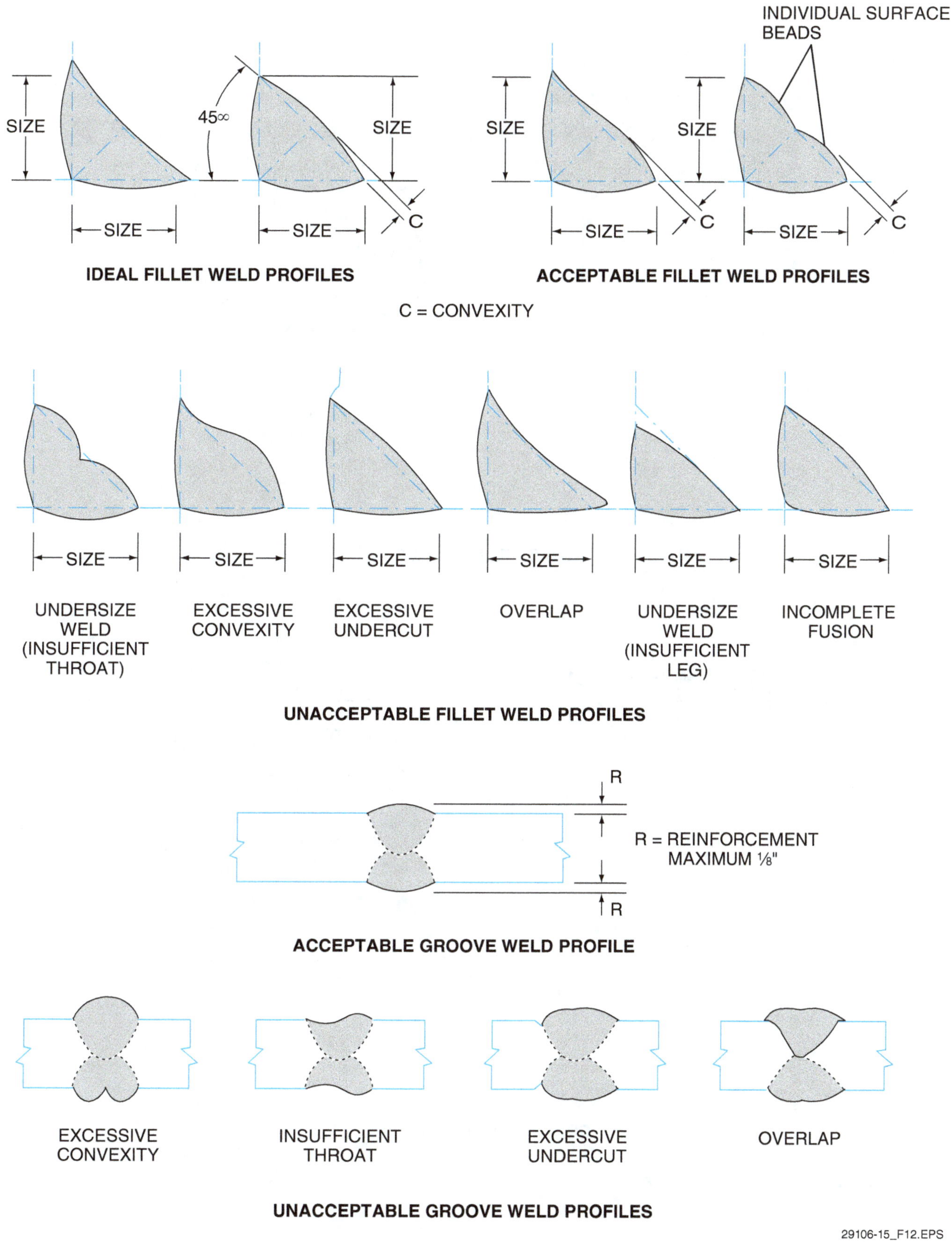

Figure 12 Acceptable and unacceptable weld profiles.

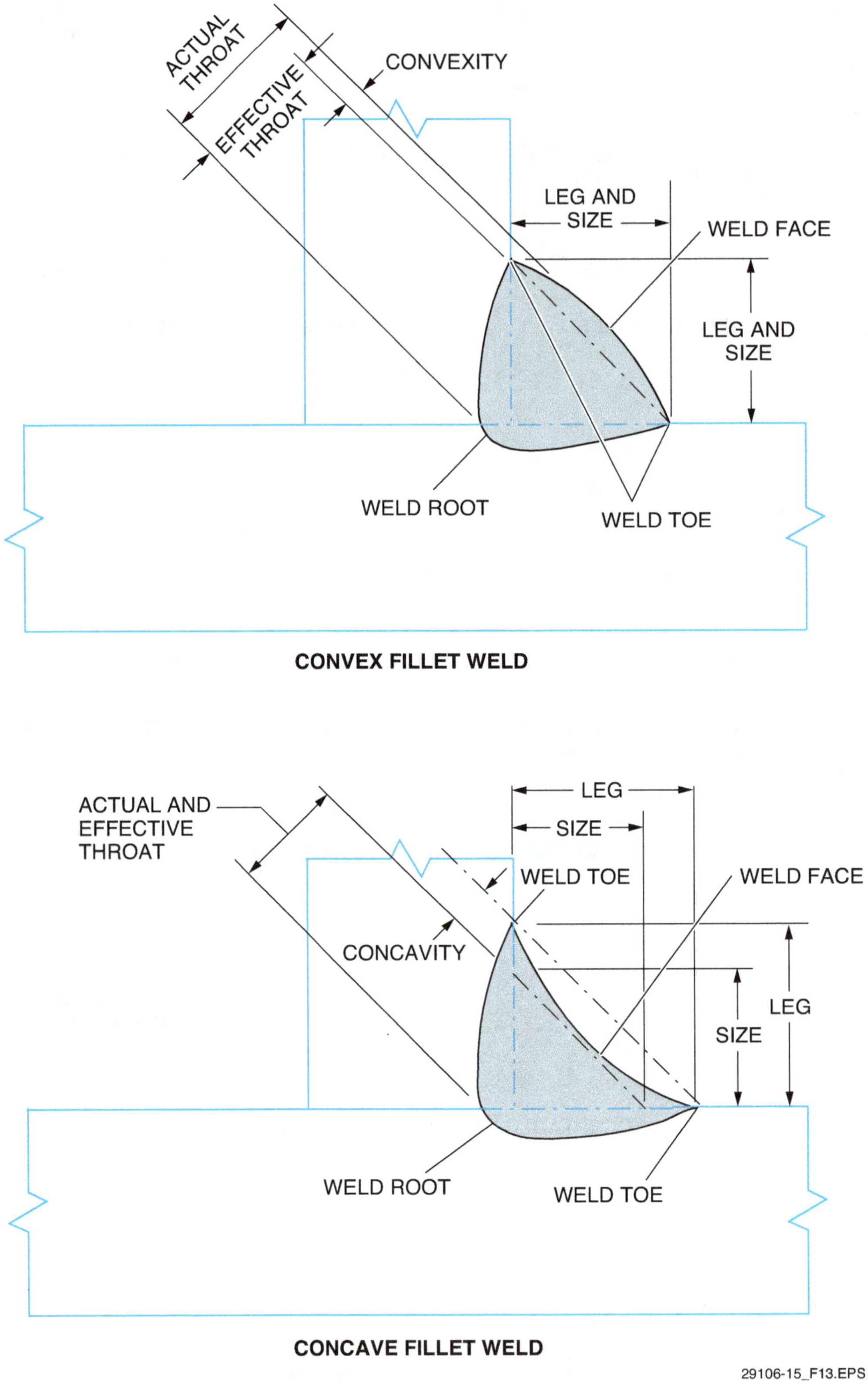

Figure 13 Convex and concave fillet welds.

2.4.2 Groove Welds

Groove welds should be made with reinforcement not exceeding ⅛" (3.2 mm) and a gradual transition to the base metal at each toe. Groove welds should not have excess reinforcement, insufficient throat, excessive undercut, or overlap. If a groove weld has any of these defects, it is considered unacceptable. The bead width should not exceed the groove width by more than ⅛" (3.2 mm).

NOTE

Refer to your site's WPS for specific requirements on fillet and butt welds. The information provided here is only a general guideline. The site WPS or quality specifications must be followed for all welds. Check with your supervisor if you are unsure of the specifications for your application.

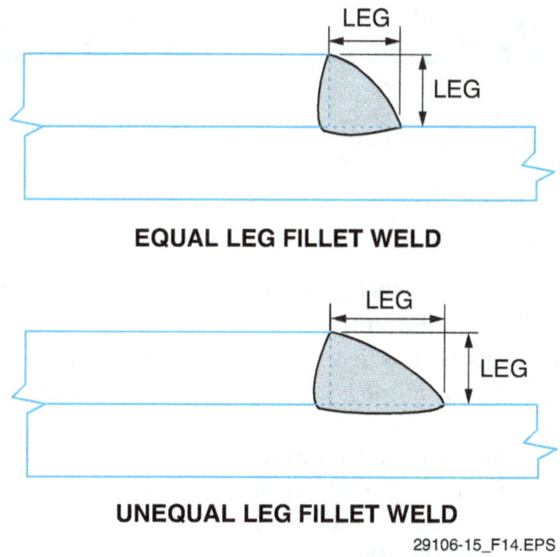

Figure 14 Equal leg and unequal leg fillet welds.

Additional Resources

Visual Inspection Workshop Reference Manual, AWS VIW-M-2008. Miami, FL: American Welding Society.

Welding Inspection Technology. Miami, FL: American Welding Society.

2.0.0 Section Review

1. The entrapment of gas pockets in a weld will cause ____.
   a. inclusions
   b. cracks
   c. porosity
   d. undercut

2. Which of these discontinuities *cannot* be detected by visual inspection?
   a. Toe cracks
   b. Underbead cracks
   c. Transverse cracks
   d. Face cracks

3. Undercut is usually caused by improper electrode manipulation.
   a. True
   b. False

4. Which of the following describes the leg of a weld?
   a. The shortest distance from the root of the weld to its face
   b. The junction between the face of the weld and the base metal
   c. The minimum distance, minus any convexity, from the root of the weld to its face
   d. The distance from the root of the joint to the toe of a fillet weld

SECTION THREE

3.0.0 WELD EXAMINATION PRACTICES

Objective

Describe various nondestructive and destructive weld examination practices.

a. Describe basic visual inspection methods including measuring devices and liquid penetrants.
b. Describe magnetic particle and electromagnetic inspection processes.
c. Describe the radiographic and ultrasonic inspection processes.
d. Describe destructive testing processes.

Performance Task

1. Perform a visual inspection (VT) on a fillet and/or groove weld and complete an inspection report.

Trade Terms

Laminations: Cracks in the base metal formed when layers separate.

Radiographic: Describes images made by passing X-rays or gamma rays through an object and recording the variations in density on photographic film.

Transducer: A device that converts one form of energy into another.

Nondestructive examination (NDE), sometimes referred to as nondestructive testing (NDT) or nondestructive inspection, is a term used for those inspection methods that allow materials to be examined without changing or destroying them. NDE methods can usually detect the discontinuities and defects previously described.

Nondestructive examination is usually performed by the site quality group as part of the site quality program. Inspectors trained in the proper test methods conduct the examinations.

The welder should be familiar with the following basic nondestructive examination practices:

- Visual inspection (VT)
- Liquid penetrant inspection (PT)
- Magnetic particle inspection (MT)
- Radiographic inspection (RT)
- Ultrasonic inspection (UT)
- Electromagnetic (eddy current) inspection (ET)
- Leak testing (LT)

There is considerable overlap in the application of nondestructive and destructive tests. Destructive tests, which destroy the weld, are frequently used on several sample weldments to supplement, confirm, or establish the limits of nondestructive tests. Destructive testing is also used to provide supporting information. Once this information has been established, nondestructive examinations can be made on similar welds to locate all discontinuities above the critical defect size that was determined by the destructive tests.

3.1.0 Visual Inspection

In visual inspection, the surface of the weld and the base metal are observed for visual imperfections. Certain tools and gauges may be used during the inspection. Visual inspection is the examination method most commonly used by welders and inspectors. It is the fastest and most inexpensive method for examining a weld. However, it is limited to what can be detected by the naked eye or through a magnifying glass.

Properly done before, during, and after welding, visual inspection can detect more than 75 percent of discontinuities before they are found by more expensive and time-consuming nondestructive examination methods.

Prior to welding, the base metal should be examined for conditions that may cause weld defects. The required dimensions of the material, including edge preparations, should also be confirmed by measurements. If problems or potential problems are found, corrections should be made before proceeding any further.

After the parts are assembled for welding, the weld joint should be visually checked for a proper root opening and any other aspects that might affect the quality of the weld. Visually check for the following conditions:

- Proper cleaning
- Joint preparation and dimensions
- Clearance dimensions for backing strips, rings, or consumable inserts
- Alignment and fit-up of the pieces being welded
- Welding procedures and machine settings
- Specified preheat temperature (if applicable)
- Tack-weld quality

Fixed Weld Fillet Gauge

This fixed weld fillet gauge can be used to measure 11 fillet weld sizes in fractions or decimals. Metric versions are also available.

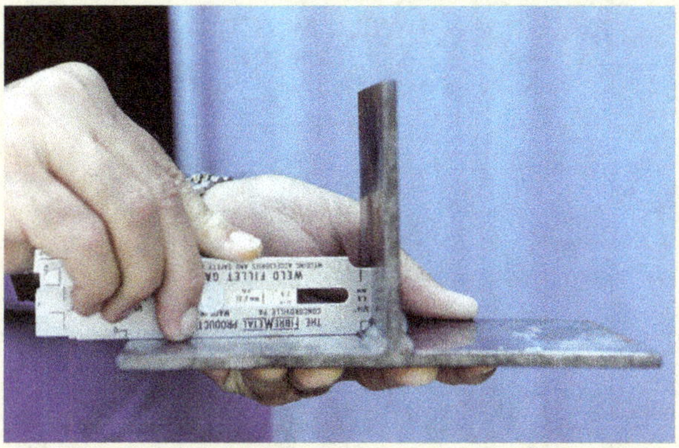

29106-15_SA01.EPS

During the welding process, visual inspection is the primary method for controlling quality. Some of the aspects that should be visually examined include the following:

- Quality of the root pass and the succeeding weld layers
- Sequence of weld passes
- Interpass cleaning
- Root preparation prior to welding a second side
- Conformance to the applicable procedure

After the weld has been completed, the weld surface should be thoroughly cleaned. A complete visual examination may disclose weld surface defects such as cracks, shrinkage cavities, undercuts, incomplete penetration, incomplete fusion, overlap, and crater deficiencies before they are discovered using other nondestructive inspection methods.

An important aspect of visual examination is checking the dimensional accuracy of the weld after it has been completed. Dimensional accuracy is determined by the use of measuring gauges. The purpose of using the gauges is to determine if the weld is within allowable limits as defined by the applicable codes and specifications.

Some of the more common welding gauges are the following:

- Undercut gauge
- Butt weld reinforcement gauge
- Fillet weld blade gauge set

3.1.1 Undercut Gauge

An undercut gauge is used to measure the amount of undercut on the base metal. Typically, codes allow for undercut to be between 0.01" and 0.031" deep. These gauges have a pointed end that is pushed into the undercut. The reverse side of the gauge indicates the measurement in either inches or millimeters. Two types of undercut gauges currently used are the bridge cam gauge and the V-WAC gauge (*Figure 15*). These gauges can be used for measuring undercut and for many other measurements.

3.1.2 Butt Weld Reinforcement Gauge

The butt weld reinforcement gauge has a sliding pointer calibrated to several different scales that are used to measure the size of a fillet weld or the reinforcement of a butt weld. To use the gauge for a fillet weld, position it as shown in *Figure 16*, and then slide the pointer to contact the base metal or weld metal. Be sure to read the correct scale for the measurement being taken. The other end of the gauge is used for butt welds.

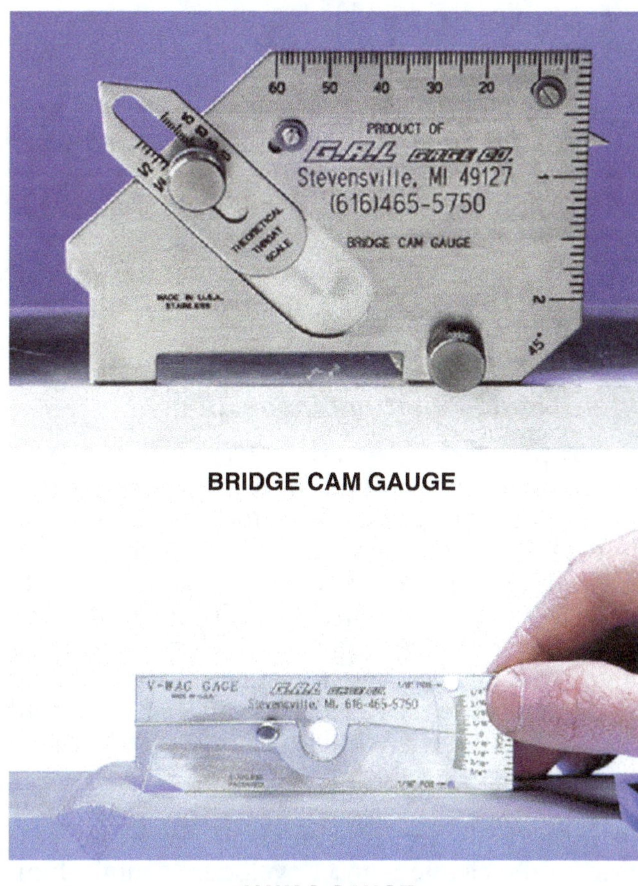

29106-15_F15.EPS

Figure 15 Undercut gauges.

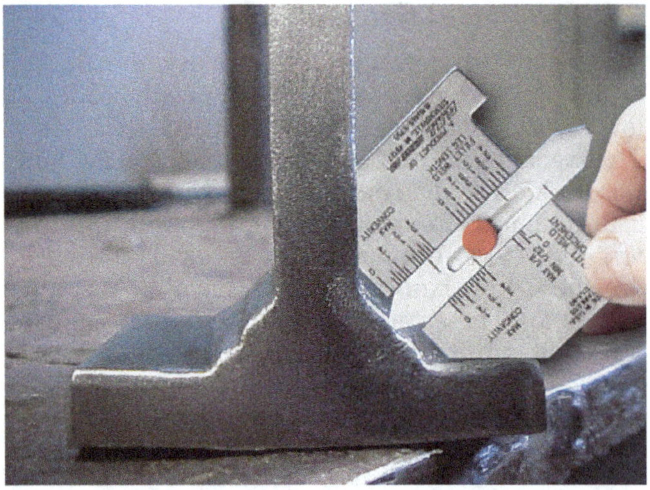

29106-15_F16.EPS

Figure 16 Automatic weld size gauge (AWS).

3.1.3 Fillet Weld Blade Gauge Set

The fillet weld blade gauge set has seven individual blade gauges for measuring convex and concave fillet welds. The individual gauges are held together by a screw secured with a knurled nut. The seven individual blade gauges can measure eleven concave and convex fillet weld sizes: 1/8", 3/16", 1/4", 5/16", 3/8", 7/16", 1/2", 5/8", 3/4", 7/8", and 1" as well as their metric equivalents.

To use the fillet weld blade gauge set, identify the type of fillet weld to be measured (concave or convex) and the size. Select the appropriate blade and position it. Be sure the gauge blade is flush to the base metal with the tip touching the vertical member. *Figure 17* shows an application of a fillet weld blade gauge.

3.1.4 Liquid Penetrant Inspection

Liquid penetrant inspection (PT) examination is a nondestructive method for locating defects that are open to the surface. It cannot detect internal defects. The technique is based on the ability of a penetrating liquid, which is usually red, to wet the surface opening of a discontinuity and be drawn into it. A liquid or dry powder developer, which is usually white, is then applied over the metal. If the flaw is significant, red penetrant bleeds through the white developer to indicate a discontinuity or defect.

The dye, cleaner, and developer are available in spray cans for convenience. Some solvents used in the cleaners and developers contain high amounts of chlorine, a known health hazard, to make the liquids nonflammable.

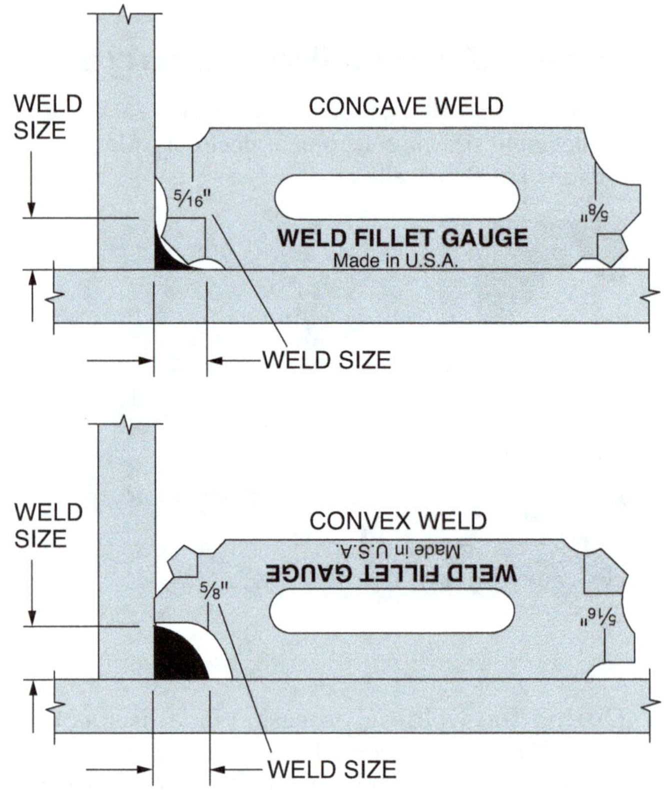

29106-15_F17.EPS

Figure 17 Fillet weld blade gauge.

WARNING!

Refer to the safety data sheet (SDS/MSDS) for hazards associated with the liquid penetrant solvent.

The most common defects found using this process are surface cracks. Most cracks exhibit an irregular shape. The width of the bleed-out (the red dye bleeding through the white developer) is a relative measure of the depth of a crack.

Surface porosity, metallic oxides, and slag will also hold penetrant and cause bleed-out. These indications are usually more circular and have less width than a crack. *Figure 18* shows liquid penetrant materials and an example of the results of liquid penetrant inspection.

The advantages of liquid penetrant inspection are that it can find small defects not visible to the naked eye; it can be used on most types of metals; it is inexpensive; and it is fairly easy to use and interpret. PT is most useful in examining welds that are susceptible to surface cracks. Except for visual inspection, it is perhaps the most commonly used nondestructive examination method for surface inspection.

29106-15_F18.EPS

Figure 18 Liquid penetrant materials and inspection example.

The disadvantages of liquid penetrant inspection are that it takes more time to use than visual inspection, and it can only find surface defects. The presence of weld bead ripples and other irregularities can also hinder the interpretation of indicators. Because chemicals are used, care must be taken when performing the inspection. When testing rough, irregular surfaces, the presence of irrelevant indicators may also make interpretation difficult.

3.1.5 Leak Testing

Leak testing (*Figure 19*) is used to determine the ability of a pipe or vessel to contain a gas or liquid under pressure. Testing methods vary depending on the application of the weldment. In some cases, the vessel is pressurized and tested by immersing it in water or by applying a soap bubble solution to the weld. An open tank can be tested using water that contains fluorescein, which can be detected by ultraviolet light.

A method called the vacuum box test is used to test a vessel when only one side of the weld is accessible. The base of a storage tank is an example. The vacuum box is a transparent box with a soft rubber seal. A vacuum pump is used to extract all the air from the box. A leak is indicated by the presence of bubbles.

In the helium spectrometer leak test, helium is used as a tracer gas. Because of the small size of helium atoms, they can pass through an opening so small that it might not be detectable by other test methods. Sensitive instruments are used to detect the presence of helium.

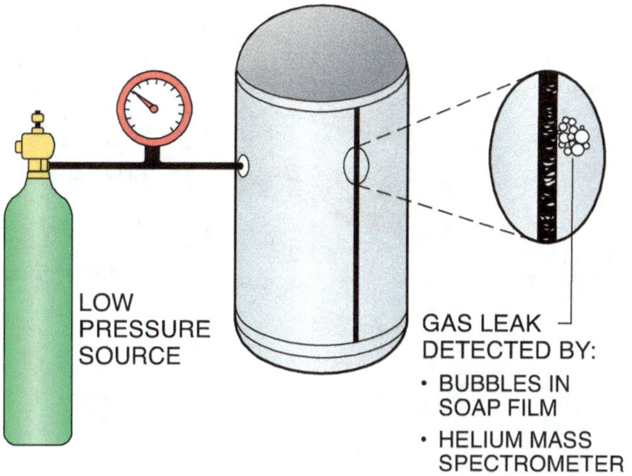

29106-15_F19.EPS

Figure 19 Example of a leak test.

3.2.0 Magnetic Particle and Electromagnetic Inspection

These non-destructive methods make use of electromagnetism to perform weld examination. Magnetic particle inspection methods are only effective when used with ferrous metals. Electromagnetic inspection can be used with non-ferrous metals as well as ferrous metals.

3.2.1 Magnetic Particle Inspection

Magnetic particle inspection (MT) is a nondestructive examination method that uses electricity to magnetize the weld that will be examined (*Figure 20*). After the metal has been magnetized, metal particles are sprinkled onto the weld surface. If there are defects in the surface or just below it, the metal particles will be grouped into a pattern around the defect. The defect can be identified by the shape, width, and height of the particle pattern.

Magnetic particle inspection is used to test welds for such defects as surface cracks, incomplete fusion, porosity, and slag inclusion. It can also be used to inspect plate edges for surface imperfections prior to welding. Defects can be detected only at or near the surface of the weld. Defects much deeper than this are not likely to be found. Certain discontinuities exhibit characteristic powder patterns that can be identified by a skilled inspector.

For magnetic particle examination, the part to be inspected must be ferromagnetic (made of steel or a steel alloy), smooth, clean, dry, and

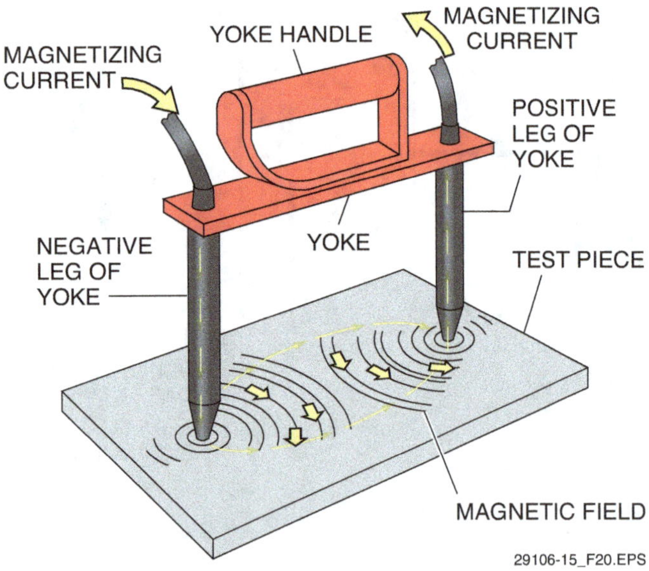

Figure 20 Magnetic particle examination with electromagnetic yokes.

free from oil, water, and excess slag. The part is magnetized by using an electric current to set up a magnetic field within the material. The magnetized surface is covered with a thin layer of magnetic powder. If there is a defect, the powder is held to the surface at the defect because of the powerful magnetic field.

When this examination method is used, there is normally a code or standard that governs both the method and the acceptance/rejection criteria of indications.

The advantages of magnetic particle inspection are that it can find small defects not visible to the naked eye, and it is faster than liquid penetrant inspection.

Disadvantages of magnetic particle inspection are that the materials must be capable of being magnetized; the inspector must be skilled in interpreting indications; rough surfaces can interfere with the results; the method requires an electrical power source; and it cannot identify internal discontinuities located deep in the weld.

3.2.2 Electromagnetic (Eddy Current) Inspection

Like magnetic particle testing, electromagnetic, or eddy current inspection (ET) uses electromagnetic energy to detect defects in the joint. A coil, which produces a magnetic field, is placed on or around the part being tested. After being calibrated, the coil is moved over the part to be inspected. The coil produces a current in the metal through induction. The induced current is called an eddy current. If a discontinuity is present in the test part, it will interrupt the flow of the eddy currents. This change can be observed on the oscilloscope display.

Eddy currents only detect discontinuities near the surface of the part. This method is suitable for both ferrous and nonferrous materials, and it is used in testing welded tubing and pipe. It can determine the physical characteristics of a material and the wall thickness in tubing. It can also check for porosity, pinholes, slag inclusions, internal and external cracks, and incomplete fusion.

The advantage of eddy current inspection is that it can detect surface and near-surface weld defects. It is particularly useful in inspecting circular parts like pipes and tubing.

The disadvantage of eddy current inspection is that eddy currents decrease with depth, so defects farther from the surface may go undetected. The accuracy of the examination depends in large part on the calibration of the instrument and the ability of the inspector. *Figure 21* shows an eddy current test being conducted.

29106-15_F21.EPS

Figure 21 Eddy current test.

3.3.0 Radiographic and Ultrasonic Inspection

These non-destructive inspection methods rely on technology that is commonly used in medical diagnostics. The ability to apply these methods and to correctly interpret the results requires extensive training and experience.

3.3.1 Radiographic Inspection

Radiography (RT) is a nondestructive examination method that uses radiation (X-rays or gamma rays) to penetrate the weld and produce an image. Like medical X-rays, this weld inspection technology originally worked by exposing treated film to the X-rays. In traditional RT work, X-ray film is placed against the weld and exposed (*Figure 22*). When a joint is radiographed, the radiation source is placed on one side of the weld and the film on the other. The joint is then exposed to the radiation source. The radiation penetrates the metal and produces an image on the film. The film is called a radiograph and provides a permanent record of the weld quality.

Modern RT offers two additional options: digital detector array (DDA) and computed radiography (CR). DDA radiography, also referred to as digital radiography, uses a semiconductor array instead of silver-based film to capture the image. Like film radiography, digital radiography is considered a form of direct radiography. Once the DDA is exposed, the image it captures can be transferred directly to a computer and enhanced for viewing. DDA radiography uses the same type of X-ray source as film radiography.

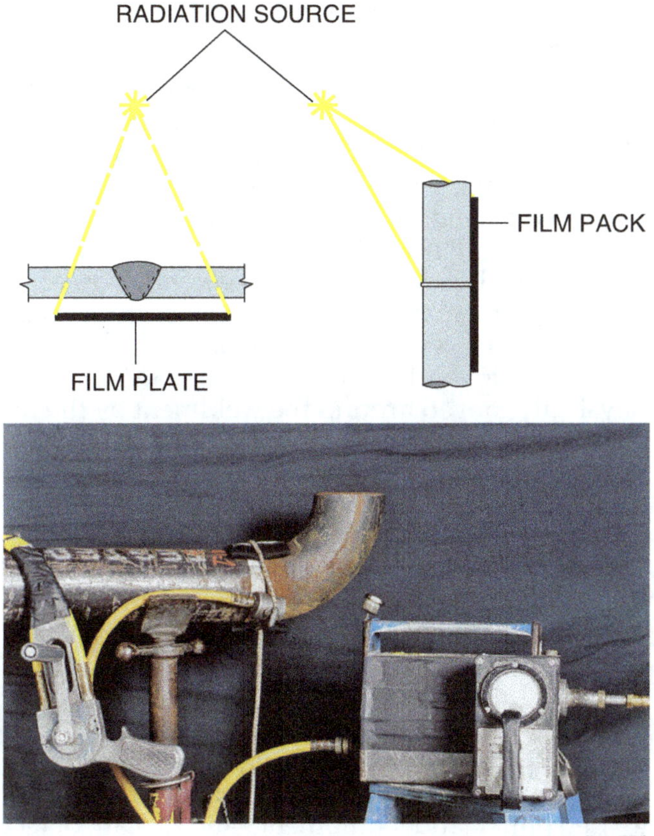

29106-15_F22.EPS

Figure 22 Weld examination using radiography.

CR generally uses a cassette that contains a reusable imaging plate. Once the plate is exposed, it is processed through a special laser scanner that captures the image and allows it to be transferred to a computer for enhancement and interpretation. *Figure 23* shows CR radiography being performed.

Radiographic inspection can produce a visible image of weld discontinuities, both surface and subsurface, when they are different in density from the base metal and different in thickness parallel to the radiation. Surface discontinuities are better identified by visual, penetrant, or magnetic particle examination. Radiographic weld inspection requires specialized skill. It should therefore be done and interpreted only by trained, qualified personnel.

The advantages of radiographic inspection are that it provides a permanent record of the weld quality, stored either on film or as a computer image. Through radiography, the entire thickness can be examined, and it can be used on all types of metals.

The disadvantages of radiographic inspection are that it is a slow and expensive method for inspecting welds; some joints are inaccessible to radiography; and excessive exposure to radiation of any type is very hazardous to humans. Cracks can frequently be missed if they are very small or are not aligned with the radiation beam.

3.3.2 Ultrasonic Inspection

Ultrasonic inspection (UT) is a relatively low-cost, nondestructive examination method that uses sound wave vibrations to find surface and subsurface defects in the weld material. Ultrasonic waves are passed through the material being tested and are reflected back by any density change caused by a defect. The reflected signal is shown on the screen display of the instrument (*Figure 24*).

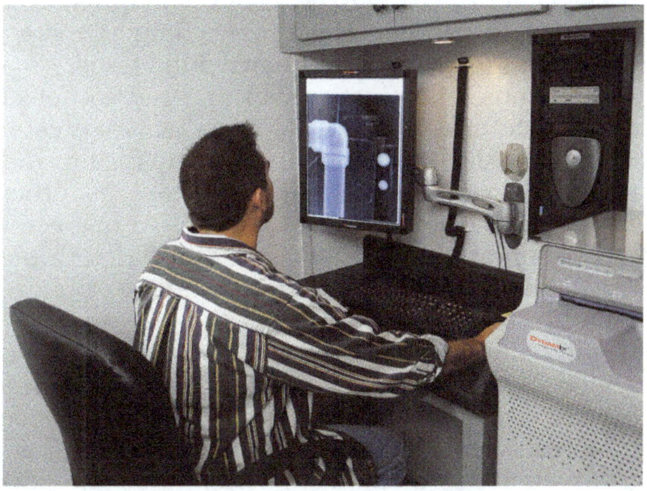

29106-15_F23.EPS

Figure 23 Computed radiography (CR) testing in process.

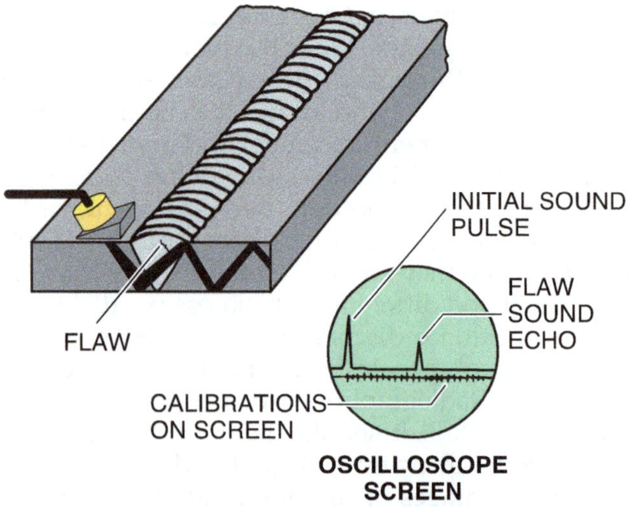

29106-15_F24.EPS

Figure 24 Portable ultrasonic device.

The term *ultrasonic* indicates frequencies above those heard by the human ear. Ultrasonic devices operate very much like depth sounders or fish finders. The key element is a transducer that is passed over the weld to be tested. The transducer converts electrical energy into mechanical energy that is emitted as high-frequency vibrations.

Ultrasonic examination can be used to detect and locate cracks, laminations, shrinkage cavities, pores, slag inclusions, incomplete fusion, and incomplete joint penetration, as well as other discontinuities in the weld. A qualified inspector can interpret the signal on a screen to determine the approximate position, depth, and size of the discontinuity.

Ultrasonic beams can deeply penetrate weld material with high sensitivity and accuracy. The output is readily digitized, and operation of the monitor is entirely nonhazardous. Some of the different types of ultrasonic techniques are the following:

Self-Propelled NDT

This ultrasonic test instrument is often used to inspect pipe and tanks. Its magnetic wheels allow it to cling to the surface as it moves along, controlled by a joystick mechanism.

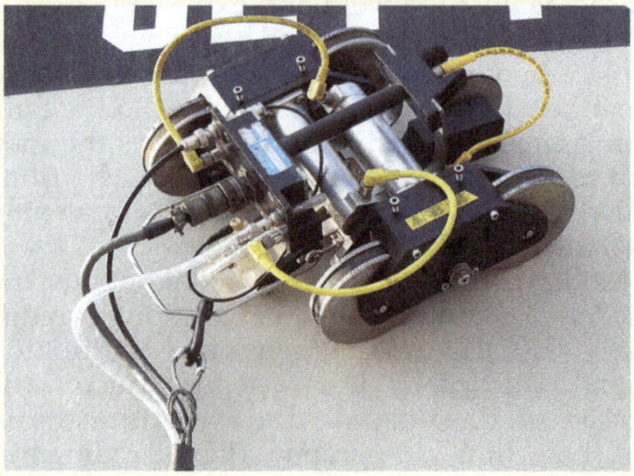

29106-15_SA02.EPS

- Pulse-echo
- Pulse-transmission
- Ultrasonic attenuation
- Continuous wave resonance
- Ultrasonic spectroscopy
- Phased array

Figure 25 shows the effects of moving an angle-beam transducer back and forth across a weld. *Figure 26* shows the effects of moving an angle-beam transducer in an arc around a weld.

Ultrasonic monitors are very portable, but they require experienced technicians to monitor the equipment. Reading an ultrasonic flaw detector requires some practice because the screen image shows the peaks of back reflection.

In traditional UT, a single-element probe is physically moved around the weldment by the inspector to create a cross-section of the joint being tested. In contrast, the newer phased array (PA) ultrasonic testing method (*Figure 27*) uses a probe containing an array of ultrasonic transducers. The beam created by these transducers is electronically steered. The data from the individual elements is combined electronically to create a cross-section of the weld.

Among the advantages of ultrasonic inspection are that it can find defects throughout the material being examined; it can be used to check materials that cannot be radiographed; it is nonhazardous to personnel and equipment; and it can detect even small defects from one side of the material.

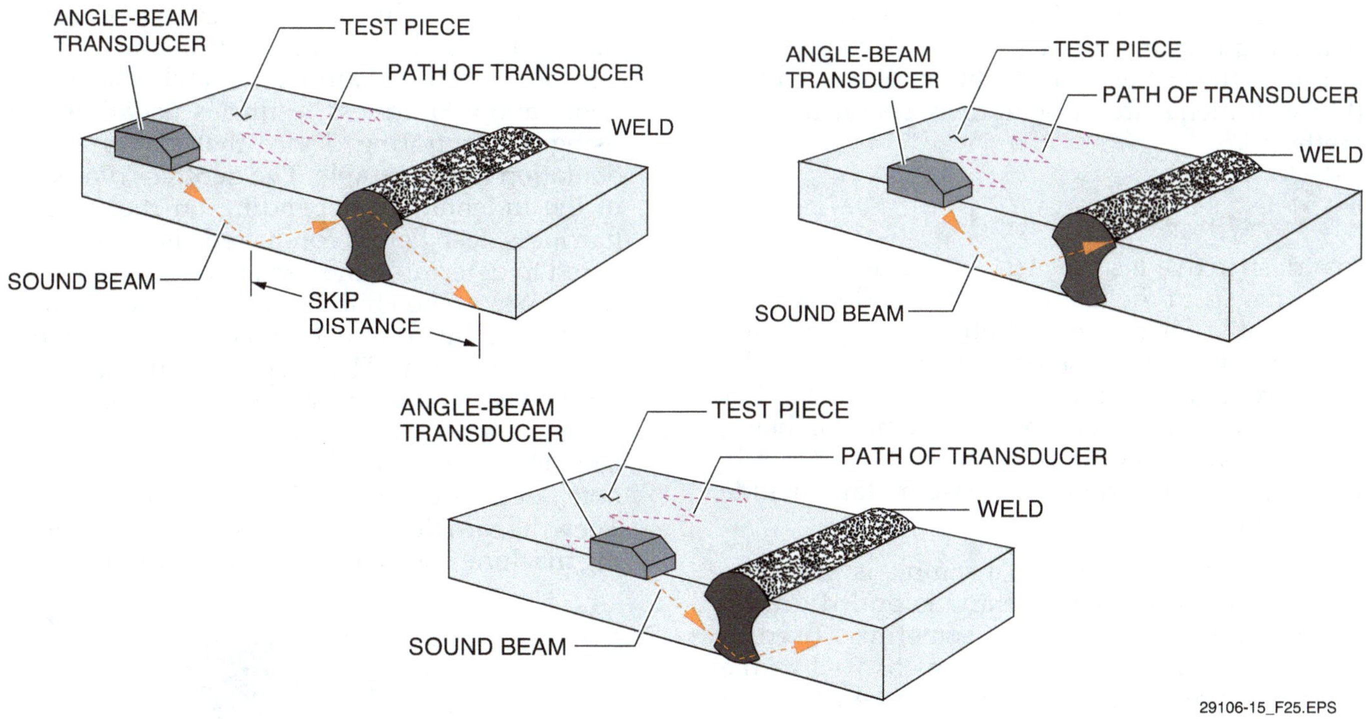

Figure 25 Moving a transducer back and forth.

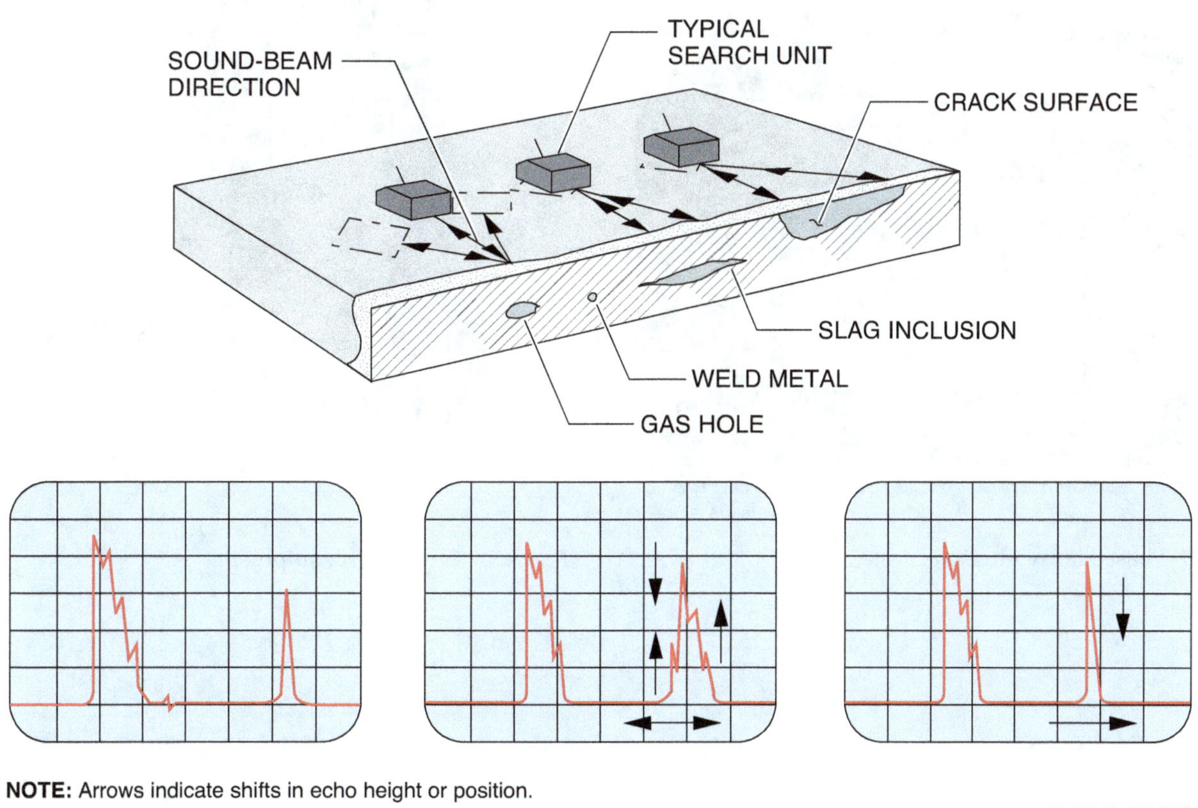

Figure 26 Moving a transducer in an arc.

The disadvantages of ultrasonic inspection are that it requires a high degree of skill to properly interpret the patterns. In addition, very small or thin weldments are difficult to inspect using this method.

3.4.0 Destructive Testing

In a destructive test, the test sample is destroyed or damaged in the testing process and is no longer suitable for use. Destructive testing is often done in a laboratory setting using machines designed to apply different stresses to the weld or the base metal, depending upon which is being tested. Examples of destructive tests commonly used to examine welds and base metals include the following:

- *Tensile test* – In this test, a sample is placed in a tensile testing machine and is pulled until it breaks. The tensile test is sometimes used to determine if the weld performs as well as the base metal. In most cases, however, the test is performed to determine specific properties of the weldment, such as strength and ductility (the amount the sample will stretch before breaking).
- *Hardness test* – While strength testing examines the ability of the sample to transmit a load, hardness testing examines its ability to resist penetration. Hardness testing is usually done using a penetrating device that leaves an indentation in the sample. The depth or diameter of the indentation, depending on the type of hardness test being conducted, is then measured to determine the hardness.
- *Impact test* – The ability of a weld to withstand an impact, such as a hammer strike, is measured by this test. The property being tested is referred to as toughness. In a laboratory environment, a pendulum-type machine that simulates a heavy hammer blow is used (*Figure 28*). A notch of a specified size is made in the sample. Then the sample is placed in the jaws of the testing machine and struck with a pendulum blow.

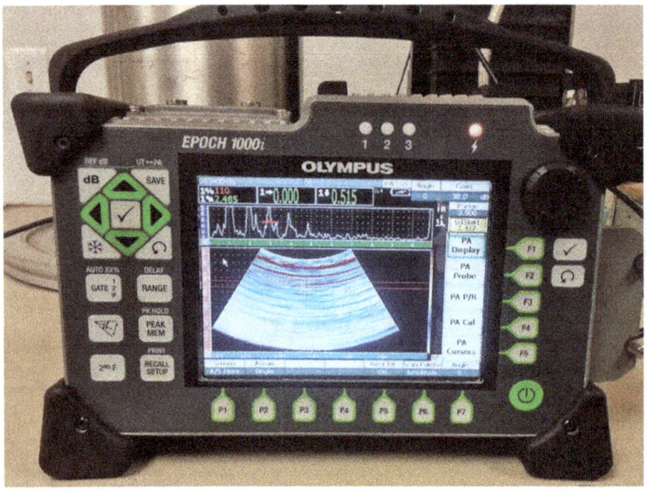

29106-15_F27.EPS

Figure 27 Phased array ultrasonic tester.

29106-15_F28.EPS

Figure 28 Impact test equipment.

- *Soundness test* – The three types of soundness tests include bend, nick-break, and fillet weld break tests.
  - Bend testing is the test most commonly used in determining the qualifications of a welder or welding procedures. Bend tests are performed by placing the sample in a special fixture and applying stress in a way that causes the sample to bend 180 degrees. The bend is then inspected for weld defects. Guided bend tests (*Figure 29*) are used to evaluate groove test welds on both plates and pipes. In this method, specimens are bent into a U-shape with a device called a jig. The bending action places stress on the weld metal and reveals any discontinuities in the weld. The jig has a plunger and die that are dimensioned for the thickness of the specimen being bent (*Figure 30*). Refer to the welding code being used for testing for the required dimensions of the bending jig's plunger and die.

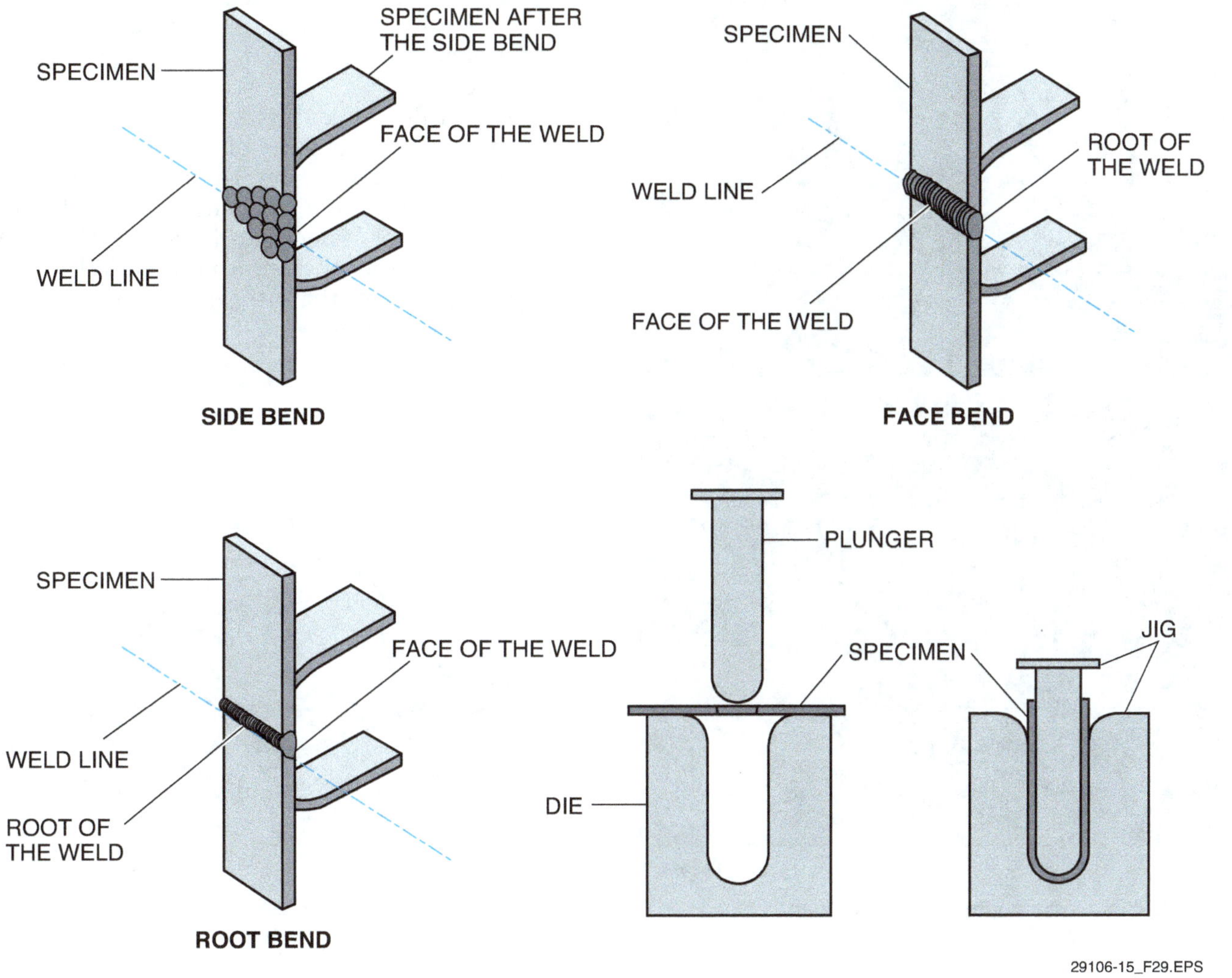

Figure 29 Guided bend test method and samples.

29106-15_F30.EPS

Figure 30 Bend test equipment and specimen.

- The three types of tests performed on the jig are root bends, face bends, and side bends. These bends test for penetration and fusion throughout the root joint. They also test for porosity, inclusions, and other defects, as well as the quality of the fusion to the side walls and the face of the weld joint. Face and root bend tests are used for materials up to ⅜" (≈10 mm metric plate) thick. For materials thicker than this, side bend testing may be used.
- Nick-break testing is used primarily in the pipeline industry. The specimen is saw-cut so it will break in a specific place. Then it is broken, and the weld zone is examined for defects.
- In a fillet weld break test, a fillet weld is made on one side of a T-joint. Stress is then applied to the T-joint until the weld fractures (*Figure 31*). The weld is then examined for defects and fusion. This is not a weld strength test; it is intended to break the weld so that it can be examined for discontinuities.

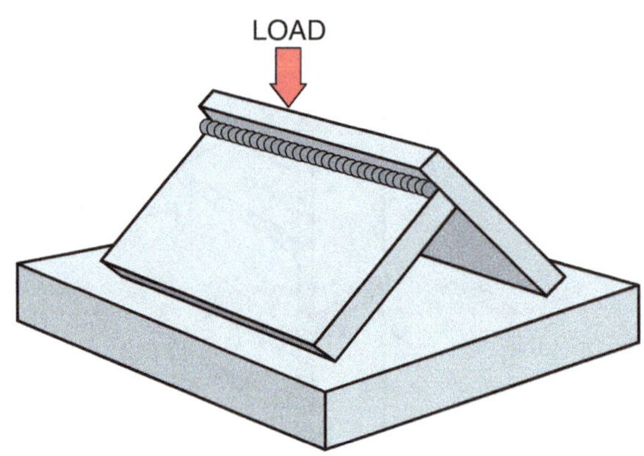

29106-15_F31.EPS

Figure 31 Fillet weld break test.

Additional Resources

Visual Inspection Workshop Reference Manual, AWS VIW-M-2008. Miami, FL: American Welding Society.

Certification Manual for Welding Inspectors. Miami, FL: American Welding Society.

Welding Inspection Technology. Miami, FL: American Welding Society.

B1.10M/B1.10:2009 Guide for the Nondestructive Examination of Welds. Miami, FL: American Welding Society.

3.0.0 Section Review

1. A bridge cam gauge is used to measure _____.
   a. undercut
   b. fillet weld depth
   c. butt weld reinforcement
   d. overlap

2. Electromagnetic inspection can be used to test welds in non-ferrous metals.
   a. True
   b. False

3. Which of these test methods uses an imaging plate placed in a cassette?
   a. Ultrasonic
   b. Phased array UT
   c. Computed radiography
   d. Digital radiography

4. A tensile test is one in which the test coupon is _____.
   a. bent
   b. pulled
   c. struck
   d. compressed

SECTION FOUR

4.0.0 WELDER PERFORMANCE QUALIFICATION TESTS

Objective

Describe the welder performance testing process.

a. Describe the qualification of welders by position.
b. Describe welder qualification testing to meet AWS and ASME requirements.
c. Describe the process for completing a weld test.

The purpose of welder performance qualification is to measure the proficiency of individual welders. As previously discussed, codes require that welders take a test to qualify for performing a welding procedure.

> **NOTE**
> Various codes and specifications often require similar methods for qualifying welders. The applicable code or specification should be consulted for specific details and requirements. Ask your supervisor if you are unsure about which codes or specifications apply to your project.

4.1.0 Welding Position Qualification

The welder or welding machine operator is qualified by welding position. Welders may be qualified to perform a welding procedure in only one position, or possibly all positions by passing a welding qualification test. The qualification tests are designed to measure the welder's ability to make groove and fillet welds in different positions on plate, pipe, or both in accordance with the applicable code. Each welding position is designated by a number and a letter, such as 1G. These designations are standard for all codes.

The letter G designates a groove weld. The letter F designates a fillet weld. For plate welding, the positions are designated by the following numbers:

- 1 – Flat position welding
- 2 – Horizontal position welding
- 3 – Vertical position welding
- 4 – Overhead position welding

Figure 32 shows the plate welding positions for both fillet welds and groove welds.

For pipe welding, there are these additional positions: 5F, 5G, 6G, and 6GR. The number indicates that multiple-position welds are required. Also, in the 1G (flat groove) and 1F (flat fillet) positions, the pipe is rotated during welding. *Figure 33* shows the pipe welding positions for both fillet welds and groove welds.

A welder who qualifies in one position does not automatically qualify to weld in all positions. However, in most cases qualification for groove welds qualify the welder for fillet welds; qualification for pipe qualify the welder for plate. Qualification in one code may or may not qualify the welder in other codes. The qualification requirements between codes may not match. Refer to your site requirements/code for qualifying requirements.

4.2.0 Code-Required Testing

AWS, ASME, and API codes require that welders be qualified through testing to perform specific types of welds based on base metal, welding position, joint design, type of electrode, and other variables.

4.2.1 AWS Structural Steel Code

The AWS structural steel code provides information concerning the qualification of welding procedures, welders, and welding operators for the types of welding done by contractors and fabricators in building and bridge construction. Qualification for plate welding also qualifies the welder for rectangular tubing.

The mild steel electrodes used with shielded metal arc welding (SMAW) are classified by F numbers: F1, F2, F3, and F4. Qualification with an electrode in a particular F-number classification will qualify the welder with all electrodes identified in that classification and in lower F-number classifications. *Table 1* shows AWS F-number electrode classifications.

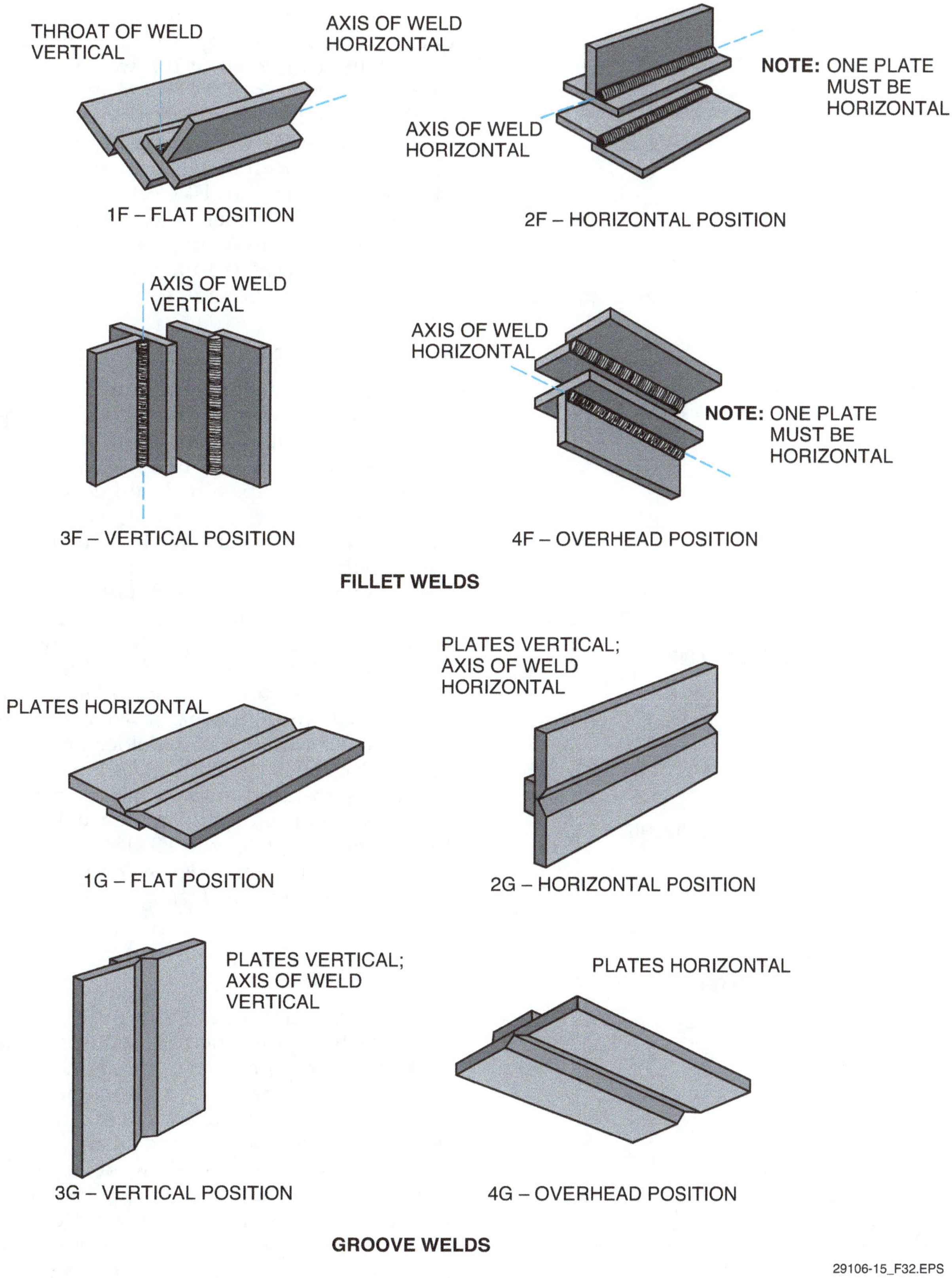

Figure 32 Welding positions for plate.

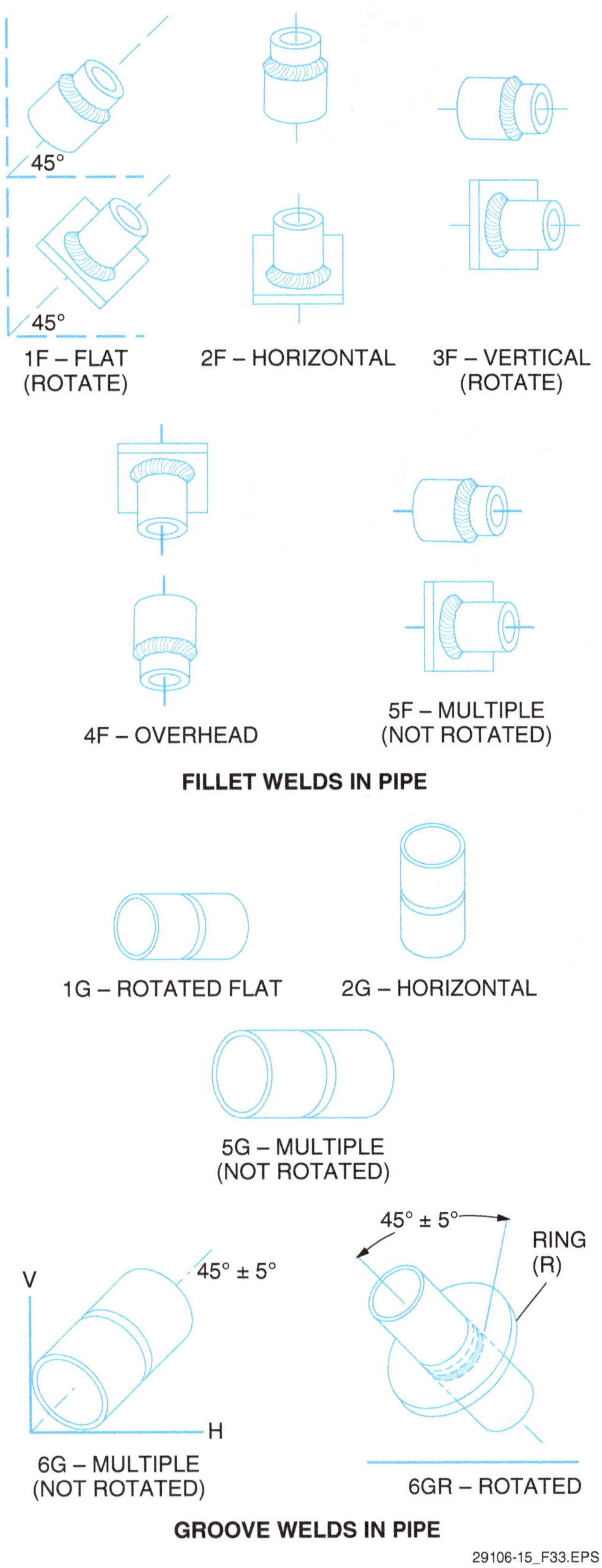

Figure 33 Welding positions for pipe.

Material thickness is an essential variable in qualification tests under AWS code. Some of the tests in the code qualify the welder only up to twice the thickness of the test piece. Others qualify the welder for unlimited thicknesses.

A typical AWS welder qualification test is a V-groove weld with metal backing in the 3G and 4G positions using an F4 electrode. Passing this test qualifies the welder to weld with F4 or lower electrodes, and to make groove and fillet welds in all positions. *Figure 34* shows an example of a fit-up for the AWS structural test.

4.2.2 ASME Code

Individual welders and welding operators who are required to weld to ASME code must qualify in accordance with Section IX of the *ASME Boiler and Pressure Vessel Code* on either plate or pipe. Qualification on pipe also qualifies the welder to weld plate, but not vice versa. Qualification with groove welds also qualifies the welder for fillet welds, but not vice versa. It is possible under the code to qualify for fillet welds only.

The typical ASME welder qualification test is to weld pipe in the 6G position using an open root. Passing this test qualifies the welder to weld pipe in all positions and plate in all positions (fillet and groove). If F3 electrodes are used all the way out, the welder only qualifies on F3 electrodes on both plate and pipe. If F3 electrodes are used for welding the root and F4 electrodes are used for filler, the welder qualifies to weld pipe or plate with F3 electrodes and pipe or plate with F4 electrodes as long as backing is used. The drawing in *Figure 35* shows a typical ASME pipe test.

4.2.3 API Code

The API requires welders to make butt or fillet welds using qualified procedures. Welds are performed on pipe nipples or segments of pipe nipples. Completed welds are subjected to visual examination and destructive or radiographic testing. Welders may receive a single qualification by performing a single weld in the fixed or rolled position. Multiple-weld qualification can be obtained by first making a butt weld in a fixed position on 6" pipe (DN150) diameter with a minimum thickness of 0.250" (6.4 mm). A second test requires the welder to cut, fit, and weld a pipe of the same size without a backing strip.

Table 1 AWS F-Number Electrode Classification

Group	AWS Electrode Classification				
F4	EXX15	EXX16	EXX18		Low-Hydrogen
F3	EXX10	EXX11			Fast-Freeze
F2	EXX12	EXX13	EXX14		Fill-Freeze
F1	EXX20	EXX24	EXX27	EXX28	Fast-Fill

29106-15_F34.EPS

Figure 34 Typical AWS plate test coupon.

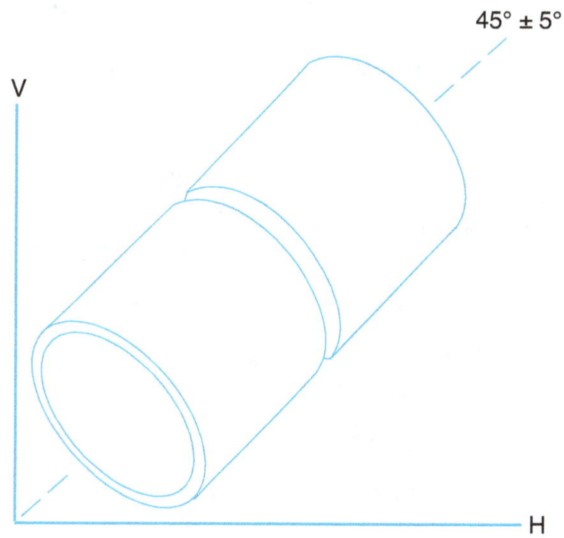

6G – MULTIPLE (NOT ROTATED)

29106-15_F35.EPS

Figure 35 Typical ASME pipe test.

4.3.0 Welder Qualification Tests

The welder becomes qualified by successfully completing a weld made in accordance with the WPS. It is general practice for code welding to qualify welders on groove weld tests. Passing these tests also permits the welder to perform fillet welds.

4.3.1 Making the Test Weld

Although qualification tests are designed to determine the capability of welders, some welders fail for reasons not related to their welding ability. This is due principally to carelessness in the application of the weld and in the preparation of the test specimen. It is important to note prior to welding where the test strips will be cut from the weld coupon. By doing this, you can avoid potential problems such as restarts in the area of the test strips. The following sections explain how to prepare a test specimen.

4.3.2 Removing Test Specimens

After making the qualification test weld, the test specimens are cut from the test pipe or plate by any suitable means. There are specific locations where the test specimen is cut from the pipe or plate. Refer to the applicable code for details related to specimen location and quantity.

Typical specimen locations for plate welds are shown in *Figure 36*. For material ⅜" (10 mm metric plate) thick, a face bend and a root bend are required. For material thicker than ⅜" (10 mm metric plate), two side bends are required.

Note that tests are usually given on plate thicknesses of ⅜" (10 mm metric plate) for limited thickness and 1" (25 mm metric plate) for unlimited thickness qualifications.

4.3.3 Preparing the Specimens for Testing

After the specimen has been cut from the test piece, it must be properly prepared for testing (*Figure 37*). Poor specimen preparation can cause a sound weld metal to fail. For example, a slight nick may open up under the severe bending stress of the test, causing the specimen to fail. To properly prepare the test specimen, do the following:

- Grind or machine the surface to a smooth finish. All grinding and machining marks must be lengthwise on the sample. Otherwise, they produce a notch effect, which may cause failure.

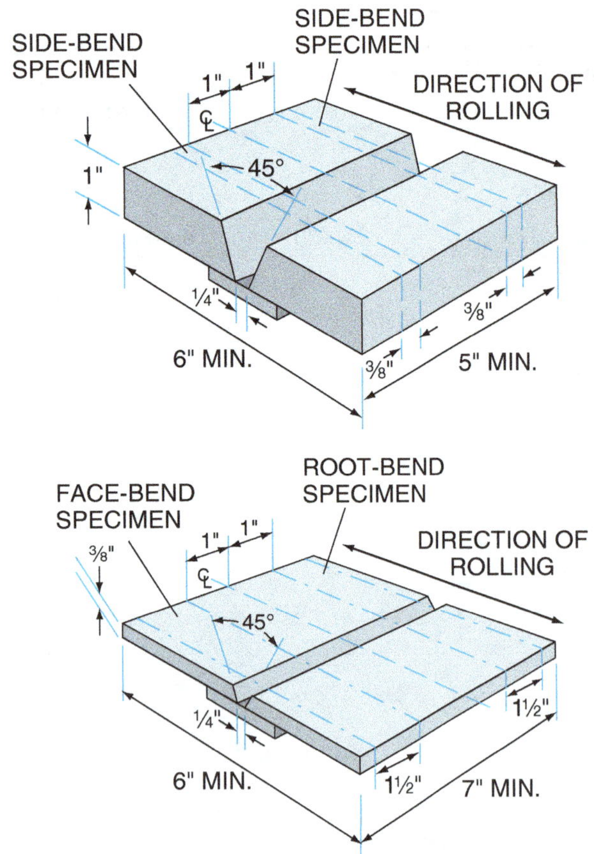

METRIC EQUIVALENTS	
¼"	6.4 MM
⅜"	10 MM
1"	25 MM
1½"	38 MM
6"	15.2 CM
7"	19.8 CM

29106-15_F36.EPS

Figure 36 Specimen locations for plate welds.

- Remove any face or root reinforcement from the weldment. This is part of the test requirement and, more important to the welder, failure to do so can cause the failure of a good weld. Grinding of the weld reinforcement may not exceed 1/32" (0.8 mm) or 5 percent of the thickness of the base material, according to AWS standards.
- Round the edges to a smooth 1/16" (1.6 mm) radius. This can be done with a file. Rounded edges help prevent failure caused by cracks starting at a sharp corner.
- Refrain from quenching specimens you are grinding when they are hot. Quenching may create small surface cracks that become larger during the bend test.

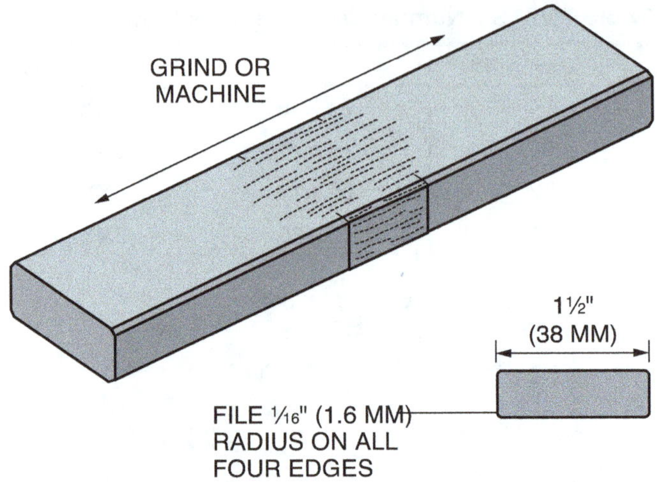

29106-15_F37.EPS

Figure 37 Example of a prepared test specimen.

After bending, the specimen is evaluated by measuring the discontinuities that are exposed. The criteria for acceptance can vary by code or site quality standards. AWS standards require that the surface shall contain no discontinuities exceeding the following dimensions:

- *⅛" (3.2 mm)* – Measured in any direction on the surface.
- *⅜" (10 mm)* – Sum of the greatest dimensions of all discontinuities exceeding 1/32" (0.8 mm) but less than or equal to ⅛" (3.2 mm).
- *¼" (6.4 mm)* – Maximum corner crack, except when the corner crack results from visible slag inclusion or other fusion-type discontinuities, then a ⅛" (3.2 mm) maximum shall apply. A specimen with corner cracks exceeding ¼" (6.4 mm), with no evidence of slag inclusions or other fusion-type discontinuities may be discarded, and a replacement test specimen from the original weldment shall be tested.

NOTE

In some cases, a radiographic inspection will be used instead of the guided bend test. This allows the entire weld to be examined and can detect small discontinuities at any location within the weld.

When the welder passes the qualification tests, the test results and the WPSs that the welder may weld are listed on a record that is kept by the company. This record becomes part of the quality documentation, and the welder becomes qualified to weld to that procedure.

AWS vs. ASME Pipe Welds

AWS and ASME codes have different requirements. The AWS version is shown on the left, while the ASME version is shown on the right.

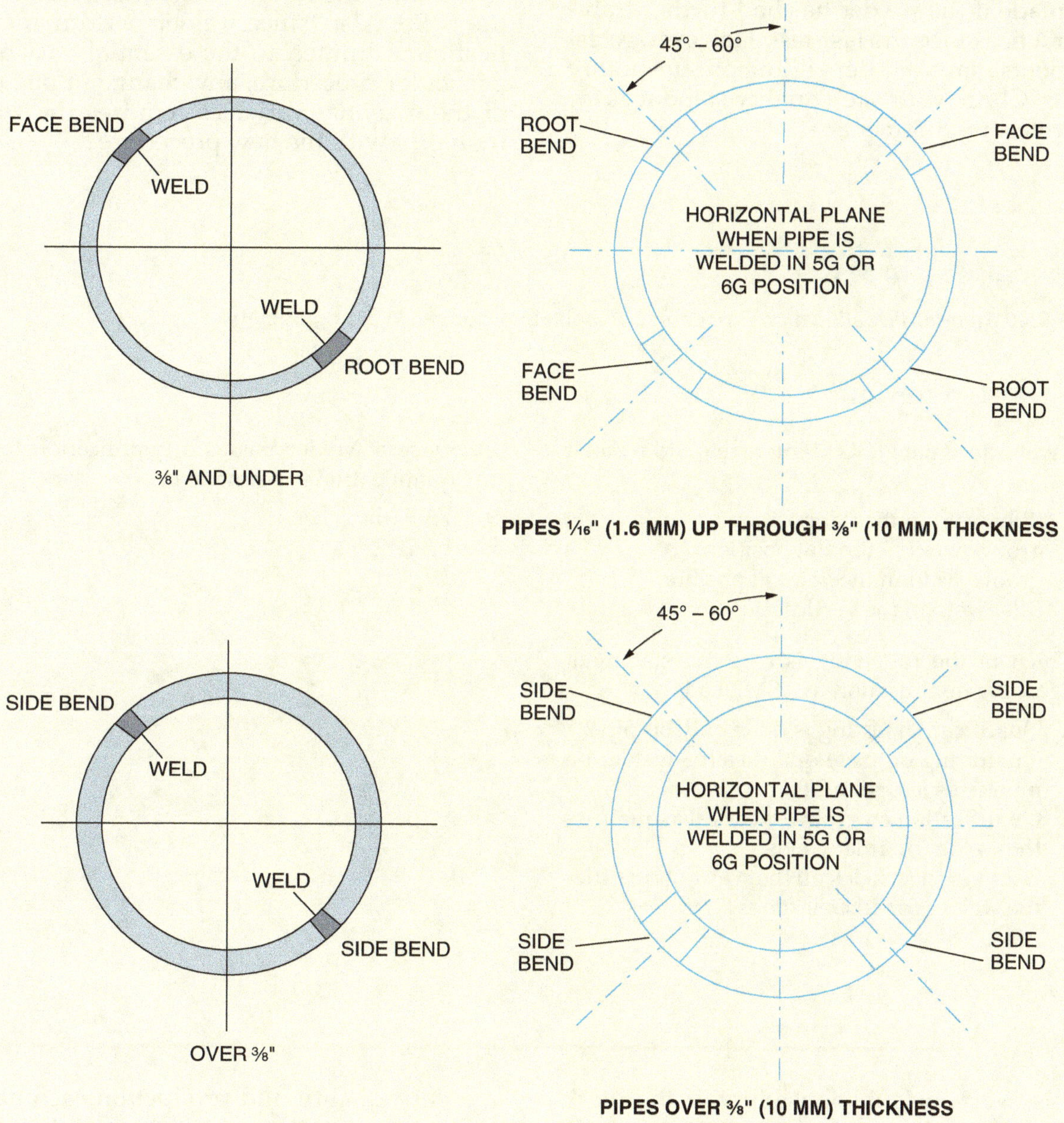

29106-15_SA03.EPS

4.3.4 Welder Qualification Limits

Welders may retest if they initially fail the test. An immediate retest consists of two test welds of each type of test that was failed. All the test specimens must pass this retest. A complete retest may be made if the welder has had further training or practice since the last test. However, retest requirements may vary depending on site quality standards. Check your site's quality standards for specific retest requirements.

After welders have qualified, they may have to requalify if they have not used the specific process for a certain time period. This time period varies depending on the codes. Welders may also be required to requalify if there is a reason to question their ability to make welds that meet the WPS. Also, since welder performance qualification is limited to the essential variables of a particular procedure, any change in one or more of the essential variables requires the welder to requalify with the new procedure.

Additional Resources

AWS B3.0 Standard Qualification Procedure. Miami, FL: American Welding Society.

4.0.0 Section Review

1. A weld designated 3G is interpreted to mean a _____.
   a. fillet weld in the flat position
   b. groove weld in the flat position
   c. groove weld in the vertical position
   d. fillet weld in the vertical position

2. Which of the following is a *correct* statement regarding qualification to ASME code?
   a. Qualification testing is done only on pipe.
   b. Qualifying on plate qualifies the welder on pipe sizes less than 24".
   c. Qualification on groove welds also qualifies the welder on fillet welds.
   d. Using F3 electrodes all the way out qualifies the welder for all electrodes.

3. Once a welder passes a qualification test, the qualification is permanent.
   a. True
   b. False

Summary

Quality is everyone's responsibility. If the work being done cannot be defined as quality work, it reflects on all those involved in the process. One essential trait of a craftsperson is a sense of quality workmanship. The craftsperson is generally closest to the work and will therefore have a major impact on product quality. Keeping quality in mind as you perform each step of your job will help you identify and correct small problems before they become major ones. This will make everyone's job easier and instill a sense of pride in what has been accomplished.

Safety, quality, and production each has a cost of its own. On the project, each of these factors should have proper guidelines. At the completion of the project, when all records for safety, cost, planning, scheduling, and effectiveness have been evaluated, the papers are usually filed away, but quality remains indefinitely for all eyes to see. Quality is, perhaps, the major reason for repeat business. How well the craftsperson performed will be noticed long after the project has been completed.

Review Questions

1. Which section of the *ASME Boiler and Pressure Vessel Code* contains the specifications for acceptable ferrous (Part A) and nonferrous (Part B) base metals?
   a. I
   b. II
   c. V
   d. IX

2. The minimum requirements for the design, materials, fabrication, erection, testing, and inspection of various types of piping systems are covered by _____.
   a. *ASME B31*
   b. *AWS D1.1*
   c. *AWS B1.11*
   d. *API Std. 1104*

3. The code that covers the requirements for welding structures or components under the surface of water is _____.
   a. *ASME B31*
   b. *AWS D1.1*
   c. *AWS D3.6M*
   d. *API 1104*

4. If you are directed to perform a weld for which you are not qualified, you should _____.
   a. bypass the chain of command
   b. report directly to the safety engineer as soon as possible
   c. try to resolve the matter with your immediate supervisor
   d. perform the weld to the best of your ability

5. Which of the following is considered a nonessential variable for a welding performance qualification?
   a. Material thickness
   b. Post-heat treatment
   c. Travel speed
   d. Welding process

6. The presence of voids or empty spots in the weld metal is called _____.
   a. porosity
   b. inclusion
   c. defect
   d. discontinuity

7. The type of porosity in which the weld exhibits a localized grouping of pores is known as _____.
   a. uniformly scattered porosity
   b. clustered porosity
   c. linear porosity
   d. piping porosity

8. The type of weld discontinuity referred to as wormholes is _____.
   a. clustered porosity
   b. piping porosity
   c. cold cracks
   d. incomplete fusion

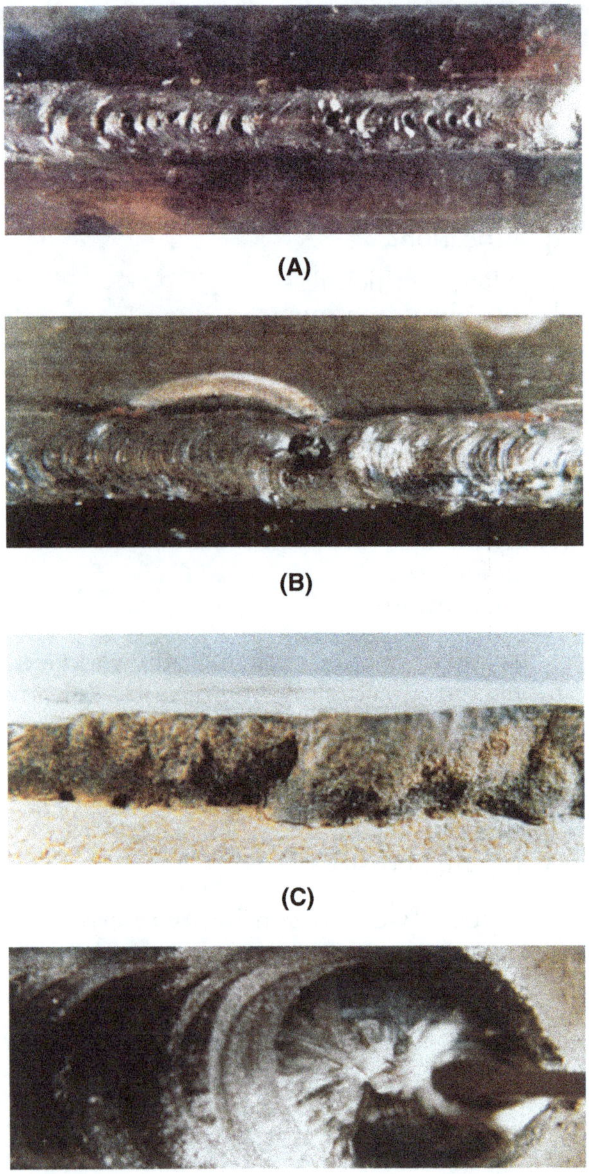

(A) (B) (C) (D)

29106-15_RQ01.EPS

Figure 1

For Questions 9 through 11, match a picture in Review Question Figure 1 to each discontinuity listed by putting the correct letter in the blank.

9. Linear porosity ________

10. Surface slag inclusion ________

11. Toe crack ________

12. One method of reducing weld metal cracking is to use _____.
    a. fast-fill electrodes
    b. fill-freeze electrodes
    c. low-hydrogen electrodes
    d. fast-freeze electrodes

13. Dross from flame cutting, if not removed, is likely to cause _____.
    a. porosity
    b. spatter
    c. undercut
    d. incomplete penetration

14. Dimensional accuracy of a weld is checked by _____.
    a. visual inspection
    b. using a tape measure
    c. radiography
    d. using welding gauges

15. A vacuum box is used to check for _____.
    a. inclusions
    b. porosity
    c. leaks
    d. incomplete fusion

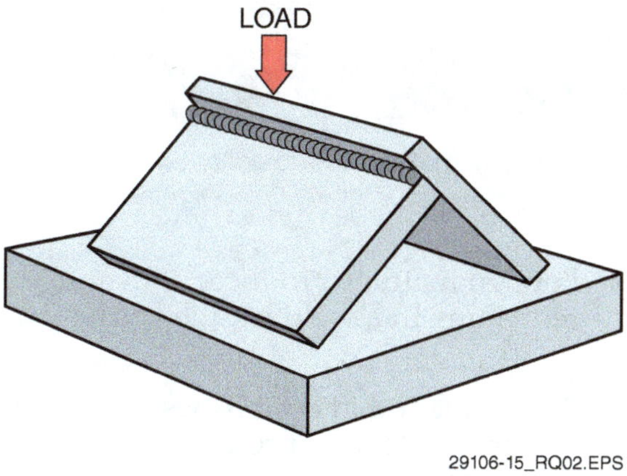

Figure 2

16. The test method shown in *Review Question Figure 2* is the _____.
    a. nick-break test
    b. fillet weld break test
    c. impact test
    d. hardness test

17. The nick-break test is a type of _____.
    a. tensile test
    b. hardness test
    c. soundness test
    d. impact test

18. A typical AWS welder qualification test is a V-groove weld with metal backing in the 3G and 4G positions using an _____.
    a. F4 electrode
    b. F3 electrode
    c. F2 electrode
    d. F1 electrode

19. An ASME welder qualification that qualifies a welder to weld pipe in all positions and plate in all positions is _____.
    a. 1F position test
    b. 6G position test
    c. 4F position test
    d. 4G position test

20. To properly prepare a test specimen, it is important to _____.
    a. quench the specimen while it is hot
    b. grind it width-wise to produce a notch effect
    c. remove any face or root reinforcement from the weldment
    d. make sure the edges are sharp

Trade Terms Quiz

Fill in the blank with the correct term that you learned from your study of this module.

1. ________ is the demonstration that welds made following a specific procedure can meet prescribed standards.
2. A(n) ________ is a discontinuity or imperfection that renders a part of the product or the entire product unable to meet minimum acceptable standards or specifications.
3. The type of discontinuity that is identified by cracking in the base metal near the weld, but under the surface is ________.
4. Metals that have the ability to be made harder by heating and then cooling are called ________.
5. The document containing the results of the nondestructive and destructive testing required to qualify a welding procedure specification (WPS) is called a(n) ________.
6. ________ is a change or break in the shape or structure of a part that may or may not be considered a defect, depending on the code.
7. The quality or state of having a uniform structure or composition throughout is known as ________.
8. The testing method is which images are made by passing X-rays or gamma rays through an object and recording the variations in density on photographic film is known as ________.
9. Cracks in the base metal formed when layers separate are called ________.
10. ________ metal has been made brittle and will tend to crack with little bending.
11. Foreign matter introduced into and remaining in a weld is called a(n) ________.
12. A document that establishes the minimum requirements for a product or process and can be adopted into law is a(n) ________.
13. A welding parameter that cannot be changed without requalifying a procedure is a(n) ________.
14. The ability of a material to absorb energy in the presence of a flaw is known as ________.
15. A welding parameter that can be changed without requalifying a procedure is a(n) ________.
16. The variables that must be considered when notch toughness requirements are invoked are called ________.
17. A document that defines how a code is to be implemented is a(n) ________.
18. A(n) ________is a document that defines in detail the work to be performed or the materials to be used in a product or process.
19. A device that converts one form of energy into another is called a(n) ________.
20. ________ is a method of joining metal using heat and a filler metal with a melting point above 842°F (450°C).
21. The deflection of an arc from its intended path by magnetic forces is known as ________.

Trade Terms

Arc blow
Brazing
Code
Defect
Discontinuity
Embrittled
Essential variables
Hardenable materials
Homogeneity
Inclusion
Laminations
Non-essential variables
Notch toughness
Procedure qualification record (PQR)
Radiographic
Specification
Standard
Supplemental essential variables
Transducer
Underbead cracking
Welding procedure qualification

Appendix A

Procedure Qualification Record

FORM QW-483 SUGGESTED FORMAT FOR PROCEDURE QUALIFICATION RECORDS (PQR)
(See QW-200.2, Section IX, ASME Boiler and Pressure Vessel Code)
Record Actual Variables Used to Weld Test Coupon

Organization Name ____________________
Procedure Qualification Record No. ____________________ Date ____________________
WPS No. ____________________
Welding Process(es) ____________________
Types (Manual, Automatic, Semi-Automatic) ____________________

JOINTS (QW-402)

Groove Design of Test Coupon
(For combination qualifications, the deposited weld metal thickness shall be recorded for each filler metal and process used.)

BASE METALS (QW-403)
Material Spec. ____________________
Type/Grade, or UNS Number ____________________
P-No. ______ Group No. ______ to P-No. ______ Group No. ______
Thickness of Test Coupon ____________________
Diameter of Test Coupon ____________________
Maximum Pass Thickness ____________________
Other ____________________

POSTWELD HEAT TREATMENT (QW-407)
Temperature ____________________
Time ____________________
Other ____________________

GAS (QW-408)

	Gas(es)	Percent Composition (Mixture)	Flow Rate
Shielding			
Trailing			
Backing			
Other			

FILLER METALS (QW-404)	1	2
SFA Specification		
AWS Classification		
Filler Metal F-No.		
Weld Metal Analysis A-No.		
Size of Filler Metal		
Filler Metal Product Form		
Supplemental Filler Metal		
Electrode Flux Classification		
Flux Type		
Flux Trade Name		
Weld Metal Thickness		
Other		

ELECTRICAL CHARACTERISTICS (QW-409)
Current ____________________
Polarity ____________________
Amps. ____________________ Volts ____________________
Tungsten Electrode Size ____________________
Mode of Metal Transfer for GMAW (FCAW) ____________________
Heat Input ____________________
Other ____________________

POSITION (QW-405)
Position of Groove ____________________
Weld Progression (Uphill, Downhill) ____________________
Other ____________________

TECHNIQUE (QW-410)
Travel Speed ____________________
String or Weave Bead ____________________
Oscillation ____________________
Multipass or Single Pass (Per Side) ____________________
Single or Multiple Electrodes ____________________
Other ____________________

PREHEAT (QW-406)
Preheat Temperature ____________________
Interpass Temperature ____________________
Other ____________________

(07/13)

29106-15_A01A.EPS

FORM QW-483 (Back)

PQR No. ____________

Tensile Test (QW-150)

Specimen No.	Width	Thickness	Area	Ultimate Total Load	Ultimate Unit Stress, (psi or MPa)	Type of Failure and Location

Guided-Bend Tests (QW-160)

Type and Figure No.	Result

Toughness Tests (QW-170)

Specimen No.	Notch Location	Specimen Size	Test Temperature	Impact Values			Drop Weight Break (Y/N)
				ft-lb or J	% Shear	Mils (in.) or mm	

Comments ____________

Fillet-Weld Test (QW-180)

Result — Satisfactory: Yes ________ No ________ Penetration into Parent Metal: Yes ________ No ________

Macro — Results ____________

Other Tests

Type of Test ____________

Deposit Analysis ____________

Other ____________

Welder's Name ____________ Clock No. ________ Stamp No. ________

Tests Conducted by ____________ Laboratory Test No. ________

We certify that the statements in this record are correct and that the test welds were prepared, welded, and tested in accordance with the requirements of Section IX of the ASME Boiler and Pressure Vessel Code.

Organization ____________

Date ____________ Certified by ____________

(Detail of record of tests are illustrative only and may be modified to conform to the type and number of tests required by the Code.)

(07/13)

29106-15_A01B.EPS

Appendix B

PERFORMANCE ACCREDITATION TASK – VISUAL WELD TEST INSPECTION

The American Welding Society (AWS) School Excelling through National Skills Standards Education (SENSE) program is a comprehensive set of minimum Standards and Guidelines for Welding Education programs. The following performance accreditation is aligned with and designed around the SENSE program.

The Performance Accreditation Tasks (PATs) correspond to and support the learning objectives in *AWS EG2.0, Guide for the Training and Qualification of Welding Personnel: Entry-Level Welder*.

Note that in order to satisfy all learning objectives in *AWS EG2.0*, the instructor must also use the PATs contained in the second level of the NCCER Welding curriculum.

PAT 1 corresponds to *AWS EG2.0, Module 9 – Welding Inspection and Testing Principles, Unit 4 Welding Inspection and Testing*, Key Indicators 1 and 2.

PATs provide specific acceptable criteria for performance and help to ensure a true competency-based welding program for students.

The following task is designed to develop your competency in performing a visual weld test inspection. Practice this task until you are thoroughly familiar with the procedure.

As you complete the task, take it to your instructor for evaluation. Do not proceed to the next task until instructed to do so by your instructor.

VISUAL WELD TEST INSPECTION

Obtain a completed fillet weld on ¼" plate (3.2 mm metric plate) a minimum of 6" (15.2 cm) long, and a complete joint-penetration groove weld minimum 6" (15.2 cm) long, and then perform a visual test inspection (VT) on each weld using the following guide and acceptance criteria.

Visual Test Inspection Report

Trainee Inspector ______________________ **Instructor** ______________________

Discontinuity Category	Acceptance Criteria	¼" Fillet Weld	CJP Groove Weld
Crack Prohibition	Any crack shall be unacceptable regardless of size or location	Acceptable ________ or Reason for rejection ________	Acceptable ________ or Reason for rejection ________
Weld/Base Metal Fusion	Complete fusion shall exist between adjacent layers of weld metal and base metal	Acceptable ________ or Reason for rejection ________	Acceptable ________ or Reason for rejection ________
Crater Cross Section	All craters shall be filled to provide the specified weld size	Acceptable ________ or Reason for rejection ________	Acceptable ________ or Reason for rejection ________
Weld Profiles	Weld profiles shall be in accordance with page 6.15 figure 14	Acceptable ________ or Reason for rejection ________	Acceptable ________ or Reason for rejection ________
Fillet Weld Size	The specified nominal fillet weld size tolerance is ±1⁄16 inch	Acceptable ________ or Reason for rejection ________	N/A
Undercut	Undercut shall not exceed 1⁄32" for any accumulated length up to 2 inches	Acceptable ________ or Reason for rejection ________	Acceptable ________ or Reason for rejection ________
Porosity	The sum of visible porosity 1⁄32" or greater shall not exceed 3⁄8" in any linear inch and shall not exceed ¾" in any linear 12 inches of weld	Acceptable ________ or Reason for rejection ________	Acceptable ________ or Reason for rejection ________
Complete Weld	Circle accept or reject for fillet and groove final inspection	Acceptable ________ or Reason for rejection ________	Accept or Reject

Trainee Inspector's Signature ______________________ **Date** __________

Instructor's Verification: PASS/FAIL Signature ______________________

Trade Terms Introduced in This Module

Arc blow: The deflection of an arc from its intended path by magnetic forces.

Brazing: A method of joining metal using heat and a filler metal with a melting point above 842°F (450°C). Unlike welding, the base metal is not melted during the brazing process.

Code: A document that establishes the minimum requirements for a product or process. Codes can be, and often are, adopted as laws.

Defect: A discontinuity or imperfection that renders a part of the product or the entire product unable to meet minimum acceptable standards or specifications.

Discontinuity: A change or break in the shape or structure of a part that may or may not be considered a defect, depending on the code.

Embrittled: Metal that has been made brittle and that will tend to crack with little bending.

Essential variable: Items in a welding procedure specification (WPS) that cannot be changed without requalifying the WPS.

Hardenable materials: Metals that have the ability to be made harder by heating and then cooling.

Homogeneity: The quality or state of having a uniform structure or composition throughout.

Inclusion: Foreign matter introduced into and remaining in a weld.

Laminations: Cracks in the base metal formed when layers separate.

Non-essential variable: Items in a welding procedure specification (WPS) that can be changed without requalifying the WPS.

Notch toughness: The ability of a material to absorb energy in the presence of a flaw such as a notch.

Procedure qualification record (PQR): The document containing the results of the nondestructive and destructive testing required to qualify a WPS.

Radiographic: Describes images made by passing X-rays or gamma rays through an object and recording the variations in density on photographic film.

Specification: A document that defines in detail the work to be performed or the materials to be used in a product or process.

Standard: A document that defines how a code is to be implemented.

Supplemental essential variable: The variables that must be considered when notch toughness requirements are invoked.

Transducer: A device that converts one form of energy into another.

Underbead cracking: Cracking in the base metal near the weld, but under the surface.

Welding procedure qualification: The demonstration that welds made following a specific process can meet prescribed standards.

Additional Resources

This module presents thorough resources for task training. The following resource material is suggested for further study.

2014 Technical Training Guide. Current Edition. Cleveland, OH, USA: The Lincoln Electric Company. **www.lincolnelectric.com**

AWS B3.0 Standard Qualification Procedure. Miami, FL: American Welding Society.

B1.10M/B1.10:2009 Guide for the Nondestructive Examination of Welds. Miami, FL: American Welding Society.

Certification Manual for Welding Inspectors. Miami, FL: American Welding Society.

Visual Inspection Workshop Reference Manual, AWS VIW-M-2008. Miami, FL: American Welding Society.

Welding Inspection Technology. Miami, FL: American Welding Society.

Figure Credits

Applied Technical Services, Module Opener, Figures 21, 23, 24 (photo), 28, SA02

AWS D1.1/D1.1M: 2008, Welding Procedure Specification, Reproduced with permission from the American Welding Society (AWS), Miami, FL, USA, Figure 1

AWS D1.1/D1.1M: 2008, Procedure Qualification Record #231, Reproduced with permission from the American Welding Society (AWS), Miami, FL, USA, Figure 2

AWS B1.11-2000, Figures 1, 2, and 3, Reproduced with permission from the American Welding Society (AWS), Miami, FL, USA, Figure 3

AWS B1.11-2000, Figure 30, Reproduced with permission from the American Welding Society (AWS), Miami, FL, USA, Figure 4

AWS B1.11-2000, Figure 20, Reproduced with permission from the American Welding Society (AWS), Miami, FL, USA, Figure 5

AWS B1.11-2000, Figures 3, 24, 25, 26, and 28, Reproduced with permission from the American Welding Society (AWS), Miami, FL, USA, Figure 6

Topaz Publications, Inc., Figures 7 (photo), 18, 22 (photo), 30, SA01

AWS B1.11-2000, Figures 7 and 8, Reproduced with permission from the American Welding Society (AWS), Miami, FL, USA, Figure 9

AWS B1.11-2000, Figure 18, Reproduced with permission from the American Welding Society (AWS), Miami, FL, USA, Figure 10

G.A.L. Gage Co, Figures 15, 16

Certification Manual for Welding Inspectors, Figure 12.11, Reproduced with permission from the American Welding Society, Miami, FL, Figure 19

Olympus NDT Inc., Figure 27

Zachry Industrial, Inc., Figure 34

Reprinted from ASME 2013 BPVC, Section IX, by permission of The American Society of Mechanical Engineers. All rights reserved.

Section Review Answer Key

Answer	Section Reference	Objective
Section One		
1. b	1.1.0	1a
2. b	1.2.1	1b
Section Two		
1. c	2.1.1	2a
2. b	2.2.2	2b
3. a	2.3.3	2c
4. d	2.4.1	2d
Section Three		
1. a	3.1.1	3a
2. a	3.2.0	3b
3. c	3.3.1	3c
4. b	3.4.0	3d
Section Four		
1. c	4.1.0	4a
2. c	4.2.2	4b
3. b	4.3.4	4c

NCCER CURRICULA — USER UPDATE

NCCER makes every effort to keep its textbooks up-to-date and free of technical errors. We appreciate your help in this process. If you find an error, a typographical mistake, or an inaccuracy in NCCER's curricula, please fill out this form (or a photocopy), or complete the online form at **www.nccer.org/olf**. Be sure to include the exact module ID number, page number, a detailed description, and your recommended correction. Your input will be brought to the attention of the Authoring Team. Thank you for your assistance.

Instructors – If you have an idea for improving this textbook, or have found that additional materials were necessary to teach this module effectively, please let us know so that we may present your suggestions to the Authoring Team.

NCCER Product Development and Revision

13614 Progress Blvd., Alachua, FL 32615

Email: curriculum@nccer.org
Online: www.nccer.org/olf

❑ Trainee Guide ❑ Lesson Plans ❑ Exam ❑ PowerPoints Other ______________

Craft / Level: Copyright Date:

Module ID Number / Title:

Section Number(s):

Description:

Recommended Correction:

Your Name:

Address:

Email: Phone:

Joint Fit-Up and Alignment

Overview

Regardless of a welder's talent and experience, a joint that is poorly fit or aligned is subject to failure and a liability. All welded joints must be properly fit according to codes, specifications, and industry standards. Due to the intense heat of the welding process, care must also be taken to ensure that a properly fitted joint is not distorted and pulled out of alignment due to expansion and contraction. This module introduces welders to the basic tools and practices related to the proper fit of both plate and pipe joints.

Module 29110

Trainees with successful module completions may be eligible for credentialing through the NCCER Registry. To learn more, go to **www.nccer.org** or contact us at **1.888.622.3720**. Our website has information on the latest product releases and training, as well as online versions of our *Cornerstone* magazine and Pearson's product catalog.

Your feedback is welcome. You may email your comments to **curriculum@nccer.org**, send general comments and inquiries to **info@nccer.org**, or fill in the User Update form at the back of this module.

This information is general in nature and intended for training purposes only. Actual performance of activities described in this manual requires compliance with all applicable operating, service, maintenance, and safety procedures under the direction of qualified personnel. References in this manual to patented or proprietary devices do not constitute a recommendation of their use.

29110 V5

29110
Joint Fit-up and Alignment

Objectives

When you have completed this module, you will be able to do the following:

1. Identify and describe various types of fit-up and alignment tools.
   a. Identify and describe various fit-up gauges and measuring devices.
   b. Identify and describe common weldment positioning equipment.
   c. Identify and describe various plate alignment tools.
   d. Identify and describe various pipe and flange alignment tools.
2. Describe techniques to avoid weldment distortion and describe the role of codes and specifications.
   a. Describe the causes of weldment distortion.
   b. Describe the techniques and tools used to control weldment distortion.
   c. Describe the role of codes and specifications in welding procedures and techniques.

Performance Tasks

Under the supervision of your instructor, you should be able to do the following:

1. Fit up joints using plate and pipe fit-up tools.
2. Check the joint for proper fit-up and alignment using gauges and measuring devices.

Trade Terms

Addendum
Ambient temperature
Coefficient of thermal expansion (linear)
Consumable insert
High-low
Level
Peening
Plumb
Residual stress
Specific heat per unit volume
Tack weld

Industry Recognized Credentials

If you are training through an NCCER-accredited sponsor, you may be eligible for credentials from NCCER's Registry. The ID number for this module is 29110. Note that this module may have been used in other NCCER curricula and may apply to other level completions. Contact NCCER's Registry at 888.622.3720 or go to **www.nccer.org** for more information.

Contents

Topics to be presented in this module include:

Figures and Tables

SECTION ONE

1.0.0 JOINT FIT-UP AND ALIGNMENT TOOLS

Objective

Identify and describe various types of fit-up and alignment tools.

a. Identify and describe various fit-up gauges and measuring devices.
b. Identify and describe common weldment positioning equipment.
c. Identify and describe various plate alignment tools.
d. Identify and describe various pipe and flange alignment tools.

Performance Tasks

1. Fit up joints using plate and pipe fit-up tools.
2. Check the joint for proper fit-up and alignment using gauges and measuring devices.

Trade Terms

Consumable insert: Preplaced filler metal that is completely fused into the root of the joint during welding, becoming part of the weld.

High-low: The discrepancy in the internal alignment of two sections of pipe.

Level: A line on a horizontal axis parallel with the earth's surface (horizon). Also refers to the tool used to check if an object is level.

Plumb: A line on a vertical axis perpendicular to the earth's surface (horizon).

Tack weld: A weld made to hold parts of a weldment in proper alignment until the final weld is made.

Joint design and setup affect the safety and quality of a completed weldment. Because joint design and setup are so important, they are covered by written codes and specifications that must be followed. Special tools to measure and support weldment fit-up are also available. This module explains the codes and introduces special tools and measuring devices that aid setup. In addition, joint inspection techniques and procedures to ensure that proper joint setup is maintained during welding will also be explained. This first section of the module focuses on the tools used in fit-up.

1.1.0 Fit-Up Gauges and Measuring Devices

Before making a weld, the joint must be fit up and checked to ensure it conforms to the Welding Procedure Specification (WPS) or site quality standards. The most common tools used to lay out and check joint fit-ups are straightedges, squares, levels, and Hi-Lo gauges.

Before proceeding, it is important to note that fit-up tools and instruments are made from a variety of metals. Of course, weldments can consist of a wide variety of metals as well. The tools and instruments used in the fit-up and preparation of a joint must be compatible with the base metal to avoid contamination. Most combinations are not a large problem, but tools and instruments made from carbon steel must not be used on stainless steel. Stainless steel devices can be used on carbon steel base metals without concern. However, it is important to note that some stainless steel tools, such as wire brushes, should not be used on stainless steel once they have been used with carbon steel. The concern is the contamination of stainless steel by carbon steel, but not the reverse.

1.1.1 Straightedges

Straightedges are used to scribe straight lines and check joint alignment. Many have calibrations along their length for measuring. Straightedges, particularly longer ones, are typically fabricated on the job from small channel or angle iron. When using a straightedge, be careful not to apply heat to it by placing it on hot metal or near the flame of a cutting torch or welding arc. Heat can cause the straightedge to become permanently distorted. Before using a straightedge, check it for straightness by visually sighting down the edge. If it is distorted and cannot be straightened, destroy it.

1.1.2 Squares

Two types of squares are commonly used for layout: pipefitter's squares and combination squares (*Figure 1*). Pipefitter's squares, which are used to measure angles and check squareness, are available in a variety of sizes. Like a carpenter's framing square, the pipefitter's square has a great deal of helpful information stamped into the metal.

Combination squares are smaller than pipefitter's squares, with blades typically 12" or 18" long (≈30 cm or 45 cm). They have replaceable attachments that slide and lock along a groove on the blade. Attachments include a combination 90-degree/45-degree level head, a centering head, and a protractor head. The combination attachment is used to check and lay out 90-degree and

45-degree angles, to check level, and to measure depth. The V-shaped centering head is used to measure round stock and to locate the center of shafts or other round objects. The protractor head is used to lay out and check angles.

When using squares, avoid overheating them as it could cause distortion. Also, protect squares from welding spatter that could stick to measuring surfaces, causing a false reading. To preserve their accuracy, squares should be stored in a dry, protected area when not in use. Carbon steel squares should be kept free of corrosion.

1.1.3 Levels

Levels come in a variety of sizes and shapes (*Figure 2*). Most levels designed for plate and piping work are made from magnesium or aluminum. Some have magnetized bases. Levels are used to check that layouts are level (horizontal) and **plumb** (vertical). Levels use a bubble in a glass vial to check level and plumb. Centering the bubble between the lines marked on the vial indicates level when the level is in the horizontal position and plumb when the level is in the vertical position.

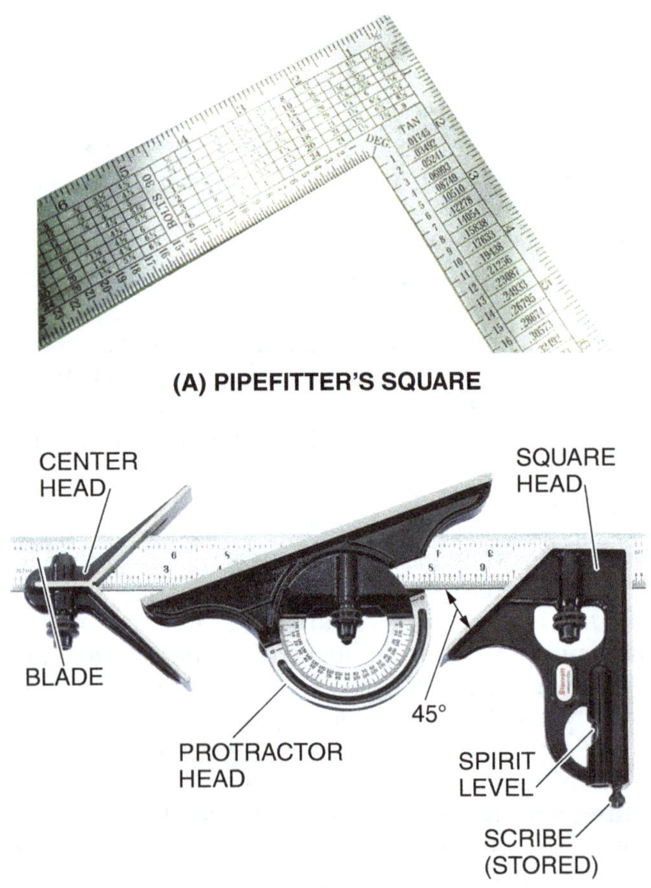

Figure 1 Pipefitter's square and combination square.

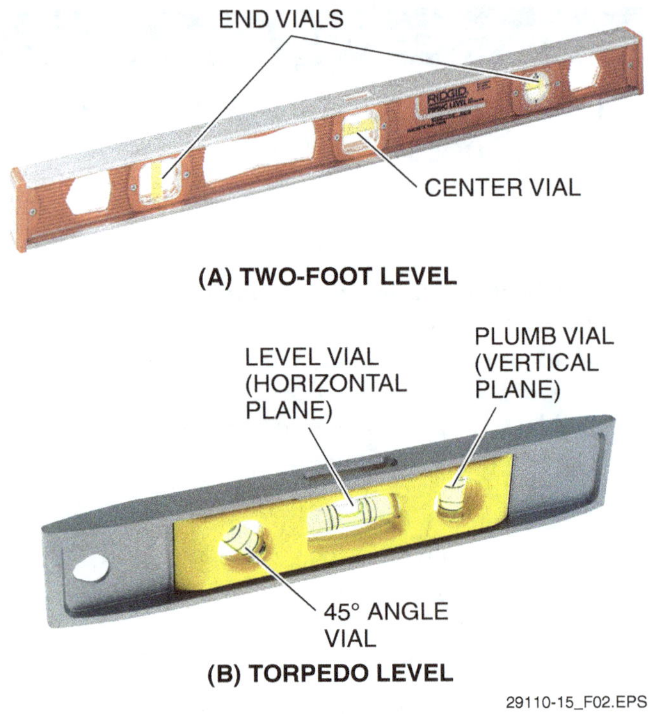

Figure 2 Common levels.

Some levels also have a vial set at 45 degrees to check 45-degree angles. There are also special levels that have an adjustable protractor scale that can be set to check any angle.

The torpedo level is about 9" (≈23 cm) long and tapered on each end. The torpedo level has three vials: one to check level, one to check plumb, and one to check 45-degree layouts.

Protect levels from heat and weld spatter, which can break the glass tube. Also, be aware that weld spatter can stick to the bottom of the level, giving a false reading. The edge of the level that is designed to be placed against the surface in question must be kept clean and free of significant damage that might alter the reading.

1.1.4 Hi-Lo Gauges

The primary purpose of a Hi-Lo gauge is to check for pipe joint misalignment, although plate joint misalignment can also be checked. The name of the gauge comes from the relationship between the alignment of one pipe or plate to the other, which is called **high-low**. When either type of weldment is misaligned, one half is considered low and the other half is high.

To check for internal misalignment of pipe, Hi-Lo gauges have two prongs, or alignment stops, that are pulled tightly against the inside diameter of the joint so that one stop is flush with each side of the joint. The variation between the two stops is read on a scale marked on the gauge.

To measure pipe misalignment with a Hi-Lo gauge, insert the prongs of the gauge into the root opening. Pull up on the gauge until the prongs are snug against both inside surfaces, and read the misalignment indicated on the scale. *Figure 3* shows checking internal misalignment with a Hi-Lo gauge.

Many Hi-Lo gauges also have the capability to check the following:

- Root opening width
- Material thickness
- Weld reinforcement
- Scribe lines for socket welds

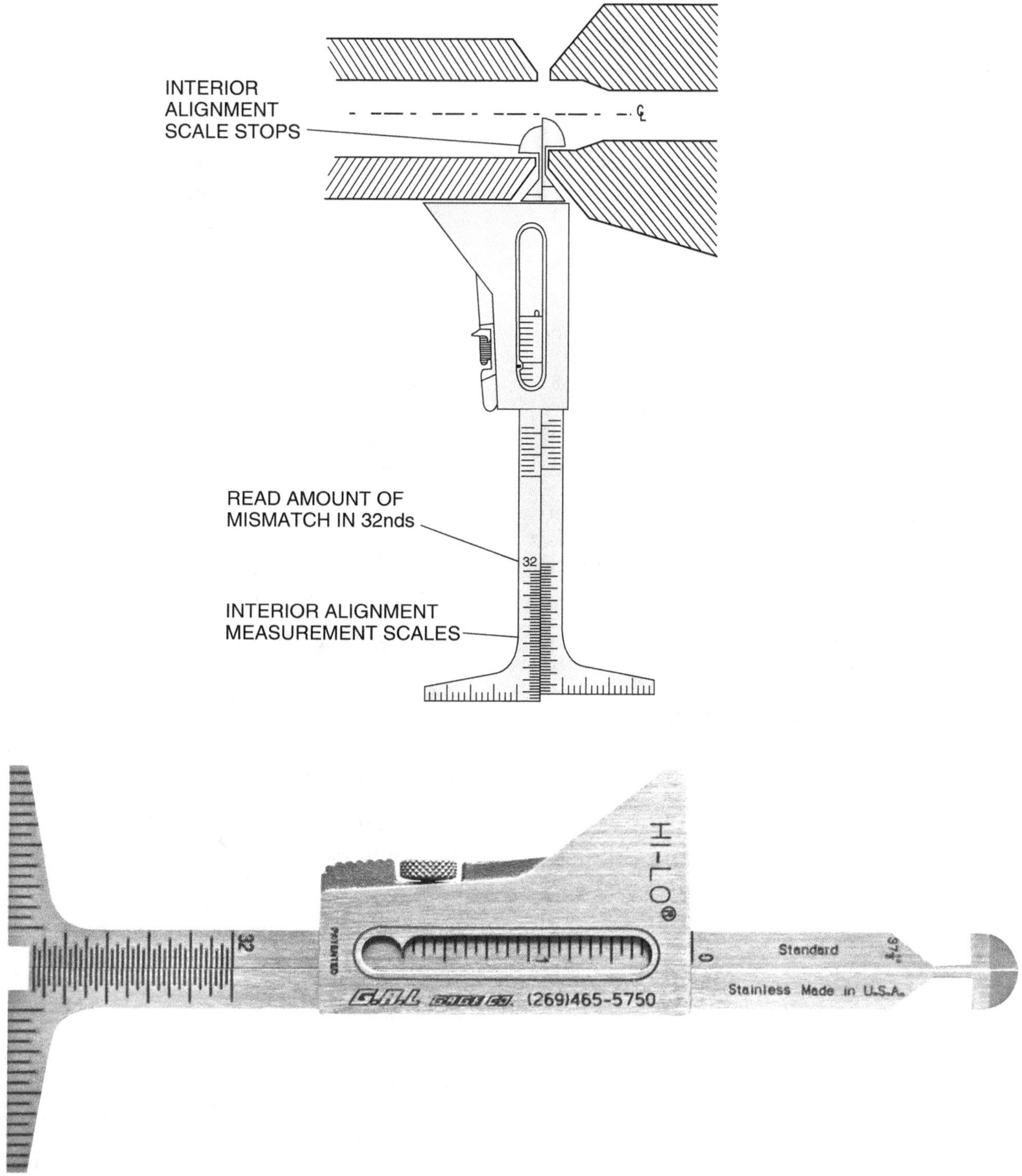

29110-15_F03.EPS

Figure 3 Checking internal misalignment.

Handy Level and Angle Finder

There might be an angle finder and level closer to you than you might think. The next time you need one, reach instead for your mobile phone. A number of smartphones, including the iPhone, offer a level/angle finder application by default. For the iPhone, access the compass app and, with the compass on screen, swipe left to reveal the level and angle finder tool. For phones not equipped with these apps by default, there are a number of them available for download. Of course, mobile phones are easily scratched and damaged, so you won't want to make a habit of using it on the job. But for leveling a picture or a shelf, it's a handy tool.

When checking for internal misalignment, take two gauge readings: one before tacking together the two pieces of pipe and one after the tack weld has been applied. The Hi-Lo gauge is designed to fit into the gap between the two pipes, even after tacking, and are available with both metric and English scales.

The alignment stops are tapered for checking the root opening (*Figure 4*). They are 3⁄32" (2.4 mm) thick and taper to 1⁄16" (1.6 mm) thick at both ends. Check the root opening by inserting the tapered head and visually checking the opening.

Weld reinforcement is another check that can be performed with the Hi-Lo gauge. It is measured by placing one foot of the gauge on the pipe and the other on the weld face, making sure the foot on the pipe surface is flush against the surface. Weld reinforcement is then read on the material thickness scale (*Figure 5*).

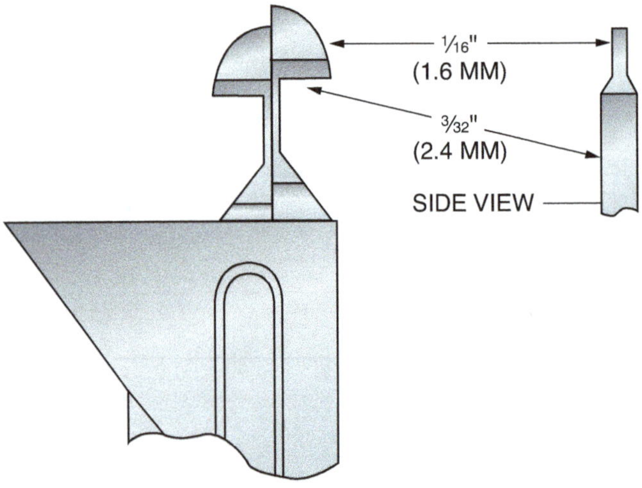

Figure 4 Tapered alignment stops.

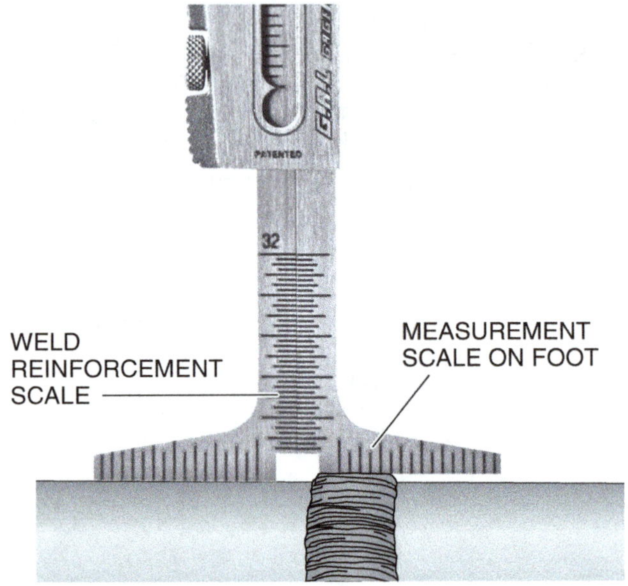

Figure 5 Measuring weld reinforcement.

CAUTION

Actual requirements for socket weld gaps must be verified for your job by checking the WPS or site quality standards. Failure to follow the WPS can cause equipment damage or result in a rejected weld.

When fitting up socket welds, the pipe is scribed so that a measurement can be made to ensure that a gap is left between the end of the pipe and the base of the socket. This gap prevents cracking when the pipe expands during the welding process; the pipe should never be bottomed-out in the socket when it is welded. The scribe-line scale on the feet of the Hi-Lo gauge can be used to measure from the socket to the scribed line to ensure the gap is as specified on the WPS or site quality specifications. Most codes require a minimum 1⁄16" (1.6 mm) gap, but actual requirements for your job may vary due to pipe size, base metal type, or a unique specification. *Figure 6* shows a Hi-Lo gauge measuring scribe lines on a socket weld.

1.2.0 Common Weldment Positioning Equipment

Many tools are available to aid plate and pipe fit-up. Some are special tools designed specifically for fit-up; others are common hand tools. Before using any tool, be sure to inspect it for damage or missing parts. Never use a tool for a job for which it was not intended. Be sure to clean and return all tools when you have finished using them.

Also remember to ensure the tool material is compatible with the base metal being fitted. Carbon steel tools, as well as some stainless steel tools that have been used with carbon steel, should not be used on stainless steel.

Hydraulic jacks, chain falls, and come-alongs are sometimes used to position parts of a weldment. These types of positioning tools are described in the following sections.

1.2.1 Hydraulic Jacks

When using hydraulic jacks, never weld directly on the jack base or ram. To secure the jack, a ring can be tack welded to act as a socket for the base and/or ram. When welding near hydraulic jacks, protect the ram from weld spatter. If the ram is off the floor, be sure to secure it with rope or chain. Also monitor the jack for oil leaks. Oil will create a fire hazard, and oil in the area to be welded will create weld defects from porosity. Oil leaks must be cleaned up before welding. *Figure 7* shows using a hydraulic jack to aid joint fit-up.

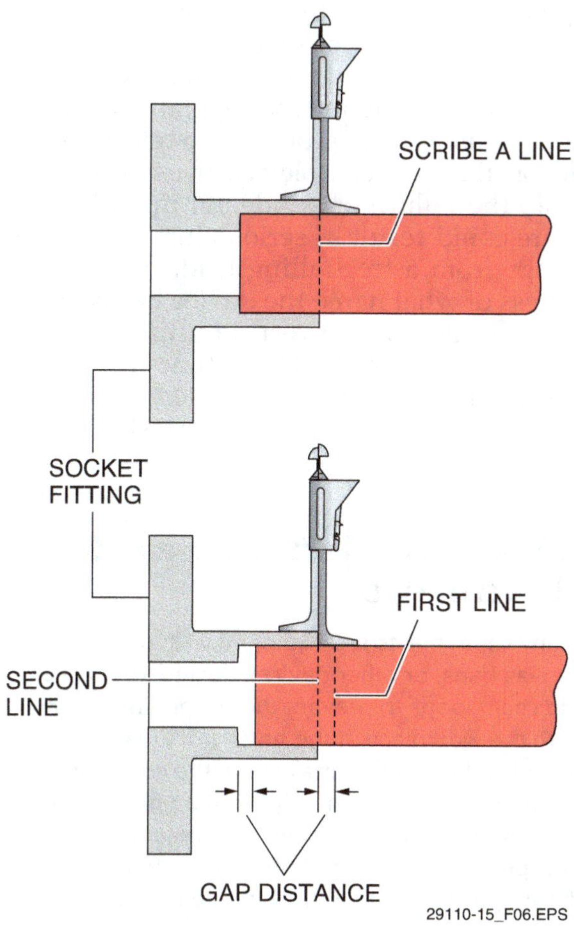

Figure 6 Measuring the distance between scribe lines on a socket weld.

WARNING!

Make sure the current path does not pass through the hydraulic jack supporting the load you are welding. Extreme heat generated by the current flow can cause the hydraulic jack to fail, releasing the load support. This can cause serious injury or even death.

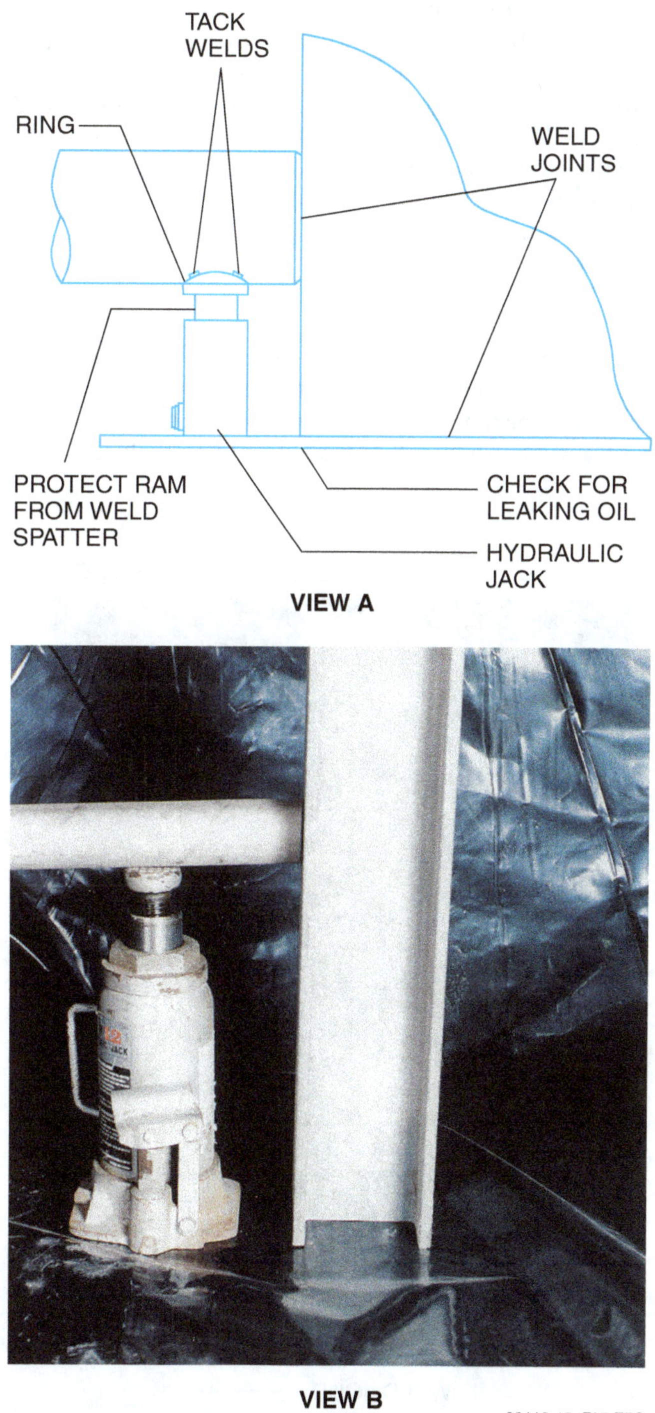

Figure 7 Proper use of a hydraulic jack for joint fit-up.

1.2.2 Chain Hoists, Chain Falls, and Come-Alongs

Chain hoists, also called chain falls, are used to lift or lower weldment parts. Always position the chain hoist directly over the center of gravity of the object being lifted. Secure the chain hoist over the weldment with an approved sling wrapped around a structural member. A suitable structural member may be a large I-beam or other significant building support component.

Never hang chain hoists from piping, ducts, conduit, or raceways. These items are not designed to carry external loads, and damage could result. The ground lead should be attached directly to the item to be welded. The chain fall or hoist should not be part of the welding current path. *Figure 8* shows using a chain hoist to aid joint fit-up.

CAUTION

Never weld a lifting eye to a structural member. If a weld is made on a structural member, it weakens the structural member, and the structural member may have to be replaced.

Figure 8 Chain hoist lifting a pipe into position.

WARNING!

Make sure the current path does not pass through the chain hoist supporting the load you are welding. Extreme heat generated by the current flow can cause the chain hoist to fail, releasing the load support. This can cause serious injury or even death.

Often, more than one come-along or chain fall may be used to precisely position a weldment. Before welding, check the ground lead connection location to be sure the welding current will not pass through the come-along. The ground lead should be attached directly to the item to be welded.

The difference in chain falls, chain hoists, and come-alongs can be a confusing topic. Some manufacturers label their products in ways that adds to the confusion. Some may call an item a chain fall while another item with the same features is referred to as a come-along. As a general rule, chain hoists and chain falls do not have a setting or feature that allows for the free-fall of the load. The device must be operated manually, by pulling the chain in one direction or another, to both lift and descend. Motorized versions also exist. Come-alongs may have chains or cables. Most have a ratcheting handle for operation, and can be switched using a lever or pawl to release the tool's hold on the load completely. If used for vertical lifting, the failure or accidental initiation of this feature could result in serious injury or property damage from a free-falling load. As a result, regardless of what name the device has been given, do not use devices for vertical lifting if they have a feature allowing the free-fall or free-release of the load. They can be safely used for horizontal positioning of a load and similar tasks.

Chain Collectors and Retainers

Many chain hoists are equipped with buckets, canvas bags, or other devices designed to collect the chain loops to prevent them from interfering with the workpiece. If the hoist you are using includes this feature, make sure the collector or retainer is aligned correctly with the drop of the chain. Otherwise, the collector or retainer may tilt, dumping the heavy load of chain on top of you. Wear appropriate PPE, including a hard hat, when using chain hoists or falls.

WARNING!

Before lifting any weldment part, add the weight of the part to the weight of the rigging (lifting) devices to determine the overall weight that will be placed on the structural member. Make sure that the structural member and all associated rigging devices are strong enough to handle the overall weight being lifted. Always inspect all rigging devices for damage and make sure that they are approved for the intended use.

1.3.0 Plate Fit-Up Tools

The most common method of holding a joint in place after it has been fitted up is to tack weld it in place. This works very well for small weldments and joints that are straight. However, large weldments or long or thick joints often require some type of mechanical means in addition to tack welding to set up and hold them in place for welding. The most common tools for plate fit-up are strongbacks, clips, yokes, and wedges. In addition, special plate alignment tools are available.

1.3.1 Strongbacks

Strongbacks (*Figure 9*) are typically made on the job site from heavy bar stock. They are notched at the weld joint to allow access to the joint so that welds can be made without interference. The strongback can be placed on the face or root side of the weldment. The plates to be joined are clamped or tack welded flush against the strongback, which holds the joint in alignment. If possible, when tack welding strongbacks, place the tack welds on only one side of the strongback. This will make the strongback easier to remove.

A good way to remove a strongback that has been tack welded is to use a grinder to remove or greatly reduce the tack welds. After the tack welds are minimized with a grinder, the tap of a hammer should break the strongback free of the base metal. After the strongback is removed, the grinder can be used to grind the tacks down flush, but care must be taken to not gouge the base metal during the grinding process.

1.3.2 Clips, Yokes, and Wedges

Clips, yokes, and wedges are also made on the job site. They can be used to align joints and then hold them in place during welding. Clips are welded to the edge of one plate and then wedges are positioned on the other plate and driven under the clips to force the joint into alignment. Yokes work in a similar manner. A yoke is welded to one plate. A slotted plate is then placed over the yoke, and a wedge is driven under the yoke to force the joint into alignment. *Figure 10* shows yokes and wedges being used to align and hold a joint.

1.3.3 Combining Alignment Tools

If the joint needs to be forced into alignment, yokes and wedges or bolts can be used with a strongback (*Figure 11*). When yokes and wedges are used, the strongback is positioned on the weldment straddling the joint. Yokes are positioned over the strongback and welded to the plates. Wedges are then driven under the yokes to force the joint into alignment. When bolts are used, the strongback must be made of channel stock or reinforced angle iron. Holes are cut in the channel or angle iron so that the bolts can pass through. The strongback is then positioned on the weldment straddling the joint. The position of the bolts is marked, and the strongback is removed. Bolts long enough to protrude through the strongback or threaded stock are then welded onto the plates.

The strongback is then positioned with the bolts protruding. Nuts are placed on the bolts and tightened with a ratchet or impact wrench, pulling the joint into position.

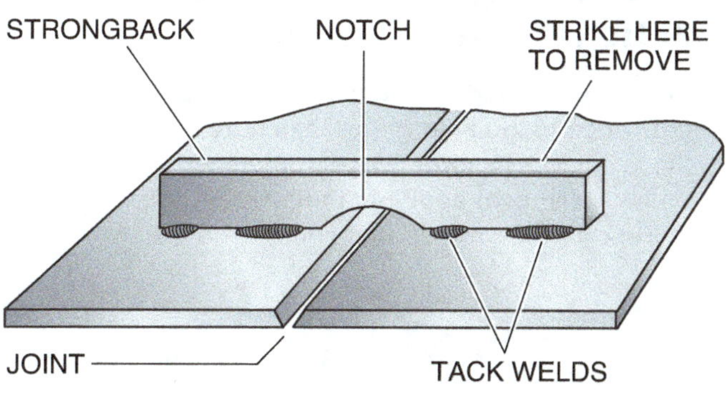

Figure 9 Strongback.

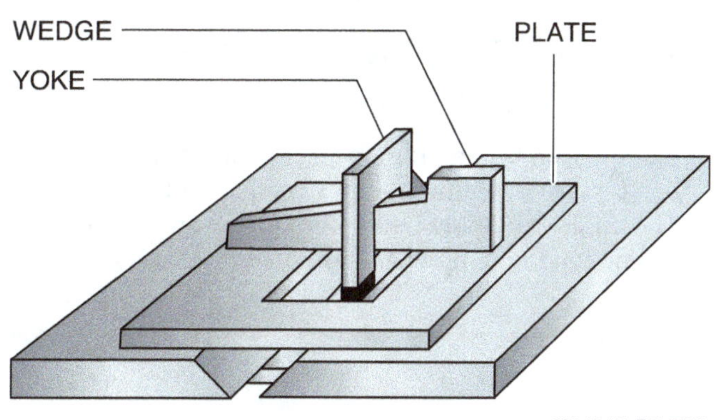

Figure 10 Yokes and wedges.

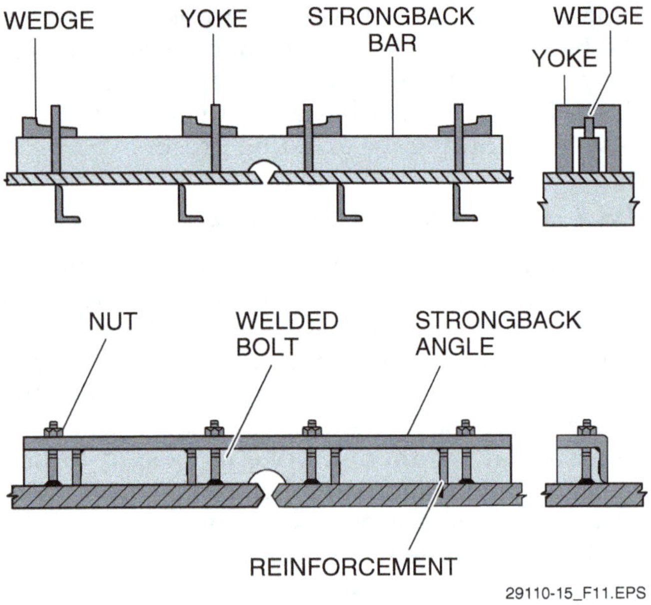

Figure 11 Aligning a joint using several tools.

1.3.4 Plate Alignment Tools

Special tools are manufactured for aligning plate that can make the task easier. A typical configuration of an alignment tool consists of a yoke, threaded adjusting rod, gap plate, and root bar (*Figure 12*). The alignment tool is used by straddling the joint opening with the yoke. The gap plate can be changed to match the specified root opening. The threaded adjusting rod is lowered until the opening in the gap plate is below the joint and the root bar can be inserted. The threaded adjusting rod is then tightened, bringing the joint into alignment. If necessary, the root bar can be adjusted to compensate for uneven joint thicknesses. Remember to properly match the tool materials with the base metal to eliminate contamination. Alternatively, isolating the two surfaces from each other so they don't touch while in use is also an option.

Using Angle Iron for a Strongback

If angle iron is used for the strongback, the bolts can be welded without removing the angle iron. The angle iron must be reinforced to prevent it from rolling when the bolts are tightened.

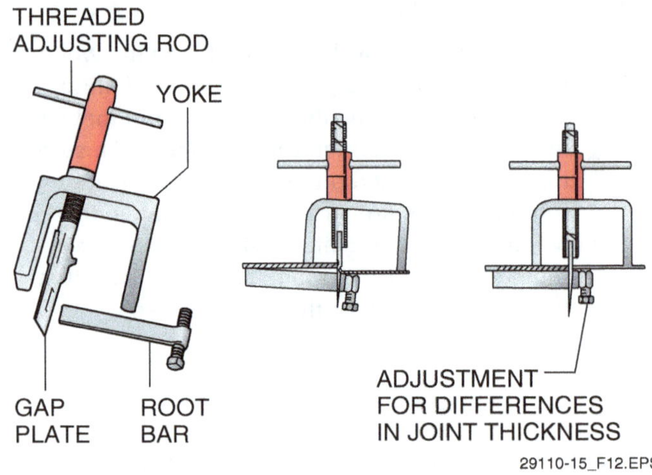

Figure 12 Plate alignment tool.

1.4.0 Pipe and Flange Fit-Up Tools

There are many different manufacturers of pipe fit-up tools. The tools they produce are similar, but before using a tool, be sure to review the guidelines provided by its manufacturer.

1.4.1 Pipe Jacks and Rollers

Pipe jack stands and rollers (*Figure 13*) are used to support pipe for fit-up and welding. Pipe jack stands typically have either a V-head or roller head and a height adjustment. Rollers, which can be floor-stand or table models, can be adjusted horizontally for various pipe diameters.

Adjustable jack stands and rollers are necessary tools for pipe fabrication, especially in the shop environment. They allow the welder to position various sizes of pipe for alignment and to manually roll the pipe as needed so the weld can be performed from the top, which is the preferred position. When fabricating piping systems in place, however, jack stands and rollers cannot

Marking Fabricated Alignment Tools

Strongbacks, wedges and similar tools are typically fabricated in the field rather than purchased. Mark these devices in some way so others won't discard them as scrap. In addition, they should be marked to indicate their material of construction.

Figure 13 Pipe jack stands and rollers.

be used because of positioning limitations. One way to deal with this is to weld straight sections of piping systems using jack stands and rollers. The straight sections can then be lifted and welded in place.

> **WARNING!**
>
> Jack stands and roller assemblies should not be field-fabricated. Load capacities and other safety considerations are built into these devices by their manufacturers, so only commercial jack stands and rollers should be used. Field-fabricated jack stands and rollers can fail and cause injury or even death. For pipe over 10" (DN250), OSHA requires that four-legged jack stands and roller assemblies be used.

> **CAUTION**
>
> When using pipe rollers to tack or weld pipe, be sure to connect the ground lead directly to the pipe to prevent arcing damage to the rollers and roller bearings.

1.4.2 Chain Clamps

Chain clamps are used to align and hold pipe for fit-up and tacking. Some manufacturers use link chain, others use chain similar to bicycle chain. Regardless of the type of chain used, the basic procedure for using chain clamps is the same. The chain, which is anchored to one side of the clamp, is passed around the pipe and secured. The slack in the chain is then taken up using a screw jack to pull the pipe tightly against the clamp.

Several types of chain clamps can be used to provide precise fit and alignment:

- Single-chain, single-jackscrew clamps
- Single-chain, double-jackscrew clamps
- Double-chain, double-jackscrew clamps
- Rim clamps

The single-chain, single-jackscrew clamp shown in *Figure 14* will align and even re-form out-of-round pipe up to Schedule 40 wall thickness. It can also be used on pipe heavier than Schedule 40 as long as no re-forming is required. Single-chain, double-jackscrew clamps have even more power to reform pipe. These clamps have a one or two jackscrews on each jack bar with optional spacing screws that allow the gap to be set quickly and precisely. A spacer may be used to maintain the required gap in the joint. Both the single- and double-jackscrew clamps are shown in *Figure 14*.

In addition to fitting up straight pipe, attachments for the precision fit-up clamps allow them to be used to fit up elbows, Ts, and flanges. The attachments connect by chain to the main block and to the elbow, T, or flange with a chain or clamp, as shown in *Figure 14*. The attachment has a fine adjustment crank to remove slack from the chain to support the elbow, T, or flange.

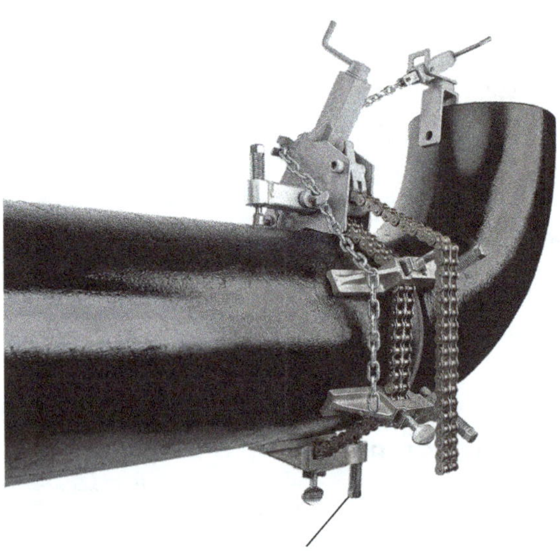

(A) SINGLE-CHAIN, SINGLE-JACKSCREW PIPE CLAMP

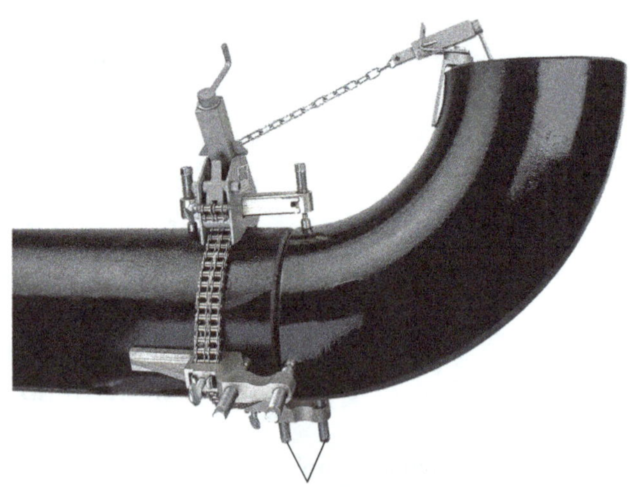

(B) SINGLE-CHAIN, DOUBLE-JACKSCREW PIPE CLAMP

29110-15_F14.EPS

Figure 14 Chain clamps.

1.4.3 *Other Pipe Alignment and Clamping Tools*

For one section of pipe to be welded end-to-end to another section of pipe, their ends must be cut or ground so that they mate properly. As you have seen in previous sections, jack stands and rollers are used to support sections of pipe to be welded. A chain clamp is one way to secure the two sections of pipe together. Cage clamps are another style of clamp that can also be used for the same purpose. Cage clamps come in a variety of styles and sizes. All must be installed manually, but the ways they are secured to the pipe sections vary. The latches on some of the old-style cage clamps had to be hammered on or off, but hammering tends to damage the clamps. Damaging the clamp can create problems joining the pipe sections.

Newer cage clamps are now available that allow the cages to be custom fitted to the pipes. *Figure 15* shows a cage clamp installed on two pipe sections, a clamp that is manually closed and opened, and a cage clamp that is tightened by a locking screw turned by a ratchet handle.

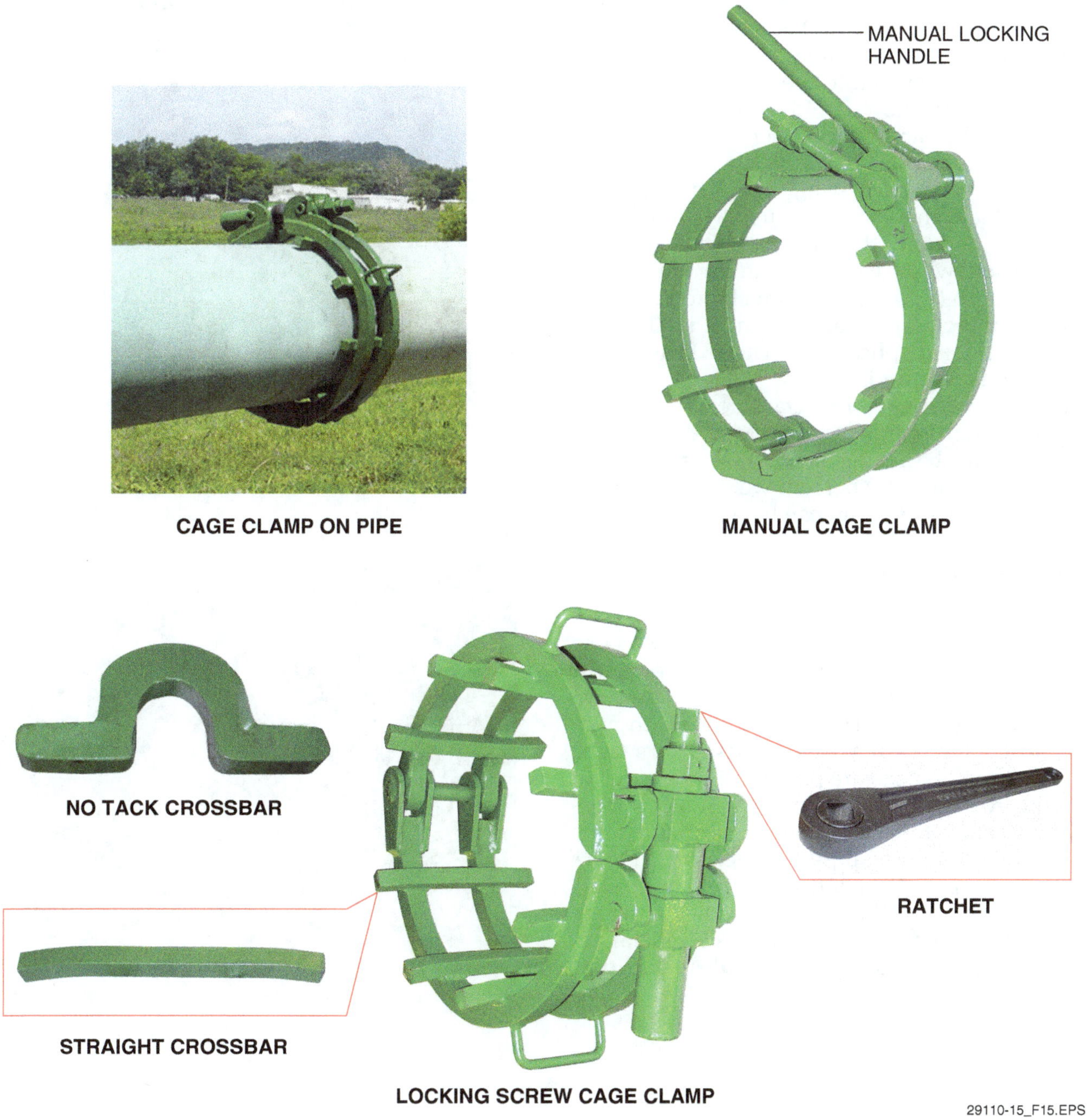

Figure 15 Cage clamps.

Nonferrous Alignment Tools

In some industries, fabricated alignment tools such as wedges, strongbacks, and yokes must be constructed of nonferrous materials to prevent sparking. Brass is a popular nonferrous metal for this purpose. This is a common requirement in some segments of the petrochemical industry.

Also shown in *Figure 15* are replaceable bars used with the cage clamps. The straight crossbars have flared ends, which allow the pipe sections to be more easily moved into the clamp. The arched no-tack crossbar allows welds to be made under the clamp or its bars without them being moved.

While cage clamps are good for aligning and holding two sections of straight pipe so that they can be welded, they are of no use when a flange must be welded onto the end of a pipe. One of the devices used when welding flanges onto pipes is a rim clamp (*Figure 16*). A rim clamp is a more versatile, non-chain clamp design. It is also recommended for applications where heavy-duty re-forming is required.

The jackscrews on a rim clamp exert pressure on specific high points of the pipe so that precise alignment can be made. These clamps are ideal for tasks in which 100 percent weld and grind is required before the clamp can be released.

1.4.4 Clamping Devices for Small-Diameter Pipe

Typical chain clamps are often too bulky to use on smaller-diameter pipe. A simple shop-built clamp for small-diameter pipe can be made from a piece of angle iron. The pipe is laid in the angle, and C-clamps or wires are used to secure the pipe. Small blocks are placed on both sides of the angle to hold it in position. When more accuracy and adjustability is required, special clamps for smaller-diameter pipe can be used (*Figure 17*). These clamps have a three-point jaw design to clamp and align the pipe when the clamping screws are turned. Special stainless steel clamps are available for working with stainless steel piping.

1.4.5 Pipe Pullers

Pipe pullers are clamps, often chain-type, that are used to pull together two pipe workpieces so that they can be joined by welding. These types of pullers are often applied in field fabrication or addition projects.

Figure 16 Rim clamp.

(A) ADJUSTABLE PIPE CLAMP OR JEWEL CLAMP

(B) ULTRACLAMP

29110-15_F17.EPS

Figure 17 Pipe clamps for smaller pipe.

In the petrochemical industry, for example, a new unit may be added to an existing facility. The piping must be joined through the use of manifolds and other piping designs. In order to make the final tie-ins from existing piping to the new piping, pipe pullers are often used to align the joining parts precisely.

CAUTION

When new piping will be coupled to machinery such as valves or pumps, it is extremely important that the flange of the pipe be perfectly aligned to the flange of the valve or pump. Using a come-along, jack, or other such device can put excessive stress on both the piping and the machine. Piping stress on pumps in particular will make them misalign, which in turn will destroy their bearings.

Do not force joints into alignment and then hold them in place with welds. Stress will be exerted at the welded joint in such an alignment, and cracking could result. Precise alignment and fitting without significant stress should be accomplished before finalizing the installation with the weld.

Pipe pullers (*Figure 18*) are frequently used to pull together sections of large-diameter pipe for welding. Fillers or consumable inserts may be installed between the faces of the joints to achieve precisely the necessary root opening between the parts. These inserts or fillers may become a permanent part of the finished weld.

CAUTION

Pipe pullers may not be permitted in some applications or projects.

29110-15_F18.EPS

Figure 18 Pipe puller.

Field-Fabricated Jack Stands

A mechanic was crushed to death when a bus he was working on fell off a set of jack stands. The stands were fabricated of plate steel by a local welding shop. The top plate of each was completely flat; they had no lips, which commercial stands always have. The front tires of the bus were not chocked, and there was nothing to prevent it from falling off the stands.

The investigation of the accident revealed that the jack stands had not been tested or certified for their rated capacity, nor were they marked with such a capacity. They were not fabricated in accordance with commercial jack stand construction. The lips on commercial jack stands cradle the area being supported. Also, commercial jack stands normally have three legs that help them compensate for irregular surfaces.

Sections of large-diameter pipe supported by jack stands may not be as large as buses, but they can cause serious or fatal injuries if they fall from their supports. Always use commercial jack stands that have been tested and certified.

1.4.6 Flange Alignment Tools

Various methods and tools can be used to make sure a flange facing aligns precisely with the pipe before it is permanently welded in place. Flange pins can be installed in the flange bolt holes to provide a reference point from which leveling can be measured for tack welds or complete welding (*Figure 19*). It is important to use a minimum of two flange pins installed in two of the bolt holes to provide a precise reference point from the center of one hole to another. From this flat, level benchmark, other tools can be used to adjust the flange to the pipe for true and precise alignment. The tool shown in *Figure 20* is also used to precisely align a flange for welding.

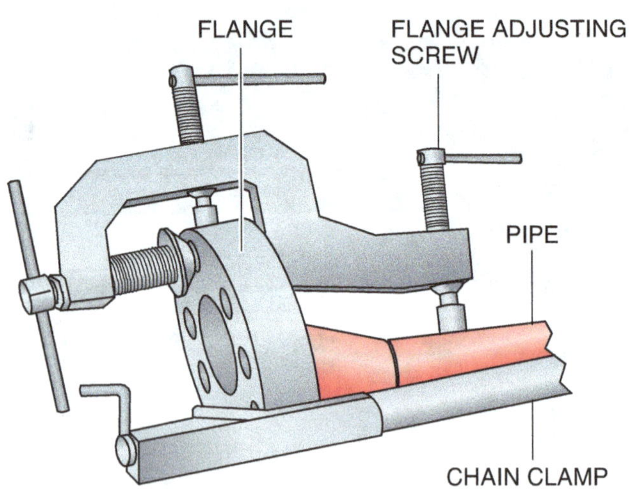

29110-15_F20.EPS

Figure 20 Flange alignment tool.

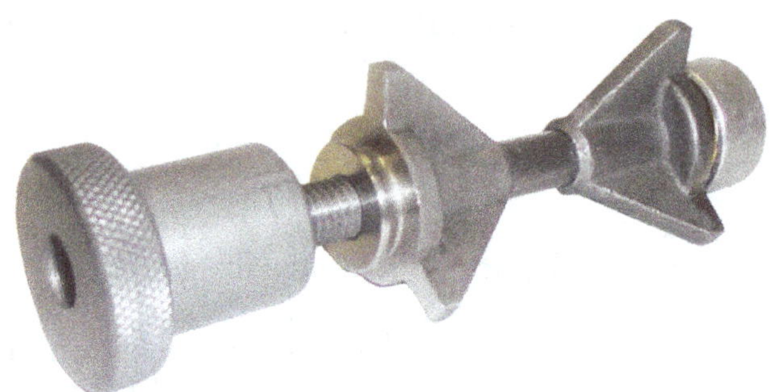

(A)

(B)

29110-15_F19.EPS

Figure 19 Using flange pins.

Types of Welded Flanges

Some of the more common flanges generally installed using welded connections are the welding neck flange, the lap joint flange, the socket weld flange, and the slip-on flange. Threaded flanges may also be back-welded, as determined by the process. Welding neck flanges are preferred for use in severe service applications, such as those involving high pressure, subzero temperatures, or elevated temperatures.

Think About It

Root Openings in Aligning and Fitting

Why is it necessary to maintain a root opening between two pieces, especially pipe pieces, when aligning and fitting prior to welding?

Additional Resources

2014 Technical Training Guide. Current Edition. Cleveland, OH. USA: The Lincoln Electric Company. **www.lincolnelectric.com**

Welding Handbook. Current Edition. Miami, FL: The American Welding Society.

Mathey Dearman Product Manuals. The Mathey Dearman Company. **www.matheydearman.com**

1.0.0 Section Review

1. Most levels designed for plate and piping work are made from _____.
   a. plastic
   b. wood
   c. aluminum or magnesium
   d. fiberglass or carbon steel

2. When lifting components into place to prepare for welding, vertical lifts should *not* be conducted using _____.
   a. a device with a free-fall mode or setting
   b. a device without a free-fall mode or setting
   c. any type of chain fall
   d. any type of synthetic lifting strap

3. Clips, yokes, and wedges are generally _____.
   a. purchased by the bag
   b. made on site
   c. shipped with the base metal
   d. made of aluminum

4. A simple shop-built clamp can be made from angle iron to help align _____.
   a. heavy plate
   b. large-diameter pipe
   c. small-diameter pipe
   d. sheet metal

Section Two

2.0.0 Weldment Distortion

Objective

Describe techniques to avoid weldment distortion and describe the roles of codes and specifications.

a. Describe the causes of weldment distortion.
b. Describe the techniques and tools used to control weldment distortion.
c. Describe the role of codes and specifications in welding procedures and techniques.

Trade Terms

Addendum: Supplementary information typically used to describe corrections or revisions to documents.

Ambient temperature: The temperature of the atmosphere surrounding an object or individual.

Coefficient of thermal expansion (linear): The change in length per unit length of material for a 1°C change in temperature.

Peening: The mechanical working of metal using repeated hammer blows, typically using a ball-shaped hammer head such as that found on a ball peen hammer. Peening may be done for both metallurgical and appearance reasons.

Residual stress: Stress remaining in a weldment as a result of heat.

Specific heat per unit volume: The quantity of heat required to raise one unit mass of a substance by one unit degree.

Besides the obvious cosmetic problems resulting from poor alignment and fit, the proper application of fit-up methods must be applied to deal with distortion and other problems. Distortion is the expansion and contraction of metal as it responds to changes in temperature. Tools such as chain clamps, rim clamps, C-clamps, strongbacks, and other devices can be used to redirect this expansion or contraction. This is referred to as controlling the distortion rather than preventing it. Distortion cannot be prevented as long as the metal experiences changes in temperatures created by the welding process. It can, however, be controlled, resulting in a weldment that is relatively free of damaging stress.

2.1.0 Causes of Distortion

Distortion is caused by the non-uniform expansion and contraction of the weld metal and adjacent base metal during the heating and cooling cycles of welding. When a metal block is heated, it expands uniformly in all directions. As the metal block cools, it contracts uniformly in all directions. If the metal block is restricted on two sides as it is heated, expansion cannot take place in the direction of restriction. Because the metal block must expand the same amount, all expansion occurs in the unrestricted directions. As the metal block cools, it contracts uniformly in all directions, leaving it narrower in the restricted direction and longer in the unrestricted direction. *Figure 21* illustrates the expansion, contraction, and distortion control of a block of steel.

During welding, these same forces act on the weld metal and base metal. As the weld metal is deposited and fuses with the base metal, it is at its maximum expanded state. Upon cooling, it attempts to contract but is restricted by the surrounding base metal. Stresses develop within the weld until the stress reaches the yield point of the base metal, causing the base metal and weld metal to stretch. These same stresses also cause the base metal to move. When the base metal is heated during the welding process, it

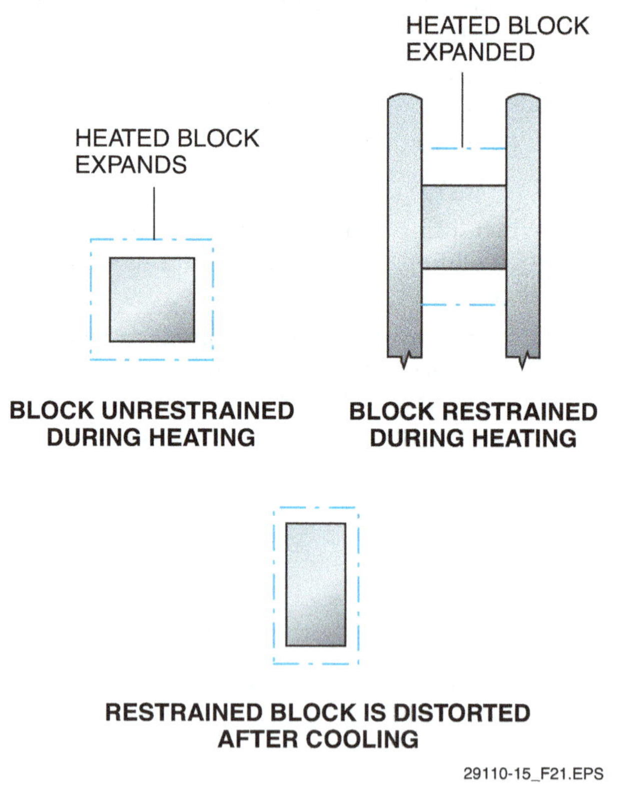

Figure 21 Controlling distortion.

will attempt to expand but will be restricted by the surrounding cooler base metal. Some movement will occur, but as the weld cools, the base metal that was welded will contract more due to the stress caused by the restriction of the surrounding base metal. This causes distortion. Even when the weld is at room temperature, stress equal to the strength of the base metal will be locked in the weldment. This is called residual stress. *Figure 22* shows how stress and distortion affect a weld.

The degree of distortion is directly related to the stresses generated during welding. Two of the metal properties that figure prominently in these stresses are the coefficient of thermal expansion (linear) and the specific heat per unit volume. Metals with higher values for these two properties will experience a higher degree of distortion than those with lower values. Stainless steel, for example, has a much higher coefficient of thermal expansion and specific heat per unit volume than carbon steel. It is therefore more likely to be affected by distortion. *Table 1* compares the coefficients of thermal expansion and the specific heats per unit volume for some common metals.

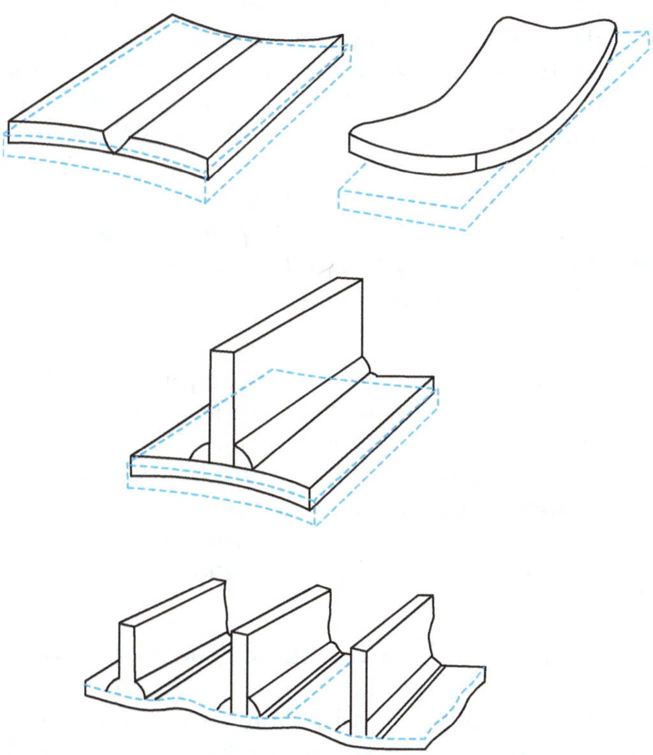

Figure 22 Stress and distortion in welds.

2.2.0 Controlling Distortion

Distortion can be controlled using a variety of techniques and tools. The following sections explain how to control distortion to minimize its effect on weldments.

2.2.1 Clamping and Bracing

Chain clamps, rim clamps, strongbacks, and other field-fabricated alignment devices are often used to hold weldments firmly in place during the welding process in order to prevent warping caused by distortion. These devices must be designed, constructed, and installed such that their overall support can withstand the stresses of welding. They must also be left in place long enough to allow the weldment to cool to ambient temperature. Some distortion may still take place once the devices are removed. This is caused by residual stress present in the weldment.

2.2.2 Tack Welding

A tack weld holds parts of a weldment in proper alignment until the finish welds are made. In order for them to be effective, pay attention to the number of tack welds, their length, and the distance between them. If too few tacks are made, the chance of the joint closing up as the welding proceeds is much greater. Clamps and other fit-up and aligning devices must be used to maintain proper alignment and the proper gap during the process.

Latent Distortion

If a weldment is clamped firmly during welding to prevent any movement, it will stay straight as long as the clamps are in place. Even after cooling, when the clamps are removed, residual stress can cause some distortion. When weldments that have residual stress are machined, they distort as the machining process removes metal, allowing the residual stress to overcome the strength of the base metal. This is referred to as latent distortion since it occurs long after the stress was created.

Table 1 Thermal Expansion and Specific Heat per Unit Volume for Common Metals

Metal	Coefficient of Thermal Expansion at 20°C	Specific Heat Per Unit Volume J/g°C
Aluminum	$22–24 \times 10^{-6}$/°C	0.900
Stainless steel (316)	$14–17 \times 10^{-6}$/°C	0.500
Steel	$12–13 \times 10^{-6}$/°C	0.448
Copper	$16–17 \times 10^{-6}$/°C	0.390

If tack welds are to be included with the main weld, they must be applied by qualified welders. They must also follow the qualifying WPS for the project, if one applies. Likewise, the filler metals must meet the same filler metal requirements as the filler metal for the finished welds. For those tack welds that require removal prior to the finish weld, use extreme care to avoid causing defects in the material.

2.2.3 Amount of Weld Material

The more weld metal placed in a joint, the greater the forces of shrinkage. Properly sizing a weld not only minimizes distortion, it also saves weld metal and time. Excess reinforcement on the face of a weld increases the forces of distortion and adds nothing to the strength of the weld. Excess face reinforcement actually reduces the strength of a weld and is therefore prohibited by welding codes. The face of fillet welds may be slightly convex, flat, or slightly concave. Groove welds should have slight reinforcement of no more than ⅛" (3.2 mm) for butt or corner welds. *Figure 23* shows acceptable weld reinforcement profiles.

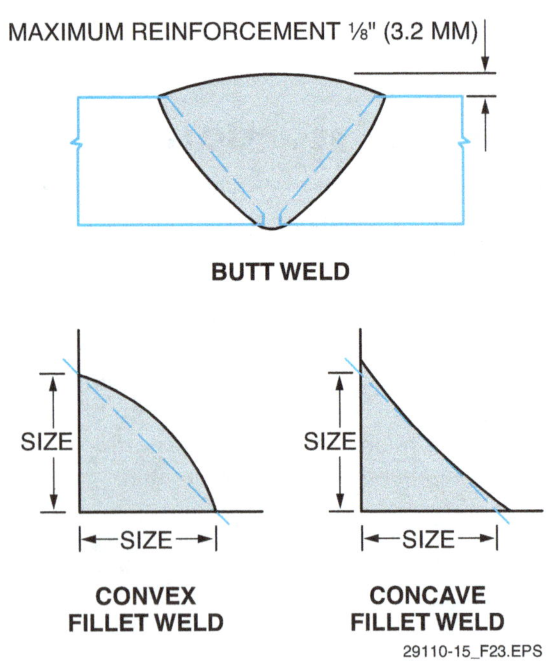

Figure 23 Weld reinforcement.

NOTE

Refer to your site's WPS for specific requirements on weld reinforcement. The information in this module is provided as a general guideline only. The site WPS must be followed for all welds covered by a procedure. Check with your supervisor if you are unsure of the specifications for your application.

Proper fit-up and edge preparation also reduce the amount of weld required. Open-root joints should have a root opening from 1⁄16" to ⅛" (1.6 to 3.2 mm). To control melt-through, a root face of 1⁄16" to ⅛" (1.6 to 3.2 mm) is used. Each side of the open-root joint is beveled from 30 to 37½ degrees (included angle of 60 to 75 degrees). The greater the angle of bevel, the more weld will be required to fill the joint. The bevel angle must be sufficient to allow access to the root but not so large that it requires excess weld metal to fill. Welds made with GMAW or GTAW generally require larger openings than welds made with SMAW because the nozzles used with GMAW and GTAW are larger in diameter than the electrodes used with SMAW. These nozzles require more room in the joint to reach the root. *Figure 24* shows an open V-groove joint preparation.

CAUTION

When preparing a joint, refer to your site's WPS for specific requirements. The information in this manual is provided as a general guideline only. The site WPS must be followed for all welds covered by a procedure. Check with your supervisor if you are unsure of the specifications for your application.

When possible, a double V-groove should be used in place of a single V-groove (*Figure 25*). The double V-groove requires half the weld metal as compared to the single V-groove. Also, welding from both sides reduces distortion since the forces of distortion will be working against each other. On thick joints, U- and J-grooves also require less weld metal than V-grooves or bevel grooves, although U- and J-grooves require more preparation time.

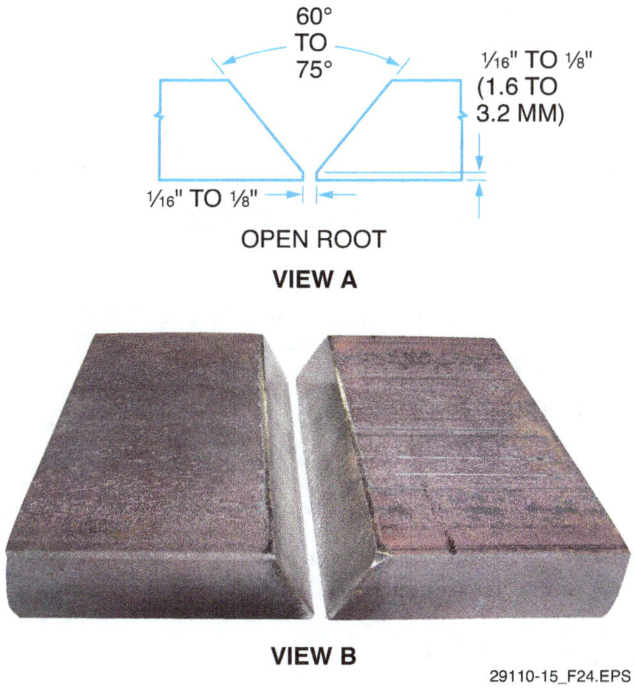

Figure 24 Open V-groove preparation.

2.2.4 Backing Strips on Groove Welds

Various forms of backing materials or strips can be applied to the back side of plate weldments. Temporary strips typically have a groove machined directly into them that controls the shape and size of the penetration material once it cools. Backing strips can be purchased in rolls from which the desired length can be cut and then taped directly to the back of the prepared joint. If the backing strip is designed to become part of the permanent weldment, it must be made of material similar to the alloy or metal being welded. If the backing strip is to be removed, it normally is made of dissimilar metal to prevent it from being welded into the joint. The backing strip can then be peeled from the back side of the finished weldment. Backing strips of copper, for example, are often used on weldments of stainless steel.

WPS

The qualifying WPS that specifies the tack-welding procedures, as well as the consumables to be used, is typically designed as part of the engineering plans. It is not normally a function or responsibility of the welder. However, the welder must be qualified to perform the tasks and follow the procedures specified in the WPS.

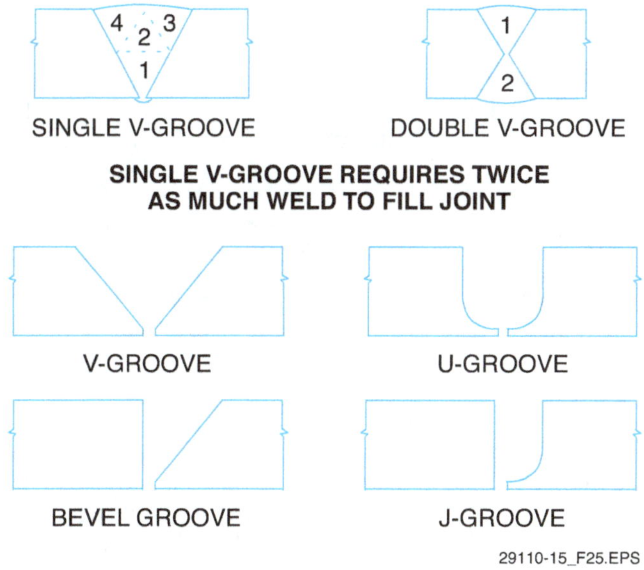

Figure 25 Types of groove styles.

When thick metal backing strips are to be used on groove welds, the root opening normally opens to ¼" (6.4 mm). The bevel is reduced to 22½ degrees (a total angle of 45 degrees) as shown in *Figure 26*. A root face is not required because the backing will prevent melt-through. Distortion can be reduced considerably through this control of the size and shape of the penetration.

2.2.5 Open Root Pipe Welds

It is important to note here that, in pipe welding, the alignment of the inside diameter is the most important, as opposed to the alignment of the outside diameter. In some cases, counter-boring of the pipe can be used to achieve alignment. However, there are limitations to counter-boring, since a minimum wall thickness must be maintained.

Molten metal that extends beyond the back or opposite side of the groove is called root reinforcement. Controlling the amount and shape of this root reinforcement is important to the function of the weldment. It is also important because controlling the root reinforcement helps reduce distortion by controlling the weld material uniformly. When welding pipe, the amount of root reinforcement on the inside of the pipe must be controlled. Excessive root reinforcement in the interior of the pipe interferes with the flow within the pipe, causing turbulence, pressure drop, and possibly other problems. Controlling this melt-through is also necessary to reduce stress on joints. However, it is difficult, if not impossible, to gain access to the interior of the pipe. Various methods are used to provide backing and to control penetration in pipe welding.

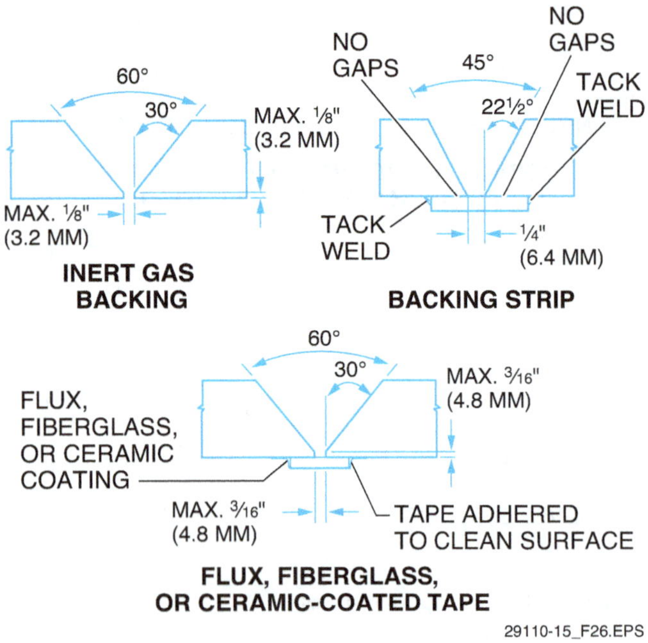

Figure 26 Groove welds with backing strip.

Gas backing using gases such as nitrogen or argon is one method used when welding pipe joints. When gas backing is used, some method of containing the gas inside the pipe must be used (*Figure 27*). For a short section of pipe, the ends of the pipe can be capped by taping or clamping a metal disk over the ends. For larger or longer pipe sections, however, this method may not be feasible due to the amount of gas that would be required to fill the pipe. For these cases, inflatable bladders, water-soluble plugs, or soft plastic bags can be inserted into the pipe near the weld joint. The inflatable bladders can be fished out of the pipe after the weld cools. The water-soluble plugs or soft plastic bags can either be blown or washed out of the pipe after the weld cools.

> **CAUTION**
>
> The pipe end opposite the end into which gas enters must have a vent hole to purge the gas and prevent pressure from building up. What is desired is an inert gas atmosphere in the pipe, rather than gas under pressure.

The type of gas used will depend on the base metal of the pipe being welded. Generally, nitrogen or carbon dioxide is used for carbon steel piping; argon is used for stainless steel, aluminum, and alloy steel piping.

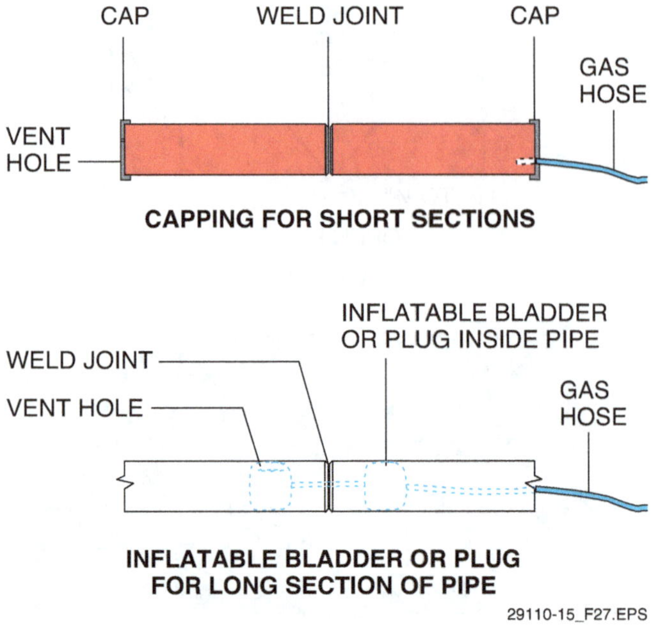

Figure 27 Gas backing.

2.2.6 Backing Rings

Backing rings (*Figure 28*) are flat metal strips that have been rolled to fit inside a pipe. They can be ordered in a variety of base metal types to match the base metal being welded. Some of these rings are split, making them easier to insert and adjust to piping that may be slightly out of round. Backing rings are available that have three or more nubs around the outside of the ring. The diameter of the nub is the root opening required for the pipe diameter being welded. Nubs can be pegs, buttons, or indents punched into the ring. Pegs must be removed after the joint is tack welded. They are removed by striking them with a chipping hammer; they snap off rather easily. Buttons can be removed or left in position to be melted during the root pass.

The joint preparation for backing rings (*Figure 29*) is similar to the preparation for open-root joints. The joint angle should be 60 or 75 degrees with a root opening and root face up to 1/8" (3.2 mm), or as specified on the WPS. If backing rings with nubs are used, the root opening will be set automatically by making sure the beveled ends of the pipe are positioned firmly against the nubs.

2.2.7 Consumable Inserts on Pipe Welds

Consumable inserts are similar to backing rings. As the name implies, they are completely consumed during welding and become part of the weld metal.

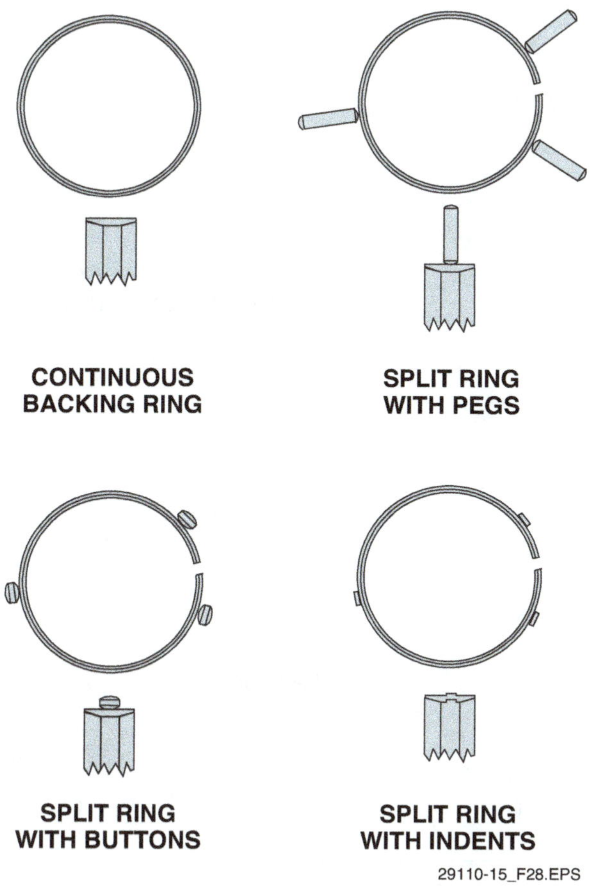

Figure 28 Examples of backing rings.

Because they are part of the finished weld, they must match the filler metal requirements for the weld being made. Gas backing often is required when using consumable inserts. Consumable inserts come in a variety of shapes and are identified by class numbers, as shown in *Figure 30*.

Consumable inserts can be purchased in coils and in preformed split rings. The rectangular Class 3 insert is also available as a solid ring.

Joint preparation for consumable inserts (*Figure 31*) can consist of a V-groove or a J-groove, with the V-groove having a joint angle of 75 degrees and a root face of 1⁄16" (1.6 mm). The J-groove joint angle is 20 degrees, also with a 1⁄16" (1.6 mm) root face. The consumable insert usually is tacked against the prepared ends of the pipe. The WPS will specify the tacking requirements.

2.2.8 Inserts on Socket Joints

Socket joints are generally used on pipe that is 5" (DN125) or smaller in diameter. A socket joint uses a prefabricated fitting containing sockets on the ends; the pipe slips into the sockets. The fitting and pipe are joined using a fillet weld.

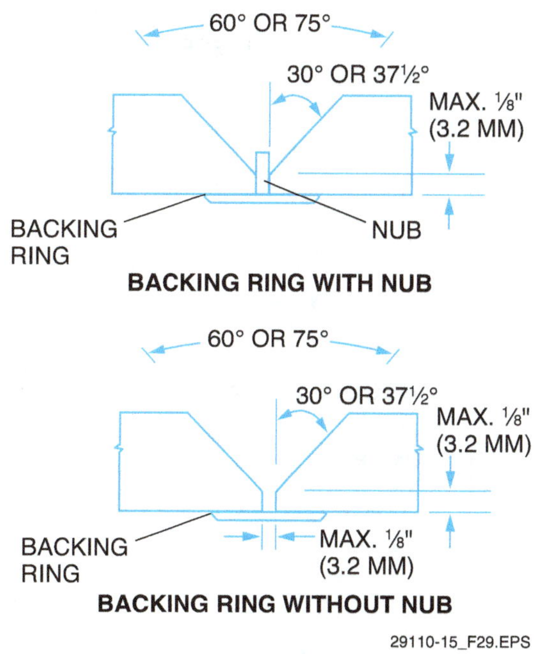

Figure 29 Pipe joint with backing ring.

All the common fittings, such as elbows, flanges, couplings, reducers, and even valves, are available as socket fittings. Using socket joints is a quick and easy method for joining pipe evenly.

In order to eliminate stress, distortion, and possible cracking by expansion, the end of the pipe must not touch the bottom of the socket. Depending on the code, socket welds require a 1⁄16" to ⅛" (1.6 to 3.2 mm) gap between the end of the pipe and the bottom of the socket fitting.

When preparing the joint, there are two ways to ensure that the gap between the end of the pipe and the socket fitting is maintained.

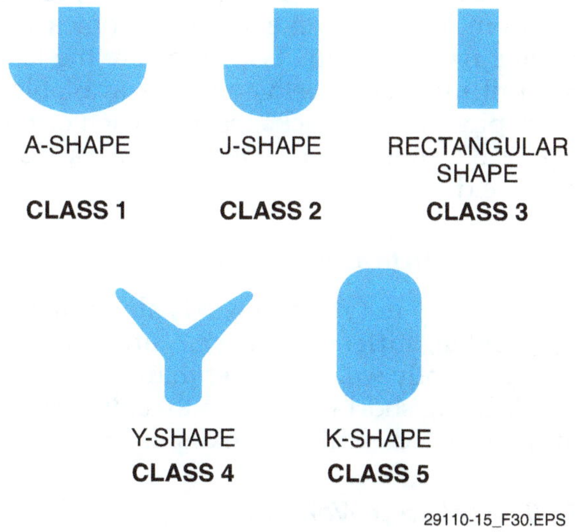

Figure 30 Consumable insert profiles.

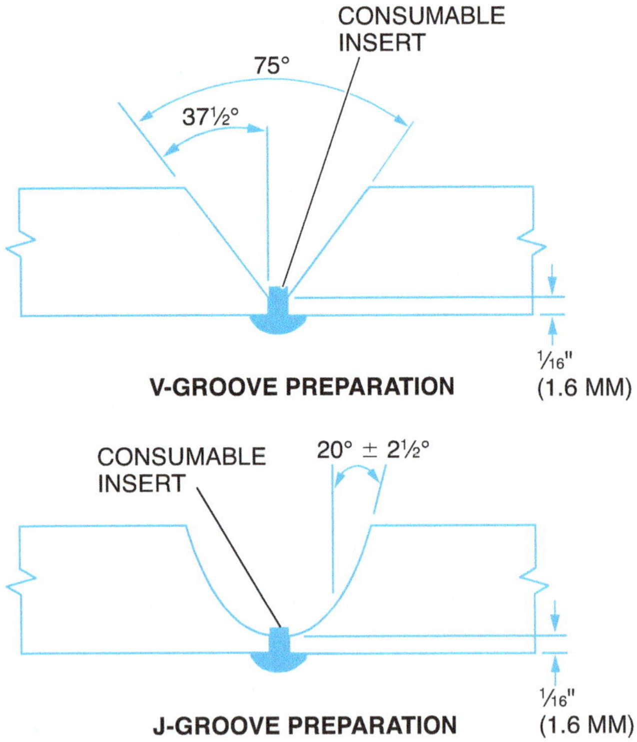

29110-15_F31.EPS

Figure 31 Joint preparation for consumable inserts.

One way, described earlier, is to scribe the pipe a preset distance from the end. The gap can be checked by measuring from the scribed line to the socket and then adding the socket depth. The scribed lines also can be used to check setup after the joint is tacked up or welded.

The second method of ensuring the correct gap is to use a gap ring. A gap ring is a split ring that is formed to the gap required for the pipe diameter being welded. The gap ring is placed in the bottom of the socket end and becomes a permanent part of the joint. Gap rings are the fastest way to fit up socket welds and are available for various pipe sizes in packages of 20 each or more.

Figure 32 shows both methods of preparing socket-weld pipe joints.

2.2.9 Intermittent Welding

Many joints do not require welds along their full length. Often, stiffeners, brackets, and braces can be intermittently welded rather than continuously welded. Using short intermittent welds instead of continuous welds will reduce distortion.

2.2.10 Backstep Welding

Backstep welding (*Figure 33*) is a welding technique in which the general progression of welding is from left to right, but the weld beads are deposited in short increments from right to left.

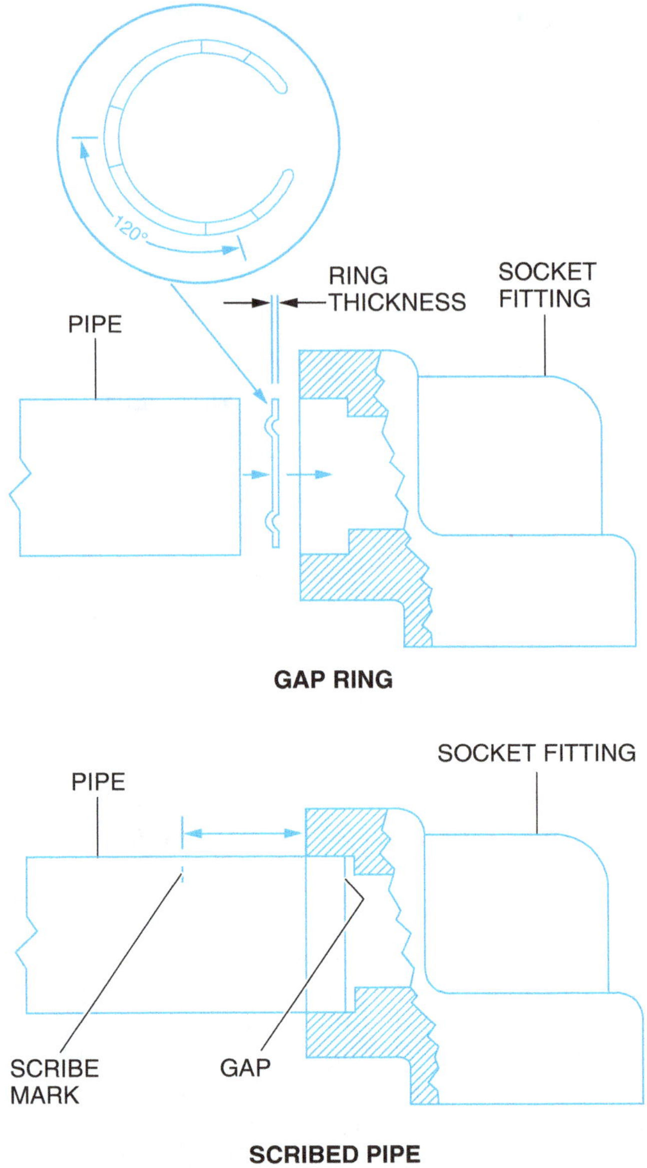

29110-15_F32.EPS

Figure 32 Socket-weld joint preparation.

This technique reduces distortion by minimizing and interrupting the heat input.

2.2.11 Welding Sequence

A welding sequence involves placing welds at different points on a weldment so that shrinkage forces in one location are counteracted by shrinkage forces in another location. Welding sequences can be quite complicated for large complex weldments or quite simple for less complicated weldments. When possible, use a welding sequence because it is a very effective means of controlling distortion. A simple welding sequence is to make short welds on alternating sides of the joint. This can be used for fillet or groove, intermittent, or continuous welding. *Figure 34* shows examples of intermittent welding sequences.

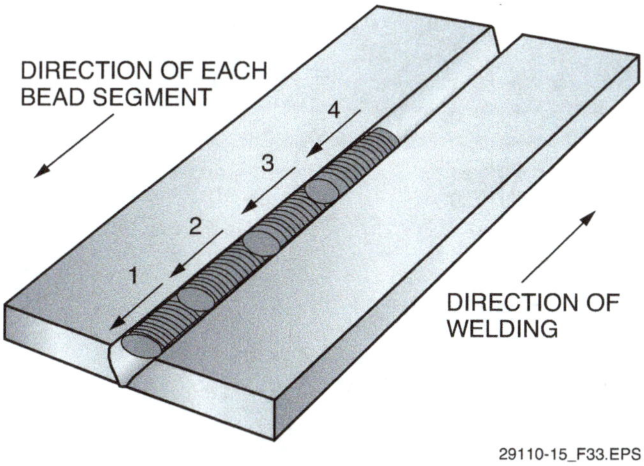

Figure 33 Backstep welding.

Welding sequences are often performed by two individuals welding on opposite sides of a joint at the same time. This type of welding is often called buddy welding and is a very effective way to control distortion.

2.2.12 Heat Treatments

Distortion can be controlled by preheating and postheating. Applying heat before welding is called preheating. Applying heat immediately after welding is called postheating. Preheating and postheating will be covered in detail in another module.

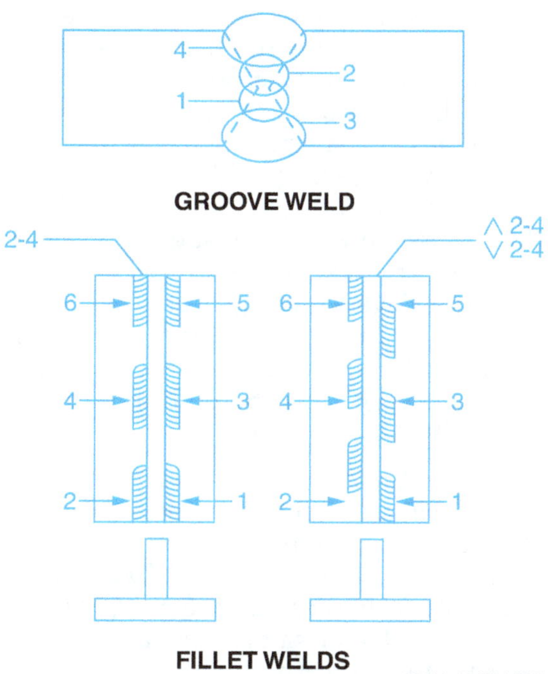

Figure 34 Intermittent welding sequences.

Welding must be stopped periodically to allow inspection of the weld in progress. These hold points should be scheduled strategically. Any work that will be obstructed after welding is complete, and therefore will not be easily inspected, must be inspected before the weld covers it. The following are among the items to be checked:

- Preheat temperature
- Postheat temperature
- Compliance with WPS requirements
- Weld root pass
- Weld layers

2.2.13 Checking Joint Misalignment and Fit-Up

The quality of joint preparation and fit-up directly affects the quality of the completed weld. After a weld is made, it is very costly and time-consuming to repair a defect. By thoroughly checking the joint fit-up, potential problems can be avoided. Follow these steps to check a fit-up:

Step 1 Determine if the weld is covered by a WPS. If the weld has a WPS, obtain a copy and follow it to make the weld.

Step 2 Check that the base metal type and grade are as specified. If either the base metal type or grade is wrong, contact your supervisor.

Step 3 Check, if required, that the welder who will perform the welding is certified/qualified to the WPS.

Alternating Welds

When possible, an alternating weld should be placed directly adjacent to or across from its matching weld on the opposite side. On lengthy welds, however, this sequence may not be advisable because it could leave long runs unwelded in between these welds. In these cases, stagger the welds from side to side.

If welding alternately on either side of the joint is not possible, or if one side must be completed first, a joint preparation may be used that deposits more weld metal on the second side. The greater contraction resulting from depositing more weld metal on the second side will help counteract the distortion on the first side.

Step 4 Check that the joint surfaces are free of contamination such as grease, oil, moisture, and rust. Also, check the surfaces parallel to the root and face of the weld. If there is contamination, clean the joint before continuing.

Step 5 Check the joint surface for cracks or laminations. If you find either, contact your supervisor.

Step 6 Check that the edge preparation is as specified for the joint (*Figure 35*). Check the following:
- Groove type
- Root opening
- Root face
- Included angle
- Bevel angle
- Base metal thickness

Step 7 Check the fit-up of backing strips, backing rings, and consumable inserts if required. Backing strips and backing rings must be tight against the joint along or around its entire length with even root spacing. Consumable inserts must fit tightly at the root of the weld. Correct any fit-up problems before proceeding.

Step 8 Check that the parts of the weldment are square, level, and plumb, and check that any angles are as specified. Correct any problems before proceeding.

Step 9 Check that the tack welds have been cleaned and feathered. Correct any problems before proceeding.

Step 10 Check that the welding process to be used is as specified on the WPS.

Step 11 Check that the consumables (type and size) are as specified on the WPS. Also check that arrangements have been made for the heated storage of consumables if required on the WPS.

Step 12 Check that the welding sequence is available if required.

Step 13 Check that the welding machine settings are as specified on the WPS.

Step 14 Check that arrangements have been made for preheating, interpass temperature control, and postheating if required on the WPS.

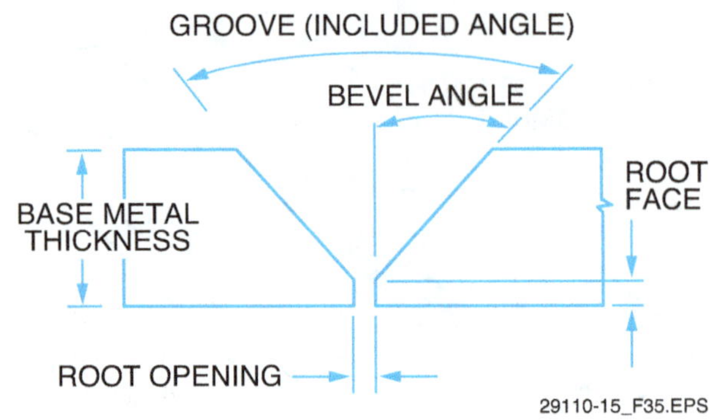

Figure 35 Edge preparation.

2.3.0 Codes and Welding Procedure Specifications

Whenever a bridge, building, ship, or pressure vessel is welded, the manufacturer and the buyer must reach agreement on how each weld will be made. To eliminate the need to write a new code for each job, government agencies, societies, and associations have developed codes. These codes are used universally to ensure safety and quality when welds are made.

2.3.1 Governing Codes and Standards

A welding code is a detailed listing of the rules and principles that apply to specific welded products. Codes ensure that safe and reliable welded products will be produced and that persons associated with the welding operation will be safe. Imagine the potential for disaster if there were no standards to control the quality and technique of welds in a nuclear plant or on an oil pipeline.

All welding should be performed following the guidelines and specifications outlined in various codes, which are specified by clients when they place orders or contract work to be done. In addition, when codes are specified, the use of these codes is mandated with the force of law by one or more government jurisdictions. They are not suggestions; they are requirements. Always check the contract, order, or project specification for the specified code(s). Some of the more common codes are the following:

- *API 1104, Standards for Welding of Pipelines and Related Facilities* – American Petroleum Institute (API); used for pipelines.
- *ASME Boiler and Pressure Vessel Code* – American Society of Mechanical Engineers (ASME); establishes the rules of safety relevant to pressure integrity that govern the design, fabrication, and inspection of boilers and pressure vessels, including nuclear power plant components.

- *ASME B31.1, Power Piping* – American Society of Mechanical Engineers (ASME); used for pressure piping.
- *ASME B31.3, 2006 Process Piping* – American Society of Mechanical Engineers (ASME); used for process piping in refineries, chemical and pharmaceutical, and similar plants.
- *ABS* – American Bureau of Shipping & Coast Guard; sets standards of excellence in Marine and Offshore Classifications. (see: **http://www.eagle.org/company/overview.html**).
- *AWS D1.1, Structural Welding Code* – American Welding Society (AWS); commonly specified for buildings and civil structures of all types for welding standard carbon and low alloy structured steel materials
- *AASHTO/AWS D1.5, 2008 Bridge Welding Code* – American Welding Society (AWS); used for welded bridges.

2.3.2 Code Changes

Periodically, codes are reviewed and changes made. Normally this occurs yearly, but some codes may go more than one year before they are updated. Most codes that are extremely broad in scope may undergo changes one section or category at a time. When a code is updated, the entire code is reissued with an updated year suffix, or addendum sheets for the areas of the code affected by the changes are issued. Changes are typically noted to alert readers to what has changed from the prior edition. The AWS code indicates additions to the code with a double vertical line in the margin alongside the changed area. Editorial changes to the AWS code are indicated by a single vertical line.

The ASME updates the entire code every three years. An addendum is issued annually. The annual addendum is identified by placing the letter A in front of the year on the cover of the code. API updates their entire code every five years.

It is important to recognize code changes. When a client specifies a code year, the code for that year, and that year only, is to be used unless otherwise specified by the client. When referring to a code, be sure the year matches the specifications for the job.

2.3.3 Welding Procedure Specifications

A WPS is a written set of instructions for producing sound welds in a prescribed manner. Each WPS is written and tested in accordance with a particular welding code or specification and must be in accordance with industry practice. Note that all welding requires that acceptable industry standards be followed, but not all welds require a WPS. If a weld does require a WPS, the WPS must be followed. The consequences of not following a required WPS are severe. Consequences include producing an unsafe weldment that could endanger life as well as the rejection of the weldment and lawsuits. Even if an unsafe condition is not known to exist, a structure known to contain numerous welds that were not done according to the WPS may have to be destroyed and redone. The economic impact can be devastating and cost welders their jobs.

When it is required, always follow the WPS. The requirement for the use of a WPS is often listed on job blueprints as a note or in the tail of the welding symbol. If you are unsure whether the welding being performed requires a WPS, do not proceed until you check with your supervisor. As a welder, you have a responsibility to follow the WPS and the applicable codes any time you are on the job. Your employer and fellow workers depend on you to carry that responsibility without deviation.

Each WPS is written by an individual who knows welding codes, specifications, and acceptable industry practices. It then becomes the responsibility of each manufacturer or contractor to test and qualify the WPS before using it. The WPS is tested by welding test coupons. Then, the coupons are tested according to the applicable code. For shielded metal arc welding (SMAW) of complete-penetration groove welds, the testing required includes nondestructive testing (NDT), tensile strength tests, and root, face, or side bend tests. The results of the testing are recorded on a Procedure Qualification Record (PQR). The WPS and PQR must be kept on file.

Many different formats are used for WPSs, but they all contain the same essential information. *Figure 36* shows a portion of a typical WPS.

Information typically found on a WPS includes the following:

- *Scope* – The welding process to which the WPS applies as well as the governing code and/or specification to be used.
- *Base metal* – The chemical composition or specification of the applicable base metal using the industry-standard identification; for example, A36 for carbon steel.
- *Welding process* – The welding process to be used to make the weld. For example, SMAW, gas metal arc welding (GMAW), flux-core arc welding (FCAW), or gas tungsten arc welding (GTAW).
- *Filler metal* – The composition, identifying type, or classification of the filler metal to be used.

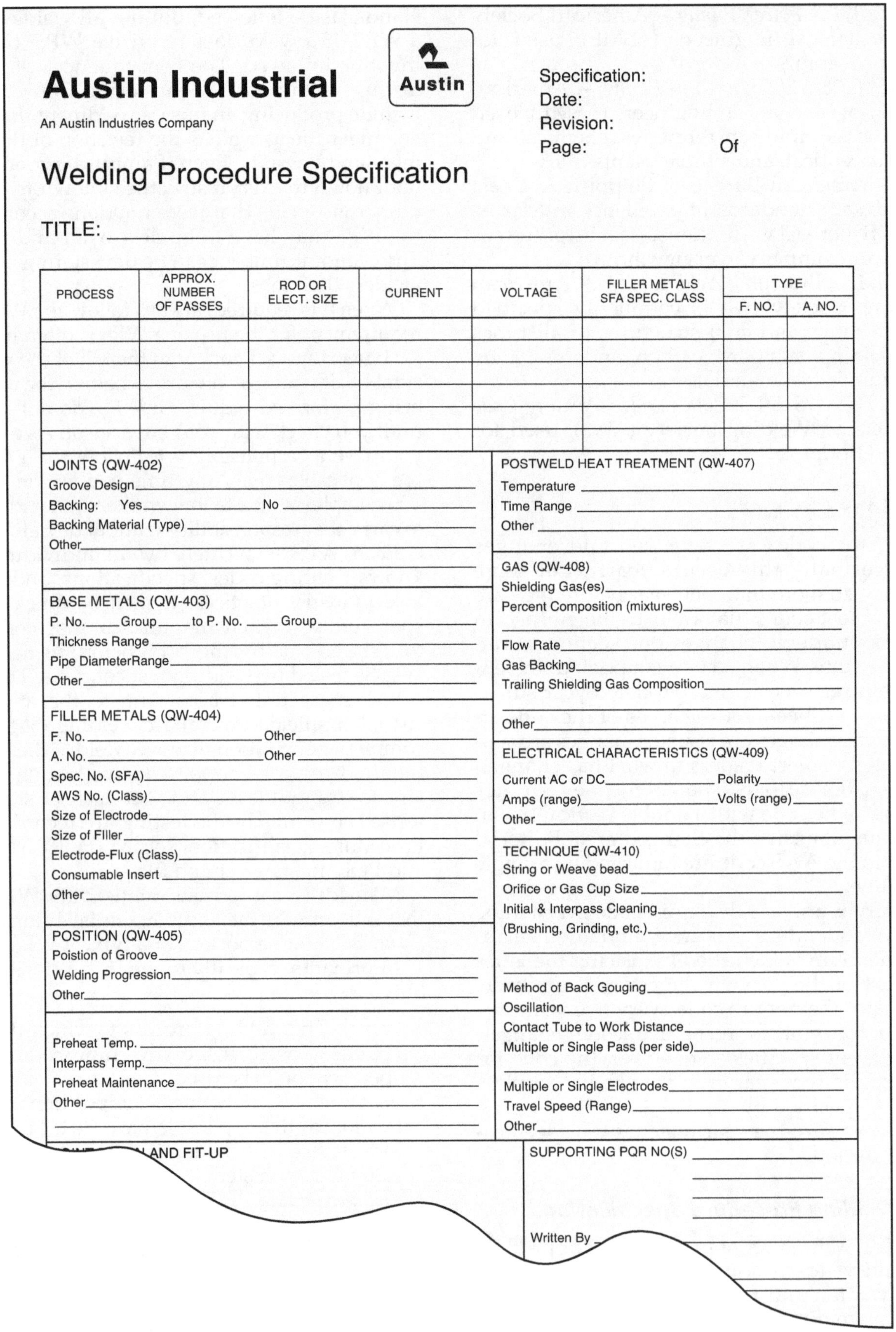

Austin Industrial

Austin

An Austin Industries Company

Specification:
Date:
Revision:
Page: Of

Welding Procedure Specification

TITLE: ____________________

PROCESS	APPROX. NUMBER OF PASSES	ROD OR ELECT. SIZE	CURRENT	VOLTAGE	FILLER METALS SFA SPEC. CLASS	TYPE	
						F. NO.	A. NO.

JOINTS (QW-402)
Groove Design ____
Backing: Yes ____ No ____
Backing Material (Type) ____
Other ____

BASE METALS (QW-403)
P. No. ____ Group ____ to P. No. ____ Group ____
Thickness Range ____
Pipe DiameterRange ____
Other ____

FILLER METALS (QW-404)
F. No. ____ Other ____
A. No. ____ Other ____
Spec. No. (SFA) ____
AWS No. (Class) ____
Size of Electrode ____
Size of Filler ____
Electrode-Flux (Class) ____
Consumable Insert ____
Other ____

POSITION (QW-405)
Poistion of Groove ____
Welding Progression ____
Other ____

Preheat Temp. ____
Interpass Temp. ____
Preheat Maintenance ____
Other ____

AND FIT-UP

POSTWELD HEAT TREATMENT (QW-407)
Temperature ____
Time Range ____
Other ____

GAS (QW-408)
Shielding Gas(es) ____
Percent Composition (mixtures) ____
Flow Rate ____
Gas Backing ____
Trailing Shielding Gas Composition ____
Other ____

ELECTRICAL CHARACTERISTICS (QW-409)
Current AC or DC ____ Polarity ____
Amps (range) ____ Volts (range) ____
Other ____

TECHNIQUE (QW-410)
String or Weave bead ____
Orifice or Gas Cup Size ____
Initial & Interpass Cleaning
(Brushing, Grinding, etc.) ____
Method of Back Gouging ____
Oscillation ____
Contact Tube to Work Distance ____
Multiple or Single Pass (per side) ____
Multiple or Single Electrodes ____
Travel Speed (Range) ____
Other ____

SUPPORTING PQR NO(S) ____

Written By ____

29110-15_F36.EPS

Figure 36 Portion of typical WPS.

- *Type of current* – Either alternating current (AC) or direct current (DC). If DC is specified, the polarity must also be given.
- *Arc voltage and travel speed* – An arc voltage range. Ranges for travel speed for automatic processes are mandatory and are recommended for semiautomatic processes.
- *Joint design and tolerances* – Joint design details, tolerances, and welding sequences that are given as a cross-sectional drawing or as a reference to drawings or specifications.
- *Joint and surface preparation* – The methods that can be used to prepare joint faces and the degree of surface cleaning required.
- *Tack welding* – Details pertaining to tack welding. Tack welders must use the WPS.
- *Welding details* – The size of electrodes to use for different portions and positions of the joint, the arrangement of welding passes to fill the joint, and the pass width and weave limitations.
- *Positions of welding* – The welding positions that can be used; for example, 1G, 3F, 5G.
- *Peening* – The details of peening and the type of tool to be used if peening is permitted.
- *Heat input* – The details to control heat input.
- *Second-side preparation* – The method used to prepare the second side when joints are welded from two sides.
- *Postheat treatment* – The details about postheat treatment or a reference to a separate document.

Essential and Nonessential Variables on a WPS

A different WPS is required for each change to an essential variable. Some essential variables for welds made with SMAW include the following:

- Base-metal thickness (within range listed)
- Composition of base metal
- Strength and composition of filler metal
- Base-metal preheat temperature

Nonessential variables for welds made with SMAW include the following:

- Type of groove used
- Omission of backing material in butt joints
- Electrode size

Changes in these nonessential variables would not normally require a different WPS.

Additional Resources

2014 Technical Training Guide. Current Edition. Cleveland, OH. USA: The Lincoln Electric Company. **www.lincolnelectric.com**

Welding Handbook. Current Edition. Miami, FL: The American Welding Society.

Mathey Dearman Product Manuals. The Mathey Dearman Company. **www.matheydearman.com**

2.0.0 Section Review

1. Stainless steel is more likely to be affected by distortion than carbon steel because it has a _____.
   a. higher coefficient of thermal expansion and a lower specific heat per unit volume
   b. lower coefficient of thermal expansion and a lower specific heat per unit volume
   c. lower coefficient of thermal expansion and a higher specific heat per unit volume
   d. higher coefficient of thermal expansion and a higher specific heat per unit volume

2. If a WPS is being followed, it also applies to the _____.
   a. tack weld
   b. welding machine model
   c. ground lead
   d. alignment tool

3. Existing welding codes eliminate the need for _____.
   a. welder training and certification
   b. creating standards for each and every job
   c. Welding Procedure Specifications (WPS)
   d. quality control inspections

Summary

It is very important to perform proper joint fit-up and alignment to ensure an acceptable weld. The proper fit-up measuring devices and tools must be used to accomplish this. Welders must know how to compensate for welding distortion as well as how to check for poor fit-up after the weld has been completed. Even though many welders are not required to fit joints to be welded, especially pipe joints, many others are able to fit as well as they weld, adding to their value and versatility. When the welder is not responsible for fitting the joint, he or she must be familiar with the techniques and tools used to create a reliable joint that meets the specifications.

Government agencies, professional societies, and associations have written guidelines for joint fit-up and alignment. These guidelines not only help ensure quality welds, but also help ensure safe welds and welding environments. All fit-up procedures must embrace these guidelines that have been proven over the years through use and repeated testing. When specified, these codes must be followed under penalty of law.

Review Questions

1. Which tool is often fabricated on the job from small channel or angle iron?
   a. Straightedge
   b. Plumb bob
   c. Square
   d. Level

2. Before welding a 45-degree joint, check the angle of the joint with a _____.
   a. straightedge
   b. Hi-Lo gauge
   c. framing square
   d. combination square

3. The primary use of a Hi-Lo gauge is to check for _____.
   a. pipe joint misalignment
   a. I-beam joint alignment
   b. 45-degree joints
   c. 90-degree joints

4. Secure a chain hoist over a weldment to be raised with an approved sling wrapped around _____.
   a. conduit
   b. piping
   c. a structural member
   d. raceways

5. When using a strongback to hold two weldment parts that have been aligned and fitted up, the tack welds used to hold the strongback to the weldment parts should be placed on only _____.
   a. opposite sides of the strongback
   b. the notched side of the strongback
   c. the vertical sides of the strongback
   d. one side of the strongback

6. If a joint needs to be forced into alignment, yokes and wedges, or bolts, can be used with _____.
   a. strongbacks
   b. cage clamps
   c. peening
   d. a come-along

7. When working with pipe over 10" (DN250), OSHA requires _____.
   a. only specific filler metals be used
   b. a WPS
   c. chain hoists with a minimum capacity of 10 tons
   d. four-legged jack stands

8. The arched crossbar available for cage clamps is designed to _____.
   a. strengthen the clamp
   b. allow welds to be made under the clamp
   c. allow the clamp to be used on a many pipe sizes
   d. hold a section of plate to the pipe

9. Which of the following devices may *not* be allowed on some applications or projects?
   a. Strongbacks
   b. Cage clamps
   c. Chain clamps
   d. Pipe pullers

10. The stress locked in a weldment when it returns to the ambient temperature, and equal to the strength of the base metal, is called _____.
    a. base stress
    b. residual stress
    c. thermal conductivity
    d. coefficient of expansion

11. The more weld metal placed into a joint, the greater the forces of _____.
    a. stress
    b. magnetism
    c. shrinkage
    d. thermal conductivity

12. Socket-welded joints are generally found on _____.
    a. lightweight plate
    b. heavy plate
    c. pipe sizes 5" (DN125) or smaller
    d. pipe sizes larger than 6" (DN150)

13. A detailed listing of the rules and principles that apply to specific welded products is called a _____.
    a. work order
    b. procedure
    c. punch list
    d. welding code

14. The written set of instructions for producing sound welds in a prescribed manner is called a _____.
    a. contract
    b. Welding Procedure Specification (WPS)
    c. welding code
    d. PQR

15. Welding test coupons prepared for a job are tested according to the _____.
    a. applicable code
    b. WPS guidelines
    c. manufacturer's guidelines
    d. PQR

Trade Terms Quiz

Fill in the blank with the correct term that you learned from your study of this module.

1. ________ refers to the difference in the internal alignment of two pipe sections.
2. The stress remaining in a weldment as a result of heating and cooling is called ________.
3. The quantity of heat needed to raise one unit mass of a substance by one unit degree is called ________.
4. Preplaced filler metal called a(n) ________ is completely fused into the root of a joint during welding.
5. The term ________ refers to a vertical line that is perpendicular to the earth's horizontal axis.
6. A(n) ________ is extra information used to identify corrections or revisions made to documents and shared with document users.
7. Mechanically working metal with repeated hammer blows is called ________.
8. The ________ refers to the temperature of air surrounding an object or individual.
9. The term ________ refers to a line that is parallel to the earth's horizontal axis.
10. A weld made to hold weldment parts in alignment until a final weld is made is called a(n) ________.
11. ________ is the change in length, per unit length of material, for each 1°C of change in temperature.

Trade Terms

Addendum
Ambient temperature
Coefficient of thermal expansion (linear)
Consumable insert
High-low
Level
Peening
Plumb
Residual stress
Specific heat per unit volume
Tack weld

Trade Terms Introduced in This Module

Addendum: Supplementary information typically used to describe corrections or revisions to documents.

Ambient temperature: The room temperature of the atmosphere completely surrounding an object on all sides.

Coefficient of thermal expansion (linear): The change in length per unit length of material for a 1°C change in temperature.

Consumable insert: Preplaced filler metal that is completely fused into the root of the joint during welding, becoming part of the weld.

High-low: The discrepancy in the internal alignment of two sections of pipe.

Level: A line on a horizontal axis parallel with the earth's surface (horizon). Also refers to the tool used to check if an object is level.

Peening: The mechanical working of metal using repeated hammer blows, typically using a ball-shaped hammer head such as that found on a ball peen hammer. Peening may be done for both metallurgical and appearance reasons.

Plumb: A line on a vertical axis perpendicular to the earth's surface (horizon).

Residual stress: Stress remaining in a weldment as a result of heat.

Specific heat per unit volume: The quantity of heat required to raise one unit mass of a substance by one unit degree.

Tack weld: A weld made to hold parts of a weldment in proper alignment until the final weld is made.

Additional Resources

This module presents thorough resources for task training. The following resource material is suggested for further study.

2014 Technical Training Guide. Current Edition. Cleveland, OH. USA: The Lincoln Electric Company. **www.lincolnelectric.com**

Welding Handbook. Current Edition. Miami, FL: The American Welding Society.

Mathey Dearman Product Manuals. The Mathey Dearman Company. **www.matheydearman.com**

Figure Credits

Topaz Publications, Inc., Module Opener, Figures 7 (photo), 8, 24 (photo)

Courtesy of Mathey Dearman, Figures 1A, 14–16, 18, 19

Courtesy of The LS Starrett Company, Figure 1B

RIDGID®, Figure 2A

© Stanley Black & Decker, Figure 2B

G.A.L. Gage Co, Figures 3 (photo), 5 (photo)

Courtesy of Sumner Manufacturing Co., Figures 13, 17B

Courtesy of H & M Pipe Beveling Machine Company, Inc., Figure 17A

Section Review Answer Key

Answer	Section Reference	Objective
Section One		
1. c	1.1.3	1a
2. a	1.2.2	1b
3. b	1.3.2	1c
4. c	1.4.4	1d
Section Two		
1. d	2.1.0	2a
2. a	2.2.2	2b
3. b	2.3.0	2c

NCCER CURRICULA — USER UPDATE

NCCER makes every effort to keep its textbooks up-to-date and free of technical errors. We appreciate your help in this process. If you find an error, a typographical mistake, or an inaccuracy in NCCER's curricula, please fill out this form (or a photocopy), or complete the online form at **www.nccer.org/olf**. Be sure to include the exact module ID number, page number, a detailed description, and your recommended correction. Your input will be brought to the attention of the Authoring Team. Thank you for your assistance.

Instructors – If you have an idea for improving this textbook, or have found that additional materials were necessary to teach this module effectively, please let us know so that we may present your suggestions to the Authoring Team.

NCCER Product Development and Revision

13614 Progress Blvd., Alachua, FL 32615

Email: curriculum@nccer.org
Online: www.nccer.org/olf

❑ Trainee Guide ❑ Lesson Plans ❑ Exam ❑ PowerPoints Other ______________

Craft / Level: Copyright Date:

Module ID Number / Title:

Section Number(s):

Description:

Recommended Correction:

Your Name:

Address:

Email: Phone:

Welding Symbols

Overview

Project drawings and specifications contain numerous symbols that communicate critical information about the welds to be used at various locations throughout the project. Welders must learn this symbolic language so they can properly interpret the symbols and make welds that meet design specifications. This module introduces a wide range of welding symbols, describes how they are structured, and explains the basic rules for applying the information that the symbols convey.

Module 29201

Trainees with successful module completions may be eligible for credentialing through the NCCER Registry. To learn more, go to **www.nccer.org** or contact us at **1.888.622.3720**. Our website has information on the latest product releases and training, as well as online versions of our *Cornerstone* magazine and Pearson's product catalog.

Your feedback is welcome. You may email your comments to **curriculum@nccer.org,** send general comments and inquiries to **info@nccer.org**, or fill in the User Update form at the back of this module.

This information is general in nature and intended for training purposes only. Actual performance of activities described in this manual requires compliance with all applicable operating, service, maintenance, and safety procedures under the direction of qualified personnel. References in this manual to patented or proprietary devices do not constitute a recommendation of their use.

29201 V5

29201
Welding Symbols

Objectives

When you have completed this module, you will be able to do the following:

1. Identify and interpret welding symbols and their structure.
   a. Describe the structure and placement of welding symbols and identify basic symbols.
   b. Identify and interpret size and dimension markings for common types of welds.
   c. Identify and interpret various supplemental symbols.
   d. Identify and interpret less common welding symbols.

Performance Task

Under the supervision of your instructor, you should be able to do the following:

1. Identify and interpret welding symbols on an instructor-provided drawing.

Trade Terms

Arrow line
Countersink
Pitch
Reference line
Weld symbol
Welding symbol

Industry Recognized Credentials

If you are training through an NCCER-accredited sponsor, you may be eligible for credentials from NCCER's Registry. The ID number for this module is 29201. Note that this module may have been used in other NCCER curricula and may apply to other level completions. Contact NCCER's Registry at 888.622.3720 or go to **www.nccer.org** for more information.

Contents

Topics to be presented in this module include:

Figures and Tables

SECTION ONE

1.0.0 WELDING SYMBOLS

Objective

Identify and interpret welding symbols and their structure.

a. Describe the structure and placement of welding symbols and identify basic symbols.
b. Identify and interpret size and dimension markings for common types of welds.
c. Identify and interpret various supplemental symbols.
d. Identify and interpret less common welding symbols.

Performance Task

1. Identify and interpret welding symbols on an instructor-provided drawing.

Trade Terms

Arrow line: The line drawn at an angle from the reference line (either end or both ends) to an arrowhead at the location of the weld.

Countersink: A hole with tapered sides and a wider opening that allows a flat-head fastener to seat flush to the surface of the material in which the hole is made.

Pitch: The center-to-center distance between welds.

Reference line: The horizontal line in the center of the welding symbol from which all elements of the welding symbol are referenced. The reference line is one of the most important elements of the welding symbol.

Weld symbol: A graphic character connected to the reference line of a welding symbol specifying the weld type.

Welding symbol: A graphical representation of the specifications for producing a welded joint; includes a reference line and arrow line and many also include a weld symbol.

Welding symbols are used on drawings, project specifications, and welding procedure specifications (WPS) to convey the design specifications for welds. A series of symbols is used to indicate the joint configuration and weld type, location, size, and length of weld required. Welders must be able to properly interpret welding symbols to ensure that the welds they make will meet the design specifications.

Most companies use the symbols that have been standardized by the American Welding Society (AWS) in *AWS A2.4, Standard Symbols for Welding, Brazing, and Nondestructive Examination.* Most companies also have a site quality standard that provides guidelines and examples of how welding symbols are to be used. Always refer to the site-specific quality standard when interpreting welding symbols.

1.1.0 Welding Symbol Construction

The base for all welding symbols is the horizontal reference line which has an arrow at one end. The arrow line, which can be on either side of the reference line, points to the location to which the welding symbol applies (*Figure 1*). The welding symbol describes the type of weld, its size, and its surface finish.

The opposite end of the reference line, called the tail, is used for information that aids in making the weld but does not have its own special place on the symbol (*Figure 2*). For instance, the tail may be used to indicate the welding and cutting processes, the reference to a note, or the welding procedure or electrode type to be used. When a reference is not required, the tail is omitted, as shown in *Figure 1.*

In the case of a T-joint, the arrow points to the two sides of the joint, and welding may be accomplished on either side. The welding symbol distinguishes between the two sides of a joint by using the arrow line and the spaces above and below the reference line. The side of the joint to which the arrow points is called the arrow side, and its weld is made in accordance with the instructions below the reference line. The opposite side of the joint is referred to as the other side,

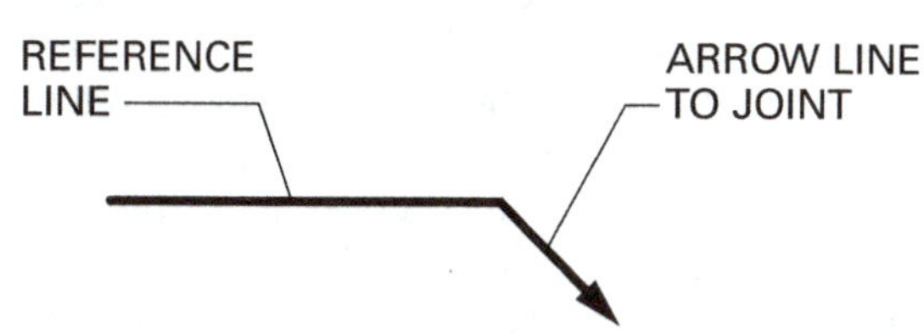

Figure 1 Horizontal reference line.

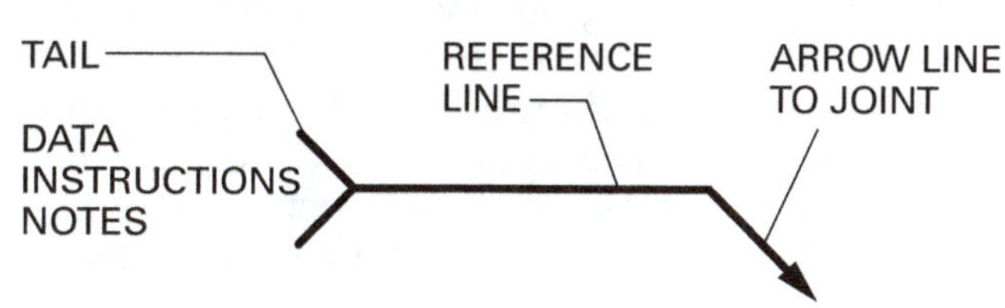

Figure 2 Tail-side information

Welding and Cutting Processes

The *Appendixes* in this module contain tables listing AWS abbreviations for various welding and cutting processes. Abbreviations from all three tables can be used in tail references, or in single-line or multiple-line welding symbols.

and its weld is made in accordance with the instructions found above the reference line. *Figure 3* shows examples of these symbols.

Regardless of which end of the reference line the arrow line is on, information on the reference line is always read from left to right. *Figure 4* shows a welding symbol base and the numerous elements of a welding symbol.

The weld symbols used to identify the type of weld to be made signify a basic type of weld or joint preparation (*Figure 5*). It is important to note the difference between welding symbols and weld symbols. A weld symbol is a graphic character connected to the reference line of a welding symbol, specifying the weld type. A welding symbol graphically represents the specifications for producing a welded joint; it includes any weld symbols as well as the additional components. All welding symbols include a reference line and arrow line, at a minimum. Many also include a weld symbol.

1.1.1 Symbols for Welds

The type of weld used is determined by the way the members to be welded are positioned to form the joint. The five basic types of weld joints are the butt, corner, tee, lap, and edge joints (*Figure 6*).

1.1.2 Location of Weld Symbols

When the weld symbol is on both the top and bottom of the reference line, the information applies to both sides of the joint or surface to which the arrow is pointing. Symbols appear on both sides of the reference line when a double-fillet or double-groove weld is specified. *Figure 7* shows symbols for arrow-side, other-side, and both sides for fillet and V-groove welds.

The shape of the groove weld symbol indicates how the groove is to be prepared. In the case of the bevel-groove and J-groove weld symbols, only one of the two members to be welded is prepared. The arrow indicates which member is to be prepared by breaking toward that member (*Figure 8*).

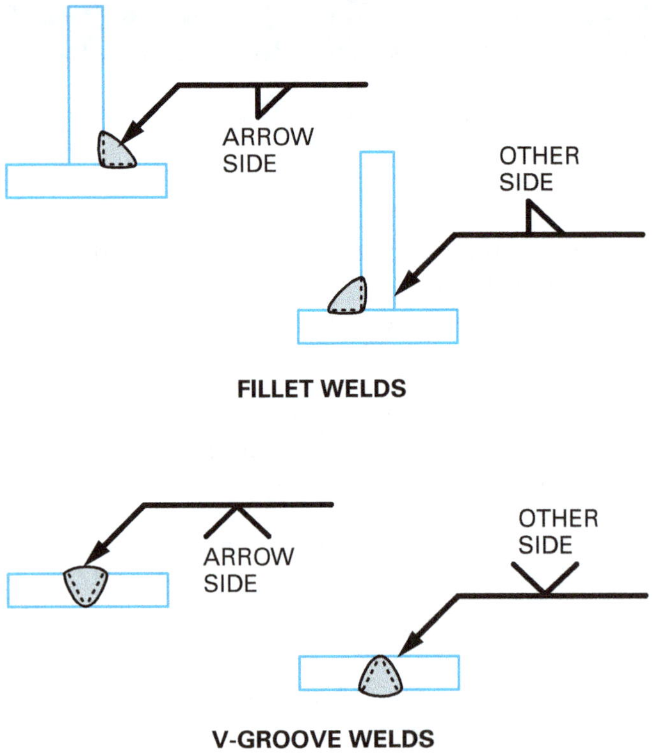

Figure 3 Arrow-side and other-side significance.

Notice also that the vertical leg of the weld symbol is shown drawn to the left of the slanted/curved leg(s). Regardless of whether the symbol is for a fillet, bevel-groove, J-groove, or flare bevel-groove weld, the vertical leg is always drawn to the left.

1.1.3 Combining Weld Symbols

When more than one type of weld is to be made on the same joint, it is necessary to combine weld symbols (*Figure 9*). A typical combination weld is made by adding a fillet weld to a single or double bevel groove weld in a T-joint. The fillet weld symbol is always placed on top of the groove weld symbol, just as it would be on the actual weld.

1.1.4 Multiple Reference Lines

When more than one welding operation must be performed in a certain sequence, two or more reference lines may be used to indicate the sequence.

Think About It

Breaking the Arrow

Are the symbol representations shown in *Figure 8* the only ways a breaking arrow can be shown?

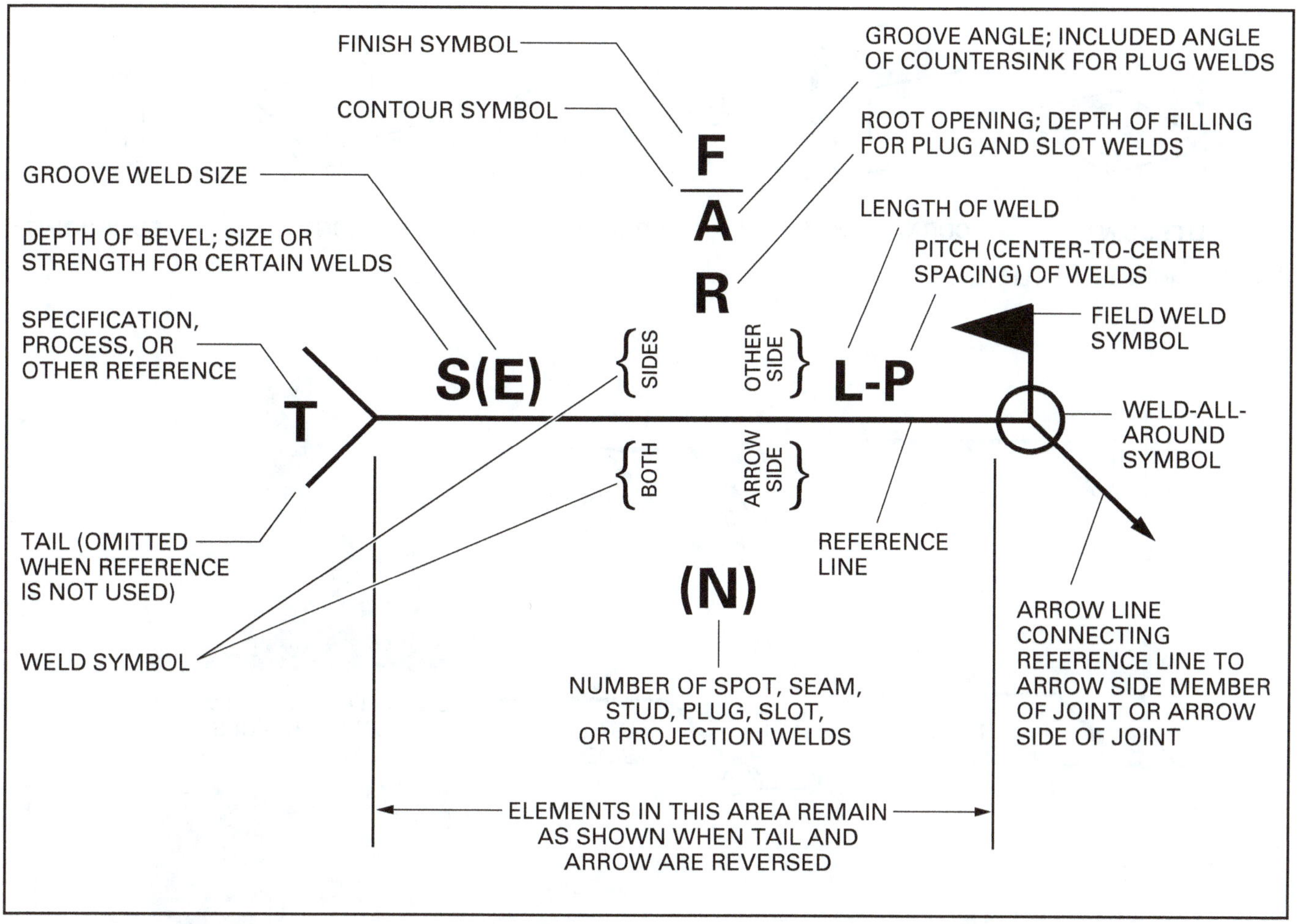

Figure 4 Standard location of elements of a welding symbol.

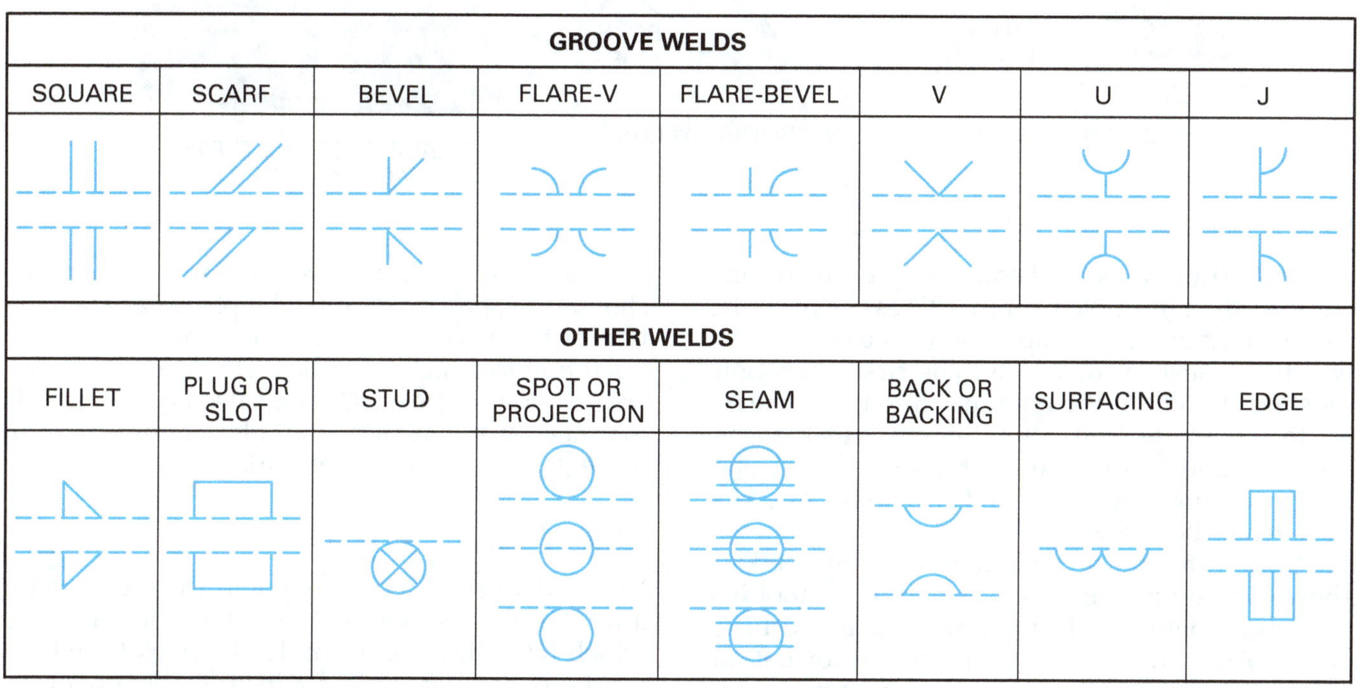

Figure 5 Basic weld symbols.

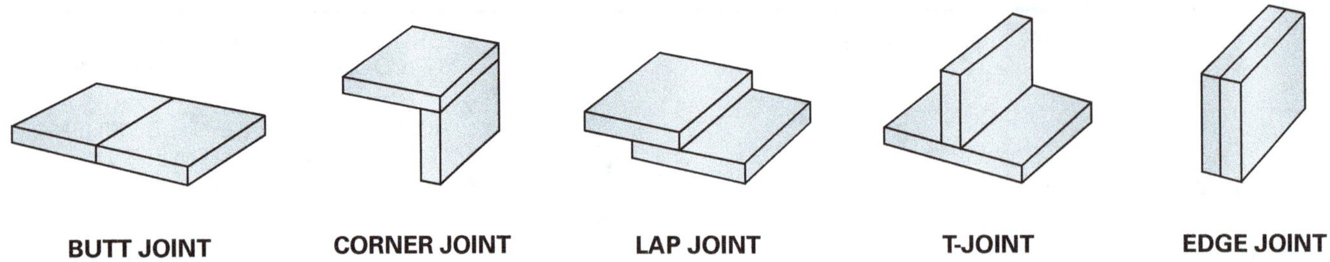

Figure 6 Basic types of weld joints.

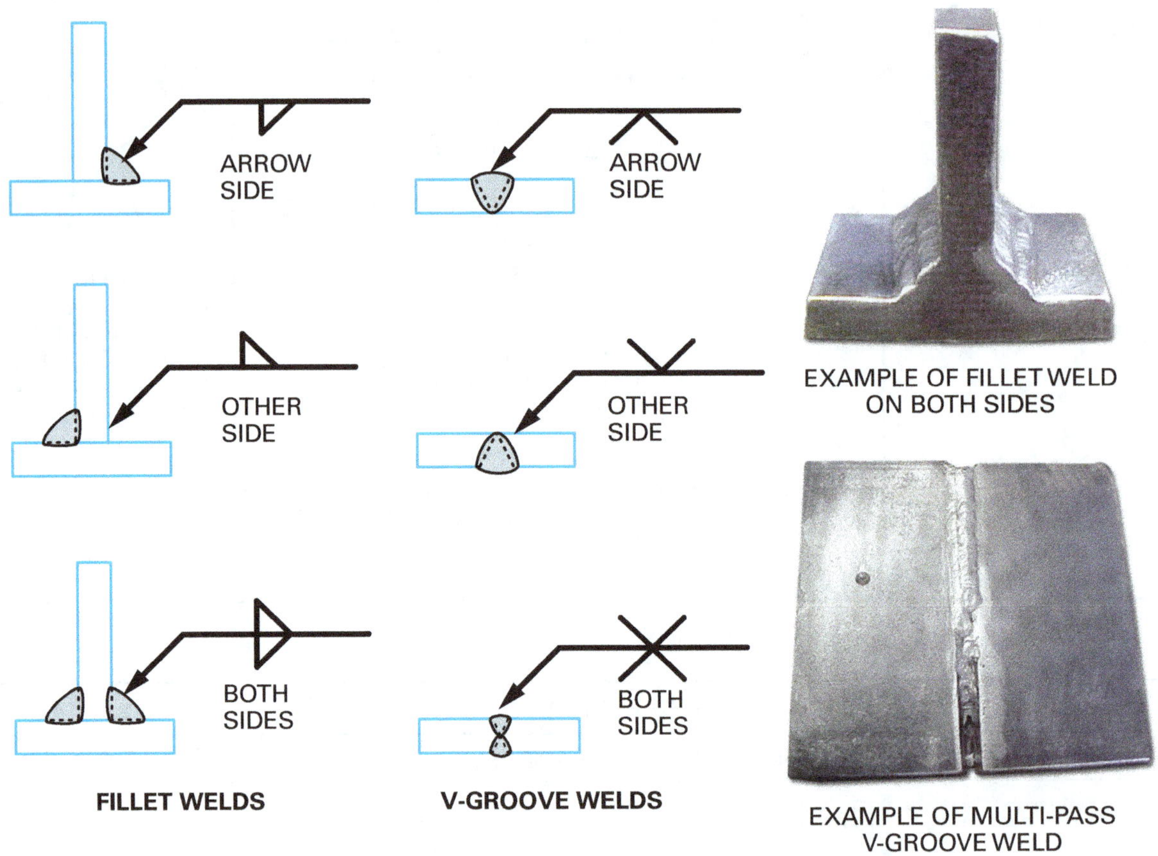

Figure 7 Fillet and V-groove welds.

Figure 10 illustrates the typical sequences of operation when multiple reference lines are used. In each sequence, the first operation to be performed will be closest to the arrow. The first operation must be shown on the reference line that is nearest to the arrow. Subsequent operations may be shown sequentially on other reference lines. The last operation to be performed will be the one farthest from the arrow.

Multiple reference lines may also be used to show data supplementary to welding symbol information that is on the reference line nearest the arrow. As shown in *Figure 11,* test information may be shown on a second or third line, away from the arrow.

When required, the weld-all-around symbol, shown in *Figure 12(A),* must be placed at the junction of the arrow line and reference line for each operation to which it applies. The field weld symbol shown in *Figure 12(B)* may also be used in this manner. Supplemental symbols are discussed in more detail later in this module.

1.2.0 Sizing and Dimensioning Welds

Unless defined in a drawing note, the size data for a weld is always shown to the left of the symbol to which it applies. For example, the size of the fillet weld shown in *Figure 13* is ¼ inch. The length of the weld is always shown to the right of the symbol to which the length applies. An example of this is seen in *Figure 14,* where the length is 6 inches.

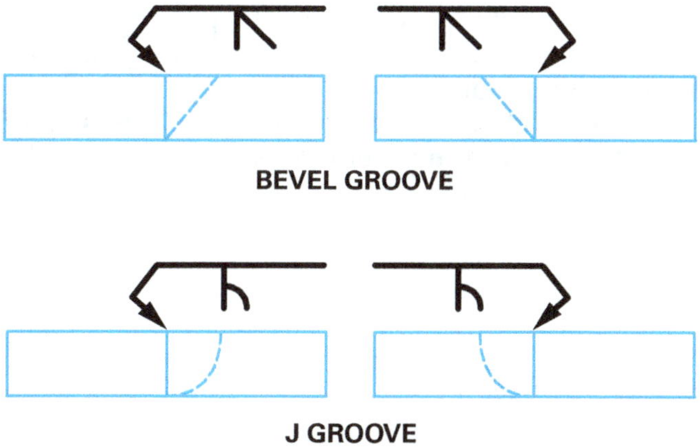

NOTE: DASHED LINES INDICATE PREPARATION SPECIFIED BY WELD SYMBOL

Figure 8 Bevel and J-groove weld symbols.

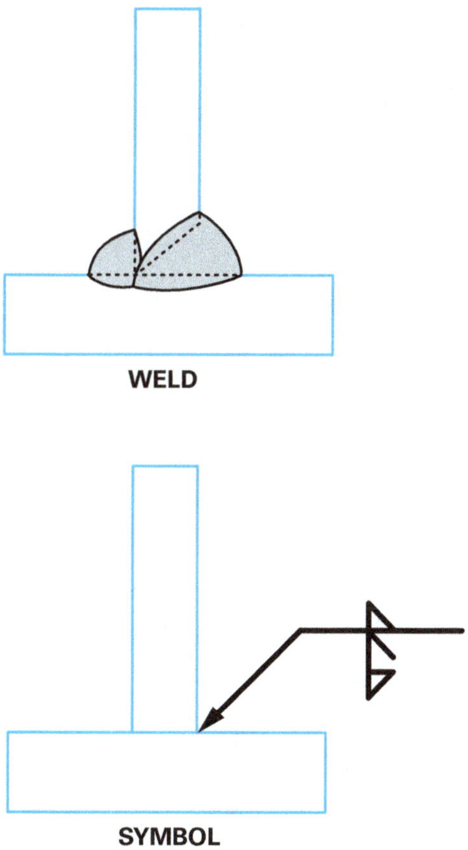

Figure 9 Combination weld symbol.

Dimensions can be given as fractions, decimals, or metric measurements. The following sections explain how to size and dimension the following weld types:

- Fillet
- Groove
- Plug
- Slot

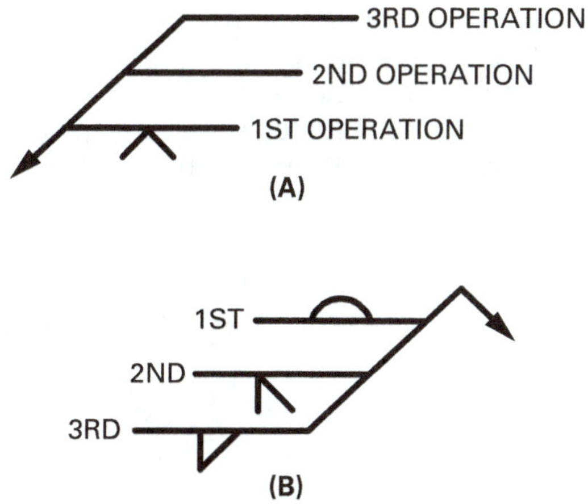

Figure 10 Multiple reference lines.

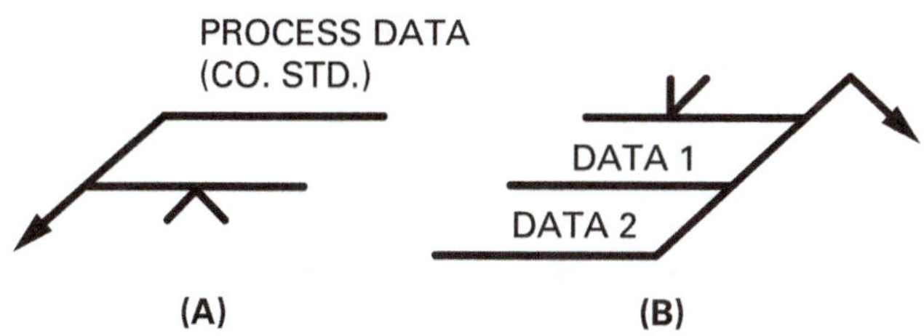

Figure 11 Supplementary data.

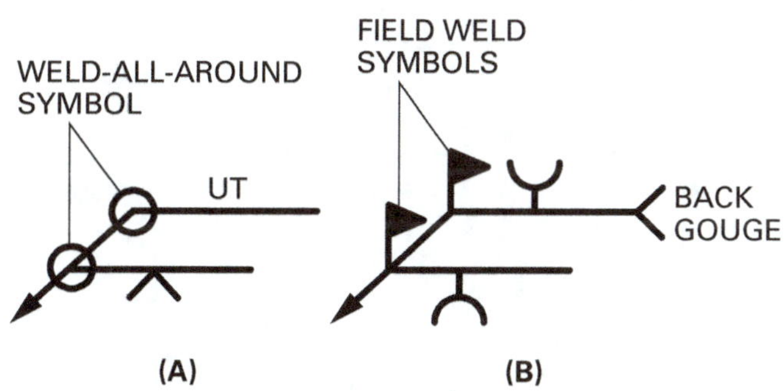

Figure 12 Typical placement of supplemental symbols.

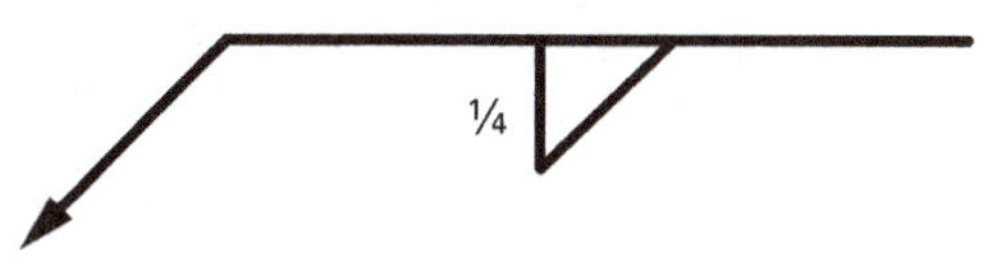

Figure 13 Sizing welds

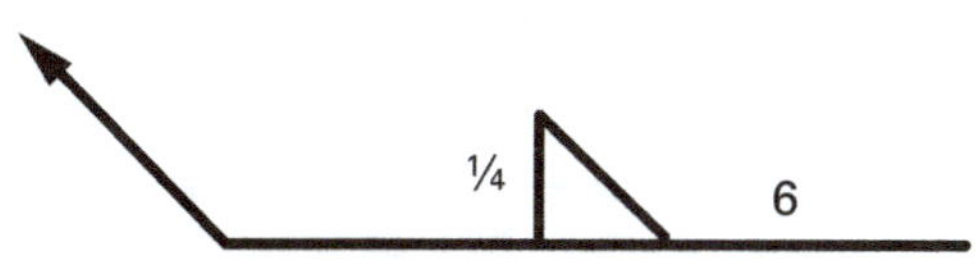

Figure 14 Dimensioning welds.

1.2.1 Sizing Fillet Welds

A fillet weld is a weld made in a corner. It is roughly triangular in shape and joins two welding surfaces that are perpendicular to each other, such as in a lap, T-, or corner joint. The size of a fillet weld is the length of the legs (sides) of the largest triangle that can be inscribed within the fillet weld cross-section. *Figure 15* shows the profiles of convex and concave welds.

The lengths of the legs on a typical fillet weld are equal, so only one dimension is given. Equal-leg fillet welds are sized by placing the size to the left of the fillet weld symbol or as defined in a drawing note.

If the fillet weld has unequal legs, the size of each leg is given in parentheses or brackets to the

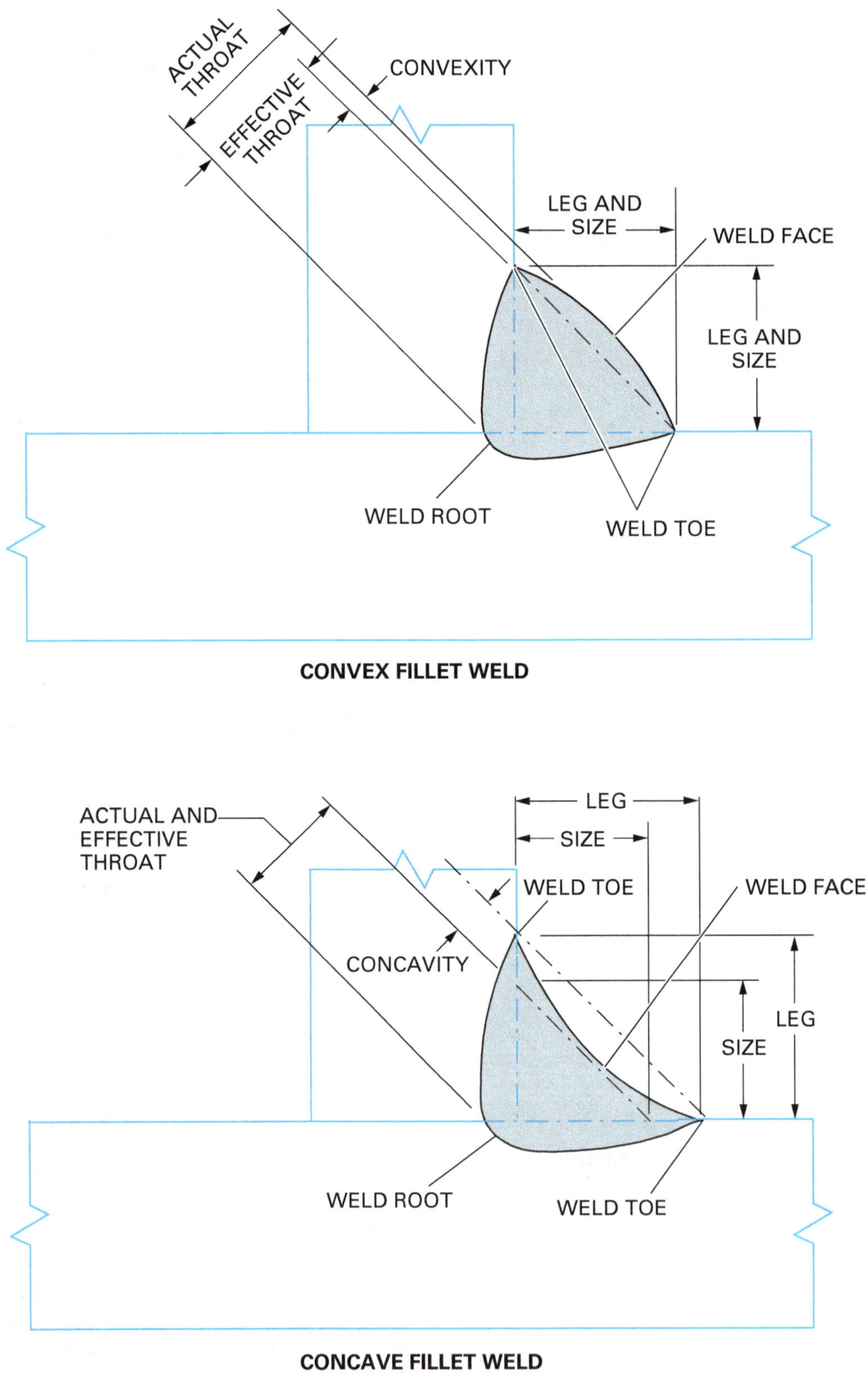

Figure 15 Profiles of fillet welds.

Nondestructive Examination Symbols

Nondestructive examination symbols may appear on the same drawings as (and look very similar to) welding symbols. Inspectors use nondestructive examination symbols to examine joints once welding is complete. The nondestructive methods of inspection are defined by letter codes on the reference line. For instance, VT indicates visual testing, UT indicates ultrasonic testing, RT indicates radiographic testing, and PT indicates penetrant testing.

NDT symbols and weld symbols may be combined by using two reference lines. In contrast to the arrow for welding symbols, the arrow for NDT always points to the weld that is to be tested, and never to a weld on the other side of the joint. If a double fillet weld were to be tested, for example, this would be shown with a separate arrow on each side of the joint.

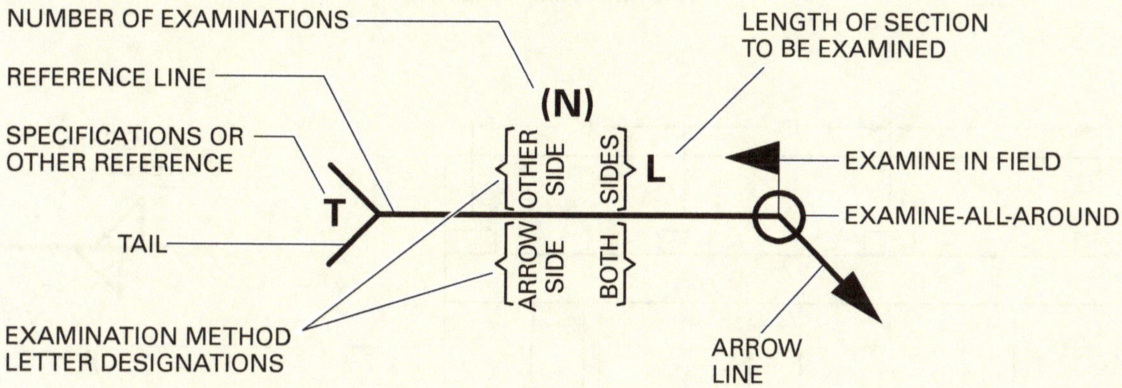

left of the fillet weld symbol with a drawing or explanation of leg orientation. An example of this is shown in *Figure 16* where the unequal legs are ¼ inch (6.4 mm) and ½ inch (12.7 mm).

1.2.2 Dimensioning Fillet Welds

The length of a fillet weld is shown to the right of the fillet weld symbol. If the fillet weld is to be an intermittent fillet weld, the length (length of each segment) and the pitch (center-to-center distance between welds) are shown with a dash. The dash appears between the length, which is always shown first, and the pitch, which is always shown second. *Figure 17* shows a length of 2 inches (5.1 cm) and a pitch of 6 inches (15.2 cm). If fillet welds are required on both sides of the joint, they can be back-to-back or staggered. If the welds are back-to-back, as shown in *Figure 17(A)*, the fillet weld symbols are aligned evenly on both sides of the reference line, as shown in *Figure 17(B)*. If the welds are to be staggered, as shown in *Figure 17(C)*, the fillet weld symbols above and below the reference line are offset, as shown in *Figure 17(D)*.

Unequal Fillet Weld Legs

The only practical use of unequal fillet weld legs is on metal of unequal thickness. Unequal legs do not make the weld stronger; the thinner material prevails. An unequal fillet leg weld helps distribute loads and eliminates incomplete fusion discontinuities that would be caused by making a small weld on heavy material without preheat.

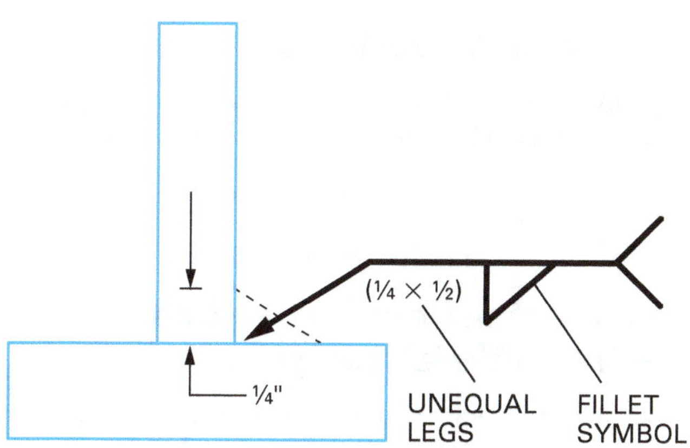

Figure 16 Sizing fillet welds.

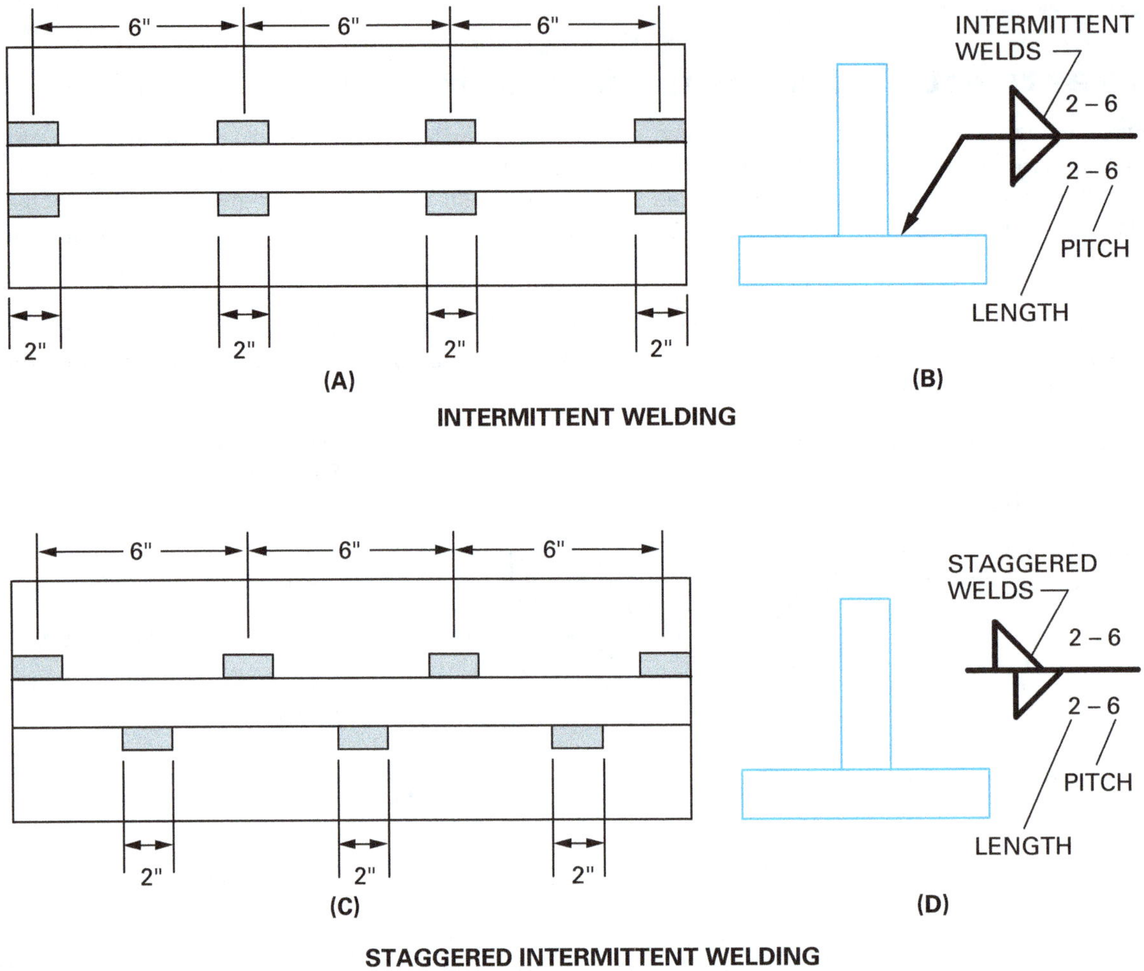

INCHES	CENTIMETERS
2	5.1
6	15.2

Figure 17 Dimensioning fillet welds.

When there is no dimension to the right of the fillet weld symbol, the weld extends the full length of the joint.

1.2.3 Sizing Groove Welds

A groove weld is a weld that is made in the groove formed by two beveled pieces of metal (*Figure 18*).

Think About It

Location of Unequal Fillet Weld Legs

How does a welder determine which side of a fillet weld has the longer leg?

Several factors must be considered when sizing groove welds. These factors apply to all groove weld types.

Groove preparation is the depth to which the groove extends into the base metal. When the groove weld extends all the way through the joint, as shown in *Figure 19(A)*, no preparation size needs to be shown. When the groove weld extends only part of the way through the joint, as shown in *Figure 19(B)*, the depth of the groove is shown to the left of the symbol. *Figure 19(B)* shows sizing a groove with a depth of ¼ inch (6.4 mm).

For partial joint penetration (PJP) groove welds, both the depth of the groove and weld size (joint penetration) are shown, as in *Figure 20*. The depth of the groove is shown first and the weld size is shown next, in parentheses. In

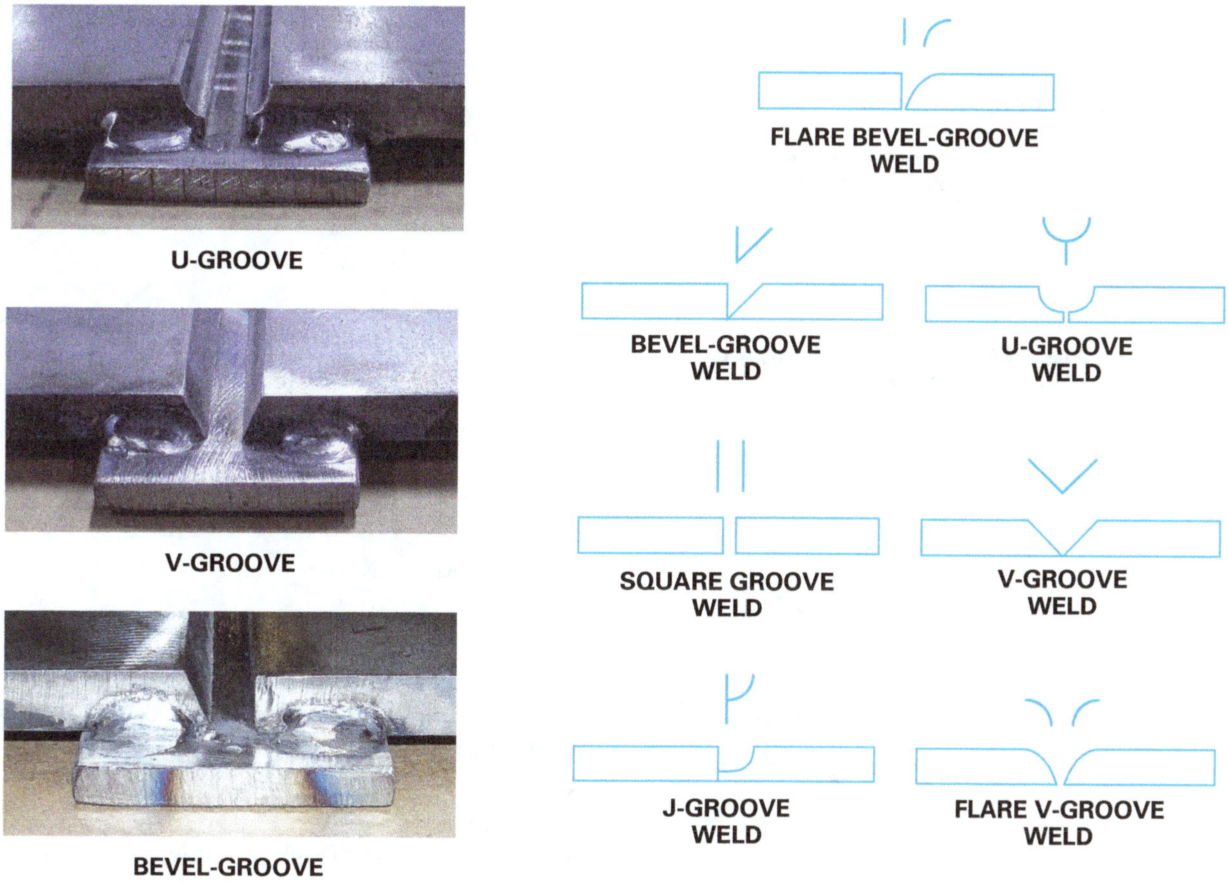

Figure 18 Types of groove welds.

this example, the groove depth is 0.25 inch (6.4 mm) and the weld size is 0.385 inch (9.8 mm). For complete joint penetration (CJP) groove welds, the weld metal extends through the joint thickness.

If a root opening is required, the root opening dimension of a groove weld will be shown inside the symbol. In *Figure 21*, the root opening is $\frac{1}{8}$ inch (3.6 mm) and is shown on only the arrow-side symbol because the opening is common to both grooves.

The angle for the type of groove is shown above or below the welding symbol, depending on whether the symbol is above or below the reference line. In *Figure 22*, the groove angle is shown as 45 degrees and is shown only by the arrow-side symbol because the groove is the same on both sides. The bevel angle for groove preparation on each workpiece is $22\frac{1}{2}$ degrees.

1.2.4 Dimensioning Groove Welds

The length of a groove weld is shown to the right of the groove-weld symbol. If there is no dimension on the right of the groove-weld symbol, the weld extends the full length of the joint.

1.2.5 Sizing and Dimensioning Plug Welds

A plug weld is a weld made in a circular hole in one member of a joint, fusing that member to another member. The size of a plug weld, which is the diameter of the plug, is shown to the left of the welding symbol. The number of plug (and slot) welds can be shown below the symbol. If the plug is not completely filled, the depth of the fill is given inside the plug weld symbol, as shown in *Figure 23(A)*. If the plug weld is to be countersunk, the angle of the countersink is shown below or above the weld symbol, depending on which side of the reference line the symbol is on, as shown in *Figure 23(B)*. The pitch, shown in *Figure 23(C)*, represents the center-to-center spacing of the plugs shown in *Figure 23(B)* to the right of the plug weld symbol. *Figure 23* shows a composite view of sizing and dimensioning plug welds.

1.2.6 Sizing and Dimensioning Slot Welds

The only time depth of fill is used with a slot weld symbol is when the slots are partially filled. The partial depth of fill is shown in the symbol (*Figure 24*). All other information, such as length, width, spacing, angle of countersink, and location of slots, is shown in a special detail on the print.

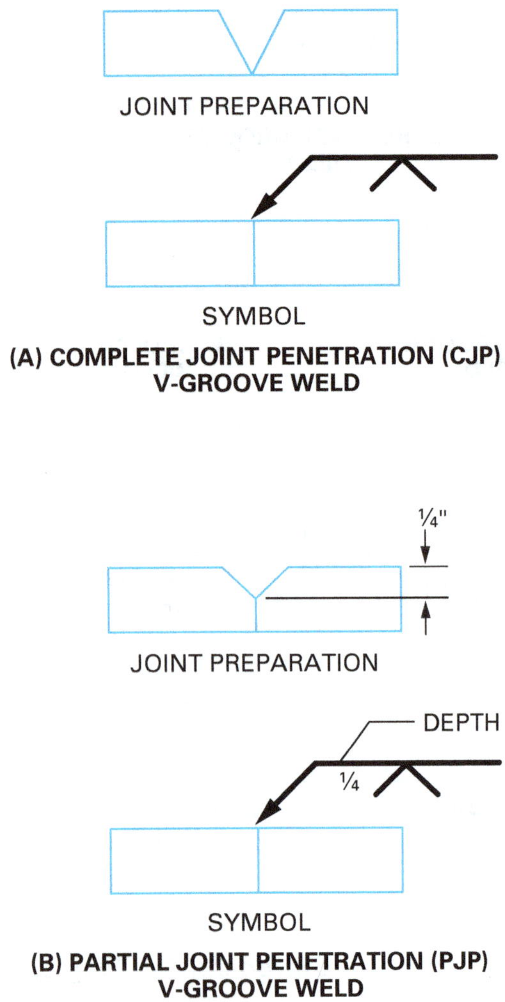

Figure 19 Sizing groove welds.

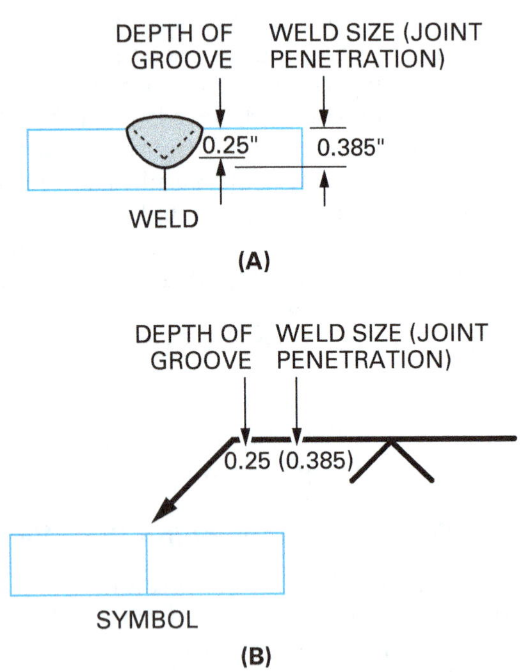

Figure 20 Sizing partial joint preparation (PJP) welds.

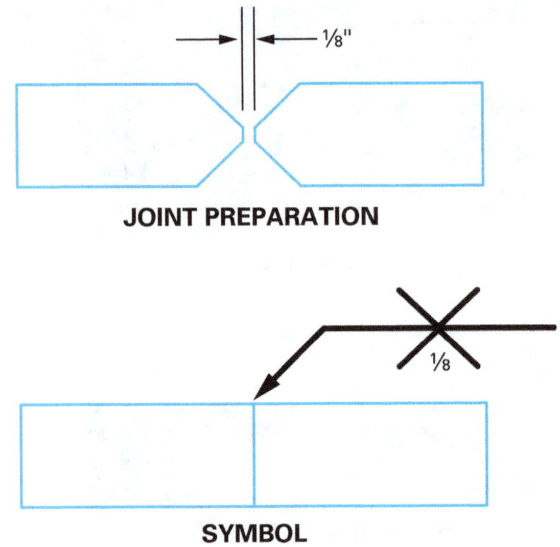

Figure 21 Sizing the root opening.

1.3.0 Supplemental Symbols

The following supplemental symbols are used to convey special instructions that apply to the welding symbol:

- Weld-all-around
- Field weld
- Contour finish (flush, convex, concave)

Figure 25 shows these and other supplemental symbols.

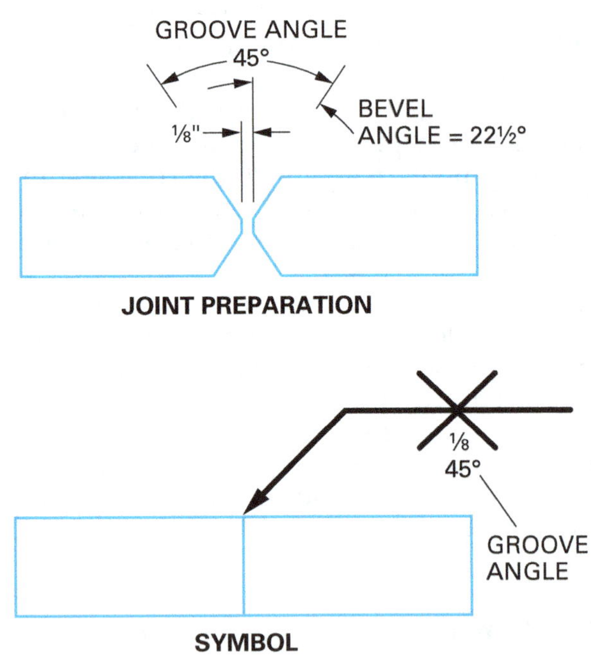

Figure 22 Specifying the groove angle.

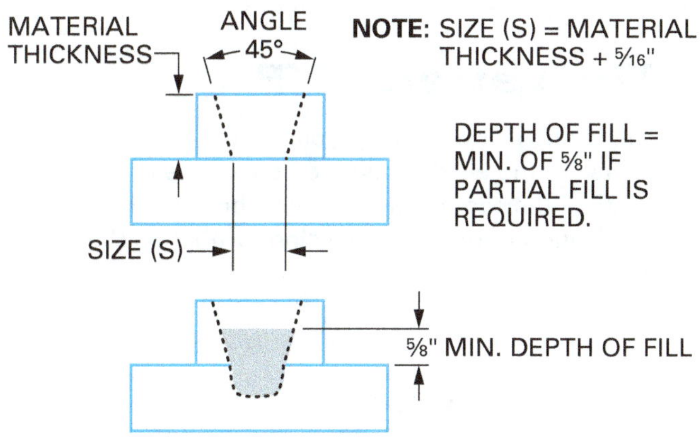

(A)

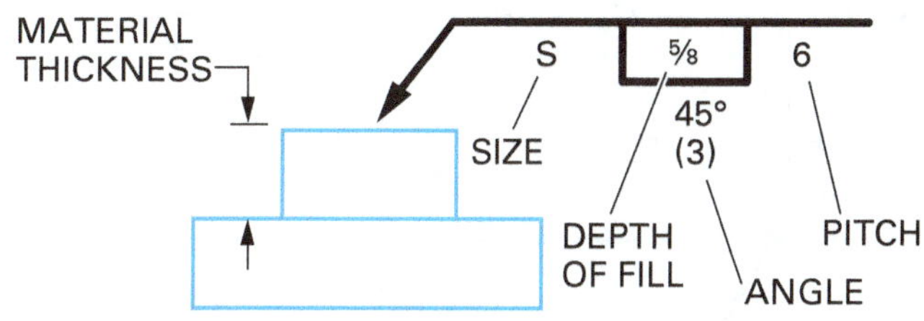

(B)

TYPICAL PLUG WELD

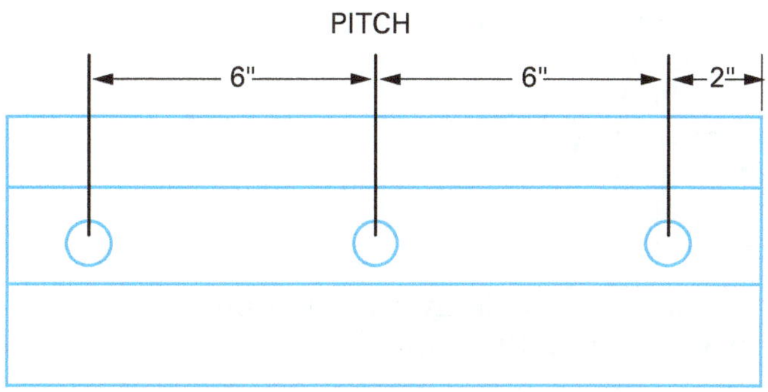

(C)

INCHES	MILLIMETERS
5/16	7.9
5/8	15.9
2	50.8
6	152.4

Figure 23 Sizing and dimensioning plug welds.

1.3.1 Weld-All-Around

When a weld is to extend completely around the joint, such as when welding a channel to a base plate, a small circle is shown where the arrow line meets the reference line. *Figures 25* and *26* both show examples of the symbol placement. *Figure 26* also shows the intended result of a typical weld-all-around symbol.

1.3.2 Field Weld

When the weld indicated by the welding symbol is to be made at a location other than that of the initial construction, a small flag is placed where the arrow line meets the reference line. The flag can be pointing either left or right. It can also point either up and down. *Figure 27* shows two examples of field welding symbols.

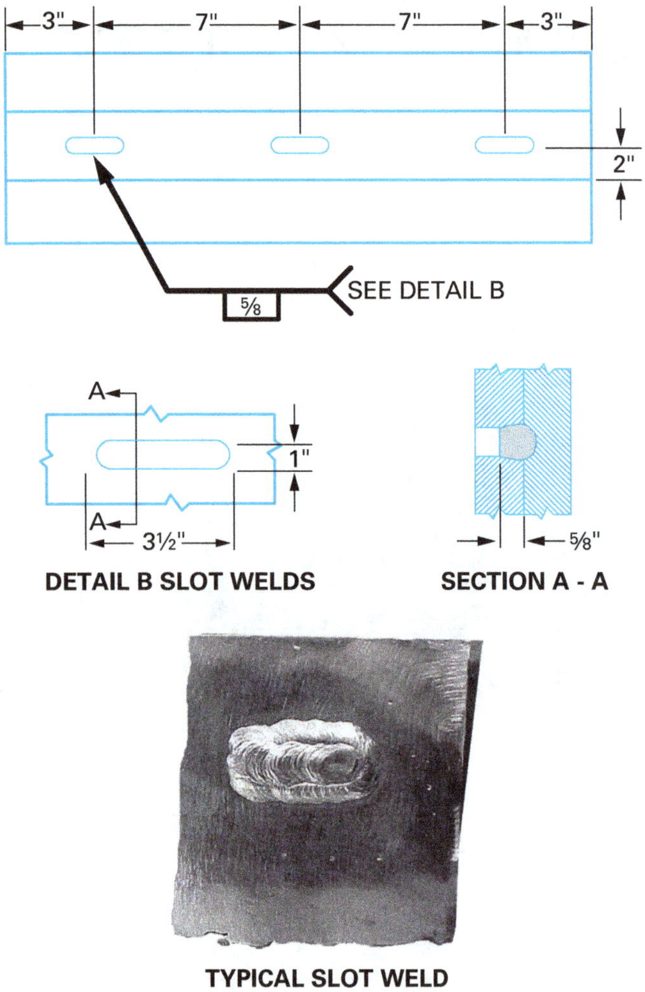

INCHES	MILLIMETERS
5/8	15.9
1	25.4
2	50.8
3	76.2

Figure 24 Sizing slot welds.

1.3.3 Contour Finish

When the face of the weld must have a finished shape that is not its normal as-welded condition, a finish symbol that indicates a flush, convex, or concave surface is placed adjacent to the welding symbol. The method used to finish the weld face is specified by a letter, such as M for machining (*Figure 28*). *Table 1* shows the related finishing methods with their letter symbols.

The Letters TYP

The letters TYP (meaning typical) are placed in the tail of the welding symbol to indicate multiple joints of the same configuration. This eliminates duplication of the same welding symbol on the drawing.

1.4.0 Other Symbols

Other, less common weld symbols are shown in *Figure 29*. They include the following:

- Backing and spacer
- Back or backing
- Melt-through
- Surfacing
- Edge
- Spot
- Seam

If a consumable insert is required, details are provided with the symbol.

1.4.1 Backing and Spacer Symbols

Backing is used at the root of the weld to control melt-through. The backing can be metal strips or shapes, flux-coated tape, fiberglass-coated tape, or ceramic tape. The backing symbol is placed opposite a groove weld symbol on the reference line. The material, size, and shape of the backing are specified in the tail of the welding symbol. Spacers are metal strips or shapes that are placed at the root of the joint to control burn-through. They are used most often on double-groove joint preparations. The spacer symbol is centered on the reference line. The joint weld symbol is modified to project from the corners of the spacer symbol. The material and size of the spacer are noted in the tail of the welding symbol. *Figure 30* includes illustrations of typical backing and spacer symbols.

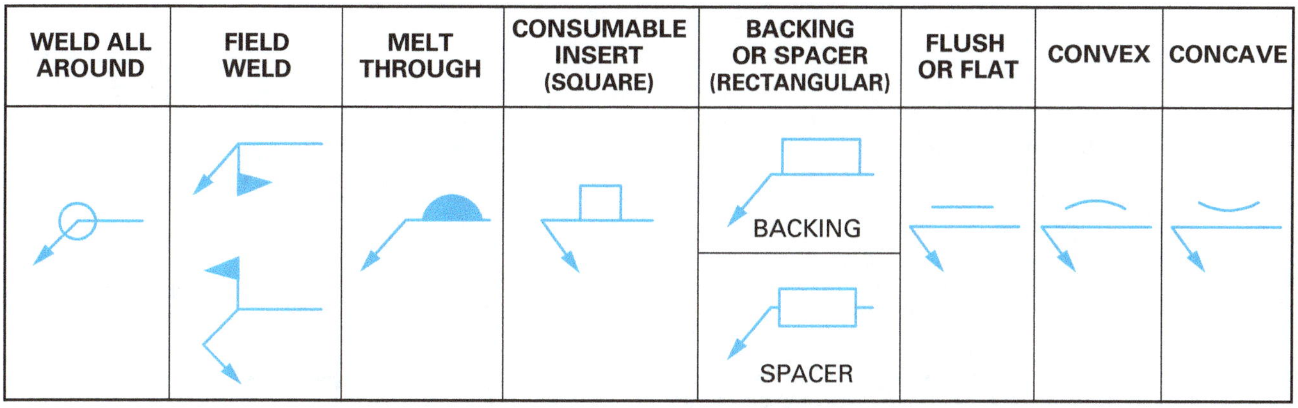

INCHES	MILLIMETERS
3 ½	88.9
7	177.8

Figure 25 Supplemental symbols.

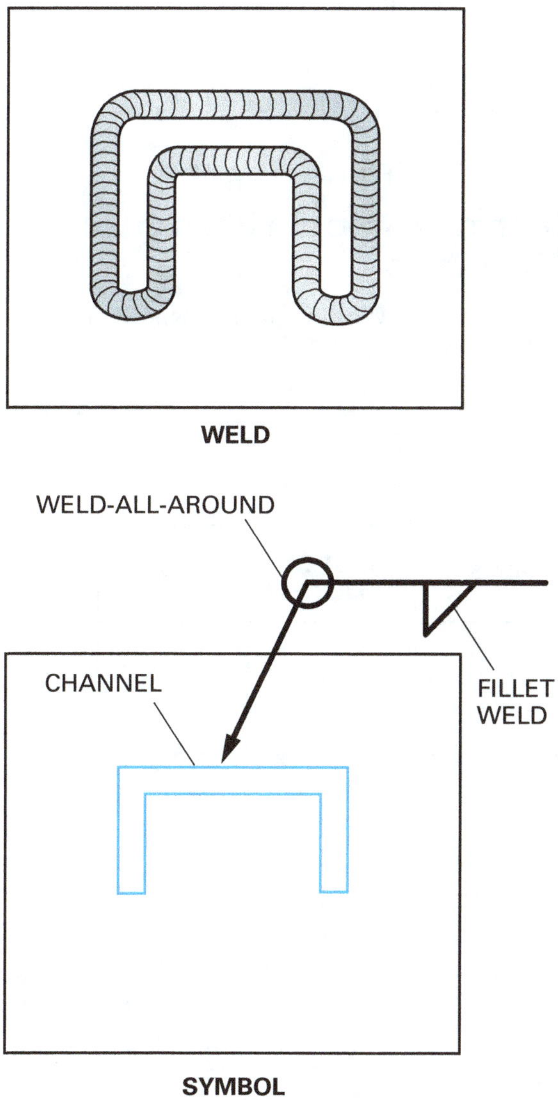

Figure 26 Weld-all-around symbol.

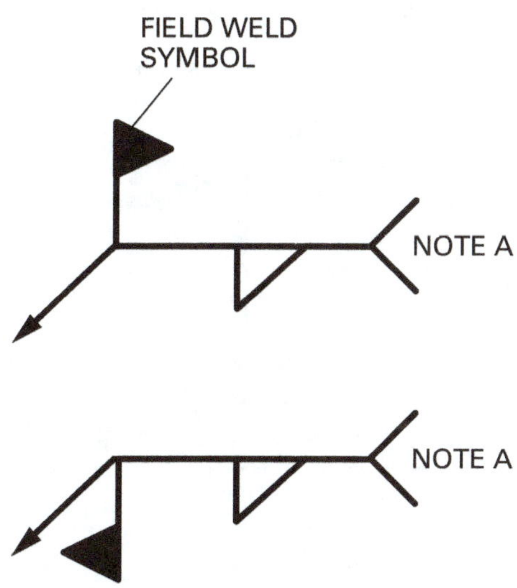

Figure 27 Field welding symbols.

1.4.2 Back or Backing Weld Symbols

A back weld is made after the groove weld. A backing weld is made before the groove weld. The same symbol is used to indicate either a back or a backing weld (*Figure 31*). The back or the backing weld symbol is placed opposite a groove weld symbol on the reference line. A note in the tail of the welding symbol indicates which weld is to be made.

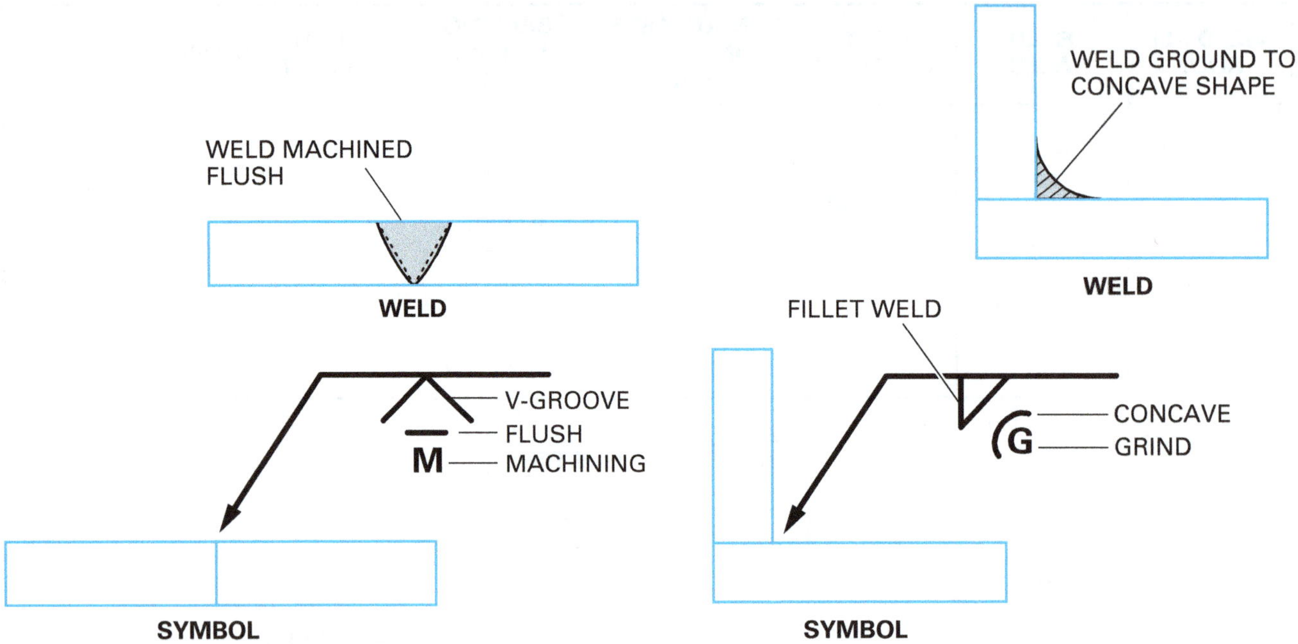

Figure 28 Examples of contour finishes and symbols.

Table 1 Finishing Methods

Method	Letter
Chipping	C
Grinding	G
Hammering	H
Machining	M
Rolling	R
Unspecified	U

Think About It

Unspecified Contour Finishes

Refer to *Table 1.* What might an unspecified (U) contour finish mean?

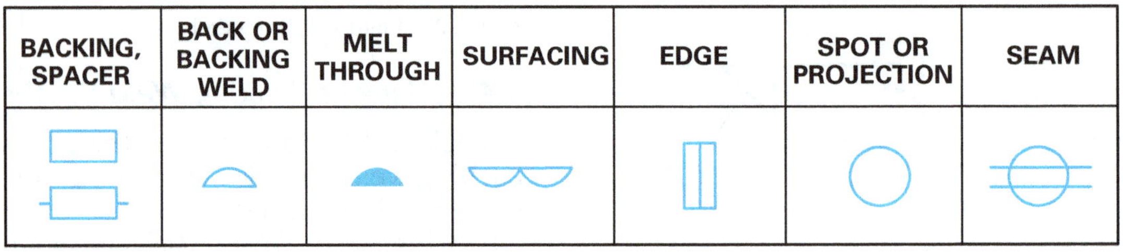

Figure 29 Other symbols.

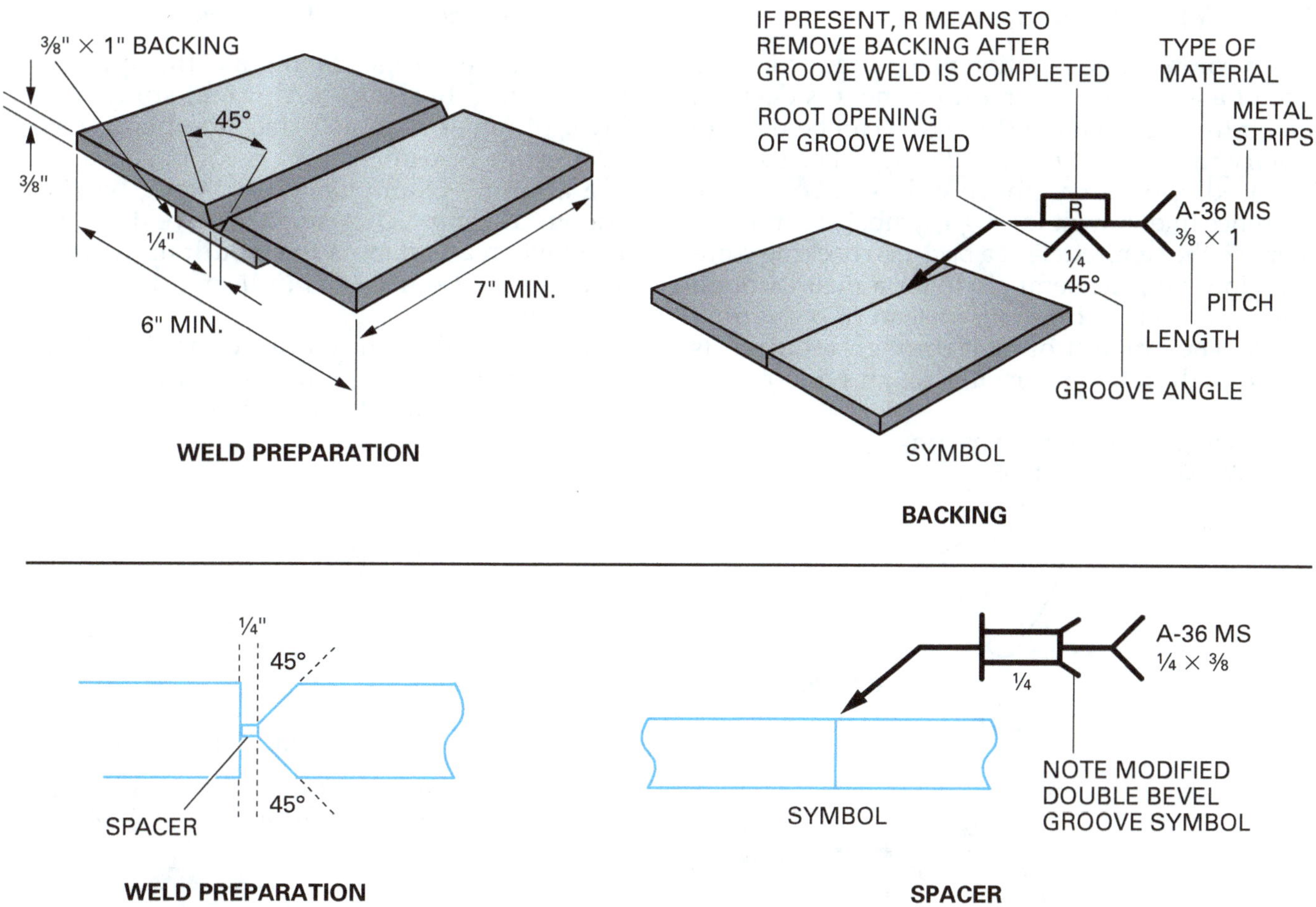

INCHES	MILLIMETERS
1/4	6.4
3/8	9.5
1	25.4
6	152.4
7	177.8

Figure 30 Backing and spacer symbols.

An Alternate Welding Symbol

The welding symbol shown in *Figure 31* could also be presented as a multiple line symbol. Using a multiple line symbol eliminates the need for the tail information showing the back weld. The multiple line symbol indicates that the first operation is the groove weld and the second operation, by definition, can only be a back weld rather than a backing weld.

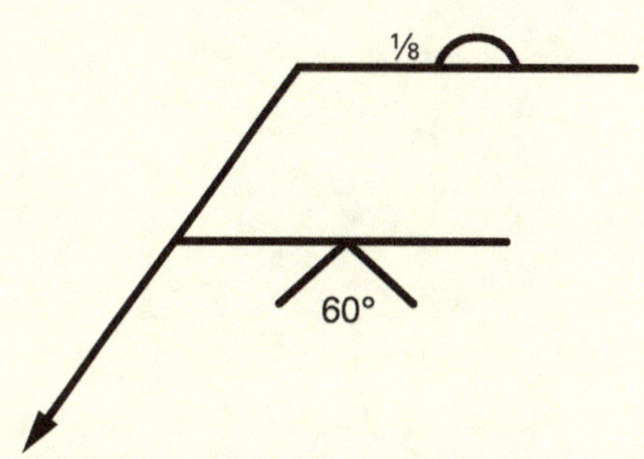

1.4.3 Melt-Through Symbols

When complete penetration is required from one side of a joint and root reinforcement is desired, a melt-through symbol (*Figure 32*) can be used to specify the amount of melt-through. The symbol is placed opposite a groove weld symbol on the reference line. A melt-through symbol is similar in shape to the semi-circle of a back or a backing symbol. However, the semi-circle for a melt-through symbol is solidly filled in, instead of remaining blank. The amount of melt-through required is placed to the left of the melt-through symbol.

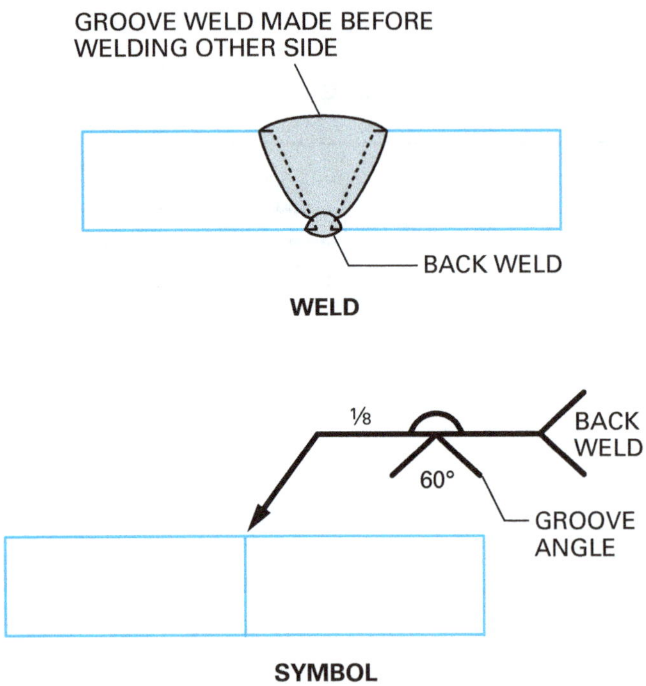

Figure 31 Back or backing weld symbol.

1.4.4 Surfacing Weld Symbols

The AWS defines surfacing as "the application by welding, brazing, or thermal spraying of a layer of material to a surface to obtain desired properties or dimensions, as opposed to making a joint." Examples of surfacing welds are shown in *Figure 33*. A surfacing weld symbol is used when a surface is to be built up with weld. The depth of buildup is indicated to the left of the symbol.

When less than the entire surface is to be built up, the surface to be built up is dimensioned on the blueprint in a detail drawing.

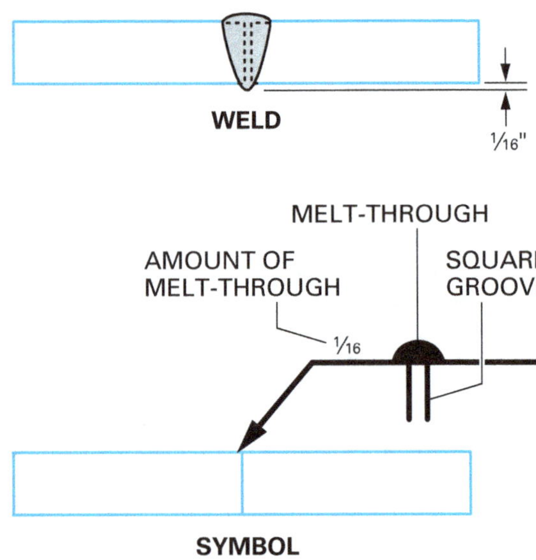

Figure 32 Melt-through symbol.

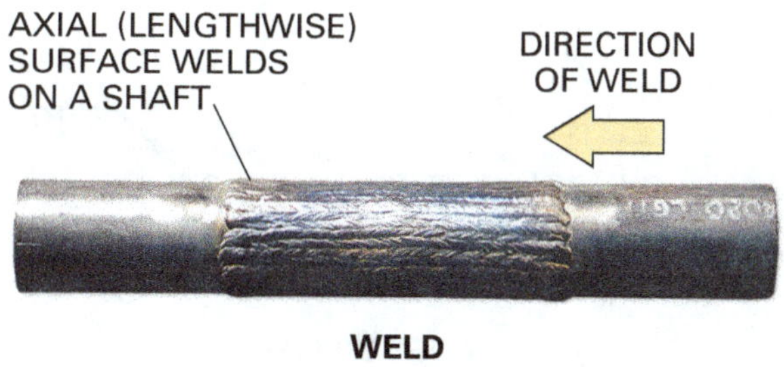

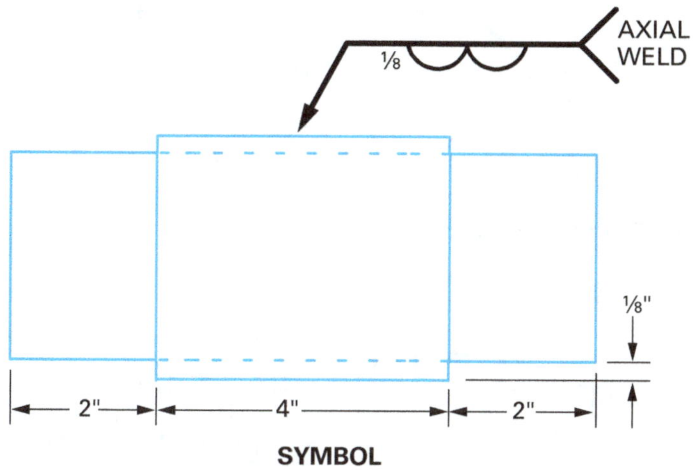

INCHES	MILLIMETERS
⅛	3.2
2	50.8
4	101.6

Figure 33 Surfacing welds and symbols.

1.4.5 Edge Weld Symbols

The AWS defines an edge weld as "a joint between the edges of two or more parallel or nearly parallel members." Edge welds are commonly used on sheet metal and can be flanged butt joints or flanged T-joints. *Figure 34* shows a variety of edge welds, symbols, and dimensioning.

1.4.6 Spot Weld Symbols

Spot welding (*Figure 35*) is a process used to join sheet metal using a series of small spots. There are two types: resistance spot welding (RSW) and arc spot welding. RSW uses electrical resistance heating and clamping pressure to fuse sheet metal panels. Arc spot welding involves either gas metal arc welding (GMAW) or gas tungsten arc welding (GTAW). The process used to make the spot weld is specified in the tail. The size of a spot weld is indicated to the left of the symbol. The pitch is indicated to the right of the symbol. The number of spot welds is defined in parentheses beneath the symbol. For arrow-side spot welds, the circle is placed on the bottom of the reference line. When the spot weld is placed on the other side, the circle is placed on top of the reference line. When there is no side significance (such as with a resistance spot), the circle is centered on the reference line. The drawings will define the starting or ending locations for the spot welds.

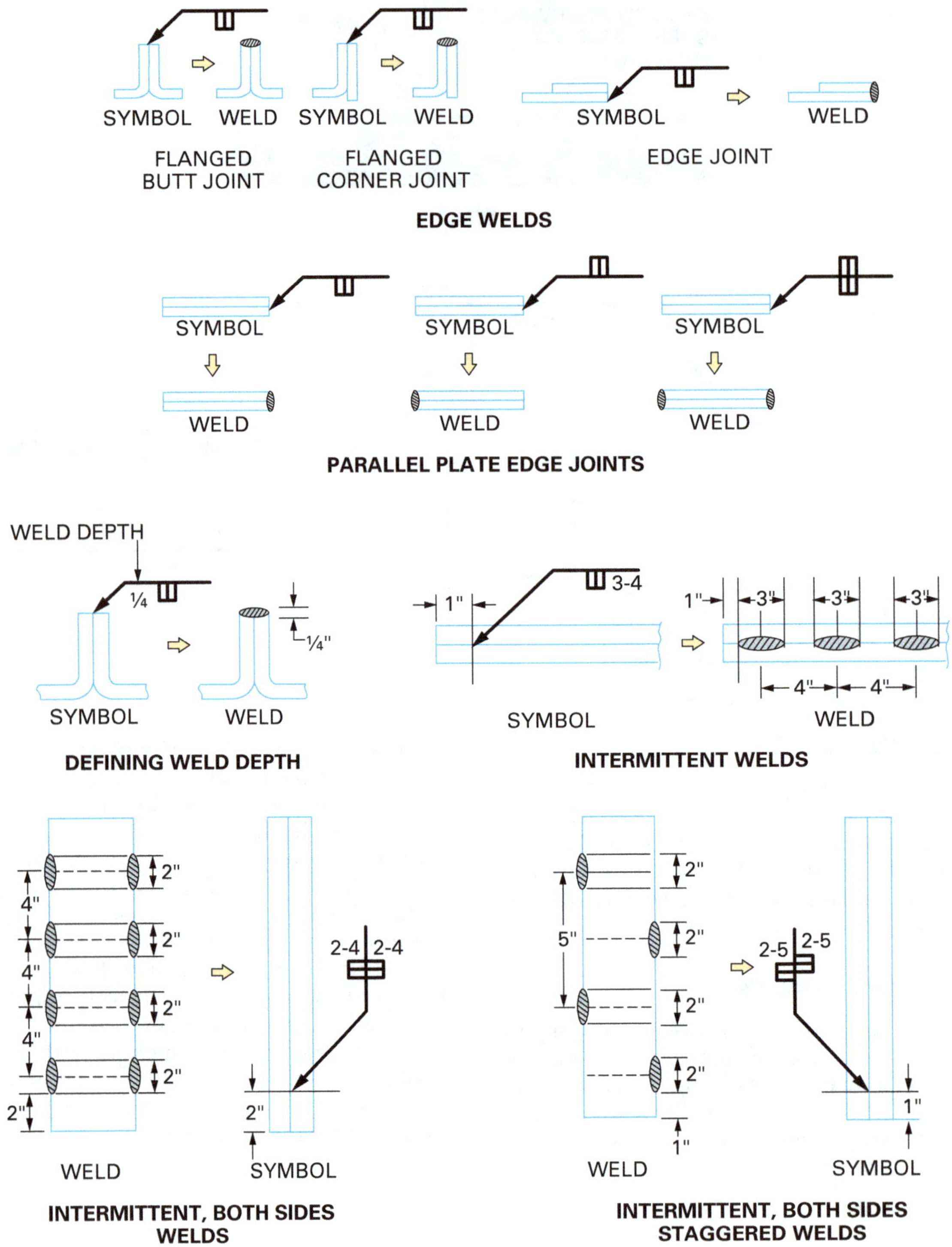

INCHES	MILLIMETERS
¼	6.4
1	25.4
2	50.8
3	76.2
4	101.6
10	254

Figure 34A Edge welds and symbols (1 of 2).

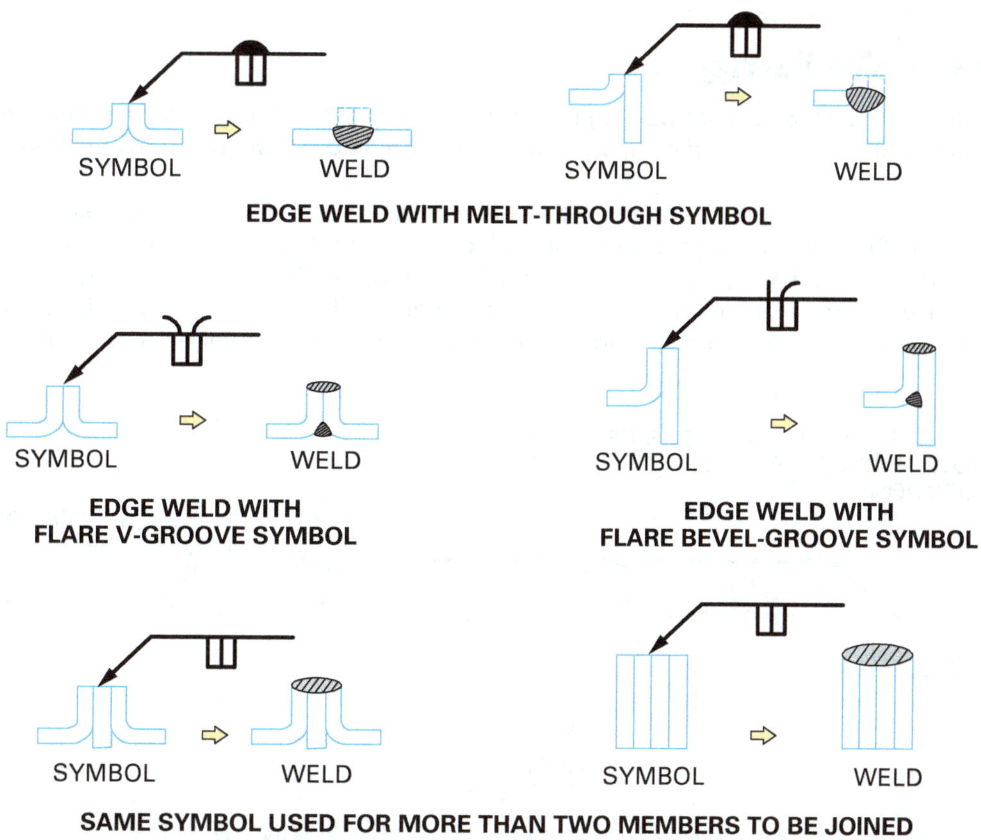

Figure 34B Edge welds and symbols (2 of 2).

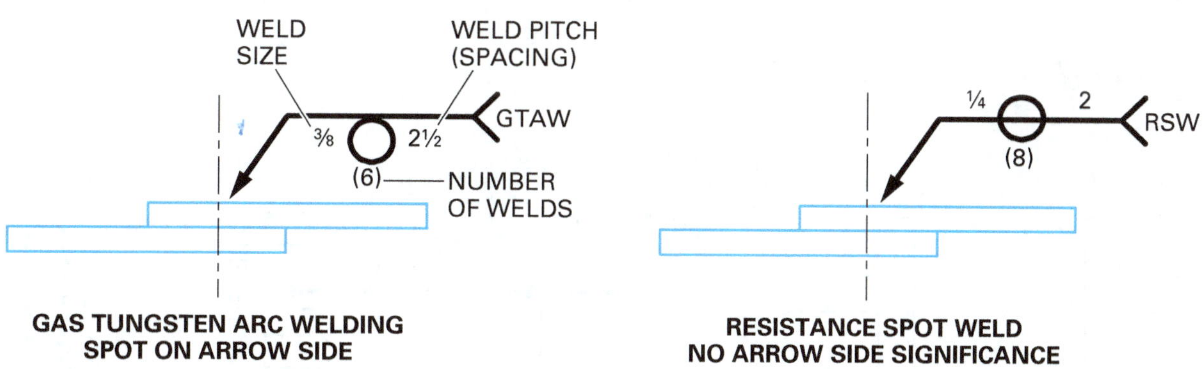

Figure 35 Spot weld symbols.

Projection Welding

Projection welding (PW) is a type of spot welding in which a number of welds are performed simultaneously with one welding machine. Because of tooling costs, projection welding is primarily used in high-production applications.

Prior to welding, one of the members to be joined is passed through a machine that presses projections at specified locations on the member. This member is joined to another member in a welding machine by pressing the two members together and applying current through the projections. The resistance at the points of contact of the projections causes the members to heat, melt, and fuse together while under pressure. This results in welded members that are fused closely together. The side without the projections has minimal surface marking as a result of the weld.

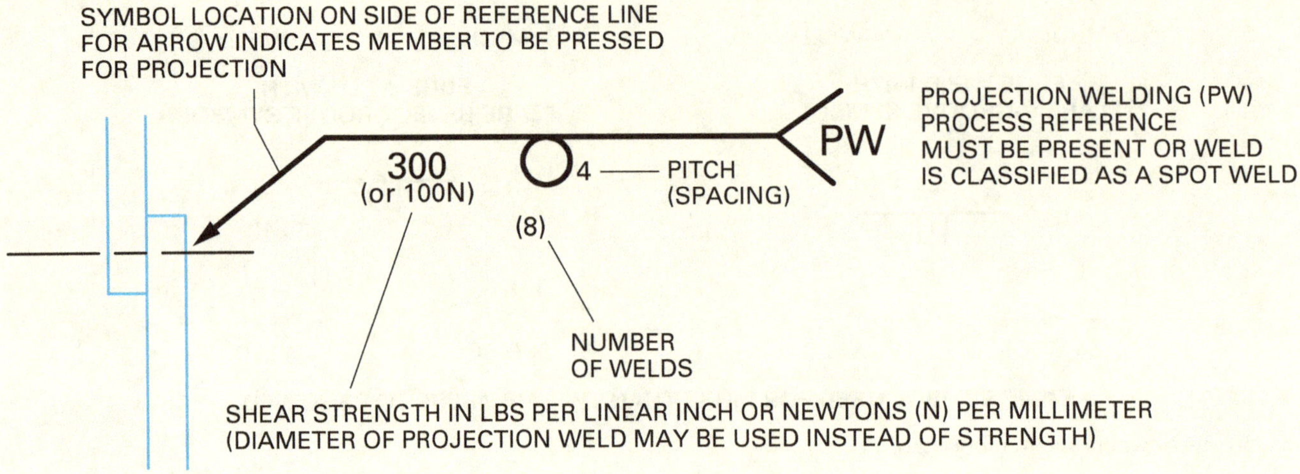

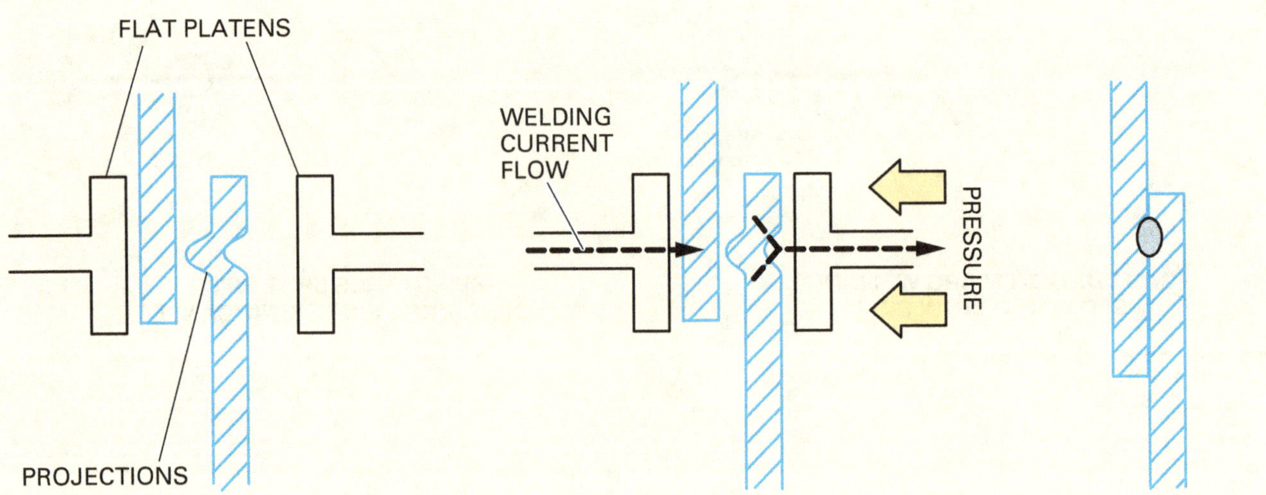

STEP 1 ONE MEMBER IS PRESSED TO CREATE PROJECTIONS AND BOTH MEMBERS ARE PLACED BETWEEN FLAT PLATENS IN WELDING MACHINE

STEP 2 MEMBERS HEAT AND MELT AT PROJECTIONS UNDER PRESSURE

STEP 3 COMPLETED WELD

1.4.7 Seam Weld Symbols

Seam welding is used to join two pieces of sheet metal in a continuous seam or in a series of overlapping spot welds. Several processes can be used to make a seam weld. The most common are GMAW, GTAW, and resistance seam welding (RSEW). The process used to make the seam weld is specified in the tail. The size of a seam weld (width) or strength of the weld is given to the left of the symbol. The length and pitch are given on the right of the symbol for intermittent welds. For welds that are less than the full length of the joined members, the drawing details the starting point and the length of the weld.

For arrow-side seam welds, the circle is placed on the bottom of the reference line, as shown in *Figure 36(A)*. When the seam weld is placed on the other side, the circle is placed on top of the reference line. When there is no side significance, such as with a resistance weld seam, the circle is centered on the reference line, as shown in *Figure 36(B)*. Flush contour symbols can be applied to seam weld symbols for other than resistance weld seams. For parallel seams, a drawing note detailing the spacing and number of seams is included in the tail.

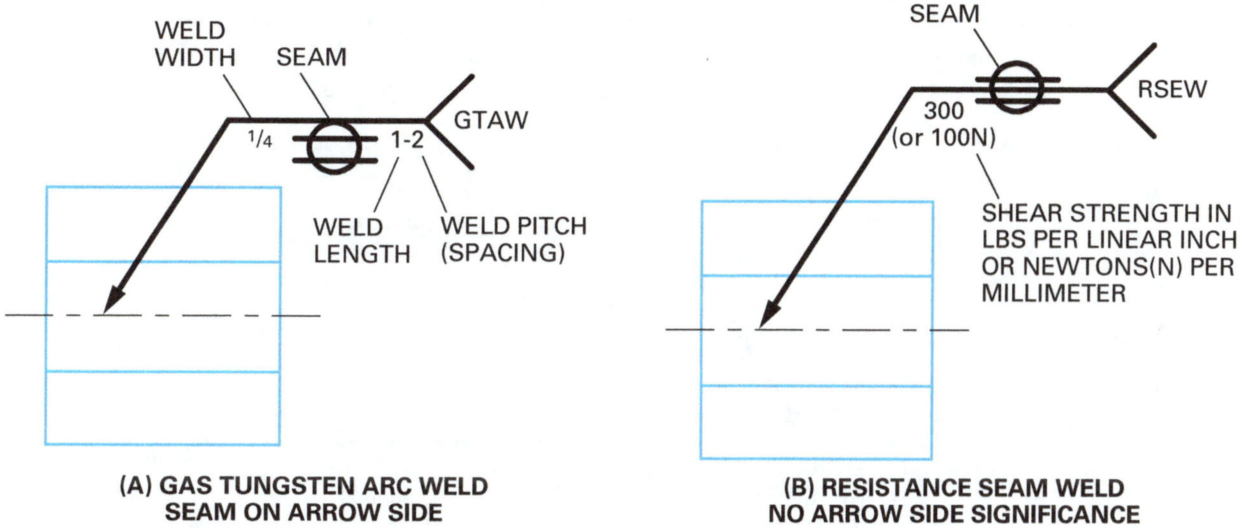

Figure 36 Seam weld symbols.

Around the World

AWS and ISO Welding Symbol Standards

AWS A2.4, Standard Symbols for Welding, Brazing, and Nondestructive Examination is the welding symbols standard used in the United States. AWS welding symbols comply with the requirements of the American National Standards Institute (ANSI) and are designated ANSI/AWS. The AWS system of welding symbols has long been used in the oil industry, which is a global enterprise. Today, the AWS system is used by approximately half of the world's welding industry.

The other widely used welding symbols standard is *ISO 2553, Welded, Brazed, and Soldered Joints – Symbolic Representation on Drawings.* The ISO system was designed by the International Organization for Standardization (ISO).

The AWS and the ISO systems are very similar. For example, the reference line and arrow system in the ISO system are the same as in the AWS system. There are, however, differences that welders need to be aware of in order to interpret the symbols correctly. The most significant difference, as shown in *Figure A*, is that the ISO system uses a dashed identification line that is not used in the AWS system. The dashed identification line, which may be drawn above or below the solid reference line, is used to indicate the other side of the joint. Information that applies to the arrow side of the joint is placed on the solid reference line. Information that applies to the other side of the joint is placed on the dashed line.

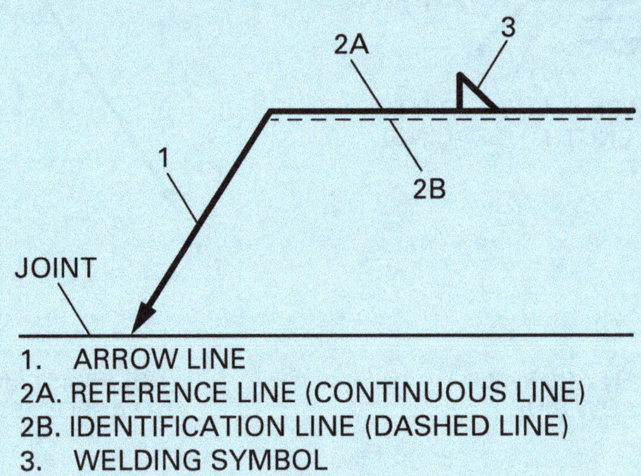

(A) ISO WELDING SYMBOL SYSTEM

Figure B uses fillet welding symbols as an example to further illustrate typical differences between AWS and ISO welding symbols.

Some countries that have extensive international trade relationships have actually used one system of welding symbols in their own country and symbols from the other system as needed to satisfy the requirements of their overseas customers. With the publication of *ISO 2553-2013*, the ISO has sought to provide a set of standardized welding symbols that recognizes both the AWS and the ISO welding symbol systems and can be applied on a worldwide basis.

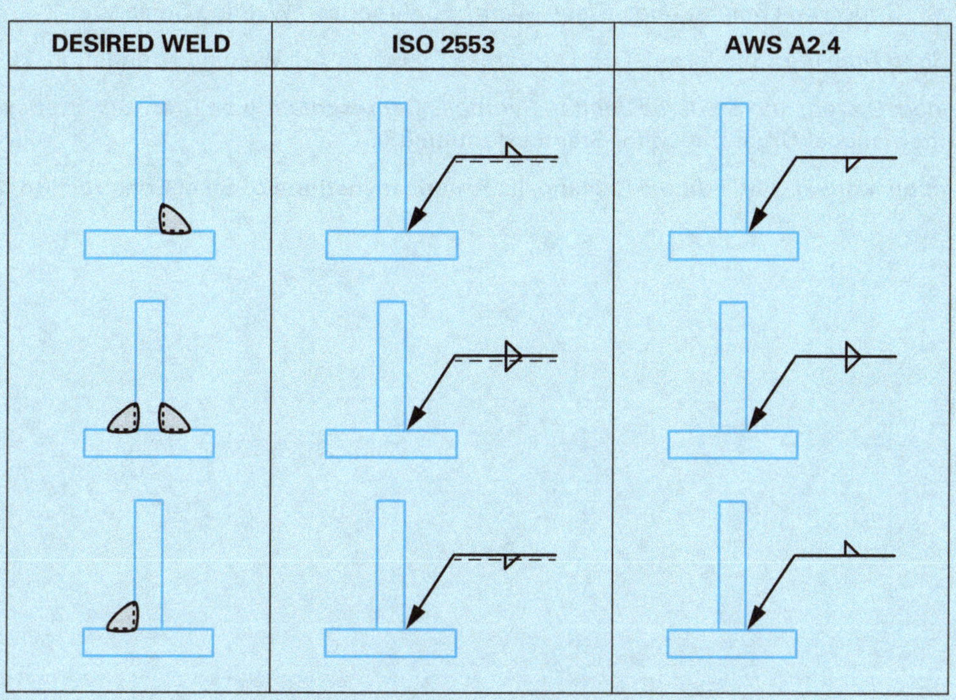

(B) COMPARISON OF ISO AND AWS SYMBOLS

Additional Resources

ASTM A325, Standard Specification for Structural Bolts, Steel, Heat Treated, 120/105 ksi Minimum Tensile Strength. West Conshohocken, PA: ASTM International.

AWS A2.4, Standard Symbols for Welding, Brazing, and Nondestructive Examination. Miami, FL: American Welding Society.

AWS A3.0, Standard Definitions; Including Terms for Adhesive Bonding, Brazing, Soldering, Thermal Cutting, and Thermal Spraying. Miami, FL: American Welding Society.

AWS D1.1/D1.1M, Structural Welding Code Steel. Miami, FL: American Welding Society.

How to Read Shop Drawings. Cleveland, OH: The James F. Lincoln Arc Welding Foundation, 2012.

ISO 2553, Welded, Brazed, and Soldered Joints—Symbolic Representation on Drawings. Geneva, Switzerland: International Organization for Standardization (ISO).

Steel Construction Manual. 14th Edition. Chicago, IL: American Institute of Steel Construction, 2011.

1.0.0 Section Review

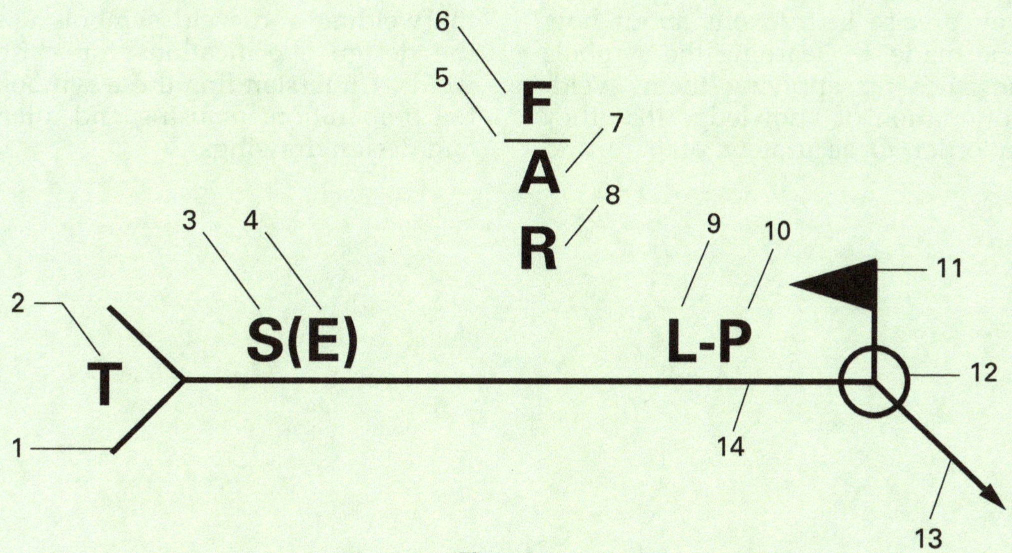

Figure 1

1. Match each of the terms listed below to the number that identifies the corresponding welding symbol shown on *Section Review Question Figure 1*.
   a. Arrow line = ____
   b. Groove weld size = ____
   c. Field weld symbol = ____
   d. Specification or process = ____
   e. Reference line = ____

2. A weld made in a corner is a ____.
   a. slot weld
   b. fillet weld
   c. plug weld
   d. groove weld

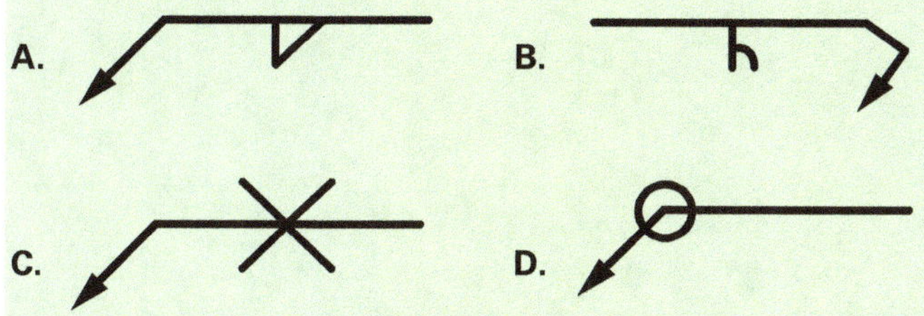

Figure 2

3. Which one of the symbols in *Section Review Question Figure 2* is a supplemental symbol?
   a. Symbol A
   b. Symbol B
   c. Symbol C
   d. Symbol D

4. The symbol in *Section Review Question Figure 3* that indicates that complete penetration from one side of a joint and root reinforcement are required is ____.
   a. Symbol A
   b. Symbol B
   c. Symbol C
   d. Symbol D

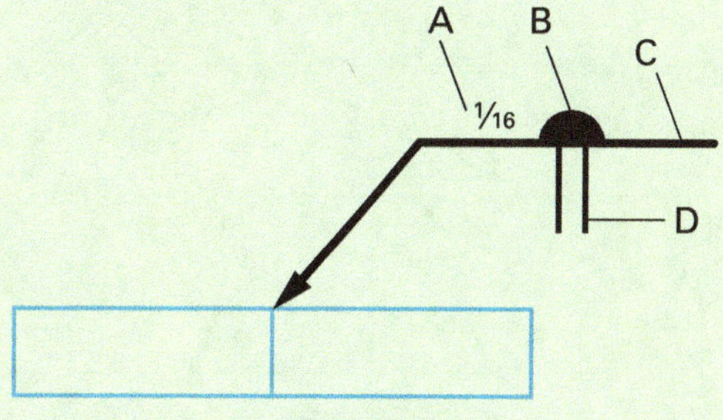

Figure 3

SUMMARY

Welding symbols are a graphic language that is used to convey precise instructions about how a weld is to be made. By learning the symbols and the basic rules for applying them, welders build a foundation of knowledge that they must have in order to accurately prepare and make the required welds. This module covered the welding and weld symbols used to convey the design specifications for various types of welds. Understanding these symbols is an essential preparation for using and interpreting WPS and design drawings.

Review Questions

1. The welding symbol describes the type of weld, its size, and its ____.
   a. surface finish
   b. tail references
   c. cutting process
   d. depth of field

2. The side of the joint or surface that is opposite the side to which the arrow of a reference line is pointing is called the ____.
   a. arrow side
   b. other side
   c. tail joint
   d. countersink

3. Regardless of which end of the reference line the arrow line is on, information on the reference line is always read from ____.
   a. top to bottom
   b. bottom to top
   c. left to right
   d. right to left

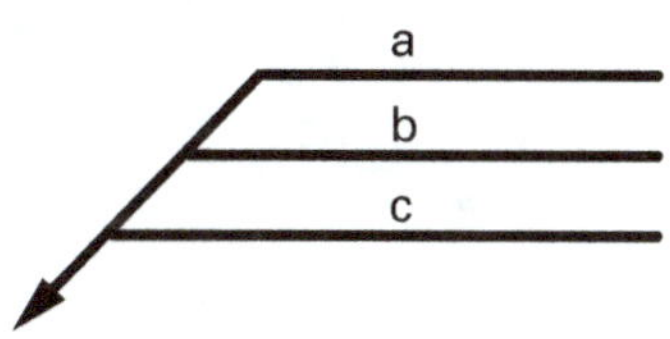

Figure 1

4. In the welding symbol shown in *Review Question Figure 1*, the operation to be conducted first is ____.
   a. Operation A
   b. Operation B
   c. Operation C
   d. not critical

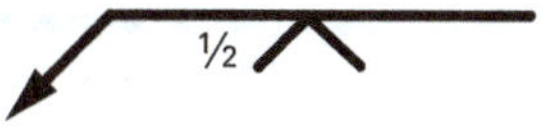

Figure 2

5. In the welding symbol shown in *Review Question Figure 2*, ½ represents the ____.
   a. size of the weld
   b. width of the weld
   c. length of the weld
   d. depth of preparation

6. In relation to the fillet weld symbol, the fillet weld size is placed ____.
   a. above the symbol
   b. below the symbol
   c. to the left of the symbol
   d. to the right the symbol

Figure 3

7. In the welding symbol shown in *Review Question Figure 3*, the number 9 represents the weld ____.
   a. pitch
   b. depth
   c. width
   d. length

8. In relation to the plug weld symbol, the pitch of a plug weld is shown ____.
   a. above the symbol
   b. below the symbol
   c. to the left of the symbol
   d. to the right of the symbol

9. The field weld symbol is a(n) ____.
   a. number
   b. letter
   c. arrow
   d. flag

10. When the face of the weld must have a finished shape, a finish symbol is placed adjacent to the welding symbol, along with a letter indicating ____.
    a. size of the weld
    b. method of finish
    c. smoothness of the finish
    d. width of the finished area

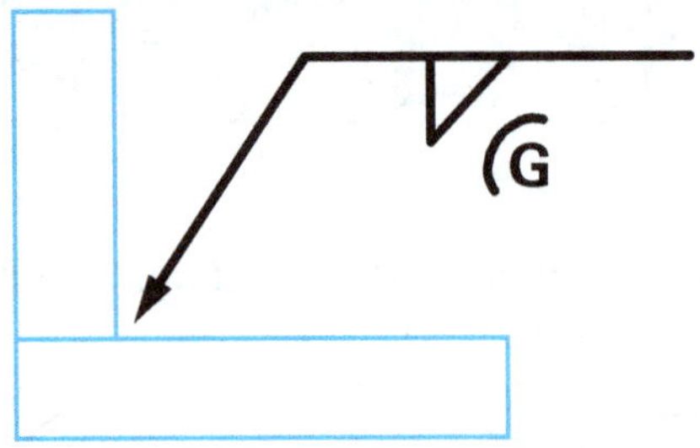

Figure 4

11. The finish of the weld face in *Review Question Figure 4* is to be achieved by ____.
    a. filleting
    b. grinding
    c. chipping
    d. reaming

12. When backing is to be used at the root of the weld to control melt-through, the backing symbol is placed opposite which weld symbol on the reference line?
    a. tail
    b. field
    c. spacer
    d. groove

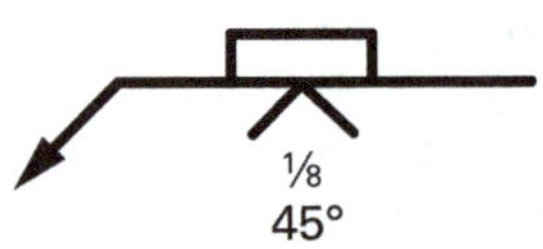

Figure 5

13. In the weld symbol shown in *Review Question Figure 5*, the numbers inside the symbol represent the ____.
    a. root opening and groove angle
    b. groove angle and depth
    c. root opening and pitch
    d. pitch and angle

Figure 6

14. The type of weld symbol shown above the reference line in *Review Question Figure 6* is a ____.
    a. spot weld
    b. back weld
    c. surfacing weld
    d. melt-through

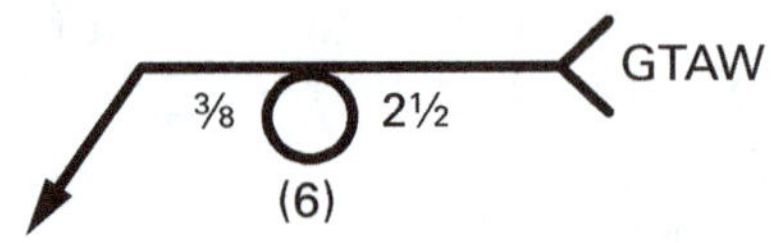

Figure 7

15. The type of weld represented by the weld symbol in *Review Question Figure 7* is a ____.
    a. spot weld
    b. backing weld
    c. surfacing weld
    d. melt-through

Appendix A

Letter Designations for Welding Processes			
AAW	Air Acetylene Welding	GTAW	Gas Tungsten Arc Welding
AB	Adhesive Bonding	GTAW-P	Gas Tungsten Arc Welding–Pulsed Arc
AHW	Atomic Hydrogen Welding	HPW	Hot Pressure Welding
AW	Arc Welding	IB	Induction Brazing
B	Brazing	IRB	Infrared Brazing
BB	Block Brazing	IW	Induction Welding
BMAW	Bare Metal Arc Welding	LBW	Laser Beam Welding
CAW	Carbon Arc Welding	OAW	Oxyacetylene Welding
CAW-G	Gas Carbon Arc Welding	OHW	Oxyhydrogen Welding
CAW-T	Twin Carbon Arc Welding	PAW	Plasma Arc Welding
CEW	Coextrusion Welding	PEW	Percussion Welding
CW	Cold Welding	PGW	Pressure Gas Welding
DP	Dip Brazing	PW	Projection Welding
DFB	Diffusion Brazing	RB	Resistance Brazing
DFW	Diffusion Welding	ROW	Roll Welding
EBW	Electronic Beam Welding	RSEW	Resistance Seam Welding
EBW-HV	Electronic Beam Welding–High Vacuum	RSEW-HF	Resistance Seam Welding–High Frequency
EBW-MV	Electronic Beam Welding–Medium Vacuum	RSEW-I	Resistance Seam Welding–Induction
EBW-NV	Electronic Beam Welding–Nonvacuum	RSW	Resistance Spot Welding
EGW	Electrogas Welding	RW	Resistance Welding
ESW	Electroslag Welding	SAW	Submerged Arc Welding
EXW	Explosion Welding	SAW-S	Series Submerged Arc Welding
FB	Furnace Brazing	SMAW	Shielded Metal Arc Welding
FCAW	Flux Cored Arc Welding	SSW	Solid-State Welding
FLB	Flow Brazing	SW	Stud Arc Welding
FLOW	Flow Welding	TB	Torch Brazing
FOW	Forge Welding	TCAB	Twin Carbon Arc Brazing
FRW	Friction Welding	TW	Thermite Welding

Appendix B

Letter Designations for Cutting Processes			
AC	Arc Cutting	LOC	Oxygen Lance Cutting
AOC	Oxygen Arc Cutting	MAC	Metal Arc Cutting
CAC	Carbon Arc Cutting	OC	Oxygen Cutting
CAC-A	Air Carbon Arc Cutting	OC-F	Flux Cutting
EBC	Electron Beam Cutting	OFC	Oxyfuel Gas Cutting
GMAC	Gas Metal Arc Cutting	OFC-A	Oxyacetylene Cutting
GTAC	Gas Tungsten Arc Cutting	OFC-H	Oxyhydrogen Cutting
LBC	Laser Beam Cutting	OFC-N	Oxynatural Gas Cutting
LBC-A	Laser Beam Cutting–Air	OFC-P	Oxypropane Cutting
LBC-EV	Laser Beam Cutting–Evaporative	OC-P	Metal Powder Cutting
LBC-IG	Laser Beam Cutting–Inert Gas	PAC	Plasma Arc Cutting
LBC-O	Laser Beam Cutting–Oxygen	SMAC	Shielded Metal Arc Cutting

Appendix C

Letter Designations for Applying Welding Processes	
AU	Automatic
ME	Mechanized
MA	Manual
SA	Semi-Automatic

Trade Terms Introduced in This Module

Arrow line: The line drawn at an angle from the reference line (either end or both ends) to an arrowhead at the location of the weld.

Countersink: A hole with tapered sides and a wider opening that allows a flat-head fastener to seat flush to the surface of the material in which the hole is made.

Pitch: The center-to-center distance between welds.

Reference line: The horizontal line in the center of the welding symbol from which all elements of the welding symbol are referenced. The reference line is one of the most important elements of the welding symbol.

Weld symbol: A graphic character connected to the reference line of a welding symbol specifying the weld type.

Welding symbol: A graphical representation of the specifications for producing a welded joint; includes a reference line and arrow line and many also include a weld symbol.

Additional Resources

This module is intended as a thorough resource for task training. The following reference works are suggested for further study.

ASTM A325, Standard Specification for Structural Bolts, Steel, Heat Treated, 120/105 ksi Minimum Tensile Strength. West Conshohocken, PA: ASTM International.

AWS A2.4, Standard Symbols for Welding, Brazing, and Nondestructive Examination. Miami, FL: American Welding Society.

AWS A3.0, Standard Definitions; Including Terms for Adhesive Bonding, Brazing, Soldering, Thermal Cutting, and Thermal Spraying. Miami, FL: American Welding Society.

AWS D1.1/D1.1M, Structural Welding Code Steel. Miami, FL: American Welding Society.

How to Read Shop Drawings. Cleveland, OH: The James F. Lincoln Arc Welding Foundation, 2012.

ISO 2553, Welded, Brazed, and Soldered Joints—Symbolic Representation on Drawings. Geneva, Switzerland: International Organization for Standardization (ISO).

Steel Construction Manual. 14th Edition. Chicago, IL: American Institute of Steel Construction, 2011.

Figure Credits

The Lincoln Electric Company, Cleveland, OH, USA, Module opener

AWS A2:2007, Figure 2, Reproduced with permission from the American Welding Society (AWS), Miami, FL, USA, Figure 4

Topaz Publications, Inc., Figures 7 (photos), 24 (photo), 33 (photo)

Specialty Welding & Fabricating of New York, Inc., Figures 18 (photos), 23 (photos)

Section Review Answer Key

Answer	Section Reference	Objective
Section One		
1. a. 13 b. 4 c. 11 d. 2 e. 14	1.1.0	1a
2. b	1.2.1	1b
3. d	1.3.0	1c
4. b	1.4.3	1d

NCCER CURRICULA — USER UPDATE

NCCER makes every effort to keep its textbooks up-to-date and free of technical errors. We appreciate your help in this process. If you find an error, a typographical mistake, or an inaccuracy in NCCER's curricula, please fill out this form (or a photocopy), or complete the online form at **www.nccer.org/olf**. Be sure to include the exact module ID number, page number, a detailed description, and your recommended correction. Your input will be brought to the attention of the Authoring Team. Thank you for your assistance.

Instructors – If you have an idea for improving this textbook, or have found that additional materials were necessary to teach this module effectively, please let us know so that we may present your suggestions to the Authoring Team.

NCCER Product Development and Revision

13614 Progress Blvd., Alachua, FL 32615

Email: curriculum@nccer.org
Online: www.nccer.org/olf

❑ Trainee Guide ❑ Lesson Plans ❑ Exam ❑ PowerPoints Other ____________

Craft / Level: Copyright Date:

Module ID Number / Title:

Section Number(s):

Description:

Recommended Correction:

Your Name:

Address:

Email: Phone:

Reading Welding Detail Drawings

Overview

Reading and interpreting the information contained on construction drawings is an essential skill for welders. Incorrectly interpreting this information can result in producing work that fails to meet design specifications, building codes, and safety standards. This module focuses on how to read and interpret assembly and detail drawings. It introduces the various object views and basic drawing elements that are used in welding detail drawings. It also explains how to interpret dimensional information, notes, and bills of materials.

Module 29202

Trainees with successful module completions may be eligible for credentialing through the NCCER Registry. To learn more, go to **www.nccer.org** or contact us at **1.888.622.3720**. Our website has information on the latest product releases and training, as well as online versions of our *Cornerstone* magazine and Pearson's product catalog.

Your feedback is welcome. You may email your comments to **curriculum@nccer.org**, send general comments and inquiries to **info@nccer.org**, or fill in the User Update form at the back of this module.

This information is general in nature and intended for training purposes only. Actual performance of activities described in this manual requires compliance with all applicable operating, service, maintenance, and safety procedures under the direction of qualified personnel. References in this manual to patented or proprietary devices do not constitute a recommendation of their use.

29202 V5

29202 READING WELDING DETAIL DRAWINGS

Objectives

When you have completed this module, you will be able to do the following:

1. Describe welding detail drawings and identify basic drawing elements and features.
   a. Describe the object views used to depict welding details.
   b. Identify basic drawing elements related to welding detail drawings.
2. Identify and explain how to interpret dimensional information, notes, and a bill of materials.
   a. Identify and explain how to interpret dimensional information.
   b. Identify and explain how to interpret notes and a bill of materials.

Performance Task

Under the supervision of your instructor, you should be able to do the following:

1. Draw or sketch a welding drawing based on an instructor-provided image or object.

Trade Terms

Blind hole
Counterbore
Elevation
Isometric view
Mark
Orthographic projection
Scale
Through hole
Tolerance

Industry Recognized Credentials

If you are training through an NCCER-accredited sponsor, you may be eligible for credentials from NCCER's Registry. The ID number for this module is 29202. Note that this module may have been used in other NCCER curricula and may apply to other level completions. Contact NCCER's Registry at 888.622.3720 or go to **www.nccer.org** for more information.

Contents

Topics to be presented in this module include:

Figures

SECTION ONE

1.0.0 BASIC WELDING DETAIL DRAWINGS

Objective

Describe welding detail drawings and identify basic drawing elements and features.

a. Describe the object views used to depict welding details.
b. Identify basic drawing elements related to welding detail drawings.

Performance Task

1. Draw or sketch a welding drawing based on an instructor-provided image or object.

Trade Terms

Isometric view: A drawing depicting a three-dimensional view of an object.

Orthographic projection: A method of developing multiple flat views of an object by projecting lines out from the object at right angles from the face.

Engineers, designers, and drafters use drawings, also called prints or blueprints, to convey information. There are many categories of drawings, including civil, structural, mechanical, piping, electrical, and instrumentation. Welders must be able to correctly interpret drawings so that the design specifications and safety factors shown in the drawings will be included in the finished product. Altering or failing to comply with a drawing could result in a violation of federal or state standards or building codes, which could lead to fines and other penalties. More importantly, a deviation from the drawings could create an unsafe structure or product, which might result in death or injury. To avoid these problems, welders must be able to read and interpret drawings accurately. This module explains how to read and interpret assembly and detail drawings.

Each category contains several types of drawings. The most common drawing used by welders is the welding detail drawing, which generally depicts the weldment that must be fabricated and the pieces to be joined by welding. The parts are drawn assembled, as they will appear after welding, except that the welds are not shown. Welding symbols show how the pieces are to be joined. *Figure 1* is an example of a small bracket that may need to be fabricated on the job.

Figure 2 shows how the same bracket would appear on a welding detail drawing.

> **NOTE**
> The performance task for this module is to draw or sketch a welding detail drawing based on a given image or object. An example of such an image and its corresponding drawing is provided in the *Appendix* of this module.

When the object to be fabricated is too complicated for one drawing, the object may be shown on an assembly drawing as well as one or more detail drawings. The assembly drawing (*Figure 3*) shows the completed and assembled object. The main purpose of the assembly drawing is to show how the parts are assembled in relation to one another. Detail drawings of each part are then provided (*Figure 4*). The detail drawings contain all the information necessary to make the part.

> **NOTE**
> Letters and leader lines identify the parts of the object in an assembly drawing. These same letters are then used to label the parts in the corresponding detail drawing.

1.1.0 Object Views

Drawings must clearly show the shape of an object. To do this, most drawings have more than one view of the object. The most commonly used views are the following:

- Isometric
- Multiview
- Section

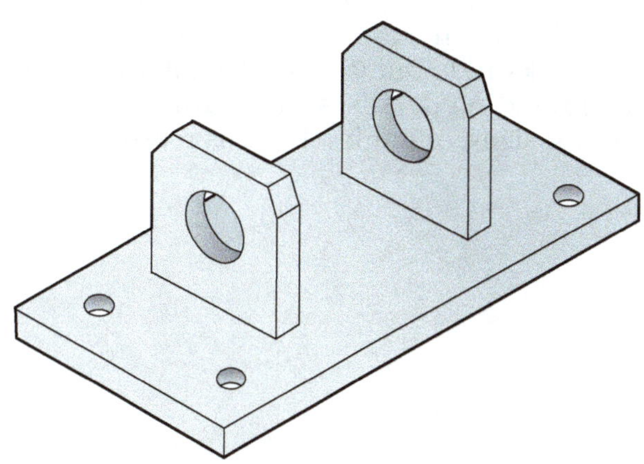

Figure 1 Bracket.

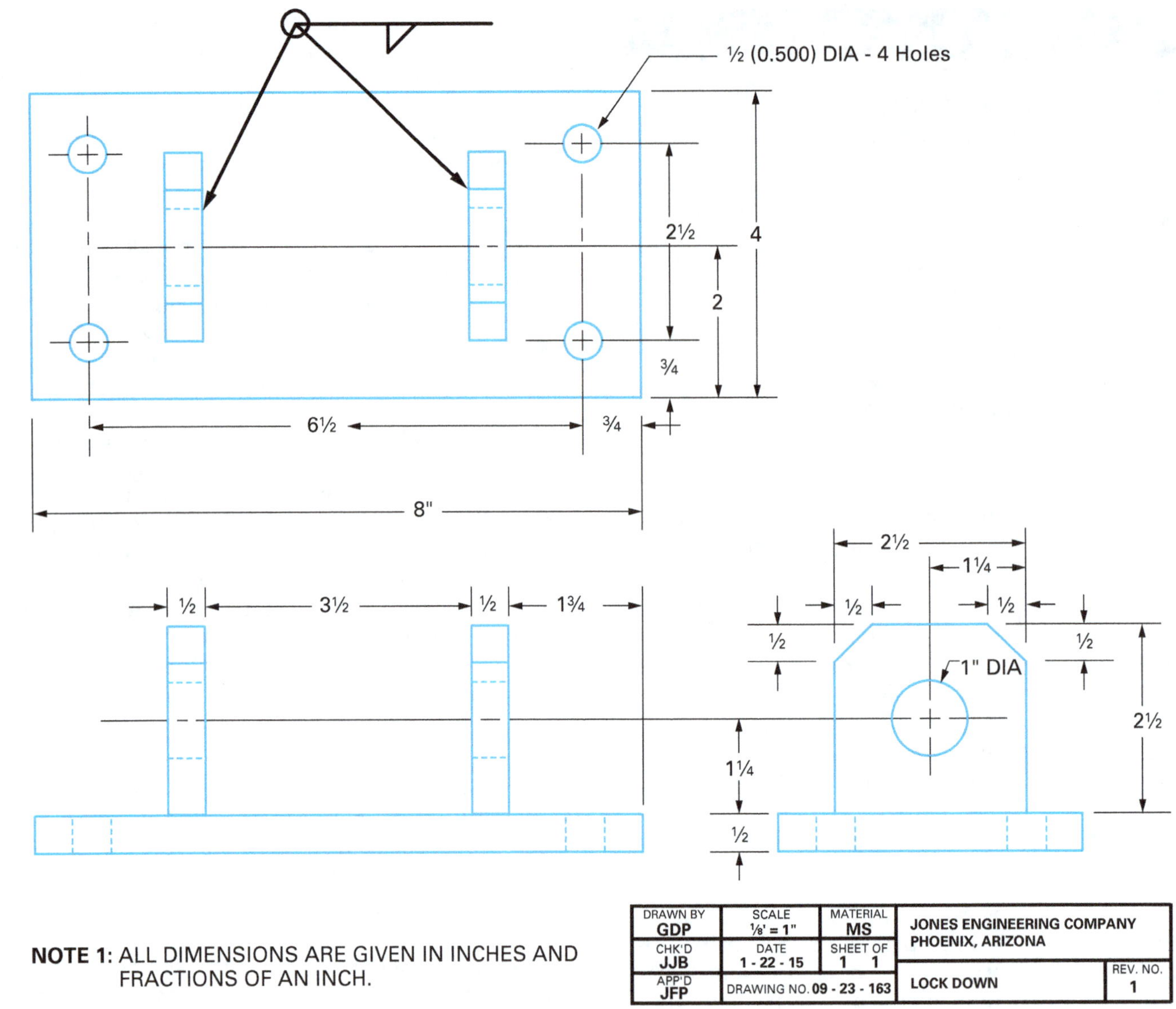

(A) IMPERIAL MEASUREMENTS

Figure 2A Welding detail drawing. (1 of 2)

1.1.1 Isometric View

The isometric view shows a three-dimensional picture of an object (*Figure 5*). The isometric (pictorial) view is an excellent way to show what an object looks like, but it is a difficult view to use when providing details, such as dimensions, hole locations, sizes, or angles. For this reason, the isometric view is generally not used on detail drawings.

1.1.2 Multiview

Multiviews are developed by a process called orthographic projection, and they are generally used on detail drawings. Several views, each from a different direction, are developed and shown on the same drawing. The views are designed as if the object had been placed inside a transparent box. Each view is then expanded by projecting points at right angles from the object onto a surface of the transparent box. If the sides of the box were folded open, a flat, multiview drawing would result, as shown in *Figure 6*.

Isometric Views

An isometric drawing of an object shows the object at 30-degree angles. This causes the object to appear on paper similar to the way the real, three-dimensional object appears.

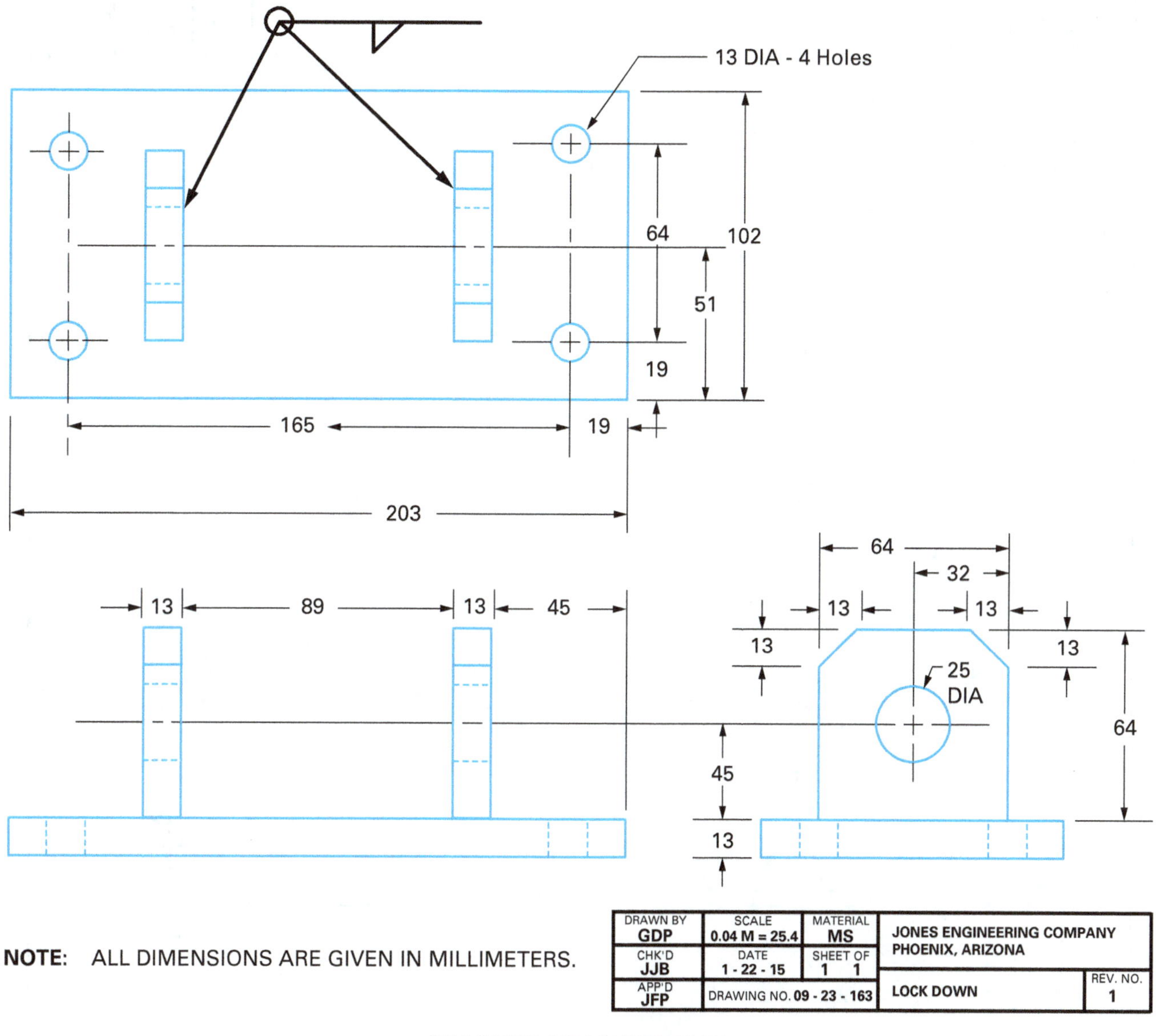

(B) METRIC MEASUREMENTS

Figure 2B Welding detail drawing. (2 of 2)

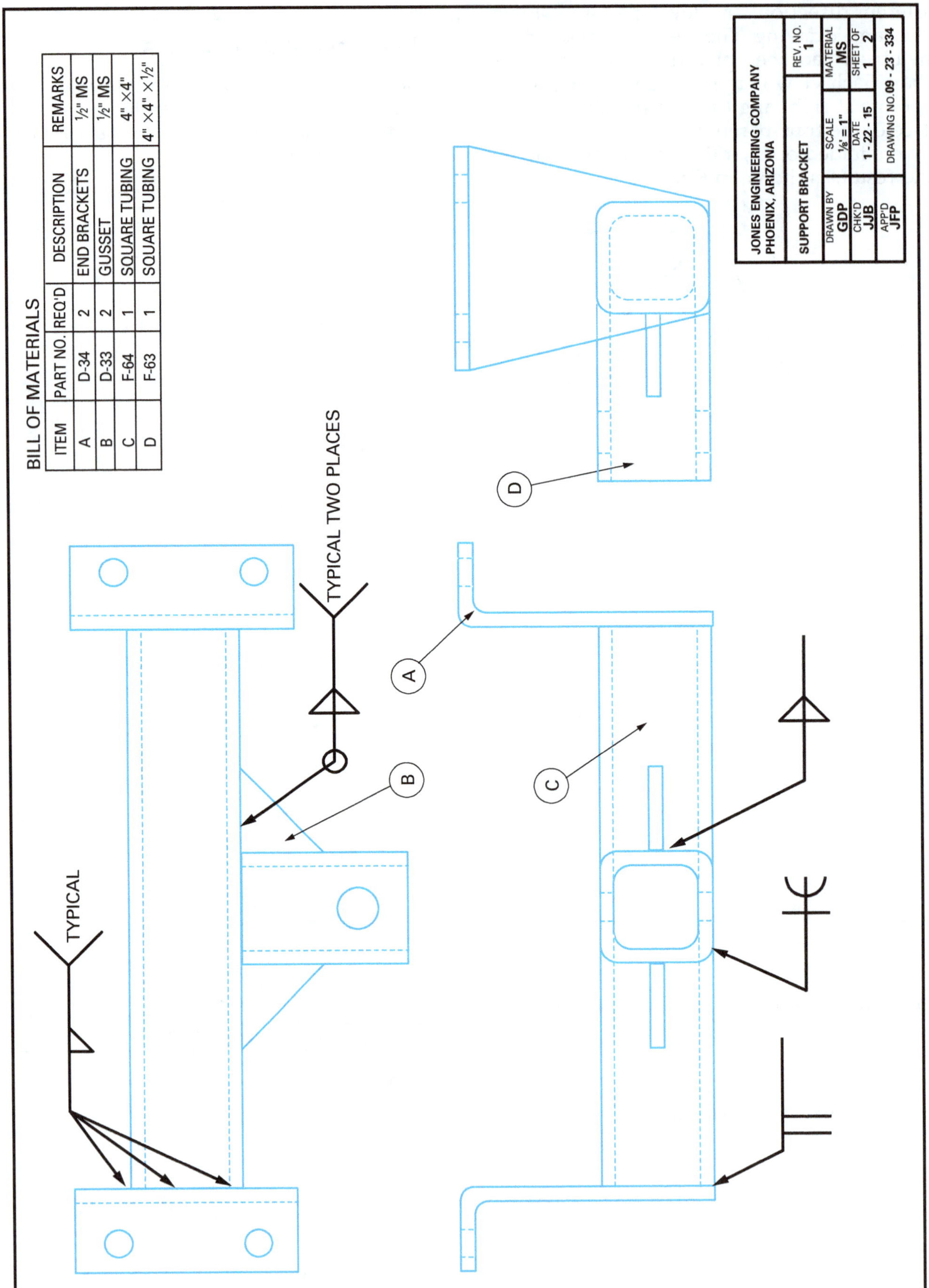

Figure 3 Assembly drawing.

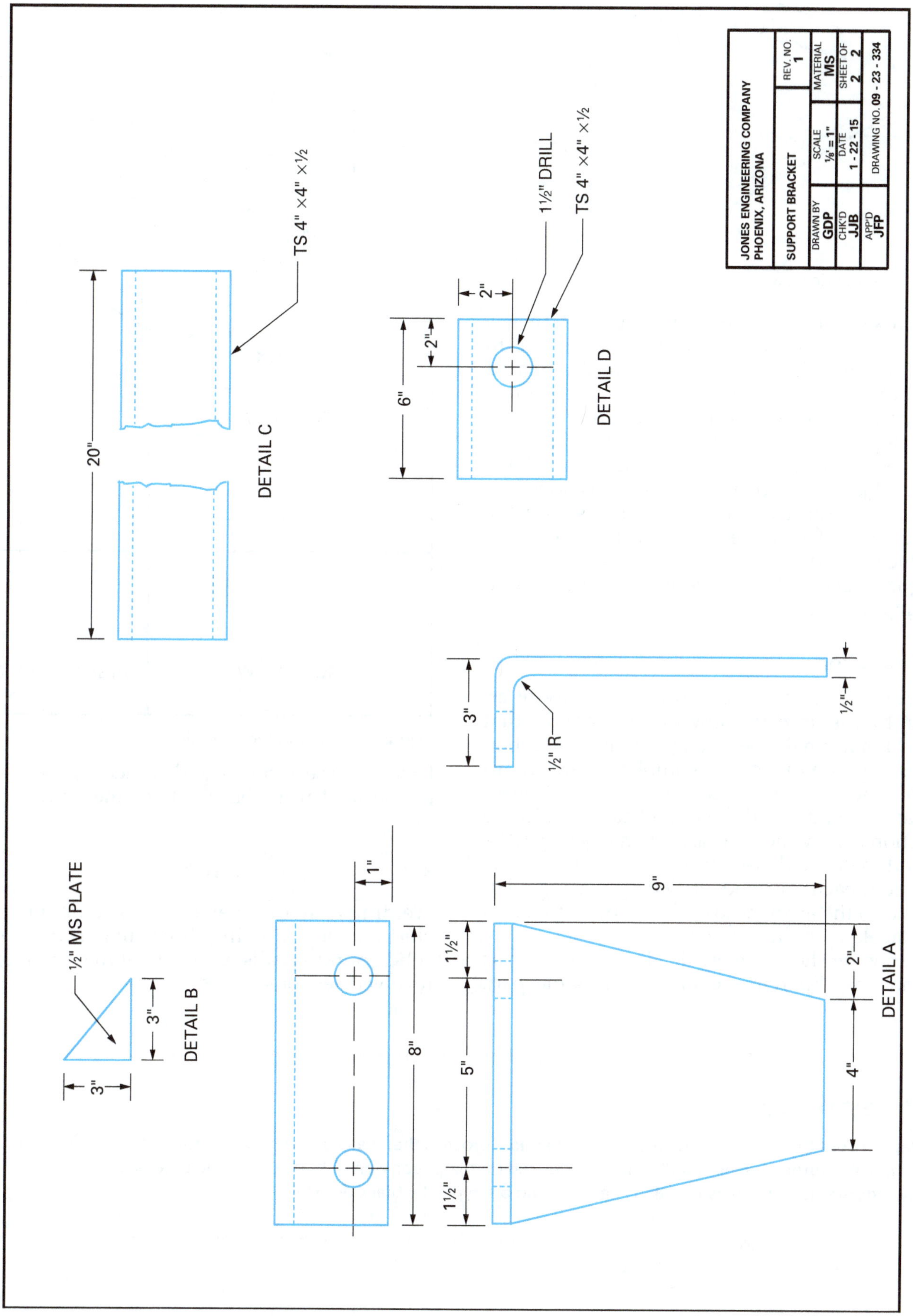

Figure 4 Detail drawing.

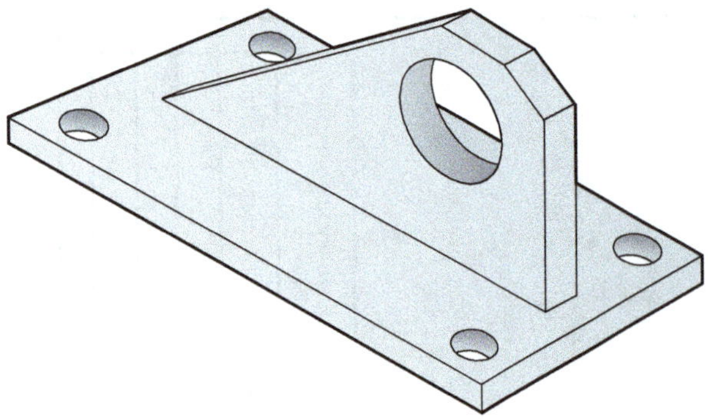

Figure 5 Isometric view

Each view in an orthographic projection has a set position in relation to the other views. In addition, the views are aligned vertically and horizontally. All sides of an object can be shown, although the views used most frequently are the front, top, and right side. The front view is selected as the view that best represents the shape of the object. All other views are positioned in reference to the front view. *Figure 7* shows the relationship of possible views used with orthographic projection.

Figure 8 shows the relationship of views on a welding detail drawing.

1.1.3 Section

A full section, or cross-section, is a view of an object that is used to show internal details that would not normally be visible (*Figure 9*). A section line is used on one of the complete views to indicate where the part has been cut or sectioned. The arrowheads on the section line point in the direction from which the sectioned object is to be viewed. Because there can be more than one section view, each one is identified by placing a letter next to the arrowheads. The sectioned view is then labeled with the letters.

A drawing shows a more detailed view of a part by showing different sections of the same part.

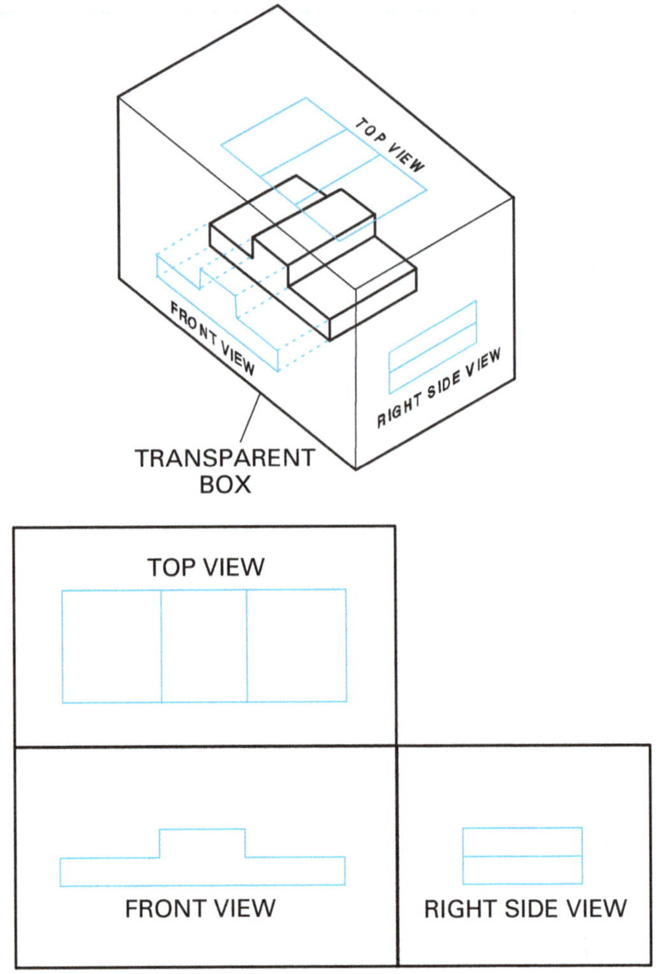

Figure 6 Orthographic projection.

Each of these sections is identified by pairs of letters located near the points of the cutting plane line.

1.2.0 Drawing Components

Welding drawings have several components, including the following: lines; material symbols; solid round breaks; pipe, or tubing, breaks; and revolved sections.

Blueprints

Blueprints got their name from an obsolete chemical process that was used to develop chemically treated print paper, which turned paper blue. The lines produced on the paper were white against the blue background. Drawings are now made using other methods and are typically black on white.

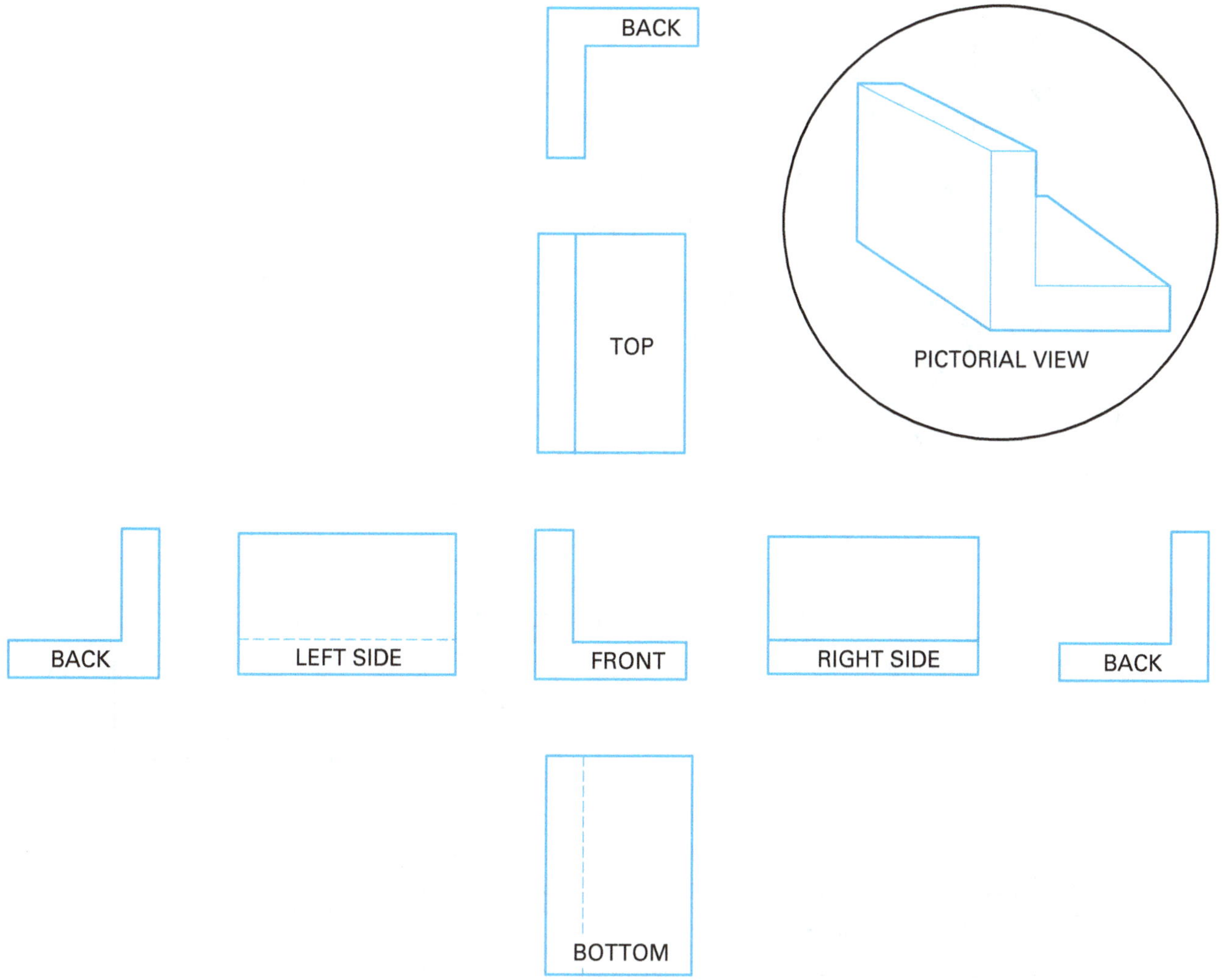

Figure 7 Relationship of possible views.

1.2.1 Lines

Many different types of lines are used on drawings, and each type has its own special meaning. The lines are drawn to a standard that specifies thickness (thick, medium, or thin) and the way the lines are interpreted or broken. The more commonly used lines are shown in *Figure 10* and are described as follows:

- *Object line* – A thick, solid line used to show all visible edges of a part; also called a visible line.
- *Hidden line* – A medium, broken line made up of evenly spaced dashes; used to show edges that are not visible.
- *Center line* – A thin, broken line made up of short and long dashes; used to show the center of holes or other objects.
- *Dimension line* – A thin line with arrowheads that shows the extent of a dimension; the line is broken by the dimension number.
- *Extension line* – A thin line used with dimension lines; it extends from the point on the object that the dimension applies to.
- *Leader line* – A thin, horizontal line with an angled arrowhead at one end; used with a note or dimension. The arrowhead points directly to the surface of the note or dimension to which the line applies.
- *Cutting plane line* – A line used to show where an imaginary cut has been made through an object. It has a heavy line that can be made up of either a series of one long and two short dashes or a series of short dashes. Short lines with arrowheads are on both ends of the line. The arrowheads point in the direction in which the section should be viewed.
- *Short break line* – A thick, solid, irregular line drawn freehand to indicate that part of the object has been removed or broken away; used to conserve space or show internal detail.
- *Long break line* – A thin, solid line with zigzags inserted at several places; used to show where a section of an object has been removed to conserve space on the drawing.

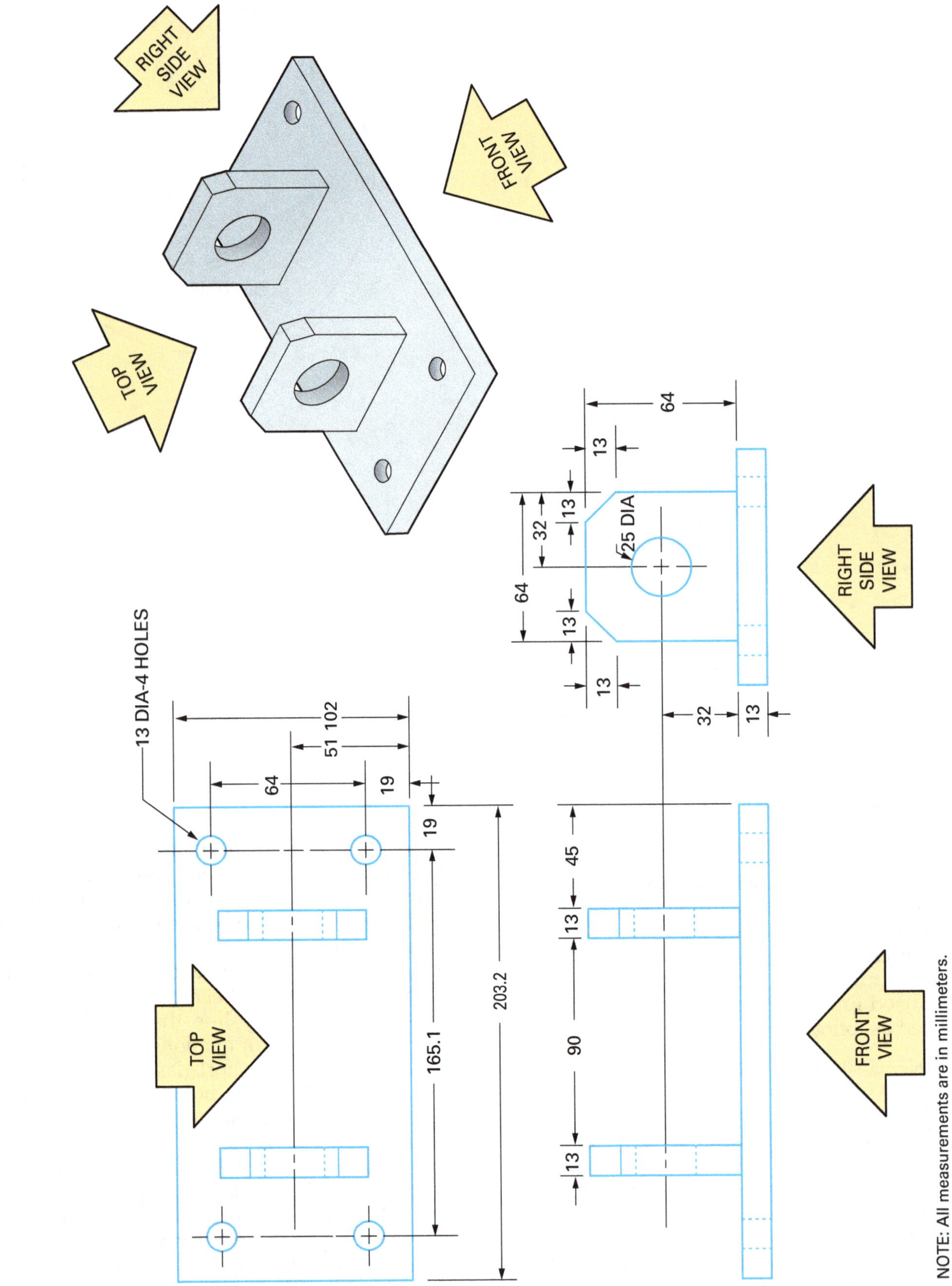

Figure 8 Relationship of views.

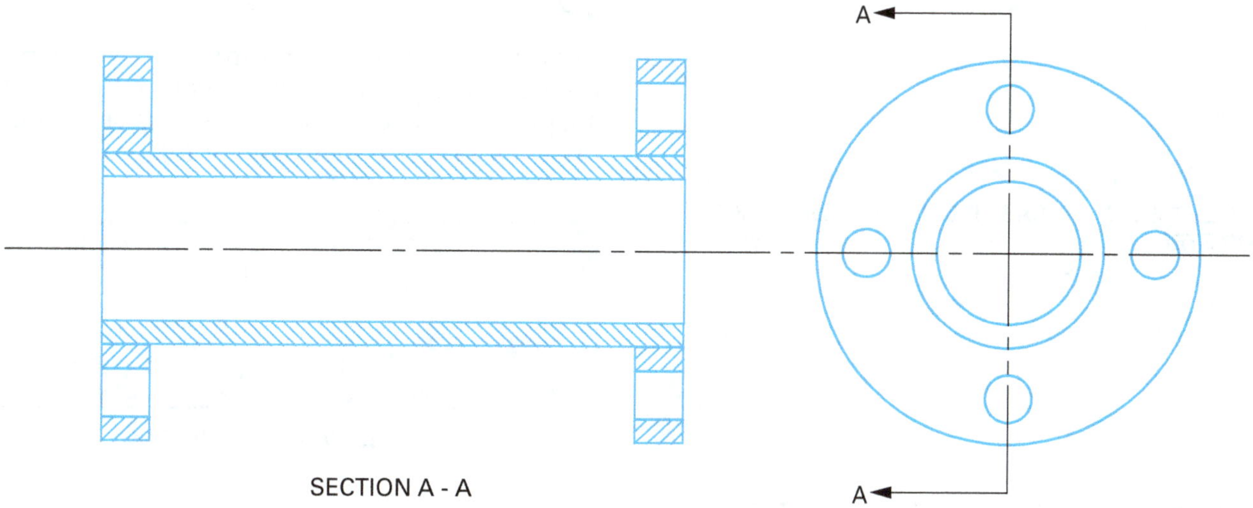

INCHES	MILLIMETERS
6 ½	165
8	204

Figure 9 Section view.

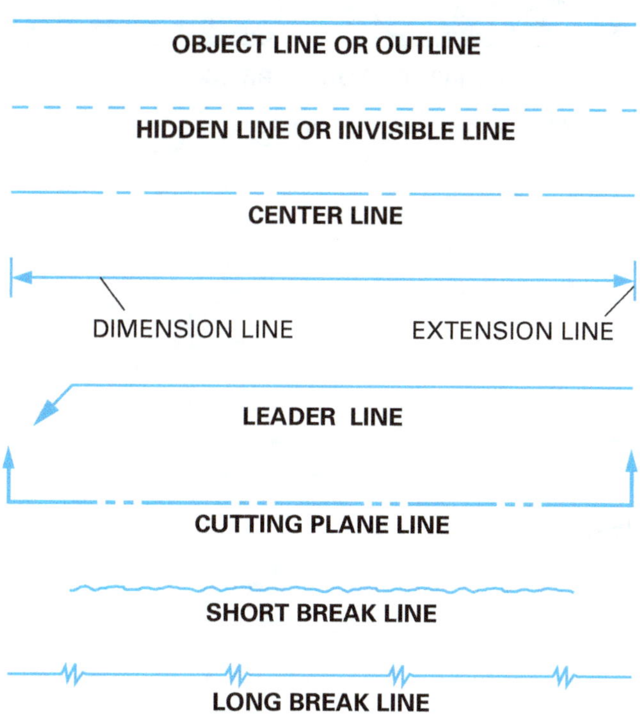

Figure 10 Common drawing lines.

1.2.2 Material Fill Symbols

Material symbols are a series of thin lines, either solid or a combination of solid and broken, that are arranged in a pattern to represent various types of materials. The symbols are used to identify internal surfaces that have been exposed by a cutting plane line or a short cutaway view. *Figure 11* shows common material symbols used to represent various metals.

1.2.3 Bar Stock and Pipe Breaks

A bar stock or pipe break is represented by thick, curved lines that loop back on themselves (*Figure 12*). The approach is the same regardless of the shape of the stock. The loop is filled with a material symbol, such as cross-hatching, to indicate an internal surface. Breaks are used to show where a section of an object has been removed to conserve space on the drawing.

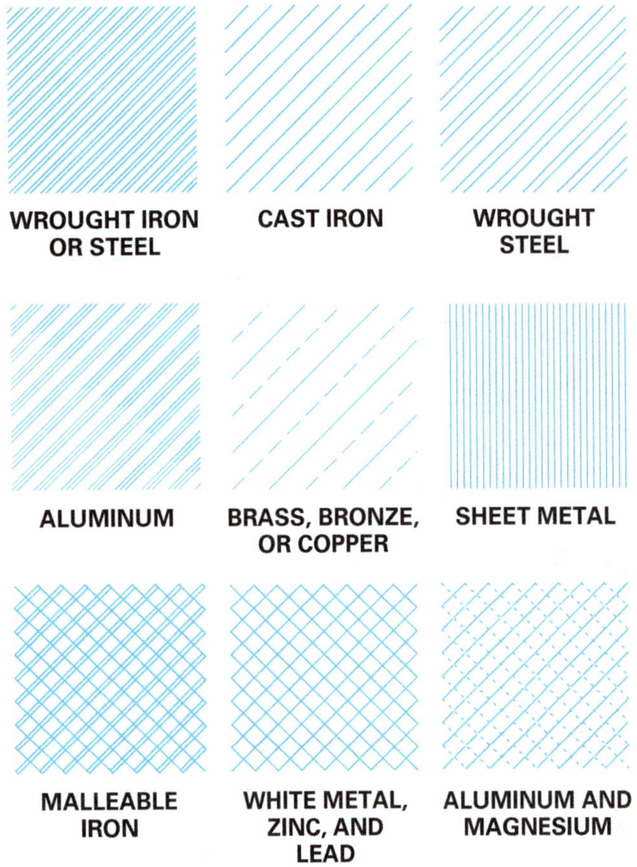

Figure 11 Material symbols.

1.2.4 Revolved Sections

Revolved sections are similar to solid round breaks and pipe, or tubing, breaks except that the shape of the object that is broken is shown in the break as a section view (*Figure 13*).

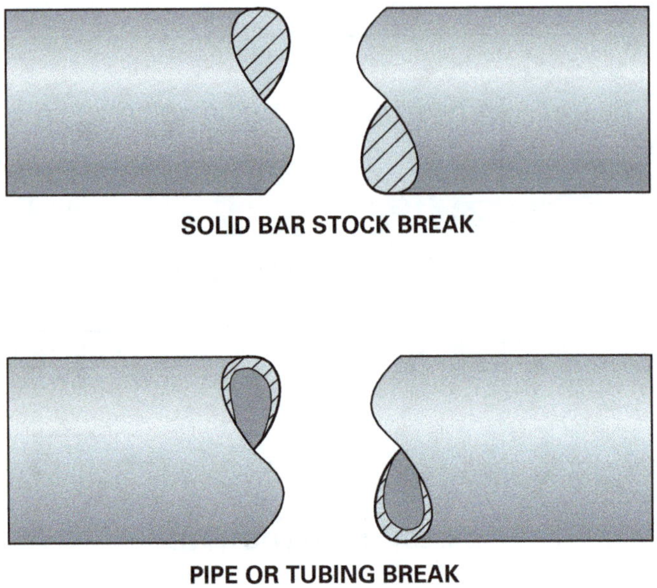

Figure 12 Bar stock and pipe breaks.

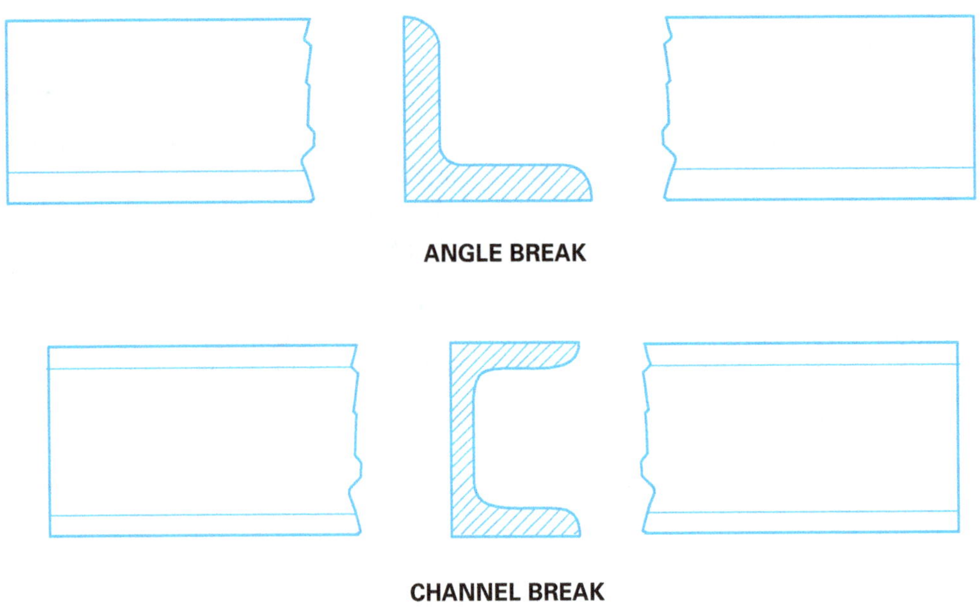

Figure 13 Revolved sections.

Material Symbols

Do not select materials based upon material symbols shown on a drawing. These symbols only represent the described material and do not provide specific information about the material. For example, there are two or three symbols that represent steel, but there are many different types of steel, each with specific characteristics. Always use the drawing notes, material list, or a bill of materials to determine the exact materials that are required.

Additional Resources

ASTM A325, Standard Specification for Structural Bolts, Steel, Heat Treated, 120/105 ksi Minimum Tensile Strength. West Conshohocken, PA: ASTM International.

AWS A2.4, Standard Symbols for Welding, Brazing, and Nondestructive Examination. Miami, FL: American Welding Society.

AWS A3.0, Standard Welding Terms and Definitions. Miami, FL: American Welding Society.

How to Read Shop Drawings. Cleveland, OH: The James F. Lincoln Arc Welding Foundation, 2012.

Steel Construction Manual. Fourteenth Edition. Chicago, IL: American Institute of Steel Construction, 2011.

1.0.0 Section Review

1. The most common drawing used by welders is the ____.
   a. section view
   b. isometric view
   c. assembly drawing
   d. welding detail drawing

2. In a drawing, a solid round break and a pipe, or tubing, break are used to show where a section of an object has been ____.
   a. broken
   b. repaired
   c. inserted
   d. removed

Section Two

2.0.0 Dimensions, Notes, and Materials

Objective

Identify and explain how to interpret dimensional information, notes, and a bill of materials.

a. Identify and explain how to interpret dimensional information.
b. Identify and explain how to interpret notes and a bill of materials.

Performance Task

1. Draw or sketch a welding drawing based on an instructor-provided image or object.

Trade Terms

Blind hole: A drilled hole that does not go all the way through the object.

Counterbore: To enlarge a portion of a drilled hole so that a socket head fastener will seat flush or below the surface.

Elevation: A height measurement above sea level or other identified point; also, the vertical sides of an object in an orthographic view.

Mark: A system of letters and numbers used with structural steel to identify the type of member and where it is placed.

Scale: A method of drawing an object larger or smaller than its actual size while keeping the proportions the same.

Through hole: A drilled hole that goes completely through an object.

Tolerance: The allowance over or under the specified dimension, value, or other parameter.

Welding detail drawings include precise dimensional information for the various pieces that make up an object. Additional details about the construction of the object are often provided in notes and a bill of materials. To fabricate an object that meets design specifications, welders must be able to read and correctly interpret the dimensional information, general or local notes, and the bill of materials.

2.1.0 Dimensions

Dimensions identify sizes. Drawings must have dimensions if the object shown is to be fabricated. The drawing must include all dimensions in correct relationship to one another in order to give a complete size description of the object. The two basic methods for dimensioning a part are conventional dimensioning and baseline dimensioning.

In conventional dimensioning (*Figure 14*), the overall dimensions of length, width, and height are given. Dimensions A and B are not given because they are easily found by adding and/or subtracting the given dimensions. If they were given, dimensions A and B would be identified in parentheses as reference dimensions. A reference dimension, such as the 1¾ inch dimension in parentheses on the drawing, is obtained by adding or subtracting other dimensions found on the drawing. If the overall size of a dimensioned part is not given, then those dimensions that make up the total must be added to find the overall size. Reference dimensions help to speed up the reading of drawings.

In baseline dimensioning (*Figure 15*), all dimensions originate from baselines. When imperial measurements are used, baseline dimensions are usually expressed as three-place decimals for accuracy. Metric baseline dimensions are also normally expressed as three-place decimals, depending upon the chosen unit of measure.

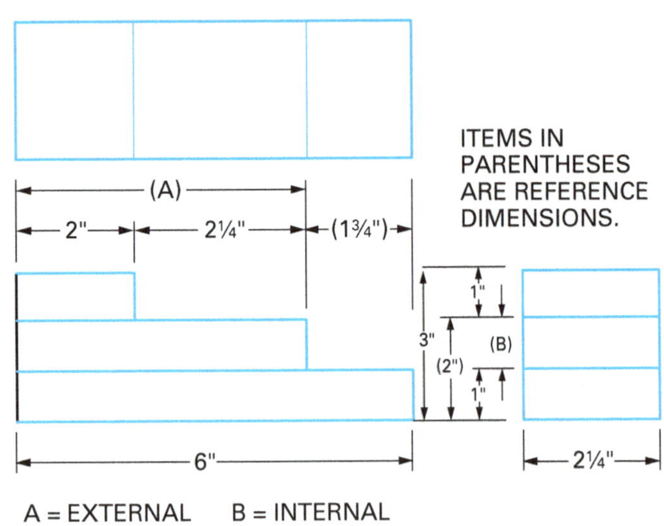

Figure 14 Conventional dimensioning.

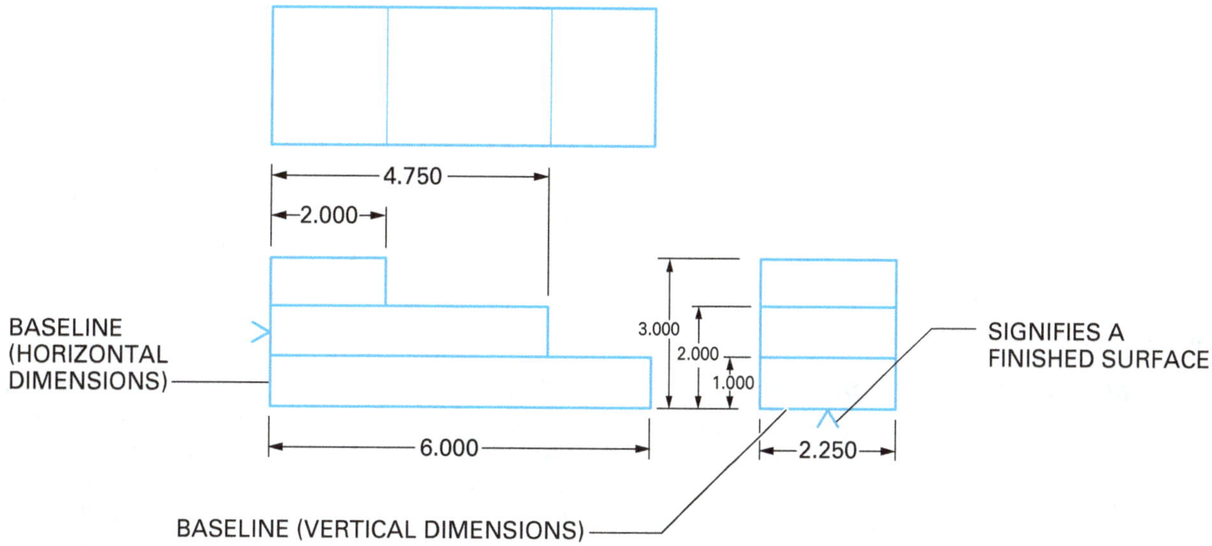

Figure 15 Baseline dimensioning.

> **NOTE**
>
> Baseline dimensions on metric drawings are never fractional because, unlike imperial measurements, metric measurements are always shown as decimals. When working with drawings that have baseline dimensions, always check to make sure which system of measurement (metric or imperial) the drawing is based on.

The baselines are typically machined surfaces that are indicated by the symbol V. Baselines may also originate from center lines or from a critical edge. The advantage of baseline dimensioning is that it eliminates the cumulative error that often occurs with conventional dimensioning. Baseline dimensioning does not require calculations to find a dimension. Most welding drawings that require hand assembly and welding do not use three-place decimal dimensioning because of the difficulty in maintaining tight tolerances.

Think About It

Dimension Calculation

What is the reference dimension of a section whose overall part dimension is 4⅞ inches? The part consists of three sections. Two of the sections are 1⅞ inches and 1⅛ inches, respectively.

2.1.1 Scale

Most drawings are made in scale. This means that the drawing can be larger, the same size, or smaller than the actual object, as long as the proportions in the drawing are the same as the proportions of the actual object. Most objects shown in welding detail drawings are too large to fit on a standard drawing, so they are drawn smaller than the actual object's size. The scale used on the drawing is given in the title block in the form of an equation. Many different scales can be used, for example:

- ¼ inch = 1 inch
- ⅛ inch = 1 inch
- ¼ inch = 1 foot

The figure to the left of the equal sign in a scale equation represents the measure used for the drawing, while the figure to the right of the equal sign corresponds to the measure used for the object. For example, if the scale ¼ inch = 1 inch is used, an actual surface 3 inches long would be represented on the drawing by a line ¾ inches long (¼ inch for each actual inch). If the scale ¼ inch = 1 foot is used, an actual surface 4 feet long would be shown on the drawing by a line 1 inch long. *Figure 16* shows how scale is used to reduce the size of a drawing.

On drawings that are dimensioned in metric units, the scale is stated as a ratio. Actual size would be stated as 1:1. A reduction ratio would be stated as 1:2, 1:5, 1:10, etc. An increasing ratio would be stated as 2:1, 5:1, etc. The drawing title block will usually specify the applicable scale, unless more than one scale is used on the drawing.

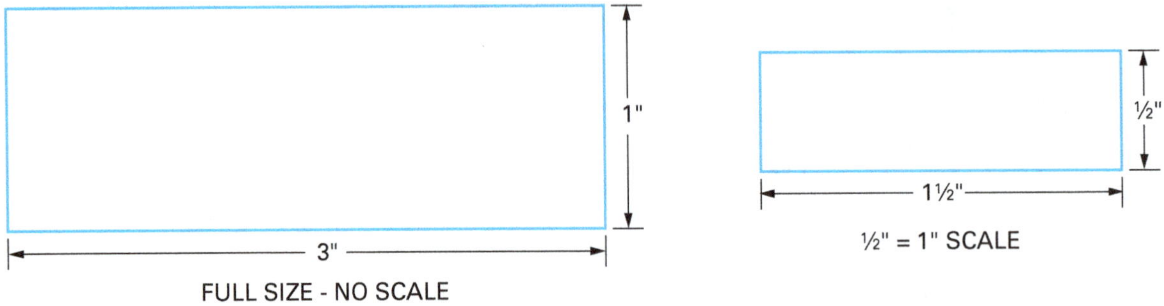

Figure 16 Drawing in scale.

2.1.2 Size and Locations Dimensions

The most common dimensions are size and location. Size dimensions provide the measurements for the overall size of the object. Location dimensions provide measurements from one feature of an object to another part, hole, break, or component. Examples of both dimension types are shown in *Figure 17*.

2.1.3 Hole Dimensions

Drilled hole dimensions are shown with a leader line and a note (*Figure 18*). The note gives the hole size and the number of holes to be drilled, if more than one. If the hole is to be drilled only partway through the object, which is called a **blind hole**, then the depth will also be given. If the depth is not given, then the hole goes completely through the object and is called a **through hole**.

Holes may also be **counterbored** or countersunk. A hole that is counterbored has a square shoulder that is used to recess socket head fasteners. A hole that is countersunk is tapered from the surface to the drilled hole size and is used for flat-head fasteners. The required dimensions for countersinking or counterboring holes are added to the note (*Figure 19*).

2.1.4 Angle and Bevel Dimensions

Angles and bevels are common on welding detail drawings because they are used with weld joint preparation. They can be specified with a leader line and a note, or they can be specified using extension and dimension lines (*Figure 20*). The size can be given as angular, linear, or a combination of angular and linear dimensions.

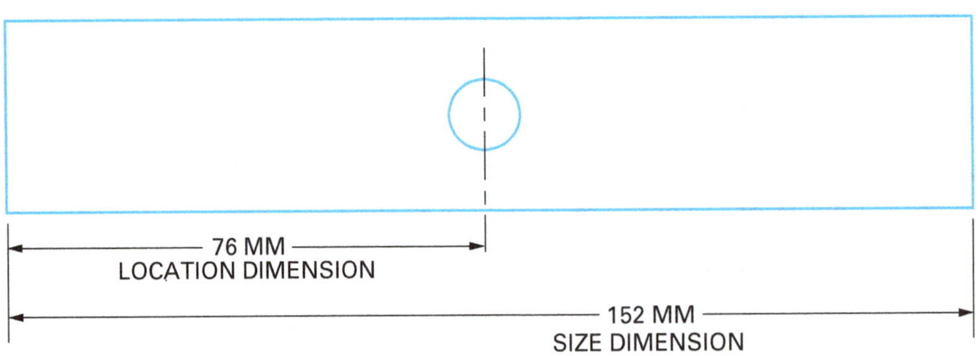

Figure 17 Size and location dimensions.

Drawing Measurements

Be sure to use only dimensions provided on the drawing. If a dimension is needed but is not provided, add or subtract the dimensions that were furnished to obtain the needed dimension. Avoid measuring directly from a drawing (called scaling a drawing). Measuring directly from a drawing almost always results in errors. If you cannot determine a measurement by adding or subtracting given dimensions, check with your supervisor.

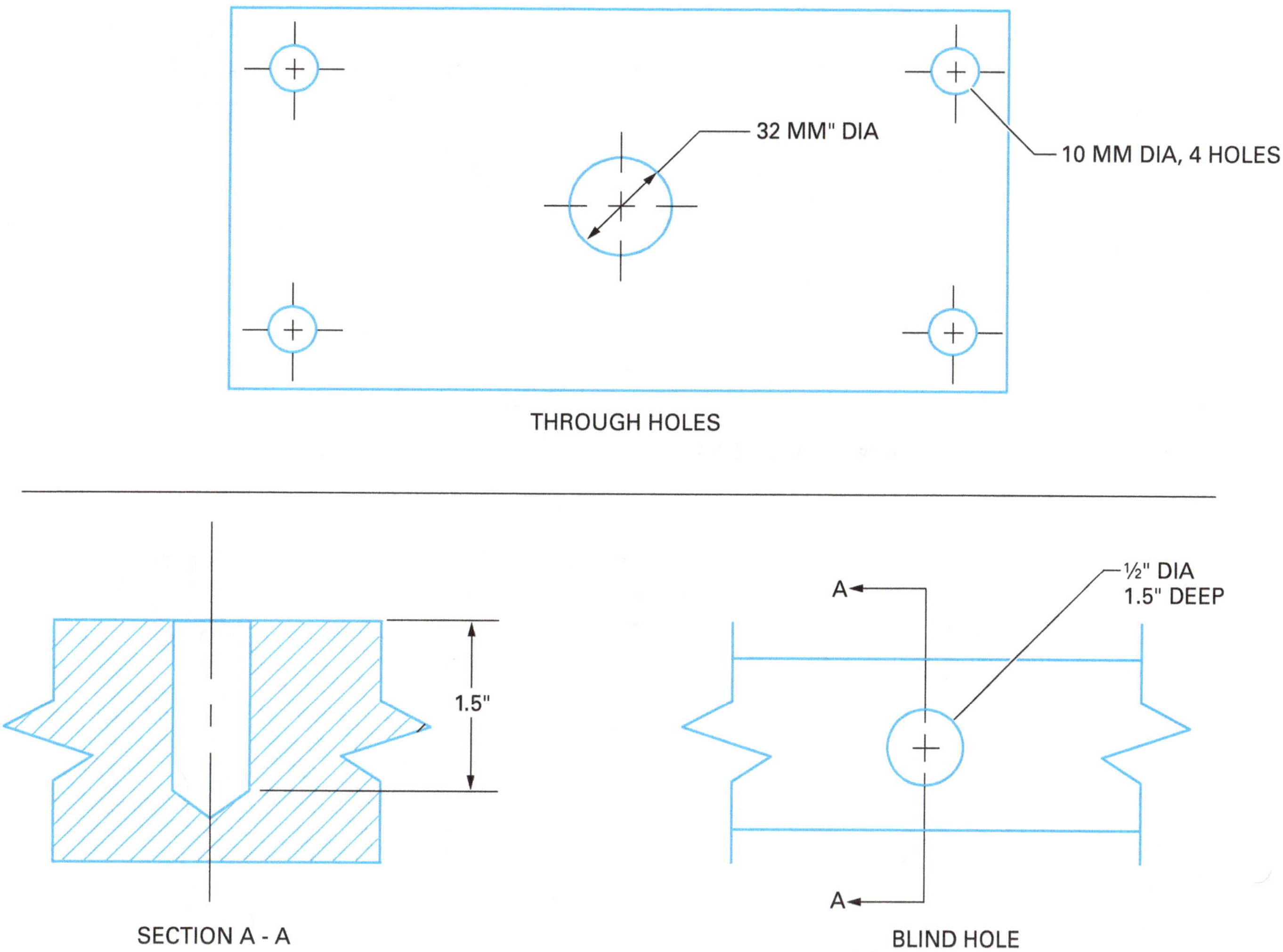

Figure 18 Dimensioning holes.

Structural Steel and Pipe Sizes

Be aware that structural steel and pipe sizes, if given on drawings, are usually nominal (approximate) sizes only. Structural steel is classified as hot rolled or cold rolled. Hot-rolled varieties are somewhat oversized and have rounded corners and edges. The sizes of cold-rolled types are more precise, and the corners and edges are square. Size specifications for some structural shapes including S-beams (formerly called I-beams), H-beams, and wide flange (WF) beams are nominal.

Pipe is sometimes sized through the application of a schedule number as part of a specification. The eleven schedules commonly used are: 5, 10, 20, 30, 40, 50, 60, 80, 100, 120, and 160. The schedule numbers, which represent the strength of the pipe, refer to wall thickness. Pipe schedule tables provide the nominal wall thickness, weight, and outside diameter for each size and type of pipe.

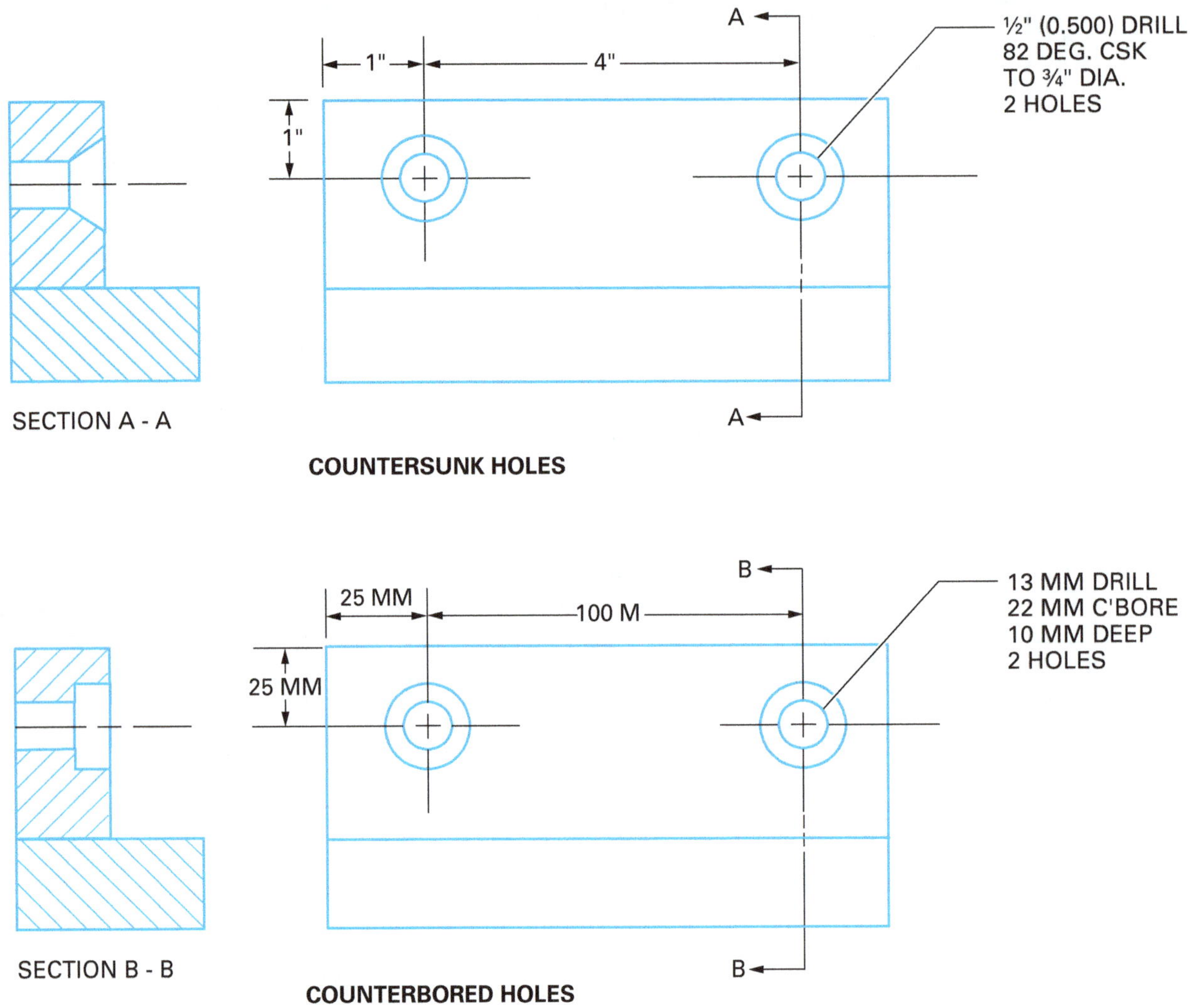

Figure 19 Dimensioning counterbores and countersinks.

2.1.5 Radius and Arc Dimensions

When the ends, sides, or corners of an object are to be rounded, a radius or arc is shown and dimensioned (*Figure 21*). Typically a combination of center lines, radius, extension, and dimension lines are used. To obtain the arc required, the radius line may start outside the physical shape of the object.

2.1.6 Tolerances

Tolerance is an important part of dimensioning because it identifies the variance in size that is allowable on a part. Depending on its intended use, a part can have a large or a small tolerance. If a part is fabricated outside the allowable tolerance, the part will have to be reworked or scrapped at an additional cost. On the other hand, if a part with a large tolerance is fabricated to very fine tolerances, this will generally add to the cost since parts with large tolerances typically can be fabricated faster than parts with smaller tolerances. For these reasons, it is important to identify the tolerances on a drawing and then work accordingly.

Tolerances can be given following the dimension, as a note, or in a table as part of the title block. They are generally given as a plus (+) or minus (−) quantity. For example, if the dimension 10 inches was followed by the tolerance, + or − ⅛ inch, the object could be 9⅞ inches through 10⅛ inches.

The plus and minus tolerances for a size can be different. For example, a dimension of 12 inches, + ⅛ inch and −0, indicates that the object can be from 12 inches through 12⅛ inches. *Figure 22* shows three ways of indicating tolerances.

Think About It

Cut Tolerance

What are the minimum and maximum allowable sizes of an 18-inch cut that has to be made with a tolerance of 6¼ inches?

2.2.0 Notes and Bill of Materials

Additional information that can be found on many drawings includes notes and a bill of materials. These forms of documentation provide data that are not given elsewhere on the drawing.

2.2.1 Notes

A note is an explanatory or critical comment concerning the details of construction. Two types of notes are found on drawings: general notes and local (specific) notes (*Figure 23*). A general note applies to the drawing in its entirety, and it is placed in an area away from the views so that it will be noticed. On large jobs, general notes are placed on a separate drawing. On smaller jobs, general notes are placed on detail drawings, where they refer to the entire drawing. General notes are grouped together and numbered.

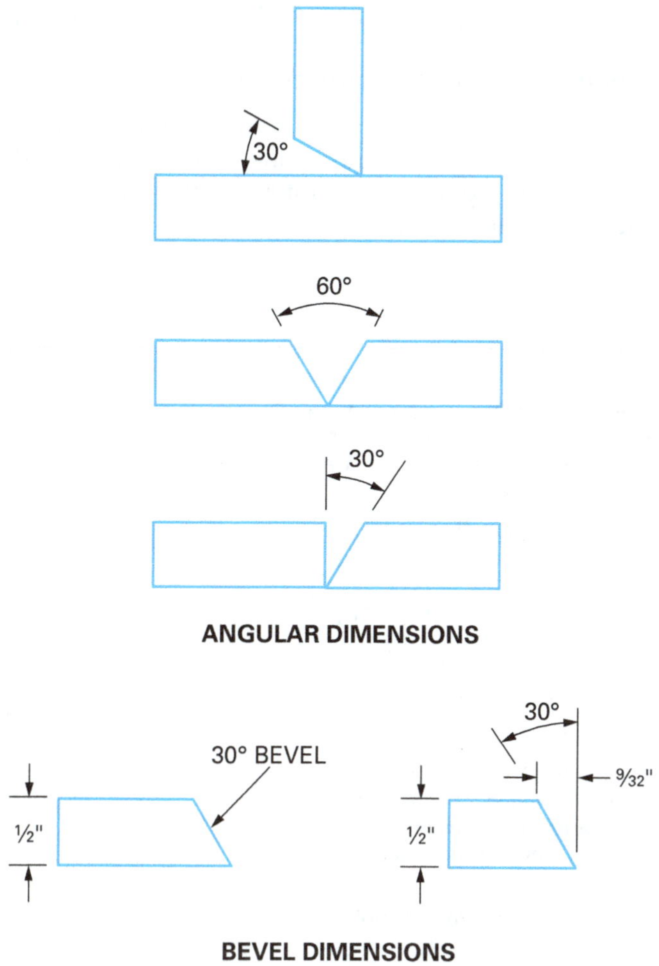

Figure 20 Specifying angles and bevels.

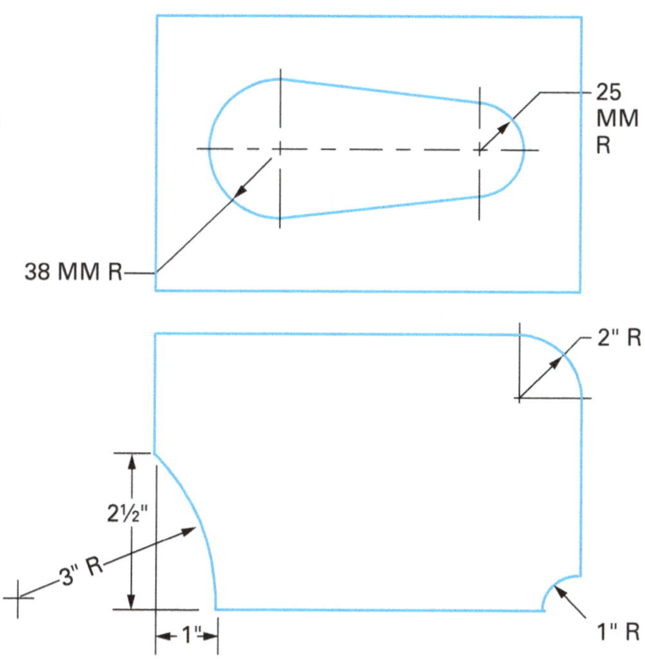

Figure 21 Dimensioning radius and arcs.

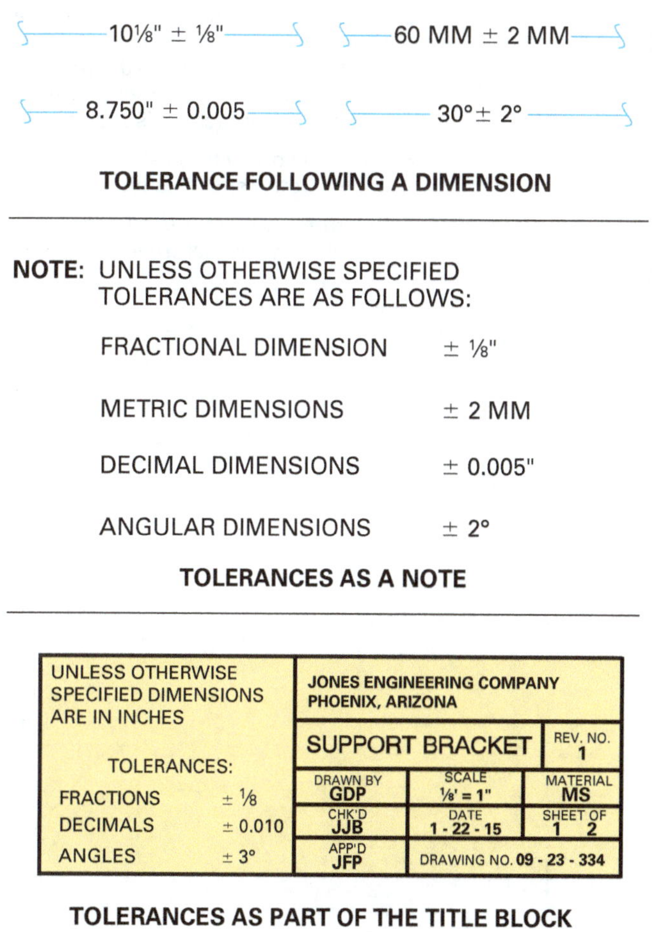

Figure 22 Dimensioning tolerances.

Items typically found in general notes include the following:

- References to specifications, procedures, or other notes
- Identification of work responsibility
- Surface finish requirements (smoothness, type of finish)
- Special handling requirements
- Material requirements such as specification, grade, and class
- Special assembly requirements

A local note applies to a particular part of an object. It is placed near one of the views that represent the part. Local notes are used for items such as the following:

GENERAL NOTES - STEEL

1. Design, fabrication, and erection of structural steel shall be in accordance with the American Institute of Steel Construction (AISC) *Steel Construction Manual*, Thirteenth Edition (2007), unless otherwise modified on the drawings or in the specifications.
2. Material shall meet the requirements of the following specifications unless otherwise noted:

STRUCTURAL STEEL	-	ASTM A 36 - 08
HIGH STRENGTH BOLTS	-	ASTM A325 - 07a, $\frac{7}{8}$ INCH DIA
WELD STEEL	-	AWS D 1.1 - 08, CLASS E70

3. Shop and field paint shall be as called for in the specifications.
4. Steel to be encased in concrete shall not be painted.
5. Hot dipped galvanized surfaces that have been damaged by welding, cutting, burning, shearing, or other damage incurred during transit or erection shall be repaired with Z.R.C. Cold Galvanizing Compound as manufactured or approved equivalent. Repair shall be in strict accordance with the manufacturer's recommendations.
6. All shop connections shall be welded or made with high strength bolts unless noted specifically.
7. Erectors shall provide all temporary shoring & bracing needed for stability until structure is completed.

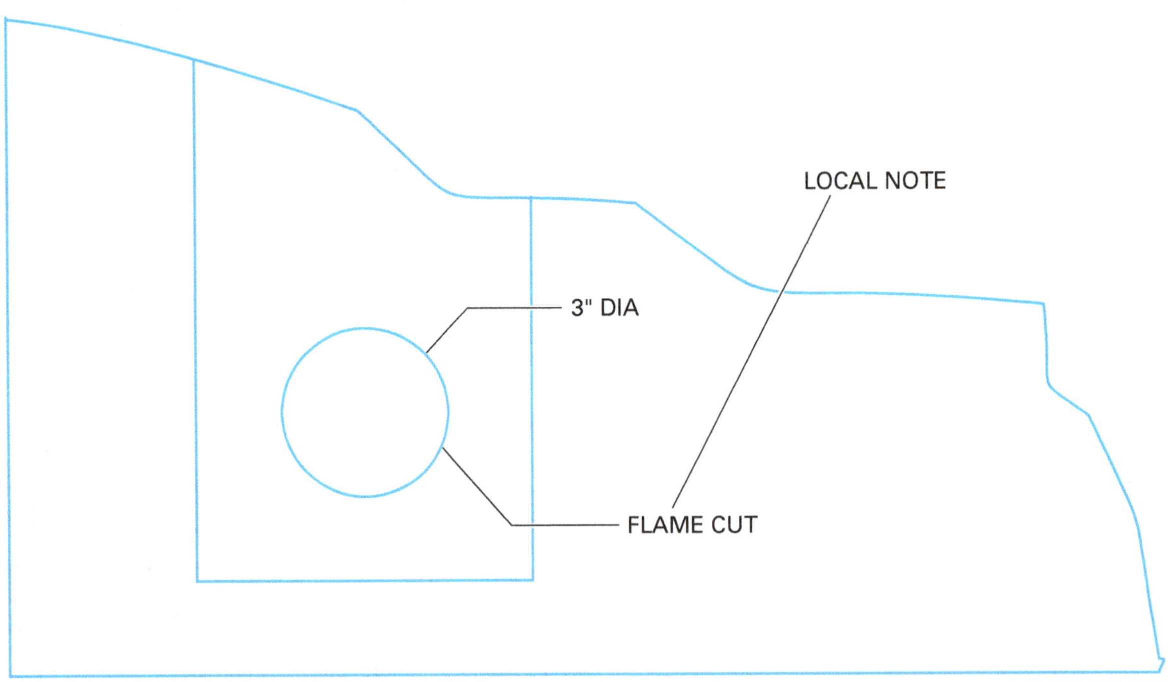

Figure 23 General notes and local notes.

- Thread fastener type and size
- Location of special items (parts to be supplied by others)
- Reference to details on another drawing
- Elevations
- Special treatments or finishes
- Drilled hole size

A specification is a note that provides a detailed description of requirements for the type and size of the parts or material to be used. For instance, it may specify the type and size of electrode and/or the welding process to be used. In *Figure 24*, the letter A refers to Specification A. In this case, Specification A in the tail of the arrow indicates that No. 30, ⅛-inch copper rod is to be used for the weld. Specifications are usually located near the views to which they refer. If there are many specifications, they are attached on a separate sheet of paper and referenced to the drawing.

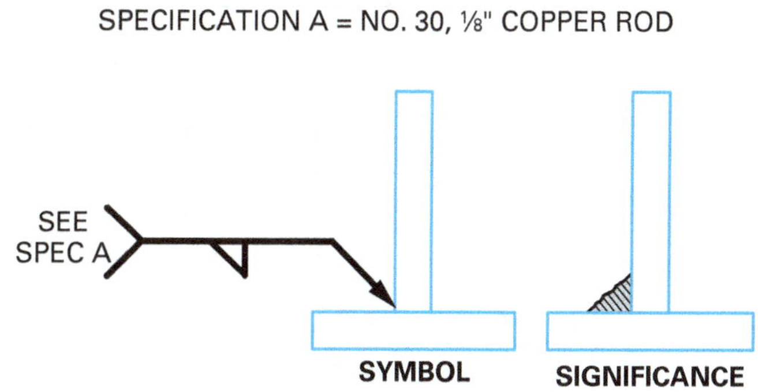

Figure 24 Application of a specification

2.2.2 Bill of Materials

A bill of materials, also called a list of material, parts list, or schedule of parts, is a list of all the parts required to fabricate an object. It generally identifies each part with a number or letter, sometimes called a mark. The bill of materials includes a description of the part, the quantity needed for each assembly, and the part number. Some bills of materials may also include size and dimensions for the part. This is often the case when the parts being described are structural shapes. *Figure 25* shows a portion of a drawing that includes a bill of materials. Not every drawing, however, will have a bill of materials. For larger projects, the bill of materials alone may require several drawing sheets.

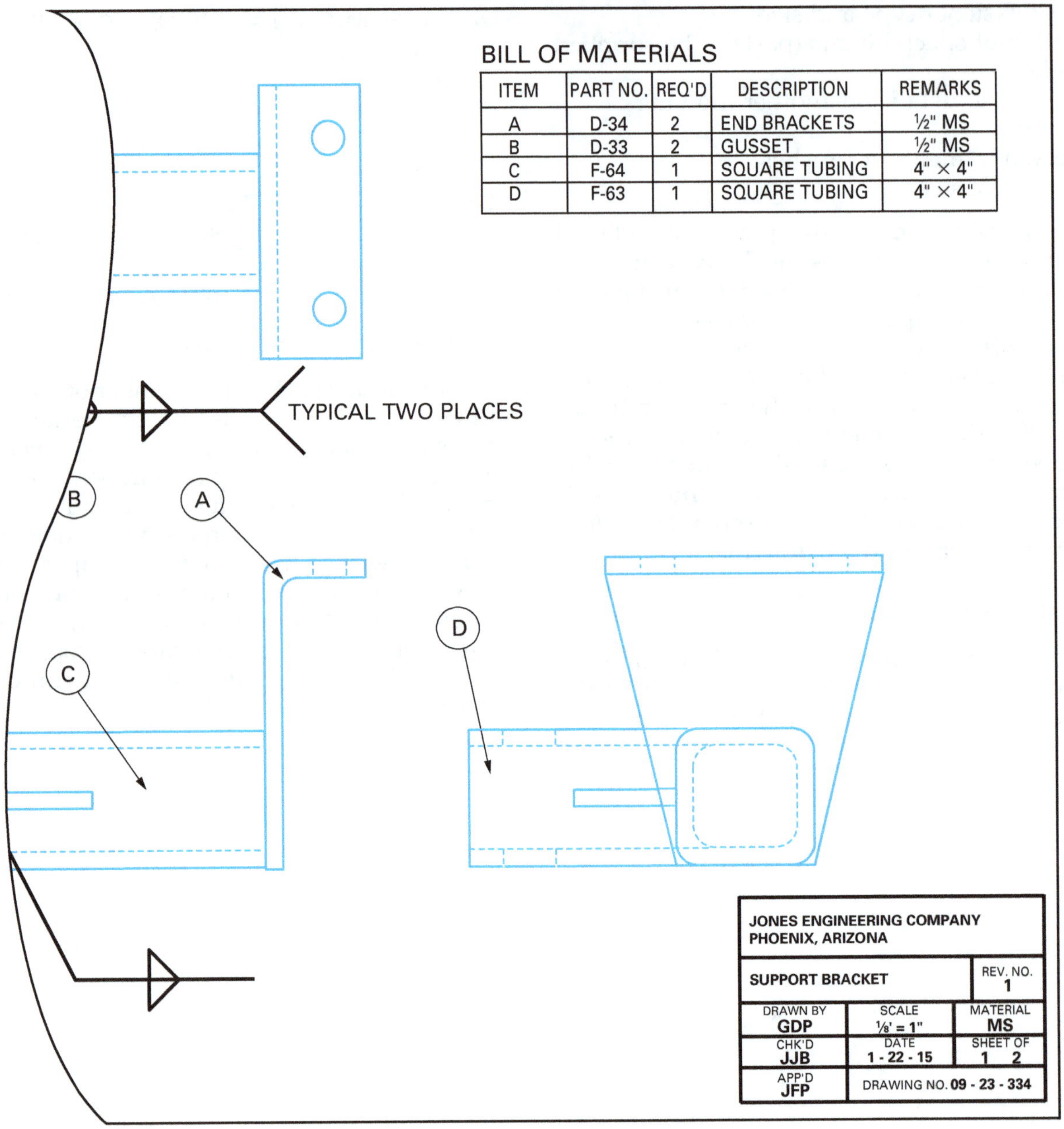

Figure 25 Bill of materials.

Additional Resources

ASTM A325, Standard Specification for Structural Bolts, Steel, Heat Treated, 120/105 ksi Minimum Tensile Strength. West Conshohocken, PA: ASTM International.

AWS A2.4, Standard Symbols for Welding, Brazing, and Nondestructive Examination. Miami, FL: American Welding Society.

AWS D1.1/D1.1M, Structural Welding Code Steel. Miami, FL: American Welding Society.

AWS A3.0, Standard Welding Terms and Definitions. Miami, FL: American Welding Society.

How to Read Shop Drawings. Cleveland, OH: The James F. Lincoln Arc Welding Foundation, 2012.

Steel Construction Manual. Fourteenth Edition. Chicago, IL: American Institute of Steel Construction, 2011.

Welding Symbols on Drawings. E. N. Gregory and A. A. Armstrong. Boca Raton, FL: CRC Press, 2005.

2.0.0 Section Review

1. A hole that will only be drilled part of the way through an object is called a ____.
   a. counterbored hole
   b. countersunk hole
   c. blind hole
   d. through hole

2. Items such as identification of work responsibility and special assembly requirements are given in ____.
   a. classification schedules
   b. specification notes
   c. welding parts lists
   d. general notes

SUMMARY

Engineers, designers, and draftsmen use welding detail drawings to precisely convey information to welders in the field. To correctly interpret this information, welders must be able to accurately read welding detail drawings.

Reading and interpreting welding detail drawings requires welders to have an understanding of lines, fills, views, and dimensioning. It is also important for welders to be able to read and interpret drawing notes and the bill of materials.

Review Questions

1. What type of weld is used to join parts 1F and 1B of the object shown in *Review Question Figure 1*?
   a. Concave
   b. Convex
   c. Groove
   d. Fillet

2. When an object to be fabricated is too complicated for one drawing, the complete and assembled object may be shown on one or more detail drawings plus a(n) ____.
   a. general note
   b. bill of materials
   c. structural print
   d. assembly drawing

3. On a typical drawing, a thick, solid line used to show all visible edges of a part is called a(n) ____.
   a. object line
   b. center mark
   c. leader line
   d. hidden section

4. A series of thin lines that are either solid, or solid and broken, and are arranged in a pattern to indicate various types of substances, such as metals, is called a(n) ____.
   a. material symbol
   b. cut-away projection
   c. cutting plane
   d. extension grid

5. How tall is part 1C of the object shown in *Figure 1*?
   a. 2"
   b. 3$\frac{13}{16}$"
   c. 4"
   d. 6"

6. Which part or parts of the object shown in *Figure 1* must be drilled?
   a. 1A
   b. 1F
   c. 1A and 1D
   d. 1B and 1C

7. Angles and bevels are common on welding detail drawings because they are used with ____.
   a. fillet welds
   b. groove welds
   c. oxyfuel cuts
   d. weld joint preparation

8. What degree of angle is used for the bevel groove weld joining 1D and 1C in *Figure 1*?
   a. 60 degrees
   b. 50 degrees
   c. 30 degrees
   d. 20 degrees

9. The tolerances used for the dimensions of a drawing can be given (shown) following the dimension, as a note, or in the ____.
   a. title block
   b. scale factor
   c. drawing title
   d. object drawing itself

10. How many pieces make up the weldment shown in *Figure 1*?
    a. 4
    b. 5
    c. 6
    d. 8

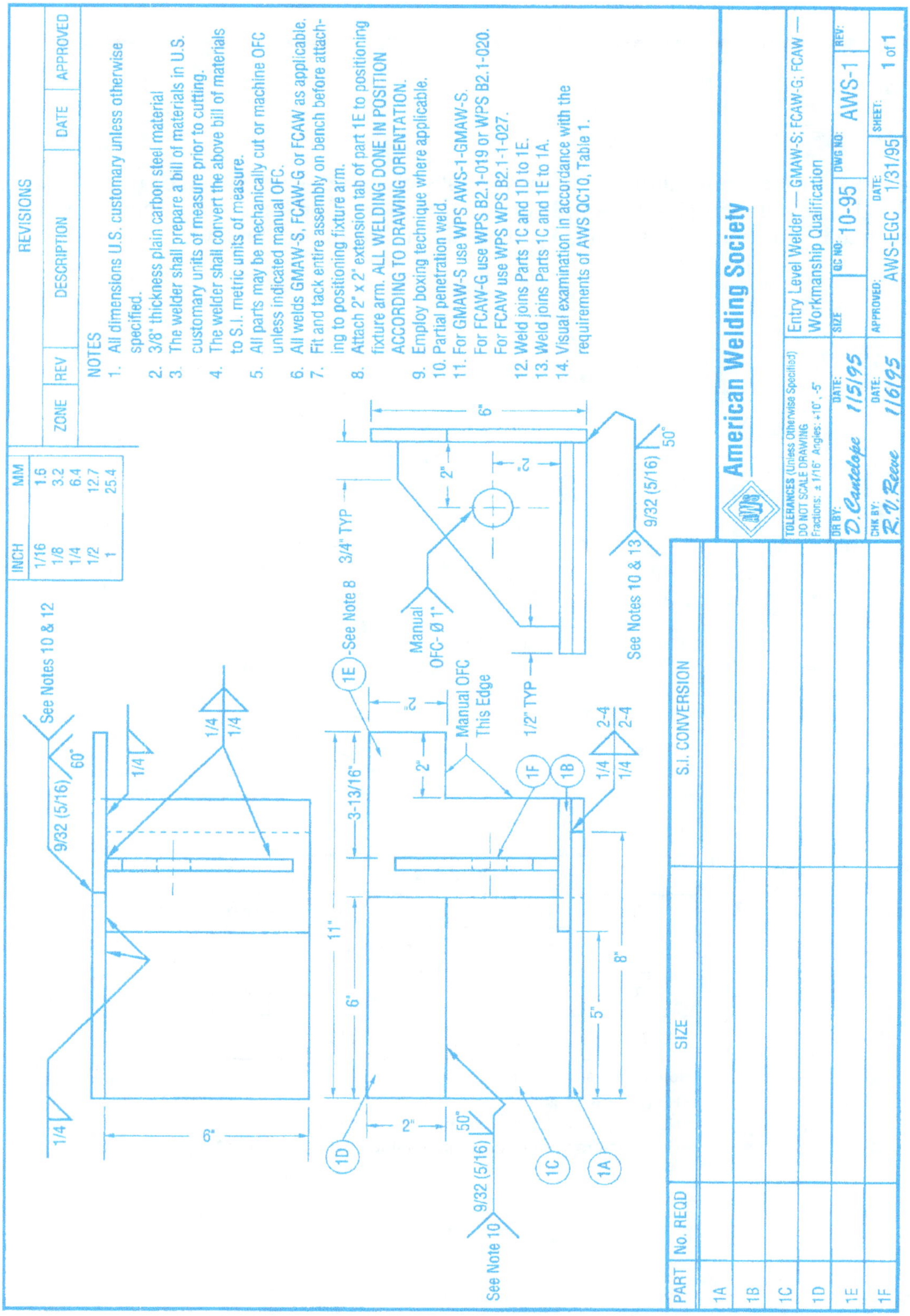
REVISIONS
ZONE
REV
DESCRIPTION
DATE
APPROVED
NOTES
1. All dimensions U.S. customary unless otherwise specified.
2. 3/8" thickness plain carbon steel material
3. The welder shall prepare a bill of materials in U.S. customary units of measure prior to cutting.
4. The welder shall convert the above bill of materials to S.I. metric units of measure.
5. All parts may be mechanically cut or machine OFC unless indicated manual OFC.
6. All welds GMAW-S, FCAW-G or FCAW as applicable.
7. Fit and tack entire assembly on bench before attaching to positioning fixture arm.
8. Attach 2" x 2" extension tab of part 1E to positioning fixture arm. ALL WELDING DONE IN POSITION ACCORDING TO DRAWING ORIENTATION.
9. Employ boxing technique where applicable.
10. Partial penetration weld.
11. For GMAW-S use WPS AWS-1-GMAW-S.
For FCAW-G use WPS B2.1-019 or WPS B2.1-020.
For FCAW use WPS WPS B2.1-027.
12. Weld joins Parts 1C and 1D to 1E.
13. Weld joins Parts 1C and 1E to 1A.
14. Visual examination in accordance with the requirements of AWS QC10, Table 1.
INCH
MM
1/16
1.6
1/8
3.2
1/4
6.4
1/2
12.7
1
25.4
See Notes 10 & 12
9/32 (5/16)
60°
1/4
1/4
1/4
6"
11"
6"
3-13/16"
2"
2"
5"
8"
1E
-See Note B
3/4" TYP
Manual OFC- Ø 1"
2"
2"
6"
Manual OFC This Edge
1/2" TYP
1F
1B
1/4
1/4
2-4
2-4
See Notes 10 & 13
9/32 (5/16)
50°
1D
1C
1A
50°
9/32 (5/16)
See Note 10
PART
No. REQD
SIZE
S.I. CONVERSION
1A
1B
1C
1D
1E
1F
American Welding Society
TOLERANCES (Unless Otherwise Specified)
DO NOT SCALE DRAWING
Fractions: ± 1/16" Angles: +10°, -5°
Entry Level Welder — GMAW-S; FCAW-G; FCAW — Workmanship Qualification
DR BY:
D. Cantelope
DATE:
1/5/95
SIZE
QC NO:
10-95
DWG NO:
AWS-1
REV:
CHK BY:
R.V. Reeve
DATE:
1/6/95
APPROVED:
AWS-EGC
DATE:
1/31/95
SHEET:
1 of 1

Figure 1

Performance Task

The Performance Task for this module is to draw or sketch a three-view detail drawing based on a given image or object. The following is an example of such an image. The three-view detail drawing representing the completed Performance Task is also provided here for trainee review.

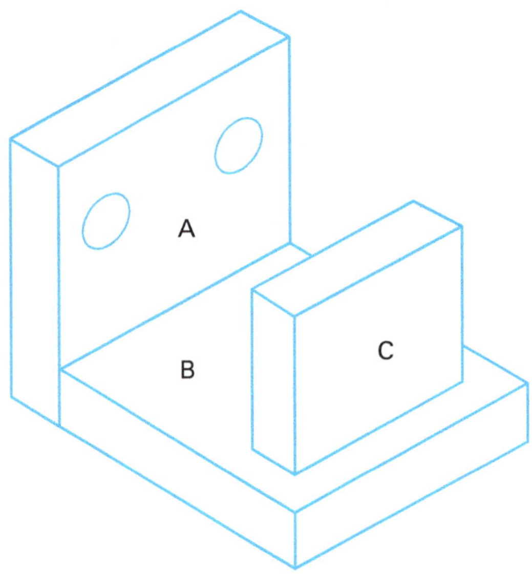

DETAILS			
Part	Thickness	Length	Width
A	¾"	3½" (Tall)	3½"
B	¾"	4" (Long)	3½"
C	¾"	2½" (Tall)	2½"

Hole Dimensions and Location

- Hole diameter is ¾"
- Hole centers are 1¼" from the top edge of block "A"
- Hole centers are ¾" from the outside edges of block "A"

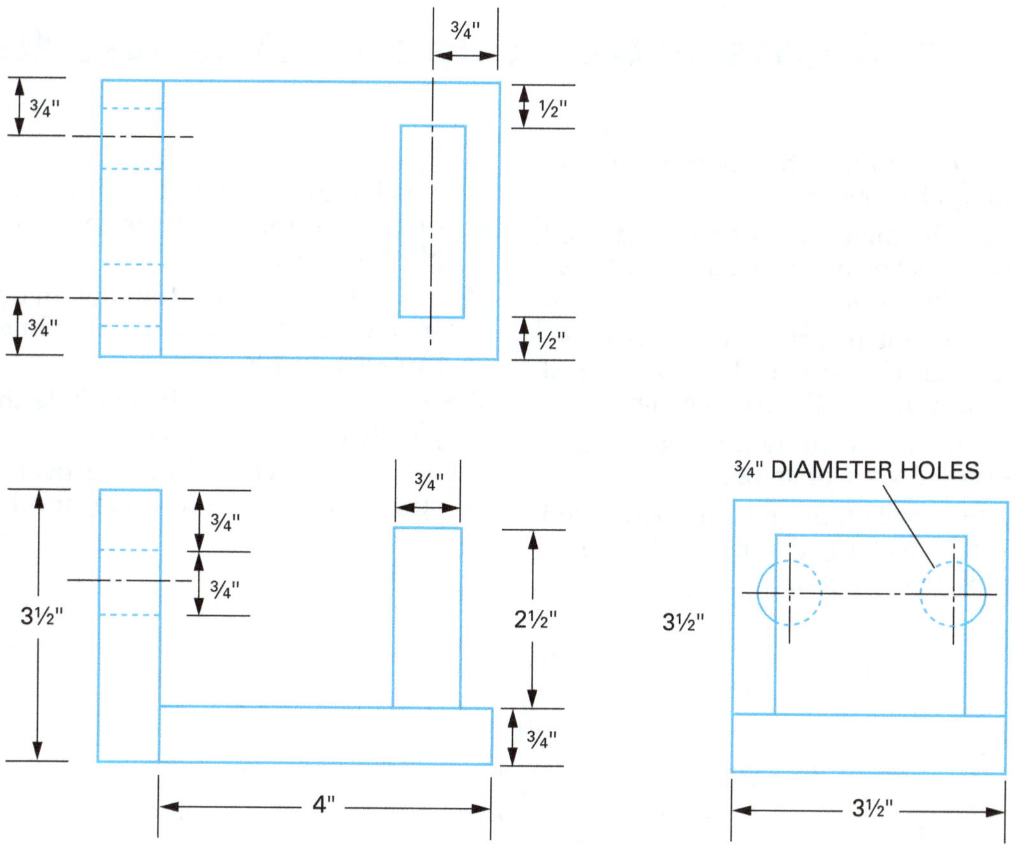

RESULT OF REDRAWING THE OBJECT USING THREE ORTHOGRAPHIC VIEWS

Trade Terms Introduced in This Module

Blind hole: A drilled hole that does not go all the way through the object.

Counterbore: To enlarge a portion of a drilled hole so that a socket head fastener will seat flush or below the surface.

Elevation: A height measurement above sea level or other identified point; also, the vertical sides of an object in an orthographic view.

Isometric view: A drawing depicting a three-dimensional view of an object.

Mark: A system of letters and numbers used with structural steel to identify the type of member and where it is placed.

Orthographic projection: A method of developing multiple flat views of an object by projecting lines out from the object at right angles from the face.

Scale: A method of drawing an object larger or smaller than its actual size while keeping the proportions the same.

Through hole: A drilled hole that goes completely through an object.

Tolerance: The allowance over or under the specified dimension, value, or other parameter.

Additional Resources

This module is intended as a thorough resource for task training. The following reference works are suggested for further study.

ASTM A325, Standard Specification for Structural Bolts, Steel, Heat Treated, 120/105 ksi Minimum Tensile Strength. West Conshohocken, PA: ASTM International.

AWS A2.4, Standard Symbols for Welding, Brazing, and Nondestructive Examination. Miami, FL: American Welding Society.

AWS A3.0, Standard Welding Terms and Definitions. Miami, FL: American Welding Society.

AWS D1.1/D1.1M, Structural Welding Code Steel. Miami, FL: American Welding Society.

How to Read Shop Drawings. Cleveland, OH: The James F. Lincoln Arc Welding Foundation, 2012.

Steel Construction Manual. Fourteenth Edition. Chicago, IL: American Institute of Steel Construction, 2011.

Welding Symbols on Drawings. E. N. Gregory and A. A. Armstrong. Boca Raton, FL: CRC Press, 2005.

Figure Credits

The Lincoln Electric Company, Cleveland, OH, USA, Module opener

AWS EG2.0-95, Figure 2, Reproduced with permission from the American Welding Society (AWS), Miami, FL, USA, RQ01

Section Review Answer Key

Answer	Section Reference	Objective
Section One		
1. d	1.0.0	1a
2. d	1.2.3	1b
Section Two		
1. c	2.1.3	2a
2. d	2.2.1	2b

NCCER CURRICULA — USER UPDATE

NCCER makes every effort to keep its textbooks up-to-date and free of technical errors. We appreciate your help in this process. If you find an error, a typographical mistake, or an inaccuracy in NCCER's curricula, please fill out this form (or a photocopy), or complete the online form at **www.nccer.org/olf**. Be sure to include the exact module ID number, page number, a detailed description, and your recommended correction. Your input will be brought to the attention of the Authoring Team. Thank you for your assistance.

Instructors – If you have an idea for improving this textbook, or have found that additional materials were necessary to teach this module effectively, please let us know so that we may present your suggestions to the Authoring Team.

NCCER Product Development and Revision

13614 Progress Blvd., Alachua, FL 32615

Email: curriculum@nccer.org
Online: www.nccer.org/olf

❑ Trainee Guide ❑ Lesson Plans ❑ Exam ❑ PowerPoints Other ____________

Craft / Level: Copyright Date:

Module ID Number / Title:

Section Number(s):

Description:

Recommended Correction:

Your Name:

Address:

Email: Phone:

Physical Characteristics and Mechanical Properties of Metals

Overview

In order for welders to accurately identify base metals and select the proper filler metals, they must have a thorough understanding of the composition, classification, physical characteristics, and mechanical properties of common metals and alloys. They must be familiar with metallurgical factors that need to be considered during welding activities and must also be able to recognize standard commercial steel shapes used for fabrication and construction.

Module 29203

Trainees with successful module completions may be eligible for credentialing through the NCCER Registry. To learn more, go to **www.nccer.org** or contact us at **1.888.622.3720**. Our website has information on the latest product releases and training, as well as online versions of our *Cornerstone* magazine and Pearson's product catalog.

Your feedback is welcome. You may email your comments to **curriculum@nccer.org**, send general comments and inquiries to **info@nccer.org**, or fill in the User Update form at the back of this module.

This information is general in nature and intended for training purposes only. Actual performance of activities described in this manual requires compliance with all applicable operating, service, maintenance, and safety procedures under the direction of qualified personnel. References in this manual to patented or proprietary devices do not constitute a recommendation of their use.

29203 V5

29203
Physical Characteristics and Mechanical Properties of Metal

Objectives

When you have completed this module, you will be able to do the following:

1. Describe the composition and classification systems for a variety of metals.
   a. Describe the composition and classification system for ferrous metals.
   b. Describe the composition and classification system for low-alloy steel.
   c. Describe the composition and classification system for common-grade stainless steel.
   d. Describe the composition and classification system for specialty-grade stainless steel.
   e. Describe the composition and classification system for nonferrous metals.
2. Describe the physical and mechanical characteristics of metals and explain how to identify base metals.
   a. Describe the physical characteristics of different metals.
   b. Describe the mechanical properties of different metals.
   c. Explain how to identify base metals in field conditions.
   d. Describe metallurgy-related considerations for welding.
3. Identify the common structural shapes of metal.
   a. Identify the most common structural steel shapes.
   b. Identify different structural beam shapes.
   c. Identify pipe and tubing types.
   d. Identify other common metal forms, including rebar.

Performance Tasks

This is a knowledge-based module; there are no performance tasks.

Trade Terms

Alloy
American Iron and Steel Institute (AISI)
American Society for Testing and Materials (ASTM International)
Austenitizing
Casting
Certified mill test report (CMTR)
Coefficient
Ductile
Ductility
Ferrous
Hardenability
Heat-affected zone (HAZ)
Hot shortness
Interpass temperature
Malleable
Mechanical properties
Metallic
Mill test report (MTR)
Nonferrous metal
Notch toughness
Postweld heat treatment (PWHT)
Quench
Residual stress
Sintered
Society of Automotive Engineers (SAE)
Stress-relief heat treatment
Tempering
Tensile strength
Underbead cracking
Unified numbering system (UNS)
Web
Wide flange (WF) beam
Wrought

Industry Recognized Credentials

If you are training through an NCCER-accredited sponsor, you may be eligible for credentials from NCCER's Registry. The ID number for this module is 29203. Note that this module may have been used in other NCCER curricula and may apply to other level completions. Contact NCCER's Registry at 888.622.3720 or go to **www.nccer.org** for more information.

Contents

Topics to be presented in this module include:

Figures and Tables

Section One

1.0.0 Metal Composition and Classification

Objective

Describe the composition and classification systems for a variety of metals.

a. Describe the composition and classification system for ferrous metals.
b. Describe the composition and classification system for low-alloy steel.
c. Describe the composition and classification system for common-grade stainless steel.
d. Describe the composition and classification system for specialty-grade stainless steel.
e. Describe the composition and classification system for nonferrous metals.

Trade Terms

Alloy: A metal that has had other elements added to it that substantially change its mechanical properties.

American Iron and Steel Institute (AISI): An industry organization responsible for preparing standards for steel and steel alloys that are based upon the Society of Automotive Engineers (SAE) code system.

American Society for Testing and Materials (ASTM International): An organization that developed a code system for identifying and labeling steels.

Austenitizing: Heating a steel up to or above the transformation temperature range to achieve partial or complete transformation to austenite grain structure.

Casting: The metal object produced by pouring molten metal into a mold.

Certified mill test report (CMTR): A quality assurance report provided to a customer by a metal manufacturer that shows the chemical content and testing results of the metal being purchased; also called a mill test report (MTR).

Coefficient: A numerical measure of a physical or chemical property that is constant for a system under specified conditions, such as the coefficient of friction.

Ductile: The term used to describe a material that can be bent or shaped without breaking.

Ductility: The characteristic of metal that allows it to be stretched, drawn, or hammered without breaking.

Ferrous: Relating to iron or an alloy that is mostly iron; a metal containing iron.

Hardenability: A characteristic of a metal that enables it to become hard, usually through heat treatment.

Heat-affected zone (HAZ): The part of the base metal that has not been melted but has been altered structurally by the weld heat.

Hot shortness: The condition of metal when it proves to be very brittle and unbendable at red heat but can be bent without showing signs of brittleness when cold or at white heat. This condition is often a result of high sulfur and phosphorus content.

Interpass temperature: In a multiple-pass weld, the temperature of the base metal at the time the next weld pass is started; it is usually the same temperature that is used for preheating.

Malleable: Able to be hammered or pressed into another shape without breaking.

Mechanical properties: The characteristics or traits that indicate how flexible and strong a metal will be under stress.

Metallic: Having characteristics of metal, such as ductility, malleability, luster, and heat and electrical conductivity.

Mill test report (MTR): A quality assurance report provided to a customer by a metal manufacturer that shows the chemical content and testing results of the metal being purchased; also called a certified mill test report (CMTR).

Nonferrous metal: A metal, such as aluminum, copper, or brass that lacks sufficient quantities of iron to be affected by its properties.

Notch toughness: The impact load required to break a test piece of weld metal measured by a notch toughness test (Charpy V-notch test). This test may be performed on metal below room temperature, at which point it may be more brittle.

Postweld heat treatment (PWHT): The process of heat treating the base metal and weld after the weld has been made.

Quench: To rapidly cool a hot metal using air, water, or oil.

Society of Automotive Engineers (SAE): An automotive society responsible for creating an early system for classifying carbon steels.

Stress-relief heat treatment: A technique used to reduce welding stresses and to increase the ductility and toughness of the weld metal and the heat-affected zone (HAZ).

Tempering: The process of reheating quench hardened or normalized steel to a temperature below the transformation range and then cooling it at any rate desired.

Tensile strength: The maximum stress or force that a material can withstand without breaking.

Unified numbering system (UNS): An industry-accepted numbering system for identifying ferrous metals and alloys.

Web: The plate joining the flanges of a girder, joist, or rail.

Wide flange (WF) beam: A metal I-beam that has an extra wide flange at the top and bottom.

Wrought: Formed or shaped by hammering or rolling.

All living and nonliving things are made up of chemical elements that can be altered by changes in temperature. A steak left on a grill will eventually burn; a piece of wood thrown into a fire will also burn. Metals are no different. When they are heated to a high enough temperature they will melt, and some will burn.

Metals, like other substances, are a collection of chemical elements bound together by natural forces or by man-made production processes. The chemical elements in some metals make them stronger than other metals. Those same chemical elements allow some metals to withstand stress or force better than others.

The composition of common welding base metals varies widely, from metals made of essentially one metallic element, to alloys (mixtures) of metallic and nonmetallic components. Metals are classified into two basic groups: ferrous metals, which are composed mainly of iron, and nonferrous metals, which contain very little or no iron. Ferrous metals include all steel, cast iron, wrought iron, malleable iron, and ductile (nodular) iron. Nonferrous metals and their alloys include the light metals (aluminum, magnesium, titanium); the heavy metals (copper, nickel, lead, tin, zinc); and the precious metals (platinum, gold, silver).

1.1.0 Ferrous Metal Composition and Classification Systems

Ferrous metals contain mostly iron and some carbon. The amount of carbon ranges from almost none in commercially pure iron to about 1.4 percent. Steel contains less than 1.7 percent carbon. Above 1.4 percent carbon, steel develops brittle properties similar to those of cast iron. Carbon steel, the largest group of the ferrous-based metals, contains much smaller amounts of carbon than cast iron. As the carbon content increases in carbon steel, the steel becomes harder and more wear-resistant. The carbon content of a particular carbon steel is the largest factor in determining its use. Steel castings usually contain a higher percentage of carbon than rolled plate and other rolled shapes. The casting foundry can usually provide the alloy formula.

Steels also contain impurities and trace amounts of elements such as manganese, silicon, phosphorus, copper, and lead. Sometimes additional amounts of these elements are intentionally added to steel to change its characteristics, such as to increase its forgeability (the ability to be heated in a forge and then beaten or hammered into shape) or to decrease a tendency toward forming blowholes (the vents that form when air or gases escape).

Steels, including carbon steels, low-alloy carbon steels, alloy steels, and stainless steels, are classified by various systems, such as specification number and grade or manufacturer's trade name and number. Steel classification systems include those developed by the American Iron and Steel Institute (AISI), the American Society for Testing and Materials (ASTM International), and the Unified Numbering System (UNS), as well as manufacturer's trade names and identification numbers. Stainless steels have their own classification systems. They are sometimes referred to by the percentages of their chromium and nickel content, such as 18/8, 25/20, and 18/10; but this system has largely been replaced by the AISI stainless steel classification system. Common classification systems for steels are explained in the following sections.

1.1.1 Carbon Steel Classification

The principal classification system for carbon steels is the AISI numerical designation of Standard Carbon and Alloy Steels. It was originally a four-digit system developed by the Society of Automotive Engineers (SAE) for carbon steels commonly used in structural shapes, plate, strip, sheet, and welded tubing. Five digits are now used for some alloys. The classification system may be referred to as the AISI, SAE, or AISI/SAE system.

The AISI and SAE designations are essentially the same, except that the AISI system sometimes uses a letter prefix to indicate the manufacturing process that produced the steel. The following are examples of letter designations:

- *A* – Open-hearth steel
- *B* – Acid Bessemer carbon steel
- *C* – Basic open-hearth carbon steel
- *D* – Acid open-hearth carbon steel
- *E* – Electric furnace steel

Carbon and Alloy Steels

Steel is an alloy of iron and carbon typically containing less than 1 percent carbon. All steels also contain varying amounts of other elements such as manganese, silicon, phosphorus, sulfur, and oxygen. In addition, some standard alloy steels can also contain elements such as nickel, chromium, and molybdenum. There are currently about 3,500 different grades of steel with many different properties. Most of the steels produced today are carbon and alloy steels. Ninety-five percent of the construction and fabrication metals used worldwide consist of these materials.

The absence of a letter prefix indicates basic open-hearth or acid-Bessemer carbon steel. The prefixing letter (if any) designates the manufacturing process. The first two numerical digits represent the series (type and class) of steel. The third and fourth (and sometimes fifth) numerical digits specify the approximate percentage of carbon content. *Table 1* shows the AISI/SAE system for designating carbon and alloy steel types and classes and includes the AISI/SAE number.

The following examples demonstrate the use of the AISI/SAE numbers for carbon steels:

- AISI number C1020:
  - C = Indicates basic open-hearth carbon steel
  - 10 = Carbon steel, nonresulfurized
  - 20 = Contains approximately 0.20 percent carbon
- AISI number E2512:
  - E = Indicates electric furnace steel
  - 25 = Designates steel alloyed with approximately 5 percent nickel
  - 12 = Designates steel containing approximately 0.12 percent carbon
- AISI number E52100:
  - E = Indicates electric furnace steel
  - 5 = Contains either approximately 0.50 percent, 1.00 percent, or 1.45 percent chromium, designated by the next digit: 0, 1, or 2, respectively
  - 2 = Designates approximately 1.45 percent chromium
  - 100 = Designates approximately 1 percent carbon

The common group classification of carbon steels is primarily based on carbon content. Plain carbon steels (AISI series 10XX through 11XX) are basically iron-carbon alloys. High-strength low-alloy (HSLA) carbon steels (AISI series 13XX through 98XX) have small amounts of alloying elements to improve strength, hardness, and toughness or to increase resistance to corrosion,

Table 1 AISI/SAE System for Designating Carbon and Alloy Steel Types and Classes

ASI/ASE Number	Carbon Steel Types and Classes
1XXX	Carbon steels
10XX	Plain (nonresulfurized) carbon steel grades
11XX	Free machining, resulfurized (screw stock)
12XX	Free machining, resulfurized, rephosphorized
13XX	Manganese 1.75 percent
15XX	High-manganese (1.00 to 1.65 percent) carburizing steels
2XXX	Nickel steels
23XX	3.50 percent nickel
25XX	5.00 percent nickel
3XXX	Nickel-chromium steels
31XX	1.25 percent nickel, 0.65 or 0.80 percent chromium
32XX	1.75 percent nickel, 1.07 percent chromium
33XX	3.50 percent nickel, 1.50 or 1.57 percent chromium
34XX	3.00 percent nickel, 0.77 percent chromium
4XXX	Molybdenum steels
40XX	0.20 or 0.25 percent molybdenum
41XX	0.50, 0.80, or 0.95 percent chromium, 0.12, 0.20, 0.25, or 0.30 percent molybdenum
43XX	1.82 percent nickel, 0.50 or 0.80 percent chromium, 0.25 percent molybdenum
46XX	0.85 or 1.82 percent nickel, 0.20 or 0.25 percent molybdenum
47XX	1.05 percent nickel, 0.45 percent chromium, 0.20 or 0.35 percent molybdenum
48XX	3.50 percent nickel, 0.25 percent molybdenum
5XXX	Chromium steels
50XX	0.27, 0.40, 0.50, or 0.65 percent chromium
51XX	0.80, 0.87, 0.92, 0.95, 1.00, or 1.05 percent chromium
5XXXX	1.00 carbon, 0.50, 1.02, or 1.45 percent chromium
6XXX	Chromium-vanadium steels
61XX	0.60, 0.80, or 0.95 percent chromium, 0.10 or 0.15 percent (minimum) vanadium
86XX	0.55 percent nickel, 0.50 percent chromium, 0.20 percent molybdenum
92XX	0.65, 0.82, or 0.85 percent manganese, 1.40 or 2.00 percent silicon, 0.00 or 0.65 percent chromium
93XX	3.25 percent nickel, 1.20 percent chromium, 0.12 percent molybdenum
94XX	0.45 percent nickel, 0.40 percent chromium, 0.12 percent molybdenum
97XX	0.55 percent nickel, 0.20 percent chromium, 0.20 percent molybdenum
98XX	1.00 percent nickel, 0.80 percent chromium, 0.25 percent molybdenum
XXBXX	Boron steels
XXLXX	Leaded steels

Steel Alloying Elements

Some alloying elements used with steel and the reasons they are added include the following:

- Manganese – Promotes forgeability
- Silicon – Prevents blowholes
- Phosphorus – Improves machinability
- Copper – Inhibits corrosion
- Lead – Improves machinability

heat, and environmental damage. The plain carbon steels are commonly grouped as follows:

- *Low-carbon* – 0.10 percent to 0.15 percent carbon, 0.25 percent to 1.50 percent manganese
- *Mild-carbon* – 0.15 percent to 0.30 percent carbon, 0.60 percent to 0.70 percent manganese
- *Medium-carbon* – 0.30 percent to 0.50 percent carbon, 0.60 percent to 1.65 percent manganese
- *High-carbon* – 0.50 percent to 1.00 percent carbon, 0.30 percent to 1.00 percent manganese

The HSLA carbon steels are alloyed with one or more of the elements manganese, nickel, chromium, or molybdenum to provide higher strength, better toughness, improved weldability, and in some cases, greater resistance to corrosion. As the percentage of these elements increases, changes in the weldability of the low-alloy steels make electrode selection and welding procedures more critical. HSLA steels include the following:

- Low-nickel steels
- Low-nickel chrome steels
- Low-manganese steels
- Low-alloy chromium steels
- Weathering steels

Another class of steel in the AISI series is sulfur steel. It is not an alloy steel but is still included in the series.

1.1.2 ASTM Steel Specification System

ASTM International develops and publishes specifications for use in material production and testing. Welding codes and procedures typically use the ASTM designation to identify metal types because it includes specifications for mechanical and physical characteristics of the material as well as chemical composition. Piping and tubing are the largest group of materials fabricated by welding covered in ASTM specifications.

Prefix letters are a general part of each specification and provide a general idea of the specification content. For example, A is used for ferrous metals, B for nonferrous metals, and E for miscellaneous subjects, such as examination and testing.

Sometimes an S is added before the prefix letter. This indicates that ASME International, publishers of the Boiler and Pressure Vessel Code, has adopted that specification.

With the exception of a few mild steel grades, such as SA-36 (mild steel plate) and SA-53 (mild steel pipe), ASTM specifications are three-digit numbers. Most specifications also have various grades, classes, or types. *Table 2* gives examples of some of the ASTM specifications for alloy steels.

NOTE

The complete collection of ASTM standards consists of 15 sections comprising 65 volumes, with a separate index. The table is provided only to illustrate sample ASTM numbers.

When companies buy steel, the manufacturer will provide the customer with a certificate indicating the quality of the steel. This document is often called a **mill test report (MTR)** or a **certified mill test report (CMTR)**. The report mainly shows the chemical content and testing results of the metal that was purchased (*Figure 1*). Such information is necessary in case the structure being built from the material fails. For liability reasons, this information must be tracked and kept on record.

Included in the mill test report are the ASTM specifications, size, grade, and length of the product, which in this case is **wide flange (WF) beams**. The report also shows the heat number, chemical analysis, and physical properties of the steel. These markings are also found on the materials covered by the report. *Figure 2* shows how the markings look on an actual beam.

UNS

The UNS used to identify metals was developed jointly by ASTM International and SAE as a way of relating the different numbering systems for metals and alloys. Its objective is to eliminate the confusion caused when more than one identification number is used to identify the same material or when the same identification number is used to identify two different materials. It is anticipated that this system will eventually replace the AISI and other systems.

Each UNS number has a letter prefix followed by five digits. Often the prefix letter relates to the family of metals. For example, UNS numbers A00001 through A99999 refer to aluminum and aluminum alloys, and UNS numbers C00001 through C99999 signify copper and copper alloys. When possible, the digits in the UNS sequence contain the same numbering sequences used in the other systems. For example, the UNS number G10200 corresponds to the AISI/SAE number 1020 for carbon steel. Most engineering and materials reference books contain tables that cross-reference AISI and SAE numbers for metals to the corresponding UNS numbers.

Table 2 ASTM Specification Standards

ASTM Number		Tensile	Type of Base Metal
A209	Grade T1	55,000	C-0.05 Mo tubes
	Grade T1a	60,000	C-0.05 Mo tubes
	Grade T1b	53,000	C-0.05 Mo tubes
A217	Grade T17	60,000	1.00 Cr-0.15 V tubes
A334	Grade 5	65,000	5.00 Ni-Steel pipe
A335	Grade P1	55,000	C-0.05 Mo pipe
	Grade P2	55,000	0.60 Cr-0.85 Mo pipe
	Grade P3	60,000	1.57 Cr-0.70 Mo pipe
	Grade P3b	60,000	2.00 Cr-0.55 Mo pipe
	Grade P11	60,000	1.25 Cr-0.55 Mo-Si pipe
	Grade P12	60,000	0.95 Cr-0.55 Mo pipe
	Grade P15	60,000	1.40 Si-0.55 Mo pipe
A517	Grade A	115,000	0.05 Cr-0.025 Mo-Si plates
	Grade B	115,000	0.05 Cr-0.02 Mo-V plates
	Grade P	115,000	0.125 Ni-0.10 Cr-0.05 Mo plates

C – Carbon Cr – Chromium Mo – Molybdenum Ni – Nickel Si – Silicon V – Vanadium

The markings show who made the steel, where it was made, its original dimensions, and its heat number or batch number. Heat or batch numbers are tracking numbers that the steel producer assigns to a particular ladle of steel. When a product, such as a beam or plate, is made from that ladle of steel, the tracking number is placed on that product in case of recall or problems with the chemistry. The following are explanations of the letters and numbers shown on the beam in *Figure 2*:

- *SDI* – Typically the name of the manufacturer.
- *USA* – Country of origin where the steel was melted.
- *A046196* – The heat number or batch number, which can be used to match the material to a material test report. Heat numbers are normally unique to each manufacturer, and they are usually meaningful only to the producer. The number provides a way to trace the product. The material test report lists the mechanical test results and chemistry. These can be checked against the material specification to ensure that the material test report results meet the requirements of the material specification.
- *W 16 × 36* – This beam has a depth of approximately 16 inches (actual depth 15⅞ inches or about 403 mm) and weighs 36 pounds per foot of length (53.6 kilograms per meter). If you looked up W 16 × 36 in the *American Institute of Steel Construction (AISC) Handbook*, you would find the beam's flange width and thickness, web thickness, beam depth, and many other design properties.

NOTE

One piece of information that is not included on the beam in *Figure 2* is the ASTM specification to which the beam was made (for example, A36, A441, A572, A588, etc.).

Plate steel also has mill markings, such as batch or heat numbers (*Figure 3*). The information at the top of the plate indicates that it was not heat treated after being rolled. The S7-09438 heat number is the batch number. The size information is self-explanatory. Grade A36 indicates that the plate was manufactured in accordance with the ASTM specifications for A36 steel plate. The other information at the bottom of the plate shows who made it and who bought it.

Steel Mills, Inc.
First Steel Mill
123 Rebar Way
Narvona, State 90426 USA

Mill Test Report

Chemical and Physical Test Report																		**Customer P.O. Number** ________
Shape & Size	**Grade**		**Specification**															**Sales Order** ________
32mm Rebar (#10)	420(60)		ASTM A615/A615M-03															
Heat Identification Number	C	Mn	P	S	Si	Cu	Ni	Cr	Mo	V	Nb	B	N	Sn	Al	Ti	Zr	Ca	**C Equivalent**
123-123	0.32	0.91	0.015	0.067	0.32	0.38	0.13	0.16	0.016	0.000	0.014	0.000	0.000	0.003	0.000	0.000	0.000	0.0000	0.502

Mechanical Test					
Yield:79800 PSI, 550.2 MPA	Tensile:102000 PSI, 703.27 MA	% EL: 15.0/8 in, 15.0/200MM	Bend: Okay	Def: .073, 1.85MM	%I/h 3.3L

This material, including the billets, was produced and manufactured in the United States of America.

____
Signature of Quality Assurance Manager

The above figures are certified extracts from the original chemical and physical test records, as contained in the permanent records of the company.

____
Signature of Mill Metallurgist

Figure 1 Example of a mill test report.

Figure 2 Mill markings on a beam.

Figure 3 Mill markings on plate steel.

There is no set format for labeling steel, so steel producers have their own way of labeling their products. The mill markings on the beams and plates shown in *Figures 1*, *2*, and *3* were stenciled or sprayed onto the steel. Other manufacturers attach paper labels to their products, as shown in *Figure 4*.

The mill markings on most steel will not be as clearly visible as those shown in these examples. Steel is often stored outside in the weather and may be exposed to other rough conditions or handling that can damage the markings.

When preparations are being made for a welding job, the job planner determines which steel to use. On some jobs, lay-down yard personnel may bring the materials to the work area. However, the fabricators (fitters and welders) must be able to ensure that the delivered steel is the same as that indicated on the welding procedure specification. *Figures 5A* and *5B* show a WPS with standard or English measurements. *Figure 6* shows a WPS with metric values.

When only a part of the original piece of steel is needed, welding personnel must transfer the heat number of the original piece to the piece they are cutting from the main piece. The transfer of numbers must be done before the part is cut away from the original piece. Transferring the heat number ensures that the removed piece of steel can be tracked when it is welded into place. Follow site-specific guidelines for transferring numbers.

On high-profile jobs, the process of tracking these numbers can be very involved because of liability concerns. Many jobs today require the welders to closely track the materials being used. The owner representatives (usually the quality control personnel) can use these heat or batch numbers to trace where the raw materials came from, the day and shift on which the material was produced, and who was on duty at the time. Most plants will not allow material to be used if the heat number has been ground off or rusted over. From a contractor's point of view, material that lacks a heat number should never be used in any critical application. If a failure occurred and the installer could not produce a valid heat number, then the contractor/installer would assume full responsibility, even though the steel was produced by someone else.

1.2.0 Low-Alloy Steel Groups

Low-alloy steels fall into the following general groups:

- High-strength low-alloy steels
- Quench-and-tempered steels
- Heat-treatable low-alloy steels
- Chromium-molybdenum steels

1.2.1 High-Strength Low-Alloy Steels

High-strength low-alloy (HSLA) steels are designed to meet specific mechanical requirements per ASTM specifications. The principal difference between plain-carbon steels and HSLA steels is the addition of alloys such as manganese, chromium, nickel, copper, molybdenum, and vanadium. Typically, these alloys total less than 2 percent. The carbon content of HSLA steels is about the same as that in low-carbon and mild steels (0.06 percent to 0.28 percent). Compared to mild steels, HSLA steels are designed to provide a combination of higher strength, better corrosion resistance, and improved notch toughness. The weldability of HSLA steel is similar to that of mild steel.

Figure 4 Paper labels attached to steel products.

HSLA steels can be divided into two general groups: Group A and Group B. Steels in Group A are designed for high strength, while steels in Group B are designed for high strength, improved atmospheric corrosion resistance, and notch toughness. Typical steels in Group A are A441, A572, and A633, which are used for structural applications such as pipelines, buildings, bridges, machinery, and railroad equipment. Other steels in Group A are A225 and A737, which are used for pressure vessels. Typical steels in Group B are A242 and A588, which are used for high-strength structural applications that may include unpainted applications. Group B steels specifically intended for low-temperature applications (the nickel content is as high as 9 percent) are A203 and A353.

CAUTION

The welding recommendations provided are general guidelines only. When welding HSLA steel, the site WPS or site quality standards must be followed. Failure to follow proper standards can result in rejected and/or defective welds.

ABC CONSTRUCTION

DATE OF ISSUE 7/20/13

WELDING PROCEDURE SPECIFICATION
IGMA-11-AF

REVISION	2 - 11/16/14	**SUPPORTING PQR NO. AND PAGE:** IGMA-11 PAGE B
PROCESS	GTAW / SMAW	
TYPE	MANUAL	

JOINTS (QW-402)

JOINT DESIGN	V OR U GROOVES AND FILLETS, PER THE PROJECT DRAWINGS AND SPECIFICATIONS
BACKING	NONE
BACKING MATERIAL	COMPATIBLE WITH BASE MATERIALS IF USED
JOINT FIT-UP REQUIREMENTS	1/8" +/- 1/16"
OTHER	NO RETAINERS (QW 402.11) ROOT SPACING PER DRAWINGS AND SPECIFICATIONS

BASE METALS (QW-403)

P-NO 1 GRP NO. ______ to P-NO 1 GRP NO. ______

BASE METAL SPEC. ______ to ______

THICKNESS RANGE QUALIFIED (QW-451)

BASE METAL GROOVE	3/16" TO 8"	FILLET	ALL
DEPOSITED WELD METAL	GTAW - 3/4" MAX (No weld pass > 1/2" thick)		
	SMAW - 8" MAX (No weld pass > 1/2" thick)		

OTHER ______

FILLER METALS/ELECTRODES (QW-404); PROGRESSION (QW-405) **POSITIONS: ALL**

LAYER	ROOT	PROCESS	GTAW	AWS CLASS	ER70S-2	PROGRESSION	UPHILL
SFA NO.	5.18	F NO.	6	A NO.	1	SIZES	1/16", 3/32", 1/8"
LAYER	WELDOUT	PROCESS	SMAW	AWS CLASS	E-7018	PROGRESSION	UPHILL
SFA NO.	5.1	F NO.	4	A NO.	1	SIZES	3/32", 1/8", 5/32"
LAYER		PROCESS		AWS CLASS		PROGRESSION	
SFA NO.		F NO.		A NO.		SIZES	

OTHER NO CONSUMABLE INSERTS (QW 404.22) NO PENETRATION FLUX (QW 404.50)

PREHEAT (QW - 406)

PREHEAT TEMPERATURE	75° F (175F when t > 1")
MAX INTERPASS TEMPERATURE	600° F
PREHEAT MAINTENANCE	NONE
OTHER	

POSTWELD HEAT TREATMENT (QW-407)

TEMPERATURE	1150° F
TIME RANGE	1HR/INCH ONE HOUR MINIMUM
MAX HEATING & COOLING RATES	PER THE APPLICABLE ASME DESIGN CODE

Figure 5A Sample WPS with standard measurements (1 of 2).

ABC CONSTRUCTION

REVISION 2 - 11/16/14

WELDING PROCEDURE SPECIFICATION
IGMA-11-AF

GAS (QW-408)

SHIELDING GAS	GTAW - 100% ARGON	FLOW RATE	15-35 CFH
PURGING GAS	NONE	FLOW RATE	N/A
TRAILING GAS	NONE	FLOW RATE	N/A
OTHER			

ELECTRICAL CHARACTERISTICS (QW-409)

PROCESS: GTAW	PROCESS: SMAW	PROCESS:
ELECTRODE: See Note 1	ELECTRODE: E-7018	ELECTRODE:
CURRENT: DC	CURRENT: DC	CURRENT:
POLARITY: SP	POLARITY: RP	POLARITY:

CHARACTERISTICS: ER70S-2				CHARACTERISTICS: E7018				CHARACTERISTICS:			
SIZE	AMPS	VOLTS	TRAVEL SPEED IPM	SIZE	AMPS	VOLTS	TRAVEL SPEED IPM	SIZE	AMPS	VOLTS	TRAVEL SPEED IPM
1/16"	50-150	N/A	Manual	3/32"	50-120	N/A	Manual				
3/32"	75-180	N/A	Manual	1/8"	70-150	N/A	Manual				
1/8"	100-250	N/A	Manual	5/32"	90-200	N/A	Manual				

GTAW - PULSED CURRENT = NONE

TECHNIQUE (QW-410)

STRING OR WEAVE BEAD	USE STRINGER BEAD FOR POSITION 2G. USE EITHER STRINGER BEAD OR WEAVE BEAD FOR POSITIONS 1G, 3G, 4G, 5G, AND 6G.
ORIFICE OR GAS CUP SIZE	4 Thru 12
INITIAL AND INTERPASS CLEANING	BRUSHING AND/OR GRINDING
METHOD OF BACK GOUGING	THERMAL OR MECHANICAL IF REQUIRED (If Thermal grind back to bright metal)
OSCILLATION	N/A
CONTACT TUBE TO WORK DISTANCE	N/A
MULTIPLE OR SINGLE PASS PER SIDE	MULTIPLE
MULTIPLE OR SINGLE ELECTRODES	SINGLE
PEENING	NONE
OTHER	

NOTES NO AUTOGENOUS WELDING

1.) A potential health risk exists from inhaling grinding dust from EWTh-2 tungsten electrodes. As an alternative, EWCe-2 and/or EWLa-2 tungsten electrodes are acceptable.

Figure 5B Sample WPS with standard measurements (2 of 2).

Weld Procedure Number	30 P1 TIG 01 Issue A
Qualifying Welding Procedure (WPAR)	WPT17/A

Manufacturer: National Fabs Ltd 25 Lane End Birkenshaw Leeds Location: Workshop Welding Process: Manual TIG Joint Type: Single Sided Butt Weld	Method Of Preparation and Cleaning: Machine and Degrease Parent Metal Specification: Grade 304L Stainless Steel Parent Metal Thickness: 3 to 8mm Wall Pipe Outside Diameter: 25 to 100mm Welding Position: All Positions Welding Progression: Upwards

Joint Design	**Welding Sequences**
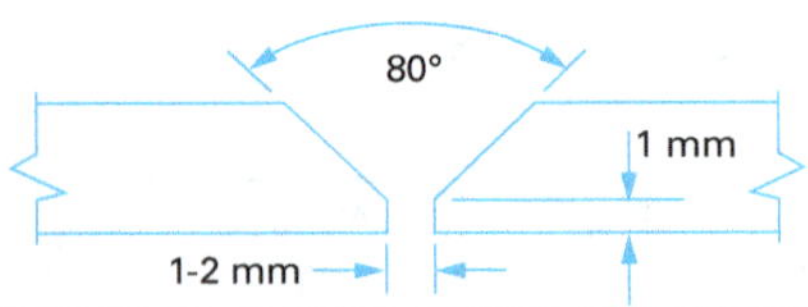	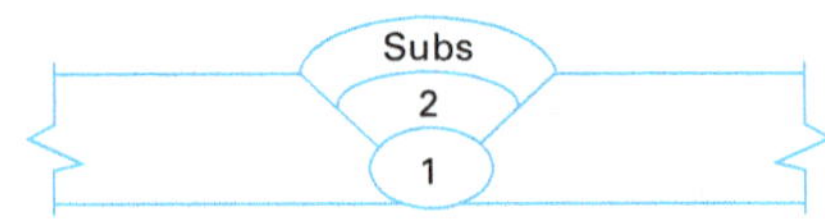

Run	Process	Size of Filler Metal	Current A	Voltage V	Type of Current/Polarity	Wire Feed Speed	Travel Speed	Heat Input
1 2 And Subs	TIG TIG	1.2 mm 1.6 mm	70 - 90 80 - 140	N/A	DC– DC–	N/A	N/A	N/A

Welding Consumables:- Type, Designation Trade Name: Any Special Baking or Drying: Gas Flux: Gas Flow Rate - Shield: - Backing: Tungsten Electrode Type/ Size: Details of Back Gouging/Backing: Preheat Temperature: Interpass temperature: Post Weld Heat Treatment Time, temperature, method: Heating and Cooling Rates*:	 BS 2901 Part 2 : 308S92 No Argon 99.99% Purity 8 - 12 LPM 5 LPM 2% Thoriated 2.4mm Dia Gas Backing 5°C Min 200°C Max Not Required	**Production Sequence** 1. Clean weld and 25mm borders to bright metal using approved solvent. 2. Position items to be welded ensuring good fit up and apply purge. 3. Tack weld parts together using TIG, tacks to at least 5mm min length. 4. Deposit root run using 1.2mm dia. wire. 5. Inspect root run internally. 6. Complete weld using 1.6mm dia wire using stringer beads as required. 7. 100% Visual inspection of completed weld

Revision History

Date	Issue	Changes	Authorization
26/11/2005	A	First changes	Jack Straw *Jack Straw*

Figure 6 Sample WPS with metric measurements.

HSLA steels can be welded using the common welding processes, although successful welding requires consideration of preheating, interpass temperature, and control of hydrogen, including the use of low-hydrogen electrodes when applicable. This information can be found on the WPS. Typically, only the root pass, and possibly the hot pass, are made with the gas tungsten arc welding (GTAW) process on HSLA steels. Fill passes and cover passes are often made with faster processes such as shielded metal arc welding (SMAW) or gas metal arc welding (GMAW). One advantage of using the GTAW process is that hydrogen is easier to control because fluxes are not used. Minimum preheat and interpass temperatures are determined based on the alloy type and the thickness of the thinner part being welded.

CAUTION

Steels that are wet or below 32°F (0°C) should always be preheated to at least 70°F (21°C) and maintained at that temperature to remove moisture and prevent its formation. Failure to preheat metals may result in rejected and/or defective welds.

Table 3 provides minimum preheat and interpass temperatures for typical HSLA steels and also includes the ASTM steel numbers and grades.

Filler metal for welding HSLA steel depends on the alloy and the application. Mild steel filler metal can be used when joint strength is not critical or when alloy steel is being joined to mild steel. Alloy steel should be used when the strength of the joint must equal or approach that of the base metal. High-alloy stainless steel is used for special applications including welding dissimilar metals or welding high-nickel steel, such as A353.

Typically, HSLA steels are not postweld heat-treated; however, this may be required to improve ductility or to maintain the dimensions of the weldment during machining.

1.2.2 Quench-and-Tempered Steels

Quench-and-tempered (Q&T) steels are furnished in the heat-treated condition, which generally consists of austenitizing, quenching, and tempering. Some Q&T steels fall into the carbon steel classification, some into the low-alloy steel classification, and others into the alloy classification of the AISI/SAE system. Q&T steels combine high yield and tensile strength with good notch toughness, ductility, corrosion resistance, or weldability, depending on their intended use. Q&T steels have less than 0.25 percent carbon, with yield strengths of 50,000 to 180,000 psi (≈344,738 to 1,241,056 kPa).

Hot Tapping

Steel pipes are commonly hot-tapped to install a bypass around a repair area or to add a branch line. With hot tapping, threaded or welded taps (called threadolets and weldolets, respectively) are welded to the outside of the pipe while material or liquid is still flowing. Special tools can then be used to work through the installed fitting, penetrate the pipe, and install a shutoff valve without the loss of the pipe contents. In other words, the complete connection is made without draining the pipe contents. The pipe remains hot.

Low-alloy pipes cannot be hot-tapped because the material flowing through the pipe will cool the pipe rapidly and prevent proper preheat, interpass, and postheat temperature maintenance. This can result in defective welds, as well as pipe failure during welding. A site engineer should be consulted before any hot tap is made due to concerns related to the contents of the pipe.

Most Q&T steels are furnished as plate, although some are available as seamless pipe.

Q&T steels are used where high strength and minimum weight are essential. Many of the Q&T steels are covered by ASTM classifications. A common group of Q&T steels known as HY steels are covered by military specifications (MIL). The most common HY steel is HY-80, which has a yield stress of 80,000 psi (≈551,581 kPa). HY-100, HY-130, HY-150, and HY-180 are also available. Types and typical uses of Q&T steels include the following:

- *A514 and A517* – Used for earth-moving equipment, pressure vessels, bridges, TV towers, and ships.
- *A533 Grade B* – Used for nuclear pressure vessels.

Table 3 Minimum Preheat and Interpass Temperatures for Typical High-Strength Low-Alloy Steels Welded with Low-Hydrogen Electrodes

ASTM Steel	Thickness (in)	Thickness (mm)	Minimum Temperature
A242	Up to 0.75	Up to 19	32°F (0°C)
A441	0.81 to 1.50	21 to 38	50°F (10°C)
A572, Gr 42, 50	1.56 to 2.50	40 to 64	150°F (66°C)
A588	Over 2.50	Over 63	225°F (107°C)
A572, Gr 60, 65	Up to 0.75	Up to 19	50°F (10°C)
A633, Gr E	0.81 to 1.50	21 to 38	150°F (66°C)
	1.56 to 2.50	40 to 64	225°F (107°C)
	Over 2.50	Over 64	300°F (140°C)

- *A543 Type B* – Used in nuclear reactor vessels, ships, and submarines. Classes 1 and 2 steels are similar to HY-80 and HY-100.
- *A553* – Used for cryogenic applications. It contains 8 to 9 percent nickel for high strength at extremely low temperatures.
- *A678* – Used for high-strength structural applications.

To prevent cracking, most Q&T steels can be welded by common welding processes using preheat and interpass temperatures as shown in *Table 4*. These temperatures vary depending on the type of steel being welded. Because Q&T steels obtain their strength from heat treatment, slow cooling can destroy the benefits of the original heat treatment. Slow cooling is generally caused by the following:

- Excess preheat
- Exceeding the recommended interpass temperature
- High heat input, such as high amperage with slow travel speed and the use of weave beads instead of stringers

CAUTION

When welding Q&T steels, pay close attention to the welding procedures concerning preheat and interpass temperature recommendations to prevent loss of strength or cracking.

On the other hand, cooling too quickly can cause a hard, brittle zone that may crack. Fast cooling is caused by the following:

- Insufficient preheat
- Insufficient interpass temperature
- Insufficient heat input during welding

When welding Q&T steels, always use low-hydrogen techniques and stringer beads to minimize heat input. Typically, the GTAW process is used only on thin sections or for the first one or two passes.

Most Q&T steels are designed to be used in the as-welded condition, with no postheat. Sometimes thick sections of A533 or A537 steels may require postheat treatment. If postheat treatment is required, use extreme care and follow the procedures exactly to prevent adversely affecting the base metal.

1.2.3 Heat-Treatable Low-Alloy Steels

Heat-treatable low-alloy (HTLA) steels are alloy steels that contain carbon in the range of about 0.25 to 0.45 percent, and small amounts of chromium, nickel, and/or molybdenum to enhance hardenability. They are generally welded in the annealed condition, and then the entire weldment is heat treated. The relatively high carbon content of HTLA steels allows them to be heat treated to a very high strength and hardness. Because HTLA steels must be heat treated after welding, they are typically used for smaller parts that can be placed in tempering ovens.

HTLA steels are produced to contain very low levels of impurities. During welding, low hydrogen techniques are used, and extreme care must be taken to keep the weld joint and filler metal absolutely clean and free of impurities. For this reason, GTAW is the preferred process for welding HTLA steels. Heat input is also very important, and welding procedures that take into account current, voltage, travel speed, and preheat temperatures must be followed.

HTLA steels generally require higher preheat and interpass temperatures than other types of alloy steels to prevent hardening of the weld zone and hot cracks. They also require postweld heat treatment (PWHT). *Table 5* shows minimum preheat and interpass temperatures for HTLA steels and also includes the AISI steel numbers and thickness ranges.

HTLA steels should receive postweld heat treatment immediately after they are welded. The type of postweld heat treatment depends on the type of steel and the preheat and interpass temperature

Table 4 Minimum Preheat and Interpass Temperatures for Q&T Steels

Thickness Range		Temperature				
Inches	**Millimeters**	**A514/A517**	**A533**	**A537**	**A543**	**A678**
Up to 0.50	Up to 13	50°F (10°C)	50°F (10°C)	50°F (10°C)	100°F (38°C)	50°F (10°C)
0.56 to 0.75	14 to 19	50°F (10°C)	100°F (38°C)	50°F (10°C)	125°F (52°C)	100°F (38°C)
0.81 to 1.00	21 to 25	125°F (52°C)	100°F (38°C)	50°F (10°C)	150°F (66°C)	100°F (38°C)
1.10 to 1.50	27 to 38	125°F (52°C)	200°F (93°C)	100°F (38°C)	200°F (93°C)	150°F (66°C)
1.60 to 2.00	40 to 51	175°F (79°C)	200°F (93°C)	150°F (66°C)	200°F (93°C)	150°F (66°C)
2.10 to 2.50	53 to 64	175°F (79°C)	300°F (149°C)	150°F (66°C)	300°F (149°C)	150°F (66°C)
Over 2.50	Over 64	225°F (107°C)	300°F (149°C)	225°F (107°C)	300°F (149°C)	—

Table 5 Minimum Preheat and Interpass Temperatures for HTLA Steels

AISI Steel	Thickness Range Inches	Thickness Range Millimeters	Minimum Preheat and Interpass Temperature
4027	Up to 0.50 0.60 to 1.0 1.1 to 2.0	Up to 13 15 to 25 27 to 51	50°F (10°C) 150°F (66°C) 250°F (121°C)
4037	Up to 0.50 0.60 to 1.0 1.1 to 2.0	Up to 13 15 to 25 27 to 51	100°F (38°C) 200°F (93°C) 300°F (149°C)
4130/ 5140	Up to 0.50 0.60 to 1.0 1.1 to 2.0	Up to 13 15 to 25 27 to 51	300°F (149°C) 400°F (204°C) 500°F (260°C)
4135/ 4140	Up to 0.50 0.60 to 1.0 1.1 to 2.0	Up to 13 15 to 25 27 to 51	350°F (177°C) 450°F (232°C) 500°F (260°C)
4340/ 8630	Up to 0.5 0.6 to 1.0 1.1 to 2.0	Up to 13 15 to 25 27 to 51	200°F (93°C) 250°F (121°C) 300°F (149°C)
8640	Up to 0.50 0.60 to 1.0 1.1 to 2.0	Up to 13 15 to 25 27 to 51	200°F (93°C) 300°F (149°C) 350°F (177°C)

maintained during welding. Allowing the weldment to cool after welding without some type of heat treatment generally results in cracks. Immediately after it is welded, HTLA steel should be stress-relieved following strict guidelines depending on the alloy and the preheat and interpass temperatures used.

1.2.4 Chromium-Molybdenum Steels

Chromium-molybdenum (Cr-Mo) steels contain 0.5 to 9 percent chromium and 0.5 to 1 percent molybdenum. The carbon content is generally less than 0.20 percent. Cr-Mo steels, which are sometimes called chrome-moly or heat-resisting alloy steels, are widely used in the petroleum industry and in steam-generating plants for elevated-temperature applications. The chromium provides improved resistance to oxidation and corrosion, and the molybdenum increases strength at elevated temperatures. Cr-Mo steels are also used in the aircraft and racing industries as tubing for highly stressed parts (*Figure 7*). Cr-Mo steels that are in the AISI/SAE 4100 series are available in forging, castings, plate, pipe, and tubing to various ASTM specifications.

Cr-Mo steels are hardenable, so a welding procedure that includes preheating, and sometimes postheating, must be used. The preheat temperature depends on the amount of alloy, the carbon content, the weld material, and base metal thickness. Generally, the higher the alloy content, carbon content, or thickness, the higher the preheat temperature. *Table 6* shows minimum preheat and interpass temperatures for Cr-Mo steels including their types and ASTM steel designations.

Once the Cr-Mo base metal has been preheated, welding should proceed without interrupting the heating cycle to avoid hard spots and cracking. If the base metal contains less than 9 percent chromium or is less than 1 inch (25 mm) thick, the heating cycle should be 800°F (427°C) per half hour per 1 inch (25 mm) of thickness. After being heated, wrap the metal and let it cool until it is at ambient temperature.

CAUTION

Check your welding procedures for any special precautions needed when interrupting a heating cycle. Failure to follow proper welding procedures will result in defective welds.

When the welding consumables are a potential source of hydrogen, such as with SMAW, it may be necessary to raise the preheat temperature and hold the weldment at this elevated temperature for a period of time. This allows the hydrogen to escape before the base metal cools, which reduces the risk of cracking. To avoid potential problems that can occur when the heating cycle of a Cr-Mo base metal is interrupted, always plan to complete the heating cycle once it has started.

Some Cr-Mo steels containing less than 1.25 percent Cr and 0.05 percent Mo can be placed in service as-welded if a high preheat temperature was used and the section is relatively thin. Other Cr-Mo steels should either be stress-relieved, annealed, or normalized and tempered.

Figure 7 Steel tubing used to construct a race car frame.

Table 6 Minimum Preheat and Interpass Temperatures for Cr-Mo Steels

ASTM Steel	Type	Up to 0.5 inch (13 millimeters)	0.5 to 1.0 inch (13 to 25 millimeters)	Over 1.0 inch (Over 25 millimeters)
A335-P2 A335-P12 A335-P11	½ Cr - ½ Mo 1 Cr - ½ Mo 1¼ Cr - ½ Mo	300°F (149°C)	300°F (149°C)	300°F (149°C)
A369-FP3b A335-P22	2 Cr - ½ Mo 2¼ Cr - 1Mo	300°F to 350°F (149°C to 177°C)	300°F to 350°F (149°C to 177°C)	350°F (177°C)
A335-P21	3 Cr - 1 Mo	300°F to 350°F (149°C to 177°C)	300°F to 350°F (149°C to 177°C)	400°F (204°C)
A335-P5 A335-P7 A335-P9	5 Cr - ½ Mo 7 Cr - ½ Mo 9 Cr - 1 Mo	400°F (204°C)	400°F (204°C)	500°F (260°C)

*Maximum carbon content of 0.15 percent. For higher content, preheat should be increased by 100°F to 200°F (38°C to 93°C)

Stress-relief heat treatment is used to reduce welding stresses and to increase the ductility and toughness of the weld metal and the heat-affected zone (HAZ). The stress-relief temperature ranges from about 1,150°F to 1,400°F (621°C to 760°C) and depends on the alloy being treated. The entire weldment or the weld zone is generally held at the stress-relief temperature for one hour per inch of thickness.

The procedures for annealing or normalizing and tempering are generally the same as those used for HTLA steels discussed earlier in this module.

1.3.0 Common Grade Stainless Steel Classifications

Stainless steels are typically classified as being common grade or specialty grade. This section covers common grade stainless steels. The specialty grades are discussed later in this module. Common grade stainless steels (corrosion-resistant steels) are iron-based alloys that normally contain at least 11 percent chromium. Other alloying elements, including nickel, carbon, manganese, and silicon, may be present in varying quantities, depending on the specific type of stainless steel, in order to enhance its physical and mechanical properties. The main physical characteristics of all stainless steels are their resistance to corrosion and heat. Some have good low- and high-temperature mechanical properties.

When compared with mild steels, stainless steels have the following characteristics:

- Lower coefficients of thermal conductivity that increase the chances of distortion
- Higher coefficients of thermal expansion that increase the chances of distortion
- Higher electrical resistances that increase the tendency to build up heat from welding current

Common grade stainless steels are classified by their grain structures. The type of grain structure is determined by the specific alloy content of the stainless steel and its heat treatment during manufacture. Based on their microcrystalline structures, the classification of common grade stainless steels is divided into these three groups:

- Austenitic
- Ferritic
- Martensitic

1.3.1 Austenitic Stainless Steels

Austenitic stainless steels are nonmagnetic in the annealed condition and non-hardenable by heat treatment. However, they can be hardened significantly by cold working. Austenitic stainless steels combine excellent corrosion and heat resistance with good mechanical properties over a broad temperature range. Austenitic stainless steels make up the largest of the three common grade stainless steel groups. For this reason, they are the type of stainless steel encountered most often by welders. They include the AISI 200 and 300 series stainless steels (*Table 7*) that all contain significant amounts of both chromium and nickel. Austenitic steels are sometimes further subdivided in two classifications based on their compositions: chromium-nickel (AISI 300 series) and chromium-nickel-magnesium (AISI 200 series). The range of applications for austenitic stainless steels includes housewares, containers, industrial piping and vessels, and architectural facades.

Austenitic stainless steels are also available in low-carbon (L-grade) and high-carbon (H-grade) types. The letter L after a stainless steel type, such as 304L, indicates a low-carbon content of 0.03 percent or under. Similarly, the letter H after a stainless steel type indicates a high-carbon content, ranging between 0.04 and 0.10 percent.

L-grades are typically used when annealing after welding is impractical, such as in the field where pipe and fittings are welded. H-grades are used when the steel will be subjected to extreme temperatures; the higher carbon content helps it retain strength.

1.3.2 Ferritic Stainless Steels

Ferritic stainless steels are always magnetic. They are hardened to some extent by cold working, not by heat treatment. Ferritic stainless steels combine corrosion resistance and heat resistance with fair mechanical properties over a narrower temperature range than austenitic steels. Ferritic stainless steels are straight chromium stainless steels containing 11.5 to 27 percent chromium, about 1 percent manganese, and little or no nickel. The carbon content is 0.20 percent or less. Examples of ferritic stainless steels include AISI types 405, 409, 430, 442, 444, and 446. They typically are used for decorative trim, sinks, and automotive applications, particularly exhaust systems.

1.3.3 Martensitic Stainless Steels

Martensitic stainless steels are magnetic and can be hardened by quenching and tempering. They are excellent for use in mild environments such as the atmosphere, freshwater, steam, and with weak acids. However, they are not resistant to severely corrosive solutions. Martensitic stainless steels comprise two groups: chromium-martensitic and martensitic. The chromium-martensitic stainless steels contain from 11.5 to 18 percent chromium, about 1 percent manganese, and in some cases 0 to 2.5 percent nickel. Examples of chromium-martensitic stainless steels are AISI types 403, 410, 414, 420, 422, 431, and 440. Martensitic stainless steels contain only 4 to 6 percent chromium and no nickel. For this reason, they are not considered true stainless steels, although their corrosion resistance is much greater than that of mild carbon steels, even at elevated temperatures. Examples of martensitic stainless steels are AISI types 501 and 502.

1.4.0 Specialty Grade Stainless Steel Classifications

In addition to the common grades of stainless steel, the following four specialty grades of stainless steel are also available:

- Precipitation-hardening
- Superaustenitic
- Superferritic
- Duplex

Table 7 AISI Classification Numbers and Composition for Selected Stainless Steels

AISI No.	C%	Mn%	Si%	Cr%	Ni%	Other Elements
Chromium-nickel-magnesium-austenitic (Nonhardenable)						
201	0.15 Max	5.5–7.5	1.0	16.0–18.0	3.0–5.5	N 0.25 Max
202	0.15 Max	7.5–10	1.0	17.0–19.0	4.0–6.0	N 0.25 Max
Chromium-nickel-austenitic (Nonhardenable)						
301	0.15 Max	2.0	1.0	16.0–18.0	6.0–8.0	—
302	0.15 Max	2.0	1.0	17.0–19.0	6.0–10.0	—
303	0.15 Max	2.0	1.0	17.0–19.0	8.0–10.0	Mo 0.0–0.6
304	0.08 Max	2.0	1.0	18.0–20.0	8.0–10.5	—
304L	0.03 Max	2.0	1.0	18.0–20.0	8.0–12.0	—
308	0.08 Max	2.0	1.0	19.0–21.0	10.0–12.0	—
309	0.20 Max	2.0	1.0	22.0–24.0	12.0–15.0	—
310	0.25 Max	2.0	1.5	24.0–26.0	19.0–22.0	—
316	0.08 Max	2.0	1.0	16.0–18.0	10.0–14.0	Mo 2.0–3.0
316L	0.03 Max	2.0	1.0	16.0–18.0	10.0–14.0	Mo 2.0–3.0
321	0.08 Max	2.0	1.0	17.0–19.0	9.0–12.0	Ti, 5xC Min
347	0.08 Max	2.0	1.0	17.0–19.0	9.0–13.0	Cb+Ta10xC Min
348	0.08 Max	2.0	1.0	17.0–19.0	9.0–12.0	Ti 0.10 Max
Chromium-martensitic (Hardenable)						
403	0.15 Max	1.0	0.5	11.5–13.0	—	—
410	0.15 Max	1.0	1.0	11.5–13.5	—	—
414	0.15 Max	1.0	1.0	11.5–13.5	1.25–2.5	—
416	0.15 Max	1.0	1.0	12.0–14.0	—	S 0.15 Min
420	0.15 Min	1.0	1.0	12.0–14.0	—	—
422	0.20–0.25	1.0	0.75	11.0–13.0	0.5–10	Mo 0.75–1.25
440A	0.60–0.75	1.0	1.0	16.0–18.0	—	W 0.75–1.25
440B	0.75–0.95	1.0	1.0	16.0–18.0	—	V 0.15–0.3
440C	0.95–1.2	1.0	1.0	16.0–18.0	—	Mo 0.75 Max
						Mo 0.75 Max
						Mo 0.75 Max
Chromium-ferritic (Nonhardenable)						
405	0.08 Max	1.0	1.0	11.5–14.5	—	Al 0.10–0.30
409	0.08 Max	1.0	1.0	10.5–11.75	—	Ti, 6xC Min
430	0.12 Max	1.0	1.0	16.0–18.0	—	—
442	0.20 Max	1.0	1.0	18.0–23.0	—	Ti, 0.20 + 4
444	0.025 Min	1.0	1.0	17.5–19.5	1.0	(%C+%N)
446	0.20 Max	1.50	1.0	23.0–27.0	1.0	Mo 1.75–2.5,
						N 0.035 Max
						(Nb+T2), 0.2 + 4
						(%C+%N) Min
						N 0.25 Max
Martensitic						
501	0.11 Max	1.0	1.0	4.0–6.0	—	Mo 0.4–0.65
502	0.10 Max	1.0	1.0	4.0–6.0	—	Mo 0.4–0.65

1.4.1 Precipitation-Hardening Stainless Steels

Precipitation-hardening (PH) stainless steels contain alloying additions, such as aluminum, that enable them to be hardened by a solution and aging heat treatment. Precipitation-hardening stainless steels were developed after World War II because the design and manufacture of jet aircraft required the use of stainless steels with a better weight-to-strength ratio. Precipitation-hardening stainless steels have designations such as 17-4PH, PH13-8MO, and AM350.

1.4.2 Superaustenitic Stainless Steels

Superaustenitic stainless steels have compositions that fall between 300-series stainless steels and nickel-alloy steels. They are highly resistant to corrosion and stress-corrosion cracking caused by chloride-containing solutions. Many of these alloys contain deliberate amounts of copper and/or nitrogen to increase their strength and corrosion resistance. These stainless steel types are widely used in process industry applications in place of common austenitic stainless steels that have a tendency to fail because of pitting, crevice corrosion, or chloride-induced stress corrosion cracking.

1.4.3 Superferritic Stainless Steels

Superferritic stainless steels differ from common-grade ferritic stainless steels in their lower carbon and nitrogen content and higher molybdenum and/or chromium content. This combination provides much higher corrosion resistance in oxidizing and chloride chemical environments and excellent resistance to corrosion in salt water. Examples of superferritic stainless steel designations include E-Brite®, Sea-Cure®, and AL 29-4-2®.

1.4.4 Duplex Stainless Steels

Duplex stainless steels are supplied with approximately equal amounts of ferrite and austenite, thus the name "duplex". They also contain about 24 percent chromium and 5 percent nickel. Developed around 1970, duplex steels are among the more recently developed stainless steels. They offer higher yield strength and superior resistance to pitting and stress cracking than the 300 series stainless steels. One example of duplex stainless steel is Type 2205, which accounts for more than 80 percent of duplex use. Duplex stainless steels are used for heat exchangers, tubes, and pipes for applications involving the production and handling of gas and oils, and for desalinization.

Highly alloyed, high-performance duplex stainless steels, called super-duplex stainless steels, are also made. These contain higher levels of chromium, molybdenum, and nitrogen. They were developed to provide outstanding resistance to acids, acid chlorides, and other caustic solutions typical of those encountered in the chemical, petrochemical, and pulp and paper industries.

1.5.0 Nonferrous Metals Classification Systems and Groups

Nonferrous metals include all metals except iron. Commonly used nonferrous metals include aluminum, magnesium, titanium, copper, nickel, zinc, tin, lead, and the precious metals. Nonferrous metals with densities lower than that of steel are considered light metals. They include, in order of increasing density, magnesium, beryllium, aluminum, and titanium. The heavier common nonferrous metals include copper, nickel, lead, tin,

Why Stainless Steel Doesn't Rust

When steel comes in contact with oxygen and water vapor in the atmosphere, a chemical reaction occurs and the steel begins to change to its original form of iron oxide. The scale that forms on the surfaces of metals when they are exposed to air is an oxide. The chromium added to the steel prevents it from rusting or staining, thus the name "stainless steel". The chromium reacts with the oxygen and water in the atmosphere to form a microscopic and transparent layer of chromium oxide called a passive film. The sizes of the chromium atoms and their oxides are about the same, so they join to form a tight, stable layer only a few atoms thick on the surface of the steel. If the stainless steel surface is cut or scratched, disrupting the passive film, it is self-healing because more chromium oxide immediately forms and covers the exposed surface, protecting it from iron oxide (rust) again.

This is why after repeated use and sharpening, stainless steel cutlery remains bright and shiny. However, if the surface of stainless steel is contaminated by a ferrous metal, the ferrous material will rust. The rust in this photo was traced back to surface preparation with tools, such as grinding wheels and points, previously used to prepare carbon steel. This rust would not be present in a properly prepared joint, nor would it be evident immediately after the weld is completed.

and zinc, as well as the precious metals platinum, gold, silver, and their alloys. Each of the nonferrous metals and alloys is classified by a different and unrelated classification system.

1.5.1 Aluminum Alloys

Pure aluminum is a soft, lightweight metal that can be used in its nearly pure form. Because aluminum is very soft and ductile and has a low tensile strength, it is often alloyed with elements such as copper, manganese, silicon, magnesium, and zinc to make different alloys with specific properties for very different purposes. When aluminum is exposed to the atmosphere, a microscopic and transparent aluminum-oxide layer forms immediately that protects the metal from further oxidation. This characteristic gives aluminum its high resistance to corrosion.

Aluminum is lightweight, has one-third the weight of steel, and can be made very strong, ductile, and malleable. It is a good conductor of heat and electricity and is highly resistive to weathering and corrosion from many acids. However, aluminum can be corroded easily by alkalis because they attack the oxide layer. Aluminum is also noncombustible and nonmagnetic and is widely used in electrical shielding and near inflammable or explosive substances. It is decorative and easily formed, cast, and machined. Polished aluminum has the highest reflectivity of any material, including mirror glass.

Aluminum alloys are broadly classified as either casting alloys or wrought alloys for rolling, forging, or extruding. The Aluminum Association, Inc. (AA) uses a three-digit plus one decimal system to designate each type of casting alloy (*Table 8*). The first digit (1 through 9) shows the main alloying element or elements. For the 1XX.X series this is essentially pure aluminum (99.00 percent aluminum). With the exception of the 1XX.X series, the second and third digits identify the specific alloys. For the 1XX.X series only, the second and third digits give the degree of aluminum purity above 99.00 percent. For example, if the second and third digits are 60, the alloy contains a minimum of 99.60 percent aluminum. The decimal indicates if the alloy composition is for the final casting (0.0) or for ingot (0.1 or 0.2, depending on impurity limits). Modifications to the alloys are indicated by the use of prefix letters such as A, B, C.

Similarly, the AA uses a four-digit numbering system for wrought aluminum alloy types (*Table 9*) that is not related to the three-digit casting alloy designation system. In this wrought aluminum system, the first digit identifies the alloy group by the major alloying element. For the 1XXX group, the second digit designates modifications and impurity limits. A zero (0) indicates no special control on individual impurities. A second digit of 1 through 9 indicates special control over one or more impurities and is assigned consecutively. For groups 2XXX through 8XXX, the second digit indicates alloy modifications and is consecutively assigned. A second digit of 0 indicates the original alloy. The last two digits indicate the minimum percentage of aluminum.

Around the World

Recycled Steel

Steel is the world's most recycled material. According to the World Steel Association, more than 650 million tons of steel were recycled annually between 2009 and 2013.

Table 8 AA System Code Designations for Basic Cast Aluminum Alloys

Designation	Major Alloying Element
1XX.X	Near pure aluminum
2XX.X	Copper
3XX.X	Silicon plus copper and/or magnesium
4XX.X	Silicon
5XX.X	Magnesium
6XX.X	Unused series
7XX.X	Zinc
8XX.X	Tin
9XX.X	Other element

Table 9 AA System Code Designations for the Basic Wrought Alloy Groups

Designation	Major Alloying Element
1XXX	Non-aluminum elements are 1 percent or less
2XXX	Copper
3XXX	Manganese
4XXX	Silicon
5XXX	Magnesium
6XXX	Magnesium and silicon
7XXX	Zinc
8XXX	Other elements
9XXX	Unused series

Recycling Aluminum

GOING GREEN

Aluminum can be recycled repeatedly with no loss of quality or its normal properties.

The following examples explain AA wrought aluminum numbers:

- AA Number 1075:
  - 1075 = Indicates aluminum content of 99.0 percent or greater
  - 1075 = Indicates no special control on the impurities
  - 1075 = Indicates an aluminum content of 99.75 percent
- AA Number 1180:
  - 1180 = Indicates aluminum content of 99.0 percent or greater
  - 1180 = Indicates impurities are limited
  - 1180 = Indicates an aluminum content of 99.80 percent

In general, aluminum and its alloys can be welded by the following methods: oxyfuel or air-fuel; electric arc; electrical resistance; and brazing.

1.5.2 Magnesium Alloys

Magnesium is mainly used in alloy form. It is the lightest of the structural metals, having less than one-quarter the density of steel. It is mainly used in the aerospace and automotive industries.

Why Aluminum Must Be Cleaned Before Welding

The aluminum-oxide coating protecting aluminum must be thoroughly cleaned from the metal before welding. The reason is that, while pure aluminum melts at about 1,200°F (about 650°C), the aluminum-oxide coating that protects the metal does not melt until it reaches the much higher temperature of 3,700°F (nearly 2,040°C).

All cleaning materials used must be contained and disposed of in accordance with local environmental laws. Cleaning something as large as a boat presents a greater challenge when it comes to containing the runoff of cleaning materials.

Magnesium is the eighth most common element in the world and the sixth most abundant metal, comprising about 2.5 percent of the earth's surface. Because seawater contains about 0.13 percent magnesium, some production facilities process seawater to obtain the magnesium. This is done after the precipitation of other salts from the seawater, which leaves a magnesium-rich brine. The uses of magnesium center on three properties of the metal: its ability to form intermetallic compounds with other metals, its high chemical reactivity, and its low density. Magnesium can be extruded, forged, rolled, and machined.

Magnesium alloys are broadly classified as casting alloys or wrought (rolled, forged, or extruded) alloys. Both of these classifications use the *ASTM Standard B951, Standard Practice for Codification of Unalloyed Magnesium and Magnesium-Alloys, Cast and Wrought.* Designations consist of one or two letters representing the alloying elements, followed by their respective percentages, and rounded off to the nearest whole number. The letters representing the alloying elements are as follows:

- *A* – Aluminum
- *B* – Bismuth
- *C* – Copper
- *D* – Cadmium
- *E* – Rare Earths
- *F* – Iron
- *G* – Magnesium
- *H* – Thorium
- *K* – Zirconium
- *L* – Beryllium
- *M* – Manganese
- *N* – Nickel
- *P* – Lead
- *Q* – Silver
- *R* – Chromium
- *S* – Silicon
- *T* – Tin
- *Y* – Antimony
- *Z* – Zinc

Magnesium alloy designations also contain information indicating the temper condition. This part of the designation, separated by a hyphen from the main designation of letters and numbers, consists of a letter or a letter and number. The letters and numbers that indicate the temper are as follows:

- *F* – As fabricated
- *O* – Annealed
- *H10, H11* – Strain hardened
- *H23, H24, H26* – Strain hardened and annealed
- *T4* – Solution heat treated
- *T5* – Artificially aged
- *T6* – Solution heat treated and artificially aged
- *T8* – Solution heat treated, cold worked, and artificially aged

In the alloy AZ92A-T6, the first A is aluminum and the Z is zinc. The 9 indicates that the aluminum percentage is between 8.6 and 9.4. The 2 shows that the zinc percentage is between 1.5 and 2.5. The final A denotes that this is the first alloy to receive the AZ92 designation. The letter and number T6 indicate solution heat treatment and artificially aged tempering.

Magnesium and its alloys are best welded by the GTAW process for thin sections and the GMAW process for thick sections. The composition of the filler metals should be compatible with the base metal. However, zinc above 1 percent increases hot shortness and can result in cracking. *Table 10* shows examples of magnesium alloys and their principal uses.

1.5.3 Titanium Alloys

Titanium is mainly used in alloy form. Its alloys are high-strength, lightweight (about one-half the density of steel), corrosion-resistant structural metals that retain their strength at elevated temperatures. It is used in aircraft engines and structures, chemical processing, surgical implants, sporting goods, and marine and other applications where corrosion resistance is required.

Titanium is a reactive metal. At high temperatures, it readily combines with oxygen, hydrogen, and nitrogen to form stable compounds. Like aluminum and magnesium, titanium forms a tight, protective, microscopic oxide film on clean surfaces at room temperature, when in contact with oxygen. This makes it highly resistant to corrosion at low temperatures. Titanium alloys are classified into three groups according to the crystal structures that are stable at room temperature.

- *Alpha alloys* – Are the usual crystalline structures of pure titanium. Aluminum is added to stabilize the crystalline structure. These alloys have excellent strength and are oxidation-resistant from 600°F to 1,100°F (about 316°C to 593°C). They have good ductility and weldability and can be hardened by working, but not by heat treatment.
- *Beta alloys* – Can contain vanadium, chromium, and aluminum. Iron and chromium are added to make them heat-treatable, high-strength alloys. They have fair weldability.
- *Alpha-beta alloys* – Contain one or more of the elements manganese, molybdenum, tin, vanadium, iron, and aluminum. They have a wide range of mechanical characteristics, but have poor weldability.

Table 10 Examples of Magnesium Alloys

ASTM Number	Nominal Alloy Composition (By %)						
	Al	Mn	Zn	Zr	Rare Earths	Th	Mg
Sand and Permanent Mold Castings							
AZ92A-T6	9.0	0.10	2.0	—	—	—	Balance
AZ63A-T6	6.0	0.15	3.0	—	—	—	Balance
AZ81A-T4	7.6	0.13 Min	0.7	—	—	—	Balance
AZ91C-T6	8.7	0.13	0.7	—	—	—	Balance
EZ33A-T5	—	—	2.6	0.8	3.3	—	Balance
HK31A-T6	—	—	0.3	0.7	—	3.3	Balance
HZ32A-T6	—	—	2.1	0.8	—	3.3	Balance
Die Castings							
AZ91A-F	9.0	0.13	0.7	—	—	—	Balance
AM60A-F	6.0	0.13	—	—	—	—	Balance
AZ91D-F	9.0	0.13	0.7	—	—	—	Balance
Extrusions							
AZ61A-F	6.5	0.15	1.0	—	—	—	Balance
MIA-F	—	1.20	—	—	—	—	Balance
AZ80A-T5	8.5	0.12	0.5	—	—	—	Balance
AZ10A-F	1.3	0.20	0.4	—	—	—	Balance
AZ31B-F	3.0	0.20	1.0	—	—	—	Balance
ZK60A-F	—	—	5.5	0.45	—	—	Balance
Sheet and Plate							
AZ31B-H24	3.0	0.20	1.0	—	—	—	Balance
HK31A-H24	—	—	—	0.7	—	3.3	Balance

Titanium is considered difficult to weld because the weld must be kept in an inert (nonreactive) gas atmosphere until it cools to below the reactive temperature. Small amounts of hydrogen or large amounts of oxygen or nitrogen can cause embrittlement in the heat-affected zone, which leads to breakability. Titanium is usually welded by the GTAW process.

About 34 different titanium alloys are currently being produced, including the ones shown in *Table 11*.

1.5.4 Beryllium Alloys

Beryllium (Be) is a high-strength, lightweight metal. It is only about two-thirds as heavy as aluminum, but is six times stiffer than steel. Beryllium's heat capacity is five times that of copper, meaning that one pound (0.45 kg) of it will absorb as much heat as five pounds (2.3 kg) of copper. For this reason, it is widely used in aerospace and nuclear reactor applications.

Titanium

Commercial production of titanium did not begin until the 1950s. At that time, titanium was recognized for its importance and was developed as a unique, lightweight, high-strength alloy for use in the engine and airframe components of high-performance jet aircraft. Today, titanium alloys are widely used in many applications that previously used metals such as stainless and specialty steels, copper alloys, and nickel-based alloys.

Table 11 Examples of Titanium Alloys

Designation	Nominal Alloy Composition (By %)				
	Al	Sn	Zr	Mo	Other
Unalloyed Grades					
ASTM Grade 1	—	—	—	—	—
ASTM Grade 7	—	—	—	—	0.2 Pd
ASTM Grade 11	—	—	—	—	0.12–0.25 Pd
ASTM Grade 12	—	—	—	0.3	0.6–0.9 Ni
Alpha and Near Alpha Alloys					
Ti-5Al-2.5Sn	5.0	2.5	—	—	—
Ti-5Al-2.5Sn-ELI	5.0	2.5	—	—	—
Ti-6Al-2Sn-4Zr-2Mo	6.0	2.0	4.0	2.0	—
Ti-6Al-2Nb01Ta-0.8Mo	6.0	—	—	1.0	2 Nb, 1 Ta
Alpha-Beta Alloys					
Ti-6Al-4V	6.0	—	—	—	4.0 V
Ti-6Al-4V-ELI	6.0	—	—	—	6.0 V
Ti-6Al-6V-2Sn	6.0	2.0	—	—	0.75 Cu, 6.0 V
Ti-8Mn	—	—	—	—	8.0 Mn
Ti-7Al-4Mo	7.0	—	—	4.0	—
Ti-6Al-2Sn-4Z-6Mo	6.0	2.0	4.0	6.0	—
Beta Alloys					
Ti-13V-11Cr-3Al	3.0	—	—	—	11.0 Cr, 13.0 V
Ti-8Mo-8V-2Fe-3Al	3.0	—	—	8.0	8.0
Ti-11.5Mo-6Zr-4.5Sn	—	4.5	6.0	11.5	—

Beryllium is a reactive metal, and at high temperatures it readily combines with oxygen, hydrogen, and nitrogen to form stable compounds. At low temperatures, it is highly resistant to corrosion. Small amounts of nitrogen and hydrogen or large amounts of oxygen cause embrittlement. *Table 12* lists beryllium alloys and some of their mechanical properties, such as tensile and yield strength.

Beryllium is considered a difficult metal to weld because it must be kept in an inert gas atmosphere while it is being welded.

WARNING!

Dust, fumes, vapors, and fine machining particles of beryllium are cancer-causing agents. If inhaled, they can cause chronic beryllium disease. An independent air supply is required when working with beryllium.

Table 12 Examples of Beryllium Alloys and Mechanical Properties

Designation and Percent Be	Tensile Strength		Yield Strength		Commercial Form	Description or Properties
	psi	kPa	psi	kPa		
HP-20 (98.0 percent)	48,000	330,948	35,000	241,317	Block	Pressed powder
HP-8 (99.0 percent)	37,000	255,106	27,000	186,158	Block	Pressed powder
HP-12 (98.5 percent)	38,000	262,200	30,000	206,843	Block	Pressed powder
HP-40 (92.0 percent)	65,000	448,159	60,000	413,685	Block	Pressed powder
C-10 (99.0 percent)	20,000	137,895	—	—	Block	Pressed powder
PR-2 (98.0 percent)	77,000	530,896	55,000	379,212	Sheet/plate	Some form at 1,350°F (732°C)
IS-2 (99.0 percent)	45,000	310,264	31,000	213,737	Sheet/plate	High form at 700°F (371°C)
XT-20 (98.0 percent)	77,000	530,896	40,000	275,790	Extrusion	Good mechanical properties
XT-40 (92.0 percent)	90,000	620,528	50,000	344,738	Extrusion	High strength
Lockalloy	60,000	413,685	50,000	344,738	Extrusion	Service at 600°F to 800°F (316°C to 427°C)

Magnesium

Magnesium is widely used in specialty racing cars and some commercial cars as a lightweight material to minimize weight and fuel consumption because of the 20 to 25 percent weight saving over aluminum. It typically is used in the construction of transmission casings, intake manifolds, and cylinder head covers. Magnesium pistons and various other engine parts are also being used in some race cars. It is difficult to visually differentiate aluminum and magnesium. Use positive metal identification (PMI) test equipment to verify these materials.

1.5.5 Copper and Copper Alloys

Copper, like aluminum, is often used in nearly pure form for electrical and heat conductors. It is alloyed with zinc to make brass, and with tin to make bronze. Other bronzes are formed by alloying copper with aluminum, silicon, or beryllium. Many variations of alloys exist within each of the brass and bronze groups, made by varying the percentages of the alloying metals and by adding other metallic and nonmetallic elements. These include zinc, tin, nickel, silicon, aluminum, cadmium, and beryllium.

Beryllium

Beryllium allows the transmission of X-rays better than other metals or glass. Because of this property, along with its high melting point, it is widely used as the window portion of high-intensity X-ray tubes, such as those used in high-resolution X-ray machines and imaging equipment.

Copper and copper alloys are classified into the following major groups:

- *Coppers* – Coppers are metals that have a designated minimum copper content of 99.3 percent or higher.
- *High-copper alloys* – For the wrought-type products, these are alloys with a copper content of less than 99.3 percent but more than 96 percent that do not fall into any other copper alloy group. Cast-type high copper alloys have a copper content in excess of 94 percent, to which silver may be added to obtain special properties.
- *Brasses* – These alloys contain zinc as the principal alloying element, with or without other alloying elements. The wrought-type alloys are grouped into three categories of brasses:

copper-zinc alloys, copper-zinc-lead alloys (leaded brasses), and copper-zinc-tin alloys (tin brasses). Cast alloys are grouped into five categories of brasses: copper-tin-zinc alloys (red, semi-red, and yellow brasses); manganese bronze alloys (high-strength yellow brasses); leaded manganese bronze alloys (leaded high-strength yellow brasses); copper-zinc-silicon alloys (silicon brasses and bronzes); and cast copper-bismuth and copper-bismuth-selenium alloys.

- *Bronzes* – Bronzes are copper alloys in which the major alloying element is not zinc or nickel. For wrought-type alloys, there are four groups of bronzes: copper-tin-phosphorus alloys (phosphor bronzes); copper-tin-lead-phosphorus alloys (leaded phosphor bronzes); copper-aluminum alloys (aluminum bronzes); and copper-silicon alloys (silicon bronzes). Cast-type alloys are divided into four groups of bronzes: copper-tin alloys (tin bronzes), copper-tin-lead alloys (leaded and high leaded tin bronzes), copper-tin-nickel alloys (nickel-tin bronzes), and copper-aluminum alloys (aluminum bronzes). Alloys known as manganese bronzes, in which zinc is the major alloying element, are included in the brasses.
- *Copper-nickels* – Copper-nickel alloys have nickel as the principal alloying element, with or without other alloying elements.
- *Copper-nickel-zinc alloys* – These alloys contain zinc and nickel as the principal and secondary alloying elements, with or without other elements. They are commonly called nickel silvers.
- *Leaded coppers* – Leaded coppers comprise a series of cast alloys of copper with 20 percent or more lead, sometimes with a small amount of silver, but without tin or zinc.
- *Special alloys* – These are alloys whose chemical compositions do not fall into any of the above categories.

Two classification systems for copper alloys are the *ASTM Standard B224, Standard Classification of Coppers,* (*Table 13*) and the Copper Development Association, Inc. designation system (*Table 14*). The Copper Development Association system has been updated to fit the UNS, and includes all commercially available metals and alloys. The format for the UNS designation is CNNN00, in which NNN designates numbers. As shown in *Table 14,* the alloys are divided into wrought and cast alloy categories. For this reason, an

Statue of Liberty

The Statue of Liberty contains 179,000 pounds (81,193 kg) of copper. When copper oxidizes, it forms a blue-green self-protective coating, which is why the copper skin of the Statue of Liberty remained virtually intact after 100 years of being subjected to high sea winds, driving rains, and scorching sun. Close analysis showed that weathering and oxidation of the copper skin had caused only 0.005 inch (0.13 mm) of wear in a century. For this reason, the copper skin did not need to be significantly rebuilt when the statue was renovated for its centennial. However, high-alloy copper saddles and rivets were used during the restoration to fasten the copper skin to the framing underneath. The outline of the copper sheets can be seen in this image. This was done to ensure the structural integrity of the statue and guard against any galvanic reaction problems. Galvanic action is corrosion caused by an electrical current produced as the result of a chemical reaction between two dissimilar metals.

Table 13 ASTM Classification System for Copper Alloys

ASTM Designation	Alloy Type	Typical Uses
Tough-Pitch Coppers		
ETP	Electrolytic, tough-pitch	Bus bars, brazing rods, wire anodes, forgings
FRHC	Fire-refined, high-conductivity, tough-pitch	Mechanical applications
FRTP	Fire-refined, tough-pitch	Sheets, strips, plate
ATP	Arsenical, tough-pitch	Roofing, radiator cores
STP	Silver-bearing, tough-pitch	Pans, printing rolls, fasteners
SATP	Silver-bearing, arsenical, tough-pitch	Pans, printing rolls, fasteners
Oxygen-Free Coppers		
OFH	Oxygen-free (without residual deoxidants)	Tubing wave guides, starting anodes, wire
OFP	Oxygen-free, phosphorus-bearing	Welding rods, forging
OFTPE	Oxygen-free, tellurium- and phosphorus-bearing	Welding rods, forging
OFS	Oxygen-free, silver-bearing	Plate, sheets, rods, forging
OFTE	Oxygen-free, tellurium-bearing	Bars, free-machining
Deoxidized Coppers		
DHP	Phosphorized, high-residual phosphorus	Tubes, pipes, anodes, projectile rotating bands
DLP	Phosphorized, low-residual phosphorus	Tubes, wave guides, general use
DPS	Phosphorized, silver-bearing	Heat exchangers, steam lines, condenser tubes, tubes for general use
DPA	Phosphorized, arsenical	Heat exchangers, steam lines, condenser tubes, tubes for general use
DPTE	Phosphorized, tellurium-bearing	Free-machining

Table 14 UNS Designations for Classifying Copper Alloys

Copper Designation	Composition
Tough-Pitch Coppers	
C11X00	99.95% oxygen-free, high-conductivity copper
C12X00	99.88+% copper (tough-pitch copper)
C19X00	96+% copper (high-copper alloys)
C2XX00	Copper-zinc alloys (brasses)
C3XX00	Copper-zinc-lead alloys (leaded brasses)
C4XX00	Copper-zinc-tin alloys (tin brasses)
C51X00	Copper-tin alloys (phosphor bronzes)
C54X00	Copper-tin-lead alloys (leaded phosphor bronzes)
C62X00	Copper-aluminum alloys (aluminum bronzes)
C65X00	Copper-silicon alloys (silicon bronzes)
C70X00	Copper-nickel alloys
Cast Alloy Groups	
C80X00	99+% copper, copper alloys
C81X00	High-copper alloys (beryllium copper)
C83X00	Red brasses and leaded red brasses
C84X00	Yellow brasses and leaded yellow brasses
C86X00	Manganese and leaded manganese-bronze alloys
C87X00	Silicon bronzes and brasses
C90X00	Tin bronzes

alloy made in both a wrought and cast form can have different UNS numbers, depending on the method of manufacture.

Tough-pitch copper is oxygen bearing, and most of the impurities have been oxidized during refining to increase conductivity. Continuous casting methods have allowed the oxygen content to be reduced to 0.03 percent or less to achieve the desired results. During fire refining, the oxygen that comes from the air injected to cause the oxidation of impurities also combines with hydrogen to form steam. The steam creates gas pockets (porosity) in the metal during solidification. (In the past, high porosity was a characteristic of tough-pitch copper.) When tough-pitch copper is reheated by welding or annealing, the hydrogen trapped in the copper is diffused into the metal and reacts with cuprous oxide to form insoluble water vapor. The expansion associated with this reaction forces the grains apart, which causes embrittlement of the metal. For applications where copper must be resistant to gassing, the oxygen is sometimes eliminated by adding a deoxidizer such as phosphorus to the molten metal before casting. Such copper is specified as deoxidized copper.

For welding copper alloys, GMAW and GTAW are most commonly used.

1.5.6 Nickel and Nickel Alloys

Nickel, a corrosion-resistant metal, is used as an alloying element as well as for plating other metals. Nickel-clad steel and nickel-based alloys containing over 50 percent nickel are used extensively in industry. They have good corrosion resistance and elasticity, and unique magnetic thermal expansion and thermal conductivity. Many high-strength, high-temperature nickel super alloys have been developed for aerospace, automotive, and hot-die applications. Alloys of nickel and chromium (Nichrome®) are used to make heating elements of high electrical resistance.

UNS and international standard designations have been assigned to many nickel-alloys. However, trade names, such as Monel 400®, Inconel 600®, and Hastelloy 22® are more commonly used to identify them. *Table 15* lists some common nickel alloys and some of their uses and properties.

In general, nickel and nickel-based alloys can be welded by SMAW, GMAW, and GTAW processes. Welding nickel alloys is very similar to welding austenitic stainless steel, with the following exceptions:

- When welding nickel alloys, the surface oxide melts at a much higher temperature than the base metal, so surfaces should be thoroughly cleaned.
- Embrittlement can be caused by lead, sulfur, phosphorus, or other low-temperature metals. The surface should be free of all contaminants.
- Weld penetration in nickel-based alloys is shallower than with other metals and requires the use of very wide bevel and groove angles.

1.5.7 Lead, Tin, and Zinc

Lead, tin, zinc, and their alloys have very low melting points. For this reason, they are usually soldered and not welded. Zinc-plated, or galvanized, steels can be welded, but the fumes are toxic and must not be inhaled.

1.5.8 Platinum, Gold, Silver, and Other Precious Metals

The precious metals are osmium, iridium, rhodium, platinum, ruthenium, gold, palladium, silver, and their alloys. They are used industrially for electrical contacts, mirror coatings, catalysts, electrodes, high-temperature crucibles, and dental structures and fillings. They are usually brazed, soldered, or welded with special equipment.

Table 15 Common Nickel Alloys and Some of Their Uses

Alloy	Ni	C	Mn	Fe	S	Si	Cu	Cr	Al	Ti	Other	Typical Properties and Uses
Commercially Pure Nickel												
A-Nickel	99.4	0.1	0.2	0.15	0.005	0.05	0.1					Chemical industry, electroplating
D-Nickel	95.2	0.1	4.5	0.15	0.005	0.05	0.05					Resistance to sulfur attack
E-Nickel	97.7	0.1	2.0	0.10	—	0.05	0.05					Similar to D-Nickel
L-Nickel	99.4	0.02	0.2	0.15	0.005	0.05	0.1					Severe plastic forming operations
Z-Nickel	94.0	0.16	0.25	0.25	0.005	0.04	0.05					Age hardenable, springs, pump rods, shafts
Nickel-Copper Alloys												
Monel	66.0	0.12	0.90	1.35	0.005	0.15	31.50		2.80	0.50		Corrosion resistance, toughness, high strength
R-Monel	66.0	0.18	0.90	1.35	0.005	0.15	31.50					Free-machining
K-Monel	65.3	0.15	0.60	1.00	0.005	0.15	29.50					Age-hardenable, nonmagnetic
H-Monel	65.0	0.1	0.9	1.5	0.015	3.0	29.50					Age-hardenable, machinable
S-Monel	63.0	0.1	0.9	2.0	0.015	4.0	30.0					Age-hardenable, anti-galling, impellers pump liners
Nickel-Iron Alloys												
Invar	36.0	*	*	63.0	*							Very low expansion coefficient length standards
Platinite	46.0											Glass seal
Permalloy	78.5											High-magnetic permeability, submarine telegraph cables
Nickel-Chromium and Nickel-Iron-Chromium Alloys												
Inconel	76.4	0.04	0.20	7.20	0.007	0.2	0.10	15.85				Heat and corrosion resisting, nonmagnetic
Inconel 702	78.0	0.02	0.05	0.30	0.007	0.15	0.05	15.85	3.0	0.60		Age-hardenable, operates up to 2,400°F (1,316°C)
S-Inconel	68.0	0.20	1.0	8.5	0.008	5.0	0.5	15.5				Age-hardenable, anti-galling
Inconel-X-750	72.9	0.04	0.70	6.8	0.007	0.30	0.05	15.0	0.80	2.5	CB-0.9	Age-hardenable, used up to 1,500°F (816°C)
Nimonic 80A	74.5	0.05	0.55	0.55	0.007	0.20	0.05	20.0	1.30	2.50		Age-hardenable, similar to Inconel X
Nimonic 90	57.0	0.05	0.50	9.50	0.007	0.20	0.05	20.45	1.65	2.60	CO-17.00	Similar to Nimonic 80A
Nichrome	80.0	0.05	0.01	0.5	—	0.20	—	20.0				Heating elements
Chromel	90.0							10.0				Heating elements, thermocouples
Nicoloy 901	37.35	0.04	0.45	33.75	0.007	0.30	1.10	13.50	0.25		MO-5.90	Age-hardenable, gas turbine wheels
Ni-O-NEL	41.35	0.03	0.65	31.65	0.007	0.35	1.80	20.2	0.40	0.50	MO-310	Resistant to certain hot acids
60-12 Ni-Cr	62.45	0.55		25.0				12.0				Resistance heating element
65-15 Ni-Cr	65.0			24.00	0.007			15.00				Resistance heating element
67-17 Ni-Cr	72.45	0.55		10.0				17.00				Heat-resisting alloy

Additional Resources

AWS B1.10M/B1.10 Guide for the Nondestructive Examination of Welds. Miami, FL: American Welding Society.

AWS B1.11 Guide for the Visual Examination of Welds. Miami, FL: American Welding Society.

AWS D3.5-93R Guide for Steel Hull Welding. Miami, FL: American Welding Society.

AWS D3.6M Specification for Underwater Welding. Miami, FL: American Welding Society.

AWS D3.7 Guide for Aluminum Hull Welding. Miami, FL: American Welding Society.

Machinery's Handbook. Erik Oberg, Franklin D. Jones, and Christopher J. McCauley. New York, NY: Industrial Press, Inc. 2004.

Metals and How to Weld Them. T. B. Jefferson. Cleveland, OH: The James F. Lincoln Arc Welding Foundation.

OSHA 1926.351, Arc Welding and Cutting. Current Edition. Washington, DC: Occupational Safety and Health Administration (OSHA).

Stick Electrode Product Catalog. Cleveland, OH: The Lincoln Electric Company, 2008.

Stick Electrode Welding Guide. Cleveland, OH: The Lincoln Electric Company, 2004.

The Procedure Handbook of Arc Welding. 14th Edition. Cleveland, OH: The James F. Lincoln Welding Foundation, 2000.

Welding Handbook, 4 vols. Miami, FL: American Welding Society.

Lincoln Electric website: **http://www.lincolnelectric.com** offers sources for products and training.

1.0.0 Section Review

1. A document that a customer receives from a steel manufacturer that indicates the quality of the steel being purchased is a(n) ____.
   a. safety data sheet (SDS)
   b. ferrous metal classification bill (FMCB)
   c. certified mill test report (CMTR)
   d. evaluated properties rating index (EPRI)

2. The amount of alloy added to make high-strength low-alloy (HSLA) steel is typically ____.
   a. less than 2 percent
   b. about 5 percent
   c. between 6 and 8 percent
   d. about 10 percent

3. Ferritic stainless steels are always ____.
   a. martensitic
   b. high in nickel content
   c. hardenable by tempering
   d. magnetic

4. A form of highly alloyed stainless steel developed to provide outstanding resistance to acids and other caustic solutions is called ____.
   a. superferritic
   b. super duplex
   c. superaustenitic
   d. super martensitic

5. What nonferrous metal has the lowest density of the structural metals and is commonly used in the aerospace industry?
   a. Magnesium
   b. Aluminum
   c. Nickel
   d. Titanium

SECTION TWO

2.0.0 METAL CHARACTERISTICS

Objective

Describe the physical and mechanical characteristics of metals and explain how to identify base metals.

a. Describe the physical characteristics of different metals.
b. Describe the mechanical properties of different metals.
c. Explain how to identify base metals in field conditions.
d. Describe metallurgy-related considerations for welding.

Trade Terms

Residual stress: The strain that occurs in a welded joint after the welding has been completed, as a result of mechanical action, thermal action, or both.

Underbead cracking: Subsurface cracking in the base metal under or near the weld.

All metals have different characteristics and mechanical properties. These physical and mechanical properties can be used to identify a particular metal. In addition, a metal's properties must be taken into consideration prior to and during welding operations.

2.1.0 Physical Characteristics of Metals

Physical characteristics are distinctive for every metal. They include the following variable traits:

- Density (specific gravity)
- Electrical conductivity
- Thermal conductivity
- Thermal expansion
- Melting point
- Corrosion resistance

2.1.1 Density

Density is the mass (weight) of a specific volume of metal. Density may also be expressed as specific gravity or specific density. Specific gravity is the weight of a metal compared to the weight of an equal volume of water. The higher the density of a material, the heavier it will be.

Light metals, such as aluminum and magnesium, have low densities (low weights for a given volume) while ferrous metals, such as carbon steel and stainless steel, have much higher densities. An iron or steel item weighs nearly three times as much as an aluminum item of the same size. *Table 16* lists the densities of some common metals in pounds per cubic inch and grams per cubic centimeter.

2.1.2 Electrical Conductivity

Electrical conductivity is a metal's ability to conduct electricity. All metals can conduct electricity to some degree. Silver is one of the best conductors, but it is too expensive to use for commercial power transmission. Copper is not quite as good a conductor as silver, but for the same diameter conductor or wire, it is a better conductor than aluminum. However, for conductors weighing the same per unit of length, aluminum will carry more current than copper because it has a larger cross-section.

The electrical conductivity of a metal is calculated from its measured resistance to current flow (its resistivity). Resistivity is the inverse (opposite) of conductivity. The lower the resistivity value, the higher the conductivity value and the more current a metal can carry for its size. The resistivity of any metal increases and its conductivity decreases as its temperature rises. This means that metals with higher electrical resistances build up heat from welding current faster. They also do not transfer heat away as efficiently, resulting in higher distortion and stresses in the weld zone. *Table 17* lists the electrical resistivity of some common metals in microhms per cubic centimeter.

Table 16 Densities of Some Common Metals

Metal	Density (lb/in^3)	Density (g/cm^3)
Aluminum	0.098	2.71
Copper	0.324	8.96
Gold	0.698	19.32
Iron or steel	0.284	7.87
Magnesium	0.063	1.74
Manganese	0.267	7.43
Molybdenum	0.369	10.20
Nickel	0.309	8.90
Tin	0.264	7.30
Titanium	0.163	4.54
Tungsten	0.697	19.30
Zinc	0.258	7.13

Table 17 Electrical Resistivity of Some Common Metals

Metal	Electrical Resistivity (microhms/cm^3)
Aluminum	2.92
Copper	1.67
Gold	2.19
Iron or steel	9.71
Magnesium	4.46
Manganese	185.00
Molybdenum	5.17
Nickel	6.84
Titanium	80.00
Zinc	5.91

2.1.3 Thermal Conductivity

Thermal conductivity is a metal's ability to conduct heat. Metals that are good conductors of electricity are also good conductors of heat. As temperature increases, a metal's resistance to the flow of heat increases. Because a metal with poor heat conductivity does not carry the weld heat away from the weld zone efficiently, the temperature rises and distortion is greater. Silver and copper are excellent heat conductors, iron and nickel are fair conductors, and titanium and manganese are poor conductors.

2.1.4 Thermal Expansion

Thermal expansion is the change in size that occurs in a material as its temperature changes. Solids expand in all dimensions when heated and contract from all dimensions when cooled. The increase in unit length when a solid is heated one degree is called its coefficient of linear expansion. The larger the coefficient of expansion, the greater the dimensional increase of the solid for each degree of temperature increase. Large dimensional changes cause distortion and stresses in weldments. *Table 18* lists the coefficients of linear expansion for a variety of metals. *Figure 8* shows the expansion of common metal pipes in inches per 100 feet for a range of temperatures in degrees Fahrenheit.

2.1.5 Melting Point

The melting point of a solid is the temperature at which the solid converts into a liquid. *Table 19* lists a number of metals and their melting points.

2.1.6 Corrosion Resistance

Metals and alloys are corroded by different agents at varying rates. Corrosion is important to consider because it can severely reduce the tensile strength of a metal. When selecting a metal for a particular use, its ability to resist corrosion by the atmosphere and by gaseous or liquid chemicals may be an important factor. Certain metals may not react to specific chemicals. For example, iron is often used to handle strong caustics. Other metals produce a protective oxide coating that retards further corrosion. Both aluminum and lead form weather-resistant oxide coatings in the atmosphere. The stainless steels and nickel-based alloys are designed to withstand corrosion from such things as the atmosphere, elevated temperatures, and strong chemicals.

2.2.0 Mechanical Properties of Metals

The mechanical properties of metals determine how they will react to or be affected by the external forces of tension, compression, torsion, shear, impacts, and cold shaping (working). The ability of metals to permanently change or deform is called plasticity. Plasticity is different from elasticity, which refers to the ability of a material to change temporarily and then revert to its original form.

2.2.1 Stress-Strain Relationship

Stress is the force or load applied to a specimen that causes it to deform. Strain is the magnitude of deformation caused by stress. Within certain limits, stress increases directly with strain. The four types of stress are as follows:

- *Tension* – Opposed, in-line forces pulling away from each other at opposite sides of a specimen
- *Compression* – Opposed, in-line forces pushing toward each other against opposite sides of a specimen
- *Torsion* – Opposed, separated, twisting forces acting on the same axis of a specimen
- *Shear* – Opposed, nonintersecting forces acting opposite each other and perpendicular to a specimen

Figure 9 shows the force directions that result from different types of stress. Often several different stress types act on a specimen at the same time.

Table 18 Coefficients of Linear Expansion of Some Common Metals

Metal	Coefficient (a) 10^{-6}/°F	Coefficient (a) 10^{-6}/°C
Aluminum	13.3	25.0
Copper	9.2	16.5
Iron	6.5	12.0
Magnesium	14.0	25.0
Manganese	12.0	22.0
Molybdenum	2.7	5.0
Nickel	7.4	12.8
Titanium	4.7	8.6
Zinc	22.1	39.7

2.2.2 Elasticity and Elastic Limit

Elasticity is the ability of a material to be deformed without the deformation becoming permanent. As long as a material remains within its elastic range, it will return to its original size and shape when the deforming force (stress) is removed. While within the elastic range, the strain will be directly proportional to the stress that produces it.

Every material has its elastic limit. When the elastic limit is exceeded, the strain is no longer proportional to the stress, and the material will not return to its original size or shape. The material has been distorted. Elastic limits are commonly measured by an extensometer.

2.2.3 Modulus of Elasticity

Every material strains (deforms) in direct proportion to the applied stress while it remains within its elastic range; but the amount each material strains for a given stress is not the same. The ratio of stress to strain is called the modulus of elasticity. It remains the same for any material under the same proportional conditions. For example, a material that will stretch 1 inch (25.4 mm) under a 10-pound (4.5-kg) tension will stretch 2 inches (50.8 mm) under a 20-pound (9-kg) tension, as long as the elastic limit is not exceeded. Similarly, a material that will stretch ¼ inch (6.4 mm) under a 10-pound (4.5-kg) tension will stretch ½ inch (12.7 mm) under a 20-pound (9-kg) tension as long as the elastic limit is not exceeded. The modulus of elasticity of a metal can be used to predict how much the metal will deform under a specified load, provided the elastic limit is not exceeded.

2.2.4 Yield Strength and Tensile Strength

Yield strength is the maximum tensile stress a material can withstand before irreversibly deforming. Once stressed beyond its yield strength, the material will not return to its original shape when the stress is removed. Tensile strength, on the other hand, is the maximum tensile stress a material can withstand without breaking. The maximum load on a tensile material when it fractures is called the ultimate load. Brittle materials tend to break at values near their yield strength, while more elastic materials may deform significantly before breaking. Both yield and tensile strength are expressed in force per cross-sectional unit area. Engineers are interested in both values during the design process.

2.2.5 Ductility

Ductility is the plasticity exhibited by a material under tension loading. Ductility is measured by dividing the change in length of a material by the original length at the point of failure. Ductility is the characteristic that allows metals to be drawn into thinner sections without breaking. Copper and aluminum are examples of common ductile metals.

2.2.6 Hardness

The hardness of a material can be defined as its resistance to indentation. It is related to its elastic and plastic properties. The greater the hardness, the greater the resistance to indentation, penetration, and wear. Hardness is measured by pressing a steel ball or diamond point against a surface with a calibrated force. The diameter or depth of the impression is then measured to determine the hardness. Standard hardness testers include the Brinell, Rockwell, and Vickers instruments. *Figure 10* shows a portable hardness tester suitable for testing on the job site.

The Brinell hardness test is performed by pressing a hardened 10-mm steel ball into the metal being tested. Force is applied by standard weights: 500 kg for aluminum, copper, and other soft materials; and 3,000 kg for iron, steel, and

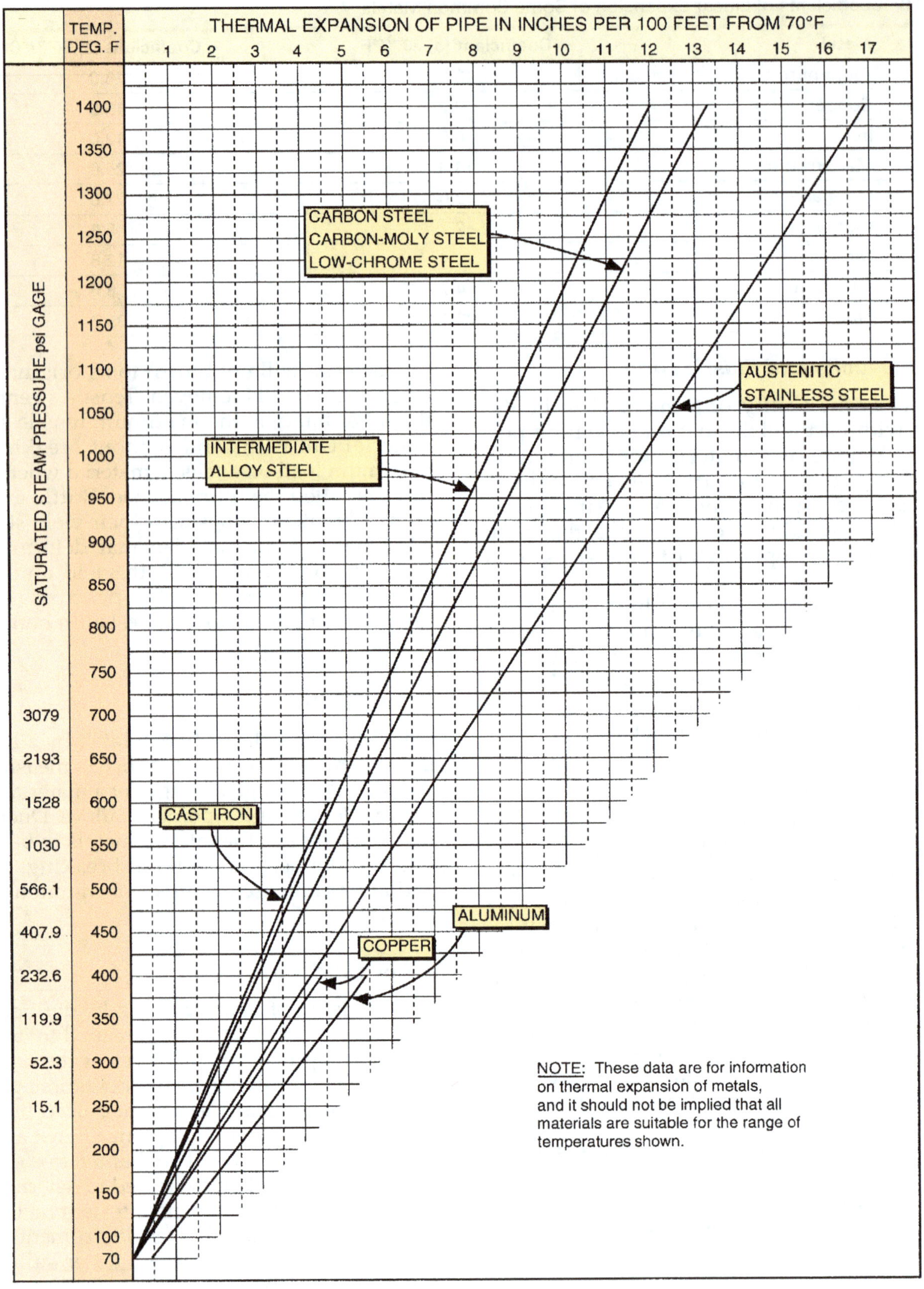

Figure 8 Expansion of different types of pipe.

Table 19 Melting Points of Some Common Metals

Metal	Melting Point
Aluminum	1,215°F (657°C)
Copper	1,981°F (1,083°C)
Iron	2,800°F (1,538°C)
Magnesium	1,202°F (650°C)
Manganese	2,273°F (1,245°C)
Molybdenum	4,760°F (2,627°C)
Nickel	2,651°F (1,455°C)
Titanium	3,300°F (1,816°C)
Zinc	787°F (419°C)

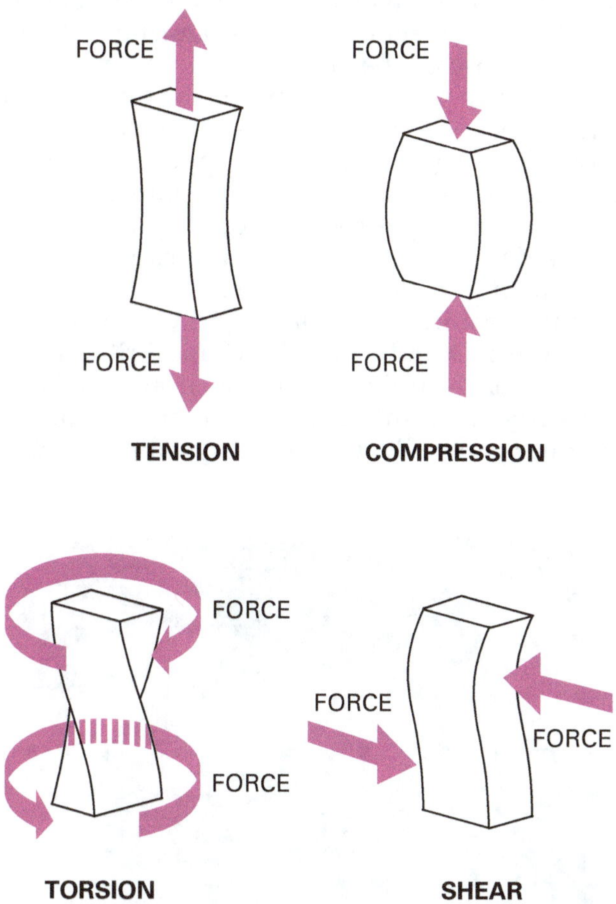

Figure 9 Force directions that result in each stress type.

other hard metals. The diameter of the indentation is measured under a microscope and then converted to the Brinell hardness number.

The Rockwell hardness tester uses a 1⁄16-inch (1.6-mm) hardened steel ball with a 100-kg weight on soft metals, and a diamond cone and 150-kg weight on hard metals. Depth of penetration is read directly from hardness scales on the tester.

The Vickers hardness tester uses a pyramid-shaped diamond penetrator with a variable 1- to 120-kg load. The hardness is read from standard

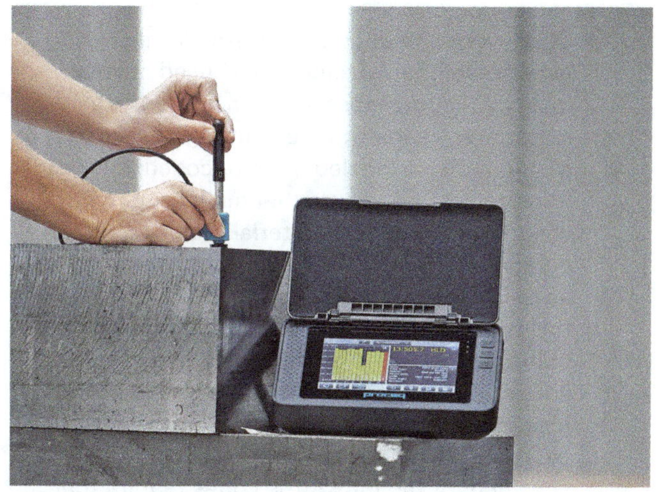

Figure 10 Portable hardness tester.

tables by comparing the width of the dent with the pressing load.

2.3.0 Field Identification of Base Metals

Many of the common base metals and alloys can be generally identified by simple tests and observation of their characteristic metallic features. This is especially true of the ferrous metals, which can be identified by magnetic attraction, surface appearance, structural form or shape, and grinding sparks.

2.3.1 Metal Labeling

Because it is difficult to identify base metal types, most sites have a system to label or color-code the various types of metals being used. Labeling is performed by printing the ASTM number on the material (shape or plate) with an indelible marker or paint stick. Color-coding involves assigning a color to a base metal type and then spraying that color on the material. When color-coding is used, a master chart is often posted near the storage area. The master chart identifies the color assigned to each base metal.

For stainless steel and nickel alloys, special care should be taken to mark only with low-chloride markers. Using markers other than the low-chloride markers will leave a residue on the metal that can contaminate the weld. Specialty markers, which usually contain white or yellow ink, are available for marking metals such as stainless steel (*Figure 11*).

Figure 11 A typical low-chloride marker for specialty metals.

NOTE

Check your site quality standards for the type of material labeling that is used at your location. Follow those standards to ensure material traceability. When cutting a piece of material that is labeled or color-coded, it is very important to make sure that the label or color code is visible on the material that is being left behind. If possible, make the cut in such a way as to leave the label or color code still visible. If this is not possible, be sure to mark the material being left behind to ensure that others can correctly identify the material. It is also a good idea to mark the material being removed so that everyone involved can identify it as the material specified for the job. Some codes require that a witness be present when marking the two pieces.

2.3.2 Identification by Magnet

All carbon steels, cast steels, and cast irons are attracted to a magnet. Of the stainless steels, only ferritic and martensitic stainless steels can be magnetized, and they have the bright, corrosion-free surface typical of stainless steel. The largest group of stainless steels—austenitic stainless steels—are nonmagnetic. Nonferrous metals and manganese steels are also nonmagnetic.

CAUTION

Magnetic testing is sometimes unreliable because of alloys or changes that can occur during welding. Do not rely solely on magnetic testing.

2.3.3 Identification by Appearance

Both hot-rolled and cold-finished carbon steel are milled in standard structural shapes, such as rods, bars, beams, channels, and angles. Unweathered, hot-rolled carbon steel has a blue-gray, smooth-to-scaly surface, but it oxidizes to a rough, red-brown rust when exposed to the weather. The corners of hot-rolled shapes are slightly rounded or chamfered.

Unweathered, cold-rolled carbon steel has a smooth, gray finish that oxidizes to a fine red-brown rust when weathered. Except for round shapes, the sides and edges are very flat and smooth, and the corners are well formed, not rounded or chamfered.

Cast steel and cast iron look similar on the surface. They both exhibit the shape characteristics typical of castings: curved surfaces and rounded edges; thin and thick regions; reinforcing ribs or raised areas; spokes; and irregular shapes. The cast surfaces are usually slightly textured and can contain mold seams. When broken, cast iron has a grainy appearance, while cast steel has a smooth texture. Cast iron is used to make everything from water-meter covers, to engine crankshafts, to fan blades (*Figure 12*).

There are three types of cast iron: gray, white, and malleable. The exterior surface of all the cast irons has a slightly irregular texture that comes from the use of sand molds. Malleable iron is sometimes forged. If it has been forged, it will have a smoother appearance and die marks.

Gray cast iron contains a high percentage of carbon and silicon. Gray cast iron is only formed by slow cooling, which forms large crystals and a soft, easily machined metal. During the slow cooling process, the high silicon content forces the carbon to separate out in the form of graphite flakes (free carbon). This gives a fresh fracture its characteristic grainy and gray appearance. If you rub a clean finger or white cloth against a freshly fractured gray cast-iron surface, the graphite will leave a dark smear. Gray cast iron can be welded by following special procedures.

White cast iron is chilled rapidly, which forms small crystals and a very hard and brittle metal. The carbon is united with the iron and is not free carbon. Freshly fractured cast iron has an irregular, fine, silvery-white, crystalline surface.

Figure 12 A compressor's cooling fan blade made of cast iron.

Identification of Nonferrous Metals

The electrical conductivity measuring instrument shown can be used to determine the composition of nonferrous metals, including titanium alloys and silver. By using eddy currents to determine the electrical conductivity of a nonferrous metal under test, conclusions can be made about the composition, structure, and condition of the material.

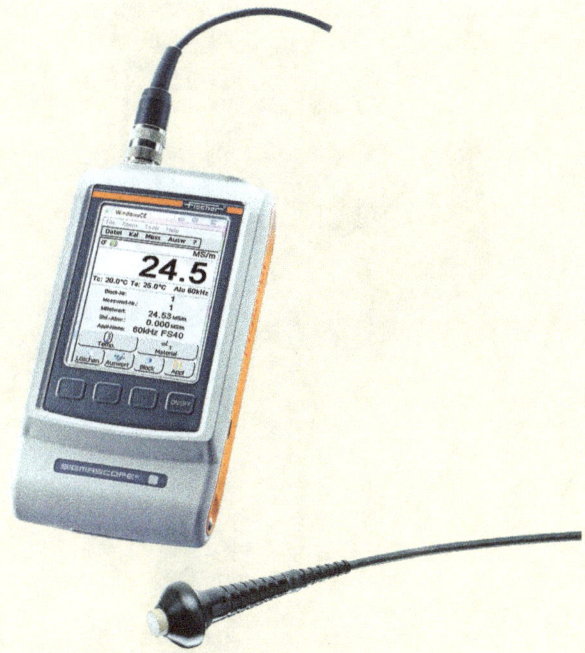

Because the weldability of white cast iron is poor, it is generally not welded.

Malleable cast iron is white cast iron that has been annealed. The annealing process results in a casting with an outer shell of white cast iron that grades into a core of gray cast iron. The resulting casting is tougher and stronger than gray cast iron and less brittle than white cast iron. Malleable cast iron can be welded, but special procedures must be used or it will revert back to white cast iron.

2.3.4 Identification by X-Ray Fluorescence Spectrometry

Portable metal analyzers that use X-ray fluorescence (XRF) spectrometry provide a fast and accurate analysis for metal and metal alloy identification. A fluorescent X-ray is created when an X-ray of sufficient energy, emitted from a miniaturized X-ray tube or from a small, sealed radioactive isotope, strikes an atom in the sample, dislodging an electron from one of the atom's inner orbital shells. The atom regains stability, filling the vacancy left in the inner orbital shell with an electron from one of the atom's higher energy orbital shells. The electron drops to the lower energy state by releasing a fluorescent X-ray, and the energy of this X-ray is equal to the specific difference in energy between two quantum states of the electron.

When a sample is measured using X-ray fluorescence, each element present in the sample emits its own unique fluorescent X-ray energy spectrum. By simultaneously measuring the fluorescent X-rays emitted by the different elements in the sample, handheld XRF analyzers can rapidly determine those elements that are present in the sample, along with their relative concentrations—in other words, the elemental chemistry of the sample. For samples with known ranges of chemical composition, such as common grades of metal alloys, the XRF analyzer also identifies most sample types by grade name, through comparing the analyzed concentrations with known values contained in an alloy library programmed within the analyzer. This entire process is typically completed in less than five seconds, and is completely nondestructive.

Thirty or more elements may be analyzed simultaneously by measuring the characteristic fluorescence X-rays emitted by a sample. *Figure 13* shows a handheld analyzer, its screen, and the analyzer in use.

2.4.0 Metallurgical Considerations for Welding

Many metallurgical variables must be considered when determining the effects of welding on both the base metal and the weld. Base metal composition and thickness, filler metal, and joint configuration all affect the weld quality. There are also several processes and design considerations used to change the metal characteristics in order

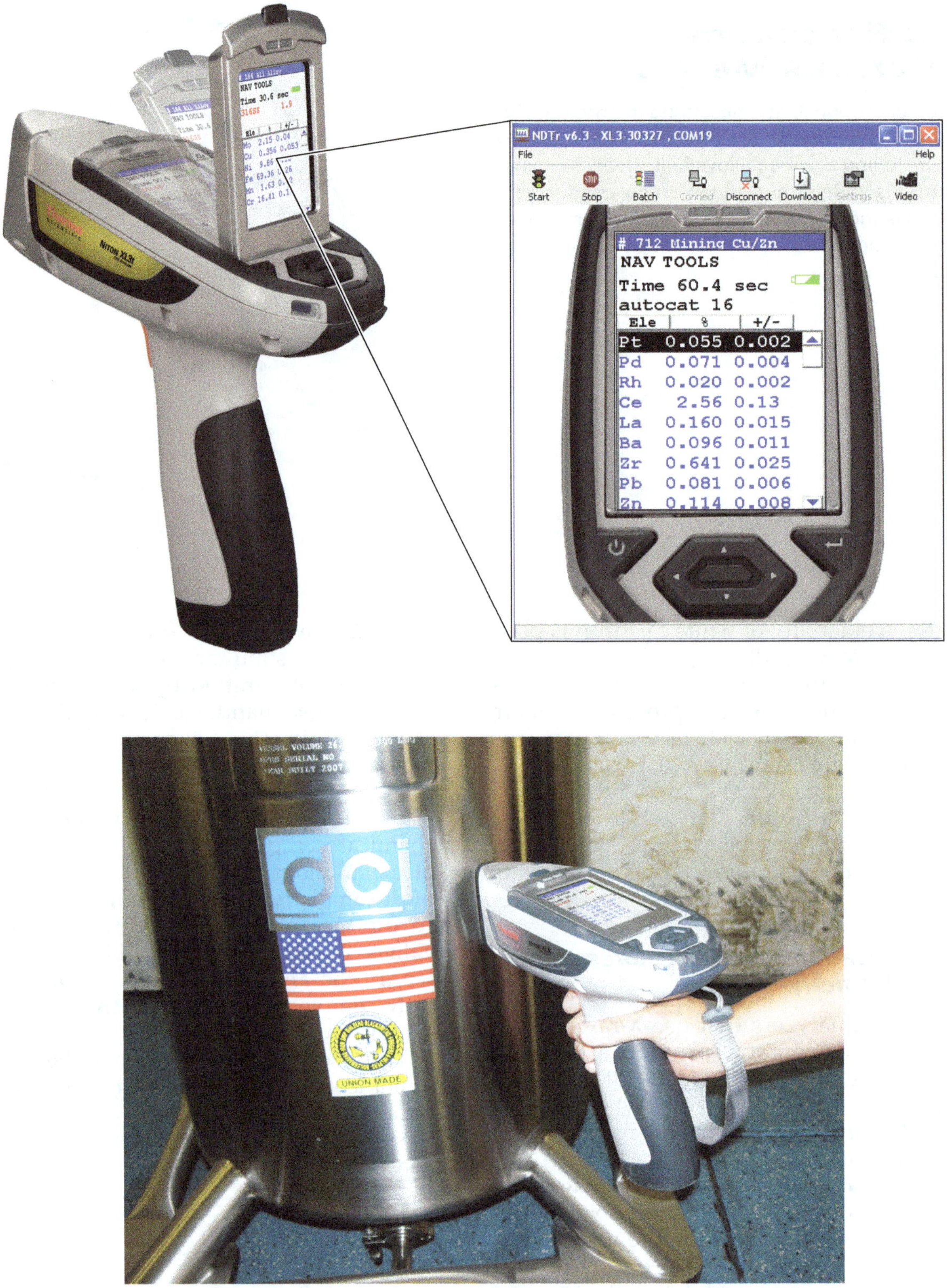

PHOTOS COURTESY OF THERMO FISHER SCIENTIFIC.

Figure 13 NITON XL3 handheld analyzer.

to produce a more desirable weld. The following sections explain the metallurgical factors that must be considered to make acceptable welds:

- Base metal preparation
- Joint design considerations
- Filler metal selection
- Filler metal and electrode selection considerations
- Preheating and interpass temperature control
- Postweld heat treatment

2.4.1 Base Metal Preparation

Because of the defects caused by surface contaminants and the possibility of toxic fumes, all welding codes require surface cleaning prior to any cutting or welding activity. In general, the codes state that the surface to be thermally cut or welded must be clean and free from paint, oil, rust, scale, and other material that would be detrimental to either the weld or the base metal when heat is applied.

The most common defect caused by surface contamination is porosity, which occurs when gas pockets or voids appear in the weld metal. When the porosity has a length greater than its width and is approximately perpendicular to the weld face, it is called piping porosity. Piping porosity is formed as the gas pocket floats toward the surface of the weld, leaving a void. The gas pocket is trapped as the weld metal solidifies, but as the next layer of weld is deposited, it continues to float up. Porosity and piping porosity may not be visible on the surface of the weld.

Removing surface contaminants from base metal is done using mechanical and/or chemical means. Mechanical cleaning is the most common method. It can involve the use of hand tools, power tools, and special sandblasting equipment.

WARNING!

When performing mechanical cleaning, be sure to wear safety glasses and a face shield for protection from the flying particles produced during the cleaning operation. In addition, special clothing is required for sandblasting. Respiratory protection may also be needed.

Several different types of chemicals can be used to clean metal before cutting or welding. Their primary purpose is to clean oil and grease from metal surfaces that require high-quality welds. The cleaning chemical to be used should be selected specifically for the type of base metal and the application. Using the wrong chemical can cause a reaction with the base metal that results in burning, pitting, or discoloration of the surface. All chemicals used for cleaning metal are very strong, which enables them to do their job quickly. Because they are so strong, they can be very hazardous if mishandled.

WARNING!

Never use a chemical without specific and recent training in its use. The mishandling of cleaning chemicals can cause severe burns to the skin, lungs, and eyes. Some chemicals can also cause death or blindness. Always use approved solvents. Procedures and methods for using, storing, and disposing of most chemicals are given in the SDS/MSDS for the specific chemical being used. These are available at the shop or job site.

2.4.2 Joint Design Considerations

Welded joints are selected primarily for the safety and strength required for the service conditions. When selecting the joint, groove, and weld to use, there are many considerations to take into account. The following sections explain those considerations.

Determining the Weight and Density of Materials

At 39°F (4°C), water is at its maximum density of 62.42 lb/ft^3 (≈1,000 kg/m^3). Since water is the standard to which other materials are compared, it is given a specific gravity value of 1.

If the specific gravity of a material is known, its weight can be calculated by multiplying its specific gravity by the weight of an equal volume of water. For example, the specific gravity of aluminum is 2.71. Its weight in pounds per cubic foot (lb/ft^3) can be found using the formula 2.71 × 62.42 lb/ft^3 = 169.16 lb/ft^3. In other words, one cubic foot of aluminum has a density (weight) of 169.16 pounds. In metric units, the calculation is 2.71 × 1,000 kg/m^3 = 2,710 kg/m^3.

Similarly, if the weight of a material is known, its specific gravity can be calculated by dividing its weight by the weight of an equal volume of water. For example, one cubic foot of aluminum weighs 169.16 pounds. The specific gravity of aluminum can then be determined using the formula 169.16 lb/ft^3 ÷ 62.42 lb/ft^3 = 2.71. In other words, aluminum has a specific gravity of 2.71. In metric units, the calculation is 2,710 kg/m^3 ÷ 1,000 kg/m^3 = 2.71.

A welding code is a detailed listing of the rules and principles that apply to specific welded products. Codes ensure that safe and reliable welded products will be produced and that the weldments will be reasonably safe. Clients specify which codes to follow when placing orders or letting contracts. These orders or contracts impose severe penalties for not conforming to the code(s) specified. In addition, when codes are specified, the use of these codes is mandated with the force of law by one or more government jurisdictions. Always check the contract, order, or project specification for the code(s) specified.

Codes require that a WPS be written for all critical welds. A WPS is a written set of instructions for producing sound welds. It includes the type of joint to use, as well as the groove preparation, if required. Each WPS is written and tested in accordance with a particular welding code or specification. All welding requires that acceptable industry standards are followed, but not all welds require a WPS. If a weld does require a WPS, it must be followed. The requirement to use a WPS is often listed on project drawings as a note or in the tail of the welding symbol. If you are unsure whether the welding being performed requires a WPS, do not proceed until you check with your supervisor.

One of the easiest joint preparations is the square groove joint because it only requires butting the edges of the plate together. Butt joints are welded with a square groove weld that can have a partial or complete joint penetration weld. A partial joint penetration weld has much less strength than a complete penetration weld. The *AWS D1.1 Structural Welding Code – Steel* imposes restrictions for complete penetration welds. With SMAW, the maximum base metal thickness is ¼ inch (6 mm plate) and welding from both sides is required. In addition, a root opening of half the thickness of the base metal is required, and the root of the first weld must be gouged before the second weld is made. For GMAW and FCAW, the maximum base metal thickness is ⅜ inch (10 mm plate) with the same requirements for welding from both sides and for back-gouging.

Fillet welds also require very little joint preparation. When using a fillet weld on outside corner welds, two types of fit-up can be used: corner to corner or half lap. The corner-to-corner joint is difficult to assemble because neither plate can support the other, and care must be taken not to burn through the corner when welding. The half-lap joint is easier to assemble, requires less weld metal, and has less danger of corner burn-through. The half-lap requires a second weld on the inside of the corner. If a half-lap fit-up is used, allowances in the plate dimensions for the lap must be made. Several methods of joint preparation are explained in the following items:

- *Single- and double-groove welds (Figure 14)* – When possible, use a double-groove joint in place of a single groove. The double groove requires half the weld metal of the single groove. Also, welding from both sides reduces distortion because the forces of distortion work against each other.
- *U- and J-groove preparation* – On thick joints, U- and J-grooves often require less weld metal than V or bevel joints. The major disadvantage of U- and J-grooves is that they require more preparation time. For this reason, V and bevel joints are used much more frequently than J- or U-grooves.
- *Groove angles and root openings (Figure 15)* – The purpose of the groove angle is to allow electrode access to the root of the weld. If the groove angle is larger than necessary, it requires additional weld metal to fill. This increases the time and cost to complete the weld and increases the chance of distortion. If the groove angle is too small, it can result in weld discontinuities. The root preparation is sized to control melt-through. Increasing or decreasing the root preparation (root opening and root face) results in excess melt-through or insufficient root penetration. As a general rule, as the groove angle decreases, the root opening increases to compensate. Root faces are used with open-root joints, but not when metal backing strips are used (except for aluminum).

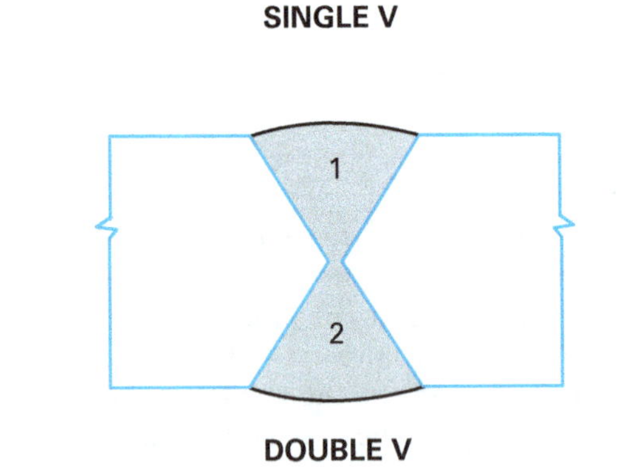

Figure 14 Single- and double-groove welds.

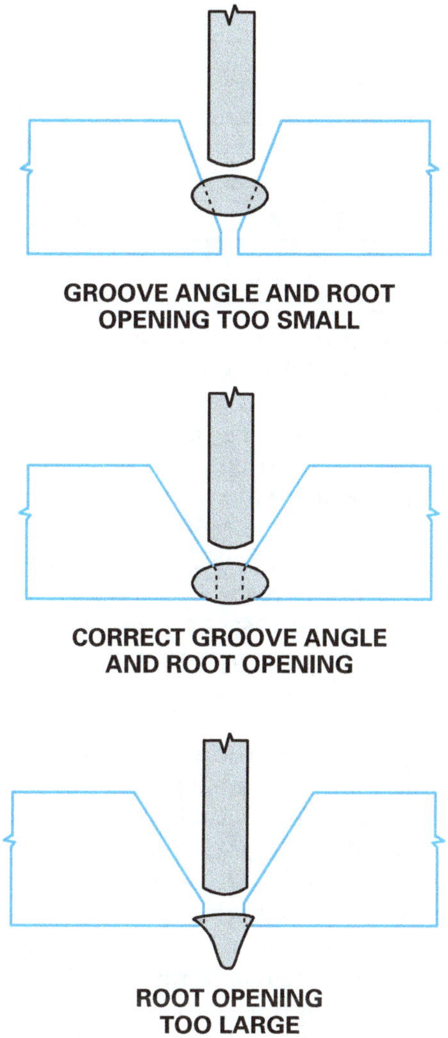

Figure 15 Proper and improper groove angles and root openings.

- *Groove welds with root openings* – For groove welds on plate, the groove angle is 60 degrees, and the maximum size of the root opening and root face is ⅛ inch (about 3.2 mm). For open root welds on pipe, the groove angle is 60 or 75 degrees, depending on the code or specifications used. The maximum size of the root opening and root faces is referred to in the WPS. *Figure 16* shows open-root joint preparation, including the groove angles.

2.4.3 Filler Metal Selection

Because so many different electrodes are available, the welder must be able to distinguish between them and select the correct electrode for the job. The following sections will explain how to select electrodes.

Electrodes are often classified into one of four groups based on their general characteristics:

- Fast-freeze

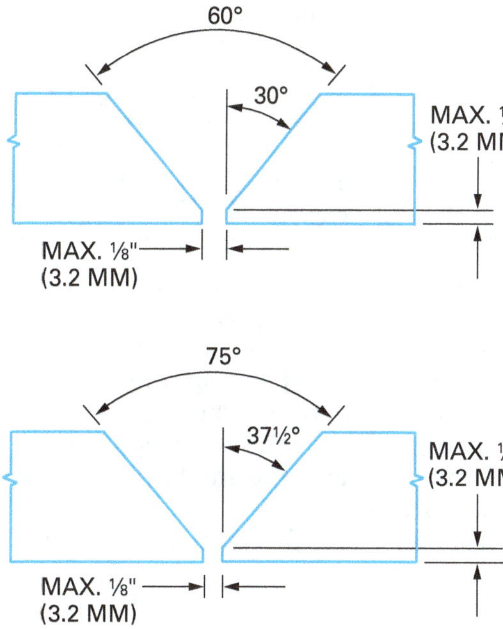

Figure 16 Open-root joint preparation.

- Fast-fill
- Fill-freeze
- Low-hydrogen

These four groups are identified by F-numbers. Refer to *Table 20* as you progress through the discussion of each group and their characteristics. The mild steel electrodes used with SMAW are classified as F1, F2, F3, and F4. Qualification with an electrode in a particular F-number classification qualifies the welder with all electrodes identified in that classification and in lower F-number classifications. For example, a typical AWS welder qualification test is a V-groove weld with metal backing in the 3G and 4G positions using an F4 electrode. Passing this test qualifies the welder to weld with F4 and all lower F-number electrodes, and to make groove and fillet welds in all positions.

Fast-fill electrodes (F1) have powdered iron in the flux, which gives them high deposition rates but requires that they only be used for flat welds and horizontal fillet welds. These electrodes have shallow penetration with excellent weld appearance and almost no spatter. The heavy slag covering is easy to remove. Electrodes in the fast-fill group will also have a 2 as the next-to-last number, as shown in *Table 20*. They include electrodes such as the following:

- E6027
- E7024
- E7028 (also in the low-hydrogen F4 group)

Typical applications for fast-fill electrodes are production welds on plate more than ¼" thick and flat and horizontal fillets and lap welds.

Table 20 AWS F-Number Electrode Classification

Group	AWS Electrode Classification				
F4	EXX15	EXX16	EXX18		Low-Hydrogen
F3	EXX10	EXX11			Fast-Freeze
F2	EXX12	EXX13	EXX14		Fill-Freeze
F1	EXX20	EXX24	EXX27	EXX28	Fast-Fill

Fill-freeze electrodes (F2), also called fast-follow electrodes, are all-position electrodes that are commonly used for vertical and flat welding. They have medium deposition rates and penetration and are excellent for sheet metal. The weld bead ranges from smooth and ripple-free to even with distinct ripples. The medium slag covering is easy to remove.

Electrodes in the fill-freeze group include the following:

- E6012
- E6013
- E7014

Typical applications for fill-freeze electrodes are the following:

- Vertical fillet or lap welds
- Sheet metal lap and fillet welds
- Joints having poor fit-up in the flat position
- General purpose welding

Fast-freeze electrodes (F3) are all-position electrodes that provide deep penetration. However, the arc creates a lot of fine weld spatter. The weld bead is flat with distinct ripples and a light slag coating that can be difficult to remove. The arc is easy to control, making fast-freeze electrodes good for overhead and vertical welding. These more challenging positions often make it harder for the welder to maintain a precise arc gap. Electrodes in the fast-freeze group include the following:

- E6010
- E7010
- E6011

Typical applications for fast-freeze electrodes are as follows:

- Vertical and overhead plate welding
- Pipe welding (cross-country and in-plant)
- Welds made on galvanized, plated, or painted surfaces
- Joints requiring deep penetration, such as square butts
- Sheet metal welds

WARNING!

Many surface coatings, such as galvanized coatings, give off very toxic fumes when heated, cut, or welded. Fumes from certain types of electrodes may also be hazardous. Breathing these fumes can cause damage to the lungs and possibly other organs. Identify any hazard for the coatings with which you are working by referring to the SDS/MSDS for that particular coating. Also, read the SDS/MSDS contained in the electrode package. Use proper ventilation and take the necessary precautions to protect yourself and other workers in the immediate area.

Low-hydrogen electrodes (F4) are designed for welding high-sulphur, phosphorus, and medium- to high-carbon steels, which have a tendency to develop porosity and underbead cracking. These problems are caused by hydrogen that is absorbed during welding. Low-hydrogen electrodes are also available with fast-fill and fill-freeze characteristics. They produce dense, X-ray-quality welds with excellent notch toughness and ductility. Electrodes in the low-hydrogen group include the following:

- E7015
- E7016
- E7018
- E7028
- E7048

Typical applications for low-hydrogen electrodes are the following:

- X-ray quality welds
- Welds requiring high mechanical properties
- Crack-resistant welds in medium- to high-carbon steels
- Welds that resist hot-short cracking in phosphorus steels
- Welds that minimize porosity in sulphur steels
- Welds in thick or highly restrained mild or alloy steels to minimize the danger of weld cracking
- Welds in alloy steels requiring a strength of 70,000 psi (≈4,826 bars) or more
- Multiple-pass welds in all positions

If hydrogen is present in the weld zone, it will create defects. The greatest source of hydrogen is airborne water vapor, which is composed of oxygen and hydrogen. Low-hydrogen electrodes are carefully manufactured to eliminate moisture in the flux coating. Low hydrogen is defined as being less than 16 milliliters (ml) per 100 grams of weld metal. The amount of hydrogen in these electrodes varies from 4 to 16 ml per 100 grams of metal weld.

Low-hydrogen electrodes have two unique suffixes. The M suffix is for compliance. Most military codes require an M suffix. When the suffix H and a number are added, the numbers relate to the amount of hydrogen in the electrodes. As a result, the electrodes will have suffixes such as H4, H8, and H16. These three suffixes indicate the dryness of the electrodes, identifying the milliliters of hydrogen per 100 grams of metal weld they contain. An additional suffix may also be added to these dryness suffixes. A suffix such as H4R would be one such additional suffix. The R means indicates they are moisture resistant. Without an R suffix, the electrodes may only be exposed to the atmosphere for a maximum of 4 hours. Those with an R suffix have an allowable exposure of up to 8 hours.

If low-hydrogen electrodes are exposed to the atmosphere, they will begin to absorb moisture and will no longer be low-hydrogen as a result. This is the reason for their short atmospheric exposure limits. For this reason, all low-hydrogen electrodes are shipped in hermetically sealed metal containers. Once opened, the electrodes must be stored in a heated oven. The temperature range of this oven is 100°F to 550°F (38°C to 228°C). However, it is important to note that placing the oven in areas below freezing may not allow it to maintain its maximum temperature, if that setting is necessary. Additional guidelines for the storage and handling of low-hydrogen electrodes are covered later in this module.

CAUTION

The welding codes establish strict guidelines for the storage, handling, and re-drying of low-hydrogen electrodes. These procedures will be referenced in your site quality standards and the WPS. Severe penalties will result from violation of these standards. Improperly storing these electrodes will lead to poor-quality welds and the possibility of having numerous welds rejected. Be sure you read and understand your site standards before working with low-hydrogen electrodes.

2.4.4 Filler Metal and Electrode Selection Considerations

If there is a WPS for the weld to be made, the filler metal will be specified. It will be given as an AWS classification and/or manufacturer's standard and will sometimes include a manufacturer's name.

When selecting filler metals and electrodes not specified in a WPS, consider the factors explained in the following sections.

The filler metal should be compatible with the base metal's chemical composition and mechanical properties. The chemical composition and mechanical properties of the base metal can be obtained from the base metal mill specifications. The mill specifications can then be compared to various tables that list base metal and filler metal acceptability.

A weld made on mild steel with a high-tensile electrode, such as E12018, is weaker than one made with E7018. The high-tensile weld is so strong that when stress is placed on the weld, it does not give. It transfers all stress to the heat-affected zone along the weld. This can cause cracking. A weld made with E7018 will give along with the base metal and distribute the stress across the weld, which makes a stronger joint. The opposite occurs when mild-steel electrodes are used to weld high-strength steel. All the stress occurs in the weld because the surrounding base metal is so much stronger. This causes the weld to fail prematurely.

Always use an electrode smaller in diameter than the thickness of the metal to be welded. For sheet metal, 3⁄32-inch (2.4 mm) E6013 is recommended because it has shallow penetration. To minimize the joint preparation, a deep penetration E6010 or E7010 electrode can be used for the root pass. Before welding, always clean the base metal surface to remove surface corrosion, plating, coatings, and paint.

The welding position is important because some electrodes, such as E7024 and E7028, can only be used in the flat and horizontal fillet weld positions. In addition, never attempt to weld in vertical or overhead positions with electrodes larger than 3⁄16 inch (4.8 mm), except for low-hydrogen electrodes. The recommended limitation for low-hydrogen is 5⁄32 inch (4 mm).

The type of joint to be welded is also important to consider when selecting an electrode. When lap and T joints can be positioned flat, they are usually welded with E7024 or E7028. Weld butt joints that need deep penetration using E6010. If the joint has poor fit-up, use a shallow-penetrating electrode such as E6012 or E6013. If low-hydrogen electrodes are used for open-root joints, back-gouging may be required. To eliminate the need for back-gouging and to make the root pass easier to run, use metal backing, or put the root in with E6010 and then fill out with low-hydrogen.

2.4.5 Preheating and Interpass Temperature Control

Preheating and interpass temperature control procedures are sometimes performed to preserve weldment strength, ductility, and weld quality. Proper preheating and interpass temperature control can reduce shrinkage, prevent excessive hardening, and prevent underbead cracking. Interpass temperature control is also required during certain welding procedures to prevent or reduce localized hardening and shrinkage. If certain metals cool too quickly between weld passes, undesirable changes occur in the base and weld metals, and the heat from successive passes will not reverse these changes. Preheating and interpass temperature heat treatment of metals is described in detail in module titled, *Preheating and Postheating of Metals.*

2.4.6 Postweld Heat Treatment

Postweld heat treatment (PWHT) is the heating of a weldment or assembly after welding has been completed. It can be used on most types of base metals. Postweld heat treatment may be necessary to reduce **residual stresses** and to improve corrosion resistance, dimensional stability, fatigue resistance, and impact and low-temperature strength. Postweld heat treatment equipment and techniques include furnaces, banding, and localized heat treatment. Postweld heat treatments include stress relieving, annealing, normalizing, and tempering. Postweld heat treatment of metals is described in detail in *Preheating and Postheating of Metals.*

Additional Resources

AWS B1.10M/B1.10 Guide for the Nondestructive Examination of Welds. Miami, FL: American Welding Society.

AWS B1.11 Guide for the Visual Examination of Welds. Miami, FL: American Welding Society.

AWS D3.5-93R Guide for Steel Hull Welding. Miami, FL: American Welding Society.

AWS D3.6M Specification for Underwater Welding. Miami, FL: American Welding Society.

AWS D3.7 Guide for Aluminum Hull Welding. Miami, FL: American Welding Society.

Lincoln Electric website: **http://www.lincolnelectric.com** offers sources for products and training.

Machinery's Handbook. Erik Oberg, Franklin D. Jones, and Christopher J. McCauley. New York, NY: Industrial Press, Inc. 2004.

Metals and How to Weld Them. T. B. Jefferson. Cleveland, OH: The James F. Lincoln Arc Welding Foundation.

OSHA 1926.351, Arc Welding and Cutting. Current Edition. Washington, DC: Occupational Safety and Health Administration (OSHA).

Stick Electrode Product Catalog. Cleveland, OH: The Lincoln Electric Company, 2008.

Stick Electrode Welding Guide. Cleveland, OH: The Lincoln Electric Company, 2004.

The Procedure Handbook of Arc Welding. 14th Edition. Cleveland, OH: The James F. Lincoln Welding Foundation, 2000.

Welding Handbook, 4 vols. Miami, FL: American Welding Society.

2.0.0 Section Review

1. The mass, or weight, of a specific volume of metal is commonly expressed as ____.
   a. purity
   b. hardness
   c. density
   d. resistivity
2. A type of stress in which opposed, in-line forces pull away from each other at opposite sides of a specimen is ____.
   a. torsion
   b. compression
   c. shear
   d. tension
3. A handheld analyzer that can be used to accurately identify most types of metal relies on the use of ____.
   a. X-ray fluorescence spectrometry
   b. magnetic resonance imagery
   c. the modulus of elasticity
   d. tension loading ductility
4. The type of electrode that is best suited for steel that has a tendency to develop porosity and underbead cracking is ____.
   a. fast-freeze
   b. low-hydrogen
   c. fill-freeze
   d. fast-fill

SECTION THREE

3.0.0 COMMON STRUCTURAL STEEL SHAPES

Objective

Identify the common structural steel shapes of metal.

a. Identify the most common structural steel shapes.
b. Identify different structural beam shapes.
c. Identify pipe and tubing types.
d. Identify other common metal forms, including rebar.

Trade Term

Sintered: A process of making parts from a powdered metal in which the metal is molded and then heated in a furnace without melting to fuse it into a solid metal.

Metals are commercially available in many shapes and sizes. They can be shaped by casting, rolling, drawing, forging, cutting, machining, extruding, and spinning. Structural steel and other common structural shapes are formed by rolling hot or cold metal between a succession of rollers. The rollers are specially configured for the shapes they will produce.

3.1.0 Common Structural Shapes

Structural steel includes a wide range of steel types, classes, and shapes. To ensure uniform standards, the ASTM specifies the properties of strength, weight, corrosion resistance, and weldability for various steel classifications. *Table 21* lists some typical structural steel types and classifications. Note that the minimum yield stress is expressed in the table as ksi. This is one of the standard notations used for structural steel. The term *ksi* is defined as kips per square inch, where kips means kilo-pounds (1,000 pounds) of dead load. Hence, a yield stress of 36 ksi means 36,000 pounds per square inch of dead load.

The shape or type of structural steel is identified on drawings by a symbol or abbreviation. Sizes and dimensions are always given in a specified order. The specification format for designating shape and size is shown with the relevant shape in the following sections.

NOTE

It is important to remember that dimensions for various structural steel shapes that are given in US customary units (such as inches) do not directly correspond to metric dimensions for those same shapes. In other words, a steel plate that measures ¼" × 4' × 10' will not have an exact metric equivalent. An actual metric steel plate size that is close to the same size might measure 6 mm × 1,250 mm × 3,000 mm.

3.1.1 Plate

Plate is rolled metal with a uniform thickness equal to or greater than 3⁄16 inch (4.8 mm), and its identifying symbol is PL. Its thickness is given in inches or millimeters, and its length and width are typically listed in feet or meters.

Example specification:

PL ¼" × 4' × 10'

3.1.2 Sheet Metal

Sheet metal is rolled metal with a uniform thickness less than 3⁄16 inch (4.8 mm). Thickness is given in inches or millimeters, or indicated by an American Wire Gauge (AWG) number. Length and width are given in feet or a metric equivalent. Sheet metal is often rolled into coils.

Example specification:

Sheet No. 12 AWG × 4' × 8'

3.1.3 Bars

Bars are rolled to a variety of cross-sectional shapes and sizes (*Figure 17*). Standard bar shapes include the following:

- *Round* – Round bar is specified by BAR, followed by the bar diameter, often with a diameter symbol (a circle with a slash through it), and then the length.
- *Square* – Square bar is specified by BAR, followed by the face width, often with a face width symbol (a square with a slash through it), and then the length.
- *Bar (flat)* – Flat bar is specified by BAR, followed by the width dimension, the thickness dimension, and then the length.
- *Z-bar* – Z-bar is specified by Z, followed by the flange width, the web depth, the flange and web thickness, and then the length. The flange and the web are always the same thickness.
- *Hexagonal* – Hexagonal bar is specified by HEX, followed by the bar thickness (measured across the flats), and then the length.

Table 21 Common Types and Classifications of Structural Steel

Steel Type	ASTM Class	Minimum Yield Stress (ksi)	Form		Remarks
Carbon	A 36	36	Plate Shapes Bars	Sheets, strips Rivets Bolts, nuts	For buildings and general structures; available in high-toughness grades
	A529	42	Plate Shapes	Bars	For buildings and similar construction
High-strength	A440	42 to 50	Plate Shapes	Bars	Lightweight and superior corrosion resistance
High-strength, low-alloy	A 441	40 to 50	Plates Shapes	Bars	Primarily for lightweight welded buildings and bridges
	A 572	42 to 65	Several types Plate	Shapes Bars	Lightweight high-toughness for buildings, bridges, and similar structures
Corrosion-resistant, high-strength, low-alloy	A 242	42 to 50	Plate Shapes	Bars	Lightweight and added durability weathering grades available
	A 588	42 to 50	Plate Shapes	Bars	Lightweight, durable in high thickness; weathering grades available
Quenched and tempered alloy	A 514	90 to 100	Several types Plate	Shapes Bars	Strength varies with thickness and type

- *Octagonal* – Octagonal bar is specified by OCT, followed by the bar thickness (measured across the flats), and then the length.

3.1.4 Angles

Angles are L-shaped bars that have either equal-sized or unequal-sized legs that are always at 90 degrees to each other. *Figure 18* shows both types of angle bar stock and the specification formats for each type. The dimensions for the angle bar stock examples are given in US customary units and metric units. Remember, though, these two units of measurement do not directly correspond. The specification format represents the size and thickness of the angle bar stock legs. For example, a specification of L2 × 2 × ⅜ would indicate a 2" × 2" angle of ⅜" thickness in US customary units. For unequal leg stock, the longest leg is listed first, as shown in *Figure 18*.

3.1.5 Channels

Channels are U-shaped forms made of two flanges connected by a common web. The flanges extend from the same side of the web. The flanges can be of uniform thickness or tapered toward the outer edges.

There are several channel variations, each having a unique specification format designation. *Figure 19* shows two different channels and their related American Standard Channels specification format designations. It also shows similar channels with metric dimensions. The channel C8 × 11.5 has a nominal depth of 8 inches and a weight of 11.5 pounds per foot of length. Similarly, the channel C8 × 18.75 has a nominal depth of 8 inches and a weight of 18.75 pounds per foot of length. The metric channels both have a nominal depth of 200 mm. The thinner channel has a weight of 22 kg per meter of length, while the thicker channel has a weight of 25.30 kg per meter of length. As shown in the figure, the channel C8 × 11.5 has a flange width of 2¼ inches and a web thickness of ¼ inch; while the C8 × 18.75 channel has a flange width of 2¼ inches and a web thickness of ½ inch. The metric channels both have a flange width of 75 mm. The thinner channel has a web thickness of 10.5 mm, while the thicker channel has a web thickness of 11.5 mm. The specific dimensions for the beam depth, flange width, and web thickness for all common channel designations can be found in related tables of dimensions and properties in most structural engineering and architectural standards reference books.

SHAPE		SPECIFICATION FORMAT
ROUND	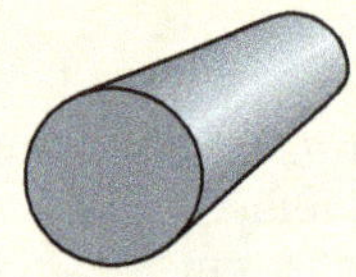	BAR (DIAMETER) ×(LENGTH) BAR ¾" ∅ × 8'3" OR BAR 19 ∅ × 3 M
SQUARE	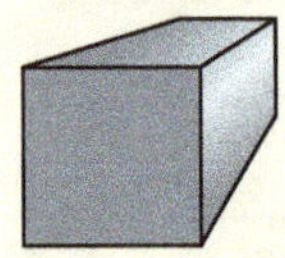	BAR (FACE WIDTH ⧄) ×(LENGTH) BAR ¾" ×8'3" OR BAR 19⧄ ×3 M
BAR	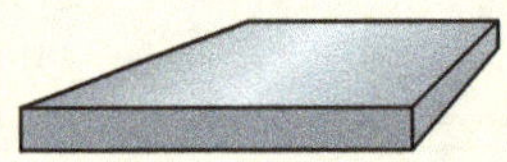	BAR (WIDTH) ×(THICKNESS) ×(LENGTH) BAR 2" ×¼"×8'3" OR BAR 50 MM ×6 MM ×3 M
Z-BAR		Z (FLANGE WIDTH) ×(WEB DEPTH) ×(FLANGE AND WEB THICKNESS) ×(LENGTH) Z 2" ×2"×½"×8'3" OR Z 50 MM ×50 MM×13 MM ×3 M
HEXAGONAL	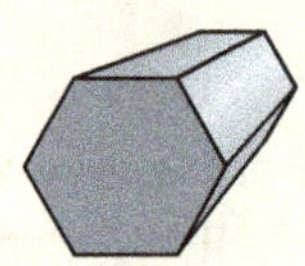	HEX (BAR THICKNESS) ×(LENGTH) HEX ⅞"×10' OR HEX 22 MM ×3 M
OCTAGONAL	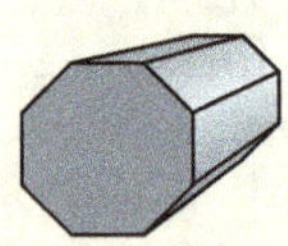	OCT (BAR THICKNESS) ×(LENGTH) OCT 1"×6' OR OCT 25 MM ×2 M

Figure 17 Bar shapes and specification formats.

3.2.0 Beams and Shapes from Beams

Beams (*Figure 20*) are made in I-shaped, H-shaped, and T-shaped cross sections, and they are made with flat or tapered flanges. In addition to different flange widths and web depths, the thickness of the flanges and webs varies with beam size. Also, beam weight (per linear foot or meter) for a given dimension can be increased by adding thickness to the web and flanges with very little change in beam width or depth. The specification designator describes the beam type, its nominal depth, and its weight per foot of length or kilogram per meter. An example specification designator for a beam is S24 × 120, where S24 represents the beam type (S) and the nominal depth in inches (24). The 120 value is the weight in pounds per foot of length. A comparable metric beam might have a 300 mm nominal depth and have a weight of 42.2 kilograms per meter. The specific dimensions for the depth, web thickness, and flange thickness for the various designations of beams are given in related tables of dimensions and properties in most structural engineering and architectural standards books.

I-Beams

The older American Standard I-beams have been replaced by the American Institute of Steel Construction (AISC) wide flange (WF) beams. These wide flange shapes are now used for most beams and columns. The wider flange provides a stronger beam than the older I-beam with the same web width.

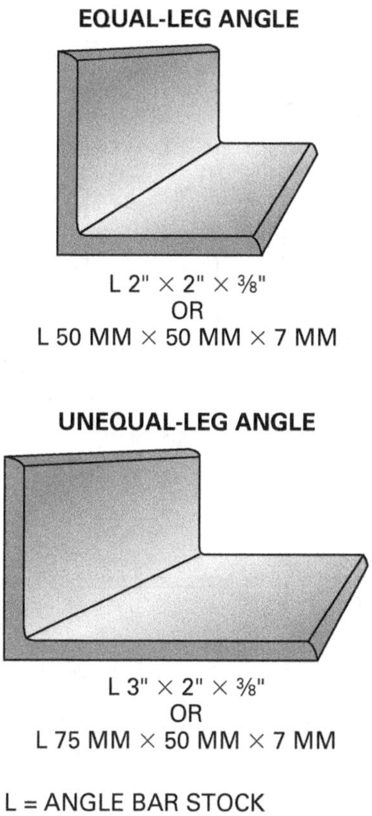

Figure 18 Angle bar stock and specification formats.

Beams are identified with a standard set of symbols. The I-beam and H-beam letter symbols include the following:

- *S (American Standard)* – I-shaped beams with tapered flanges
- *WF (Wide flange)* – Wider flanges than S beams with thinner webs and nontapered flanges
- *M* – Similar to WF beams but with short flanges
- *HP* – H-shaped beams with nontapered flanges

T-beams include structural tees and tee shapes. Structural tees are made by cutting S, W, and M beams down the center, usually by shearing. The conventional symbol for a structural tee is the original beam letter followed by a T. For example, a tee cut from an S8 × 18.4 beam would be identified as ST4 × 9.2 in US customary units.

T-shapes are rolled into their final tee shape. The symbol for tee shapes is a T, without any other letter. The specification gives the nominal depth, flange width, thickness, and length. An example specification is T3 × 7.8. T-shapes are short pieces, commonly used as connections or supports.

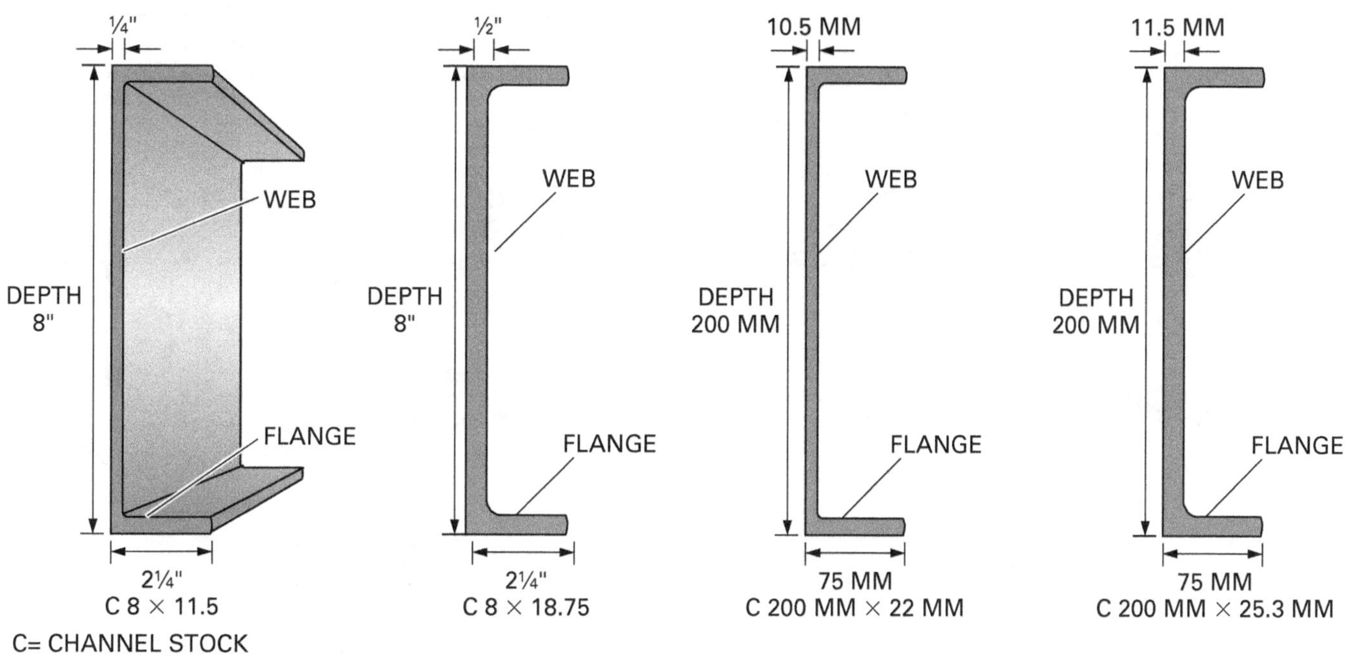

Figure 19 Channels and specification formats.

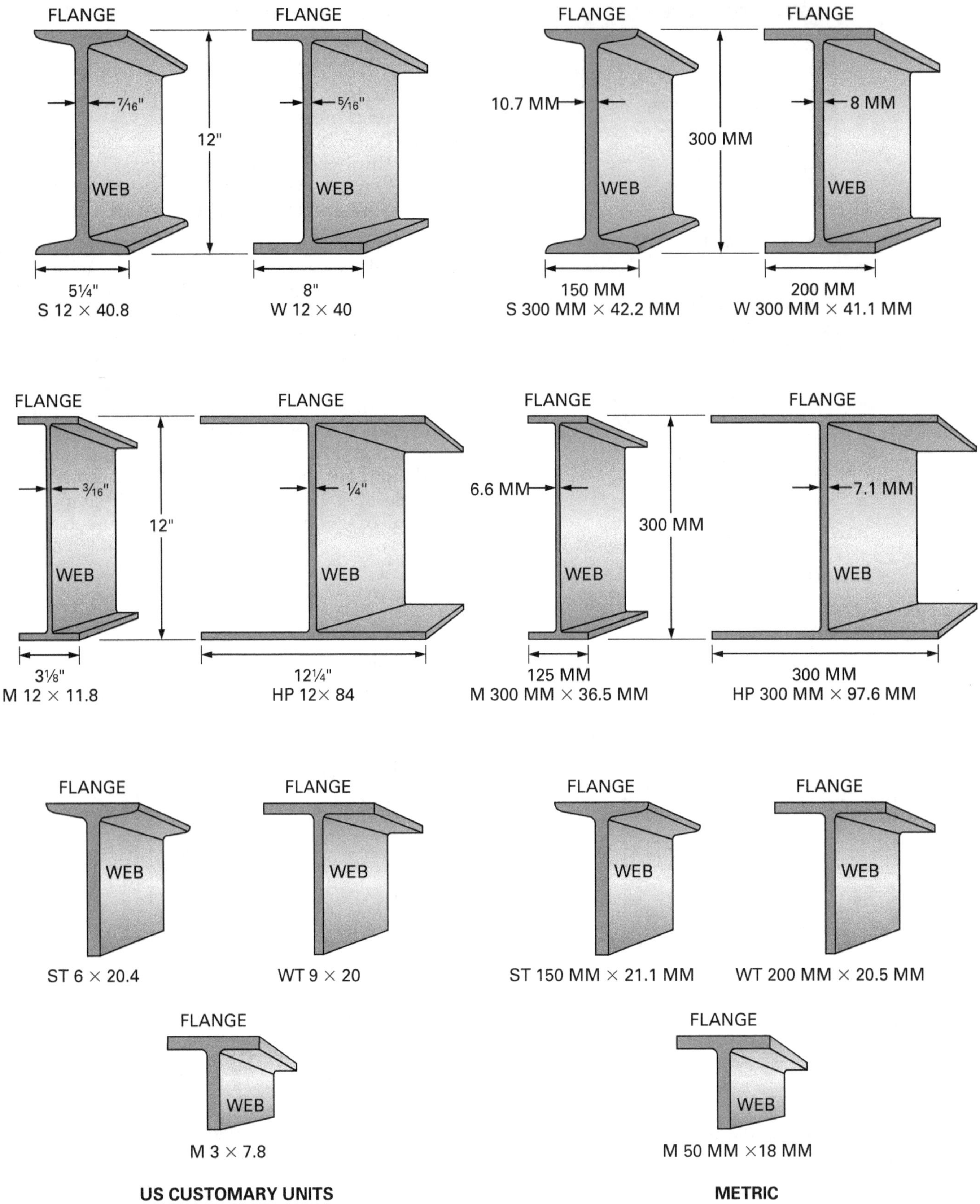

Figure 20 Structural beam and T-beam shapes.

3.3.0 Pipe and Tubing

Pipe and tubing share some basic similarities, but they are identified using different measuring methods and standards. This section focuses on various sizes of pipe, some common tubing shapes, and different methods for producing pipe and tubing.

3.3.1 Pipe

Steel pipe size is specified as a nominal inside diameter (ID) up through 12 inches (or 300 mm). The word *nominal* is used to identify the design size, but the actual inside diameter will not be exactly that given size. Pipe 14 inches (350 mm) and larger is specified by the actual outside diameter (OD). Pipe weight, which is determined by the wall thickness, is specified by a schedule number or strength. Common schedule numbers are 5, 10, 20, 30, 40, 60, 80, 100, 120, 140, and 160. Strengths are specified as standard (STD), extra strong (XS), and extra-extra strong (XXS). Standard falls between Schedules 30 and 40. Extra-strong falls between Schedules 60 and 80. Extra-extra strong is Schedule 160 or higher.

Because the OD of a given pipe size is always the same, the ID becomes smaller as the pipe schedule increases. The additional wall thickness is taken from the ID while the OD remains fixed. *Table 22* shows common pipe sizes and schedules with nominal wall thicknesses in US customary units and similarly sized metric pipe measurements. As with other steel shapes, nominal size dimensions for pipe that are given in US customary units do not directly correspond to metric pipe. For instance, a 12-inch pipe is not identical to a 300-millimeter pipe.

3.3.2 Tubing

Tubing is manufactured in square, rectangular, and round cross sections. Round tubing can be distinguished from pipe by its dimensions. Unlike pipe, standard round tubing is measured by its actual OD.

The cross sections of square and rectangular tubing have even outside dimensions and slightly rounded corners. *Figure 21* shows three standard tubing shapes and their specification formats in inches, as well as three similarly sized metric tubing shapes.

3.3.3 Seamed and Seamless Tubing and Pipe

Both standard and nonstandard shapes can also be produced by other processes. For example, tubing can be produced by extrusion or by rolling and welding. Extrusions can be produced in an almost limitless variety of shapes. When tubing is produced by rolling and welding, the most common welding process used is electronic resistance welding (ERW).

3.4.0 Other Common Forms

In addition to common structural steel shapes, beams, pipe, and tubing, there are other forms of steel that are commonly used in industry. These other forms include reinforcing bars, forged objects, cast objects, and powdered metals.

Why Doesn't the Pipe Size Match Pipe ID or OD?

When pipe was first made, it was made from lead, which required a thick wall to hold the specific pressure for a given ID. When the switch was made from lead pipe to steel pipe, a decision had to be made whether to maintain the OD so the steel pipe would fit existing fittings, or to change the OD and keep the ID size. The decision was made to maintain the OD to provide direct replacement with the existing fittings. At that point, neither nominal ID nor OD corresponded to size.

Table 22 Standard Pipe Sizes and Schedules with Nominal Wall Thicknesses

		Nominal Wall Thickness (in)												
Nom. Pipe Size	Outside Diam.	Sched. 10	Sched. 20	Sched. 30	STD	Sched. 40	Sched. 60	XS	Sched. 80	Sched. 100	Sched. 120	Sched. 140	Sched. 160	XXS
⅛	0.405	–	–	–	0.068	0.068	–	0.095	0.095	–	–	–	–	–
¼	0.540	–	–	–	0.088	0.088	–	0.119	0.119	–	–	–	–	–
⅜	0.675	–	–	–	0.091	0.091	–	0.126	0.126	–	–	–	–	–
½	0.840	–	–	–	0.109	0.109	–	0.147	0.147	–	–	–	0.188	0.294
¾	1.050	–	–	–	0.113	0.113	–	0.154	0.154	–	–	–	0.219	0.308
1	1.315	–	–	–	0.133	0.133	–	0.179	0.179	–	–	–	0.250	0.358
1¼	1.660	–	–	–	0.140	0.140	–	0.191	0.191	–	–	–	0.250	0.382
1½	1.900	–	–	–	0.145	0.145	–	0.200	0.200	–	–	–	0.281	0.400
2	2.375	–	–	–	0.154	0.154	–	0.218	0.218	–	–	–	0.344	0.436
2½	2.875	–	–	–	0.203	0.203	–	0.276	0.276	–	–	–	0.375	0.552
3	3.500	–	–	–	0.216	0.216	–	0.300	0.300	–	–	–	0.438	0.600
3½	4.000	–	–	–	0.226	0.226	–	0.318	0.318	–	–	–	–	–
4	4.500	–	–	–	0.237	0.237	–	0.337	0.337	–	0.438	–	0.531	0.674
5	5.563	–	–	–	0.258	0.258	–	0.375	0.375	–	0.500	–	0.625	0.750
6	6.625	–	–	–	0.280	0.280	–	0.432	0.432	–	0.562	–	0.719	0.864
8	8.625	–	0.250	0.277	0.322	0.322	0.406	0.500	0.500	0.594	0.719	0.812	0.906	0.875
10	10.750	–	0.250	0.307	0.365	0.365	0.500	0.500	0.594	0.719	0.844	1.000	1.125	1.000
12	12.750	–	0.250	0.330	0.375	0.406	0.562	0.500	0.688	0.844	1.000	1.125	1.312	1.000
14 OD	14.000	0.250	0.312	0.375	0.375	0.438	0.594	0.500	0.750	0.938	1.094	1.250	1.406	–
16 OD	16.000	0.250	0.312	0.375	0.375	0.500	0.656	0.500	0.844	1.031	1.219	1.438	1.594	–
18 OD	18.000	0.250	0.312	0.438	0.375	0.562	0.750	0.500	0.938	1.156	1.375	1.562	1.781	–
20 OD	20.000	0.250	0.375	0.500	0.375	0.594	0.812	0.500	1.031	1.281	1.500	1.750	1.969	–
22 OD	22.000	0.250	0.375	0.500	0.375	–	0.875	0.500	1.125	1.375	1.625	1.875	2.125	–
24 OD	24.000	0.250	0.375	0.562	0.375	0.688	0.969	0.500	1.218	1.531	1.812	2.062	2.344	–
26 OD	26.000	0.312	0.500	–	0.375	–	–	0.500	–	–	–	–	–	–
28 OD	28.000	0.312	0.500	0.625	0.375	–	–	0.500	–	–	–	–	–	–
30 OD	30.000	0.312	0.500	0.625	0.375	–	–	0.500	–	–	–	–	–	–
32 OD	32.000	0.312	0.500	0.625	0.375	0.688	–	0.500	–	–	–	–	–	–
34 OD	34.000	0.312	0.500	0.625	0.375	0.688	–	0.500	–	–	–	–	–	–
36 OD	36.000	0.312	0.500	0.625	0.375	0.750	–	0.500	–	–	–	–	–	–
42 OD	42.000	–	–	–	0.375	–	–	0.500	–	–	–	–	–	–

3.4.1 Reinforcing Bars

Reinforcing bars (rebars), sometimes called re-rods, are used for concrete reinforcement. They are available in several grades. These grades vary in yield strength, ultimate strength, percentage of elongation, bend-test requirements, and chemical composition. Reinforcing bars can be coated with different compounds, such as epoxy, for use in concrete where corrosion can be a problem.

The ASTM International has established standard specifications for reinforcing bars. These grades appear on bar-bundle tags, in color coding, in rolled-on markings on the bars, and/or on bills of materials. The ASTM specifications are as follows:

- *A615, Standard Specification for Deformed and Plain Billet-Steel Bars for Concrete Reinforcement*
- *A706, Standard Specification for Low-Alloy Steel Deformed and Plain Bars for Concrete Reinforcement*
- *A996, Standard Specification for Rail-Steel and Axle-Steel Deformed Bars for Concrete Reinforcement*

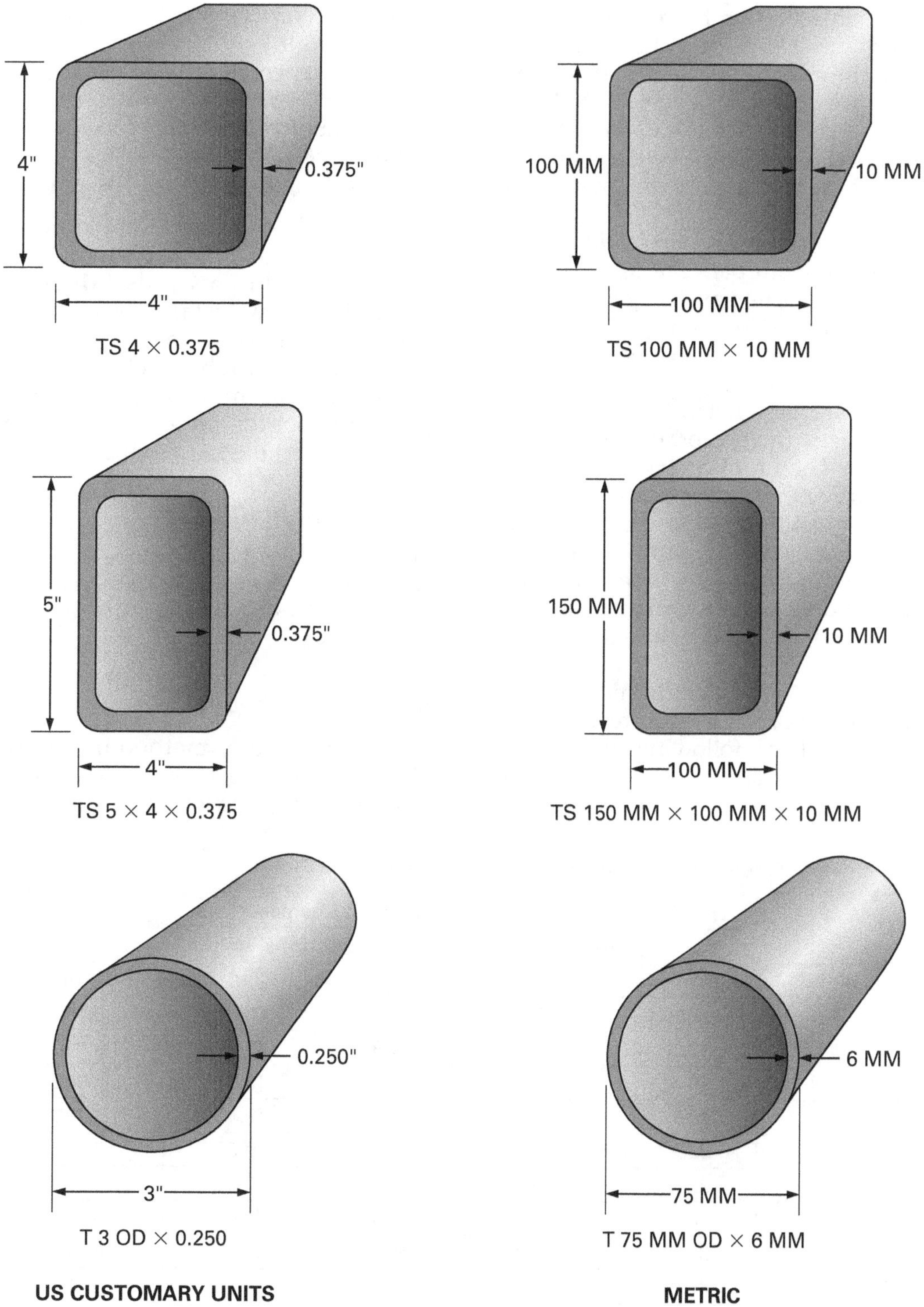

Figure 21 Standard tubing shapes and specification formats.

The standard configuration for reinforcing bars is the deformed bar. Different patterns may be impressed upon the bars depending on the manufacturer, but all are rolled to conform to ASTM specifications. The purpose of the deformation is to improve the bond between the concrete and the bar and to prevent the bar from moving within the concrete.

Plain bars are smooth and round without deformations on them. Plain bars are used for special purposes: dowels at expansion joints, where the bars must slide in a sleeve; for expansion and contraction joints in highway pavement; and for column spirals.

Deformed bars are designated by a number in eleven standard sizes (metric or inch-pound) as shown in *Table 23*. The number indicates the approximate diameter of the bar in eighths of an inch or millimeters. For example, a #16 metric bar has an approximate diameter of 16 mm. The nominal dimension of a deformed bar (the nominal diameter does not include the deformation) is equivalent to that of a plain bar having the same weight per foot.

As shown in *Figure 22*, bar identification is made by ASTM specifications; these require that each bar manufacturer roll the following information onto the bar:

- A letter or symbol to indicate the manufacturer's mill
- A number corresponding to the size number of the bar (*Table 23*)
- A symbol or marking to indicate the type of steel (*Table 24*)
- A marking to designate the grade (*Table 25*)

The grade represents the minimum yield or tension strength, measured in kips per square inch (ksi) or megapascals (MPa), that the type of steel will withstand before it permanently stretches (elongates). Today, Grade 420 is the most commonly used rebar. Rebars are normally supplied from the mill bundled in 60-foot lengths (12-meter lengths are common for metric rebar) that are cut in the field to the required length.

3.4.2 Forged Shapes

Forging is used to produce specialized shapes where high strength is required. An example is forged high-pressure pipe fittings such as flanges and elbows.

3.4.3 Cast Shapes

Casting is another method of producing shapes. Cast iron and steel are used for machine parts, automobile components, and large castings.

Table 23 ASTM Standard Metric and Inch-Pound Reinforcing Bars

Bar Size		Nominal Characteristics					
		Diameter		Cross-Sectional Area		Weight	
Metric	[in-lb]	mm	[in]	mm	[in]	kg/m	[lbs/ft]
#10	[#3]	9.5	[0.375]	71	[0.11]	0.560	[0.376]
#13	[#4]	12.7	[0.500]	129	[0.20]	0.944	[0.668]
#16	[#5]	15.9	[0.625]	199	[0.31]	1.552	[1.043]
#19	[#6]	19.1	[0.750]	284	[0.44]	2.235	[1.502]
#22	[#7]	22.2	[0.875]	387	[0.60]	3.042	[2.044]
#25	[#8]	25.4	[1.000]	510	[0.79]	3.973	[2.670]
#29	[#9]	28.7	[1.128]	645	[1.00]	5.060	[3.400]
#32	[#10]	32.3	[1.270]	819	[1.27]	6.404	[4.303]
#36	[#11]	35.8	[1.410]	1006	[1.56]	7.907	[5.313]
#43	[#14]	43.0	[1.693]	1452	[2.25]	11.38	[7.65]
#57	[#18]	57.3	[2.257]	2581	[4.00]	20.24	[13.60]

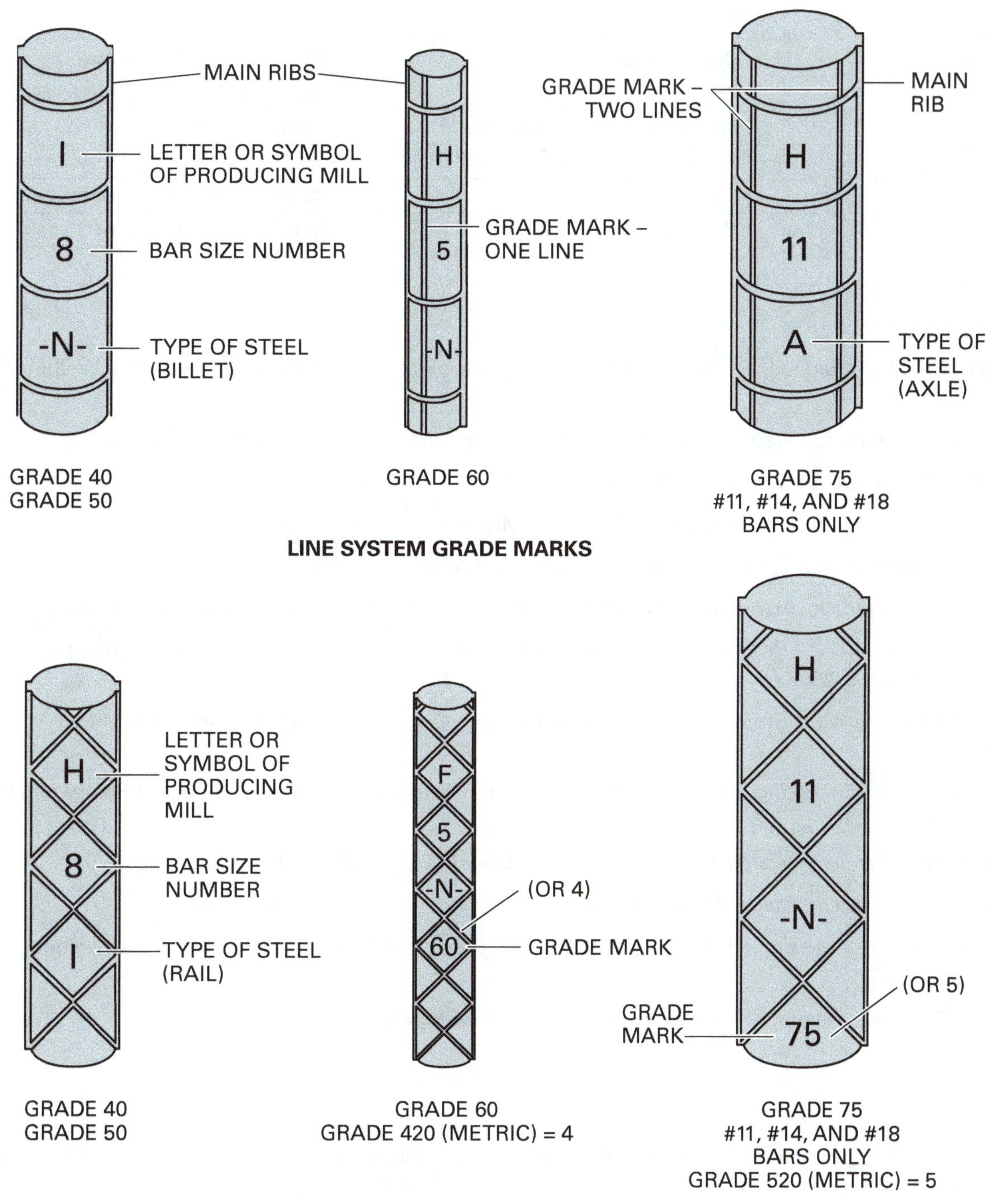

Figure 22 Reinforcement bar identification.

Table 24 Reinforcement Bar Steel Types

Symbol/Marking	Type of Steel
A	Axle *(ASTM A996)*
S or N	Billet *(ASTM A615)*
I or IR	Rail *(ASTM A996)*
W	Low-alloy *(ASTM A706)*

3.4.4 Powdered Metals

Commercially prepared powdered metals are used for welding fillers, surface coatings, and mold castings. When cast to make parts, the molded powder is usually sintered (heated in a furnace without melting) to fuse it into a solid metal. This is a common technique for manufacturing complex-shaped or porous parts. It is also used with high-temperature metals that cannot be liquid-cast.

Table 25 Reinforcement Bar Grades

Grade	Identification	Mimimum Yield Strength
40 and 50	None	40,000 to 50,000 psi (40 to 50 ksi)
60	One line or the number 60	60,000 psi (60 ksi)
75	Two lines or the number 75	75,000 psi (75 ksi)
420	The number 4	420 MPa (60,000 psi or 60 ksi)
520	The number 5	520 MPa (75,000 psi or 75 ksi)

Additional Resources

AWS B1.10M/B1.10 Guide for the Nondestructive Examination of Welds. Miami, FL: American Welding Society.

AWS B1.11 Guide for the Visual Examination of Welds. Miami, FL: American Welding Society.

AWS D3.5-93R Guide for Steel Hull Welding. Miami, FL: American Welding Society.

AWS D3.6M Specification for Underwater Welding. Miami, FL: American Welding Society.

AWS D3.7 Guide for Aluminum Hull Welding. Miami, FL: American Welding Society.

Lincoln Electric website: **http://www.lincolnelectric.com** offers sources for products and training.

Machinery's Handbook. Erik Oberg, Franklin D. Jones, and Christopher J. McCauley. New York, NY: Industrial Press, Inc. 2004.

Metals and How to Weld Them. T. B. Jefferson. Cleveland, OH: The James F. Lincoln Arc Welding Foundation.

OSHA 1926.351, Arc Welding and Cutting. Current Edition. Washington, DC: Occupational Safety and Health Administration (OSHA).

Stick Electrode Product Catalog. Cleveland, OH: The Lincoln Electric Company, 2008.

Stick Electrode Welding Guide. Cleveland, OH: The Lincoln Electric Company, 2004.

The Procedure Handbook of Arc Welding. 14th Edition. Cleveland, OH: The James F. Lincoln Welding Foundation, 2000.

Welding Handbook, 4 vols. Miami, FL: American Welding Society.

3.0.0 Section Review

1. Steel plate is rolled metal with a uniform thickness equal to or greater than ____.
   a. ⅛" (0.04 mm)
   b. $^3/_{16}$" (0.06 mm)
   c. ¼" (0.08 mm)
   d. $^5/_{16}$" (0.10 mm)

2. The three steel beam cross-section shapes are ____.
   a. A, H, and I
   b. I, L, and T
   c. H, T, and V
   d. I, H, and T

3. Pipe that is 14" (350 mm) and larger is specified by its ____.
   a. actual ID
   b. nominal ID
   c. actual OD
   d. nominal OD

4. The steel form that is most likely to be used for expansion and contraction joints in highway pavement is ____.
   a. forged high-pressure pipe
   b. plain reinforcing bars
   c. sintered powdered metal
   d. deformed rebar rods

SUMMARY

Being successful as a welder depends not only on the ability to make satisfactory welds, but also on having a thorough understanding of base metals, alloys, and filler metals. Welders must know about the composition, classification, physical characteristics, and mechanical properties of metals. They must also understand the numerous metallurgical variables that must be considered during welding activities, and they need to be able to identify different forms and shapes of structural metal. The information presented in this module should provide an adequate foundation for career advancement in welding.

Review Questions

1. The largest group of the ferrous-based metals includes the ____.
   a. cast irons
   b. alloy steels
   c. heavy metals
   d. carbon steels

2. Quench-and-tempered steels and chromium-molybdenum steels are examples of ____.
   a. common-grade stainless steels
   b. low-alloy steels
   c. high-alloy steels
   d. specialty-grade stainless steels

3. A lower coefficient of thermal conductivity along with a higher coefficient of thermal expansion and a higher electrical resistance are all characteristics of ____.
   a. nickel alloy
   b. carbon steel
   c. stainless steel
   d. aluminum alloy

4. A type of stainless steel that provides excellent resistance to salt water corrosion is ____.
   a. superferritic
   b. duplex
   c. superaustenitic
   d. precipitation-hardening

5. The principal alloying element in brasses is ____.
   a. aluminum
   b. zinc
   c. beryllium
   d. bronze

6. Corrosion resistance is an important characteristic of a metal because corrosion can severely reduce a metal's ____.
   a. tensile strength
   b. interpass ductility
   c. thermal expansion
   d. residual stress

7. The ability of a material to be strained (deformed) without permanent deformation is known as ____.
   a. ductility
   b. modulus
   c. elasticity
   d. tensile strength

8. The resistance to indentation of a material is called ____.
   a. ductility
   b. hardness
   c. tensile strength
   d. modulus of elasticity

9. The three types of cast iron are malleable, white, and ____.
   a. ductile
   b. green
   c. blue
   d. gray

10. Malleable cast iron is white cast iron that has been ____.
    a. annealed
    b. cooled slowly
    c. forged
    d. chilled rapidly

11. The most common defect caused by surface contamination of a metal is ____.
    a. rust
    b. porosity
    c. corrosion
    d. toxic fumes

12. A channel that has a designation of C12 × 30 is a (n) ____.
    a. American Standard Channel with a depth of 12 inches
    b. Miscellaneous Channel with a depth of 30 inches
    c. Miscellaneous Channel with a weight of 12 pounds per foot
    d. American Standard Channel with a weight of 12 pounds per foot

13. Shearing an S, W, or M beam down the center creates a(n) ____.

a. wide-flange beam
b. H beam
c. structural tee
d. seamed I

14. Tubing can be produced by rolling and welding or by ____.

a. casting
b. compression
c. annealing
d. extrusion

15. The standard configuration for reinforcing bars is the ____.

a. dowel bar
b. deformed bar
c. column bar
d. plain bar

Trade Terms Introduced in This Module

Alloy: A metal that has had other elements added to it that substantially change its mechanical properties.

American Iron and Steel Institute (AISI): An industry organization responsible for preparing standards for steel and steel alloys that are based upon the Society of Automotive Engineers (SAE) code system.

American Society for Testing and Materials (ASTM International): An organization that developed a code system for identifying and labeling steels.

Austenitizing: Heating a steel up to or above the transformation temperature range to achieve partial or complete transformation to austenite grain structure.

Casting: The metal object produced by pouring molten metal into a mold.

Certified mill test report (CMTR): A quality assurance report provided to a customer by a metal manufacturer that shows the chemical content and testing results of the metal being purchased; also called a mill test report (MTR).

Coefficient: A numerical measure of a physical or chemical property that is constant for a system under specified conditions, such as the coefficient of friction.

Ductile: The term used to describe a material that can be bent or shaped without breaking.

Ductility: The characteristic of metal that allows it to be stretched, drawn, or hammered without breaking.

Ferrous: Relating to iron or an alloy that is mostly iron; a metal containing iron.

Hardenability: A characteristic of a metal that enables it to become hard, usually through heat treatment.

Heat-affected zone (HAZ): The part of the base metal that has not been melted but has been altered structurally by the weld heat.

Hot shortness: The condition of metal when it proves to be very brittle and unbendable at red heat but can be bent without showing signs of brittleness when cold or at white heat. This condition is often a result of high sulfur and phosphorus content.

Interpass temperature: In a multiple-pass weld, the temperature of the base metal at the time the next weld pass is started; it is usually the same temperature that is used for preheating.

Malleable: Able to be hammered or pressed into another shape without breaking.

Mechanical properties: The characteristics or traits that indicate how flexible and strong a metal will be under stress.

Metallic: Having characteristics of metal, such as ductility, malleability, luster, and heat and electrical conductivity.

Mill test report (MTR): A quality assurance report provided to a customer by a metal manufacturer that shows the chemical content and testing results of the metal being purchased; also called a certified mill test report (CMTR).

Nonferrous metal: A metal, such as aluminum, copper, or brass, that lacks sufficient quantities of iron to be affected by its properties.

Notch toughness: The impact load required to break a test piece of weld metal measured by a notch toughness test (Charpy V-notch test). This test may be performed on metal below room temperature, at which point it may be more brittle.

Postweld heat treatment (PWHT): The process of heat treating the base metal and weld after the weld has been made.

Quench: To rapidly cool a hot metal using air, water, or oil.

Residual stress: The strain that occurs in a welded joint after the welding has been completed, as a result of mechanical action, thermal action, or both.

Sintered: A process of making parts from a powdered metal in which the metal is molded and then heated in a furnace without melting to fuse it into a solid metal.

Society of Automotive Engineers (SAE): An automotive society responsible for creating an early system for classifying carbon steels.

Stress-relief heat treatment: A technique used to reduce welding stresses and to increase the ductility and toughness of the weld metal and the heat-affected zone (HAZ).

Tempering: The process of reheating quench hardened or normalized steel to a temperature below the transformation range and then cooling it at any rate desired.

Tensile strength: The maximum stress or force that a material can withstand without breaking.

Underbead cracking: Subsurface cracking in the base metal under or near the weld.

Unified numbering system (UNS): An industry-accepted numbering system for identifying ferrous metals and alloys.

Web: The plate joining the flanges of a girder, joist, or rail.

Wide flange (WF) beam: A metal I-beam that has an extra wide flange at the top and bottom.

Wrought: Formed or shaped by hammering or rolling.

Additional Resources

This module is intended as a thorough resource for task training. The following reference works are suggested for further study.

AWS B1.10M/B1.10 Guide for the Nondestructive Examination of Welds. Miami, FL: American Welding Society.

AWS B1.11 Guide for the Visual Examination of Welds. Miami, FL: American Welding Society.

AWS D3.5-93R Guide for Steel Hull Welding. Miami, FL: American Welding Society.

AWS D3.6M Specification for Underwater Welding. Miami, FL: American Welding Society.

AWS D3.7 Guide for Aluminum Hull Welding. Miami, FL: American Welding Society.

Machinery's Handbook. Erik Oberg, Franklin D. Jones, and Christopher J. McCauley. New York, NY: Industrial Press, Inc. 2004.

Metals and How to Weld Them. T. B. Jefferson. Cleveland, OH: The James F. Lincoln Arc Welding Foundation.

OSHA 1926.351, Arc Welding and Cutting. Current Edition. Washington, DC: Occupational Safety and Health Administration (OSHA).

Stick Electrode Product Catalog. Cleveland, OH: The Lincoln Electric Company, 2008.

Stick Electrode Welding Guide. Cleveland, OH: The Lincoln Electric Company, 2004.

The Procedure Handbook of Arc Welding. 14th Edition. Cleveland, OH: The James F. Lincoln Welding Foundation, 2000.

Welding Handbook, 4 vols. Miami, FL: American Welding Society.

Lincoln Electric website: **http://www.lincolnelectric.com** offers sources for products and training.

Figure Credits

Robert I. Carr, Ph.D., P.E., Module opener

Concrete Reinforcing Steel Institute, Figure 1

Topaz Publications, Inc., Figures 2–4, 11

Zachry Industrial, Inc., Figures 5A, 5B

The Lincoln Electric Company, Cleveland, OH, USA, Figure 6, SA03, SA04

Photo Courtesy of Proceq SA Schwerzenbach, Switzerland, Figure 9

W.W. Grainger, Inc., Figure 10

Thermo Fisher Scientific, Figure 12

Photo Courtesy of U.S. Army, SA01

American Welding Society®, SA02

U.S. National Park Service, SA05

Fischer-Technology, Inc., SA06

Section Review Answer Key

Answer	Section Reference	Objective
Section One		
1. c	1.1.2	1a
2. a	1.2.1	1b
3. d	1.3.2	1c
4. b	1.4.4	1d
5. a	1.5.2	1e
Section Two		
1. c	2.1.1	2a
2. d	2.2.1	2b
3. a	2.3.4	2c
4. b	2.4.3	2d
Section Three		
1. b	3.1.1	3a
2. d	3.2.0	3b
3. c	3.3.1	3c
4. b	3.4.1	3d

NCCER CURRICULA — USER UPDATE

NCCER makes every effort to keep its textbooks up-to-date and free of technical errors. We appreciate your help in this process. If you find an error, a typographical mistake, or an inaccuracy in NCCER's curricula, please fill out this form (or a photocopy), or complete the online form at **www.nccer.org/olf**. Be sure to include the exact module ID number, page number, a detailed description, and your recommended correction. Your input will be brought to the attention of the Authoring Team. Thank you for your assistance.

Instructors – If you have an idea for improving this textbook, or have found that additional materials were necessary to teach this module effectively, please let us know so that we may present your suggestions to the Authoring Team.

NCCER Product Development and Revision

13614 Progress Blvd., Alachua, FL 32615

Email: curriculum@nccer.org
Online: www.nccer.org/olf

❑ Trainee Guide ❑ Lesson Plans ❑ Exam ❑ PowerPoints Other ____________

Craft / Level: Copyright Date:

Module ID Number / Title:

Section Number(s):

Description:

Recommended Correction:

Your Name:

Address:

Email: Phone:

Preheating and Postheating of Metals

Overview

Welders need to understand how heat affects metals and how to control heating during each stage of a welding operation in order to make strong, enduring welds. This module explains the relationship between heat and metal and describes various devices and methods that welders use to heat metals and measure temperature.

Module 29204

Trainees with successful module completions may be eligible for credentialing through the NCCER Registry. To learn more, go to **www.nccer.org** or contact us at **1.888.622.3720**. Our website has information on the latest product releases and training, as well as online versions of our *Cornerstone* magazine and Pearson's product catalog.

Your feedback is welcome. You may email your comments to **curriculum@nccer.org,** send general comments and inquiries to **info@nccer.org**, or fill in the User Update form at the back of this module.

This information is general in nature and intended for training purposes only. Actual performance of activities described in this manual requires compliance with all applicable operating, service, maintenance, and safety procedures under the direction of qualified personnel. References in this manual to patented or proprietary devices do not constitute a recommendation of their use.

29204 V5

29204
Preheating and Postheating of Metals

Objectives

When you have completed this module, you will be able to do the following:

1. Describe the relationship between heat and metal and identify preheating methods.
   a. Describe the relationship between heat and metal.
   b. Identify and describe methods used to preheat metal prior to welding.
   c. Identify and describe devices and products used to measure temperature.
2. Describe interpass temperature control and postheating processes.
   a. Describe interpass temperature control.
   b. Describe various postheating processes.

Performance Task

Under the supervision of your instructor, you should be able to do the following:

1. Preheat base metal to 350°F (177°C) and verify preheat using a temperature-indicating device.

Trade Terms

Annealing
Constituents
Critical temperature
Crystalline
Discontinuity
Heat-affected zone (HAZ)
Interpass temperature control
Martensite
Preheat weld treatment (PHWT)
Strain hardening
Work hardening

Industry Recognized Credentials

If you are training through an NCCER-accredited sponsor, you may be eligible for credentials from NCCER's Registry. The ID number for this module is 29204. Note that this module may have been used in other NCCER curricula and may apply to other level completions. Contact NCCER's Registry at 888.622.3720 or go to **www.nccer.org** for more information.

Contents

Topics to be presented in this module include:

Figures and Tables

SECTION ONE

1.0.0 METAL AND PREHEATING

Objective

Describe the relationship between heat and metal and identify preheating methods.

a. Describe the relationship between heat and metal.
b. Identify and describe methods used to preheat metal prior to welding.
c. Identify and describe devices and products used to measure temperature.

Performance Task

1. Preheat base metal to 350°F (177°C) and verify preheat using a temperature-indicating device.

Trade Terms

Constituents: The elements and compounds, such as metal oxides, that make up a mixture or alloy.

Critical temperature: The temperature at which iron crystals in a ferrous-based metal transform from being face-centered to body-centered. This dramatically changes the strength, hardness, and ductility of the metal.

Crystalline: Like a crystal; having a uniform atomic structure throughout the entire material.

Heat-affected zone (HAZ): The portion of the base metal workpiece that is not melted but has had its properties affected by welding or other heat-intensive operation such as oxyfuel cutting.

Interpass temperature control: The process of maintaining the temperature of the weld zone within a given range during welding.

Martensite: A solid solution of carbon in alpha-iron that is formed when steel is cooled so rapidly that the change from austenite to pearlite is suppressed; responsible for the hardness of quenched steel.

Preheat weld treatment (PHWT): The process of heating the base metal before making a weld.

Preheat weld treatment (PHWT) is the controlled heating of the base metal immediately before welding begins. The main purpose of PHWT is to keep the weld and the base metal along the weld line in the heat-affected zone (HAZ) from cooling too fast. In some cases, the whole structure is preheated. It may not always be possible or practical to heat the whole structure. When this is the case, only the section near the weld is preheated.

1.1.0 Temperature and Metal Relationship

At room temperature, most metals are solid and have a crystalline structure (*Figure 1*). They become liquids if heated to a high enough temperature. Temperature is a measure of molecular activity. When the temperature is high enough, the forces that hold the molecules in their crystalline structures break down, and the metal becomes a liquid. At even higher temperatures, metals become gases. The temperature at which a metal becomes liquid is its melting point. The melting point is different for every metal. In the case of alloys (metal mixtures), the various constituents melt at different temperatures. Some, like carbon, may not melt at all.

Below the melting point, there is a temperature zone at which crystalline changes take place. Depending on the metal or alloy, these changes can seriously affect the metal's physical structure and mechanical characteristics. Quick cooling (quenching) of a metal while it is in this temperature zone can freeze some of the modified crystalline structure. As a result, the mechanical properties and physical traits of the cooled metal will be very different from those of the same metal if it had been cooled more slowly. Quenching does not have much of an effect on low-carbon and mild steels because they do not contain much carbon. However, quenching medium- and high-carbon steels produces a hard and brittle crystalline structure, which reduces the metal's toughness. *Table 1* shows the minimum preheat temperatures for various base materials, and it includes their welding processes and material thickness.

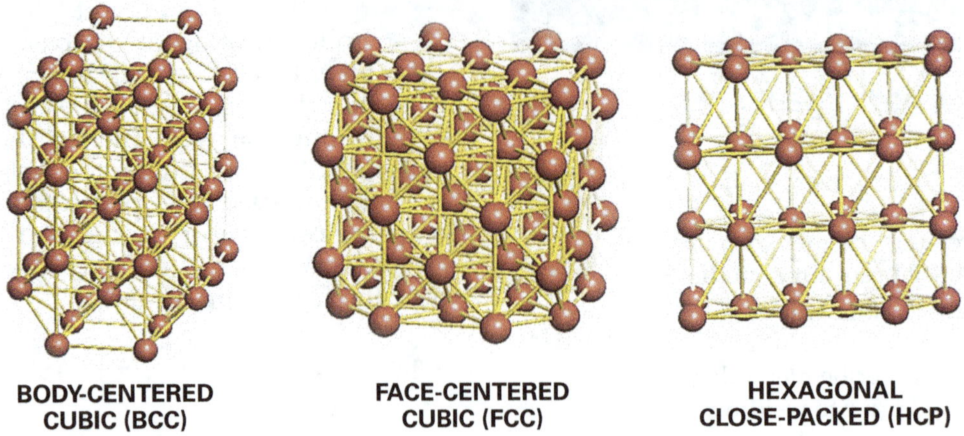

Figure 1 Crystal structures of iron.

Table 1 Minimum Preheat Temperature for Various Base Metals

Base Materials	Welding Processes	Material Thickness	Mimimum Preheat Temperature
ASTM A36 ASTM A53 GR. B ASTM A106 GR. B API 5L GR. B API 5L GR. X42	Shielded Metal Arc Welding With Other Than Low Hydrogen Electrodes	≤¾″ (19 mm) Between ¾″ and 1½″ (19 mm and 38 mm) Between 1½ and 2½″ (38 mm and 64 mm) >2½″ (64 mm)	None (See Note) 150°F (66°C) 225°F (107°C) 300°F (149°C)
ASTM A36 ASTM A53 GR. B ASTM A106 GR. B ASTM A572 GR. 42 ASTM A572 GR. 50 ASTM A586 API 5L GR. B API 5L GR. X42	Shielded Metal Arc Welding With Low Hydrogen Electrodes, Submerged Arc Welding, Gas Metal Arc Welding, and Flux-Cored Arc Welding	≤¾″ (19 mm) Between ¾″ and 1½″ (19 mm and 38 mm) Between 1½″ and 2½″ (38 mm and 64 mm) >2½″ (64 mm)	None (See Note) 50°F (10°C) 150°F (66°C) 225°F (107°C)
ASTM A572 GR. 60 ASTM A572 GR. 65 ASTM 5L GR. X52	Shielded Metal Arc Welding With Low Hydrogen Electrodes, Submerged Arc Welding, Gas Metal Arc Welding, and Flux-Cored Arc Welding	≤¾″ (19 mm) Between ¾″ and 1½″ (19 mm and 38 mm) Between 1½″ and 2½″ (38 mm and 64 mm) >2½″ (64 mm)	50°F (10°C) 150°F (66°C) 225°F (107°C) 300°F (149°C)
ASTM A514 ASTM A517	Shielded Metal Arc Welding With Low Hydrogen Electrodes, Submerged Arc Welding, Gas Metal Arc Welding, and Flux-Cored Arc Welding	≤¾" (19 mm) Between ¾" and 1½" (19 mm and 38 mm) Between 1½" and 2½" (38 mm and 64 mm) >2½" (64 mm)	50°F (10°C) 125°F (52°C) 175°F (79°C) 275°F (135°C)

NOTE: If below 32°F (0°C), preheat to 70°F (21°C)

NOTE

Tables in the Trainee Guide are for educational purposes only. These tables should not be relied upon as a resource for work outside of the learning environment. Always consult job specifications and other named project resources for current and accurate guidance.

When medium-carbon and high-carbon steel are welded, the temperature in the weld zone always reaches the melting point (above the critical temperature or transformation temperature), and the crystalline structure in the weld zone changes. If the metal has not been preheated enough, the large mass of the cool base metal will quench the weld zone and cause localized

Heat Treatments

Different forms of heat treatments are used to alter the properties of metals. Soft metals can be made hard, brittle, or strong using the correct heat treatment. They can be returned to their original soft form with heat treatment. Welding specifications often require that welded joints be heat-treated before welding or after fabrication. To avoid mistakes, heat treatments made to metals of any kind must be done in a planned and systematic way. This is because the use of temperatures that are too low or too high can result in doing more harm than good.

hardness and brittleness. If the metal has been adequately preheated, the weld zone will cool much more slowly, which will allow the crystalline structure to return to the crystalline form of the surrounding base metal.

Preheating may be required to do the following:

- Reduce shrinkage in the weld and the adjacent base metal
- Prevent excessive hardening and reduced ductility of the weld filler and base metals in the heat-affected zone
- Reduce hydrogen gas in the weld zone
- Ensure compliance with required welding procedures

The critical temperature of a metal is somewhere below its melting point. The temperature range that exists between a metal's critical temperature and its melting point is called the transformation temperature range. This range, generally considered to lie between 932°F and 1,040°F (500°C and 560°C), is the lowest temperature point at which the molecular structure of the metal will change during preheating, welding, and postheating. The melting point, strength, and hardness of steel are determined by the amount of carbon in the steel. *Table 2* shows the classifications of steel based on carbon content.

When preheating metal, apply heat to the region of the intended weld long enough to sufficiently heat the area. Insufficient heating will allow a relatively large weld to cool too quickly. To ensure proper preheating, measure the temperature of the preheated region again after waiting 15 to 20 minutes.

Table 2 Classification of Carbon Steel

Common Name	Carbon Content
Low-carbon steel	0.15 percent maximum
Mild steel	0.15 to 0.30 percent
Medium-carbon steel	0.30 to 0.50 percent
High-carbon steel	0.50 to 1.00 percent

1.1.1 Reducing Shrinkage

Because restrained assemblies are held firmly in position to maintain their shape, they can build up residual stresses following welding. The section that is welded expands due to the heat. When it cools, it shrinks, and stresses may occur if the surrounding area does not shrink along with it. With steel weldments, this shrinkage can cause distortion, warping, and even cracking. With cast-iron weldments, shrinkage usually causes the base metal to crack or the weld to fail. For example, if a broken cast-iron pulley spoke is being repaired, the rim and other spokes must be adequately preheated or they will be placed under heavy strain by the thermal expansion of the welded spoke. When the weld cools, the residual strains could cause the rim to crack or the welded spoke to crack again. If the rim and spokes near the weld are adequately preheated, they will shrink along with the welded spoke, reducing or eliminating most of the stress on the rim. *Figure 2* shows the areas of heavy strain resulting from inadequate preheating of a cast-iron pulley.

1.1.2 Preventing Excessive Hardening and Reduced Ductility

All welding processes use or generate temperatures higher than the melting point of the base

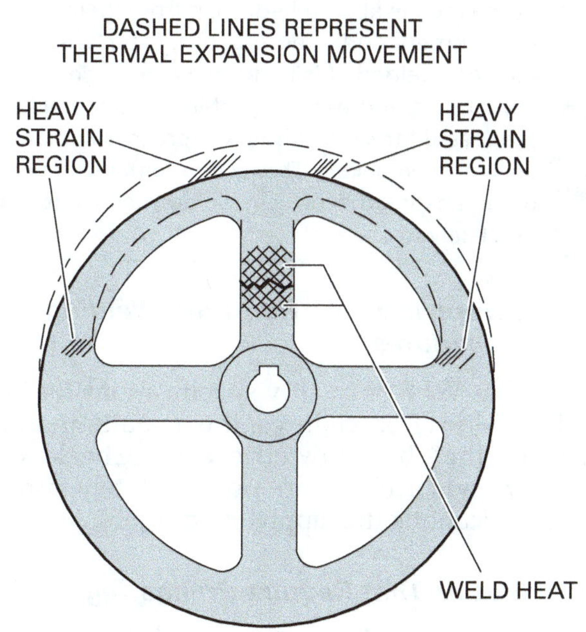

Figure 2 Strain caused by inadequate preheating of a cast-iron pulley.

metal. If the welded metal is a high-carbon alloy steel or cast iron that has been quenched, a hard, brittle metal called **martensite** will form in the weld zone. This causes the entire heat-affected zone parallel to the weld fusion line to become hard and brittle (*Figure 3*). The large temperature difference also causes differential thermal expansion and creates stresses in the weld region. Heavy plate and castings have a high heat absorption capacity that will cause the weld metal and adjacent base metal to quench unless they are adequately preheated.

1.1.3 Reducing Hydrogen Gas in the Weld Zone

The welding arc can cause moisture in the weld zone to separate into hydrogen and oxygen. The hydrogen dissolves in the molten base metal, and the oxygen combines with other elements to form slag. As the metal begins to cool, the hydrogen forms gas bubbles in the base metal along the weld boundary. These bubbles create pressure that strains the metal and causes underbead cracking along the weld boundary and at the toe. Preheating helps to remove moisture from the weld zone. This results in slower cooling, which allows the hydrogen more time to diffuse up through the filler metal and down into the adjacent base metal. The more time the hydrogen has to diffuse, the less pressure it causes, and the less chance there is for underbead cracking to occur. *Figure 4* is a simplified diagram showing underbead cracking caused by hydrogen bubbles.

> **NOTE**
> Some metals absorb hydrogen from their environment during normal use. Later, when they are welded, this hydrogen can cause cracking. To address this problem, the metal is preheated for several hours, a process known as hydrogen baking. This action removes much of the absorbed hydrogen, so normal welds are less likely to crack.

1.1.4 Complying with Required Welding Procedures

If a site has WPS or quality specifications for particular welding procedures, it is usually mandatory that they be followed exactly. Check with your supervisor if you are not absolutely sure of the specifications that apply to your job.

1.1.5 Metals That Require Preheating

Determining whether metal assemblies and conditions require preheating can be very difficult.

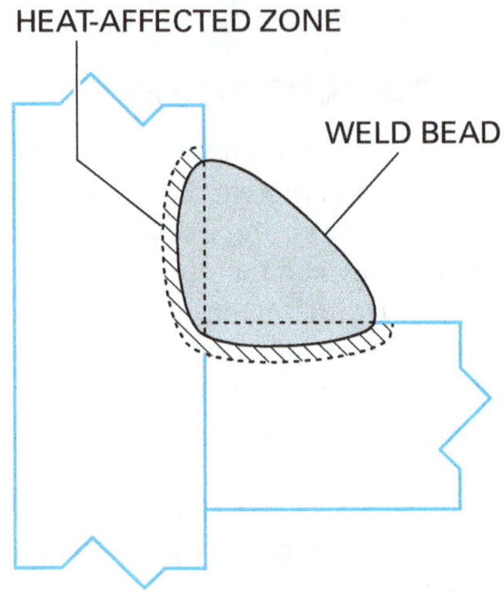

Figure 3 Heat-affected zone (HAZ).

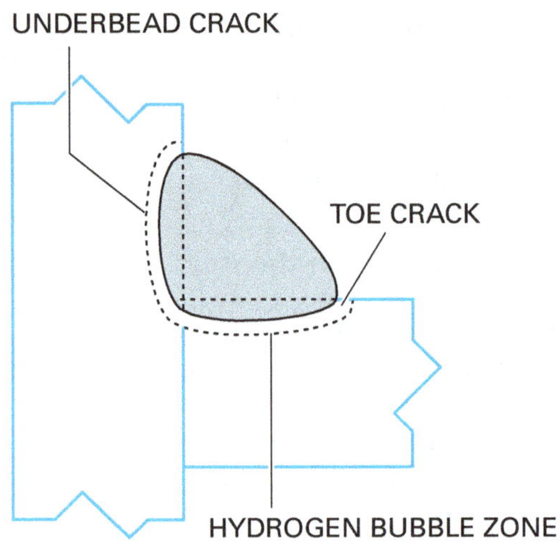

Figure 4 Underbead cracking caused by hydrogen bubbles.

Preheating and **interpass temperature control** depend on the base metal composition, thickness, and degree of restraint. Mild carbon steel, for example, requires preheating if its carbon content exceeds 0.30 percent. If the carbon content is lower than 0.30 percent, it must be heated to a temperature above 70°F (21°C). Some base metals require preheating only when they exceed a certain thickness or are too cold. For example, metals exposed to freezing winter temperatures usually require some preheating. Welding assembles and castings with complex shapes usually require preheating to avoid severe stresses or warping. Some alloys, such as high-carbon or alloy steels, require preheating to avoid brittleness and cracking. Not all metals require preheating. Low-carbon steels rarely need preheating. Always consult the WPS for specifics. *Table 3* shows the metals that usually require preheating and their conditions and forms.

Preheating requirements are affected by the alloy composition and thickness of the base metal,

Table 3 Metals That Usually Require Preheating

Metal or Alloy	Conditions or Forms
Aluminum	Large or thick section castings
Copper	All (prevents too-rapid heat loss)
Bronze, copper-based	All (prevents too-rapid heat loss)
Mild carbon steels	Restricted joints, complex shapes, freezing temperatures, and carbon content over 0.30 percent
Cast irons	All types and all shapes
Cast steels	Higher carbon content or complex shapes
Low-alloy steels	Thicker sections or restricted joints
Low-alloy steels	Heating temperature dependent on carbon or alloy content
Manganese steels	Heating temperature dependent on carbon or alloy content
Martensitic stainless steels	Carbon content over 0.10 percent
Ferritic stainless steels	Thick plates

the welding process to be used, and the ambient temperature. The hardenability of a steel is directly related to its carbon content and alloying elements. Because different types of steel have different amounts of carbon and alloying elements, they have different preheat requirements. As a general rule, the higher the carbon content, the higher the preheating temperature required. *Table 4* lists preheat requirements for steels, including the steel type, shapes or conditions, and preheat temperature.

1.2.0 Preheating Methods

Preheating is most effective when the entire welding assembly is preheated (general heating) in an oven or preheating furnace because the heating is even. However, because of site conditions or the

Underbead Cracking

Always consider preheating and postheating a weld when there is the danger of underbead cracking, also called hydrogen cracking. The risk of underbead cracking increases when welding carbon and low-alloy steels. Underbead cracking can occur when there is a combined presence of the following conditions:

- Diffusible (atomic) hydrogen in the steel
- A steel with a partly or wholly martensitic microstructure
- Tensile stress at the weldment and heat-affected zone
- A temperature below 300°F (149°C)

Proper preheat, interpass temperature maintenance, and postweld heat treatment, together with the proper electrode/shielding gas combination, can eliminate this problem.

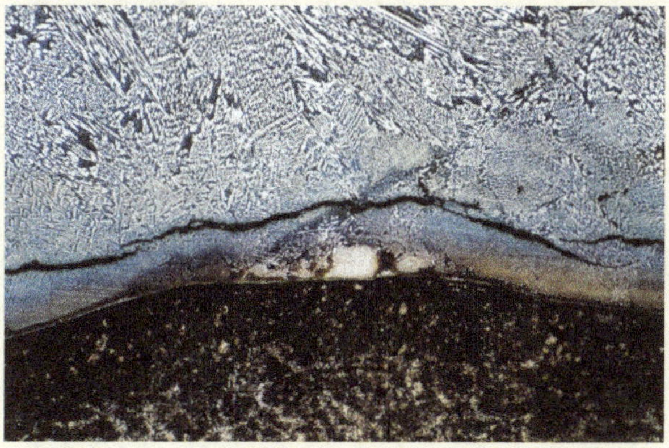

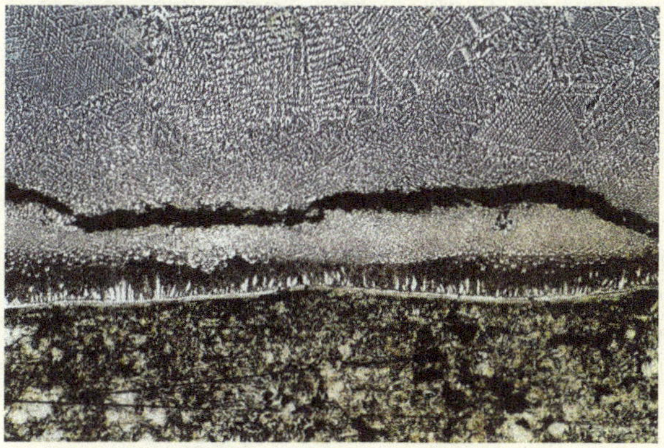

Table 4 Preheat Requirements for Steels

Steel Type	Shapes or Conditions	Preheat Temperature
Low-carbon steel (0.10–0.30 percent C)	Freezing temperatures	Above 70°F (21°C)
Low-carbon steel (0.10–0.30 percent C)	Complex large shapes	100°F–300°F (38°C–149°C)
Medium-carbon steel (0.30–0.45 percent C)	Complex large shapes	300°F–500°F (149°C–250°C)
High-carbon steel (0.45–0.80 percent C)	All sizes and shapes	500°F–800°F (250°C–427°C)
Low-alloy steel	Lower C or alloy percent	100°F–500°F (38°C–250°C)
Low-alloy steel	Higher C or alloy percent	500°F–800°F (250°C–427°C)
Manganese steel	Lower C or alloy percent	100°F–200°F (38°C–93°C)
Manganese steel	Higher C or alloy percent	200°F–500°F (93°C–250°C)
Cast iron	All shapes	200°F–400°F (93°C–204°C)
Martensitic stainless steel	0.10–0.20 percent C	500°F (250°C)
Martensitic stainless steel	0.20–0.50 percent C	500°F (250°C) and postheat
Ferritic stainless steel	¼" (6.4 mm) and thicker	200°F–400°F (93°C–204°C) and postheat

size or shape of the welded assembly, preheating by general heating is not always possible. Therefore, a variety of heating devices can be used for preheating. Localized heating of specific regions can be done with gas torches, blowtorches, electric resistance heaters, radio frequency induction heaters, or partial ovens that are built only around the region to be heated. Parts of the oven or enclosure can be removed so that the welder can have access and still maintain the preheating temperature.

Most welding shops have one or more types of preheating devices, such as the following:

- Oxyfuel torches
- Portable preheating torches
- Open-top preheaters
- Electric resistance heaters
- Induction heaters
- Enclosed furnaces

1.2.1 Oxyfuel Torches

If properly monitored, oxyfuel torches equipped with specialized heating tips can be used for localized preheating. *Figure 5* shows two types of torch handles to which heating tips can be mounted.

The specialized heating tips are designed to produce blowtorch flame patterns. Heating tip attachments with multiflame heating heads are often referred to as rosebuds. They are designed to mount directly to torch handles like welding tips. Heating tips designed to replace the cutting tips on straight cutting torches or combination cutting torch cutting attachments are also available. *Figure 6* shows several types of heating tips.

WARNING!

Always use the correct type of tip for the fuel gas being used. Using a tip with the wrong gas can cause the tip to explode. Observe the pressures and flow rates recommended by the manufacturer.

1.2.2 Portable Preheating Torches

Portable preheating torches are designed to burn either oil or gas combined with compressed air. Some use only gas and have an air blower to increase the intensity of the flame. *Figure 7* shows an air torch suitable for preheating large weldments. The combustion chamber head can easily be removed and an extension added to increase the length. This torch operates on various fuel gases combined with compressed air.

Preheating Steel in Low-Temperature or High-Humidity Conditions

Steel should not be welded if the steel temperature is below 32°F (0°C). If below 32°F (0°C), steel should be heated to at least 70°F (21°C) prior to welding. Under humid conditions, steel should be heated to a higher preheat temperature in order to dry any moisture from its surface.

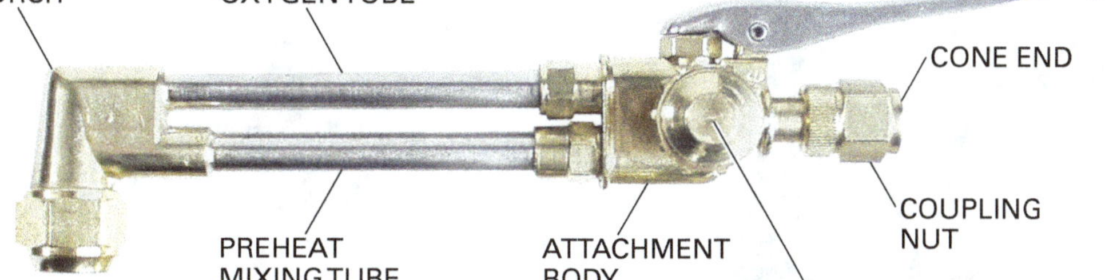

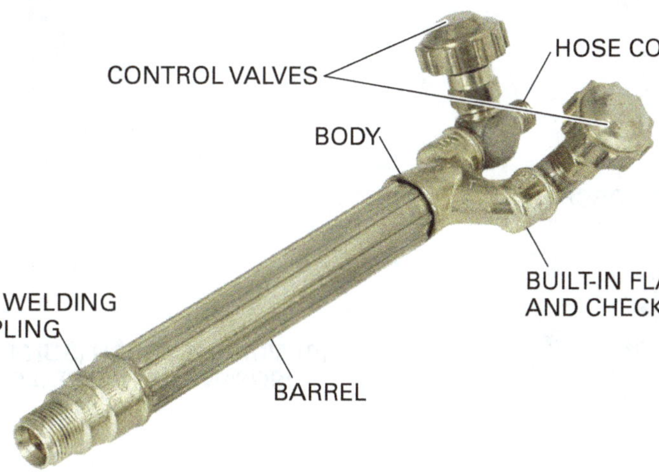

Figure 5 Combination torch handle and cutting torch attachment.

1.2.3 Open-Top Preheaters

Open-top preheaters come in several basic designs. The simplest is a large, horizontal gas burner with an adjacent or overhead support structure. The object to be heated is mounted on the support structure directly over the gas burners. A variation is the open flat-top preheater, which consists of a horizontal grate or series of bars with gas burners underneath (*Figure 8*). The item to be preheated is placed on the grating. This type of preheater works similar to a gas-powered kitchen stove.

Air Tube Burner

The gas fuel/compressed air tube burner can be used for the heat treatment of large weldments. The tube section has evenly spaced multiple heating tips along its length. A burner of this type provides a more even heat distribution to a large weldment.

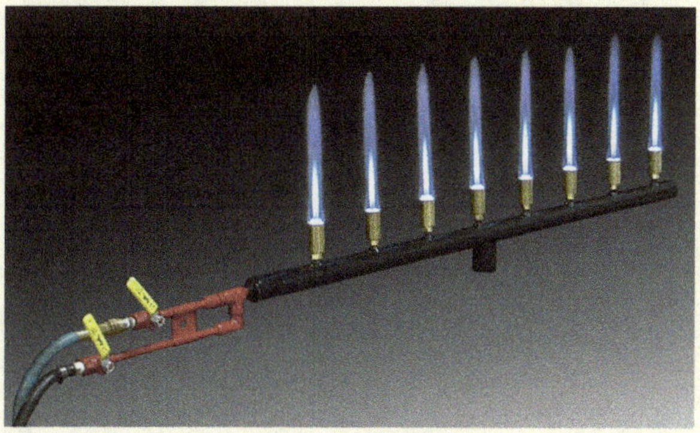

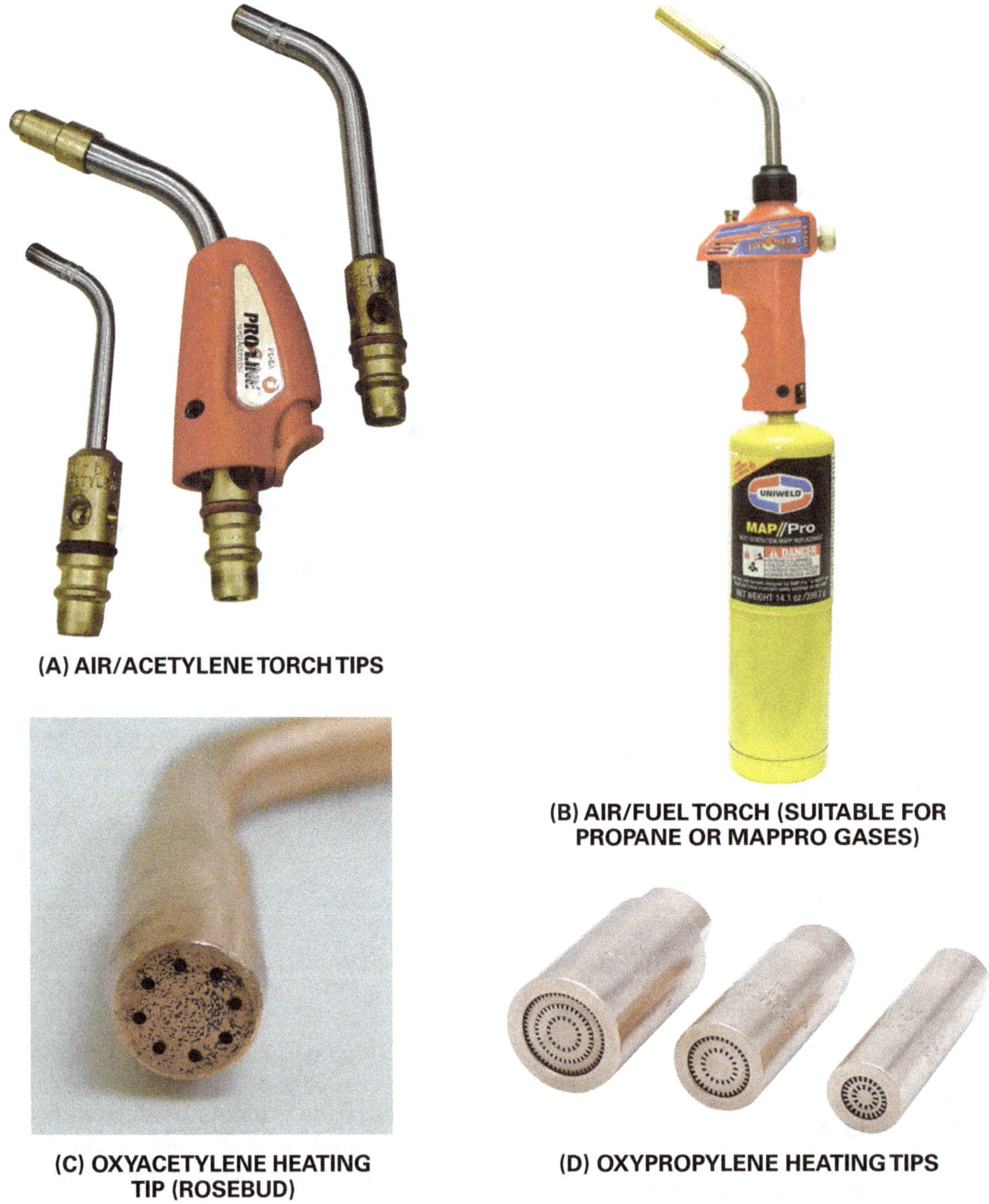

Figure 6 Types of heating tips.

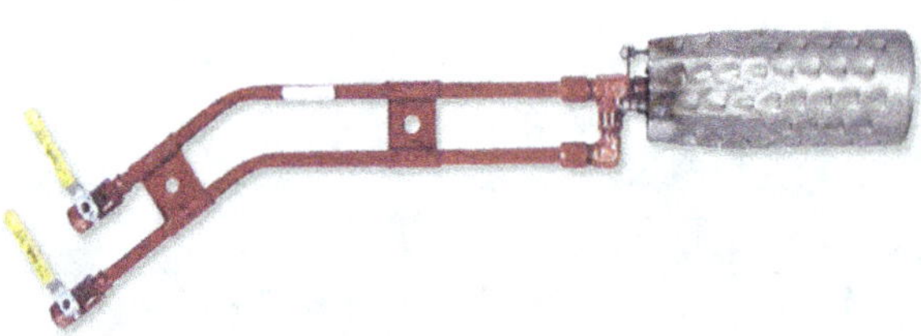

Figure 7 Gas preheating torch.

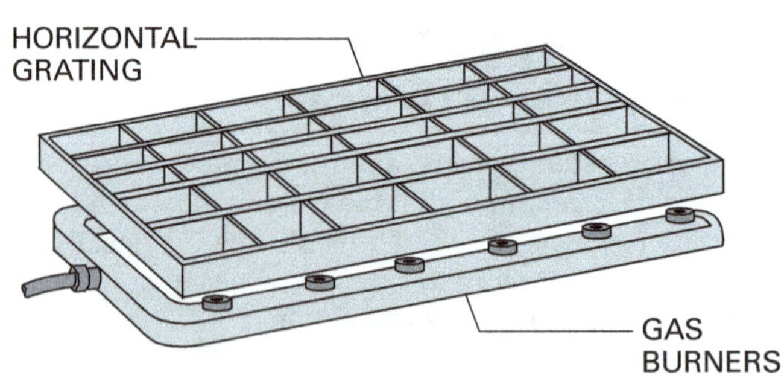

Figure 8 Open flat-top preheater.

1.2.4 Electric Resistance Heaters

Electric resistance heaters have tubular resistance elements that are formed in several shapes or styles to fit various applications (*Figure 9*). These elements are temporarily fastened against the surface of the metal to be heated. Electricity passing through the elements creates heat, which is transferred to the weldment.

1.2.5 Induction Heaters

In induction heating, an alternating magnetic field is generated within the metal body being heated. A magnetic coil is placed around the metal object, and a strong alternating current is passed through the coil. The resulting alternating magnetic field generated within the base metal creates circulating electrical currents known as eddy currents. The interaction of these currents with the electrical resistance of the base metal produces an electrical heating effect, raising the temperature of the base metal. *Figure 10* shows an air-cooled induction heating system consisting of a power supply, induction blanket, and the associated power and control cables. Air-cooled systems are typically used for preheat applications up to 400°F (204°C). This type of system typically uses a controller to monitor and automatically control the temperature.

1.2.6 Enclosed Furnaces

Enclosed furnaces or ovens include both permanent and temporary structures or partial enclosures that must be made of firebrick and clay, sheet metal, or some other noncombustible material. The item to be heated may be totally or partially enclosed. With very large items, it may not be practical or possible to enclose the entire item. Furnaces and ovens are heated by gas or oil burners, torches, or electric resistance heaters. *Figure 11* shows an example of a gas-fired box furnace.

1.3.0 Measuring Temperature

Control of preheating, interpass, and postheating temperatures depends on accurately determining the temperature of the weld zone or assembly. Both underheating and overheating can negatively affect a metal's physical and mechanical characteristics. There are a number of ways to determine the surface temperature of an item being heated. Some involve complex manufactured products, while others use less sophisticated methods. Devices and materials that are commonly used to determine surface temperatures include thermometers and temperature-sensitive indicators.

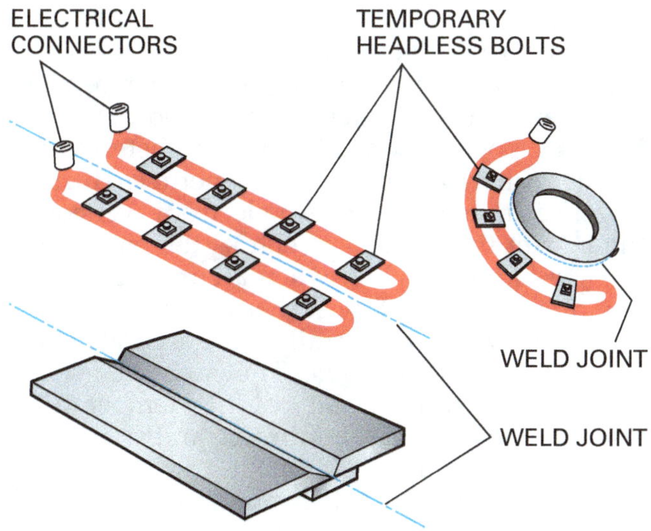

Figure 9 Resistance heating elements.

1.3.1 Thermometers

Electronic thermometers (*Figure 12*) now dominate industrial temperature measurement. These thermometers display temperature on either an analog meter or a digital liquid crystal display (LCD) readout. They generally use either a thermocouple or thermistor type of temperature probe, or both, to sense the heat and generate a temperature indication.

Thermocouple probes use a sensing device made of two dissimilar wires welded together at one end called a junction. When the junction is heated, it generates a low-level DC voltage that produces a proportional temperature reading on the electronic thermometer indicator. Thermocouple probes tend to be rugged and less expensive than thermistor probes. The probe shown in *Figure 12* is designed to clamp securely to small pipe; many probe styles are available. Thermistor probes use a semiconductor electronic element in which resistance changes in proportion to the change in temperature.

Often, several different probes are used with the same electronic thermometer to allow temperature measurement in different applications. Many electronic thermometers can also accommodate two or more probes at the same time, so that simultaneous measurements can be made at several locations. Most thermometers of this type can calculate and display the difference in temperature between the locations being measured.

Electronic thermometers are precise measuring instruments. Be sure to read and follow the manufacturer's instructions for operating electronic thermometers. Also, be sure to follow the manufacturer's instructions for calibration of the instrument.

Dial thermometers (*Figure 13*) are available in various forms. They are popular because they are rugged, small, and typically inexpensive. They come in many stem lengths and dial sizes. Because pocket-style dial thermometers have smaller calibrated scales, it is sometimes difficult to get accurate readings. An electronic version of the pocket thermometer, such as the one shown in *Figure 13*, works best when accuracy is important. Remember that most thermometers, regardless of the type, are generally more accurate in the mid-section of their range. They are often somewhat less accurate when reading temperatures that are close to the minimum and maximum values in their range.

Handheld infrared thermometers (*Figure 14*) are also very popular. Infrared thermometers allow the user to measure temperature without making contact with the surface of an object. For this reason, they are also called non-contact thermometers. They work by detecting infrared energy emitted by the object at which they are pointing. Laser sighting devices allow the user to aim the thermometer like a pistol at the surface to be measured. The temperature is displayed instantly. This type of thermometer is useful for taking temperature readings in hard-to-reach areas. However, it may not be effective on highly reflective surfaces. The unit may have an adjustable setting which can be set to allow the measurement of reflective surfaces. A thermometer with no such adjustment can still be used to measure the temperature of a shiny surface by applying a matte paint or non-reflective tape to the surface.

Induction Heaters

Because water cools more efficiently than air, water-cooled induction heating systems can be used for high-temperature preheating and stress relieving applications. This type of system is normally equipped with a temperature controller and temperature recorder. The controller and recorder are important components in stress-relieving applications.

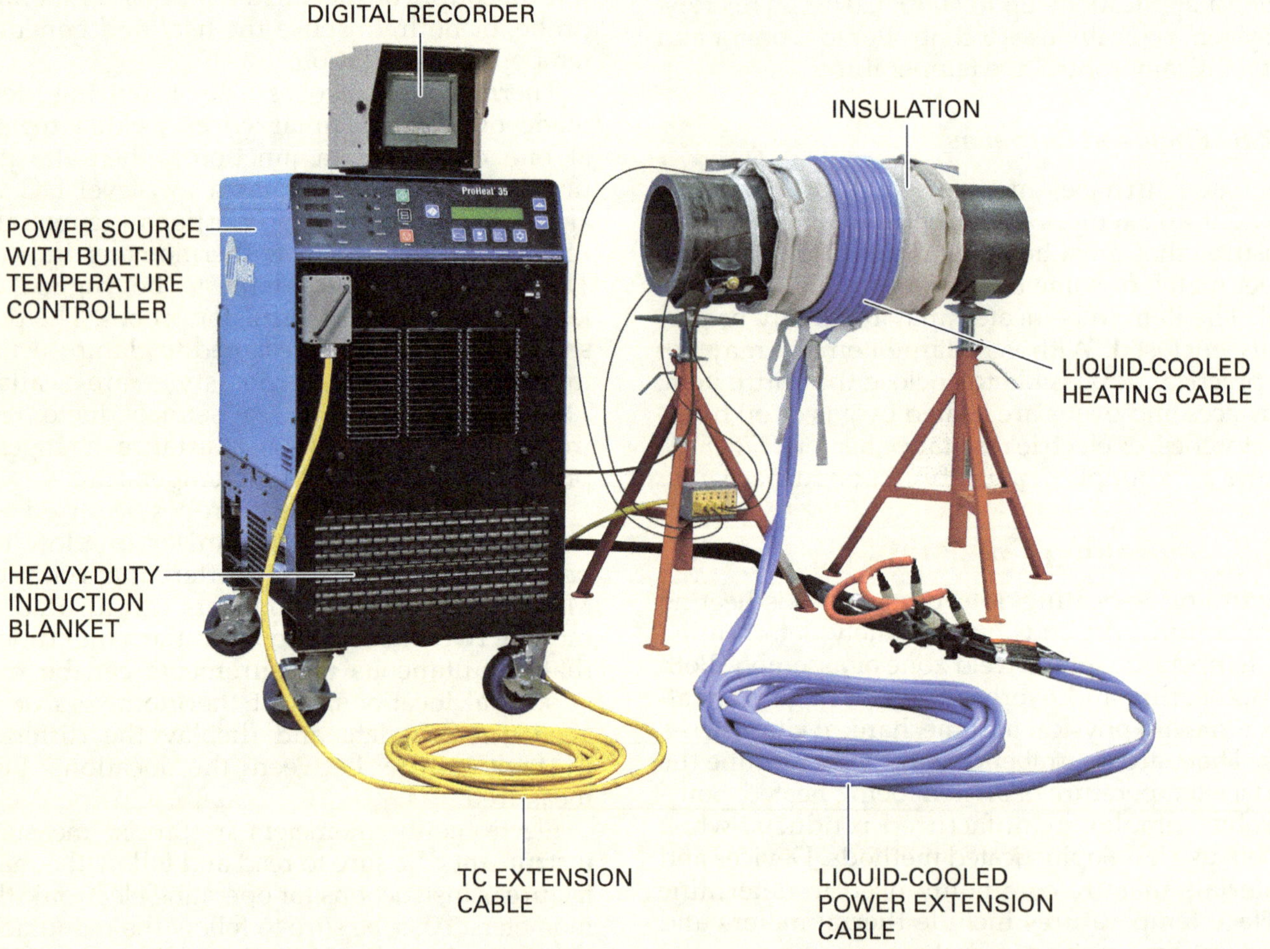

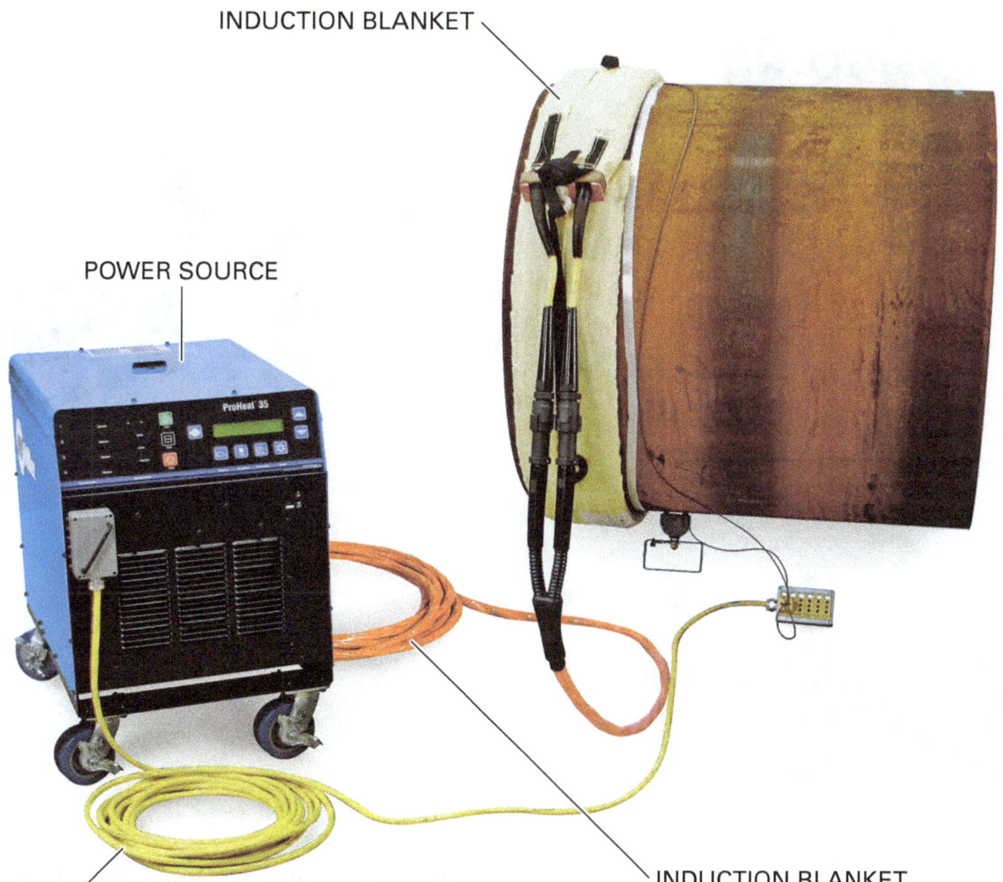

Figure 10 Induction heating system.

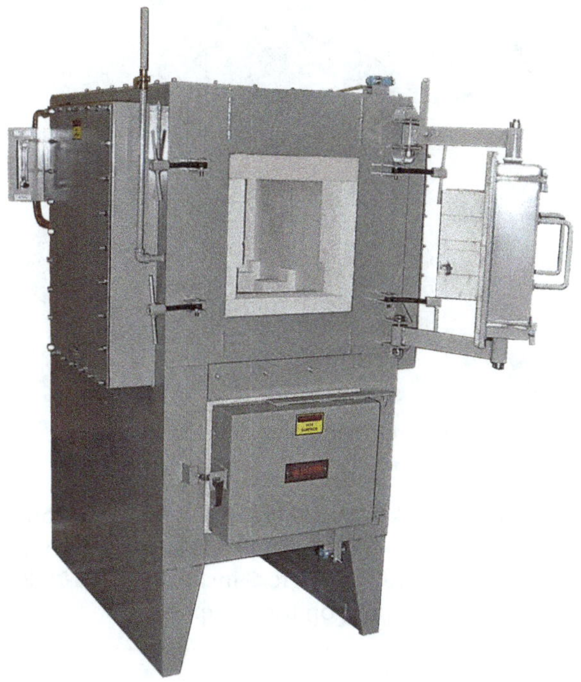

Figure 11 Box furnace.

It is important to understand the distance-to-spot ratio (D:S) of infrared thermometers. The D:S is the ratio of the distance to the object being measured compared to the diameter of the temperature measurement area. For example, if the D:S ratio is 12:1, taking a measurement from 12 inches (30.5 cm) away will give the average temperature of a spot 1" (25 mm) in diameter (*Figure 15*). The greater the distance, the larger the spot becomes. When using an infrared unit with a 12:1 D:S from 12 feet away, the spot becomes 1 foot in diameter. As a result, the temperature measurement may be inaccurate, unless the surface being measured is at least as large as the spot. Otherwise, the device will average the temperature of two or more unrelated surfaces.

1.3.2 Temperature-Sensitive Indicators

Temperature-sensitive indicators are devices that change form or color at specific temperatures or temperature ranges. These include commercially made materials and devices such as crayon sticks

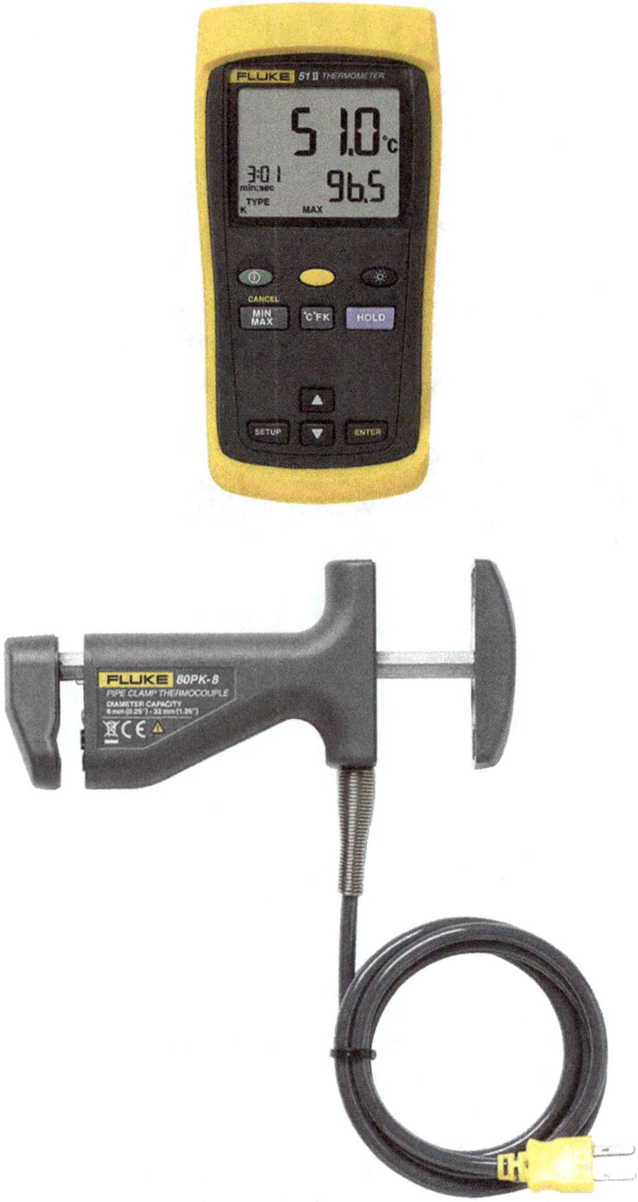

Figure 12 Electronic thermometer.

(*Figure 16 [A]*); liquids (*Figure 16 [B]*); and chalks, powders, pellets, and labels (*Figure 16 [C]*). These indicators are designed to change color or form at precise temperatures, often within a 1 percent tolerance.

Temperature sticks are among the most widely used temperature-sensitive indicators. The sticks consist of a series of graduated temperature-indicating crayons that cover the temperature range from 100°F to 1,150°F (38°C to 621°C). These sticks are used to mark on the base metal. At a specific temperature, the mark melts or changes color. *Table 5* is an example of a preheating chart that could be used when selecting temperature-indicating crayons for various metals.

Figure 13 Dial and electronic pocket thermometers.

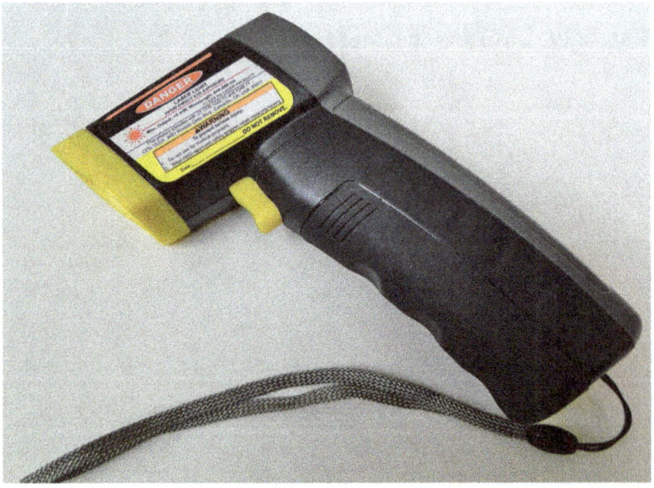

Figure 14 An infrared non-contact thermometer.

> **CAUTION**
>
> Do not mark with the crayon in the weld zone. The recommended practice is to mark from the toe of the weld out 1 to 3 inches (25.4 to 76.2 mm), depending on the code.

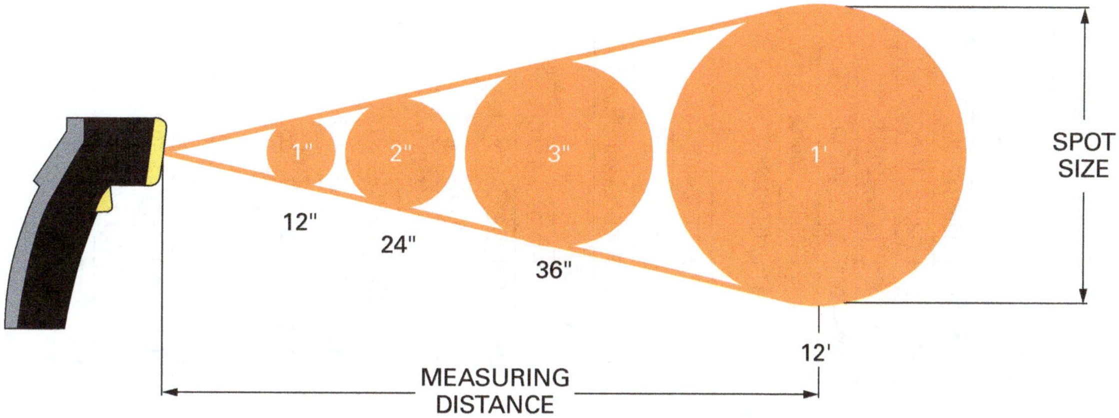

Figure 15 The D:S ratio illustrated.

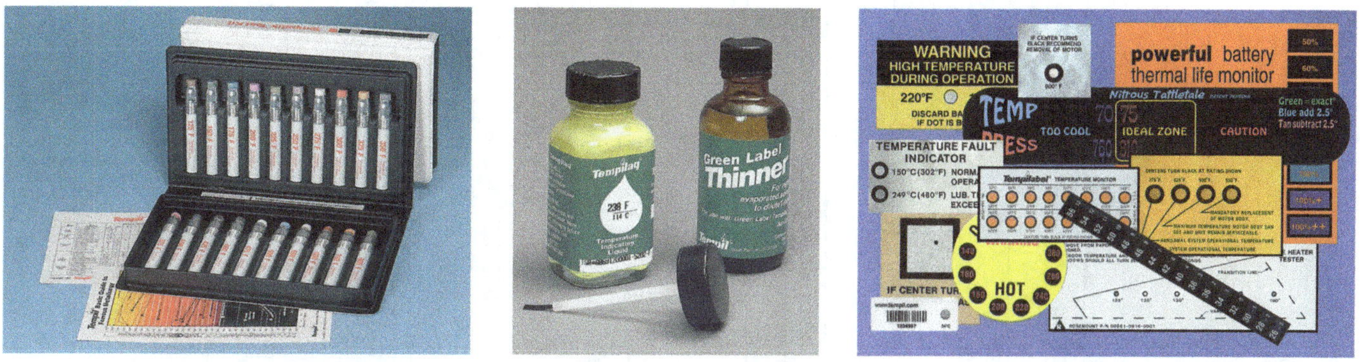

(A) TEMPERATURE STICKS (B) TEMPERATURE LIQUID (C) TEMPERATURE LABELS

Figure 16 Temperature-sensitive indicators.

Furnace Thermocouples

The rate of temperature rise, holding time at temperature, and rate of cooling are critical to proper heat treatment. For this reason, thermocouples used in furnaces must measure the temperature of the metal being heat-treated, not the ambient temperature inside the furnace. To do this, the thermocouples must, of course, be in contact with the metal.

Table 5A Preheating Chart for Selecting Temperature-Indicating Crayons (1 of 2)

Metal Group	Metal Description	Approximate Composition Percentage								Recommended Preheat Temperature
		C	Mn	Bi	Cr	Ni	Mo	Cu	Other	
Plain Carbon Steels	Plain Carbon Steel	Below 0.20								Up to 150°F (Up to 65°)
	Plain Carbon Steel	0.20–0.30								150°F–300°F (65°C–150°C)
	Plain Carbon Steel	0.30–0.45								250°F–450°F (120°C–230°C)
	Plain Carbon Steel	0.45–0.80								450°F–750°F (230°C–400°C)
Plain Moly Steels	Carbon Moly Steel	0.10–0.20					0.50			150°F–250°F (65°C–120°C)
	Carbon Moly Steel	0.20–0.30					0.50			200°F–400°F (95°C–205°C)
Manganese Steels	Silicon Structural steel	0.35	0. 80	0.25						300°F–500°F (150°C–260°C)
	Medium Manganese Steel	0.20–0.25	1.0–1.75							300°F–500°F (150°C–260°C)
	SAE 1330 Steel	0.30	1.75							400°F–800°F (205°C–425°C)
	SAE 1340 Steel	0.40	1.75							500°F–800°F (260°C–425°C)
	SAE 1045 Steel	0.50	1.75							800°F–900°F (425°C–480°C)
	12% Manganese Steel	1.25	12.0							Usually not Required
	Manganese Moly Steel	0.20	1.65	0.20			0.35	0.40		300°F–500°F (150°C–260°C)
	Jalten Steel	0.35 max	1.50	0.30				0.20		400°F–800°F (205°C–425°C)
	Manten Steel	0.30 max	1.35	0.30				0.35 min		400°F–800°F (205°C–425°C)
High Tensile Steels	Armco High Tensile Steel	0.12 max				0.50 min	0.50 min	0.50–1.50		Up to 200°F (Up to 95°C)
	Double Strength #1 Steel	0.12 max				0.50–1.25	0.10 min	0.50–1.50		300°F–800°F (150°C–425°C)
	Double Strength #1A Steel	0.30 max				0.50–1.25	0.10 min	0.80		400°F–700°F (205°C–370°C)
	Mayarir Steel	0.12 max	1.25	0.35	0.2–1.0	0.25–0.75		0.50 max		Up to 300°F (Up to 150°C)
	Otiscoloy Steel	0.12 max		0.10 max	0.10 max			0.25 max		200°F–400°F (95°C–205°C)
	Nax High Tensile Steel	0.15–0.25		0.75	0.60	0.17			Zr. 12	Up to 300°F (Up to 150°C)
	Cromansil Steel	0.14 max		0.75	0.50		0.15 max	0.40		300°F–400°F (150°C–205C°)
	A.W. Dyn-el Steel	0.11–0.14						0.40		Up to 300°F (Up to 150°C)
	Corten Steel	0.12 max		0.25–1.0	0.5–1.5	0.55 max				200°F–400°F (95°C–200°C)

Table 5B Preheating Chart for Selecting Temperature-Indicating Crayons (2 of 2)

Metal Group	Metal Description	Approximate Composition Percentage								Recommended Preheat Temperature
		C	Mn	Bi	Cr	Ni	Mo	Cu	Other	
High Tensile Steels (Cont.)	Chrome Copper Nickel Steel	0.12 max	0.75		0.75	0.75		0.55		400°F–600°F (205°C–315°C)
	Chrome Manganese Steel	0.40	0.90		0.40			1.0		200°F–400°F (95°C–205°C)
	Yoloy Steel	0.05–0.35	0.3–1.0	0.3 max		1.75		0.9–1.25		200°F–600°F (95°C–315°C)
	Hi-Steel	0.12 max	0.6			0.55				200°F–500°F (95°C–260°C)
Nickel Steels	2¼% Nickel Steel	0.25				2.25				200°F–400°F (95°C–205°C)
	3½% Nickel Steel	0.23				3.50				200°F–400°F (95°C–205°C)
Moly Bearing Chromium and Nickel Steels	SAE 4140 Steel	0.40			0.90		0.20			600°F–800°F (315°C–425°C)
	SAE 4340 Steel	0.40			0.80	1.85	0.25			700°F–900°F (370°C–480°C)
	SAE 4615 Steel	0.15				1.80	0.25			400°F–800°F (205°C–425°C)
	SAE 4820 Steel	0.20				3.50	0.25			600°F–800°F (315°C–425°C)
Low Chrome Moly Steels	1¼% Cr – ½% Mo Steel	0.17 max			1.25		0.50			250°F–400°F (120°C–205°C)
	2¼% Cr – 1% Mo Steel	0.15 max			2.25		1.0			300°F–500°F (150°C–260°C)
Medium Chrome Moly Steels	5% Cr – ½% Mo Steel	0.15 max			5.0		0.5			400°F–800°F (205°C–425°C)
	7% Cr – ½% Mo Steel	0.15 max			7.0		0.5			400°F–800°F (205°C–425°C)
	9% Cr – 1% Mo Steel	0.15 max			9.0		1.0			400°F–800°F (205°C–425°C)
Plain High Chromium Steels	11½-13% Cr Type 410	0.15 max			12.0					400°F–800°F (205°C–425°C)
	16-18% Cr Type 430	0.12 max			17.0					300°F–500°F (150°C–260°C)
	23-27% Cr Type 448	0.20 max			25.0					300°F–500°F (150°C–260°C)
	18% Cr 8% Ni Type 304	0.07			18.0	8.0				Usually do not require preheat but it may be desirable to remove chill
Chrome Nickel Stainless Steels	25-12 Type 309	0.07			25.0	12.0				
	25-20 Type 310	0.10			25.0	20.0				
	18-8 Cb Type 347	0.07			18.0	8.0			Cb 10xc	
	18-8 Mo Type 316	0.07			18.0	8.0	2.5			
	18-8 Mo Type 317	0.07			18.0	8.0	3.5			
Irons	Cast Iron									700°F–900°F (370°C–480°C)
Nonferrous	Ni Resist									500°F–1,000°F (260°C–540°C)
	Nickel, Monel, Inconel					Preheat not usually required for thin sections. 70°F–400°F				
	Aluminum-Copper					(21°C–205°C) preheat may be desirable for thick sections.				

Determining Metal Temperatures by Color

The approximate temperature of a metal heated in a furnace can be determined by the different colors exhibited by the metal. Reference charts are available that welders can use to interpret metal temperatures based on observing the color of the metal.

The chart shown below is an example only; it should not be relied upon for actual temperature measurement. Also note that the colors shown here will appear differently in various environments. Sunlight, for example, will cause the colors to appear differently than they would in a darkened room.

COLOR		APPROXIMATE TEMPERATURE	
		°F	°C
FAINT RED		930	500
BLOOD RED		1075	580
DARK CHERRY		1175	635
MEDIUM CHERRY		1275	690
CHERRY		1375	745
BRIGHT CHERRY		1450	790
SALMON		1550	845
DARK ORANGE		1630	890
ORANGE		1725	940
LEMON		1830	1000
LIGHT YELLOW		1975	1080
WHITE		2200	1205

Additional Resources

The Procedure Handbook of Arc Welding, 14th Edition. Cleveland, OH: The James F. Lincoln Arc Welding Foundation, 2000.

Galvery, William L. Jr. and Frank M. Marlow. *Welding Essentials: Questions and Answers.* 2nd Edition. New York, NY: Industrial Press, 2007.

Welding Handbook. Vol. 2, Part 1: Welding Processes. Miami, FL: American Welding Society, 2004.

Welding Handbook. Vol. 3, Part 2: Welding Processes. Miami, FL: American Welding Society, 2004.

1.0.0 Section Review

1. The temperature range that exists between a metal's critical temperature and its melting point is called its ____.
   a. transformation temperature range
   b. preheat weld treatment span
   c. heat-affected zone (HAZ)
   d. discontinuity arc

2. Open-top preheaters work similarly to a(n) ____.
   a. convection oven
   b. oxyfuel torch
   c. box furnace
   d. kitchen gas stove

3. Handheld infrared thermometers are also called ____.
   a. temperature sticks
   b. finger elements
   c. non-contact thermometers
   d. optical thermocouples

Section Two

2.0.0 Temperature Maintenance and Postheating Processes

Objective

Describe interpass temperature control and postheating processes.

a. Describe interpass temperature control.
b. Describe various postheating processes.

Trade Terms

Annealing: A postweld heat treatment for stress relieving weldments that is done at temperatures approximately 100°F (38°C) above the critical temperature with a prolonged holding period.

Discontinuity: The deformation, or interruption, of the crystal structure inside a metal.

Strain hardening: A method of increasing the strength of a metal by cold-working it.

Work hardening: The increase in strength that sometimes accompanies plastic deformation of a metal.

In addition to proper preheating, it is equally important to maintain proper temperature of a weldment or assembly both during welding operations and after the welding has been completed. This section introduces several postheating processes that help prevent weld failures by relieving stresses, improving corrosion resistance, restoring ductility, and reducing brittleness.

2.1.0 Interpass Temperature Control

Interpass temperature control is used to maintain the temperature of the weld zone during welding. It is used when the cooling rate is too high or too low to maintain the correct temperature of the weldment between weld passes. The minimum interpass temperature is usually the same as the preheat temperature. If welding is ever interrupted, the temperature of the weldment must be brought back to the interpass temperature before welding is continued. Usually, there is also a maximum interpass temperature, which is especially important in preventing overheating from heat buildup in smaller weldments.

When preheating is important, interpass temperature for multiple-pass welds is usually also important. Interpass temperature control is often simply a continuation of preheating. Heating equipment and techniques are the same.

With large, high-mass weldments and infrequent weld passes, the base metal can cool below the required preheat temperature if additional heat is not supplied. When the base metal cools below the specified minimum preheat temperature, it must be reheated to the preheat temperature before welding continues.

2.2.0 Postheating

Postweld heat treatment is the heating of a weldment or assembly after welding has been completed. It can be performed on most types of base metals. Postheat equipment and techniques are the same as those used for preheating.

Postweld heat treatments include the following the following:

- Stress relieving
- Annealing
- Normalizing
- Tempering

2.2.1 Stress Relieving

Stress relieving is the most commonly used postheat treatment. It is done below the lower transformation temperature, usually in the range of 1,050°F to 1,200°F (566°C to 649°C).

Stress relief treatment is done for the following reasons:

- To reduce residual (shrinkage) stresses in weldments and castings. This is especially important with highly restrained joints.
- To improve resistance to corrosion and stress corrosion cracking.
- To improve dimensional stability during machining operations.
- To improve resistance to impact loading and low-temperature failure.

> **Importance of Maximum Interpass Temperatures**
>
> Austenitic stainless steel typically has a 350°F (177°C) maximum interpass temperature because it is subject to intergranular migration of alloying elements (sensitization) between 800°F and 1,600°F (427°C and 871°C). Migration decreases the corrosion resistance of stainless steel.

Some codes cover stress relieving. These codes specify the heating rate, holding time, and cooling rate. The heating rate is usually 300°F to 400°F (149°C to 204°C) per hour. The holding time is typically one hour for each inch (25.4 millimeters) of thickness. The cooling rate is usually 300°F to 500°F (149°C to 260°C) per hour. Stress relieving requires the use of temperature indicators and temperature-control equipment.

2.2.2 Annealing

Annealing relieves much more stress than a stress relieving postheat treatment can, but it results in a steel of lower strength and higher ductility. Annealing is done at temperatures approximately 100°F (38°C) above the critical temperature, with a prolonged holding period. The heating is followed by slow cooling in the furnace, or by covering, wrapping, or burying the item. Austenitic stainless steel is an exception because it requires rapid cooling. Annealing is used to relieve the residual stresses associated with welds in carbon-molybdenum pipe and to relieve stresses in welded castings that contain casting strains. Generally, annealing is considered to leave the metal in its softest condition, with good ductility.

2.2.3 Normalizing

Normalizing is a process in which heat is applied to remove strains and reduce grain size. Normalizing is done at temperatures and holding periods comparable to those used in annealing. However, the cooling is usually done in still air outside the furnace, and the cooling rate is slightly faster than the slow cooling used in annealing. Normalizing usually results in higher strength and less ductility than annealing. It can be used on mild steel weldments to form a uniform austenite solid solution, to soften the steel, and to make it more ductile. Normalized metal is not as soft and free of stresses as fully annealed metal.

2.2.4 Tempering

Tempering, also called drawing, increases the toughness of quenched steel and helps avoid breakage and failure of heat-treated steel. It reduces both the hardness and the brittleness of hardened steel. Tempering is done at temperatures below the critical temperature, much lower than those used for annealing, normalizing, or stress relieving. Tempering is commonly used on tool steel.

2.2.5 Hardening

The hardness of a material is its resistance to indentation. In metals, hardness is related to their elastic and plastic properties. These properties can be altered when mechanical forces or temperature changes exert stresses or strains on the metal. The deformation, or interruption, of the crystal structure inside the metal is called **discontinuity**. Discontinuity is also called dislocation.

When a piece of metal is hit with a hammer, deformation occurs with each stroke. Hammering metal can form it into a new shape, but the amount of metal will still be the same. The reshaping of metal at ambient temperature is called cold forming or cold working. During cold forming, the metal is strained beyond its yield point, which is the point at which plastic deformation occurs and it will not return to its original form. When a metal is strained or stressed beyond its yield point at ambient temperature, it is said to have been work hardened. **Work hardening**, also called **strain hardening**, tends to make the component stronger (harder) and less prone to failure. Parts that have been hardened using the cold forming process may look brighter in appearance.

Plastic deformation is the permanent distortion of a metal after its elastic limits have been exceeded because of stress. Metals that have been cold worked have undergone plastic deformation. The ability of metals to flow in a plastic manner without breaking apart is the basis for all metal

Stress Relief

The most commonly used method for stress relief of weldments is postweld heat treatment. Stress relief can also be accomplished mechanically. One method involves attaching a mechanical vibrating device to the weldment during or immediately after welding. When activated, the device vibrates the weldment at a specific resonant frequency. The sustained force of the vibrations deforms but does not fracture the shape of the metal's grains. This is called plastic deformation. The plastic deformation of the metal's grains evens out the stress distribution within the weldment. The weight of the weldment generally determines the length of time that the weldment will be subjected to the vibrations.

Tempering Temperatures

If carbon steel is heated in an oxidizing atmosphere, a film of oxide forms on the clean surface, and the color of the surface changes as the temperature increases. The colors can be seen with the unaided human eye under low light conditions. Each color or shade is related to a specific temperature. The colors for plain carbon steel as described in the example chart provided here have been widely used in the trade in the past to determine the correct amount of temper. These colors are affected to some extent by the composition of the metal being heated; therefore, they should be used as a guide only.

COLOR		APPROXIMATE TEMPERATURE	
		°F	°C
VERY PALE YELLOW		430	221
LIGHT YELLOW		440	227
PALE STRAW-YELLOW		450	232
STRAW-YELLOW		460	238
DEEP STRAW-YELLOW		470	243
DARK YELLOW		480	249
YELLOW-BROWN		490	254
BROWN-YELLOW		500	260
SPOTTED RED-BROWN		510	266
BROWN PURPLE		520	271
LIGHT PURPLE		530	277
FULL PURPLE		540	282
DARK PURPLE		550	288
FULL BLUE		560	293
LIGHT BLUE		640	338

forming processes. Plastic flow is the process in which metals are stretched or compressed permanently without breaking or cracking. The resistance to continuing plastic flow as a metal is cold worked is also called work hardening.

When metal is heated to a point just below its melting point, the crystals in it are forced to deform to adapt to the strain caused by the heat. If the crystals in a metal are deformed during the welding process, the weldment will gain strength and become harder. In some situations, metals are intentionally heated again after they have been welded to give them more strength.

2.2.6 Heat Treatment Devices

Metals can be heated by a variety of methods and devices before or after welding. Properly controlling heat is a major consideration. Water-cooled induction heaters are used in some applications, but they are not always available or may not be the process required for the application. *Figure 17* shows a portable induction heater on top of its power supply/control box. A rear view of the power supply is also shown.

Depending on the length of the leads between the power supply and the handheld unit (work head), the operator can move the induction coils to the object rather than move the object to the coils. These portable induction units can be used to heat or cool the object. They can also be programmed to operate at different power levels and times. This variety of options allows better control of the preheat or postheat treatment.

Ceramic heating pads are electrical devices that are often used to heat and relieve stress in metals. In a ceramic heating pad (*Figure 18*), the individual ceramic components are threaded together as needed to cover the metal being heated. The heating pads can be placed on plate steel or laid over pipe. Electrical wires are threaded through the individual ceramic pieces, generating heat when power is applied. Each individual piece is rated at a specific amperage level. The number of pieces used is dependent upon the desired heating output and physical size of pad needed.

The wires from the heating pads must be connected to an electrical device that will control the amount of electrical current flowing through

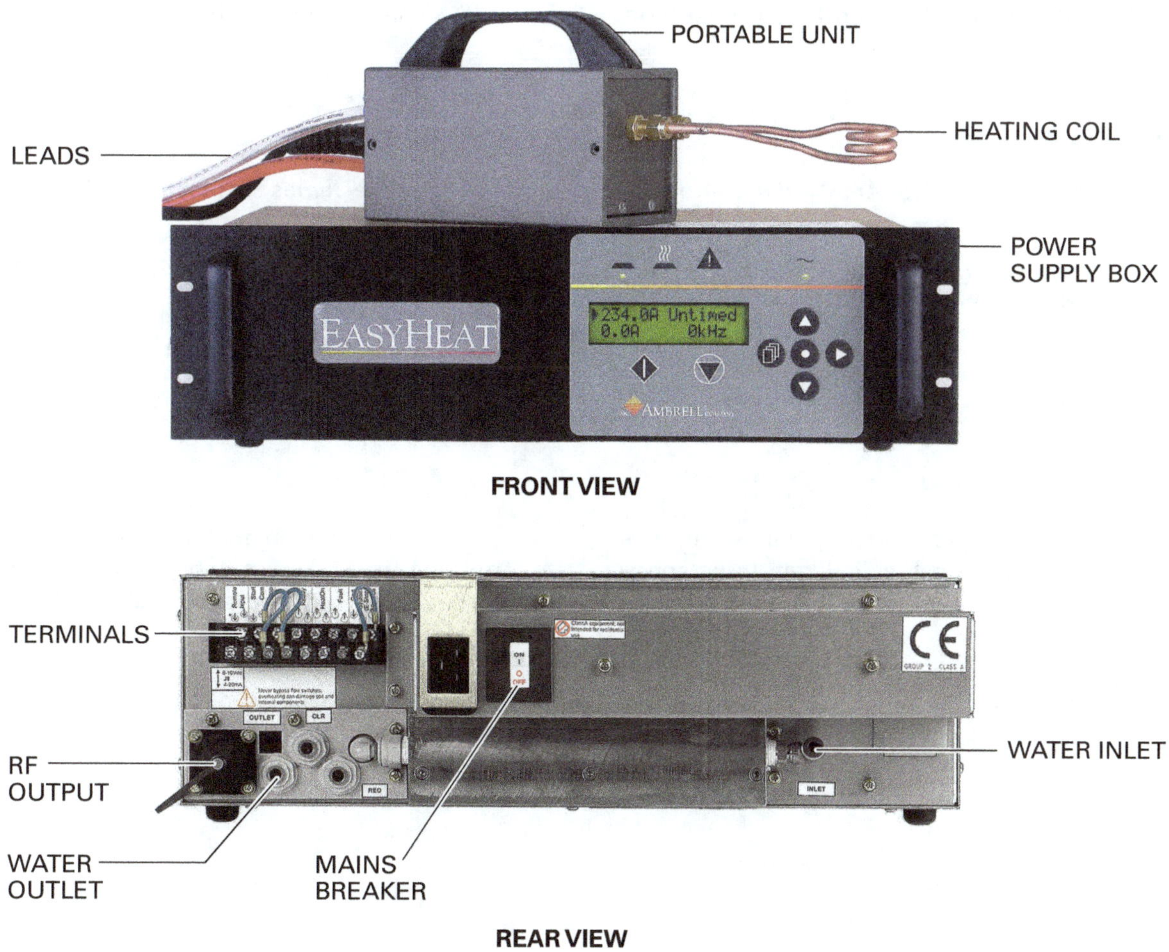

Figure 17 Portable induction heater and its power supply/control box.

the wires. Temperature-sensing thermocouples attached to the metal let the controlling device detect the temperature of the object being heated. The controller compares the thermocouple's signal against a preprogrammed temperature setting and determines if the metal needs to be warmer or cooler. If the controller senses that the metal is getting too hot or getting too hot too quickly, it can reduce the electrical power flowing through the ceramic pads of the heating device. On the other hand, if the controller determines that the metal is not heating fast enough, the controller can increase the current flow through the pads. This causes the pads to become warmer and the metal to heat more quickly. Ceramic heating pads are electrical, so they may be plugged into the AC outlet of a welding power supply.

2.2.7 Time-at-Temperature Considerations

In most heat treatment procedures, temperature and time-at-temperature are both specified by code. Time is very important because metallurgical changes are sluggish at temperatures below the critical temperatures. The maximum temperature is determined by the metal alloy. The holding time at the maximum temperature is based on the metal's thickness, usually one hour for each inch (25.4 mm) of thickness. The cooling rate is determined by the specific treatment or by any applicable code. Temperature control must also be used for postheating. Postheating temperatures are determined in the same manner as preheating temperatures.

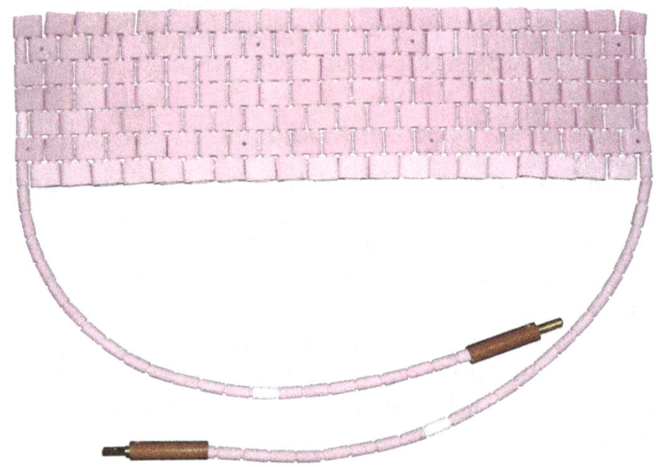

Figure 18 Ceramic heating pads.

Additional Resources

The Procedure Handbook of Arc Welding, 14th Edition. Cleveland, OH: The James F. Lincoln Arc Welding Foundation, 2000.

Galvery, William L. Jr. and Frank M. Marlow. *Welding Essentials: Questions and Answers*. 2nd Edition. New York, NY: Industrial Press, 2007.

Welding Handbook. Vol. 2, Part 1: Welding Processes. Miami, FL: American Welding Society, 2004.

Welding Handbook. Vol. 3, Part 2: Welding Processes. Miami, FL: American Welding Society, 2004.

2.0.0 Section Review

1. The interpass temperature for small weldments and frequent weld passes is likely to be specified as a ____.
   a. modal value
   b. maximum temperature
   c. calculated distribution
   d. minimum temperature

2. For stress relief on metals being welded, the holding time is typically how many hour(s) per inch (2.5 cm) of thickness of the metal?
   a. 14
   b. 12
   c. 1
   d. 2

SUMMARY

With certain alloy steel weldments and castings, preheating, interpass temperature control, and postheating are very important for relieving residual stresses and avoiding weld failures. Heating requirements vary with alloy composition, metal thickness, ambient temperature, and weldment configuration or shape. Proper preheating will preserve the weldment's strength, ductility, and weld quality. Postheating relieves residual stresses, improves corrosion resistance, restores ductility, and reduces brittleness. These processes result in stronger, tougher metal. Temperature control during preheating and postheating is very important because a temperature that is too high can weaken the base metal, and a temperature that is too low will not meet the heating objectives.

Review Questions

1. The controlled heating of the base metal immediately before welding begins is called _____.
   a. progressive base quenching
   b. thermal stress conditioning
   c. interpass temperature control
   d. preheat weld treatment

2. The temperature at which a metal turns liquid is its _____.
   a. melting point
   b. preheating point
   c. welding temperature
   d. postheating temperature

3. Hydrogen gas in the weld zone will cause _____.
   a. slow melting of the base metal
   b. underbead cracking
   c. hardening
   d. slag

4. The hardenability of steel is directly related to its alloying elements and _____.
   a. thickness
   b. carbon content
   c. preheat temperature
   d. welding temperature

5. Preheating is most effective when the welding assembly is heated _____.
   a. rapidly
   b. with a torch
   c. in an oven
   d. with induction heaters

6. Air-cooled induction heating systems are typically used for preheat applications up to _____.
   a. 300°F (149°C)
   b. 400°F (204°C)
   c. 600°F (316°C)
   d. 1,000°F (538°C)

7. Temperature-indicating crayons can be used to indicate temperatures ranging from 100°F (38°C) to _____.
   a. 400°F (204°C)
   b. 980°F (527°C)
   c. 1,150°F (621°C)
   d. 1,350°F (732°C)

8. What temperature is usually the same as the interpass temperature?
   a. Cutting
   b. Annealing
   c. Welding
   d. Preheat

9. For stress relief on metals being welded, the heating rate is usually how many degrees per hour?
   a. 100°F to 200°F (38°C to 93°C)
   b. 200°F to 300°F (93°C to 149°C)
   c. 300°F to 400°F (149°C to 204°C)
   d. 400°F to 500°F (204°C to 260°C)

10. Annealing is done at temperatures that are approximately how many degrees above the critical temperature?
    a. 100°F (38°C)
    b. 200°F (93°C)
    c. 300°F (149°C)
    d. 400°F (204°C)

Trade Terms Introduced in This Module

Annealing: A postweld heat treatment for stress relieving weldments that is done at temperatures approximately 100°F (38°C) above the critical temperature with a prolonged holding period.

Constituents: The elements and compounds, such as metal oxides, that make up a mixture or alloy.

Critical temperature: The temperature at which iron crystals in a ferrous-based metal transform from being face-centered to body-centered. This dramatically changes the strength, hardness, and ductility of the metal.

Crystalline: Like a crystal; having a uniform atomic structure throughout the entire material.

Discontinuity: The deformation, or interruption, of the crystal structure inside a metal.

Heat-affected zone (HAZ): The portion of the base metal workpiece that is not melted but has had its properties affected by welding or other heat-intensive operation such as oxyfuel cutting.

Interpass temperature control: The process of maintaining the temperature of the weld zone within a given range during welding.

Martensite: A solid solution of carbon in alpha-iron that is formed when steel is cooled so rapidly that the change from austenite to pearlite is suppressed; responsible for the hardness of quenched steel.

Preheat weld treatment (PHWT): The process of heating the base metal before making a weld.

Strain hardening: A method of increasing the strength of a metal by cold-working it.

Work hardening: The increase in strength that sometimes accompanies plastic deformation of a metal.

Additional Resources

This module is intended as a thorough resource for task training. The following reference works are suggested for further study.

The Procedure Handbook of Arc Welding, 14th Edition. Cleveland, OH: The James F. Lincoln Arc Welding Foundation, 2000.

Welding Essentials: Questions and Answers. 2nd Edition. William L. Galvery, Jr. and Frank M. Marlow. New York, NY: Industrial Press, 2007.

Welding Handbook. Vol. 2, Part 1: Welding Processes. Miami, FL: American Welding Society, 2004.

Welding Handbook. Vol. 3, Part 2: Welding Processes. Miami, FL: American Welding Society, 2004.

Figure Credits

The Lincoln Electric Company, Cleveland, OH, USA, Module opener, Figure 6D

Carnegie Institution of Washington, Figure 1

Courtesy of Uniweld Products, Figures 5, 6B

Topaz Publications, Inc., Figures 6A and C, 14

Belchfire Corp., Figure 7, SA02

Miller Electric Mfg. Co., Figure 10, SA03

Lucifer Furnaces, Inc., Figure 11

Fluke Corporation, reproduced with permission, Figure 12

Courtesy of SPX Service Solutions, Figure 13

Tempil, Inc., Figure 16

Ameritherm Inc., Figure 17

Bolttech Mannings Inc., Figure 18

Alloy Steel North America, SA01

Section Review Answer Key

Answer	Section Reference	Objective
Section One		
1. a	1.1.0	1a
2. d	1.2.3	1b
3. c	1.3.1	1c
Section Two		
1. b	2.1.0	2a
2. c	2.2.7	2b

NCCER CURRICULA — USER UPDATE

NCCER makes every effort to keep its textbooks up-to-date and free of technical errors. We appreciate your help in this process. If you find an error, a typographical mistake, or an inaccuracy in NCCER's curricula, please fill out this form (or a photocopy), or complete the online form at **www.nccer.org/olf**. Be sure to include the exact module ID number, page number, a detailed description, and your recommended correction. Your input will be brought to the attention of the Authoring Team. Thank you for your assistance.

Instructors – If you have an idea for improving this textbook, or have found that additional materials were necessary to teach this module effectively, please let us know so that we may present your suggestions to the Authoring Team.

NCCER Product Development and Revision

13614 Progress Blvd., Alachua, FL 32615

Email: curriculum@nccer.org

Online: www.nccer.org/olf

❑ Trainee Guide ❑ Lesson Plans ❑ Exam ❑ PowerPoints Other ______________

Craft / Level: Copyright Date:

Module ID Number / Title:

Section Number(s):

Description:

Recommended Correction:

Your Name:

Address:

Email: Phone:

GMAW and FCAW – Equipment and Filler Metals

Overview

Gas metal arc welding (GMAW) and flux-cored arc welding (FCAW) are two common welding processes. This module describes GMAW and FCAW processes and basic safety practices that are required for each type. It also provides an overview of GMAW and FCAW equipment and how it is set up for use. Various types of filler metals used in GMAW and FCAW are also examined.

Module 29205

Trainees with successful module completions may be eligible for credentialing through the NCCER Registry. To learn more, go to **www.nccer.org** or contact us at **1.888.622.3720.** Our website has information on the latest product releases and training, as well as online versions of our *Cornerstone* magazine and Pearson's product catalog.

Your feedback is welcome. You may email your comments to **curriculum@nccer.org,** send general comments and inquiries to **info@nccer.org**, or fill in the User Update form at the back of this module.

This information is general in nature and intended for training purposes only. Actual performance of activities described in this manual requires compliance with all applicable operating, service, maintenance, and safety procedures under the direction of qualified personnel. References in this manual to patented or proprietary devices do not constitute a recommendation of their use.

29205 V5

29205

GMAW and FCAW – Equipment and Filler Metals

Objectives

When you have completed this module, you will be able to do the following:

1. Describe basic GMAW/FCAW processes and related safety practices.
   a. Describe basic GMAW/FCAW processes.
   b. Identify GMAW/FCAW-related safety practices.
   c. Describe the various GMAW metal transfer modes.
   d. Describe the FCAW metal transfer process.
2. Describe GMAW and FCAW equipment and explain how to prepare for welding.
   a. Identify common GMAW/FCAW welding equipment.
   b. Describe power source control considerations.
   c. Identify and describe welding cables and terminations.
   d. Identify and describe external wire feeders and their controls.
   e. Identify and describe GMAW and FCAW guns, contact tips, and nozzles.
   f. Identify various shielding gases and their related equipment.
   g. Explain how to set up welding equipment for GMAW and FCAW welding.
3. Identify various GMAW and FCAW filler metals.
   a. Identify various GMAW filler metals.
   b. Identify various FCAW filler metals.

Performance Task

Under the supervision of your instructor, you should be able to do the following:

1. Set up GMAW and FCAW equipment with appropriate shielding gases and filler metals.

Trade Terms

Axially
Composite metal-core wire electrode
Globular transfer
Inductance
Pinch effect
Pulsed transfer
Short-circuiting transfer
Spray transfer
Stickout

Industry Recognized Credentials

If you are training through an NCCER-accredited sponsor, you may be eligible for credentials from NCCER's Registry. The ID number for this module is 29205. Note that this module may have been used in other NCCER curricula and may apply to other level completions. Contact NCCER's Registry at 888.622.3720 or go to **www.nccer.org** for more information.

Contents

Topics to be presented in this module include:

Figures and Tables

Section One

1.0.0 GMAW and FCAW Welding

Objective

Describe basic GMAW/FCAW processes and related safety practices.

a. Describe basic GMAW/FCAW processes.
b. Identify GMAW/FCAW-related safety practices.
c. Describe the various GMAW metal transfer modes.
d. Describe the FCAW metal transfer process.

Performance Task

1. Set up GMAW and FCAW equipment with appropriate shielding gases and filler metals.

Trade Terms

Axially: In a straight line along an axis.

Composite metal-cored wire electrode: A filler wire electrode with a hollow core containing powdered materials, primarily metals.

Globular transfer: A GMAW process in which the electrode wire does not touch the base metal, and the filler metal is transferred randomly in the arc as large, irregular globules.

Pulsed transfer: A GMAW welding process in which the welding power is cycled from a low level (background level) to a higher level (welding level), resulting in lower average voltage and current; this is a variation of the spray transfer process.

Short-circuiting transfer: A GMAW welding process in which the electrode wire shorts against the base metal. The welding current repeatedly and rapidly melts the wire and establishes an arc, which is extinguished when the wire again touches the base metal.

Spray transfer: A GMAW welding process in which the electrode wire does not touch the base metal, and the filler metal is transferred axially (in a straight line) in the arc as fine droplets.

In order to perform gas metal arc welding (GMAW) and flux-cored arc welding (FCAW) safely and effectively, a good understanding of how both processes work is required. It is also critical to understand the safety practices that are necessary for all welding operations.

1.1.0 GMAW and FCAW Processes

Gas metal arc welding (GMAW), shown in *Figure 1*, uses an electric arc to melt a consumable wire electrode and fuse it with the base metal. The arc generates an intense heat of 6,000°F to 10,000°F (3,316°C to 5,538°C). The arc is stabilized and the molten filler and base metals are protected from oxidation by a shielding gas dispensed from a GMAW gun nozzle. The wire is automatically fed into the weld puddle as it melts. The process usually operates with direct current electrode positive (DCEP).

Flux-cored arc welding (FCAW), shown in *Figure 2*, uses an electric arc to melt a flux-cored wire electrode and fuse it with the base metal to form a weld. The wire is tubular, allowing a flux material to be inserted inside the electrode. A wire feeder automatically feeds the flux-cored wire as it is being consumed in the arc. The equipment used for FCAW is basically the same as that used for GMAW. The main difference between the two processes is that GMAW uses solid or composite metal-cored wire electrodes with a shielding gas, while FCAW uses a tubular flux-cored wire electrode with or without a shielding gas.

In the two basic FCAW processes, one uses an external shielding gas while the other is self-shielding. Whether a shielding gas is used is determined by the type of flux-cored wire that is used. When the flux-cored wire does not require an external shielding gas, the process is called self-shielding FCAW, or FCAW-S. When an external shielding gas is required, the process is called gas-shielded FCAW, or FCAW-G. Equipment specifically designed for self-shielding FCAW does not have provisions for shielding gas (gas

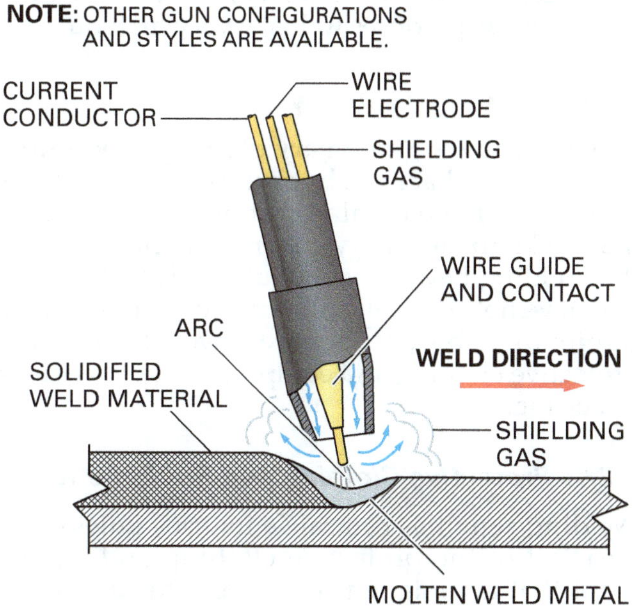

Figure 1 Basic GMAW process.

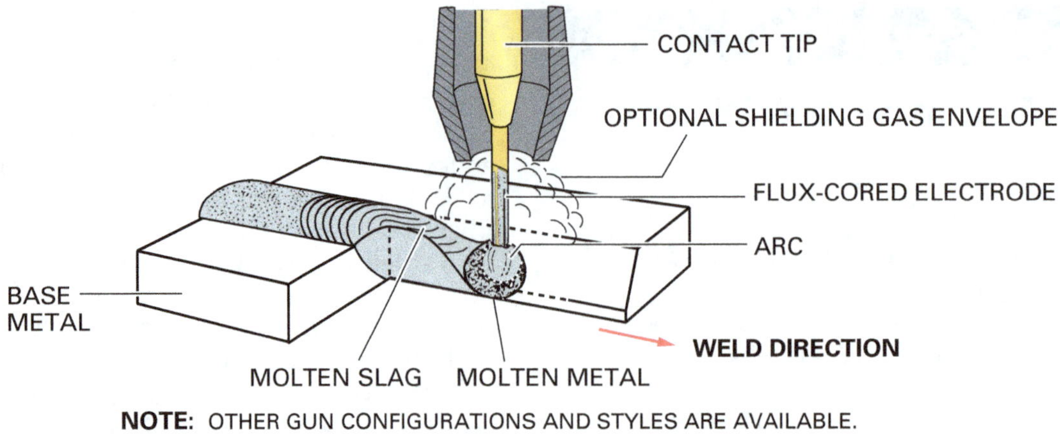

Figure 2 Basic FCAW process.

connections, solenoids, and gas preflow and postflow timers), and the guns do not have gas nozzles. Equipment used for gas-shielded FCAW welding is the same as that used for GMAW except that the wire feed drive assemblies have to be able to handle the cored wire, which is softer and generally larger in diameter than solid wire.

CAUTION

A shielding gas should not be applied to a self-shielding, flux-cored electrode. Doing so will destroy the mechanical properties of the weld. Even though the weld may actually run smoother, it will not be a good weld.

FCAW is presently limited to the welding of ferrous metals. These include low- and medium-carbon steels, some low-alloy steels, cast irons, and some stainless steels. When critical welds are required, the gas-shielded FCAW-G process is often used because the combination of flux and separate shielding gas produces welds of very high quality.

1.2.0 Safety Summary

The following is a summary of safety procedures and practices that must be observed while cutting or welding. Keep in mind that this is only a summary. Complete safety coverage is provided in the Welding Level One module, *Welding Safety*. If you have not completed that module, do so before continuing. Above all, be sure to wear appropriate protective clothing and equipment when welding or cutting.

1.2.1 Protective Clothing and Equipment

Welding activities can cause injuries unless you wear all of the protective clothing and equipment that is designed specifically for the welding industry. The following information includes important safety guidelines about protective clothing and equipment:

- Always use safety glasses with a full face shield or a welding helmet. The glasses, face shield, or helmet lens must have the proper light-reducing shade for the type of welding or cutting to be performed. Never directly or indirectly view an electric arc without using a properly tinted lens.
- Wear proper protective leather and/or flame-retardant clothing along with welding gloves that will protect you from flying sparks, molten metal, and heat (*Figure 3*).

Figure 3 A fully protected welder wearing leathers, gloves, and hood.

- Wear high-top safety shoes or boots. Make sure that the tongue and lace area of the footwear will be covered by a pant leg. If the tongue and lace area is exposed or the footwear must be protected from burn marks, wear leather spats under the pants or chaps over the front and top of the footwear.
- Wear a solid material hat, preferably 100-percent cotton, with or without a bill. Avoid hats with a fabric-covered metal button on the top. The bill can be pointed to the rear or over one shoulder to help keep sparks from going down inside the shirt when overhead cutting or welding is required. If overhead welding or welding in awkward positions is required, a full leather hood with a welding face plate and the correctly tinted lens is recommended. If a hard hat is required, use one that allows the attachment of rear deflector material and a face shield.
- Wear earmuffs or at least earplugs to protect ear canals from sparks.

1.2.2 Fire/Explosion Prevention

Welding activities usually involve the use of fire or extreme heat to melt metal. Whenever fire is used, it must be controlled and contained. Welding or cutting activities are often performed on vessels that may once have contained flammable or explosive materials. Residues from those materials can catch fire or explode when a welder begins work on such a vessel. The following are fire and explosion prevention guidelines associated with welding:

- Never carry matches or gas-filled lighters in your pockets. Sparks can cause the matches to ignite or the lighter to explode, causing serious injury.
- Never use oxygen to blow dust or dirt from clothing. The oxygen can remain trapped in the fabric for some time. If a spark hits the clothing during this time, the clothing can burn rapidly and violently out of control.
- Make sure that any flammable material in the work area is moved or shielded by a fire-resistant covering. Approved fire extinguishers must be available before attempting any heating, welding, or cutting operations. If a hot-work permit and a fire watch are required, be sure those items are in place before beginning and ensure that all site requirements have been observed.
- Never release a large amount of oxygen or use oxygen in place of compressed air. Its presence around flammable materials or sparks can cause rapid and uncontrolled combustion. Keep oxygen away from oil, grease, and other petroleum products.
- Never release a large amount of fuel gas, especially acetylene. Methane and propane tend to concentrate in and along low areas and can ignite at a considerable distance from the release point. Acetylene is lighter than air, but is even more dangerous than propane. When mixed with air or oxygen, acetylene will explode at much lower concentrations than any other fuel.
- To prevent fires, maintain a neat and clean work area, and make sure that any metal scrap or slag is cold before disposal.
- Before cutting containers such as tanks or barrels, check to see if they have contained any explosive, hazardous, or flammable materials, including petroleum products, citrus products, or chemicals that decompose into toxic fumes when heated. As a standard practice, always clean and then fill any tanks or barrels with water, or purge them with a flow of inert gas such as nitrogen to displace any oxygen. Containers must be cleaned by steam cleaning, flushing with water, or washing with detergent until all traces of the material have been removed.

WARNING!

Welding or cutting must never be performed on drums, barrels, tanks, vessels, or other containers until they have been emptied and cleaned thoroughly, eliminating all flammable materials and all substances (such as detergents, solvents, greases, tars, or acids) that might produce flammable, toxic, or explosive vapors when heated. Clean containers only in well-ventilated areas. Vapors can accumulate during cleaning, causing explosions or injury.

Proper procedures for cutting or welding hazardous containers are described in the *American Welding Society (AWS) F4.1, Safe Practices for the Preparation of Containers and Piping for Welding and Cutting.*

1.2.3 Work Area Ventilation

Vapors and fumes tend to rise in the air from their sources. Welders often have to work above the welding area where the fumes are being created. Welding fumes can cause personal injuries. Good work area ventilation helps to remove the vapors and to protect the welder. The following is a list of work area ventilation guidelines to consider before and during welding activities:

- Make sure confined space procedures are followed before conducting any welding or cutting in the confined space.

- Never use oxygen for ventilation in confined spaces.
- Always perform cutting or welding operations in a well-ventilated area. Cutting or welding operations involving zinc or cadmium materials or coatings result in toxic fumes. For long-term cutting or welding of such materials, always wear an approved full-face, supplied-air respirator that uses breathing air supplied from outside the work area. For occasional, very short-term exposure, a high-efficiency, particulate arresting (HEPA)-rated or metal-fume filter may be used on a standard respirator.
- Make sure that confined spaces are properly ventilated for cutting or welding purposes. Use powered extraction systems when available. *Figure 4* shows a welder working with a portable extraction (vacuum) system designed to remove and filter welding fumes and vapors.

1.3.0 GMAW Metal Transfer Modes

GMAW can use any of several methods of metal transfer to conduct the filler metal to the molten weld pool. The principal modes include the following:

- Spray transfer
- Globular transfer
- Short-circuiting transfer
- Pulsed transfer

Figure 4 Welder working with a portable extraction system.

1.3.1 Spray Transfer

In the spray transfer mode (*Figure 5*), the welding voltage is set in the higher ranges to keep the end of the wire electrode away from the work and to maintain a continuous arc. The high heat of the arc melts the end of the wire electrode, and fine droplets are carried by the arc axially to the molten weld puddle on the base metal. The spray transfer deposits filler metal at a high rate and produces very little spatter. However, the weld puddle is relatively large, and the mode works best in the flat or horizontal fillet positions. Also, only certain shielding gases or gas combinations can be used with spray transfer. Shielding gases are covered later in this module.

1.3.2 Globular Transfer

In the globular transfer mode (*Figure 6*), the welding voltage is set lower than it is for spray transfer. The end of the wire electrode remains a short distance from the work to maintain a continuous arc. Because of its weight, the electrode wire tip periodically melts off in a large glob, which the

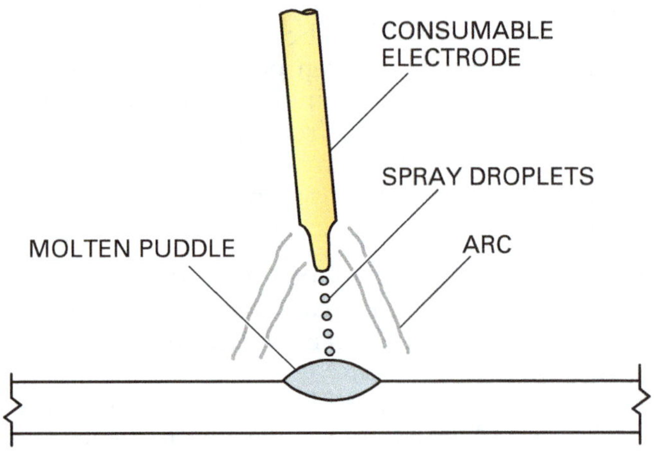

Figure 5 Spray transfer.

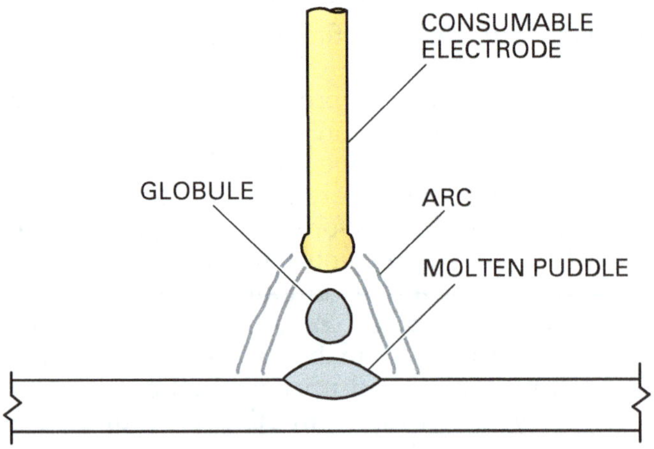

Figure 6 Globular transfer.

arc then carries to the workpiece, where it mixes with the weld puddle. Globular transfer can be used to deposit filler metal at high rates, and it is capable of producing high quality welds in all positions. However, bead appearance can be rough, and spatter may be heavy in some applications.

1.3.3 Short-Circuiting Transfer

Short-circuiting transfer (GMAW-S) uses the lowest voltage settings of the three modes (*Figure 7*). It is the best mode for welding light-gauge sheet metal. The wire electrode is fed against the workpiece. When contact is made, current flow heats the wire electrode until it melts enough for a droplet to separate and mix with the weld puddle. When the droplet separates, a gap is formed, and an arc is initiated. The continuously feeding wire closes the gap and contacts the workpiece, shorting out the arc. The current again heats the wire until another droplet separates, and the arc is again established. This process repeats many times each second. The short-circuiting process can be used to weld thin materials, and it is suitable for all positions. However, it does not penetrate deeply and may produce incomplete fusion on thicker materials because of the low heat input.

A number of manufacturers have introduced proprietary versions of short-circuiting transfer that rely on intelligent inverter power sources. These adjust the voltage and current moment-by-moment throughout each stage of the short-circuit transfer process so the droplets form and then transfer to the weld puddle optimally. The result is minimal spatter, excellent fusion, and higher-quality welds. These new technologies are especially popular in settings involving stainless steel pipe welds.

Adaptive Control Welding

Some newer inverter power sources are capable of using adaptive control welding during pulse transfer. These units sense and alter the frequency of the pulse, its width, the peak current, and wire speed automatically to maintain optimum welding characteristics during welding.

1.3.4 Pulsed Spray Transfer

Pulsed-transfer (GMAW-P) welding is a variation of spray transfer welding (*Figure 8*). The process transfers molten metal in droplets while maintaining an arc with the base metal. The pulsed GMAW process also allows the transfer action to occur while lower current levels are used. The advantages of the lower current levels are less spatter and better penetration without melting through the base metal. Pulsed transfer is especially good for aluminum welding because it results in a smaller heat-affected zone (HAZ).

In pulsed-transfer welding, the welding power is electronically controlled by a special power source that pulses the output power to the consumable electrode. Although different GMAW machines may have different types of power sources, the results are the same for each. The welding current is generated in pulses of high current at specified intervals of time. The frequency of these pulses varies from one power source to the next, but some power sources are capable of producing pulses over a wide time range. During the period of time between each pulse, the welding current is reduced, but it is still maintained at a high enough level to keep the

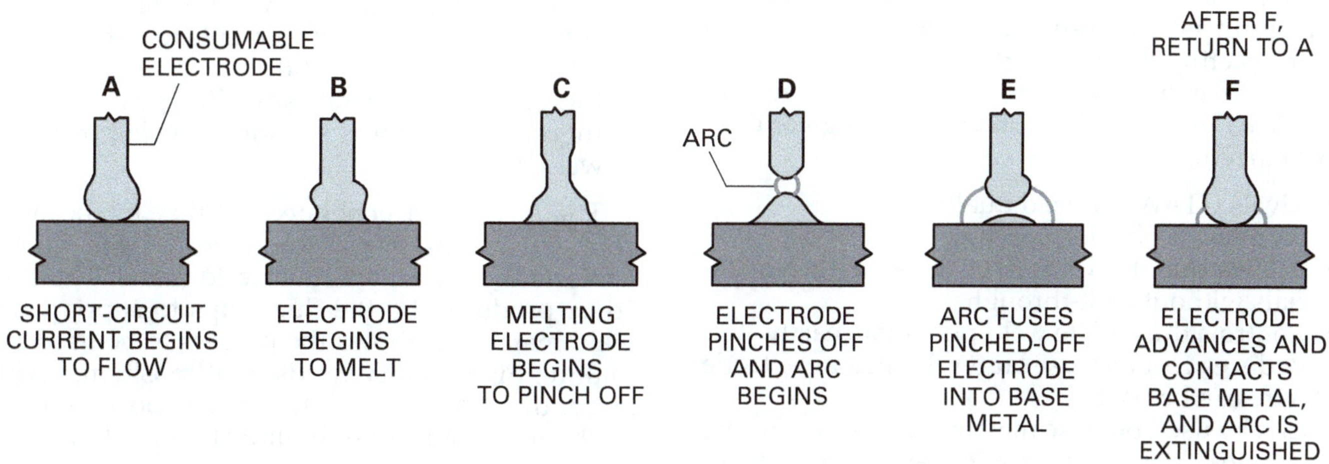

Figure 7 Short-circuiting transfer.

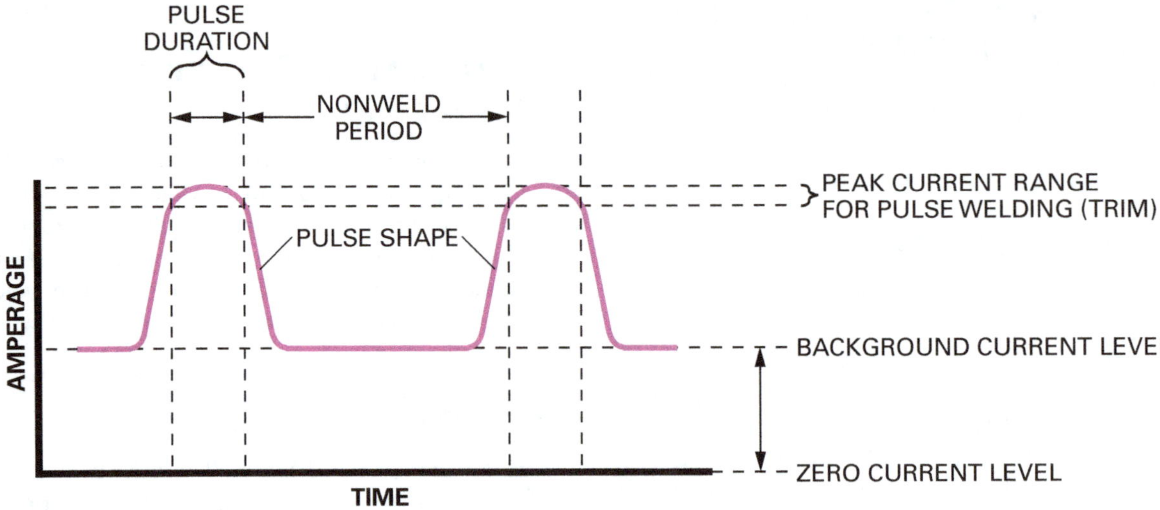

NOTE: DIFFERENT WAVEFORMS AND DURATIONS ARE USED FOR VARIOUS APPLICATIONS.

Figure 8 Output pulses of high welding current occurring over time.

wire heated and an arc established. This lowered current level is called the background current.

After an arc has been established and maintained between the consumable electrode and the base metal, the welding current to the electrode is momentarily increased to form a pulse, and then it returns to its lower background level. Each pulse of welding current must peak strongly enough and have enough of a duration to make a single drop of molten metal separate from the wire that is maintaining an arc with the base metal.

Figure 9 shows the GMAW-P process relative to one of the pulse shapes. Points A through E on the diagram correspond to the electrode arc illustrations, labeled A through E, above the diagram.

When the single drop of molten metal separates from the electrode, it is transferred across the arc to form the molten puddle on the base metal. This transfer process creates less spatter and reduces the chance of melt-through. When these pulses are repeated many times per second, the result is a high-quality weld with minimal heat affecting the welding zone.

The following is a summary of advantages of GMAW-P over welding processes using continuous currents:

- Rivals GTAW for weld quality
- Permits all-position welding
- Allows smooth welds to be made on thin materials without melt-through
- Controls thick plate weld pool more easily
- Reduces cost because larger diameter electrode wire can be used
- Can be used on base metals such as aluminum, stainless steel, copper alloy, and nickel steels, as well as carbon steels
- Reduces weld contamination from surface oxides because less surface area is exposed during the weld
- Creates a smaller heat-affected zone (HAZ), which favors its use with aluminum

1.4.0 FCAW Metal Transfer Process

The FCAW process (*Figure 10*) uses an electric arc to melt the filler and base metals and to transfer fine droplets of filler metal to the weld puddle. Globular transfer is commonly used in FCAW processes, but axial spray can be achieved if the correct electrode and shielding gas are used.

The electromagnetic force that propels the metal spray is strong, but the process produces very little spatter. The flux inside the core contains various ingredients that perform the following functions:

- Ionizers to stabilize the arc
- Deoxidizers to purge the weld of gases
- Additional metals and elements to enhance the quality of the weld metal
- Ingredients to generate shielding gas
- Ingredients to form a protective slag over the weld bead

The gas shield generated by flux in the wire core, as well as that of the external gas shield, protects the arc and molten weld metal. The slag cover produced by the flux helps to support the weld metal and allows the process to be used for making high-quality welds in all positions. With larger diameter flux-cored wires, deposition rates of about 25 pounds per hour (≈11 kg per hour) can be achieved.

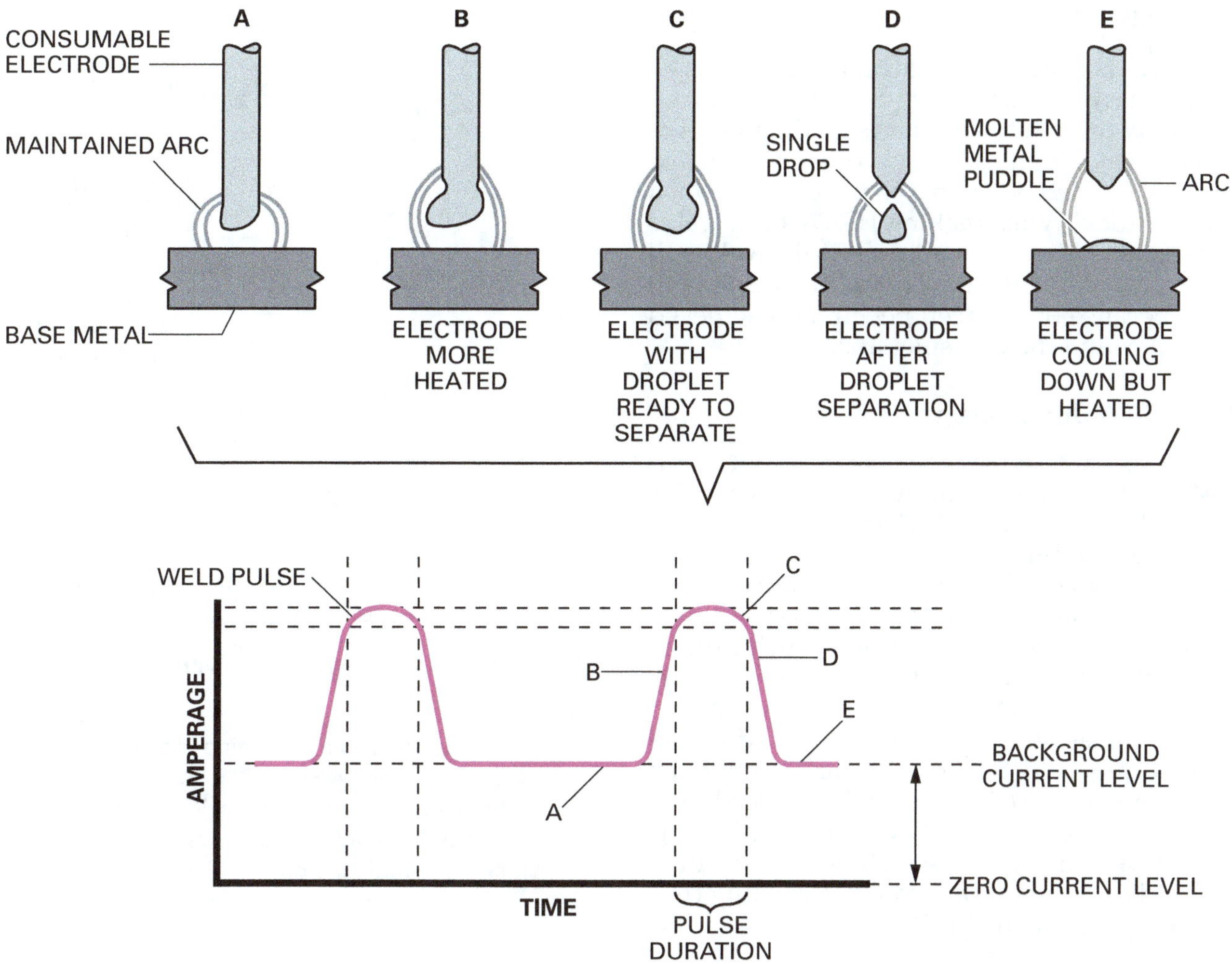

Figure 9 Typical pulsed welding process associated with one pulse shape.

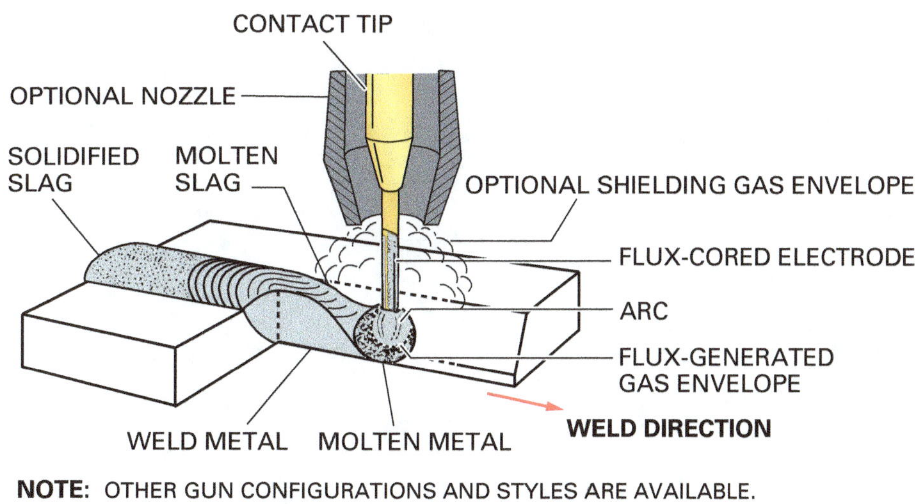

Figure 10 FCAW process.

1.4.1 FCAW Weld Penetration

FCAW welds made with carbon dioxide (CO_2) shielding have a much deeper penetration than either SMAW or FCAW without a shielding gas. For this reason, edge preparation is often not needed for double-welded butt joints on plate up to ⅜-inch (≈10 mm) thick. When making FCAW fillet welds using CO_2 shielding, a smaller size fillet weld will have as much strength as a larger fillet made with SMAW because of the deep penetration achieved with FCAW and the CO_2 shielding.

1.4.2 FCAW Joint Design

FCAW uses the same basic joint designs (*Figure 11*) as SMAW. However, when making groove welds with FCAW using a CO_2 shielding gas, the square groove (square butt) joint can be used for plate thicknesses up to ⅜ of an inch (10 mm). Above this thickness, beveled joints should be used.

Because FCAW electrode wire is much thinner than comparable SMAW electrodes and FCAW penetration with CO_2 is superior to SMAW, the included angles of V-joints can be reduced to half the angle required for SMAW. For example, an included angle of 60 degrees can be reduced to 30 degrees when using FCAW with CO_2 shielding. This can save approximately 50 percent on filler metal and a considerable amount of welding time when compared with SMAW.

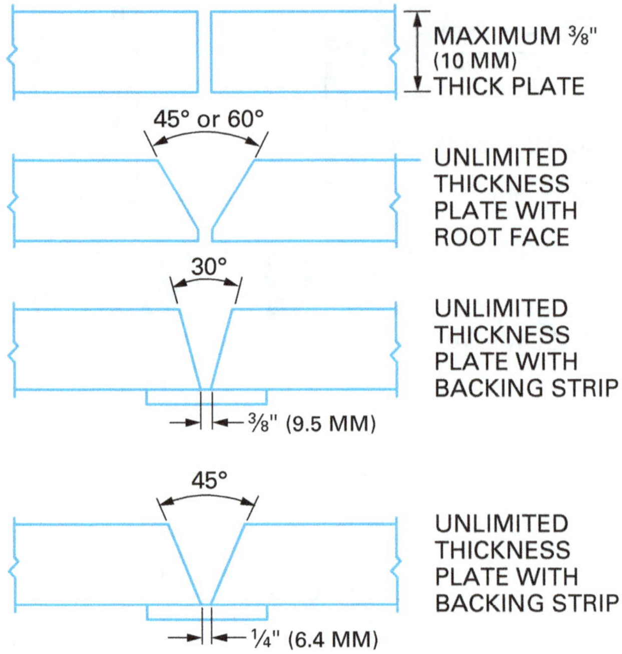

Figure 11 Joint designs for FCAW with CO_2 shielding.

CAUTION

Always refer to the Welding Procedure Specification (WPS) or site quality standards for specific information on joint requirements. Information in this module is provided as a general guideline only.

Additional Resources

AWS B1.10M/B1.10 Guide for the Nondestructive Examination of Welds. Miami, FL: American Welding Society.

AWS B1.11 Guide for the Visual Examination of Welds. Miami, FL: American Welding Society.

Modern Welding Technology. Howard B. Cary. Englewood Cliffs, NJ: Prentice Hall, Inc.

OSHA 1926.351, Arc Welding and Cutting. Current Edition. Washington, DC: Occupational Safety and Health Administration (OSHA).

Welding Handbook. Volume 1. Welding Science & Technology. Miami, FL: American Welding Society, 2001.

Welding Handbook. Volume 2. Part 1: Welding Processes. Miami, FL: American Welding Society, 2004.

Lincoln Electric website: **http://www.lincolnelectric.com** offers sources for products and training.

1.0.0 Section Review

1. Flux-cored arc welding (FCAW) uses a tubular flux-cored wire electrode with or without a(n) ____.
   a. oxidizer tip
   b. flux insulator
   c. shielding gas
   d. pulse generator

2. One good way to minimize fire hazards and vent a confined space before a cutting or welding procedure is to ____.
   a. spray the entire space with water
   b. use a powered extraction system
   c. blow compressed oxygen into the area
   d. purge the area with methane

3. The GMAW metal transfer method that uses the lowest voltage settings is the ____.
   a. globular-transfer method
   b. pulsed-transfer method
   c. spray-transfer method
   d. short-circuiting transfer method

4. In an FCAW metal transfer process, the flux inside the core of the electrode contains ionizers that are used to ____.
   a. purge the weld of gases
   b. stabilize the electric arc
   c. generate a shielding gas
   d. form a protective slag

Section Two

2.0.0 GMAW and FCAW Equipment

Objective

Describe GMAW and FCAW equipment and explain how to prepare for welding.

a. Identify common GMAW/FCAW welding equipment.
b. Describe power source control considerations.
c. Identify and describe welding cables and terminations.
d. Identify and describe external wire feeders and their controls.
e. Identify and describe GMAW and FCAW guns, contact tips, and nozzles.
f. Identify various shielding gases and their related equipment.
g. Explain how to set up welding equipment for GMAW and FCAW welding.

Performance Task

1. Set up GMAW and FCAW equipment with appropriate shielding gases and filler metals.

Trade Terms

Inductance: An electrical circuit property, adjustable on some FCAW power supplies, that influences the pinch effect.

Pinch effect: The tendency of the end of the wire electrode to melt into a droplet during short-circuit GMAW welding.

Stickout: The length of electrode that projects from the contact tube, measured from the arc tip to the contact tube.

Typical GMAW and FCAW systems (*Figure 12*) are made of four interconnected pieces of equipment, including the following:

- DC power source (welding machine)
- Wire feeder
- Welding gun
- Shielding gas supply (if required)

In some systems, equipment functions may be integrated or combined. For example, the wire feeder can be built into the same cabinet as the power source, or it may be incorporated into the welding gun. Welding guns are shown in a later section of this module.

Most systems have a separate wire feeder (push system) that pushes the wire from the feeder through the gun cable and gun. Other systems may incorporate the wire feeder into the welding gun itself (pull system) or do both (push-pull system). There are even systems that have synchronized wire feeders located at intervals along very long feeder conduits. Various types of wire feeders are covered in a later section. Heavy duty, water-cooled guns may also be equipped with a closed-loop water-cooling system. Systems that are used for FCAW-S do not require gas cylinders, since the shielding gas comes from the flux-core electrode.

2.1.0 Types of Welding Power Sources

Direct current (DC) power sources, or welding machines, are usually designed to produce either constant-current (CC) or constant-voltage (CV) DC power. A constant-current machine produces a constant-current output over a wide voltage range; this is typical of a SMAW welding machine. A constant-voltage machine maintains a constant voltage as the output current varies; this is the typical power source used in GMAW or FCAW welding machines. The output voltages and output currents of a welding machine can be plotted on a graph to form a curve. These curves show how the output voltage relates to the output current as either changes (*Figure 13*).

Different types of welding machines (power sources) are available for GMAW, including the following:

- Transformer-rectifier power sources
- Engine-driven generator power sources
- Inverter power sources

Constant Current/ Constant Voltage Power Sources

Transformer-rectifiers can be designed to produce constant-current or constant-voltage DC. Some will produce either variable or constant voltage by setting a switch. These machines are sometimes referred to as multiprocess power sources, since they can be used for any welding process, including GMAW and FCAW.

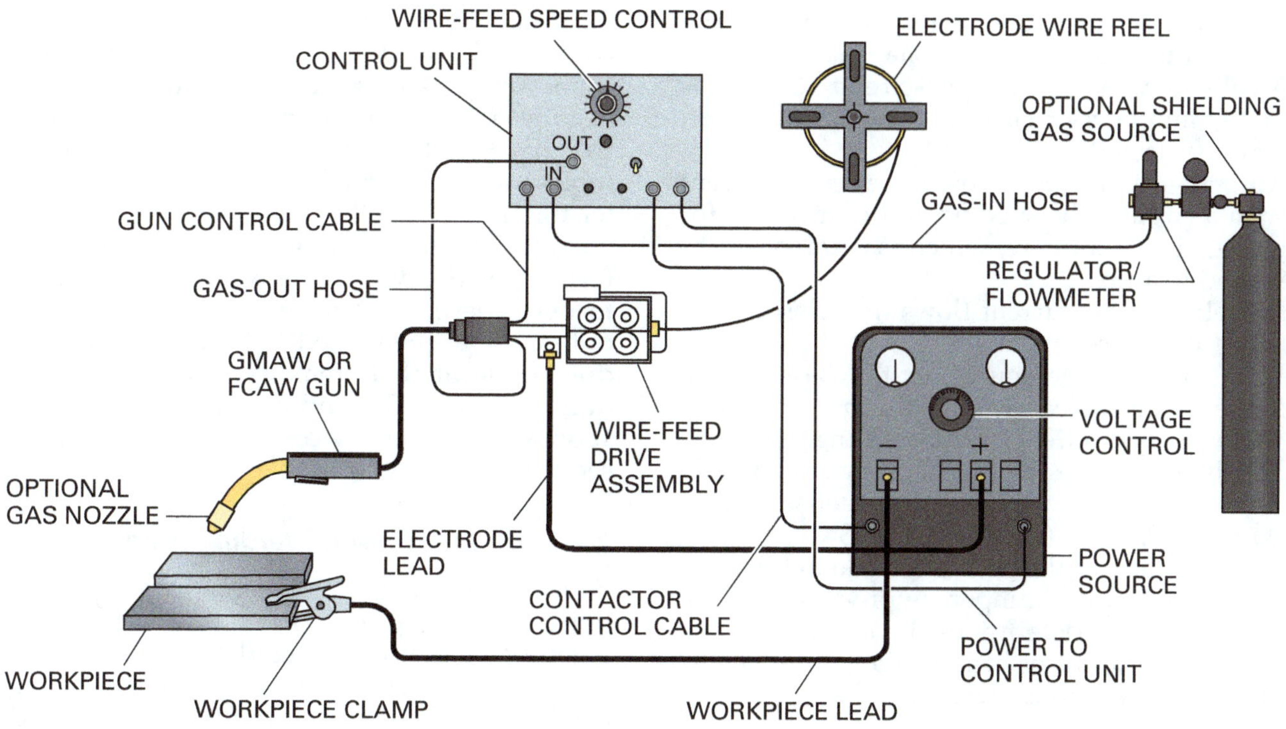

NOTE: The polarity of the gun and workpiece leads is determined by the type of filler metal and application.

Figure 12 Simplified GMAW/FCAW system diagram.

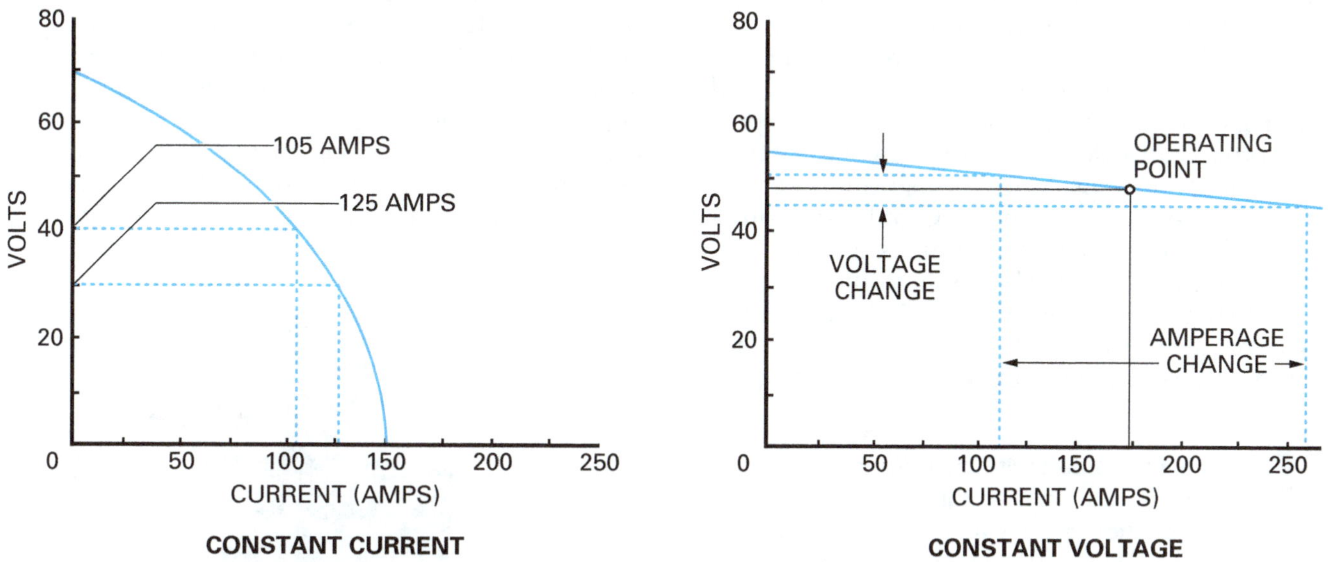

Figure 13 Constant-current and constant-voltage output curves.

2.1.1 Characteristics of Welding Current

The current produced by a welding machine has different characteristics from the current flowing through utility power lines. Welding current has low voltage and high amperage, while the power line current has high voltage and low amperage.

Voltage is the measure of the electromotive force or pressure that causes current to flow in a circuit. The specific unit of measure is the volt.

There are two types of voltage associated with welding current: open-circuit voltage and operating voltage. Open-circuit voltage is the voltage present when the machine is on but no welding is being done. There are usually ranges of open-circuit voltages that can be selected, up to about 40 volts. Operating voltage, or arc voltage, is the voltage after the arc has been struck.

Arc voltage is generally slightly lower than the open-circuit voltage. The arc voltage is typically 2 to 3 volts lower than the open-circuit voltage for each 100 amperes of current, but it depends on the range selected.

Amperage is the measure for electric current flow in a circuit. The specific unit of measure for amperage is the ampere, which is commonly abbreviated to amp.

In welding, the current flows in a closed loop through two welding cables; the workpiece lead connects the power source to the base metal, and the other cable connects the power source to the wire electrode. During welding, an arc is established between the end of the wire electrode and the work. The arc generates intense heat at 6,000°F to 10,000°F (3,316°C to 5,538°C), melting the base metal and the wire electrode to form the weld. The amount of amperage produced by the welding machine determines the intensity of the arc and the amount of heat available to melt the work and the electrode.

2.1.2 Transformer-Rectifier Power Sources

The transformer-rectifier power source uses a transformer to convert the primary current to welding current and a rectifier to change the current from AC to DC. Transformer-rectifier power sources can be designed to produce either AC and DC, or DC only. Transformer-rectifiers that produce both AC and DC are usually lighter duty than those that produce DC only. Transformer-rectifier power sources that produce DC only are sometimes called rectifier power sources or just rectifiers. Depending on the size, transformer-rectifier power sources may require 120/240-volt single-phase input power, 208/230/240-volt three-phase input power, or 480-volt three-phase input power.

WARNING!

Coming into contact with the primary voltage of a welding machine can cause major shock or electrocution. Make sure that all welding machines are properly grounded to prevent injury.

Transformer-rectifiers used for GMAW and FCAW have an On/Off switch and a voltage control. If the machine has a CC/CV or mode switch, it also has an amperage control for SMAW or GTAW. When the switch is set to CV for either GMAW or FCAW, the amperage control is disabled or becomes the voltage control.

The welding cables (electrode cable and workpiece cable) are connected to terminals marked Electrode and Ground, or Positive (+) and Negative (–). Welding machines often have selector switches to select direct current electrode negative (DCEN) or direct current electrode positive (DCEP). If there is no selector switch, the cables must be manually changed on the machine terminals to select the type of current desired.

Figure 14 shows a typical industrial transformer-rectifier CC/CV power source with a wire feeder. These units usually have a 100 percent duty cycle at their rated output and a 60 percent duty cycle at their maximum output. They require a wire feeder and a GMAW/FCAW gun or spool gun. These devices are described in a later section.

2.1.3 Engine-Driven Welding Machines

Welding machines can also be powered by gasoline or diesel engines. The engine is connected to a generator that produces the power needed for the welding machine. The size and type of engine used depends on the requirements of the welding machine.

To produce sufficient electrical power, the engine generator must turn at a required number of revolutions per minute (rpm). The engines of power generators have governors to control the engine speed. Most governors have a welding speed switch. The switch can be set to idle the engine when no welding is taking place. When the electrode is touched to the base metal, the governor automatically increases the speed of the engine to the required speed for welding. If no

Figure 14 Industrial transformer-rectifier CC/CV power source with wire feeder.

welding takes place for about 15 seconds, the engine will automatically return to idle. The switch can also be set for the engine to run continuously at the welding speed.

Figure 15 shows a diesel engine-driven welding machine. This particular unit can produce CC/AC or DC and CV/DC outputs up to 400 amps at a 100 percent duty cycle for SMAW, GTAW, GMAW, and FCAW. For GMAW and FCAW, it is available with a portable wire feeder that provides a constant wire-feed speed for use with CV power sources. When used with CV power sources, the wire feeder provides wire as needed when it senses that an arc is struck. The wire feeder is shown with an FCAW-S gun, but other guns can also be used.

Engine-driven units often have an auxiliary power unit that produces 120 volts alternating current (VAC) for lighting, power tools, and other electrical equipment. When 120 VAC is required, the engine-driven generator or alternator must run continuously at its normal welding speed.

Engine-driven units have both engine controls and welding machine controls. The engine controls vary with the type and size of the unit but normally include the following:

- Starter
- Voltage meter
- Temperature gauge
- Fuel gauge
- Hour meter

Engine-driven welding machines have a voltage/amperage control. They may have a polarity or mode selector switch, or the welding cables at the welding current terminals might have to be changed to change the polarity. If the machine has a CC/CV or mode selector switch, it also has an amperage control. When the switch is set to CV or GMAW/FCAW mode, the amperage control is disabled and becomes a voltage control.

Figure 15 Diesel engine-driven welding machine with wire feeder and FCAW-S gun.

The advantage of engine-driven generators is that they are portable and can be used in the field where electricity is not available for other types of welding machines. The disadvantage is that engine-driven generators are more costly to purchase, operate, and maintain.

2.1.4 Inverter Power Sources

Inverter welding machines (*Figure 16*) increase the frequency of the incoming primary power. This provides a smaller, lighter power source with a faster response time and more control for pulse welding. Inverter welding machines are used where space is limited and portability is important. The controls on these welding machines vary according to size and application. Smaller inverter machines are often mounted on lightweight carts for greater portability.

Typical industrial advanced inverter welding machines designed specifically for GMAW or FCAW operations usually have controls similar to those shown in *Figure 17*. For specific control functions for the power supply you are using, refer to the operator's manual.

A typical program sequence for one of the internal welding programs available in an advanced inverter power source is shown in *Figure 18*.

2.1.5 Welding Machine Ratings

The rating of a welding machine is determined by the amperage output of the machine at a given duty cycle. The duty cycle of a welding machine is based on a 10-minute period and specified by a percentage. It is the percentage of 10 minutes during which the machine can continuously produce its rated amperage without overheating. For example, a machine with a rated output of 300 amps at a 60-percent duty cycle can deliver 300 amps of welding current for 6 minutes out of every 10 (60 percent) without overheating.

The duty cycle of a welding machine will be 10, 20, 30, 40, 60, or 100 percent. A welding machine with a duty cycle of 10 to 40 percent is considered a light- to medium-duty machine. Most industrial, heavy-duty machines for manual welding have 60 to 100 percent duty-cycle ratings. Machines designed for automatic welding operations have 100 percent duty-cycle ratings.

With the exception of 100 percent duty-cycle machines, the maximum amperage that a welding machine can produce is always higher than its rated capacity. A welding machine rated at 300 amps and a 60-percent duty cycle will generally put out a maximum of 375 to 400 amps. But since the duty cycle is a function of its rated capacity, the duty cycle will decrease as the amperage is raised over 300 amps. Welding at 375 amps with a welding machine rated 300 amps at 60-percent duty cycle will lower the duty cycle to about 30 percent. If welding continues for more than 3 out of 10 minutes, the machine will overheat. Most welding machines have a heat-activated circuit breaker that shuts off the machine automatically when it overheats. Leave the machine plugged in and turned on to allow the cooling fans to continue running when this occurs.

Figure 16 Inverter welding machine.

On the other hand, if the amperage is set below the rated amperage, the duty cycle increases. Setting the amperage at 200 amps for a welding machine rated at 300 amps and a 60-percent duty cycle will increase the duty cycle to 100 percent. *Figure 19* shows the relationship between amperage and duty cycle.

2.2.0 Power Source Control

As described in earlier sections, GMAW and FCAW are best performed with a constant-voltage

GMAW/FCAW Feeders for Engine-Driven Welding Machines

If possible, select wire feeders or spool guns that operate from 24VDC battery power and arc voltage. If AC-powered feeder or spool gun controls are used, the power source will have to operate in the continuous run mode to supply AC power to operate the feeder or spool gun.

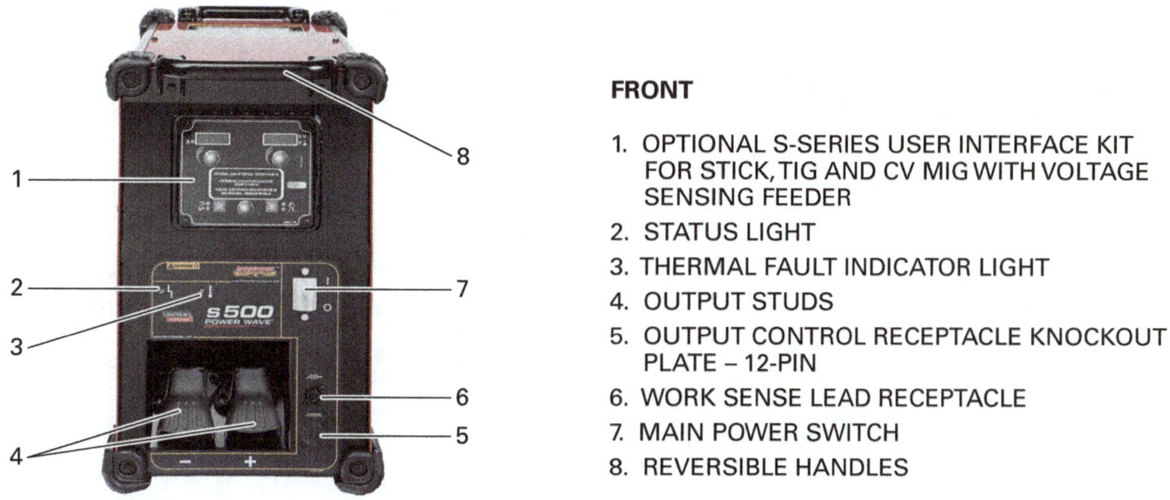

Figure 17 Typical industrial inverter unit controls.

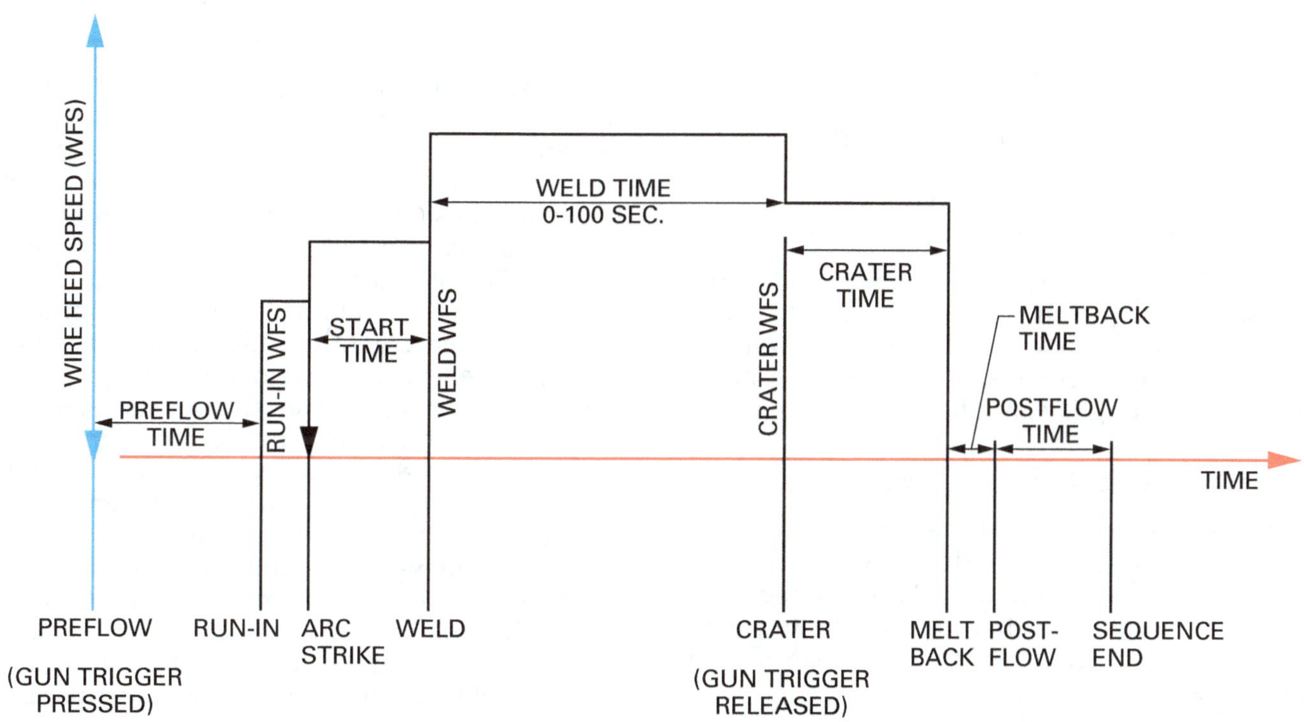

Figure 18 Typical welding program sequence.

or constant-voltage/constant-current (CV/CC) DC welding machine. In constant-voltage welding machines, the open-circuit voltage and the welding voltage are nearly the same. Constant-current variable-voltage DC welding machines, the type used for SMAW in which the output voltage varies between high open-circuit voltage and lower arc voltage, are sometimes used. However, they require specially designed voltage-sensing wire feeders. The use of CC machines for GMAW and FCAW is prohibited by some codes. CV/CC welding machines are required for GMAW-P operations.

Welding machines usually contain controls for adjusting arc voltage, slope, and inductance, and they have meters to monitor amperage and voltage. The simplest machines are power sources with built-in wire feeders. A simple combination welder and wire feeder may have only a wire-feed speed control and a voltage control.

2.2.1 Slope

The constant-voltage power sources used for GMAW and FCAW are not precisely constant in their voltage output, because the voltage always drops a little as the current increases. This voltage-to-amperage relationship forms a slight curve when plotted as a graph. The general angle

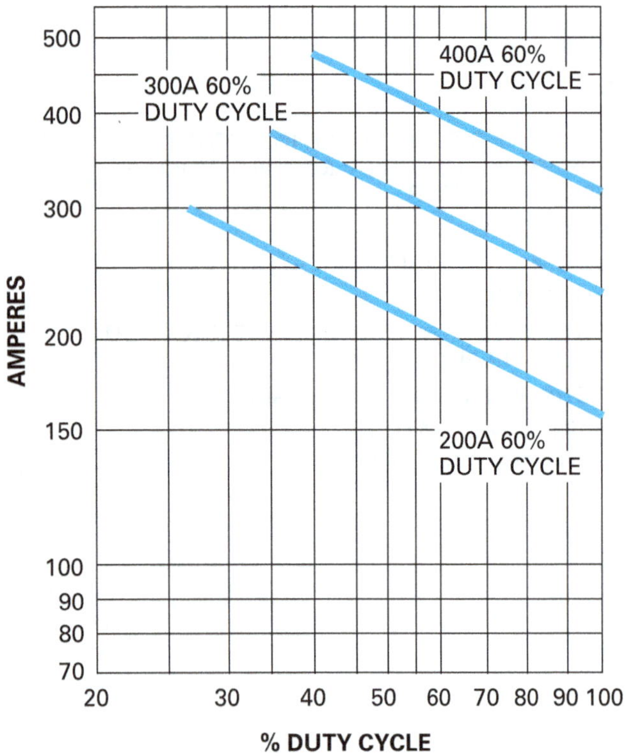

Figure 19 Relationship between amperage and duty cycle.

of the volt-ampere curve is known as the slope. This slope is adjustable on some units. A flatter (more horizontal) slope is better for spray-transfer welding, while a steeper slope is better for short-circuiting welding (*Figure 20*).

2.2.2 Pinch Effect and Inductance

When using short-circuiting welding in GMAW, the current passing through the wire electrode heats and then melts the wire. This is sometimes called the pinch effect because the molten wire appears to be pinched off, forming a droplet of molten metal (*Figure 21*).

An electrical circuit property called inductance influences the pinch effect. Low inductance produces high pinch effect. Conversely, high inductance produces low pinch effect. With low circuit inductance, the current rise is very fast, and the high pinch effect can cause the wire droplet to explode or spatter. When the circuit inductance is high, the current rises more slowly, decreasing the number of short circuits per second and extending the arc duration. The pinch effect forms the droplet more gently, which results in a more fluid puddle, smoother weld, and less spatter.

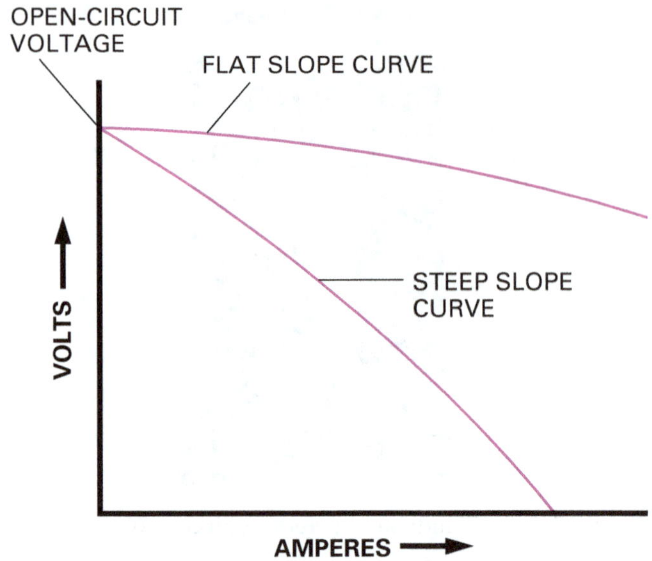

Figure 20 Volt-ampere curve slope profiles for GMAW.

Constant-voltage power sources are self-regulating and automatically produce current as it is required. Inductance controls the rate at which the current output rises. Some inductance is desirable in FCAW because it prevents explosive arc starts by slowing down the rate of current rise. Many FCAW power sources have a control to adjust the inductance. In general, lower thermal conductivity materials use a higher inductance, and higher thermal conductivity materials use a lower inductance.

All-in-One Welding Machines

Portable, light industrial GMAW/FCAW welding machines are also available. The MIG unit shown is equipped with an internal wire feeder and has dual voltage inputs of 120 or 240VAC. With a 120VAC input, it is rated at 100 amps with a 40 percent duty cycle. With a 240V input, it is rated at 200 amps with a 25 percent duty cycle. This machine can also be used to perform TIG welding.

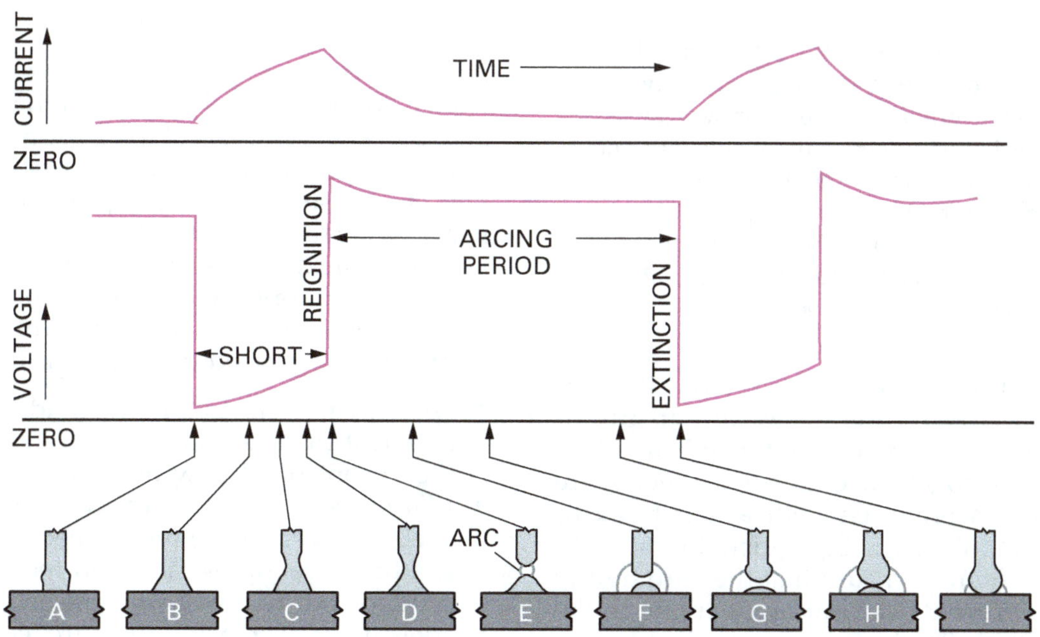

Figure 21 Pinch effect.

2.2.3 Arc Blow

When operating in GMAW spray-transfer mode or in FCAW, arc blow can sometimes be a problem. Arc blow is the deflection of the arc from its normal course because of the arc magnetic field's attraction to or repulsion from the weld current's magnetic field in the base metal. The amount and direction of arc deflection depends upon the relative position, direction, and density of the base metal's magnetic field. Arc blow can result in excess spatter and weld defects. It can be minimized by relocating the workpiece clamp or changing the weld angle of the gun.

2.3.0 Welding Cable

Cables used to carry welding current are designed for maximum strength and flexibility. The conductors inside the cable are made of fine strands of copper wire. The copper strands are covered with layers of rubber reinforced with nylon or Dacron® cord (*Figure 22*).

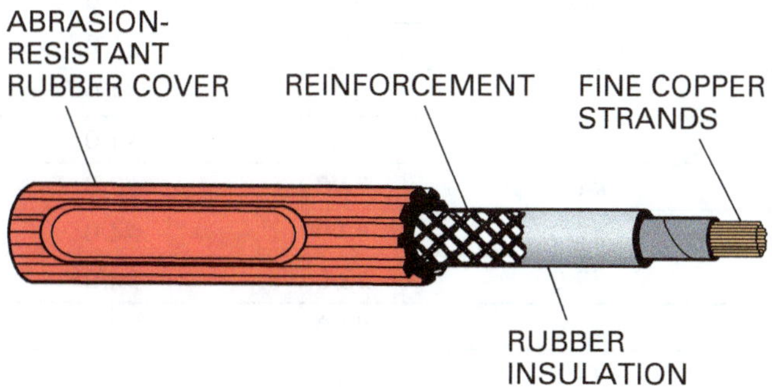

Figure 22 Cut-away section of welding cable.

The size of a welding cable is based on the number of copper strands it contains. Large-diameter cable has more copper strands and can carry more welding current. Typical welding cable sizes range from #4 to #3/0 (spoken as "3 aught"), as shown in *Table 1*.

When selecting the welding cable size, the current flow (amperage) as well as the distance the current will travel must be considered. The longer the distance the current has to travel, the larger the cable must be. Cable that is too small will become hot, and the voltage provided under load will drop due to the cable's electrical resistance. Excessive voltage drop may affect the performance of the welding machine. When selecting welding cable, use the rated capacity of the welding machine for the cable amperage requirement. For the distance, measure both the electrode and the workpiece leads, and add the two lengths together. To identify the welding cable size required, refer to a recommended welding cable size table, furnished by most welding cable manufacturers.

2.3.1 Welding Cable End Connections

Welding cables must be equipped with the proper end connections or terminals.

Lugs are used at the end of the welding cable to connect the cable to the welding machine terminals. The lugs come in various sizes to match the welding cable size, and are mechanically secured to the welding cable.

CAUTION

If the end connection is not tightly secured to the cable, the connection will overheat and oxidize. An overheated connection will cause variations in the welding current and permanent damage to the connector and/or cable. Check connections for tightness and repair loose or overheated connections. If quick-disconnects are used, they must be properly sized to the cable or overheating may occur.

Quick-disconnects are also mechanically connected to the cable ends. They are insulated and serve as cable extensions for splicing two lengths of cable together. Quick-disconnects are connected or disconnected with a half twist or similar motion. When using quick-disconnects, care must be taken to ensure that they are tightly connected to prevent overheating or arcing in the connector. *Figure 23* shows some typical quick-disconnects and lugs used as welding cable connectors.

The workpiece clamp (*Figure 24*) provides the connection between the end of the workpiece lead and the workpiece. Workpiece clamps are mechanically connected to the welding cable and come in a variety of shapes and sizes. The size of a workpiece clamp is the rated amperage that it can carry without overheating. Workpiece clamps must at least have the same rating as the rated capacity of the power source on which they will be used.

2.4.0 External Wire Feeders

An external wire feeder pulls the electrode wire from a spool and pushes it through the gun cable and the gun. It consists of a wire spool holder with a drag brake, an electric motor that drives either one or two sets of opposing slotted rollers, and various controls. The wire feeder also usually contains the shielding gas connections, the gas control solenoid, and (optionally) the cooling-water flow-control solenoid. Some of the controls that may be located on a wire feeder include the following:

- *Spool drag* – This is a manual adjustment used to adjust the drag on the wire spool to prevent the uncontrolled unwinding of the wire.
- *Wire feed speed control* – This is a variable control used to set the wire feed speed. Since GMAW welding machines are self-regulating (they produce the required current on demand), increasing or decreasing the wire feed speed will cause the welding machine to automatically increase or decrease its current output.
- *Voltage control* – This control is used to set the welding voltage.
- *Wire jog* – The manual wire jog switch is used to start and stop the wire feeder without energizing the welding current.
- *Wire meltback timer (burnback)* – This is an adjustable time delay that causes the welding current to continue after the trigger has been released. This consumes the length of wire that continues to feed after the trigger has been released so that the proper stickout is maintained to start the next weld.

Table 1 Welding Cable Sizes

Machine Size in Amperes	Duty Cycle (%)	Copper Cable Sizes for Combined Lengths of Electrodes Plus Workpiece Lead				
		Up to 50 feet (15 m)	50–100 feet (15–30 m)	100–150 feet (30–46 m)	150–200 feet (46–61 m)	200–250 feet (61–76 m)
100	20	#8	#4	#3	#2	#1
180	20	#5	#4	#3	#2	#1
180	30	#4	#4	#3	#2	#1
200	50	#3	#3	#2	#1	#1/0
200	60	#2	#2	#2	#1	#1/0
225	20	#4	#3	#2	#1	#1/0
250	30	#3	#3	#2	#1	#1/0
300	60	#1/0	#1/0	#1/0	#2/0	#3/0
400	60	#2/0	#2/0	#2/0	#3/0	#4/0
500	60	#2/0	#2/0	#3/0	#3/0	#4/0
600	60	#3/0	#3/0	#3/0	#4/0	* * *
650	60	#3/0	#3/0	#4/0	* *	* * *

** Use double strand of #2/0
*** Use double strand of #3/0

Figure 23 Quick-disconnects, lugs, and DINSE connectors.

- *Gas purge* – The manual gas purge switch is used to operate the gas solenoid in order to purge the shielding gas system prior to actual welding. It is also used to adjust the flowmeter.
- *Trigger hold* – This switch enables or disables a trigger hold function activated by the gun trigger. It enables welding without holding the trigger.
- *Gas preflow timer* – This is an adjustable time delay that prevents the welding current contactor from closing until the shielding gas is flowing.
- *Gas postflow timer* – This is an adjustable time delay that is used to keep shielding gas flowing for a short period after the gun trigger has been released.

Figure 25 shows some basic wire feeder units, while *Figure 26* shows other types of wire feeders.

Two critical properties of electrode wire must be considered before selecting the wire to be used with a wire feeder. Those two properties are cast and helix (*Figure 27*). Cast and helix are determined by cutting a strand of the wire several feet long and laying it on a flat surface. The cast of the wire is the diameter of the circle formed by the strand of wire. The helix is the height, or vertical distance, of the end of the wire from the flat surface. The cast and helix of electrode wire primarily affect the wire feeder's ability to move the wire through the gun cable and gun. The cast and helix can also cause excessive wear to the contact tip and misguide the wire, causing it to be diverted away from the joint. These factors can affect the quality of the weld.

During operation, a wire feeder grips the wire between the grooves of opposing grooved rollers. There can be one or two sets of opposed rollers (*Figure 28*). The grooves can be V-shaped or U-shaped in their cross section.

V-grooves are generally used for hard wire. One roller slot in each V-groove set may be serrated for better gripping. U-groove rollers are used for soft wire or flux-cored wire, as they crush more easily. For extremely soft wire or flux-cored wire, cogged U-groove rollers are used.

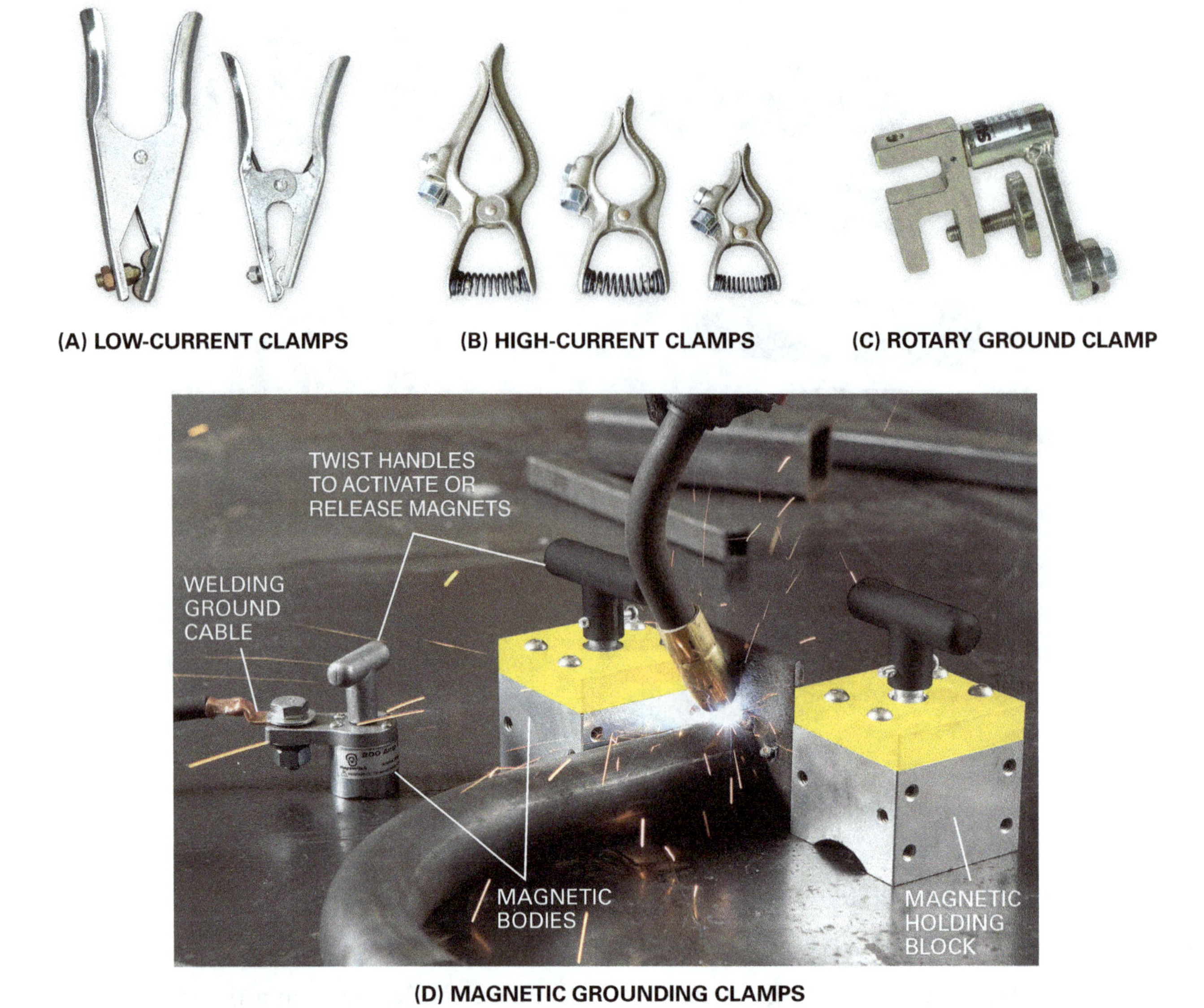

Figure 24 Workpiece clamps.

Some rollers are made in two pieces with different-size half-grooves formed on the outside edges of each half-roller. The half-rollers are assembled to fit two different ranges of wire size. *Figure 29* shows several different groove styles and combination rollers.

When dust or dirt is a problem, a felt wiper can be added to the wire between the spool and the inlet guide to the feed rollers. The felt wipers are treated with a silicone lubricant that cleans and lubricates the wire. Special liquid silicone cleaner/lubricants can occasionally be added to the felt wiper.

CAUTION

Use only special cleaner/lubricants made specifically for wire feed wipers. Use of other cleaners or lubricants may contaminate and cause porosity in the weld.

2.5.0 GMAW/FCAW-G Guns

The GMAW/FCAW-G gun (*Figure 30*) supports and guides the wire electrode and provides the electrical contact between the wire and the welding machine electrode lead. The gun also contains a remote switch to start and stop the welding current and the shielding gas flow. The tip of the gun has a nozzle to direct the shielding gas around the arc. Within the nozzle, the wire passes through a contact tip that provides the actual connection between the welding current and the electrode wire. Nozzles and contact tips are subject to wear and electrical erosion, so both components are replaceable.

The duty cycle of GMAW guns is determined and rated with CO_2. The rating is reduced from 10 to 50 percent when mixed gases are used, depending on gas mixture and welding parameters. The chosen mode of transfer (short-circuit,

Figure 25 Basic GMAW/FCAW wire feeder units.

spray, or pulsed) may further derate their duty cycle.

Some types of guns contain motor-driven wire pullers within the gun handle (*Figure 31*). These work in conjunction with a push wire feeder located with the power source. This system was developed for feeding soft wire, such as aluminum, that tends to bend easily. With the wire being both pushed and pulled, neither feeder mechanism needs to grip the wire quite as tightly.

In addition, there are spool guns designed for aluminum wire. These guns contain both a wire feeder and a small spool of wire. In some applications, an auxiliary gun control unit is required for proper operation. *Figure 32* shows one style of spool gun.

Light-duty guns are gas-cooled by the shielding gas flow. Heavy-duty guns are water-cooled. The cooling-water flow may be controlled by a solenoid in the wire feeder or by a separate dedicated cooling system. Cooling water is piped to and from the gun through flexible tubing integrated into the gun cable. Some cooling systems use domestic water to cool the gun. Others use a closed-loop system (*Figure 33*) that circulates demineralized water or special cooling fluid that will not corrode internal gun surfaces or plug the cooling passages with mineral scale.

Figure 26 Various types of wire feeders.

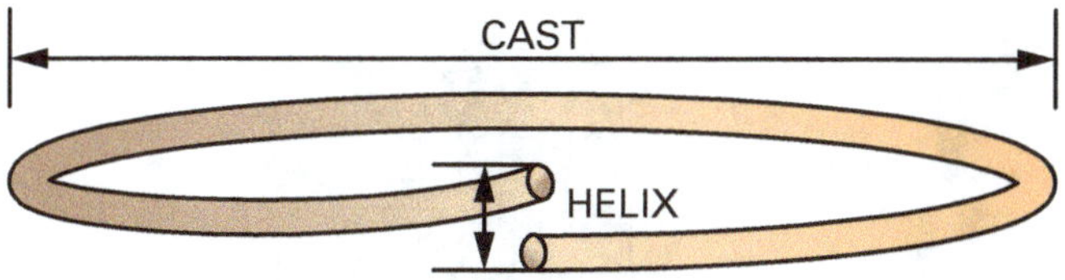

Figure 27 Cast and helix of a wire electrode.

2.5.1 FCAW-S Guns

FCAW-S guns are similar to GMAW/FCAW-G guns, but they are often designed to handle larger-diameter wire and they do not provide shielding gas flow. The FCAW gun supports and guides the wire electrode. It also contains a remote switch for starting and stopping the welding current. The wire electrode passes through the contact tip at the end of the gun, which provides the electrical connection from the welding machine electrode lead to the electrode wire.

Guns designed for self-shielded FCAW do not have provisions for delivering shielding gas. This makes the gun nozzles smaller in diameter. To protect the contact tip, these guns have a fiber insulator that threads onto the end of the gun nozzle.

FCAW-S generates smoke, which can be removed at the gun tip with a special smoke extraction nozzle. The nozzle is connected to a vacuum pump or blower by a flexible tube that draws off smoke, which improves visibility.

2.5.2 Contact Tips and Nozzles

Contact tips and nozzles are a vital part of effective welding. The contact tip provides the connection between the welding current and the electrode wire. The nozzle directs shielding gas around the arc. Both of these components are

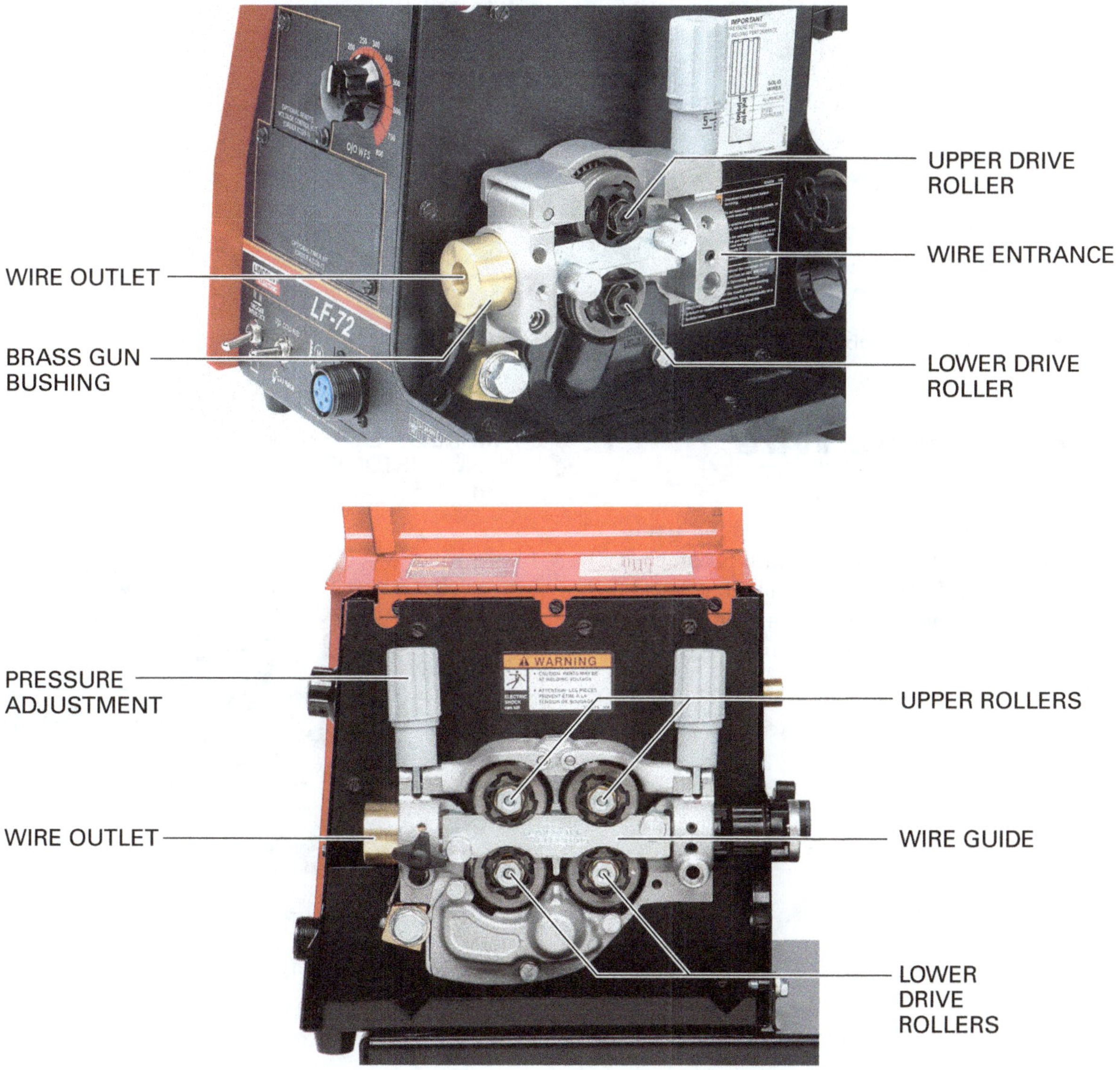

Figure 28 Wire feeder drive roll systems.

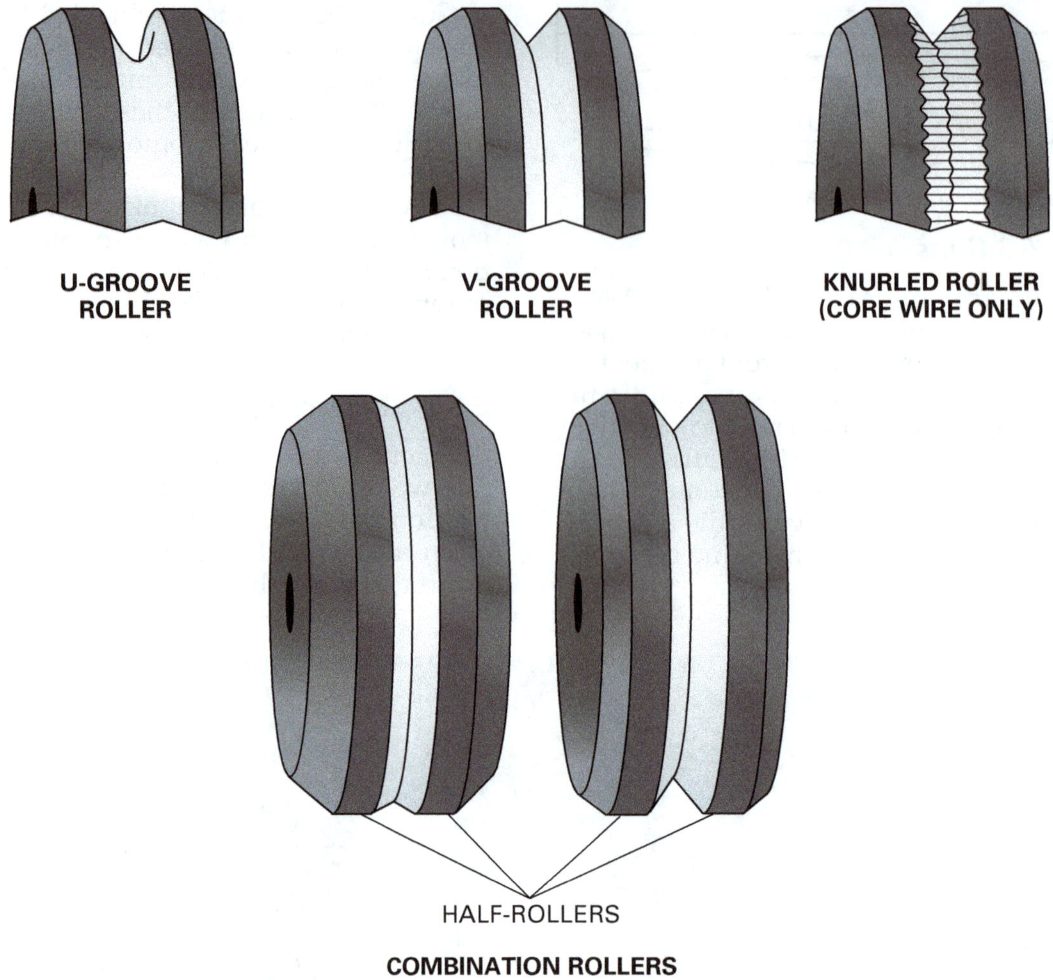

Figure 29 Groove styles and combination rollers.

Programmable Wire Feeders

Programmable wire feeders are available for many power sources. Custom welding programs can be created, or preset welding programs can be stored in the feeder or in the power source. The program can be activated by the various controls and displays on the feeder.

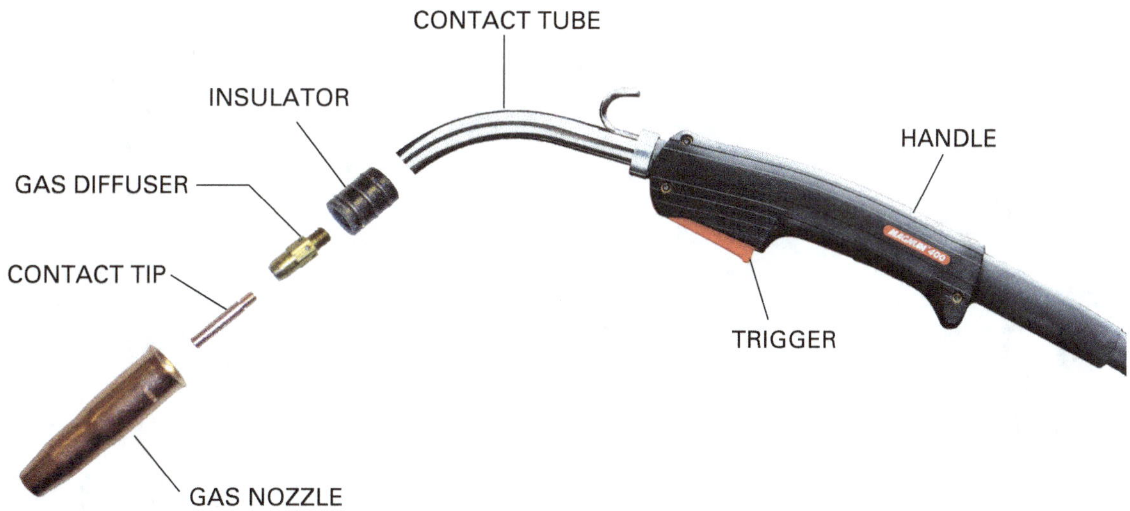

Figure 30 Construction of a typical GMAW/FCAW-G gun.

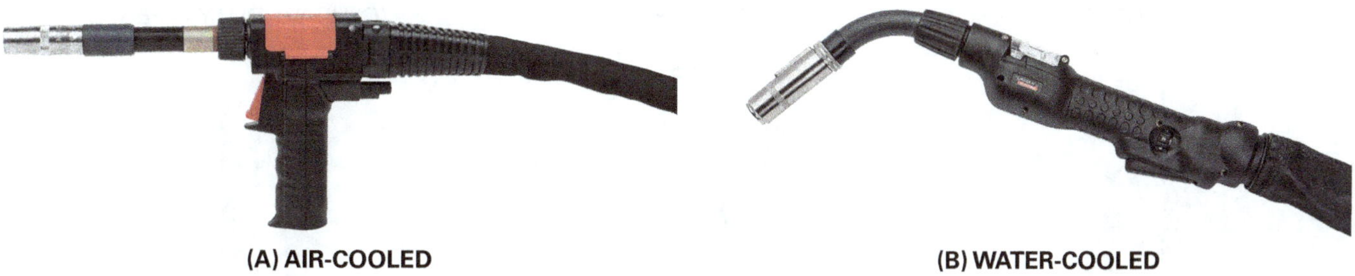

Figure 31 Pull guns for push-pull wire feeders.

Figure 32 Spool gun.

Figure 33 Typical closed-loop cooling unit.

available in many different shapes and sizes (*Figure 34*), depending on the application requirements. For that reason, a number of factors must be taken into consideration when selecting contact tips and nozzles. The chosen tip and nozzle must be compatible with the equipment in use.

A contact tip should always correspond to the wire electrode size. For example, if a wire electrode with a diameter of 0.125 inches (3.2 mm) is being used, the contact tip should also be 0.125 inches (3.2 mm). If a contact tip is too small for the wire being used, the wire will jam. A contact tip that is larger than necessary may work, but it is not the ideal setup. If the contact tip is too large, poor electrical contact will cause damage from arcing between the tip and the wire.

Other factors to consider when selecting a contact tip include the following:

- *Tapered vs. non-tapered* – Tapered tips are usually longer to allow for welding in areas with restricted access. Non-tapered tips have more mass and tend to last longer than tapered ones.
- *Threaded vs. non-threaded* – Threaded tips are screwed into the gas diffuser. These are the most common contact tips. Non-threaded tips are dropped into the gas diffuser.

Figure 34 Various contact tips and nozzles.

- *Standard vs. heavy duty* – Contact tips can be small standard-duty types or larger heavy-duty types. Most low-amperage applications can be handled with a standard contact tip. High-amperage or extensive welding applications are best handled with a heavy-duty contact tip.

Like contact tips, there are a wide variety of nozzles. They can be tapered or non-tapered, threaded or non-threaded, have small or large inner diameters, and come in numerous shapes and sizes. Ideally, it is best to choose the largest nozzle possible that still allows access to the weld joint. This way the greatest amount of shielding gas coverage is provided. If the welding joint has limited access, a smaller tapered nozzle will need to be used.

Nozzles are available in copper or brass. Copper nozzles are able to withstand extreme heat, so they are best suited for heavy-duty applications. Brass nozzles are subject to melting and burning if they are used in high-temperature applications, so they are more suited for light-duty applications. However, brass nozzles resist welding spatter better than copper nozzles.

2.6.0 Shielding Gases

Shielding gases are used to displace atmosphere from the weld zone and to prevent contamination of the weld puddle by oxygen, nitrogen, and moisture. The three principal shielding gases are argon, helium, and CO_2. These gases are often mixed in various proportions for specific applications.

The GMAW/FCAW-G process always uses shielding gas to protect the electrode and the weld from contamination. The shielding gas is directed around the end of the electrode and the weld zone by the nozzle of the gun.

Some FCAW core wires require the use of a shielding gas. The shielding gas is directed around the electrode end and the weld zone by the nozzle of the gun. The shielding gas used depends on the filler wire and base metal being welded. The most common shielding gases for FCAW carbon and low-alloy steels are CO_2 or an argon-CO_2 mixture. For FCAW high-alloy or stainless steels, argon or an argon-CO_2 mixture is generally used. These gases are available separately or premixed in pressurized cylinders. They are also available separately in bulk liquid tanks, so they can be mixed at the point of use.

CAUTION

Check your WPS or site quality standards for the type of shielding gas to use. If these are not available, refer to the filler metal supplier's specifications.

Each shielding gas has distinctive performance characteristics. Each gas performs differently in the various transfer modes and with different base metals. Mixtures of gases often have the best features of the individual constituents. The following sections explain some of the features of the principal gases and their primary uses.

Gas flow rate from the gun tip is important because it affects the quality and the cost of a weld. Too low a flow rate will not shield the weld zone adequately, which will result in a poor-quality weld. Excessive gas flow wastes expensive gases and can cause turbulence at the weld. The turbulence can pull atmosphere into the weld zone, causing porosity in the weld. Welding specifications indicate appropriate shielding gas flow rates.

2.6.1 Argon

Argon (Ar), is the least expensive and the most commonly used inert shielding gas. Argon can be used alone or in combination with other shielding gases because of its current transfer and ionization properties. It forms a tight arc column with high current density, which concentrates the arc in a small area and produces deep penetration.

Pure argon is used with aluminum, nickel-based alloys, copper alloys, and the reactive metals zirconium, titanium, and tantalum. It provides excellent arc stability, penetration, and bead shape. For ferrous metals, argon is usually mixed with oxygen, helium, hydrogen, carbon dioxide, and/or nitrogen.

2.6.2 Helium

Helium is likely the most expensive of the inert gases and supplies are diminishing. Because of thermal conductivity (heat transfer), helium (He) is used when deeper penetration and higher travel speed are required.

Arc stability with helium is not as good as it is with argon. The helium arc column is wider than the argon arc column, resulting in reduced current density. Helium is better for welding aluminum, magnesium, and copper alloys because its use results in better puddle fluidity and bead wetting. Some mixtures of helium and argon have the beneficial characteristics of both gases. Helium mixtures of 90 percent helium, 7.5 percent argon, and 2.5 percent CO_2 are widely used for short-circuit GMAW of stainless steel in all positions.

2.6.3 Carbon Dioxide

CO_2 is commonly used to weld mild steel because of its weld performance and low cost. In the welding arc, it is not an inert gas because it breaks down into carbon monoxide (CO) and free oxygen. The oxygen combines with elements in the weld puddle to form slag. CO_2 does not support spray transfer, so short-circuiting transfer and globular transfer must be used. CO_2 provides good penetration. However, in GMAW the weld surface is heavily oxidized and has high weld spatter. This is not true in FCAW. Normally, argon is mixed with no more than 25 percent CO_2 to improve the performance characteristics. *Figure 35* shows the effects of shielding gas on the welding arc in GMAW and FCAW.

2.6.4 Gas Mixtures

The principal shielding gases are often mixed to improve their overall welding characteristics. For steel, small amounts of oxygen added to argon stabilize the arc, increase filler metal deposition rate, and improve wetting and bead shape. Small percentages of oxygen added to CO_2 improve arc char-

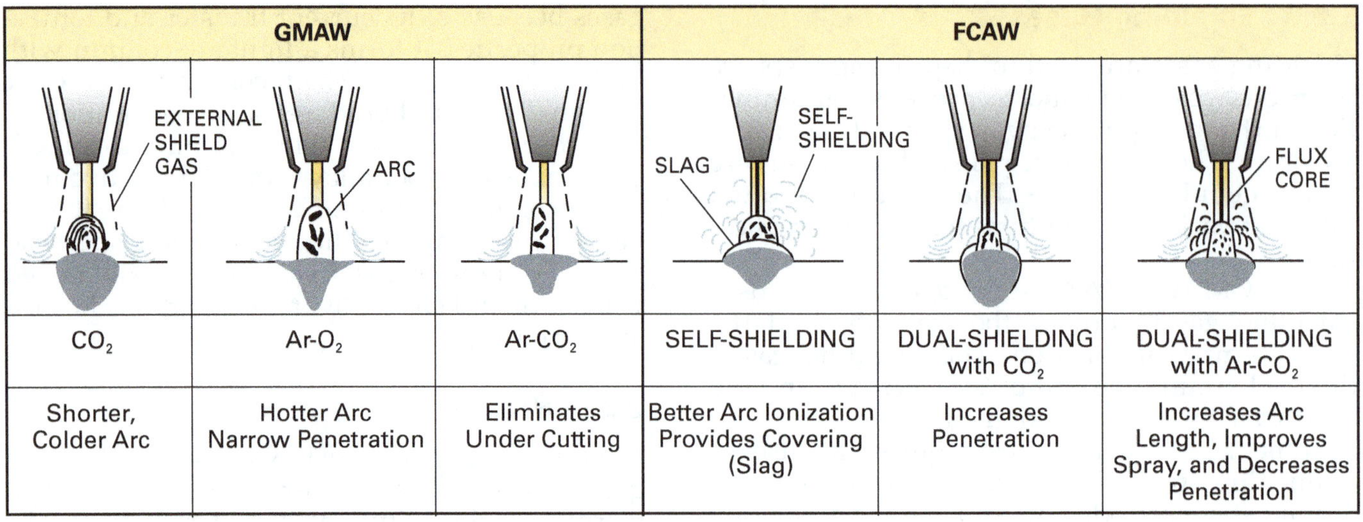

Figure 35 Effects of shield gas on the welding arc.

acteristics. Helium is added to argon to increase the arc voltage and heat for welding nonferrous metals. Small amounts of nitrogen and oxygen are added to argon for welding stainless steel when the austenitic characteristics should be preserved. Helium and CO_2 are added to argon to improve arc stability, wetting, and bead profile. Some examples and uses of gas mixtures are shown in *Table 2*. Many other typical gas mixtures are available.

2.6.5 Shielding Gas Selection

Selecting the best shielding gas for a particular welding task involves considering many factors. The shielding gas selected will affect arc shape, arc density, arc temperature, arc stability, rate of filler metal transfer, degree of spatter, weld penetration, weld bead shape, weld bead appearance, weld porosity, weld chemistry, and weld quality.

Shielding gases affect the welding process in three ways:

- Gas thermal conductivity affects arc voltage and heat delivered to the weld. The thermal conductivity of both helium and CO_2 is much higher than that of argon; both helium and CO_2 supply more heat to the weld. As a result, they also require more voltage and power to maintain a stable arc.
- Some gases can react with the filler and base metals and can also affect arc stability. CO_2 and most oxygen-bearing gases cannot be used with aluminum because they form oxides. However, they provide better fusion and arc stability when used with steels.
- Gases affect the mode of metal transfer and penetration depth. Mixtures containing more than 15 percent CO_2 do not allow true spray transfer.

Common short-circuiting shielding gases and their applications in short-circuiting transfer welding are shown in *Table 3*. Common shielding gases and their applications in spray transfer welding are shown in *Table 4*.

2.6.6 Gas Regulators/Flowmeters

A gas pressure regulator and a flowmeter are required to meter the shielding gas to the gun at the proper flow rate and pressure (*Figure 36*). The two devices can be matched to each other in a single assembly, or they may be separate devices. A typical cylinder pressure regulator includes two pressure gauges—one for cylinder pressure and another for the outlet pressure. The flowmeter may also have a gauge. The flow rate indicator is a clear cylinder with a ball that floats freely inside the housing to indicate the actual flow rate. Increased flow rates cause the ball to rise higher in the housing. The metering, or flow, valve on the flowmeter is used to adjust the gas flow rate to the gun nozzle. The flow indicator shows the gas flow rate in cubic feet per hour (cfh) or liters per minute (lpm).

Gas flow to the gun is started and stopped by an electric solenoid valve that is usually located inside the control unit or the wire feeder unit. The solenoid is controlled by the operator with the trigger switch on the gun. The same trigger switch simultaneously starts and stops the wire feeder and welding current.

2.6.7 Cylinder Safety

Cylinders must always be handled with great care because of the extremely high-pressure gases they contain. When a valve is broken off a high-pressure cylinder, the gas escapes with explosive

Table 2 Examples and Uses of Gas Mixtures

Gas Mixture	Use
Argon with 1% oxygen	Spray transfer on stainless steels
Argon with 2% oxygen	Spray transfer on carbon steels, low-alloy steels, and stainless steels
Argon with 5% oxygen	General carbon steel welding using highly deoxidized wires
Argon with 8% to 12% oxygen	Single-pass spray transfer welds with high puddle fluidity
Argon with 3% to 10% carbon dioxide	Spray transfer and short-circuiting transfer on carbon steels
Argon with 11% to 20% carbon dioxide	High-speed and out-of-position welds on carbon steels and low-alloy steels
Argon with 21% to 25% carbon dioxide	Short-circuiting transfer on mild steels
Argon with 25% helium	Sometimes used for aluminum where bead appearance is important
Argon with 75% helium	Welding heavy aluminum
Argon with 90% helium	Welding aluminum over 3" thick and copper over 1" thick
Argon with 5% to 10% carbon dioxide and 1% to 3% oxygen	Welding carbon steel, low-alloy steel, and stainless steel with all transfer modes
Argon with 10% to 30% helium and 5% to 15% carbon dioxide	Pulse spray transfer on heavy out-of-position carbon and low-alloy steel
60% to 70% helium with 20% to 35% argon and 4% to 5% carbon dioxide	Short-circuiting transfer of high-strength steels in all positions
90% helium with 7.5% argon and 2.5% carbon dioxide	Short-circuiting transfer of stainless steels in all positions
Argon, helium, carbon dioxide, and oxygen	High-rate deposition of low-alloy, high-tensile metals, and mild steels

Table 3 Short-Circuiting Shielding Gases and Their Applications

Gas Mixture	Shielding Gas	Thickness	Advantages
Carbon steel	75% Argon + 25% Carbon Dioxide	Less than ⅛" (3.2 mm)	High welding speeds without burn-through; minimum distortion and spatter
	75% Argon + 25% Carbon Dioxide	More than ⅛" (3.2 mm)	Minimum spatter; clean weld appearance; good puddle control in vertical and overhead positions
	50% Argon + 50% Carbon Dioxide	—	Deeper penetration; faster welding speeds
Stainless steel	90% Helium + 7.5% Argon + 2.5% Carbon Dioxide	—	No effect on corrosion resistance; small heat-affected zone; no undercutting; minimum distortion
Low-alloy steel	60 to 70% Helium + 25 to 35% Argon + 4.5% Carbon Dioxide	—	Minimum reactivity; excellent toughness; excellent arc stability, wetting characteristics, and bead contour; little spatter
	75% Argon + 25% Carbon Dioxide	—	Fair toughness; excellent arc stability, wetting characteristics, and bead contour; little spatter
Aluminum, copper, magnesium, nickel, and their alloys	Argon and Argon + Helium mixtures	Over ⅛" (3.2 mm)	Argon satisfactory on sheet metal; argon-helium preferred base material

Table 4 Spray Transfer Shielding Gases and Their Applications

Gas Mixture	Shielding Gas	Thickness	Advantages
Aluminum	100% Argon	0 to 1" (0 to 25 mm)	Best metal transfer and arc stability, least splatter
	65% Argon + 35% Helium	1 to 3" (27 to 76 mm)	Higher heat input than straight argon; improved fusion characteristics with 5XXX series Al-Mg alloys
	75% Helium + 25% Argon	Over 3" (76 mm)	Highest heat input; minimizes porosity
Magnesium	100% Argon	–	Excellent cleaning action
Carbon steel	Argon + 3 to 5% Oxygen	–	Improves arc stability; produces a more fluid and controllable weld puddle; good coalescence and bead contour; minimizes undercutting; permits higher speeds than pure argon
	Argon + 8 to 10% Carbon Dioxide	–	High-speed mechanized welding; low-cost manual welding
Low-alloy steel	98% Argon + 2% Oxygen	–	Minimizes undercutting; provides good toughness
Stainless steel	99% Argon + 1% Oxygen	–	Improves arc stability; produces a more fluid and controllable weld puddle; good coalescence and bead contour; minimizes undercutting on heavier stainless steels
	98% Argon + 2% Oxygen	–	Provides better arc stability, coalescence, and welding speed than 1% oxygen mixture for thinner stainless steel materials
Nickel, copper, and their alloys	100% Argon	Up to ⅛" (3.2 mm)	Provides good wetting; decreases fluidity of weld material
	Argon + Helium mixtures		Higher heat inputs of 50 and 75% helium mixtures offset high heat dissipation of heavier gauges
Titanium	100% Argon		Good arc stability; minimum weld contamination; inert gas backing is required to prevent air contamination on back of weld area

force and can severely injure personnel from the blast and blown debris. The cylinder itself can become a powerful, uncontrolled projectile. When transporting and handling cylinders, always observe the following rules:

WARNING!

Do not remove the protective valve cap unless the cylinder is secured. If a cylinder falls over and the valve breaks off, the cylinder may rocket away uncontrollably, causing severe injury or death to anyone in its way.

- The safety cap should always be installed over the valve except when the cylinder is connected for use.
- When in use, the cylinder must always be secured to prevent it from falling. It should be chained or clamped to the welding machine or to a post, beam, or pipe.
- A cylinder cart should be used to transport a cylinder. Cylinders should be secured in the cart.
- Cylinders should never be hoisted by slings or magnets. They can slip out of the sling or fall from the magnets. Do not route a sling through the cap openings. Always use a hoisting basket or similar device.
- Always open the cylinder valve slowly. Once pressure has been applied to the system, the valve should be completely opened to its backseated position to prevent leakage from around the valve stem.

2.7.0 Welding Equipment Setup

To weld safely and efficiently, the welding equipment must be properly set up (*Figure 37*). The following sections explain the steps for setting up welding equipment.

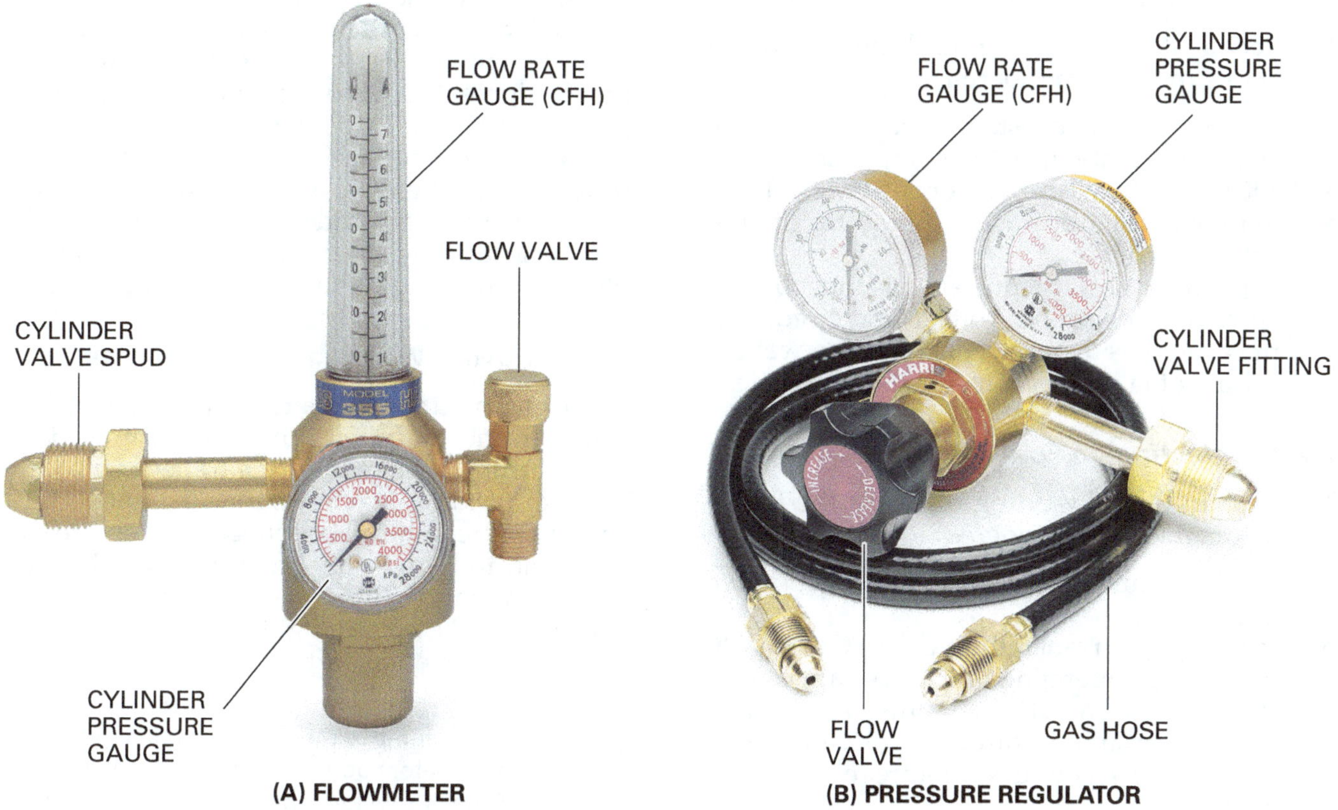

Figure 36 Shielding-gas flowmeter and pressure regulator.

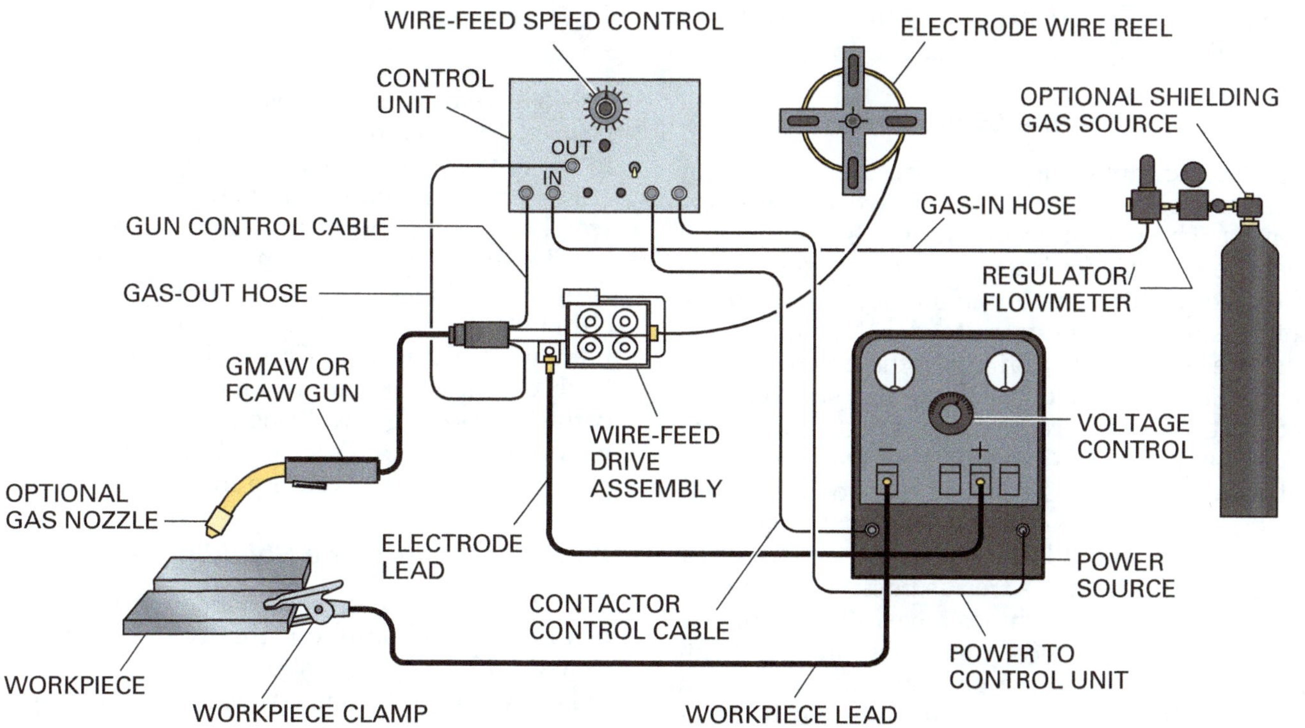

NOTE: The polarity of the gun and workpiece leads is determined by the type of filler metal and application.

Figure 37 Diagram of a GMAW/FCAW system

2.7.1 Selecting a GMAW Welding Machine

To select a GMAW welding machine, the following factors must be considered:

- GMAW requires a constant-voltage or constant-voltage/constant-current welding machine.
- A DCEP-type welding current is required.
- The maximum amperage required for the task must be available.
- The primary power requirements must be met. Ensure that the proper voltage and circuit current capacity is available, and that the welding machine plug and receptacle are compatible.

2.7.2 Selecting an FCAW Welding Machine

To select an FCAW welding machine, the following factors must be considered:

- FCAW requires a constant-voltage power source.
- The welding current required for FCAW is DC.
- The required operating amperage ranges from 150 amps with a 0.045" (1.2 mm) electrode to 650 amps or more with a ⅛" (3.2 mm) electrode; larger electrodes require even higher amperage.
- The primary power requirements must be met. Ensure that the proper voltage and circuit current capacity is available, and that the welding machine plug and receptacle are compatible.

2.7.3 Positioning the Equipment

Because the gun cables are short, the wire feeder must be positioned close to the work to be performed. The power source must also be located near the wire feeder because of the limited length of the contactor/control cable that runs from the power source to the wire feeder. Normally, the wire feeder is located on top of the power source, on the floor, or a cart near the power source.

Select a site where the equipment will not be in the way but will be protected from welding, cutting, or grinding sparks. There should be good air circulation to keep the machine cool. The environment should be free from explosive or corrosive fumes and as free as possible from dust and dirt. Welding machines have internal cooling fans that will pull these materials into the welding machine if they are present. Also, dust and dirt can collect on the wire, causing weld contamination and clogging of the liner in the gun cable. The site should be free of standing water or water leaks. If an engine-driven unit is used, position it so that it can be easily refueled and serviced.

There should be easy access to the site so that the equipment can be started, stopped, or adjusted as needed. If the machine will be plugged into an outlet, be sure that the outlet has been properly installed by a qualified electrician to ensure that it is grounded. Also, be sure to identify the location of the electrical disconnect before plugging the welding machine into the outlet.

2.7.4 Moving Welding Power Sources

Large engine-driven generators are mounted on a trailer frame, and they can easily be moved by a pickup truck or tractor with a trailer hitch. Other types of welding machines may have a skid base or be mounted on steel or rubber wheels. Be careful when moving wheel-mounted welding machines by hand. Some machines are top-heavy and could fall over in a tight turn or if the floor or ground is uneven or soft.

WARNING!

Secure or remove the wire feeder and gas cylinder before attempting to move a welding machine. If a welding machine starts to fall over, do not attempt to hold it. Welding machines are very heavy, and severe crushing injuries can occur if a welding machine falls on you.

Most welding machines have a lifting eye that is used to move machines mounted on skids or to lift any machine. Before lifting a welding machine, check the equipment specifications for the weight. Be sure the lifting device and tackle can handle the weight of the machine. Always use a sling and/or a shackle. Never attempt to lift a machine by placing the lifting hook directly in the machine's lifting eye. This is because the safety latch on the hook cannot be closed. Also, before lifting or moving a welding machine, be sure the welding cables are secure. *Figure 38* shows the proper way to lift a welding machine.

2.7.5 Connecting the Shielding Gas

The shielding gas connects to the wire feeder unit or to the power source if the wire feeder is integrated with the power source.

Follow these steps to connect the shielding gas:

Step 1 Identify the required shielding gas by referring to the WPS or site quality standards.

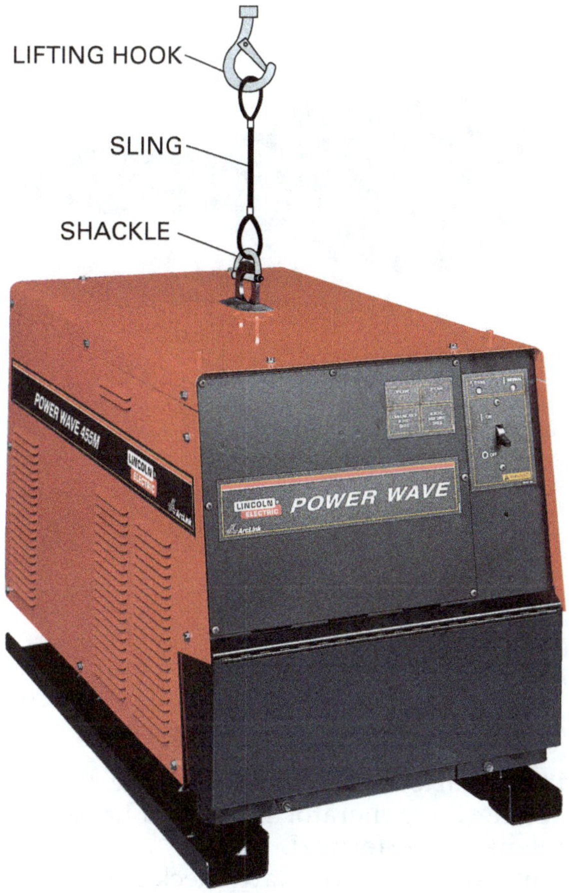

Figure 38 Lifting a welding machine.

Step 2 Locate a cylinder of the correct gas or mixture, and secure it near the wire feeder. Be sure to secure the cylinder so that it cannot fall over.

Step 3 Remove the cylinder's protective cap, momentarily crack the cylinder valve open to blow out any dirt, and then close it.

WARNING! Even though they are available, do not use an adapter to connect a regulator/flowmeter equipped with one type of Compressed Gas Association (CGA) connection to a gas cylinder that has a different CGA connection. The CGA connections are specific to the types of gas and cylinder pressures permitted for the connection.

Step 4 Using a regulator/flowmeter with the correct CGA connection for the cylinder, install the pressure regulator/flowmeter on the cylinder.

Step 5 Connect the gas hose to the flowmeter and to the gas connection on the wire feeder (or welding machine).

Step 6 Slowly crack open the cylinder valve, and then open it completely.

WARNING! Do not open the valve quickly or the flowmeter gauges may break. If a gauge breaks, the glass may shatter and the gas may escape, potentially injuring the operator.

Step 7 Adjust the gas flow to the specified flow rate using the flowmeter valve.

Some flowmeters are equipped with several scales of different calibrations around the same sight tube, for monitoring the flows of different gases. These scales differ due to the different densities of shielding gases. Be sure to rotate the scales or read the correct side for the type of gas being used.

2.7.6 Selecting and Installing Filler Wire

Follow these steps to select and install the filler wire:

Step 1 Identify the filler wire required by referring to the WPS or site quality standard.

Step 2 Locate a spool of the wire, and mount it on the wire feeder spool holder. Lock the holder, and set the drag brake for a slight drag. The brake control is usually located on the end of the spool axle.

Step 3 Check (and change if necessary) the wire feed drive wheels, liner, and gun contact tube to make sure that they are the correct size for the wire selected. Adjust the wheel tension if necessary. For detailed instructions on changing or adjusting the wire feeder drive wheels or changing the gun's contact tube or liner, refer to the specific manufacturer's instructions.

Step 4 Feed the wire into the feeder wheels, and then use the JOG control to feed it through the gun cable and the gun contact tube.

2.7.7 Placing the Workpiece Clamp

The workpiece clamp must be properly placed to prevent damage to surrounding equipment. If the electrical welding current travels through a bearing, seal, valve, or contacting surface, it could cause severe damage from heat and arcing, which would require that these items be replaced. Carefully check the area to be welded, and position the workpiece clamp so the welding current will not pass through any contacting surface. When in doubt, ask your supervisor for assistance before proceeding.

> **CAUTION**
>
> Welding current passing through electrical or electronic equipment will cause severe damage to the affected equipment. Also, before welding on any type of equipment with a battery, the ground lead at the battery must be disconnected to protect the electrical system.

> **WARNING!**
>
> If welding is to be done near a battery, the battery must be removed. Batteries produce hydrogen gas, which is extremely explosive. A welding spark could cause the battery to explode, showering the area with battery acid.

Workpiece clamps must never be connected to pipes carrying flammable or corrosive materials. The welding current could cause overheating or sparks, resulting in an explosion or fire.

The workpiece clamp must make a good electrical contact when it is connected. Dirt and paint will inhibit the connection and cause arcing, resulting in overheating of the workpiece clamp. Dirt and paint also make the welding current unstable and can cause defects in the weld. Clean the surface before connecting the workpiece clamp. If the workpiece clamp is damaged or does not close securely onto the surface, replace it.

2.7.8 Energizing the Power Source

Electrically powered welding machines are energized by plugging them into an electrical outlet. The electrical requirements will be located on the equipment specification tag that is prominently displayed on the machine. Most machines requiring single-phase, 240VAC power have a three-pronged plug. Machines requiring three-phase, 208 or 460VAC power have a four-pronged plug. Both types of plugs are shown in *Figure 39*.

If a welding machine does not have a power plug, an electrician must connect it. The electrician will add a plug or hard-wire the machine directly into an electrical box.

> **WARNING!**
>
> Never use a welding machine until you identify the location of the electrical disconnect switch for that specific machine or receptacle. In the event of an emergency, you must be able to turn off the power quickly to the welding machine at the disconnect switch.

2.7.9 Starting Engine-Driven Generators

Before welding can take place with an engine-driven generator, the engine must be checked and then started. As with a car engine, the engine powering the generator must also have routine maintenance performed.

Figure 39 Typical three- and four-pronged plugs.

Many sites have prestart checklists that must be completed and signed prior to starting or operating an engine-driven generator. Check with your supervisor. If your site has such a checklist, complete and sign it. If your site does not have a prestart checklist, perform the following checks before starting the engine:

- Check the oil, using the engine oil dipstick. If the oil is low, add the appropriate grade of oil for the conditions, according to the manufacturer's maintenance instructions.
- Check the coolant level in the radiator if the engine is liquid-cooled. If the coolant level is low, add the proper coolant.

> **CAUTION**
>
> Do not add plain water to radiators that contain antifreeze. Antifreeze not only protects radiators from freezing in cold weather, it also has rust inhibitors and additives to protect heat-transfer surfaces. If the antifreeze is diluted, it will not provide the proper protection.

- Check the fuel. The unit may have a fuel gauge or a dipstick. If the fuel is low, add the correct fuel (diesel or gasoline) to the fuel tank. The type of fuel required should be marked on the fuel tank. If it is not marked, contact your supervisor to verify the fuel required, and have the tank marked.

CAUTION

Adding gasoline to a diesel engine or diesel to a gasoline engine will cause severe engine problems, and may create a fire hazard. Always be sure to add the correct fuel to the fuel tank.

- Check the battery water level unless the battery is sealed. Add demineralized water if the battery water level is low.
- Check the electrode holder to be sure it is not grounded. If the electrode holder is grounded, it will arc and overheat the welding system when the welding machine is started. This may cause damage to the equipment.
- Open the fuel shut-off valve if the equipment has one. If there is a fuel shut-off valve, it should be located in the fuel line between the fuel tank and the carburetor.
- Record the hours from the hour meter. An hour meter records the total number of hours that the engine runs. This information is used to determine when the engine needs to be serviced. The hours will be displayed on a gauge similar to an odometer.
- Clean the unit. Use a compressed air hose to blow off the engine and generator or alternator. Use a rag to remove heavier deposits that cannot be removed with the compressed air.

WARNING!

Always wear eye protection when using compressed air to blow dirt and debris from surfaces. Never point the compressed air nozzle at yourself or anyone else.

Most engines have an On/Off ignition switch and a starter. They may be combined into a key switch similar to the ignition on a car. To start the engine, turn on the ignition switch and press the starter. Release the starter when the engine starts. The engine speed is controlled by the governor. If the governor switch is set for idle, the engine will slow to an idle after a few seconds. If the governor is set to welding speed, the engine will continue to run at the welding speed.

Small engine-driven generators may have an On/Off switch and a pull cord. These are started by turning on the ignition switch and pulling the cord, in much the same way as a lawn mower is started. Engine-driven generators should be started about 5 to 10 minutes before they are needed for welding. This will allow the engine to warm up before a welding load is applied.

If no welding is required for 30 minutes or more, stop the engine by turning off the ignition switch. If you are finished with the welding machine for the day, close the fuel valve if there is one.

Engine-driven generators require regular preventive maintenance to keep the equipment operating properly. Most sites have a preventive maintenance schedule based on the hours that the engine operates. In severe conditions, such as in very dusty environments or cold weather, maintenance may have to be performed more frequently.

CAUTION

To prevent damage to the equipment, perform preventive maintenance as recommended by the site procedures or by the manufacturer's maintenance schedule in the equipment manual.

The responsibility for performing preventive maintenance varies by site. Check with your supervisor to determine who is responsible for performing preventive maintenance. Even though the welder may not be responsible for conducting preventive or periodic maintenance, all welders share an obligation to monitor and protect the equipment, and to report problems. As is true for most crafts, a welder cannot produce welds without functional and reliable equipment.

Adjusting Flowmeters

In most cases, the top of the floating flowmeter ball is used as the reference point for setting the gas flow. However, always check the manufacturer's instructions to determine the correct adjustment method. Use the lowest flow rate as often as possible to conserve gas and protect the environment.

Gas Flow Rates

More gas flow is not always better. Too much flow can deflect off the workpiece. The resultant swirling gas can create turbulence that draws in air, contaminating the weld and creating pinholes. If black flecks occur in the weld puddle or pinholes occur in the weld, try reducing the gas flow.

When performing preventive maintenance, follow the manufacturer's guidelines in the equipment manual. Typical procedures to be performed as a part of preventive maintenance include the following:

- Changing the oil
- Changing the gas filter
- Changing the air filter
- Checking/changing the antifreeze
- Greasing the undercarriage
- Repacking the wheel bearings

Dust Removal

Blow out the welding machine and drive unit with compressed air occasionally to remove dust. A significant buildup of dust inside the machine can provide an alternate path for the electricity and could result in some of the components shorting out.

Additional Resources

AWS B1.10M/B1.10 Guide for the Nondestructive Examination of Welds. Miami, FL: American Welding Society.

AWS B1.11 Guide for the Visual Examination of Welds. Miami, FL: American Welding Society.

Lincoln Electric website: **http://www.lincolnelectric.com** offers sources for products and training.

Modern Welding Technology. Howard B. Cary. Englewood Cliffs, NJ: Prentice Hall, Inc.

OSHA 1926.351, Arc Welding and Cutting. Current Edition. Washington, DC: Occupational Safety and Health Administration (OSHA).

Welding Handbook. Volume 1. Welding Science & Technology. Miami, FL: American Welding Society, 2001.

Welding Handbook. Volume 2. Part 1: Welding Processes. Miami, FL: American Welding Society, 2004.

2.0.0 Section Review

1. The two types of voltage associated with welding current are ____.
   a. open-circuit voltage and operating voltage
   b. operating voltage and arc voltage
   c. closed-circuit voltage and welding voltage
   d. operating voltage and closed-circuit voltage

2. The deflection of an arc caused by the attraction or repulsion of the arc's magnetic field with the weld current's magnetic field in the base metal is known as ____.
   a. pinch effect
   b. slope curvature
   c. arc blow
   d. inductance bending

3. Quick disconnects used on welding cable are connected and disconnected with a ____.
   a. terminal lug
   b. screwing action
   c. pronged plug
   d. half twist

4. A device that can be used to start and stop the wire feeder on a welder without energizing the welding current is a ____.
   a. spool drag brake
   b. wire jog switch
   c. feed speed control
   d. wire meltback timer

5. The part of a GMAW/FCAW gun that provides the electrical connection from the welding machine electrode to the electrode wire is the ____.
   a. gun handle
   b. fiber insulator
   c. contact tip
   d. extraction nozzle

6. The most commonly used shielding gas for GMAW/FCAW-G processes is ____.
   a. argon (Ar)
   b. CO_2
   c. helium (He)
   d. carbon monoxide (CO)

7. What rigging device should never be placed into a welding machine's lifting eye when the machine is being lifted or moved?
   a. A synthetic sling
   b. The lifting hook
   c. A wire rope sling
   d. Any type of shackle

SECTION THREE

3.0.0 FILLER METALS

Objective

Identify various GMAW and FCAW filler metals.

a. Identify various GMAW filler metals.
b. Identify various FCAW filler metals.

Performance Task

1. Set up GMAW and FCAW equipment with appropriate shielding gases and filler metals.

GMAW and FCAW filler metals are explained in this section. The WPS specifies the filler metals to be used. The improper selection of filler metal can cause the weld to fail.

Regardless of the type of filler metal being used, there is often a need to cut and manipulate the wire. *Figure 40* shows an example of a special tool that is used for this purpose. Commonly called MIG welding pliers or wire feed pliers, this tool is designed for drawing out wire, cutting wire, removing and installing contact tips and nozzles, removing welding spatter from nozzles, and light hammering.

3.1.0 GMAW Filler Metals

GMAW uses a continuous solid wire electrode for filler material. This filler wire is drawn from high-grade pure alloys compounded for different applications. During manufacture, the wire is drawn, cleaned and inspected for defects several times. Standard wire diameters are 0.023" (0.6 mm), 0.030" (0.7 mm), 0.035" (0.9 mm), 0.045" (1.2 mm), 0.052" (1.4 mm), and 0.062" (1.5 mm). Note that the metric equivalents used here represent actual metric wire sizes, and are not direct equivalents. The finished wire is wound on spools 4" (102 mm), 8" (204 mm), 12" (30 cm), or 14" (36 cm) in diameter. The most common spool is 12" (30 cm) in diameter and weighs 30 pounds (13.6 kg). Reels or drums up to 30" (76 cm) in diameter and weighing up to 1,000 pounds (454 kg) are also available for stationary production welding.

Figure 40 Wire feed pliers.

> **Filler Wire and Filler Metal**
>
> The terms *filler wire* and *filler metal* are both used to describe the electrode used in GMAW and FCAW. Filler metal is also referred to as filler rod when used for GTAW.

Several industry organizations and the United States government publish specification standards for welding wire. The most common are the AWS specifications. The purpose of the AWS specifications is to set standards that all manufacturers must follow when manufacturing welding consumables. This ensures consistency for the user regardless of who manufactured the product. The specifications set standards for the following:

- Classification system, identification, and marking
- Chemical composition of the deposited weld metal
- Mechanical properties of the deposited weld metal

The following AWS specifications refer to gas metal arc welding wires and rods:

- *A5.7: Specification for Copper and Copper Alloy Bare Welding Rods and Electrodes*
- *A5.9/A5.9M: Specification for Bare Stainless Steel Welding Electrodes and Rods*
- *A5.10/A.510M: Specification for Bare Aluminum and Aluminum Alloy Welding Electrodes and Rods*
- *A5.14/A5.14M: Specification for Nickel and Nickel Alloy Bare Welding Electrodes and Rods*
- *A5.16/A5.16M: Specification for Titanium and Titanium Alloy Welding Electrodes and Rods*
- *A5.18/A5.18M: Specification for Carbon Steel Electrodes and Rods for Gas Shielded Arc Welding*
- *A5.19: Specification for Magnesium Alloy Welding Electrodes and Rods*
- *A5.28/A5.28M: Specification for Low-Alloy Steel Electrodes and Rods for Gas Shielded Metal Arc Welding*

NOTE

An AWS specification number is generally followed by a four-digit number that indicates the year that the specification was last revised.

Filler wire is graded for three major areas of use:

- *General use* – Wire meets specifications. No record of chemical composition, strength, or similar factors is supplied to the user with the wire purchase.
- *Rigid control fabrication* – A Certificate of Conformance is supplied with the wire at purchase. The stock is identified by code numbers located on the roll package.
- *Critical use* – A Certified Chemical Analysis report is supplied, and records are kept by the fabricator of welds and processes for later reference on aircraft, nuclear reactors, and pressure vessels.

The following factors affect the selection of a GMAW filler wire:

- Base metal chemical composition
- Base metal mechanical properties
- Weld joint design
- Service or specification requirements
- Shielding gas used

Commonly used filler wire types include the following:

- Carbon steel
- Stainless steel
- Aluminum and aluminum alloy
- Copper and copper alloy
- Low-alloy steel

Settings for electrode wire of the same classification can vary widely depending upon the manufacturer. It is important to check the manufacturer's recommended settings for each specific wire used. This information can typically be found on the wire manufacturer's website or catalog.

Rod and Electrode Designations

The letter R that follows the E in an AWS classification number indicates that the classification is used as a metal filler rod as well as an electrode. GMAW and GTAW filler metals are both labeled the same way when used for gas tungsten arc welding (GTAW).

3.1.1 Carbon Steel Filler Metals

Carbon steel filler metals are identified by specification *AWS A5.18*. The filler metal is available in wire reels and rod form. The wire classification number is located on a label on the side of the reel or container (*Figure 41*).

All carbon steel filler metals contain alloys such as silicon, manganese, aluminum, and carbon. Other alloys such as nickel, chromium, and molybdenum are also often added. The purpose of the alloys is as follows:

- *Silicon (Si)* – Concentrations of 0.40 to 1.00 percent are employed to deoxidize the puddle and to strengthen the weld. Silicon above 1 percent may make the welds crack-sensitive. Silicon also lowers the melting temperature of the wire and promotes wetting.
- *Manganese (Mn)* – Concentrations of 1 to 2 percent are also employed as a deoxidizer and to strengthen the weld. Manganese also decreases hot-crack sensitivity.
- *Aluminum (Al), titanium (Ti) and zirconium (Zr)* – One or more of these may be added in very small amounts for deoxidizing. Strength may also be increased.
- *Carbon (C)* – Concentrations of 0.05 to 0.12 percent are used to add strength without adversely affecting ductility, porosity, or toughness.
- *Nickel (Ni), chromium (Cr), and molybdenum (Mo)* – May be added in small amounts to improve corrosion resistance, strength, and toughness.

In many cases, carbon steel filler metal wire is copper coated. The copper coating protects the filler metal wire from corrosion and makes electrical contact easier as the wire passes through the copper contact tube at the end of a gun.

AWS carbon steel filler metal wire classifications and their uses are listed in *Table 5*.

3.1.2 Low-Alloy Steel Filler Metals

Low-alloy steel filler metals are identified by *AWS Specification A5.28*. The filler metal is available in wire reel and rod form. The wire classification number is located on a label on the side of the reel or container. The classification is similar to the carbon steel filler metal classification except that there is a letter in the suffix at the end of the classification (*Figure 42*). The letter suffix indicates that the filler metal contains alloys (nickel, manganese, molybdenum, etc.) that give the deposited weld metal special properties. Also, the tensile strength of low-alloy filler metals is often higher than that of carbon steel filler metals. For specific

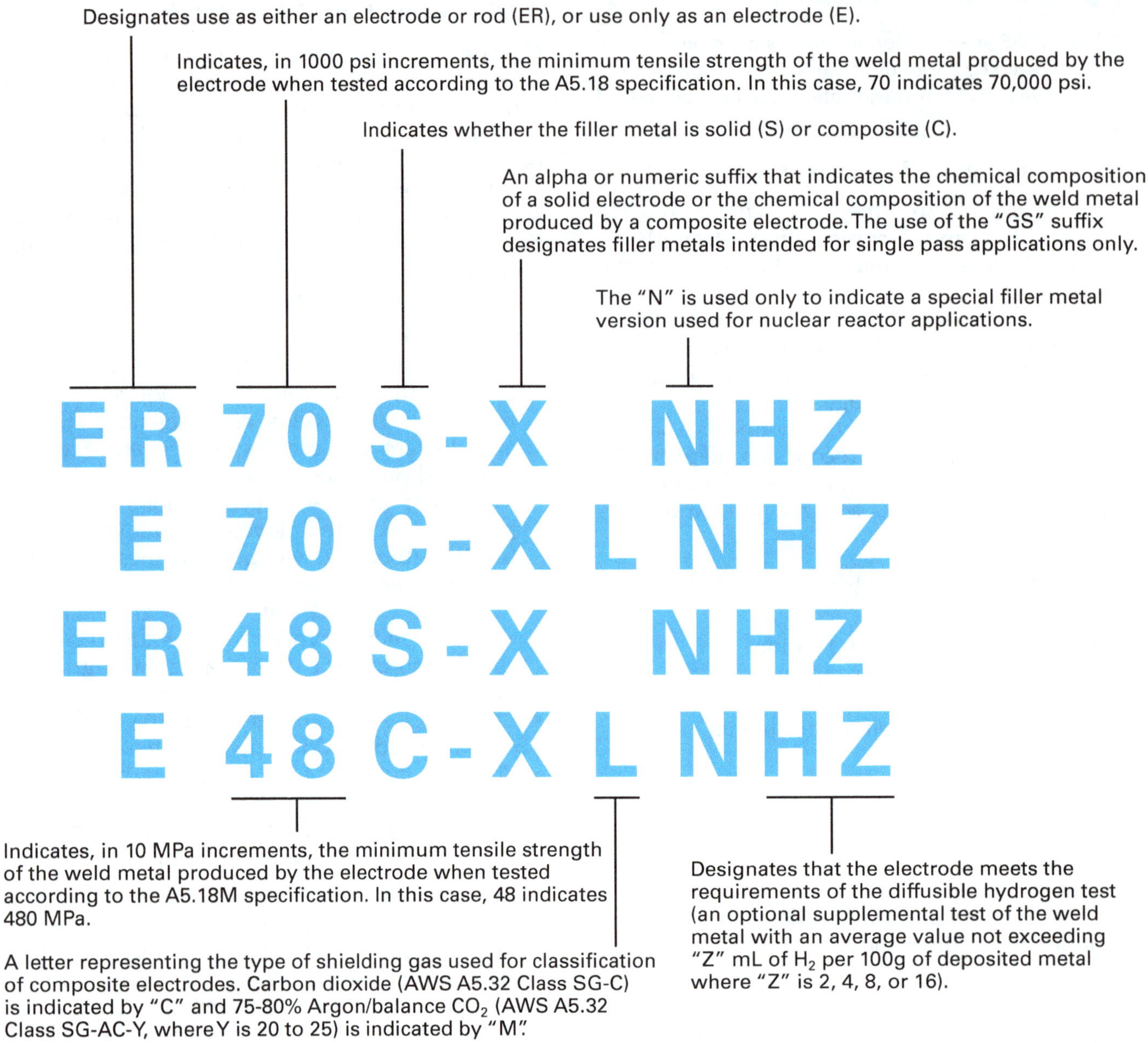

Figure 41 AWS classification for carbon steel filler metal.

information on low-alloy filler metal composition and intended use, refer to the manufacturer's specifications.

3.1.3 Stainless Steel Filler Metals

Stainless steel filler metals are identified by specification *AWS A5.9*. The filler metal is available in wire reel, flat strip, and rod form (*Figure 43*).

Stainless steel filler metal should be selected to closely match the alloy composition of the base metal. Stainless steel filler metals also require specific shielding gases or gas mixtures.

3.1.4 Aluminum and Aluminum Alloy Filler Metals

Aluminum filler metals are covered by *AWS Specification A5.10*. Aluminum filler metals usually contain magnesium, manganese, zinc, silicon, or copper for increased strength. Corrosion resistance and weldability are also considerations. Aluminum filler metals are designed to weld specific types of aluminum and should be selected for compatibility. The most widely used aluminum filler metals are ER4043 (contains silicon) and ER5356 (contains magnesium). The filler metal classification number is located on a label on the side of the reel or container (*Figure 44*).

Table 5 AWS Carbon Steel Wire Classifications and Uses

Wire Classification	Use
ER70S-2 ER48S-2	Welds all grades of carbon steel, including killed (highly deoxidized by adding aluminum powder or ferrosilicon), semi-killed, and rimmed (slightly deoxidized by adding a small amount of aluminum powder). Also welds carbon steel with rusted surfaces. Acceptable shielding gases include Ar-O_2, Ar-CO_2, and CO_2. The viscous weld puddle is well suited to out-of-position welding. Wetting is helped by keeping the oxygen or CO_2 percentage relatively high.
ER70S-3 ER48S-3	Welds killed and semi-killed steels (used in appliances, automobiles, and farm equipment) with CO_2, Ar-O_2, or Ar-CO_2 shielding gases.
ER70S-4 ER48S-4	Welds semikilled and rimmed steels (used in boilers and pressure vessels, piping, ship steels, structural steels, and A515 grades 55-70 steels) with CO_2, Ar-O_2, or Ar-CO_2 shielding gases.
ER70S-6 ER48S-6	Welds many carbon steels with CO_2 and high welding current. High-speed welding can be done with Ar-O_2 with 5% or more O_2. Either spray transfer or short-circuiting transfer can be used.
ER70S-7 ER48S-7	Welds a wide variety of steels with superior weld quality and appearance. Porosity and weld defects caused by mill scale or rust are minimized.
ER70S-G ER48S-G E70C-G E48C-G	Filler metals not included in the preceding classes that only have certain mechanical properties specified. Absence of a C or M suffix means that the manufacturer should be consulted for the recommended shielding gas and intended applications.
E70C-GS E48C-GS	Composite stranded or metal cored electrodes intended for only single pass applications. These electrodes may have higher alloy content and the manufacturer should be consulted for specific applications.
E70C-3 E48C-3 E70C-6 E48C-6	Composite stranded or metal cored electrodes intended for both single and multipass applications on steel. They are characterized by a spray arc and excellent bead wash characteristics.

3.1.5 Copper and Copper Alloy Filler Metals

Copper filler metals are covered by *AWS Specification A5.7.* Most copper filler metals contain other elements that increase strength, deoxidize the weld metal, and match the base metal composition. *Figure 45* shows the AWS classification for copper filler metal.

3.1.6 Nickel and Nickel Alloy Filler Metals

Nickel-based filler metals (*Figure 46*) are covered by *AWS Specification A5.14*. These filler metals contain other elements to match base metal applications and to increase the strength and quality of the weld metal. For GTAW, DCEN is used with high-purity argon, helium, or both as a shielding gas. For GMAW, DCEP is used. Argon shielding gas is often used, as are mixtures of argon and helium.

3.1.7 Magnesium Alloy Filler Metals

Magnesium alloy filler metals (*Figure 47*) are covered by *AWS Specification A5.19.* These filler metals are usually used with GTAW, GMAW, and plasma arc welding (PAW) processes. Oxyfuel welding should be used only for temporary repair work. For GTAW welding, the techniques and equipment are similar to those for aluminum. Argon, helium, or both are used for shielding. Alternating current is preferred for arc cleaning and penetration. Direct current is also employed: DCEP is used for thin materials, and DCEN is used for mechanized welding with helium for deep penetration. GTAW is recommended for the defect repair of clean magnesium castings. GMAW of magnesium alloys is the same as for other metals. Argon is generally used for shielding; however, mixtures of argon and helium are occasionally used. Pulsed and short-circuit GMAW are used for magnesium alloys, and spray transfer without pulsing is also used; however, globular transfer is not suitable.

3.1.8 Titanium and Titanium Alloy Filler Metals

Titanium and titanium alloy filler metals (*Figure 48*) are covered by *AWS Specification A5.16*. These filler metals are generally used with GTAW, GMAW, submerged arc welding (SAW), and PAW

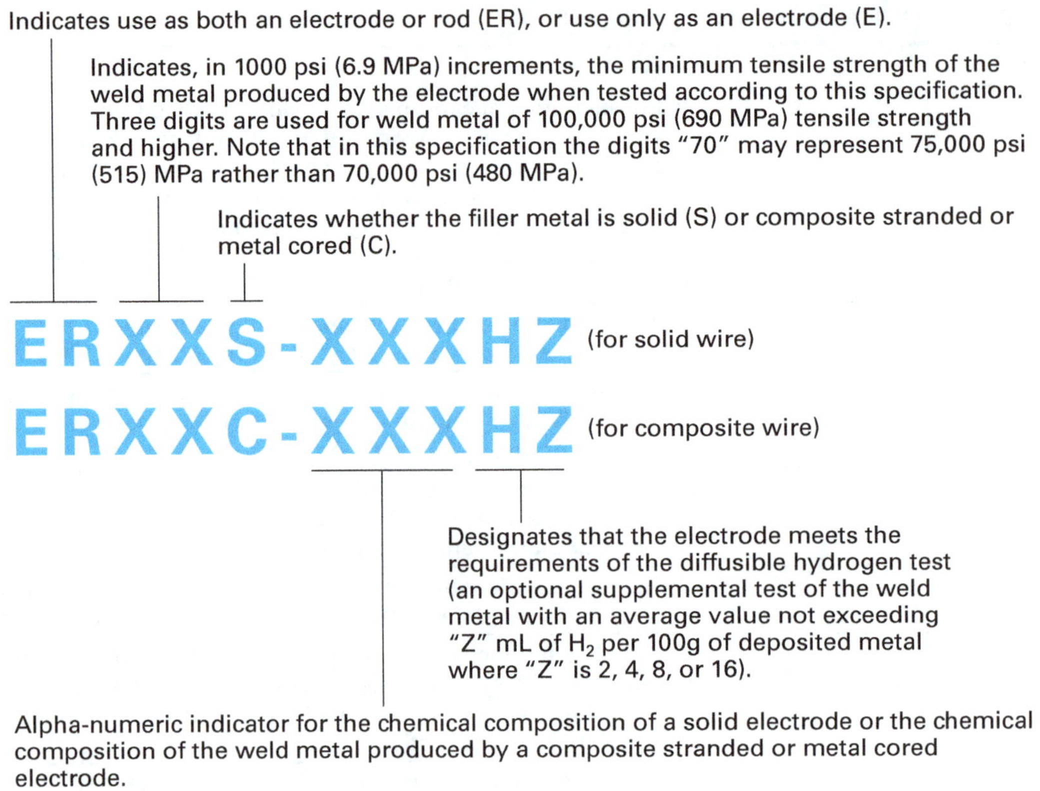

Figure 42 AWS classification for low-alloy steel filler metal.

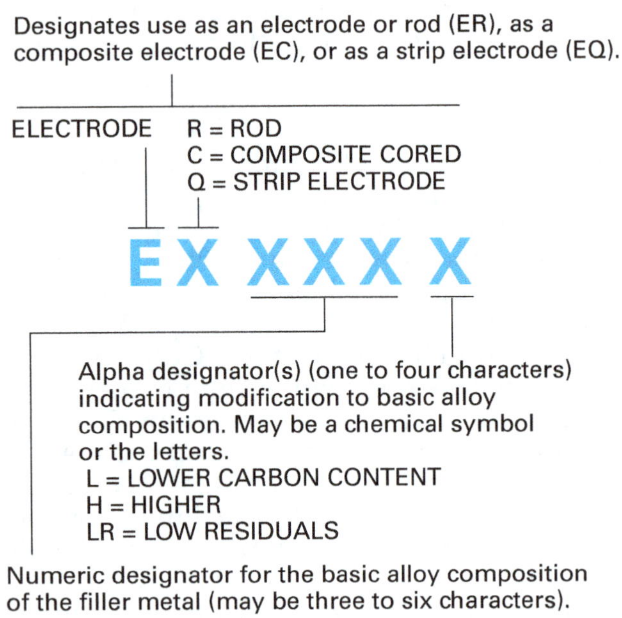

Figure 43 AWS classification for stainless steel filler metal.

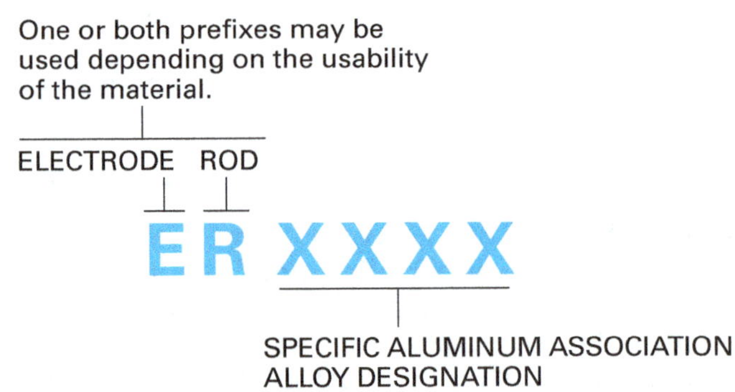

Figure 44 AWS classification for aluminum filler metal.

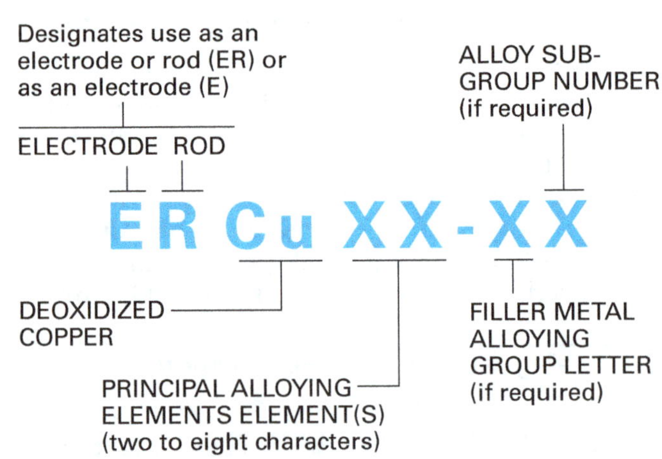

Figure 45 AWS classification for copper filler metal.

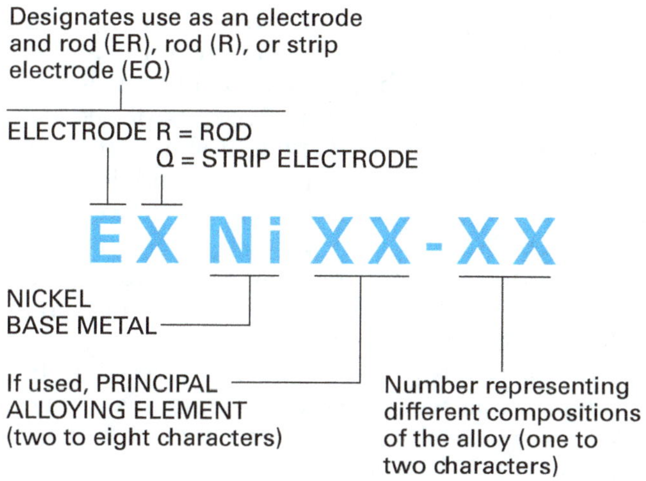

Figure 46 AWS classification for nickel filler metal.

processes. Titanium is sensitive to embrittlement by oxygen, nitrogen, and hydrogen at temperatures above 500°F (260°C). Like aluminum, titanium requires weld cleaning and high-purity gas shielding, and an especially adequate post-flow shielding to avoid embrittlement. Titanium can be successfully fusion welded to zirconium, tantalum, niobium, and vanadium. Titanium should not be welded to copper, iron, nickel, and aluminum. As with magnesium, titanium fines are also flammable, so don't allow them to accumulate.

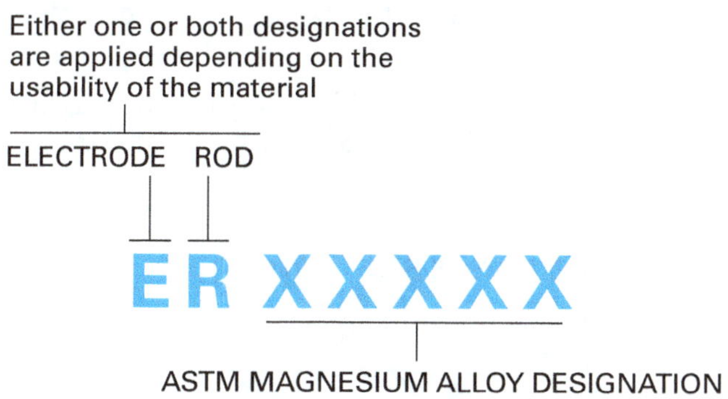

Figure 47 AWS classification for magnesium filler metal.

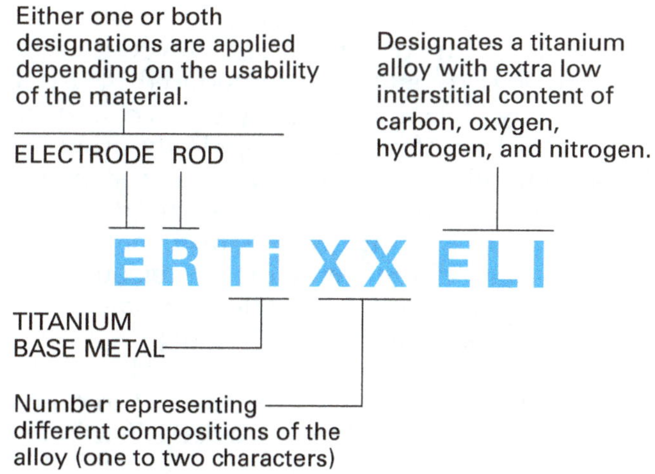

Figure 48 AWS classification for titanium filler metal.

3.2.0 FCAW Filler Metals

FCAW is primarily used for the welding of ferrous metals. These include low- and medium-carbon steels, some low-alloy steels, cast irons, and some stainless steels.

The FCAW electrode is a continuous flux-cored wire electrode. The wire is made by forming a thin filler metal strip into a U-shape, filling the U with the flux material, squeezing the strip shut around the flux, and then drawing the tube through dies to size the wire and compact the flux. During manufacture, the wire is carefully inspected for defects. Flux-cored wires are manufactured in a range of sizes beginning at 0.023" (0.6 mm). The finished flux-cored wire is coiled on all the standard spool sizes and forms.

Keep in mind that all flux-cored wires are considered to be low hydrogen. For this reason, proper electrode control procedures must be followed to keep the wire dry.

By varying the power settings on the welding machine, many smaller sizes of electrode wire can be used both for thin metals and for multiple-pass welds on metals of unlimited thickness in all positions.

Magnesium Fires

While the ignition of magnesium is a very remote possibility when welding, the ignition will cease when the heat source is removed. Ignition of the weld pool is also prevented by the gas shielding used in GTAW, GMAW, and PAW processes. Most magnesium fires occur with accumulations of fines (powder-sized granules) from grinding or filing, or chips from machining. Accumulations of fines on clothing should be avoided. Graphite-based or salt-based powders recommended for extinguishing magnesium fires should be stored in the work area. If large amounts of fines are produced, they should be collected in a water-wash dust collector designed for use with magnesium. Special precautions for the handling and final disposal of wet magnesium fines must then followed.

There have been several AWS standards related to flux-cored wire over the years. However, a new standard has recently been introduced by AWS. *AWS A5.36/A5.36M* is presently in effect and incorporates the information historically provided in several other standards. The new standard covers carbon and low-alloy carbon steel flux-cored products. Manufacturers may continue to label packaging in reference to the previous standards until January 1, 2016. After this date, labeling per the new standard will be required. As a result, both the old and new standards will be seen until all of the products manufactured and labeled before the deadline have been distributed and consumed.

Common AWS specifications for flux-cored electrodes that will be discussed briefly in this section include the following:

- *AWS A5.20/A5.20M: Specification for Carbon Steel Electrodes for Flux Cored Arc Welding*
- *AWS A5.22: Specification for Stainless Steel Electrodes for Flux Cored Arc Welding and Stainless Steel Flux Cored Rods for Gas Tungsten Arc Welding*
- *AWS A5.29/A5.29M: Specification for Low-Alloy Steel Electrodes for Flux Cored Arc Welding*

Note that these standards are discussed below, even though the *AWS A5.36/A5.36M* standard is already in effect. However, a great deal of product awaiting consumption is labeled according to the previous standards. Standards *AWS A5.20/A5.20M* and *AWS A5.29/A5.29M* will be archived and their contents are now reflected in the new standard. The standard related to flux-cored stainless steel electrodes, *AWS A5.22*, will remain in effect.

Each standard defines specifications for the following:

- Classification, identification, and marking of the electrode wire
- Chemical composition of the deposited weld metal
- Chemical composition of the flux filling
- Mechanical properties of the deposited weld metal

FCAW flux-cored electrodes are now classified by the AWS in the new standard using either a fixed or an open classification system. The fixed classification system has been carried over from an earlier system that provided limited descriptors for the welding characteristics of the electrode. As an example, *Figure 49* shows the AWS fixed classification for mild steel and low-alloy steel FCAW electrodes under the newer *AWS A5.36* standard.

The open classification system allows more designators to be defined, which provides more information to personnel for selecting electrodes based on process, welding characteristics, strengths, and weld positions. *Figure 50* shows the AWS open classification for mild steel and low-alloy steel FCAW electrodes.

3.2.1 Carbon Steel Flux-Cored Electrodes

Figure 51 shows the AWS A5.20 specification format for carbon steel flux-cored electrodes.

The carbon steel flux-cored electrode welding characteristics as indicated by the AWS classification are shown in *Table 6*. Carbon steel flux-cored electrode classes and uses are shown in *Table 7*.

3.2.2 Low-Alloy Steel Flux-Cored Electrodes

Figure 52 shows the *AWS A5.29* specification format for low-alloy steel flux-cored electrodes.

3.2.3 Stainless Steel Flux-Cored Electrodes and Rods

Figure 53 shows the *AWS A5.22* specification format for stainless steel flux-cored electrodes and rods.

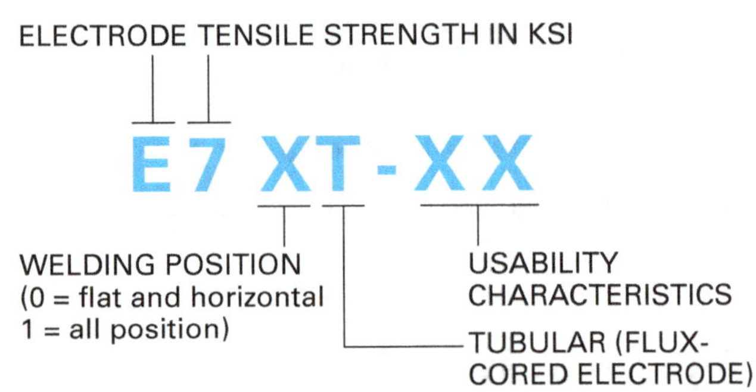

Figure 49 AWS fixed classification for mild steel and low-alloy steel FCAW electrodes.

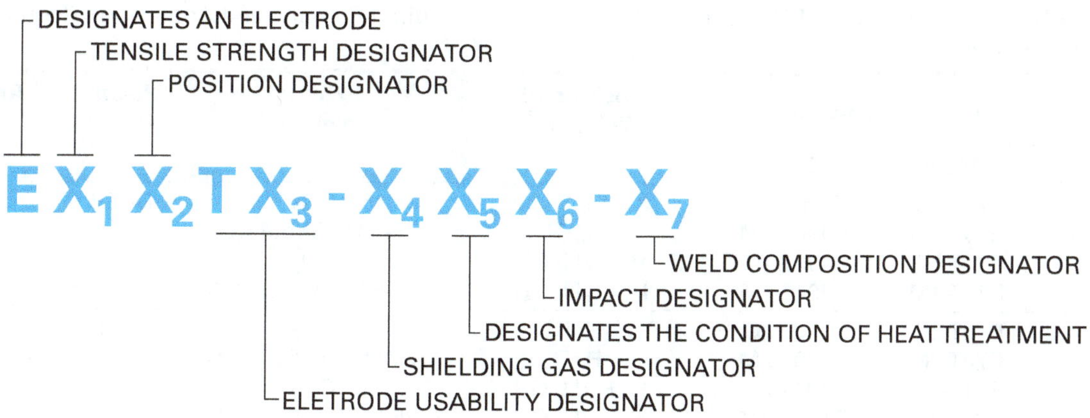

Figure 50 AWS open classification for mild steel and low-alloy steel FCAW electrodes.

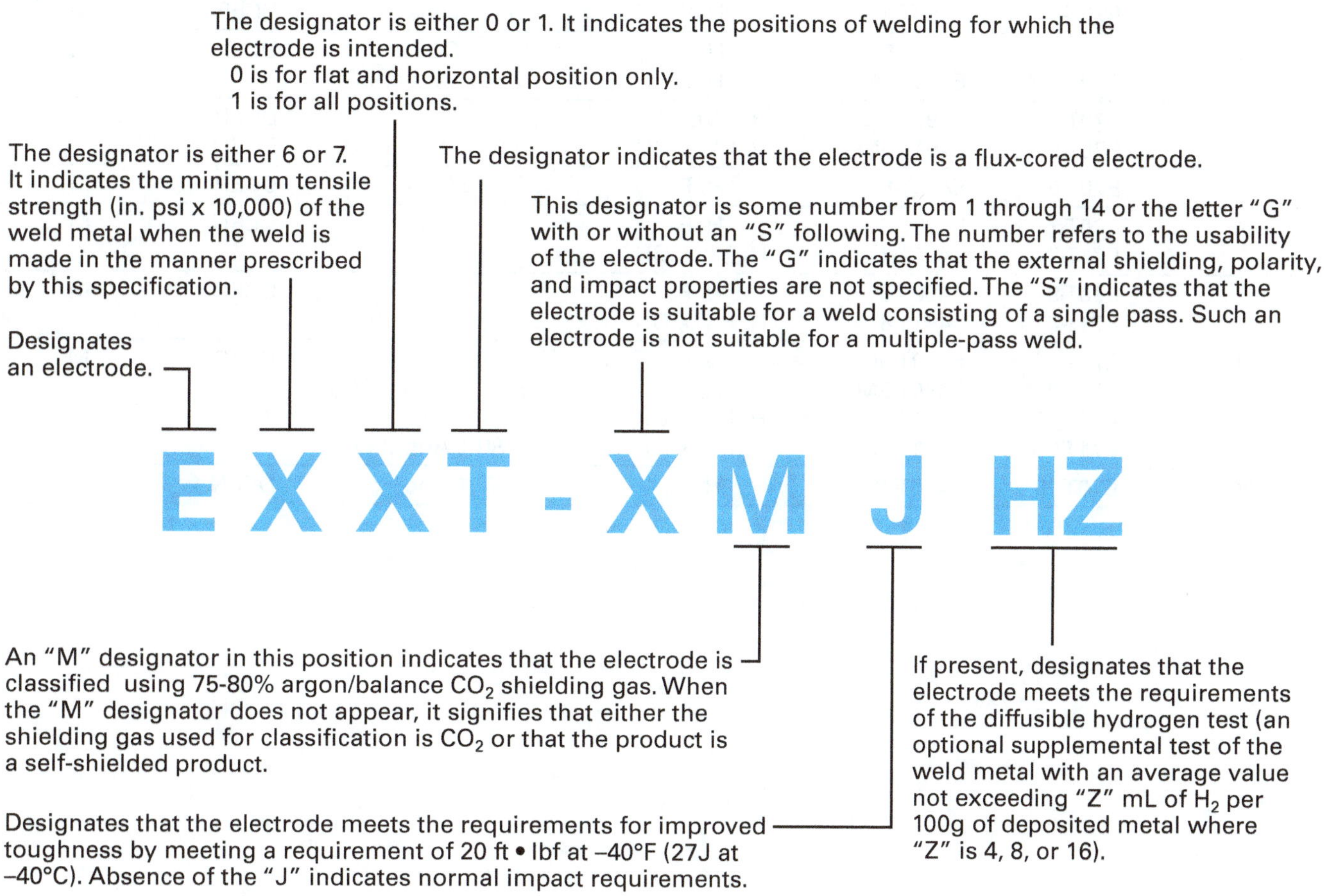

Figure 51 AWS classification for carbon steel flux-cored electrodes.

Table 6A Carbon Steel Flux-Cored Electrode Welding Position, Shielding, Polarity, and Application Requirements (Sheet 1 of 2)

Usability Designator	AWS Classification		Position of Welding[a,b]	External Shielding[c]	Polarity[d]	Application[e]
	A5.20	A5.20M				
1	E70T-1C	E490T-1C	H, F	CO_2	DCEP	M
	E70T-1M	E490T-1M	H, F	75-80% Ar/bal CO_2		
	E71T-1C	E491T-1C	H, F, VU, OH	CO_2		
	E71T-1M	E491T-1M	H, F, VU, OH	75-80% Ar/bal CO_2		
2	E70T-2C	E490T-2C	H, F	CO_2	DCEP	S
	E70T-2M	E490T-2M	H, F	75-80% Ar/bal CO_2		
	E71T-2C	E491T-2C	H, F, VU, OH	CO_2		
	E71T-2M	E491T-2M	H, F, VU, OH	75-80% Ar/bal CO_2		
3	E70T-3	E490T-3	H, F	None	DCEP	S
4	E70T-4	E490T-4	H, F	None	DCEP	M
5	E70T-5C	E490T-5C	H, F	CO_2	DCEP	M
	E70T-5M	E490T-5M	H, F	75-80% Ar/bal CO_2		
	E70T-5C	E491T-5C	H, F, VU, OH	CO_2	DCEP or	
	E70T-5M	E491T-5M	H, F, VU, OH	75-80% Ar/bal CO_2	DCEN[f]	
6	E70T-6	E490T-6	H, F	None	DCEP	M
7	E70T-7	E490T-7	H, F	None	DCEN	M
	E71T-7	E491T-7	H, F, VU, OH			
8	E70T-8	E490T-8	H, F	None	DCEN	M
	E71T-8	E491T-8	H, F, VU, OH			
9	E70T-9C	E490T-9C	H, F	CO_2	DCEP	M
	E70T-9M	E490T-9M	H, F	75-80% Ar/bal CO_2		
	E71T-9C	E491T-9C	H, F, VU, OH	CO_2		
	E71T-9M	E491T-9M	H, F, VU, OH	75-80% Ar/bal CO_2		
10	E70T-10	E490T-10	H, F	None	DCEN	S

Table 6B Carbon Steel Flux-Cored Electrode Welding Position, Shielding, Polarity, and Application Requirements (Sheet 2 of 2)

Usability Designator	AWS Classification A5.20	AWS Classification A5.20M	Position of Welding[a,b]	External Shielding[c]	Polarity[d]	Application[e]
11	E70T-11	E490T-11	H, F	None	DCEN	M
	E71T-11	E491T-11	H, F, VU, OH			
12	E70T-12C	E490T-12C	H, F	CO_2	DCEP	M
	E70T-12M	E490T-12M	H, F	75-80% Ar/bal CO_2		
	E71T-12C	E491T-12C	H, F, VU, OH	CO_2		
	E71T-12M	E491T-12M	H, F, VU, OH	75-80% Ar/bal CO_2		
13	E61T-13	E431T-13	H, F, VU, OH	None	DCEN	S
	E71T-13	E491T-13				
14	E71T-14	E491T-14	H, F, VU, OH	None	DCEN	S
	E60T-G	E430T-G	H, F	Not Specified	Not Specified	M
	E70T-G	E490T-G				
	E61T-G	E431T-G	H, F, VD or VU, OH	Not Specified	Not Specified	M
	E71T-G	E491T-G				
G	E60T-GS	E430T-GS	H, F	Not Specified	Not Specified	S
	E70T-GS	E490T-GS				
	E61T-GS	E431T-GS	H, F, VD or VU, OH	Not Specified	Not Specified	S
	E71T-GS	E491T-GS				
	EX0T-Gg	EX0T-Gg	H, F	Not Specified	Not Specified	S
	EX1T-Gg	EX1T-Gg	H, F, VD or VU, OH	Not Specified	Not Specified	
	EX0T-GSg	EX0T-GSg	H, F	Not Specified	Not Specified	
	EX1T-GSg	EX1T-GSg	H, F, VD or VU, OH	Not Specified	Not Specified	

Notes:

[a] H = horizontal position; F = flat position; OH = overhead position; VD = vertical position with downward progression; VU = vertical position with upward progression.

[b] Electrode sizes suitable for out-of-position welding, i.e., welding position other than flat or horizontal, usually are those sizes that are smaller than the 3⁄32" (2.4 mm) size. For that reason, electrodes meeting the requirements for the groove weld tests and the fillet weld tests may be classified as EX1T-X or EX1T-XM (where X represents the tensile strength and usability designators) regardless of their size.

[c] Properties of weld metal from electrodes that are used with external gas shielding (EXXT-1, EXXT-1M, EXXT-2, EXXT-2M, EXXT-5, EXXT-5M, EXXT-9, EXXT-9M, EXXT-12, and EXXT-12M) vary according to the shielding gas employed. Electrodes classified with the specified shielding gas should not be used with other shielding gases without first consulting the manufacturer of the electrode.

[d] The term "DCEP" refers to direct current electrode positive (DC, reverse polarity). The term "DCEN" refers to direct current electrode negative (DC, straight polarity).

[e] M = single or multiple pass; S = single pass only. (Single-pass only classes are highly oxidizing and are used on metals that are rusty or have mill scale.)

[f] Some E71T-5 and E71T-5M electrodes may be recommended for use on DCEN for improved out-of-position welding. Consult the manufacturer for the recommended polarity.

[g] The letter "X" can be replaced with either a "6" or "7" to designate the weld metal's tensile strength.

Table 7A Carbon Steel Flux-Cored Electrode Classes and Uses (1 of 2)

AWS Classification	Intended Uses
EXXT-1 EXXT-1M	Electrodes designated as –1 use CO_2 shielding. However, other gas mixtures can be used to improve arc stability if recommended by the manufacturer. Electrodes designated as –1M use an argon-CO_2 mixture to increase manganese and silicon in the weld metal to increase yield and tensile strengths. Reducing argon in a gas mixture or using CO_2 shielding only with –1M electrodes may cause deterioration of arc and out-of-position welding characteristics and loss of strength. Both electrodes have spray transfer characteristics and low spatter. Most electrodes of this class have rutile-base slag and produce high disposition rates.
EXXT-2	These electrodes are similar to –1 and –1M types with higher manganese or silicon, or both. The higher levels of deoxidizers allow single-pass welding of heavily oxidized or rimmed steel. The –2M electrodes have an even higher manganese content and can be used on metals with heavier mill scale, rust, or foreign matter that cannot be tolerated by some –1 or –1M electrodes.
EXXT-3	These self-shielding electrodes are designed for single-pass, very high-speed welding and use spray transfer. Besides the listed positions, they can be used on 20-degree inclined downhill welds of sheet metal. They are not recommended for T or lap joints in materials thicker than 3⁄16" or for butt, edge, or corner joints in materials thicker than ¼".
EXXT-4	These self-shielding electrodes use globular-type transfer and are designed for very high deposition rates. They produce low-sulfur welds that resist hot cracking. They are also designed for low penetration beyond the root of the weld because of poor joint fitup.
EXXT-5 EXXT-5M	These electrodes have a globular transfer and a lime-fluoride base slag that produces welds that have better impact strength and hot and/or cold cracking resistance than other similar classes with rutile-based slags. The –5 electrodes are used with a CO_2 shielding gas, however, argon-CO_2 mixes can be used to reduce spatter if recommended by the manufacturer. Reducing argon in a gas mixture or using CO_2 shielding only with -5M electrodes may cause deterioration of arc an out-of-position welding characteristics and loss of strength. Operator appeal for both types of electrodes is not as good as other similar classes with rutile slag.
EXXT-6	These self-shielding electrodes use spray-type transfer. The slag system is designed for good penetration into the root of the weld and excellent slag removal, even in deep grooves.
EXXT-7	These self-shielding electrodes use a droplet/spray-type transfer and produce a low-sulfur weld that is very resistant to cracking. The slag system is designed to produce high deposition rates for larger size electrodes in the flat or horizontal positions.
EXXT-8	These self-shielded electrodes have small droplet/spray-type transfer. They are suitable for all positions. The weld metal has good toughness, low temperature performance, and crack resistance.
EXXT-9 EXXT-9M	These electrodes are essentially the same as the –1 and –M electrodes except the deposit weld metal with improved impact properties. Some electrodes in this class require joints that are relatively clean of oil, excess oxide, and scale.
EXXT-10	These self-shielding electrodes have small droplet transfer. They are used for high speed, single-pass welds on any thickness material in the flat, horizontal, and 20-degree inclined vertical positions.
EXXT-11	These self-shielding electrodes have a smooth spray-type transfer and are general-purpose electrodes for all positions. Their use on material thicknesses greater than ¾" is not generally recommended unless preheat and interpass temperatures are maintained. Consult the manufacturer for specific recommendations.

Table 7B Carbon Steel Flux-Cored Electrode Classes and Uses (2 of 2)

AWS Classification	Intended Uses
EXXT-12 EXXT-12M	These electrodes are essentially the same as –1 and –1M electrodes except they are modified to improve impact toughness and to lower manganese requirements of the A-1 Analysis Group in the *ASME Boiler and Pressure Code Section IX*. As a result, tensile strength and hardness will be decreased from –1 and –1 electrodes. Users should check hardness when a required hardness level is specified or required.
EXXT-13	These self-shielding electrodes are usually welded with a short-circuiting transfer. The slag system allows an all-position root pass on pipe welds. They are recommended only for the root pass and are not generally recommended for multiple passes.
EXXT-14	These self-shielding electrodes have a smooth spray-type transfer. The slag system allows high speed welds in all positions. They are used to make welds on sheet metal up to 3⁄16" thick and are often designed especially for galvanized, aluminized or other coated steels. They are generally not recommended for T or lap joints in materials thicker than 3⁄16" or butt, edge, or corner joints in materials thicker than ¼". The manufacturer should be consulted for specific recommendations.
EXXT-G	These are multipass electrodes not covered by other classifications. Except for requirements to assure carbon steel deposits and tensile strengths, other requirements are agreed upon by the purchaser and manufacturer.
EXXT-GS	These are single-pass electrodes not covered by other classifications. Except for tensile strength, other requirements are agreed upon by the purchaser and the manufacturer.

FCAW of Structural Steel

For large structural welding projects such as bridges and skyscrapers, many construction companies prefer carbon or low-alloy steel FCAW to meet the toughness requirements for new construction using the newer self-shielding E71T-8 or gas-shielded E71T-1 electrodes. Many ironworkers who use SMAW are reluctant to use FCAW, and those who have used FCAW predominantly relied upon E71T-11 (a limited-thickness electrode) for structural welds because it was easy to use in all positions. However, welds with –11 wire sometimes had cracking problems and had to be rewelded. When welders are properly retrained for FCAW using the newer electrodes, the cracking problems are eliminated and the deposition rates are higher because the newer electrodes run hotter and faster than the –11 electrodes or SMAW electrodes.

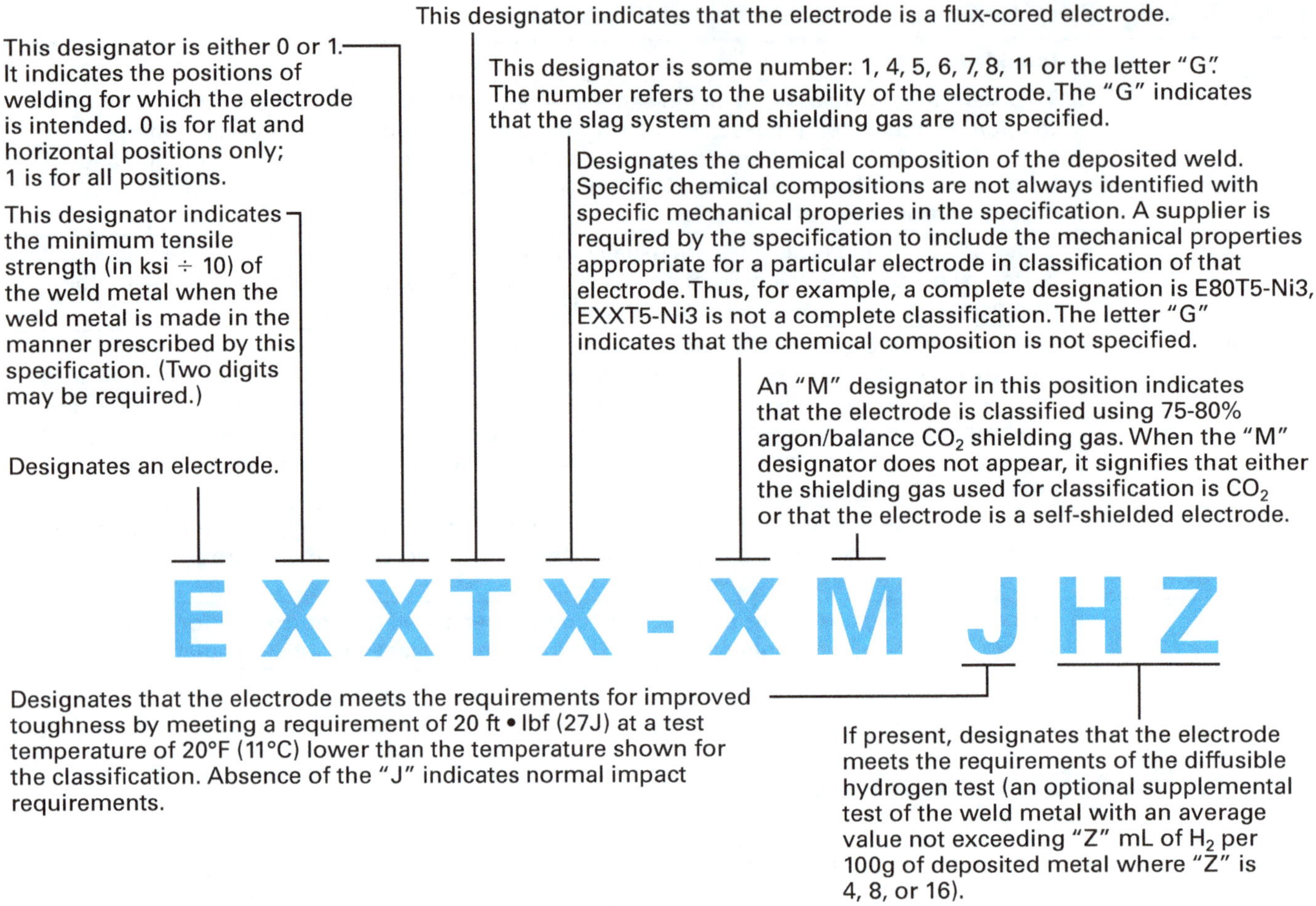

Figure 52 AWS classification for low-alloy steel flux-cored electrodes.

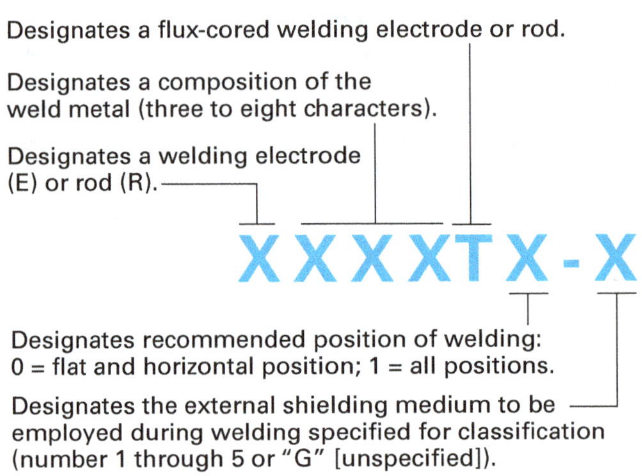

Figure 53 AWS specification for stainless steel flux-cored electrodes and rods.

Additional Resources

AWS B1.10M/B1.10 Guide for the Nondestructive Examination of Welds. Miami, FL: American Welding Society.

AWS B1.11 Guide for the Visual Examination of Welds. Miami, FL: American Welding Society.

Lincoln Electric website: **http://www.lincolnelectric.com** offers sources for products and training.

Modern Welding Technology. Howard B. Cary. Englewood Cliffs, NJ: Prentice Hall, Inc.

OSHA 1926.351, Arc Welding and Cutting. Current Edition. Washington, DC: Occupational Safety and Health Administration (OSHA).

Welding Handbook. Volume 1. Welding Science & Technology. Miami, FL: American Welding Society, 2001.

Welding Handbook. Volume 2. Part 1: Welding Processes. Miami, FL: American Welding Society, 2004.

3.0.0 Section Review

1. Which of these filler metals is sensitive to embrittlement by oxygen, nitrogen, and hydrogen at temperatures above 500°F (260°C)?
   a. Nickel
   b. Magnesium
   c. Titanium
   d. Aluminum

2. In FCAW operations, it is possible to use a smaller size of electrode wire for thin metals and multiple-pass welds by varying the welding machine's ____.
   a. pinch
   b. polarity cycling
   c. feed drag
   d. power settings

SUMMARY

Performing gas metal arc welding and flux-cored arc welding safely and effectively requires a good understanding of the welding equipment, the filler metals that are used, and the proper safety precautions that should be followed. There are many different types of GMAW and FCAW equipment and components. Before any GMAW or FCAW can begin, workers must first select the appropriate safety and welding equipment, as well as the proper filler metals. They must also know how to safely move and set up the equipment. This module has provided enough information about these basic responsibilities to enable workers to use GMAW and FCAW equipment safely and efficiently.

Review Questions

1. In GMAW, the arc is stabilized and the molten filler and base metals are protected from oxidation by the use of ____.
   a. differential pressure
   b. self-fluxing alloy
   c. intense heat
   d. shielding gas

2. Of the two basic FCAW processes, one is self-shielding while the other one ____.
   a. requires no shielding gas
   b. uses ferrous core vapors
   c. uses external shielding gas
   d. utilizes an oxygen generator

3. Areas where oil, grease, and other petroleum products are present, should be kept clear of accumulated ____.
   a. oxygen
   b. helium
   c. argon
   d. nitrogen

4. Which metal transfer mode uses a high welding voltage that ultimately causes fine droplets of electrode wire to be carried axially to the molten weld puddle?
   a. Globular transfer
   b. Inverter transfer
   c. Short-circuiting transfer
   d. Spray transfer

5. Short-circuiting metal transfer is the best mode to use for welding ____.
   a. cast iron blocks
   b. light-gauge sheet metal
   c. heavy steel beams
   d. aluminum plates

6. FCAW welds made with CO_2 shielding gas produce ____.
   a. deep penetration
   b. a slag-free weld
   c. shallow penetration
   d. a lot of spatter

7. Depending on the range selected, for every 100 amperes of current the arc voltage of a welding machine is typically lower than the open-circuit voltage by ____.
   a. 0 to 1 volts
   b. 1 to 2 volts
   c. 2 to 3 volts
   d. 4 to 5 volts

8. The welding arc generates intense heat that ranges from ____.
   a. 1,000°F to 5,000°F (530°C to 2,760°C)
   b. 6,000°F to 10,000°F (3,316°C to 5,538°C)
   c. 5,000°F to 12,500°F (2,760°C to 6,927°C)
   d. 8,500°F to 12,500°F (4,704°C to 6,927°C)

9. One advantage of using engine-driven generators as a way to power welders is that the equipment is ____.
   a. less expensive than single-phase welders
   b. easier to operate than all other welders
   c. free of most maintenance requirements
   d. portable so it can be used in the field

10. Which welding power source increases the frequency of the incoming primary power and is well suited for applications where space is limited?
    a. Inverter
    b. Engine-driven generator
    c. Axial
    d. Transformer-rectifier

11. When plotted on a graph, the general angle showing the voltage-to-amperage relationship for GMAW is known as the ____.
    a. pinch
    b. arc blow
    c. slope
    d. hypotenuse

12. Two factors that must be considered when selecting the size of a welding cable are the amperage load and the ____.
    a. type of copper strands in the cable
    b. distance the current will travel
    c. speed of the wire electrode feed
    d. pinch effect imparted on the cable

13. The duty cycle of a GMAW gun is determined and rated with _____.
    a. carbon dioxide (CO_2)
    b. helium (He)
    c. argon (Ar)
    d. oxygen (O)

14. Light-duty GMAW guns are cooled by _____.
    a. domestic water flow
    b. special cooling fluid
    c. shielding gas flow
    d. demineralized water

15. Because of its current transfer and ionization properties, the most common shielding gas that can be used alone or in combination with other gases is _____.
    a. carbon dioxide (CO_2)
    b. helium (He)
    c. nitrogen (N)
    d. argon (Ar)

16. What is required to meter shielding gas to a welding gun at the proper flow rate?
    a. A gas mixer/stabilizer
    b. A variable vacuum tip
    c. A gas regulator/flowmeter
    d. An auxiliary gun controller

17. After checking the equipment specifications to determine the weight of a welding machine, what is the proper way to lift and move the machine?
    a. Use a wire rope basket hitch under the welder.
    b. Attach a sling and/or shackle to the welder lift eye.
    c. Use numerous synthetic slings attached to the corners of the welder.
    d. Attach the lifting hook from the lifting device to the welder lift eye.

18. A tool designed for cutting and manipulating wire, removing and installing contact tips and nozzles, and cleaning welding spatter from nozzles is a pair of _____.
    a. cast and helix straightener bars
    b. knurled combination rollers
    c. push-pull wire straighteners
    d. MIG welding, or wire feed, pliers

19. Many types of GMAW carbon steel filler wire are protected from corrosion by a(n) _____.
    a. copper coating
    b. aluminum core
    c. mercury alloy
    d. nonferrous skin

20. The primary use of FCAW is for the welding of _____.
    a. nonferrous metals
    b. aluminum
    c. ferrous metals
    d. titanium

Trade Terms Introduced in This Module

Axially: In a straight line along an axis.

Composite metal-core wire electrode: A filler wire electrode with a hollow core containing powdered materials, primarily metals.

Globular transfer: A GMAW process in which the electrode wire does not touch the base metal, and the filler metal is transferred randomly in the arc as large, irregular globules.

Inductance: An electrical circuit property, adjustable on some FCAW power supplies, that influences the pinch effect.

Pinch effect: The tendency of the end of the wire electrode to melt into a droplet during short-circuit GMAW welding.

Pulsed transfer: A GMAW welding process in which the welding power is cycled from a low level (background level) to a higher level (welding level), resulting in lower average voltage and current; this is a variation of the spray transfer process.

Short-circuiting transfer: A GMAW welding process in which the electrode wire shorts against the base metal. The welding current repeatedly and rapidly melts the wire and establishes an arc, which is extinguished when the wire again touches the base metal.

Spray transfer: A GMAW welding process in which the electrode wire does not touch the base metal, and the filler metal is transferred axially (in a straight line) in the arc as fine droplets.

Stickout: The length of electrode that projects from the contact tube, measured from the arc tip to the contact tube.

Additional Resources

This module is intended as a thorough resource for task training. The following reference works are suggested for further study.

AWS B1.10M/B1.10 Guide for the Nondestructive Examination of Welds. Miami, FL: American Welding Society.

AWS B1.11 Guide for the Visual Examination of Welds. Miami, FL: American Welding Society.

Lincoln Electric website: **http://www.lincolnelectric.com** offers sources for products and training.

Modern Welding Technology. Howard B. Cary. Englewood Cliffs, NJ: Prentice Hall, Inc.

OSHA 1926.351, Arc Welding and Cutting. Current Edition. Washington, DC: Occupational Safety and Health Administration (OSHA).

Welding Handbook. Volume 1. Welding Science & Technology. Miami, FL: American Welding Society, 2001.

Welding Handbook. Miami, FL: American Welding Society, 2004.

Figure Credits

The Lincoln Electric Company, Cleveland, OH, USA, Module opener, Figures 3, 4, 14–17, SA01, 23C, 25, 26, 28, 30, SA02, 31–34, 36, 38 (photo)

Miller Electric Mfg. Co., Figure 18

AWS C5.6:1989, Figure 6, Reproduced with permission from the American Welding Society (AWS), Miami, FL, USA, Figure 21

Topaz Publications, Inc., Figures 23A and B, 24A and B, 39

Courtesy of Sumner Manufacturing Co., Figure 24C

Magswitch Technology Pty Ltd, Figure 24D

Courtesy of Irwin Tools, Figure 40

AWS A5.18/A5.18M:2005, Figure A.1, Reproduced with permission from the American Welding Society (AWS), Miami, FL, USA, Figure 41

AWS PHB-4:2000, Page 9, Reproduced with permission from the American Welding Society (AWS), Miami, FL, USA, Tables 3, 4

AWS A5.20/A5.20M:2005, Figure A.1, Reproduced with permission from the American Welding Society (AWS), Miami, FL, USA, Figure 51

AWS A5.20/A5.20M:2005, Table 2, Reproduced with permission from the American Welding Society (AWS), Miami, FL, USA, Table 6

AWS A5.29:2005, Figure A.1, Reproduced with permission from the American Welding Society (AWS), Miami, FL, USA, Figure 52

Section Review Answer Key

Answer	Section Reference	Objective
Section One		
1. c	1.1.0	1a
2. b	1.2.3	1b
3. d	1.3.3	1c
4. b	1.4.0	1d
Section Two		
1. a	2.1.1	2a
2. c	2.2.3	2b
3. d	2.3.1	2c
4. b	2.4.0	2d
5. c	2.5.2	2e
6. a	2.6.1	2f
7. b	2.7.4	2g
Section Three		
1. c	3.1.8	3a
2. d	3.2.0	3b

NCCER CURRICULA — USER UPDATE

NCCER makes every effort to keep its textbooks up-to-date and free of technical errors. We appreciate your help in this process. If you find an error, a typographical mistake, or an inaccuracy in NCCER's curricula, please fill out this form (or a photocopy), or complete the online form at **www.nccer.org/olf**. Be sure to include the exact module ID number, page number, a detailed description, and your recommended correction. Your input will be brought to the attention of the Authoring Team. Thank you for your assistance.

Instructors – If you have an idea for improving this textbook, or have found that additional materials were necessary to teach this module effectively, please let us know so that we may present your suggestions to the Authoring Team.

NCCER Product Development and Revision

13614 Progress Blvd., Alachua, FL 32615

Email: curriculum@nccer.org
Online: www.nccer.org/olf

❑ Trainee Guide ❑ Lesson Plans ❑ Exam ❑ PowerPoints Other ______________

Craft / Level: Copyright Date:

Module ID Number / Title:

Section Number(s):

Description:

Recommended Correction:

Your Name:

Address:

Email: Phone:

GTAW – Equipment and Filler Metals

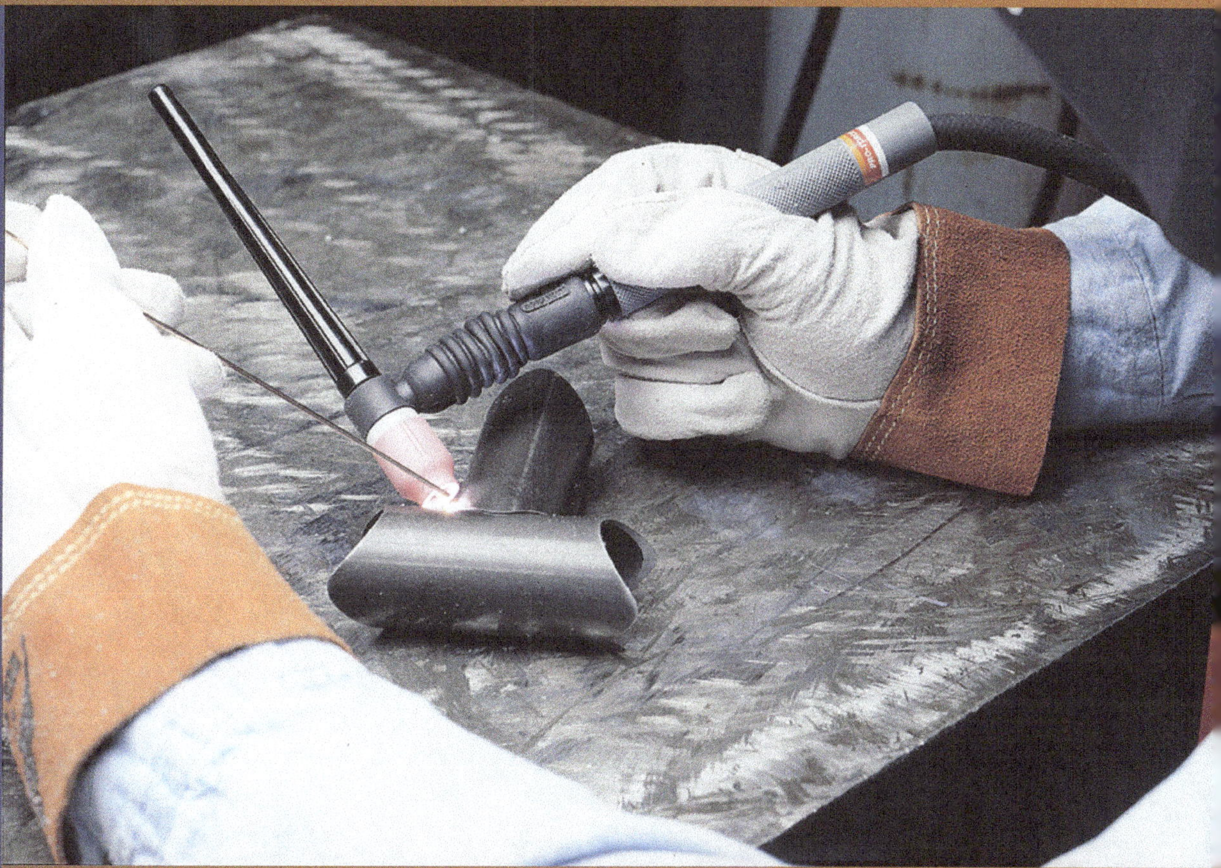

Overview

To produce consistent, high-quality welds, welders must select, install, and use the equipment and materials involved in the GTAW process properly. This module provides an overview of the equipment and safety concerns that are associated with GTAW. It describes GTAW equipment and consumables, including power sources, torches, nozzles, electrodes, shielding gases, and filler metals. It also explains how to select, prepare, and set up GTAW equipment safely and efficiently.

Module 29207

Trainees with successful module completions may be eligible for credentialing through the NCCER Registry. To learn more, go to **www.nccer.org** or contact us at **1.888.622.3720**. Our website has information on the latest product releases and training, as well as online versions of our *Cornerstone* magazine and Pearson's product catalog.

Your feedback is welcome. You may email your comments to **curriculum@nccer.org**, send general comments and inquiries to **info@nccer.org**, or fill in the User Update form at the back of this module.

This information is general in nature and intended for training purposes only. Actual performance of activities described in this manual requires compliance with all applicable operating, service, maintenance, and safety procedures under the direction of qualified personnel. References in this manual to patented or proprietary devices do not constitute a recommendation of their use.

29207 V5

29207

GTAW – Equipment and Filler Materials

Objectives

When you have completed this module, you will be able to do the following:

1. Identify GTAW-related safety practices and describe the electrical characteristics that affect GTAW.
   a. Identify GTAW-related safety practices.
   b. Describe the electrical characteristics that affect GTAW.
2. Identify and describe GTAW equipment and consumables.
   a. Identify and describe GTAW welding machines.
   b. Identify and describe GTAW torches.
   c. Identify and describe GTAW torch nozzles and electrodes.
   d. Identify and describe GTAW shielding gases.
   e. Identify and describe GTAW filler metals.
3. Explain how to set up for GTAW welding.
   a. Explain how to select and position the welding machine.
   b. Explain how to connect and set up the shielding gas flow rate.
   c. Explain how to select and prepare the tungsten electrode.
   d. Explain how to select and install the nozzle along with the tungsten electrode.

Performance Tasks

Under the supervision of your instructor, you should be able to do the following:

1. Select a GTAW shielding gas.
2. Select a GTAW filler metal.
3. Connect the shielding gas and set the flow rate.
4. Select and prepare the tungsten electrode.
5. Break down and reassemble a GTAW torch.

Trade Terms

Lift-arc technology
Potentiometer
Truncated
Tungsten

Industry Recognized Credentials

If you are training through an NCCER-accredited sponsor, you may be eligible for credentials from NCCER's Registry. The ID number for this module is 29207. Note that this module may have been used in other NCCER curricula and may apply to other level completions. Contact NCCER's Registry at 888.622.3720 or go to **www.nccer.org** for more information.

Contents

Topics to be presented in this module include:

Figures and Tables

Figures and Tables (Continued)

SECTION ONE

1.0.0 GTAW

Objective

Identify GTAW-related safety practices and describe the electrical characteristics that affect GTAW.

a. Identify GTAW-related safety practices.
b. Describe the electrical characteristics that affect GTAW.

Trade Terms

Lift-arc technology: A type of GTAW technology in which the tungsten electrode may be touched to the work, but the arc will not strike until the tip is lifted away to an appropriate distance.

Truncated: Having the pointed top (vertex) of a conical shape cut off, leaving only a plane section parallel to the base of the cone.

Tungsten: A nonstandard common name for the tungsten electrode used in GTAW; also, the principal metal from which the electrodes are constructed.

Gas tungsten arc welding (GTAW) is an arc welding process that uses a welding power source to produce an electric arc between a tungsten electrode and the base metal. The arc melts and fuses the base metal to form the weld. A filler metal is typically added. In processes where a filler metal is not added, it is referred to as autogenous welding. In manual welding (*Figure 1*), the filler metal is usually a handheld metal rod or wire that is fed into the leading edge of the weld metal pool. An inert gas or gas mixture is used to shield the electrode and molten weld metals to prevent oxidation and contamination from the atmosphere. The finished weld has no slag because no flux is used. Also, since no filler metal is transported within the arc, the process produces little or no spatter.

GTAW is used to weld a variety of metals. The process is slower than gas metal arc welding (GMAW); however, it produces high-quality welds. GTAW has been commonly used for root pass welds on pipe where the subsequent fill and cover passes are performed with SMAW, GMAW, or FCAW. In the past, GTAW was called tungsten inert gas (TIG) welding. Today, the nonstandard term TIG is still used, including in the control nomenclature marked on some welding machines.

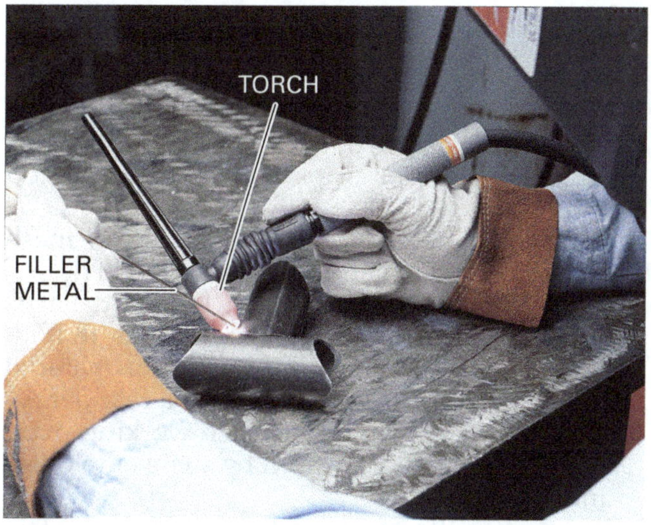

Figure 1 Performing manual GTAW.

1.1.0 Safety Summary

The following is a summary of safety procedures and practices that welders must observe when cutting or welding. Keep in mind that this is only a summary. Complete safety coverage is provided in the Welding Level One module, *Welding Safety*. That module should be completed before continuing with this one. Above all, be sure to wear appropriate protective clothing and equipment when welding or cutting.

1.1.1 Protective Clothing and Equipment

Welding activities can cause injuries unless you wear all of the protective clothing and equipment that is designed specifically for the welding industry. The following information includes important safety guidelines about protective clothing and equipment:

- Wear a face shield over snug-fitting cutting goggles or safety glasses for gas welding or cutting. Either the face shield or the lenses of the welding goggles must be an approved shade for the application. A welding hood equipped with a properly tinted lens is best for all forms of welding.
- Wear proper protective leather and/or flame-retardant clothing along with welding gloves that protect the welder from flying sparks, molten metal, and heat.

- Wear high-top safety shoes or boots. Make sure that the tongue and lace area of the footwear will be covered by a pant leg. If the tongue and lace area is exposed or the footwear must be protected from burn marks, wear leather spats under the pants or chaps over the front and top of the footwear.
- Wear a 100-percent cotton cap with no mesh material included in its construction. The bill of the cap points to the rear. If a hard hat is required for the environment, use one that allows the attachment of rear deflector material and a face shield. A hard hat with a rear deflector is generally preferred when working overhead, and may be required by some employers and job sites.
- Wear earmuffs or at least earplugs to protect your ear canals from sparks.

WARNING!

Using proper personal protective equipment (PPE) for the hands and eyes is particularly important. The most common injuries that welders experience during GTAW operations are injuries to the fingers and eyes. Welders who fail to use proper gloves, safety glasses, and face shields when they are holding and grinding the tungsten electrode by hand can cut or burn their fingers, have splintered tungsten electrode lodge in their hands or fingers, or get small slivers of the electrode stuck in their eyes.

1.1.2 Fire/Explosion Prevention

Welding activities usually involve the use of fire or extreme heat to melt metal. Whenever fire is used, it must be controlled and contained. Welding or cutting activities are often performed on vessels that may once have contained flammable or explosive materials. Residues from those materials can catch fire or explode when a welder begins work on such a vessel. The following are fire and explosion prevention guidelines associated with welding:

- Never carry matches or gas-filled lighters in your pockets. Sparks can cause the matches to ignite or the lighter to explode, causing serious injury.
- Never use oxygen to blow dust or dirt from clothing. The oxygen can remain trapped in the fabric for a time. If a spark hits clothing during this time, the clothing can burn rapidly and violently out of control.
- Make sure that any flammable material in the work area is moved or shielded by a fire-resistant covering. Approved fire extinguishers must be available before attempting any heating, welding, or cutting operations. If a hot-work permit and a fire watch are required, be sure those items are in place before beginning and ensure that all site requirements have been observed.
- Never release a large amount of oxygen or use oxygen as compressed air. The presence of oxygen around flammable materials or sparks can cause rapid and uncontrolled combustion. Keep oxygen away from oil, grease, and other petroleum products.
- Never release a large amount of fuel gas, especially acetylene. Propane tends to concentrate in and along low areas and can ignite at a considerable distance from the release point. Acetylene is lighter than air, but is even more dangerous than propane. When mixed with air or oxygen, acetylene will explode at much lower concentrations than any other fuel.
- To prevent fires, maintain a neat and clean work area, and make sure that any metal scrap or slag is cold before disposing of it.
- Before cutting containers such as tanks or barrels, check to see if they have contained any explosive, hazardous, or flammable materials, including petroleum products, citrus products, or chemicals that decompose into toxic fumes when heated. As a standard practice, always clean and then fill any tanks or barrels with water, or purge them with a flow of inert gas such as nitrogen to displace any oxygen. Containers must be cleaned by steam cleaning, flushing with water, or washing with detergent until all traces of the material have been removed.

WARNING!

Welding or cutting must never be performed on drums, barrels, tanks, vessels, or other containers until they have been emptied and cleaned thoroughly, eliminating all flammable materials and all substances (such as detergents, solvents, greases, tars, or acids) that might produce flammable, toxic, or explosive vapors when heated. Clean containers only in well-ventilated areas. Vapors can accumulate during cleaning, causing explosions or injury.

Proper procedures for cutting or welding hazardous containers are described in the *American Welding Society (AWS) F4.1, Safe Practices for the Preparation of Containers and Piping for Welding and Cutting.*

1.1.3 Work Area Ventilation

Vapors and fumes tend to rise in the air from their sources. Welders often have to work above the welding area where the fumes are being created. Welding fumes can cause personal injuries. Good work area ventilation helps to remove the vapors and to protect the welder. The following is a list of work area ventilation guidelines to consider before and during welding activities:

- Follow confined space procedures before conducting any welding or cutting in a confined space.
- Never use oxygen for ventilation in confined spaces.
- Always perform cutting or welding operations in a well-ventilated area. Cutting or welding operations involving zinc or cadmium materials or coatings result in toxic fumes. For long-term cutting or welding of such materials, always wear an approved full-face, supplied-air respirator that uses breathing air supplied from outside of the work area. For occasional, very short-term exposure, a HEPA-rated or metal-fume filter may be used on a standard respirator.
- Make sure confined spaces are properly ventilated for cutting or welding purposes. Use powered extraction systems when available.

1.1.4 GTAW-Specific Safety

GTAW, also referred to as heliarc or TIG welding, is a process that is different from the processes used in shielded metal arc welding (SMAW), GMAW, and flux-cored arc welding (FCAW). In other forms of welding, the electrode or wire is consumed in the welding process. In GTAW, the electrode itself is not consumed, but a separate rod of filler metal is consumed.

Vapors from the melted base metal, the filler metal, and from the shielding gases used in the GTAW process can create respiratory issues. As always, welders performing any kind of work must be aware of the vapors generated from the welding processes. Argon gas is often used for shielding in the GTAW process. A combination of the arc and the gas generates substantial levels of ultraviolet (UV) light, which reacts with the oxygen in the area of the weld. UV radiation can ionize oxygen in the atmosphere near the weld, which can lead to a rise in ozone and nitrogen dioxide levels. As long as the work is being performed in well-ventilated areas, the risks are minimal. If work must be performed in confined spaces, consider using additional ventilation or respirators. Follow your employer's policies in this matter.

Because GTAW work is often performed on delicate welds, the welder is frequently positioned very close to the work. The risk of UV radiation being emitted from the electrical arc is higher because the shielding gas used with GTAW is more transparent than gases used with other forms of welding. This transparency is due to the minimal smoke produced from the weld. Because of the brightness of the GTAW process, safety representatives recommend that the welder wear darker clothing to minimize UV reflections under the welding helmet. They also recommend that the walls of the work area be painted with a paint that reduces reflections from the UV light generated by the GTAW process. Because aluminum and stainless steel reflect more light than common carbon steels, welders need to take extra precautions to protect their vision when performing GTAW work on these metals.

The GTAW process requires the base metal and the filler metal to be extremely clean. Cleaning solvents are used to prepare the metals. Welders must read all Safety Data Sheet (SDS) information about any cleaning agent before using it. Some cleaning agents are flammable, while others are nonflammable. Make sure that you know which kind you are using. Pay special attention to any storage requirements for such cleaning agents.

WARNING!

Chlorinated cleaning solvents can produce toxic gases. For this reason, they should never be used for cleaning metals in a welding setting.

The tungsten electrodes used for GTAW become contaminated over time and with use. When electrodes lose their original shapes, welders must grind the tungsten tips to reshape them and make them suitable for continued use. The tungsten dust and flying particles from these grinding activities can damage the welder's vision and present a respiratory hazard. Read and understand the SDS that comes with each batch of electrodes.

1.2.0 Welding Current

The welding arc required for GTAW is produced by electrical current from a constant-current (CC) welding machine. Electrical current is the flow of electrons along a conductor. The current flows through two welding cables, one connected to the workpiece and the other connected to the torch, which holds a tungsten electrode.

An arc is established in the gap between the end of the tungsten electrode and the work. The gap has a high electrical resistance. When the electrical current meets resistance, heat is generated. When the arc is struck across the gap, the resistance generates intense heat of 6,000°F to 10,000°F (≈3,300°C to 5,500°C), melting the base metal and forming a weld. The area surrounding the arc is flooded with a shielding gas to prevent oxidation of the tungsten electrode, filler rod, and base metal surface during the welding process. *Figure* 2 shows the GTAW welding process.

1.2.1 Characteristics of Welding Current

The current produced by a welding machine to perform welding has different characteristics than the current flowing through utility power lines. Welding current has low voltage and high amperage, while the power line current has high voltage and low amperage.

Voltage is the measure of the electromotive force or pressure that causes current to flow in a circuit. There are two voltage levels associated with welding current: open-circuit voltage and operating voltage. Open-circuit voltage is the voltage that is present when the machine is on but no welding is taking place. For GTAW, there are usually ranges of open-circuit voltages that can be selected, up to about 80V. Operating voltage, or arc voltage, is the voltage after the arc has been struck. With GTAW, arc voltage is generally much lower than the open-circuit voltage. The arc voltage drops rapidly with increasing current flow.

Amperage is the electrical current flow in a circuit. The unit of measurement for amperage is the ampere, or amp. The number of amps produced by the welding machine determines the intensity of the arc and the amount of heat generated. The higher the amperage, the hotter the arc. Usually, welders use a finger control or a foot pedal to control the GTAW welding machine current.

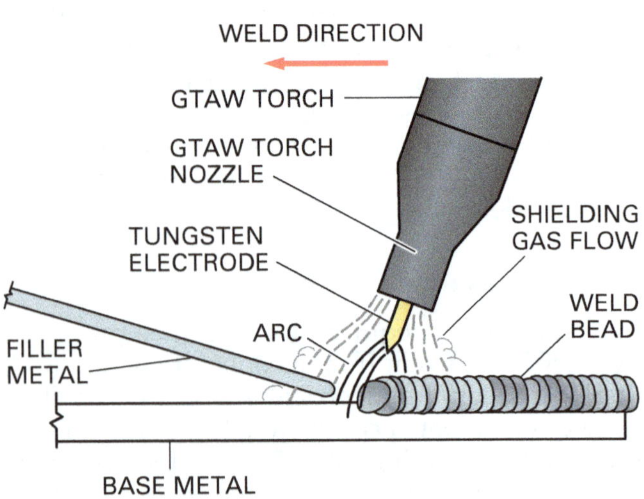

Figure 2 Diagram of the GTAW welding process.

1.2.2 DC Welding Current

Direct current (DC) is electrical current that flows in one direction only. The direction in which the current is flowing is called the polarity.

The direction in which the current flows in a circuit determines polarity. Direct current electrode negative (DCEN), also referred to by the nonstandard term *straight polarity*, is almost always used for performing GTAW on ferrous and nonferrous metals.

When the welding circuit is set up as DCEN, the tungsten electrode of the GTAW torch is connected to the negative terminal of the welding machine and the workpiece lead is connected to the positive terminal. DCEN produces the greatest amount of heat in the base metal, compared to other types of welding current. About two-thirds of the arc heat goes to the base metal and one-third goes to the tungsten electrode. Welders usually use a tungsten electrode with a pointed end that is slightly **truncated**, or blunted, for DCEN. The weld bead produced is narrow but with deep penetration. DCEN is used to weld steels, stainless steels, nickel, and titanium. Because DCEN provides limited cleaning action, the base metal surfaces must first be thoroughly cleaned of any contaminants such as rust, mill scale, or oxides.

Direct current electrode positive (DCEP), also referred to by the nonstandard term *reverse polarity*, produces the least amount of heat in the base metal. About one-third of the arc heat goes to the base metal and two-thirds goes to the tungsten electrode. Because of this, DCEP usually requires a large tungsten electrode with a rounded end. Unlike DCEN, the weld bead produced is wide but with shallow penetration. This current type also has a strong cleaning action on the surface of oxidized base metals. DCEP is only used for special applications in which shallow penetration is required. *Figure* 3 shows typical DCEN and DCEP welding circuits.

Polarity Switches

Many welding machines have a polarity switch. Instead of physically having to disconnect the welding leads and reconnect them to change polarity, the welder changes polarity by turning a switch.

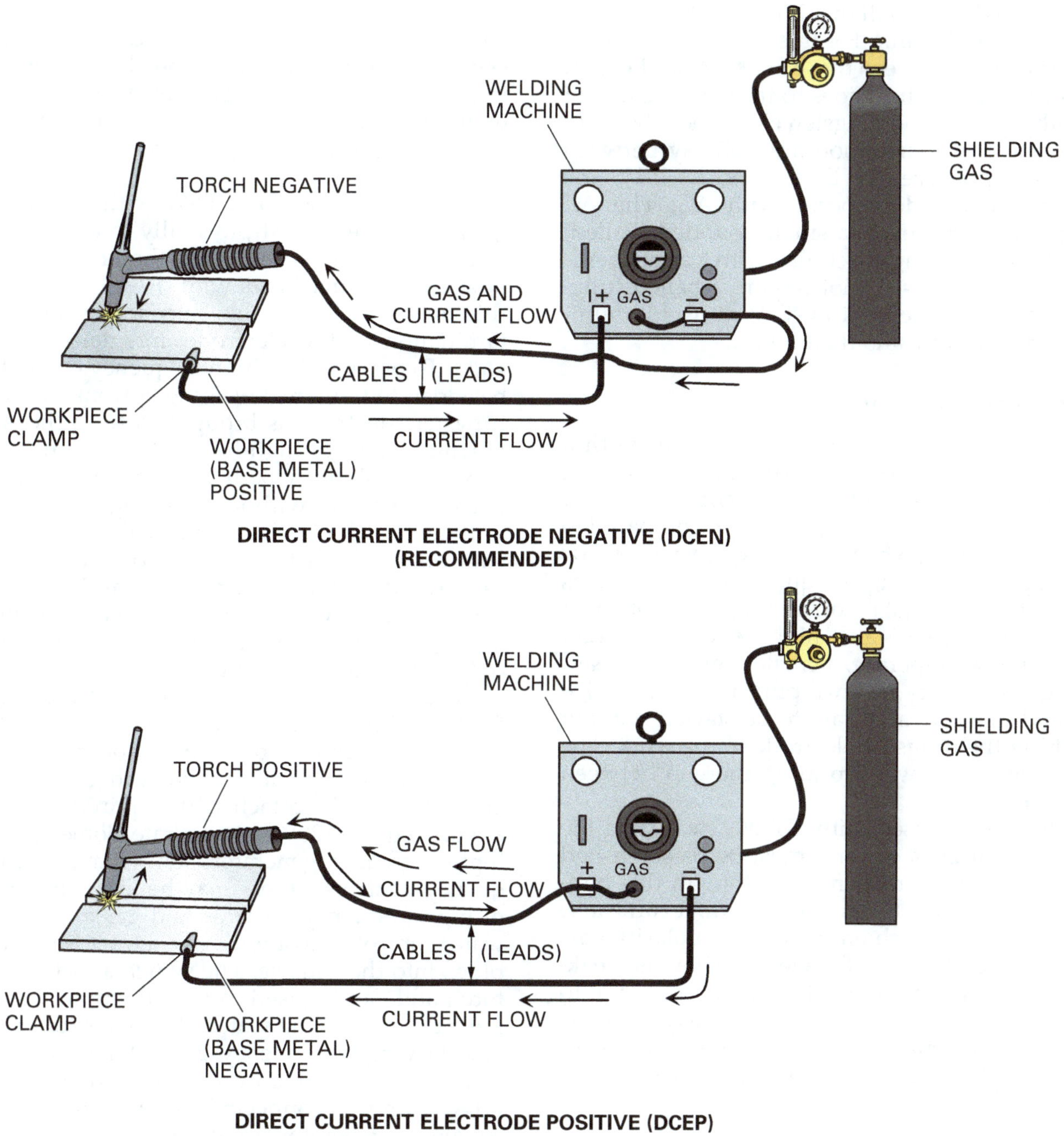

Figure 3 Representative DCEN and DCEP welding circuits.

1.2.3 AC Welding Current

Alternating current (AC) is electrical current that alternates between positive and negative values. In the positive half of the cycle, the current flows in one direction; during the negative half, the current reverses itself. The number of cycles completed in one second is called the frequency. In the United States, AC is almost always 60 cycles per second, or 60 Hertz (Hz). In other countries, the frequency is often 50 Hz. *Figure 4* shows one cycle of AC.

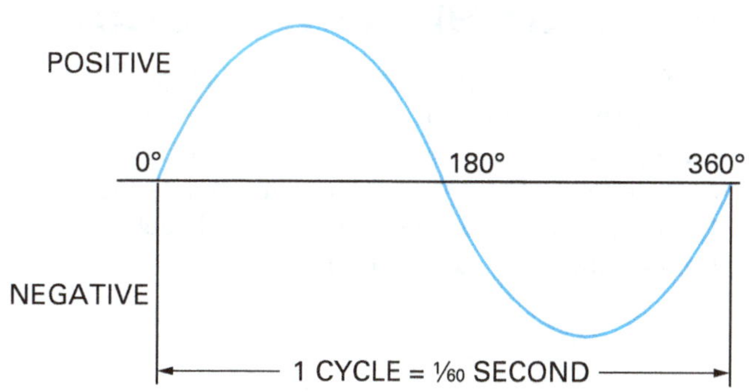

Figure 4 AC cycle.

AC is DCEN for half the time and DCEP for the remainder of the time. For this reason, AC characteristics fall in between those of DCEN and DCEP. About half of the heat goes to the base metal and the other half to the tungsten electrode. The weld bead size and penetration are midway between the beads produced by DCEN and DCEP. Some cleaning action does occur with AC. The AC tungsten electrode has a small, rounded (balled) tip. Newer welding machines with variable AC-frequency controls do not require a balled tungsten tip. AC is used with GTAW to weld aluminum, magnesium, and their alloys.

1.2.4 High Frequency

High frequency is an electrical current with a very high voltage (3,000V to 5,000V) and very low amperage. It is called high frequency because it cycles at millions of times per second compared to standard current, which has a frequency of 60 Hz. This high-voltage, high-frequency current creates a very stable arc that is able to jump a gap of about ½ inch (≈13 mm). Because high-frequency current has such low amperage, it creates very little heat.

High frequency has two purposes: to stabilize AC and to allow a DC arc to be started without touching the tungsten electrode to the work. The high voltage allows it to jump the gap between the two.

When AC is used with GTAW, each time the arc cycles, it goes out and must be reestablished in the opposite direction. Even though this happens very fast (60 times per second), it can cause the arc to extinguish on the reverse polarity part of the AC cycle when the electron flow is weakest. If this happens, oxides are absorbed into the weld, causing porosity, which weakens the weld. To prevent the AC arc from extinguishing, a high-frequency current is added to the alternating welding current. The high-frequency current, with its high voltage and low amperage, establishes a path on which the alternating welding current can travel, preventing the arc from extinguishing as it cycles. When non-squarewave AC is being used, the high frequency must be on continuously.

When using DC for GTAW without high frequency, welders must physically touch the tungsten to the work to establish the arc. Except for newer GTAW machines with lift-arc technology, this may not be desirable since it can cause contamination of the electrode and defects in the weld. To prevent this, high-frequency current can be added to the direct welding current. When the tungsten electrode is brought close to the work, the high-voltage, high-frequency arc jumps the gap, establishing a path for the direct welding current to travel without touching the tungsten to the work. Typically, a timer will automatically shut off the high frequency about two to three seconds after the welding arc has been established. Since DC flows in only one direction, it is very stable. The high voltage and frequency is only needed for starting the arc.

Welding power sources specifically designed for GTAW have a high-frequency generator built into them. If conventional power sources are used for GTAW, a separate auxiliary high-frequency generator can be attached if required. The generators require 120VAC to operate. Some auxiliary high-frequency generators only have high-frequency controls; others may have a combination of high-frequency controls and gas and cooling-water controls. Most will have a torch switch that plugs into the console. The switch attaches to the torch and can be used to turn the power, shielding gas, and cooling water (if used) on or off. The auxiliary high-frequency generator is connected to the power source by the electrode cable. The GTAW torch is connected to the auxiliary high-frequency generator (*Figure 5*).

Starting DC GTAW without High Frequency

DC GTAW is often performed without a high-frequency start for noncritical welds. The arc is touch-started in the weld zone or on a piece of copper. Copper is less likely to contaminate the tungsten. Some machines have a built-in selectable no high-frequency start mode.

High-Frequency Operation Precaution

When operating most transformer and transformer-rectifier machines in SMAW mode, welders must turn off the high-frequency generators.

REMOTE CONTROLLED ARC START SWITCH ON GTAW TORCH

GTAW WELDING MACHINE

GTAW MACHINE ON TOP OF HIGH-FREQUENCY GENERATOR

Figure 5 Auxiliary high-frequency generator with a GTAW torch.

Additional Resources

AWS B1.10, Guide for the Nondestructive Examination of Welds. Miami, FL: American Welding Society.

AWS B1.11, Guide for the Visual Examination of Welds. Miami, FL: American Welding Society.

AWS D3.5, Guide for Steel Hull Welding. Miami, FL: American Welding Society.

AWS D3.7, Guide for Aluminum Hull Welding. Miami, FL: American Welding Society.

Lincoln Electric website: **http://www.lincolnelectric.com**

Modern Welding Technology. 6th Edition. Howard B. Cary. Englewood Cliffs, NJ: Prentice Hall, Inc.

OSHA 1910.269, Appendix *C, Protection from Step and Touch Potentials.* Current Edition. Washington, DC: Occupational Safety and Health Administration (OSHA).

OSHA 19260.351, Arc Welding and Cutting. Current Edition. Washington, DC: Occupational Safety and Health Administration (OSHA).

Welding Handbook. Vol. 1. Welding Science & Technology. 9th Edition. Miami, FL: American Welding Society, 2001.

Welding Handbook. Vol. 2, Part 1: Welding Processes. Miami, FL: American Welding Society, 2004.

1.0.0 Section Review

1. When performing long-term cutting or welding of zinc or cadmium materials or coatings, always _____.
   a. ventilate the area with a continuous oxygen supply
   b. connect the welding nozzle to a nitrogen purge
   c. wear an approved full-face, supplied-air respirator
   d. use a water-cooled, multi-flame (rosebud) torch

2. Direct current electrode negative (DCEN) produces the greatest amount of heat in the _____.
   a. base metal
   b. filler metal
   c. electrode
   d. torch

Section Two

2.0.0 GTAW Equipment

Objective

Identify and describe GTAW equipment and consumables.

a. Identify and describe GTAW welding machines.
b. Identify and describe GTAW torches.
c. Identify and describe GTAW torch nozzles and electrodes.
d. Identify and describe GTAW shielding gases.
e. Identify and describe GTAW filler metals.

Performance Tasks

1. Select a GTAW shielding gas.
2. Select a GTAW filler metal.
5. Break down and reassemble a GTAW torch.

Trade Terms

Potentiometer: A three-terminal resistor with an adjustable center connection; used for variable voltage control.

GTAW operations involve the use of a number of different types of equipment and consumables. Chief among these are GTAW welding machines, torches, nozzles, electrodes, shielding gases, and filler metals.

2.1.0 Welding Power Source Types

Welders use CC welding machines to perform GTAW. The type of welding current (DCEN, DCEP) depends on the type and thickness of the base metal.

Many types of welding machines may supply welding current for GTAW. Usually, any welding machine that produces constant current, such as those used for SMAW, can be used. In addition, there are welding machines specially designed for GTAW. These special machines have built-in high frequency and controls for shielding gas and cooling water. Types of welding machines that can be used for GTAW include the following:

- Transformer welding machines
- Transformer-rectifier welding machines
- Engine-driven generator and alternator welding machines
- Inverter power sources

DC welding machines are usually designed to produce either constant-current or constant-voltage direct welding current. Constant current means that welding current is produced over a wide voltage range regardless of arc length. This is typical of an SMAW or GTAW welding machine. A constant-voltage machine maintains a constant voltage as the output current varies. This is typical of a GMAW or FCAW welding machine. The output voltages and output currents of a welding machine can be plotted on a graph to form a curve. These curves show how the output voltage relates to the output current as either of these factors changes (*Figure 6*).

2.1.1 Transformer Welding Machines

Transformer welding machines produce alternating welding current only. They use a voltage step-down transformer, which converts high-voltage, low-amperage current from commercial power lines to low-voltage, high-amperage welding current. The primary electrical power input required for a transformer welder can be 120VAC, 208-230VAC single- or three-phase, or 460V three-phase. Special light-duty transformer welding machines used for sheet metal work are designed to be plugged into a 120VAC outlet. However, most light-duty transformer welding machines require 230VAC primary current. Heavy-duty industrial transformer welders require 460V three-phase primary current.

WARNING!

Coming in contact with the primary current of a welding machine can cause electrocution. Make sure that welding machines are properly grounded to prevent injury.

Transformer welders are not as common as other types of welding machines on the job site, but they are used for special jobs. A transformer welder has an On/Off switch, amperage control, and terminals for connecting the electrode lead and the workpiece lead. Transformer welding machines used for GTAW must have high-frequency capabilities built in or they must use an auxiliary high-frequency generator.

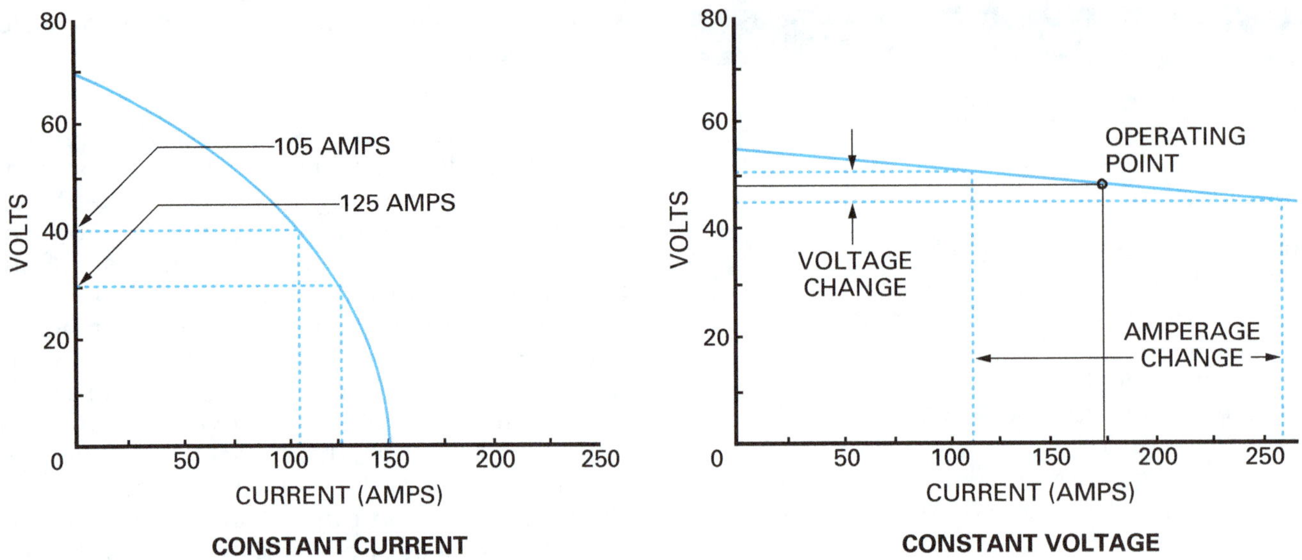

Figure 6 Constant-current and constant-voltage output curves.

2.1.2 Transformer Rectifier Welding Machines

The transformer-rectifier welding machine uses a transformer to transform the high-voltage, low-amperage primary current to low-voltage, high-amperage welding current and a rectifier to convert the AC to DC. Transformer-rectifier welding machines are designed to produce either direct welding current only or both alternating and direct welding currents. Small transformer-rectifier machines that produce both alternating and direct welding currents are usually lighter-duty than those that produce DC only. Transformer-rectifier welding machines that produce direct welding current only are sometimes simply referred to as rectifiers. Depending on their size, transformer-rectifier welding machines may require 120VAC, 208-230VAC single- or three-phase, or 460V three-phase power to operate.

Transformer-rectifiers used for GTAW have an On/Off switch and an amperage control. The welding cables (electrode lead and workpiece lead) are connected to terminals. They often have selector switches to select DCEN or DCEP. If there is no selector switch, welders must change the cables on the machine terminals manually to select the type of current desired. Most high-amperage machines designed specifically for GTAW are the transformer-rectifier type. Transformer rectifiers designed specifically for GTAW generally have the following:

- High-frequency generator
- High-frequency selector switch for DC Start, Continuous (for AC), or Off (no high frequency)
- Shielding gas preflow timer
- Control to automatically start cooling water flow when an arc is struck
- Shielding gas postflow timer control to prevent weld contamination (keeps shielding gas flowing for a set time after the arc has been terminated)
- Hand- or foot-operated remote current control
- Remote/local selector switch for remote control
- Start amperage control

Welders often use GTAW for critical welds on difficult-to-weld materials. For this reason, many special advanced features and controls have been developed and are available on machines designed for GTAW. These advanced features and controls include the following:

- High-frequency intensity control for better arc starting
- High-frequency stabilizer control to control tungsten spitting
- Crater fill timer to automatically taper the welding current from the selected setting to a minimum current at the end of a weld
- Balanced wave adjustment for penetration/cleaning control (sets percentage of time AC remains on the straight or reverse side of the cycle)

Regardless of the machine used, welders should always consult the manufacturer's documentation for specific operating information. *Figure 7* shows a typical industrial, high-amperage, heavy-duty transformer-rectifier welding machine.

KEY CONTROLS

A. MINIMUM OUTPUT PRESET CONTROL & DISPLAY SWITCH
B. SET-UP MENU
C. DIGITAL AMPERAGE METER
D. LOCAL/REMOTE CURRENT SWITCH
E. MAXIMUM OUTPUT PRESET CONTROL
F. POST FLOW TIME
G. THERMAL SHUTDOWN LIGHT
H. DOWNSLOPE TIME
I. PULSE BACKGROUND CURRENT CONTROL
J. PULSE % ON TIME CONTROL
K. SPOT TIME
L. PULSE FREQUENCY CONTROL
M. POLARITY SWITCH
N. POWER SWITCH
O. PULSE/SPOT TIME MODE SWITCH
P. TRIGGER SWITCH
Q. AC BALANCE CONTROL
R. MODE SWITCH
S. REMOTE RECEPTACLE (NOT SHOWN)

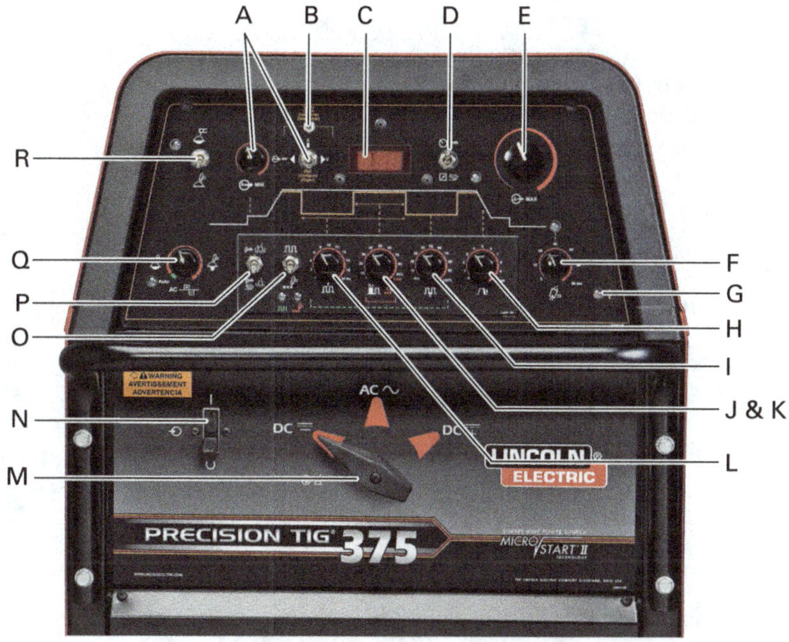

SETUP MENU

- DC TIG START MODES: HIGH FREQUENCY, SCRATCH START, AND TOUCH START TIG™
- ADJUSTABLE PREFLOW TIME
- ADJUSTABLE START PULSE FOR SOFT OR FORCEFUL STARTS
- ADJUSTABLE TIG HOT START
- ADJUSTABLE UPSLOPE TIME
- ADJUSTABLE STICK HOT START
- ADJUSTABLE STICK ARC FORCE

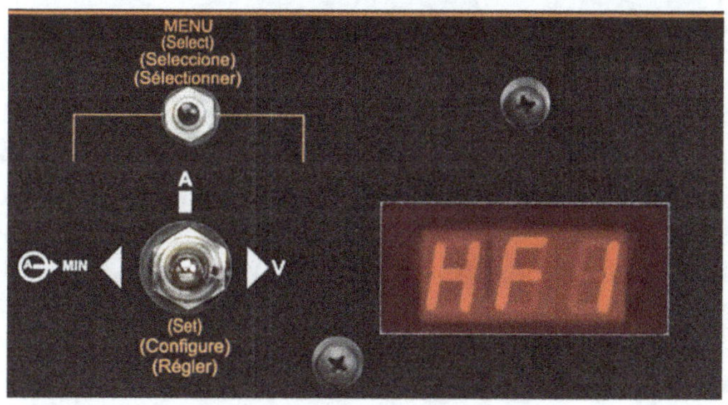

Figure 7 Transformer-rectifier welding machine designed for SMAW, GTAW, and pulsing.

2.1.3 Inverter Power Sources

Inverter power sources (*Figure 8*) increase the frequency of the incoming primary power. This provides a smaller, lighter power source with a faster response time and much more waveform control for pulse welding. Welders generally use inverter power sources in applications where space is limited and portability is important. The controls on these machines vary according to size and application.

The power source shown in *Figure 8* weighs less than 30 pounds (14 kg), and is a full-function inverter intended for alloy fabrication or critical maintenance work. It produces up to 90 amps with 100 percent duty cycle for GTAW, and 60 amps with 100 percent duty cycle for SMAW on an input power supply of 120VAC. However, it can produce up to 130 amps for both SMAW and GTAW when the input power is 208–230VAC.

Many production inverter welding machines used today are referred to as advanced inverter welding machines. They usually include all the controls listed for transformer-rectifier machines and basic inverter machines plus the following:

- Pulse arc control that allows the welding current to be pulsed between a high-current setting and a low-current setting for puddle control
- Hot-start current control to enable a controlled surge of welding current to establish a puddle quickly

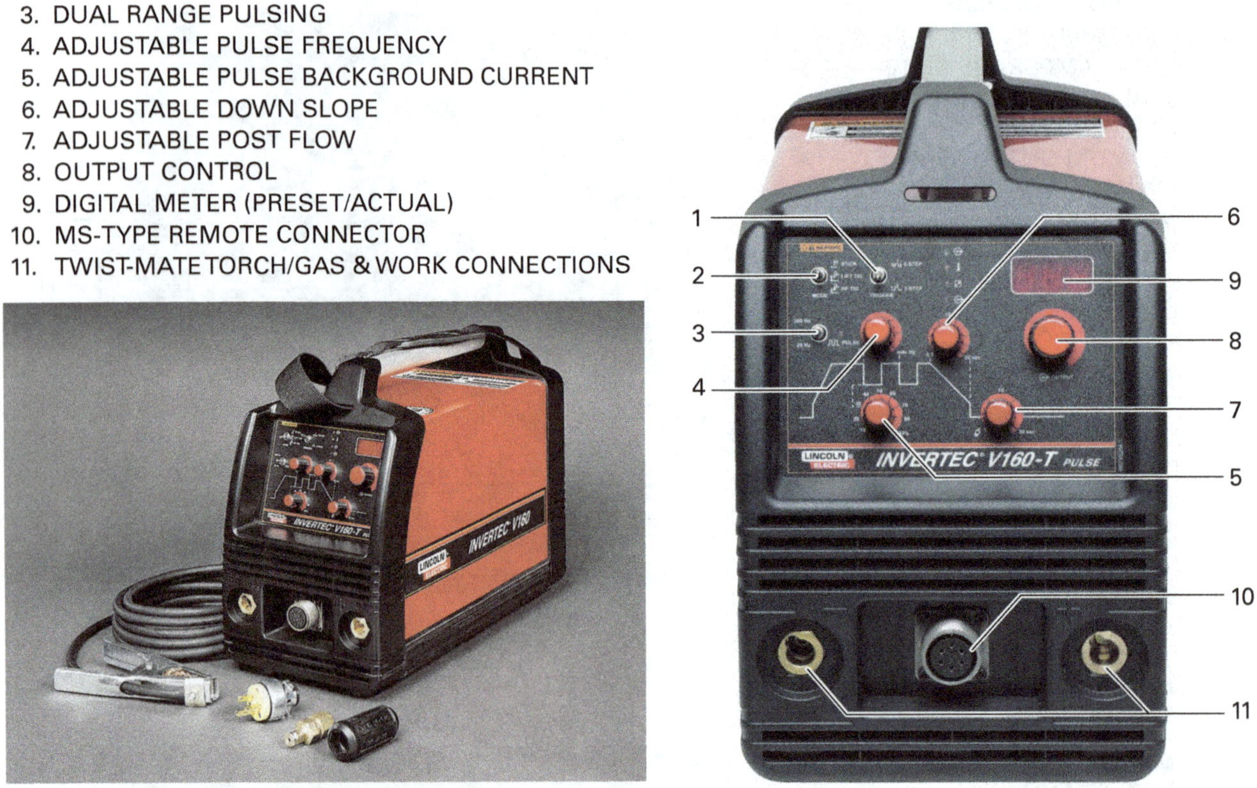

Figure 8 Basic GTAW inverter welding machine.

- Slope-up control for direct hot-start current, which controls how quickly the current rises to the hot-start setting
- Optional spot arc timer control for GTAW spot welding
- Variable AC-frequency control
- AC balance-wave control

Advanced inverter power sources are available in capacities of up to 500 amps that weigh approximately 100 pounds (45 kg).

2.1.4 Engine-Driven Generator Welding Machines

Welding machines can also be powered by gasoline or diesel engines. The engine can be connected to a generator or to an alternator. Engine-driven generators (*Figure 9*) produce direct welding current. Engine-driven alternators produce AC that is fed through a rectifier to produce direct welding current. The size and type of engine used depend on the size of the welding machine.

To produce welding current, the generator must turn at a required number of revolutions per minute. The engines powering the generators have governors that control the engine speed. Most governors have a welding speed switch that can be set to idle the engine when no welding is taking place. When the electrode is touched to the base metal, the governor automatically increases the speed of the engine to the required rpm for welding. If no welding takes place for about 15 seconds, the engine automatically returns to idle. The switch can also be set for the engine to run continuously at the welding speed.

Figure 9 Engine-driven generator welding machine.

Engine-driven generators often have an auxiliary power unit that produces 120VAC current for lighting, power tools, and other common equipment. When 120VAC current is required, the engine-driven generator must run continuously at the welding speed.

Engine-driven generators have engine controls and welding current controls. The engine controls vary with the type and size of the unit, but normally include the following:

- Starter
- Voltage gauge
- Temperature gauge
- Fuel gauge
- Hour meter
- Oil pressure indicator

The following are the common welding current controls:

- Amperage control
- Current range switch
- Amperage and voltage gauge
- Polarity switch

The advantage of engine-driven generators is that they are portable and can be used in the field where electricity is not available to power other types of welding machines. The disadvantage is that engine-driven generators are more costly to purchase, operate, and maintain.

2.1.5 Power Source Ratings

The rating of a welding machine is determined by the amperage output of the machine at a given duty cycle. The duty cycle of a welding machine is based on a 10-minute period. It is the percentage of ten minutes during which the machine can continuously produce its rated amperage without overheating. For example, a machine with a rated output of 300 amps with a 60-percent duty cycle can deliver 300 amps of welding current for 6 minutes out of every 10 without overheating.

The duty cycle of a welding machine will generally be 20, 30, 40, 60, or 100 percent. A welding machine having a duty cycle of 20 to 40 percent is considered a light- to medium-duty machine; welding can only be maintained for 2 to 4 minutes out of each 10-minute period. Most industrial, heavy-duty machines for manual welding have a 60- or 100-percent duty cycle rating. Machines designed for automatic welding operations have a 100-percent duty cycle rating.

With the exception of the 100-percent duty cycle machines, the maximum amperage that a welding machine can produce is always higher than its rated capacity. A welding machine rated for 300 amps with a 60-percent duty cycle generally puts out a maximum of 375 to 400 amps. However, since the duty cycle is a function of its rated capacity, the duty cycle will decrease as the amperage is raised over 300 amps. Welding at 375 amps with a welding machine rated for 300 amps with a 60-percent duty cycle will lower the duty cycle to about 30 percent. If welding continues for more than 3 out of 10 minutes under these conditions, the machine will overheat. Note that most welding machines have a heat-activated circuit breaker that will shut off the machine automatically when it overheats. The machine cannot be turned back on until it has cooled below a preset temperature.

Conversely, if the amperage is set below the rated amperage, the duty cycle increases. Setting the amperage at 200 amps for a welding machine rated at 300 amps with a 60-percent duty cycle will increase the duty cycle to 100 percent.

Figure 10 is a graph that shows the relationship between amperage and duty cycle.

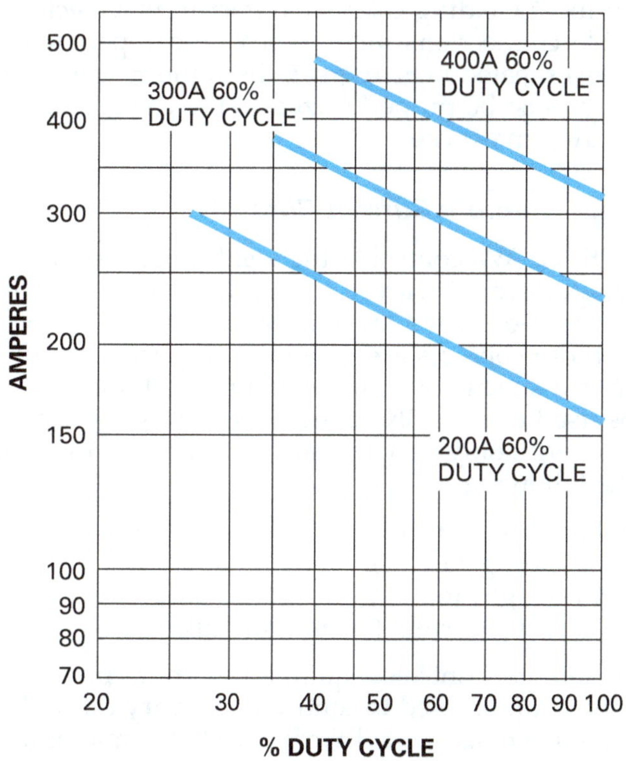

Figure 10 Relationship between amperage and duty cycle.

2.1.6 Welding Cable

Cables used to carry welding current are designed for maximum strength and flexibility. The conductors inside the cable are made of fine strands of copper wire. The copper strands are covered with layers of rubber reinforced with nylon or Dacron® cord. *Figure 11* shows the construction of a welding cable.

The size of a welding cable is based on the number of copper strands it contains. Large-diameter cable has more copper strands and can carry more welding current. Typically, the smallest cable size is number 4 and the largest is number 3/0 (spoken as "3 aught").

When selecting a welding cable size, consider the amperage load as well as the distance the current will travel. The longer the distance the current has to travel, the larger the cable must be to reduce voltage drop and heating caused by the electrical resistance in the welding cable. When selecting welding cable, use the rated capacity of the welding machine for the cable amperage requirement. For the distance, measure both the electrode and workpiece leads, and add the two lengths together. To identify the welding cable size required, refer to a table of recommended welding cable sizes. These tables are furnished by most welding cable manufacturers. Check all cable connections for tightness and repair loose or overheated lugs. If quick-disconnects are used, they must be properly sized to the cable or overheating may occur.

2.1.7 Remote Current Control

With GTAW, control of the welding current is often performed with remote controls such as foot- or hand-operated switches and potentiometers. Remote controls are generally designed for use with a specific welding machine, so they cannot be used universally on all machines. Remote controls with only a switch can control one or more of the following:

- High frequency
- Shielding gas
- Cooling water
- Welding current (on and off only)

Remote controls equipped with a potentiometer can be used to continuously vary the welding current as needed, from zero to the maximum current set on the power source.

Figure 11 Construction of a welding cable.

The most useful remote control has both switches and a potentiometer, and it controls the following:

- High frequency
- Shielding gas
- Cooling water
- Welding current (on and off)
- Welding current up or down as needed (zero to maximum setting on the power source)

Figure 12 shows various remote control units.

2.2.0 GTAW Torches

In addition to a CC welding machine (and a high-frequency generator for AC welding), GTAW requires the following:

- GTAW torch
- Shielding gas
- Optional torch cooling system

In addition, a remote amperage control is often used with GTAW. *Figure 13* shows a typical GTAW welding system.

Aluminum Welding with a Balled-Tip versus a Pointed-Tip Electrode

Traditional AC welding of aluminum using a balled-tip electrode is required for most applications using conventional power sources. However, for inverter power sources that use squarewave technology with AC balance and variable AC frequency control, welding thinner materials and creating smaller beads is accomplished much more easily using the same pointed and blunted 2 percent thoriated or ceriated electrodes as used for DCEN welding along with increased electrode-negative time.

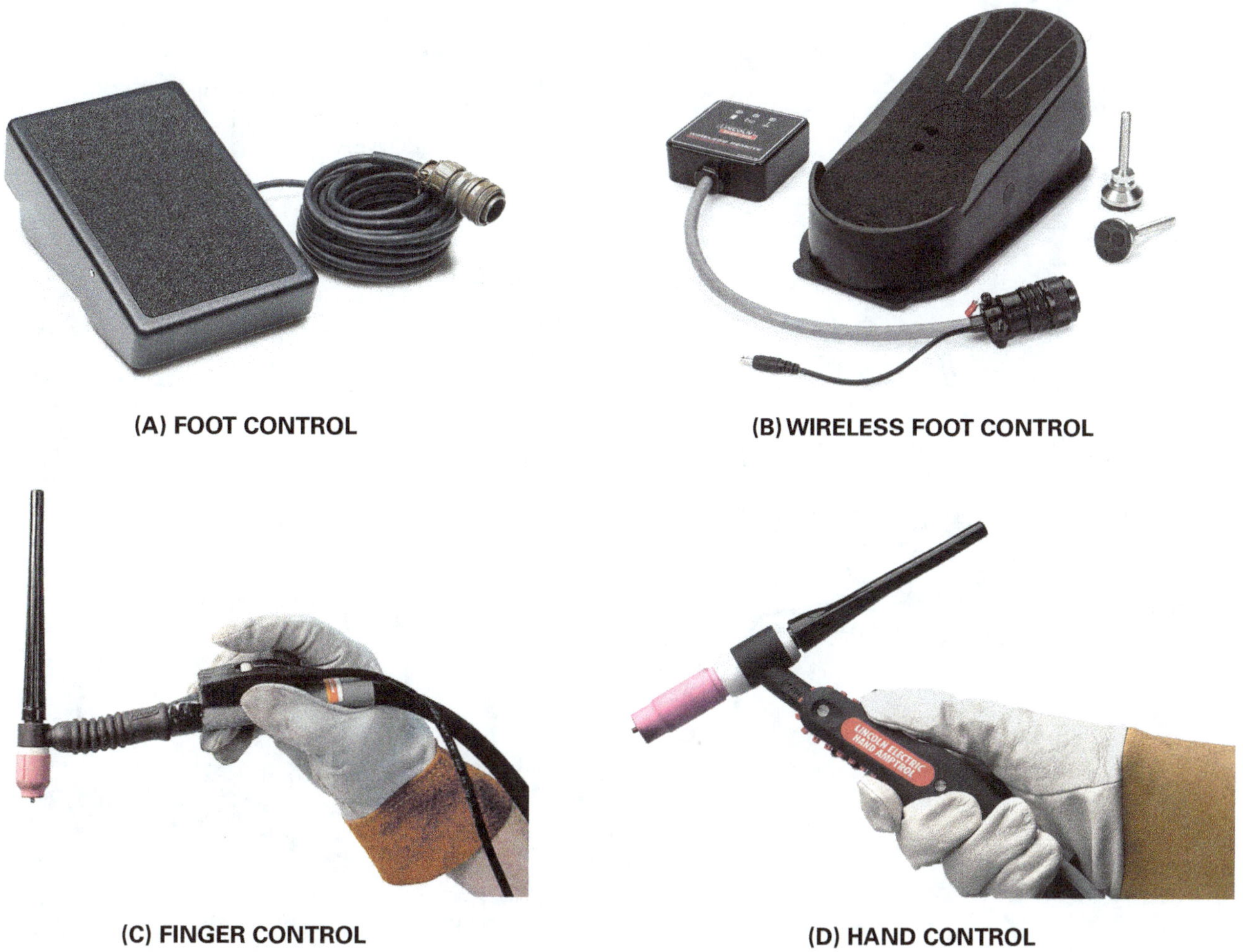

Figure 12 Typical foot-, hand-, and finger-operated remote controls.

The GTAW torch provides an electrical path for the welding current between the welding current lead and the tungsten electrode. There are many different manufacturers of torches, but all torches are basically the same even though their parts are not often interchangeable. All torch bodies have a built-in or replaceable collet body and electrode collet. The electrode collet and collet body come in a variety of sizes to match the size of the electrode being used. The collet holds the tungsten electrode in place and is secured by a back cap. The back cap, which is threaded onto the torch, is loosened to change the electrode or to adjust its stickout. The reserve length of electrode that extends from the back of the collet is insulated and protected by the back cap. Back caps are made in several lengths to cover tungstens of various lengths. The shorter back caps are used with short tungstens in confined spaces where torch clearance is inadequate.

The torch also contains passages for the shielding gas or cooling water, if needed. The shielding gas exits the torch through an insulated gas nozzle, sometimes called a cup, which surrounds the tungsten. *Figure 14* provides some GTAW torch terminology.

Heat is removed from GTAW torches either by air cooling with a shielding gas or by water cooling. *Figure 15* shows examples of both air-cooled and water-cooled GTAW torches. Two styles of air-cooled torches are common. One style uses a single plastic hose that has a power cable inside it, as shown in *Figure 15*. The shielding gas flows around the power cable, cooling it and the torch before being discharged to shield the electrode and weld. A special power cable adapter attaches to the power terminal on the power source.

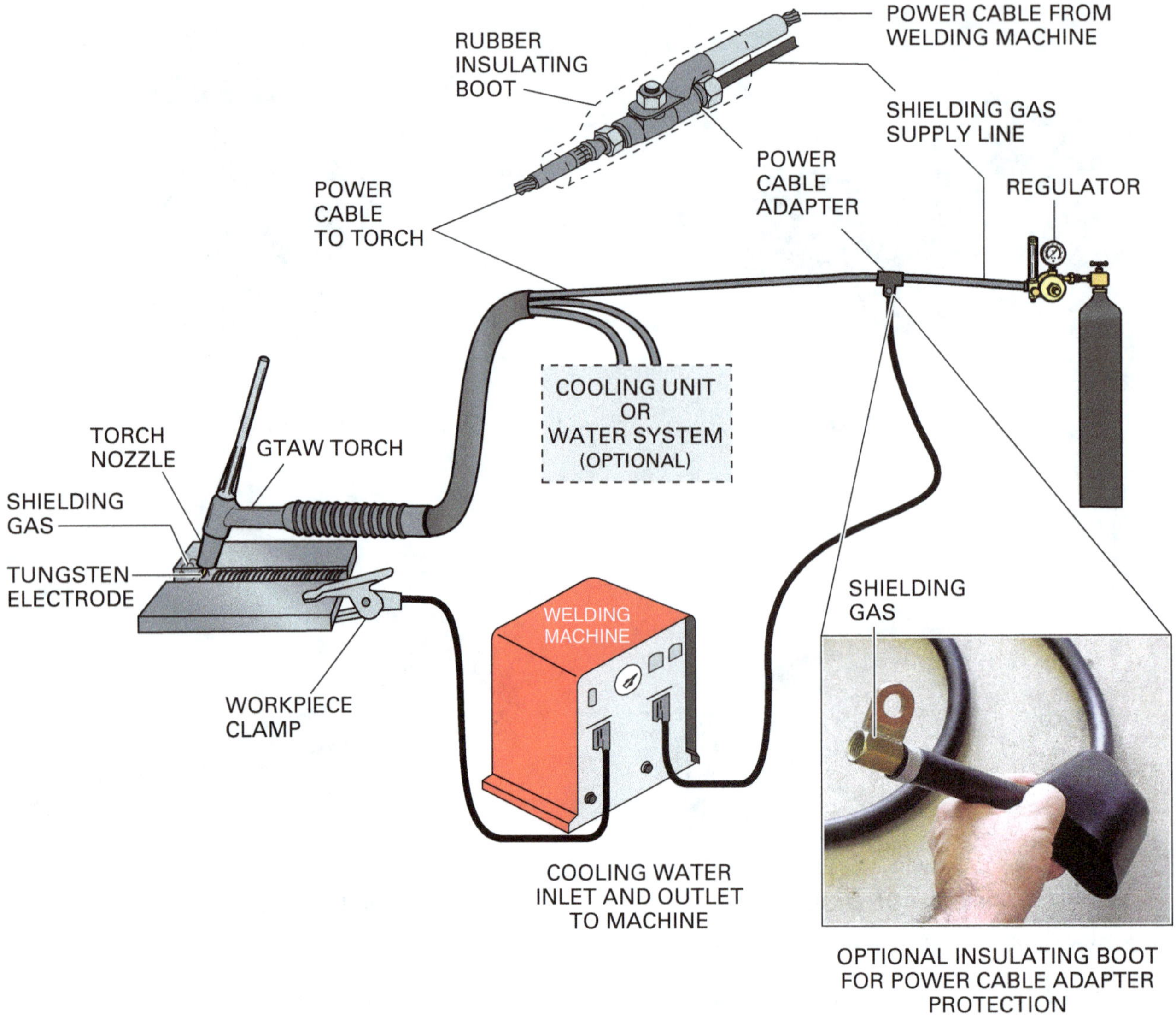

Figure 13 Basic GTAW welding system.

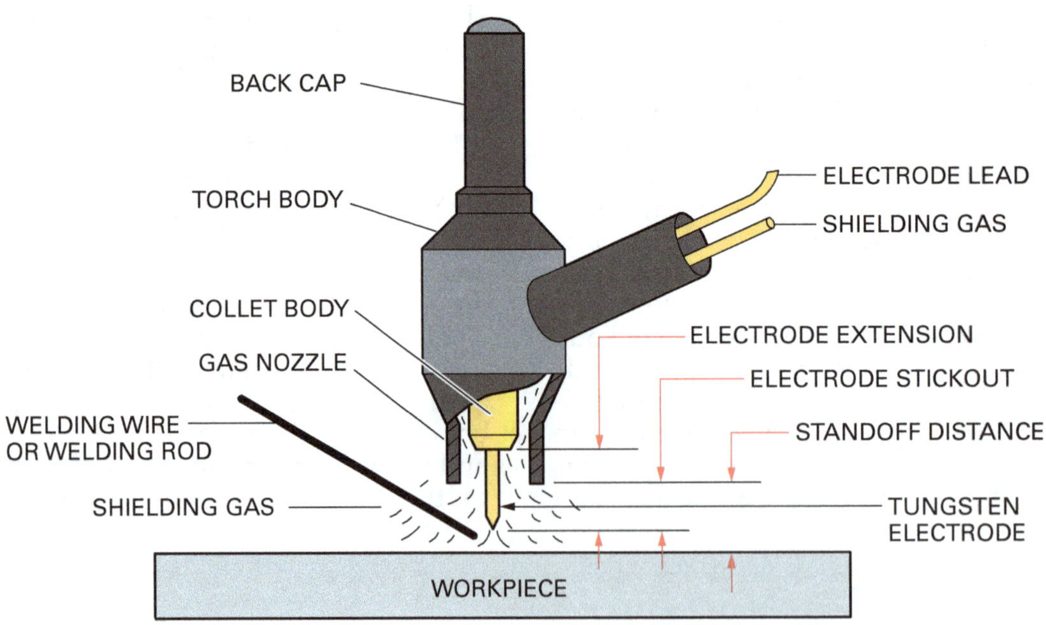

Figure 14 GTAW torch terminology.

This connector needs to be covered by a rubber insulating boot to prevent shorts. The single hose connects to one side of the adapter. This connection also makes the electrical connection. The shielding gas supply hose connects to the other side of the adapter.

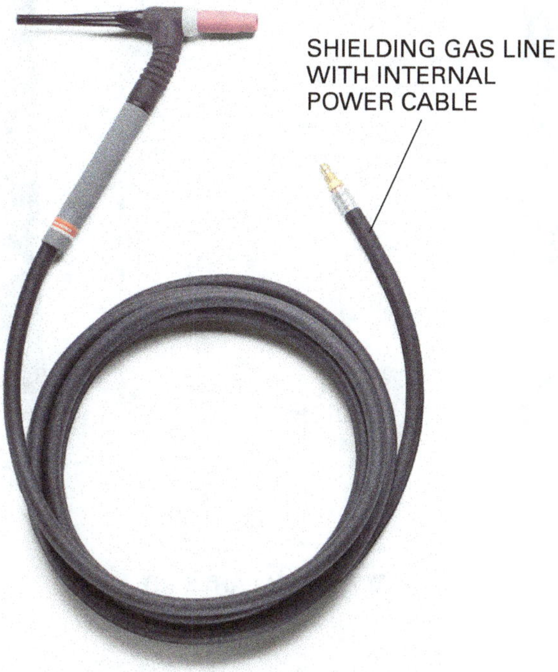

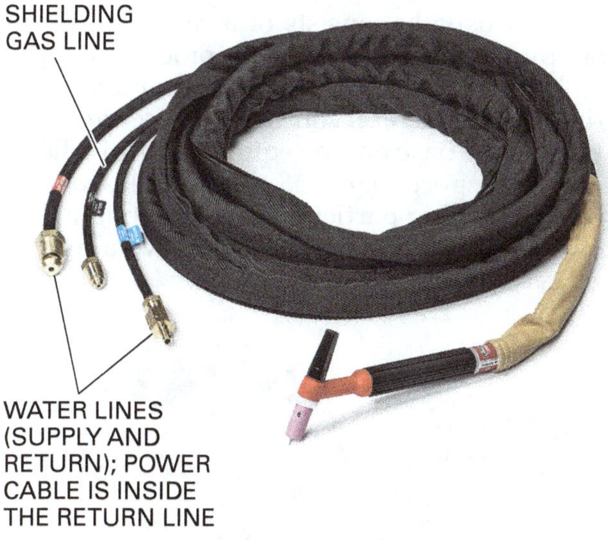

Figure 15 GTAW torches.

The second type of air-cooled torch has a separate power cable and shielding gas hose. Either the single- or double-hose torch can be purchased with a gas control valve in the torch to manually control the shielding gas flow. This is necessary when the power source being used does not have a solenoid valve to automatically start and stop the shielding gas flow. *Figure 16* shows the various parts of an air-cooled GTAW torch.

Water-cooled torches have three hoses. One hose supplies shielding gas; one delivers cooling water to the torch; the third hose contains the power cable and the cooling water return or discharge. The cooling water may be supplied from the domestic water system or from a closed-loop cooling system. Regardless of the system used, it should include a fusible link in the torch cooling line. A fusible link is a temperature-based safety device that will open and stop the welding current flow if the coolant flow to the torch is interrupted. *Figure 17* shows the parts of a water-cooled GTAW torch.

Cable Covers

To prevent damage from hot or molten metal to GTAW torch cables containing hoses for gas or water, leather or fire-resistant cable covers can be added as protection. These covers zip or snap around the torch cable for a distance of 10" to 20" (25 cm to 50 cm) behind the torch handle.

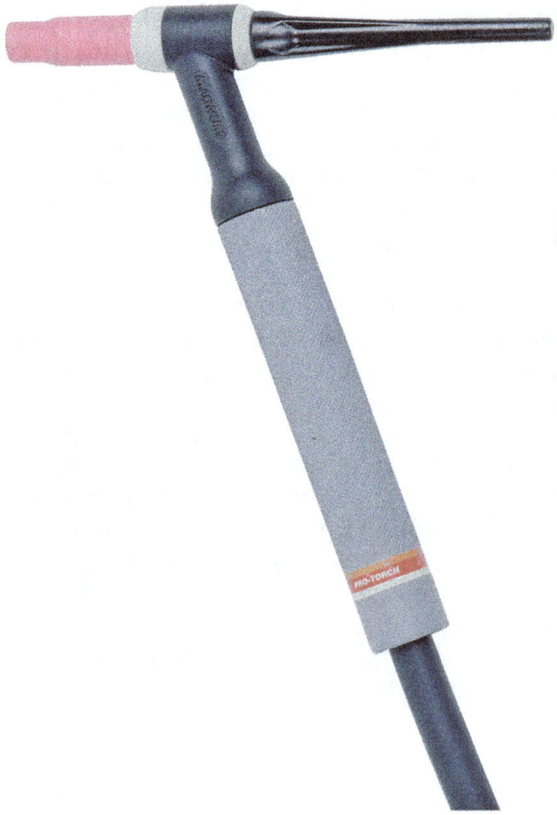

AIR-COOLED GTAW TORCH

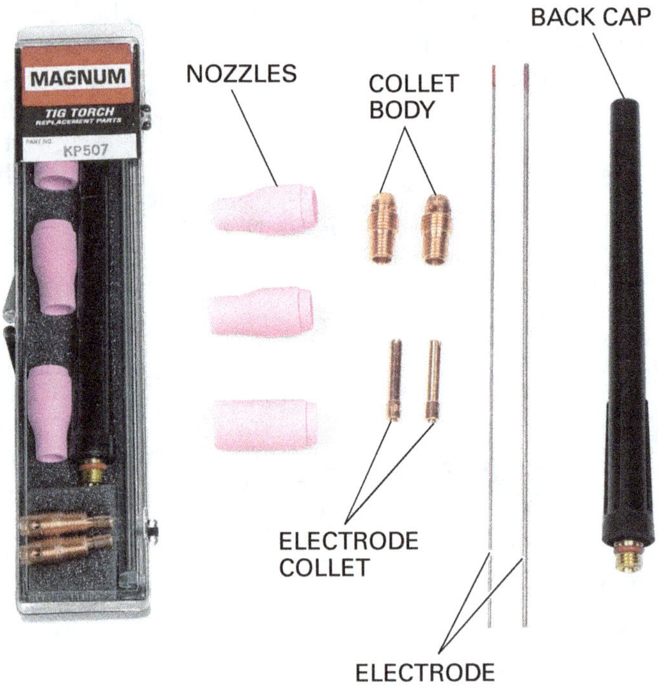

GTAW TORCH PARTS KIT

Figure 16 Air-cooled GTAW torch.

Closed-loop cooling systems recirculate demineralized water or special cooling fluids that will not corrode internal torch surfaces or plug the cooling passages with mineral scale. After cooling the torch, the water is typically returned to the system through a hose that also contains the torch power cable. A typical closed-loop cooling system usually consists of a reservoir, circulating pump, heat-transfer equipment to reject any collected heat to the atmosphere, and the connecting lines. The reservoir may also serve as the portable unit's base and be equipped with wheels for easy transport. *Figure 18* shows a closed-loop cooling unit. This particular unit is designed to stack with welding machines that have a matching chassis.

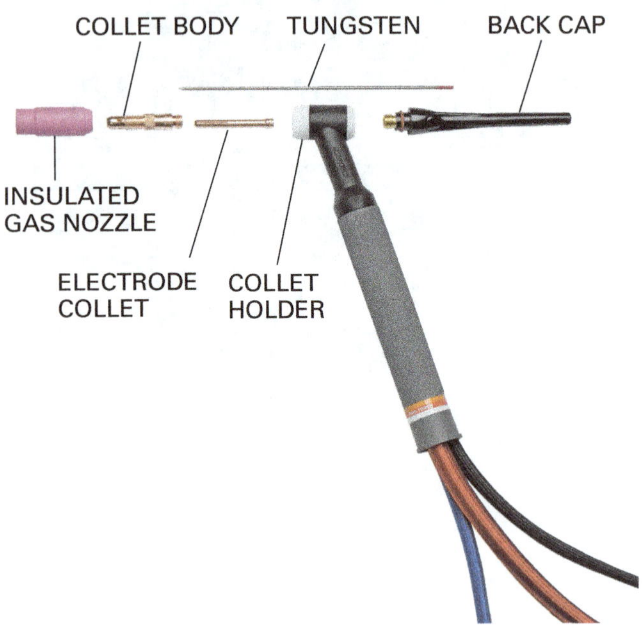

Figure 17 Water-cooled GTAW torch.

Figure 18 Typical closed-loop cooling unit.

2.3.0 Nozzles, Collets, and Electrodes

GTAW collets and electrodes are designed to work together, with the tungsten electrode fitting precisely into the collet. For this reason, welders must take both of the devices into consideration when selecting either the proper collet or electrode for a particular job.

2.3.1 Gas Nozzles

The GTAW gas nozzle shapes and directs the shielding gas flow as it exits the torch. Nozzles, sometimes called cups, are usually threaded onto the torch to form a gas-tight joint. They are made of ceramic material, chrome-plated steel, Pyrex®, or glass. Ceramic nozzles can be used up to about 300 amperes. Above that amperage, water-cooled, metal-coated ceramic nozzles or water-cooled ceramic nozzles must be used.

Gas nozzles are available in different lengths and diameters (refer to *Figures 16* and *17*). Nozzle length is determined by the job requirements, such as limited clearance or deep grooves. Nozzle size is the exit orifice inside diameter; it depends on the torch type and size and the electrode diameter. Nozzle size may be specified in fractions of an inch (¼", ⅜", $\frac{7}{16}$", ½", ¾") or in millimeters (6 mm, 10 mm, 11 mm, 13 mm, 19 mm). Metric size nozzles are not, however, designed to be equivalent in size to imperial versions, and metric size and imperial size nozzles cannot be substituted for each other. Sometimes nozzle diameter is given as a size number, such as 4, 6, 7, 8, or 12. These numbers are the nozzle opening diameters in multiples of $\frac{1}{16}$". For example, a Number 6 nozzle is $6 \times \frac{1}{16}$", or ⅜" in diameter.

A GTAW torch may also be equipped with a gas lens collet body. A gas lens collet body contains an assembly of fine screens that straightens the gas flow from the nozzle and eliminates turbulence, which causes the gas to flow smoothly past the tungsten electrode. This reduces the chance that the turbulence will pull atmosphere into the weld zone and cause contamination and weld defects. The use of a gas lens allows longer tungsten stickout. This is useful for welding in tight quarters. Increasing tungsten stickout decreases the current-carrying capacity of the tungsten. *Figure 19* shows a typical gas lens collet body and torch with a nozzle, collets, and electrodes.

Figure 19 Typical gas lens collet body and torch with nozzle.

2.3.2 Tungsten Electrodes

Tungsten electrodes used for GTAW are manufactured in different formulations and sizes to meet the requirements of *AWS Specification A5.12*. Electrode diameters range from 0.01" (0.3 mm) up to ¼" (6.4 mm) for current ratings of 5 to 1,000 amperes. Electrodes are made in lengths of 3" (76 mm), 6" (152 mm), 7" (175 mm), 12" (305 mm), 18" (457 mm), and 24" (610 mm). Manual welding is usually done with 7" (175 mm) or shorter electrodes.

Electrodes are manufactured with two different finishes—chemically cleaned or ground. The ground finish is the more expensive option.

CAUTION

Always handle tungsten electrodes with clean gloves to prevent contamination from body oils and other materials.

A number of different tungsten electrodes are manufactured. They include the following:

- Pure tungsten
- Zirconiated tungsten
- Thoriated tungsten
- Ceriated tungsten
- Lanthanated tungsten
- Rare earth tungsten
- Cryogenically treated thoriated tungsten

Pure tungsten electrodes, often simply referred to as tungstens, are used for AC welding. They provide good arc stability and good resistance against contamination. When used for DC welding, they are easily contaminated. Although relatively inexpensive in terms of per unit price, pure tungsten is more difficult to start, less able to maintain a stable arc, and has a shorter service life than other types of tungsten electrodes.

Zirconiated tungstens are also used for AC welding. They have a small percentage of zirconium added to the tungsten. They are used for AC welding where tungsten inclusions cannot be tolerated. The zirconium also gives these electrodes easy arc-starting characteristics.

Thoriated tungsten electrodes are used for DC welding. They have a small percentage of thorium added to do the following:

- Make the arc easier to start
- Increase the current range
- Help prevent tip melt
- Reduce the tendency to stick or freeze to the work
- Increase resistance to contamination, when properly used

Thoriated tungsten electrodes are most commonly manufactured in these two concentrations of thorium dioxide (thoria): EWTh-1 (approximately 1 percent thoria) and EWTh-2 (approximately 2 percent thoria).

WARNING!

Thoriated tungsten electrodes contain low-level radioactive thorium. The grinding dust from these electrodes is considered a hazard. *AWS Safety and Health Fact Sheet 27* provides further information about this subject. If possible, always use tungsten electrodes containing cerium or lanthanum instead of thorium.

Ceriated and lanthanated tungsten electrodes have the advantage of not being radioactive. Ceriated tungsten electrodes can be balled and used for AC welding.

EWG electrodes that combine three nonradioactive rare earth materials into one electrode are increasingly used as replacements for 2-percent thoriated electrodes. Besides their advantage of not being radioactive, these mixed tungsten electrodes start and re-ignite well and have very good service life for applications in which welding cycles of at least 15 minutes are used.

Thoriated electrodes that have been treated with a multi-step cryogenic process are another good replacement for 2 percent thoriated electrodes. The cryogenic treatment maximizes grain structure and improves electron flow. The cryogenically treated electrodes are easier to ignite and have longer tip life, longer overall electrode life, and higher amperage tolerance than nontreated 2-percent thoriated tungsten.

Gas Nozzles

Some manufacturers recommend that the inside diameter of the nozzle should be a minimum of three times the electrode diameter.

Think About It

Nozzle Size

What size nozzle would a welder select if a ½" diameter opening was needed?

Other tungsten electrodes continue to be developed. For example, tungsten electrodes with less common oxides, such as yttrium and magnesium, are also becoming available.

Tungsten electrodes are identified by a color band at one end. *Table 1* shows GTAW electrode color codes and AWS classifications.

2.4.0 Shielding Gas

The GTAW process always uses inert shielding gases to displace the atmosphere from the weld zone and prevent oxidation and contamination of the tungsten electrode, weld puddle, and filler metal. The two principal shielding gases used for GTAW are argon (Ar) and helium (He). The use of helium was what prompted the naming of GTAW initially as heliarc welding. These gases may be used alone or as mixtures. Occasionally, small percentages of other gases are added for special conditions.

Each shielding gas and shielding gas mixture has distinct performance characteristics. Each affects the arc differently and produces different weld characteristics. Mixtures of gases often have the best features of the individual gases within them. The following sections explain shielding gas characteristics and their principal uses.

Table 1 GTAW Electrode Color Codes and AWS Classifications

Electrode	Color Band	Electrode AWS Classification
Pure Tungsten	Green	EWP (Minimum 99.5% Tungsten)
1% Thoriated	Yellow	EWTh-1 (1% Thoria)
2% Thoriated	Red	EWTh-2 (2% Thoria)
Zirconiated	Brown	EWZr (0.15%–0.40% Zirconium Oxide)
Zirconiated	White	EWZr (8% Zirconium Oxide)
Ceriated	Gray	EWCe-2 (2% Ceria)
Lanthanated	Black	EWLa-1.0 (1% Lanthana)
Lanthanated	Gold	EWLa-1.5 (1.5% Lanthana)
Lanthanated	Blue	EWLa-2.0 (2% Lanthana)
EWG	*Not Standard	Undefined (*e.g.,* various combinations of rare earth alloys, yttrium oxide, magnesium oxide, etc.)

*Manufacturers may select any color not already in use.

2.4.1 Argon

Argon is the most common shielding gas used with GTAW. Argon provides a smooth, quiet arc that requires a lower arc voltage than other shielding gases for a given arc length. This means that it gives the welder the greatest tolerance for arc gap variation. Argon also works well with AC, and it provides better base metal cleaning than helium. Argon is ten times heavier than helium, so it forms a better gas shield than helium, which tends to rise at the same flow rates. Argon is used both alone and in combination with other shielding gases.

When compared with helium, argon has the following advantages:

- A smoother, quieter arc
- Lower arc voltage for a given arc length
- Easier arc starting
- Better cathode cleaning on aluminum and magnesium with AC
- Lower shielding flow rate
- Better shielding in cross drafts
- Least expensive shielding gas

Tungsten Electrode Packaging

Most tungsten electrodes are packaged as shown to prevent contamination or damage to the electrodes. An inner divider containing the electrodes slides out of the package. Note the orange color band (may appear reddish due to printing color variations) on the end of each electrode, identifying the type in agreement with *Table 1*. All color bands typically discolor after use due to the heat produced. The color band end is not used when sharpening an electrode.

Small amounts of hydrogen are sometimes mixed with argon for deeper penetration in welding stainless steel. Hydrogen cannot be used with aluminum or carbon steels because it produces porosity and underbead cracking. The most common argon-hydrogen mixture is 85 percent argon and 15 percent hydrogen.

Nitrogen is sometimes added to argon to stabilize austenitic stainless steel and to increase penetration in welding copper.

2.4.2 Helium

Helium is used when deeper penetration and higher travel speed are required. Arc stability is not as good as it is with argon. However, helium is capable of delivering more heat on the base metal than is possible with argon, although it does require a higher voltage. This makes helium better for welding thick sections of highly heat-conductive metals such as aluminum and copper.

When compared with argon, helium has the following advantages:

- Deeper penetration into the weld joint
- Increased welding speed
- Welding of highly heat conductive metals

Helium is often mixed with argon. A common mixture is 75 percent helium and 25 percent argon. This mixture gives good cathodic cleaning action and deep weld penetration, which are good characteristics of both gases.

2.4.3 Cylinder Safety

Shielding gases may be supplied in bulk liquid tanks, in liquid cylinders, or in high-pressure cylinders of various sizes. The most common container is the high-pressure cylinder, which is portable and can be easily moved as needed.

When transporting and handling cylinders, observe the following rules:

- Always install the safety cap over the valve, except when the cylinder has been connected for use.
- Secure the cylinder to prevent it from falling when it is in use. Chain or clamp it to the welding machine or to a post, beam, or pipe.
- Always use a cylinder cart to transport cylinders.
- Never hoist cylinders with a sling or magnets. Cylinders can slip out of the sling or fall from the magnets. Always use a hoisting basket or similar device.
- Always open the cylinder valve slowly, and open it completely to prevent leakage around the valve stem.

WARNING!

Do not remove the valve's protective cap until the cylinder has been secured. If the cylinder falls over and the valve breaks off, the cylinder will shoot like a rocket, causing severe injury or death to anyone in its way.

Around the World

Tungsten Standards

The United States, Canada, Europe, and Japan each have their own published standards for tungsten, as shown in the table below. These standards specify the dimensions, packaging, and manufacturing requirements that tungsten manufacturers must meet for the given market area.

MARKET AREA	STANDARD NAME
UNITED STATES	ANSI/AWS A5.12
CANADA	ASME/SFA
EUROPE	ISO 6848
JAPAN	JIS

Always Use the Correct Gas

Make sure to always use the correct gas or gas mixture. Using an incorrect gas mixture, like a common one used for GMAW, which is 75 percent argon/25 percent CO^2, will immediately cause the tungsten electrode to be consumed and deposited in the weld as a contaminant. This is definitely not the kind of result you are looking for.

2.4.4 Gas Regulators/Flowmeters

A gas regulator and a flowmeter are required to supply the shielding gas to the torch at the proper pressure and flow rate (*Figure 20*). A typical regulator/flowmeter has a preset pressure regulator with a cylinder valve spud, a flow-metering needle valve, and a flow-rate gauge. The pressure regulator is usually equipped with a cylinder pressure gauge to indicate the cylinder gas pressure. The metering valve is used to adjust the gas flow to the torch nozzle. The flow gauge indicates the gas flow rate in cubic feet per hour (cfh) or liters per minute (L/min) as shown by a ball-type flow indicator. Although the regulator and flowmeter are shown separately in *Figure 20,* their functions may also be combined into a single device.

Gas flow to the torch is started and stopped either by a manually operated valve or by an electric solenoid valve. The manual valve may be on the torch, or it may be in the gas line to the torch. An electric solenoid valve is controlled automatically by the welding current flow or by a manual switch.

2.5.0 GTAW Filler Metals

Filler metal for manual GTAW is generally supplied in 36" (914.4 mm) lengths in diameters from 1⁄16" (1.6 mm) to ¼" (6.4 mm). There are some automatic and manually operated wire feeders used with manual GTAW, but they are usually found only in high-production facilities. Filler metals for GTAW are drawn from high-grade pure alloys compounded for specific applications. Generally, the rods are not coated (bare, without a flux component added) except for a corrosion-resistant copper electroplating on some carbon steel rods. However, some special-purpose rods may have flux cores or coatings.

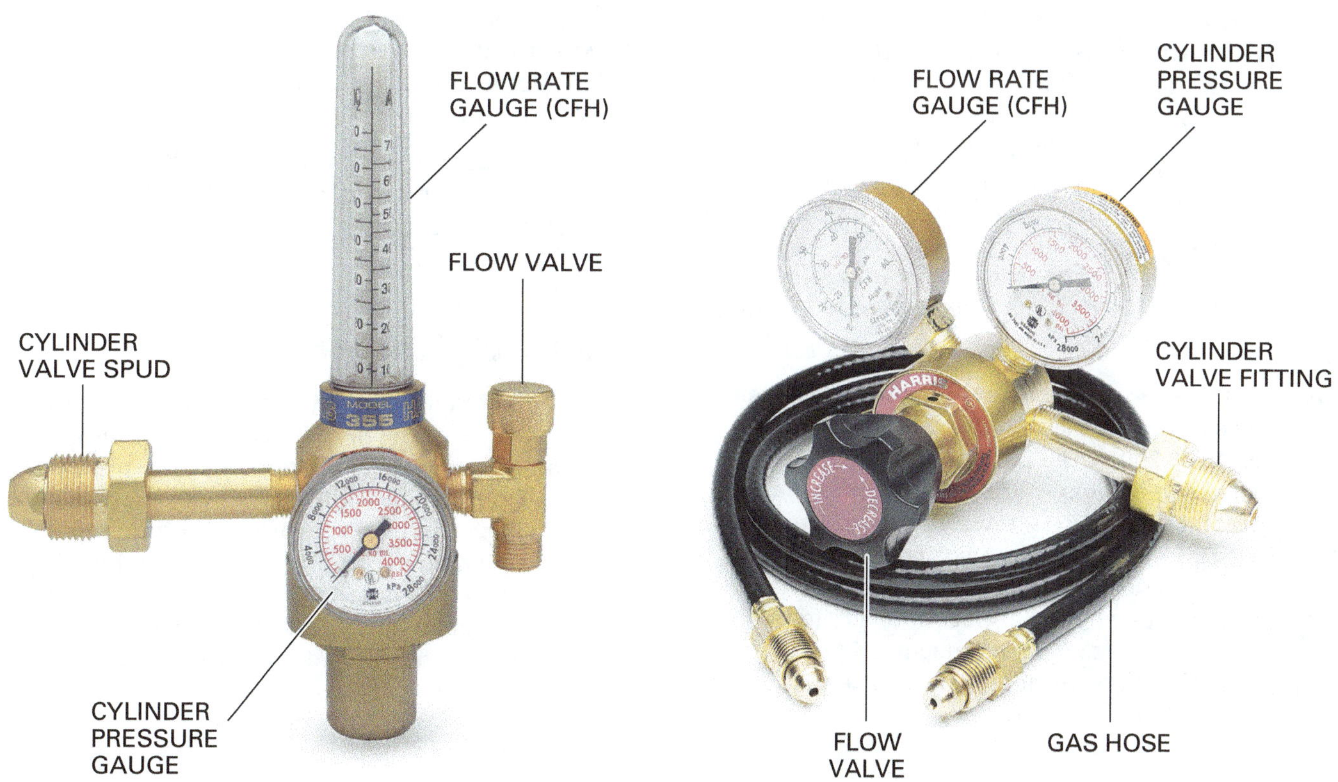

Figure 20 Shielding-gas flowmeter and gas regulator.

Several industry organizations and the United States government publish specification standards for filler metals. The most common standards are those published by the American Welding Society. The purpose of the AWS specifications is to set standards that all manufacturers must follow when manufacturing welding consumables. This ensures consistency for the user regardless of who manufactured the product. The specifications set standards for the following:

- Classification system, identification, and marking
- Chemical composition of the deposited weld metal
- Mechanical properties of the deposited weld metal

Examples of AWS specifications that apply to GTAW filler metals are the following:

- *AWS A5.7, Specification for Copper and Copper Alloy Bare Welding Rods and Electrodes*
- *AWS A5.9/A5.9M, Specification for Bare Stainless Steel Welding Electrodes and Welding Rods*
- *AWS A5.10/A5.10, Specification for Bare Aluminum and Aluminum Alloy Welding Electrodes and Rods*
- *AWS A5.14/A5.14M, Specification for Nickel and Nickel Alloy Base Welding Electrodes and Rods*
- *AWS A5.16/A5.16M, Specification for Titanium and Titanium Alloy Electrodes and Rods*
- *AWS A5.18/A5.18M, Specification for Carbon Steel Electrodes and Rods for Gas Shielded Arc Welding*
- *AWS A5.19, Specification for Magnesium Alloy Welding Electrodes and Rods*
- *AWS A5.22, Specification for Stainless Steel Electrodes for Flux Cored Arc Welding and Stainless Steel Flux Cored Rods for Gas Tungsten Arc Welding*
- *AWS A5.28/A5.2, Specification for Low Alloy Steel Electrodes and Rods for Gas Shielded Arc Welding*

NOTE

The specification number is generally followed by a colon and then a four-digit number, such as A5.18:2005 or A5.28:2015. The colon and following number indicate the year that the specification was last revised. These documents are revised periodically, usually every five years. For that reason, the year numbers have been omitted from the specifications cited in this module. When referring to an AWS specification on the job, always make sure to use the current version.

Filler metals and wire are graded for three major areas of use:

- *General use* – The wire or rod meets specifications. No record of chemical composition, strength, etc., is supplied to the user with the wire purchase.
- *Rigid control fabrication* – A Certificate of Conformance is supplied with the wire or rod at purchase. The stock is identified by heat numbers or code numbers located on the roll package.
- *Critical use* – A Certified Chemical Analysis report is supplied, and records are kept by the fabricator of welds and processes for later reference on aircraft, nuclear reactors, and pressure vessels.

Factors that affect the selection of a GTAW filler metal include the following:

- Base metal chemical composition
- Base metal mechanical properties
- Weld joint design
- Service or specification requirements
- Shielding gas used

The following are commonly used filler metal types:

- Carbon steel
- Stainless steel
- Aluminum and aluminum alloy
- Copper and copper alloy
- Low-alloy steel

Table 2 summarizes the GTAW filler metal specifications and the major AWS class covered by each specification.

Gas Cylinders

Gas cylinders must always be handled with great care because of the high-pressure gases they contain. If the valve is broken off a high-pressure gas cylinder, the gas will escape with explosive force, which can cause severe injury from the blast and blown debris. The cylinder itself may become a deadly missile. The tanks must always be chained in an upright position when being transported or positioned in their storage or use locations.

Table 2 GTAW Filler Metal Specifications and Classification System

Material	Filler Metal		X Designator Description	Example
	AWS Spec.	AWS Class*		
Carbon steel	A5.18	ERXXS-Y EXXC-Y	Tensile strength × 1,000 (psi)	ER70S-3 E70C-3
Low-alloy steel	A5.28	ERXXS-Y EXXC-Y	Tensile strength × 1,000 (psi)	ER80S-B2 E80C-B2
Stainless steel	A5.9	ERXXXY	Stainless alloy (308, 410, etc.)	ER308L EC308L
Stainless steel (flux-cored)	A5.22	RXXXTi-5	Stainless alloy (308, 309, etc.)	R309LT1-5
Aluminum	A5.10	ERXXXX-Y	Aluminum alloy (4043, 5083, etc.)	ER4043
Nickel	A5.14	ERNiXX-Y	Major alloying elements (Cr, Fe, Mo, etc.)	ERNiCr-3
Copper	A5.7	ERCuXX-Y	Major alloying elements (Al, Ni, Si, etc.)	ERCuAl-A2
Magnesium	A5.19	ERXXYYY	Major alloying elements (Al, Zn, etc.)	ERAZ92A
Titanium	A5.16	ERTi-Y		ERTi-5

*Legend

E – Filler metal may be used as an electrode.
R – Filler metal may be used as a rod.
S – Solid filler metal.
C – Composite or stranded filler metal.
Y – Designator (or combination of designators) that describes specific alloy, shielding gas to be used, diffusible hydrogen limit, etc. Refer to the appropriate AWS Filler Metal Specification shown in table for explanation.

The AWS classification of metal filler rod is identified on its container. In addition, each rod is usually marked either by stamping the AWS classification into the rod or by a paper or plastic tab wrapped around the rod (*Figure 21*). Consume filler metal rods so that the ID marker is not destroyed. Markers are located at each end. Cutting a rod in half allows you to retain the marker on the used stub. Retaining the stubs with the markings until the weld is complete provides evidence that the correct filler metal was used.

2.5.1 Carbon Steel and Low-Alloy Steel Filler Metals

Carbon steel filler metals are identified by *AWS Specification A5.18*. Low-alloy steel filler metals are identified by *AWS Specification A5.28*. The rod classification number is found on a label on the packaging. *Figures* 22 and 23 show the AWS classifications for carbon steel and low-alloy steel filler metals.

All steel filler metals contain alloys such as silicon, manganese, aluminum, and carbon. Other alloys such as nickel, chromium, and molybdenum are also often added. The purpose of the alloys is as follows:

- *Silicon (Si)* – Concentrations of 0.40 to 1.00 percent are used to deoxidize the puddle and to strengthen the weld. Silicon above 1 percent may make the welds crack-sensitive.
- *Manganese (Mn)* – Concentrations of 1 to 2 percent are also used as a deoxidizer and to strengthen the weld. Manganese also decreases hot-crack sensitivity.
- *Aluminum (Al), Titanium (Ti), and Zirconium (Zr)* – One or more of these elements may be added in very small amounts for deoxidizing. These elements may also increase strength.
- *Carbon (C)* – Concentrations of 0.05 to 0.12 percent are used to add strength without adversely affecting ductility, porosity, or toughness.
- *Nickel (Ni), Chromium (Cr), and Molybdenum (Mo)* – These elements may be added in small amounts to improve corrosion resistance, strength, and toughness.

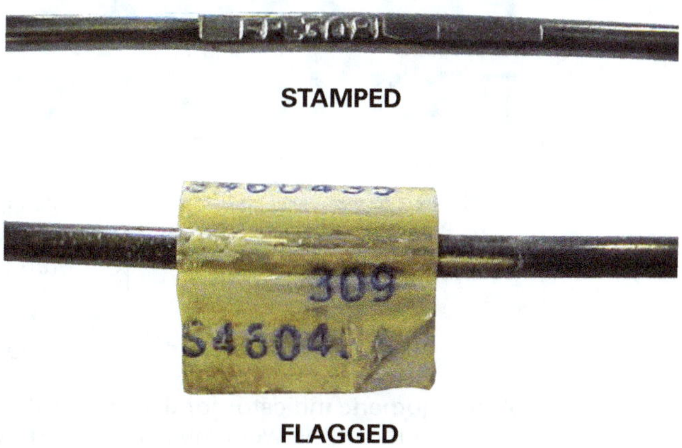

Figure 21 Filler rod markings.

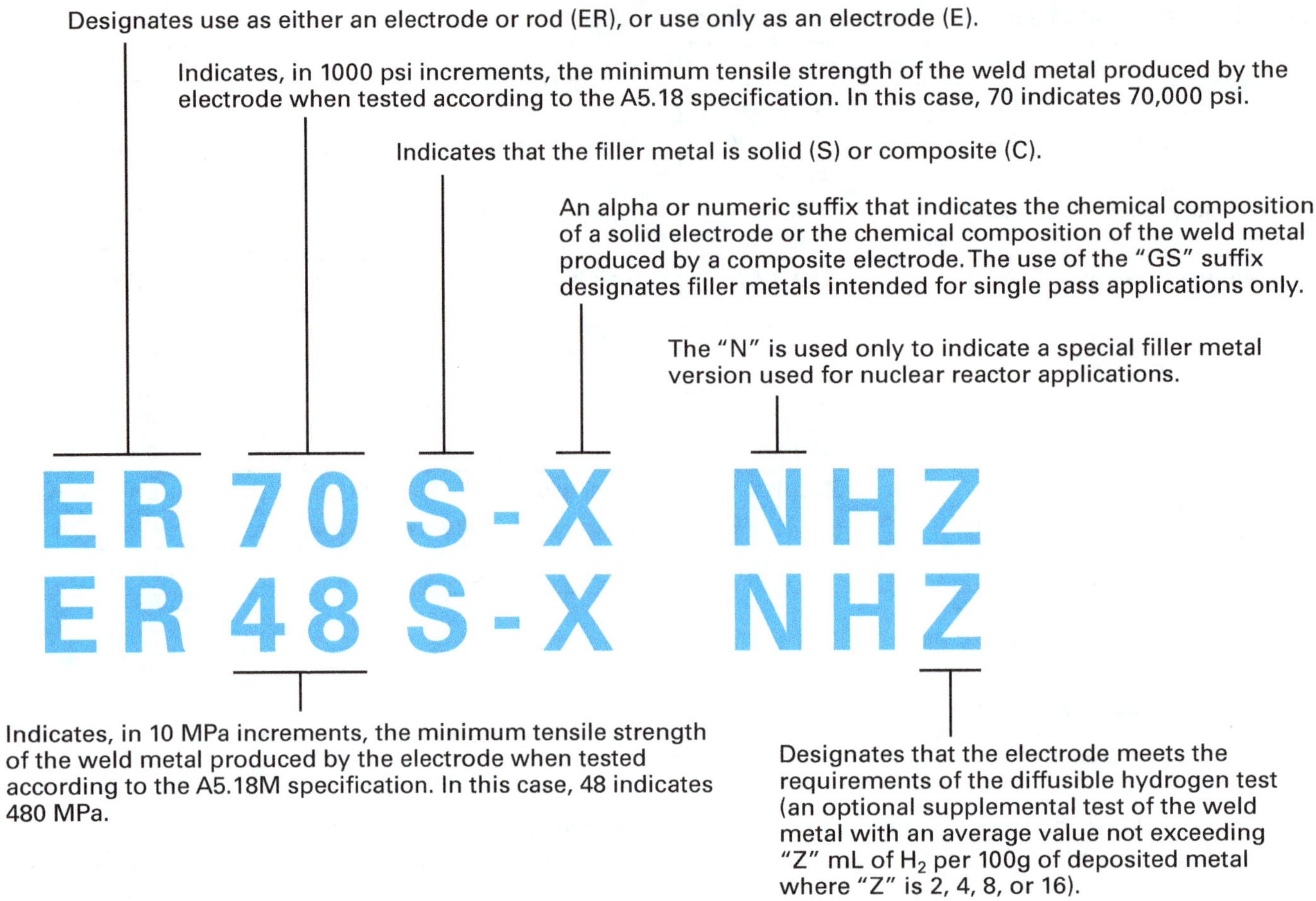

Figure 22 AWS classification for carbon steel filler metals.

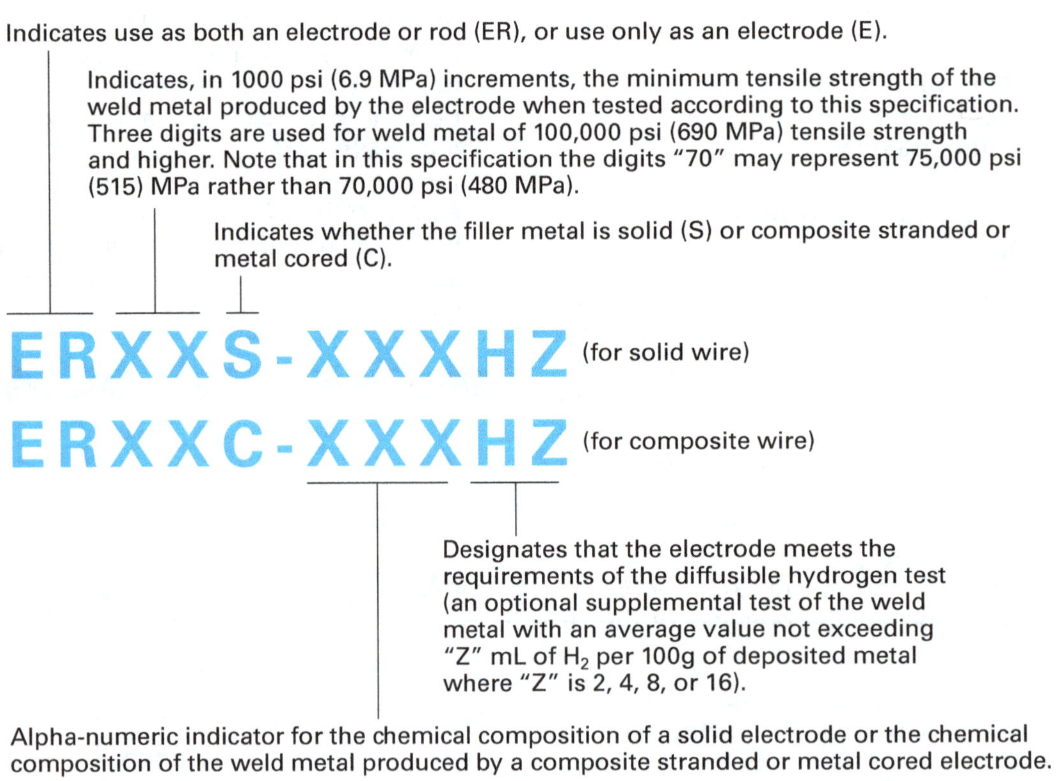

Figure 23 AWS classification for low-alloy steel filler metals.

2.5.2 Stainless Steel Filler Metals

Stainless steel filler metals are identified by *AWS Specification A5.9*. Designations for a typical AWS stainless steel electrode rod classification are shown in *Figure 24*.

Select stainless steel filler metal to closely match the alloy composition of the base metal.

2.5.3 Aluminum and Aluminum-Alloy Filler Metals

Aluminum electrode rods are covered by *AWS Specification A5.10* as shown in *Figure 25*. Aluminum filler metals usually contain magnesium, manganese, zinc, silicon, or copper for increased strength. Corrosion resistance and ease of being welded are also considerations. Aluminum filler metals are designed to weld various types of aluminum and should be selected for compatibility. The most widely used aluminum rods are ER4043 (contains silicon) and ER5356 (contains magnesium).

2.5.4 Copper and Copper Alloy Filler Metals

Copper electrode rods are covered by *AWS Specification A5.7*. Most copper filler metals contain other elements to increase strength, deoxidize the weld metal, and match the base metal composition (*Figure 26*).

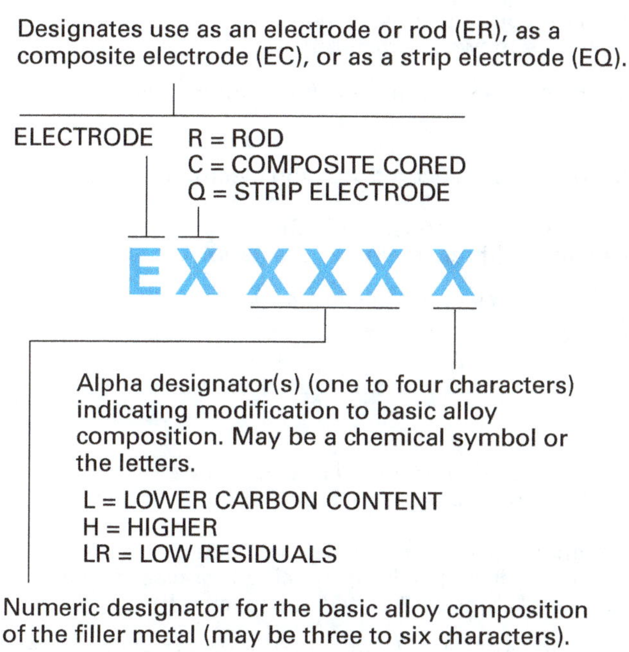

Figure 24 Typical AWS stainless steel rod classification.

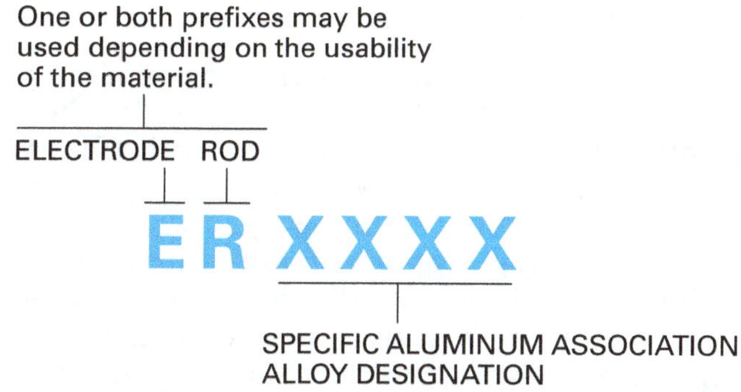

Figure 25 Typical AWS aluminum filler metal classification.

2.5.5 Nickel and Nickel Alloy Filler Metals

Nickel-based filler metals are covered by *AWS Specification A5.14*. These filler metals contain other elements to match base metal applications and to increase the strength and quality of the weld metal. For GTAW, DCEN is used with high-purity argon, helium, or both argon and helium used as a shielding gas. *Figure 27* shows the AWS classification for nickel filler metals.

2.5.6 Magnesium Alloy Filler Metals

Magnesium alloy filler metals (*Figure 28*) are covered by *AWS Specification A5.19*. These filler metals are usually used with GTAW, GMAW, and plasma arc welding (PAW) processes. For GTAW welding, the techniques and equipment are similar to those for aluminum. Argon, helium, or both are used for shielding. Alternating current is preferred for arc cleaning and penetration. Direct current is also employed with DCEP, used for thin materials; and DCEN, used for mechanized welding with helium, for deep penetration. GTAW is recommended for the defect repair of clean magnesium castings.

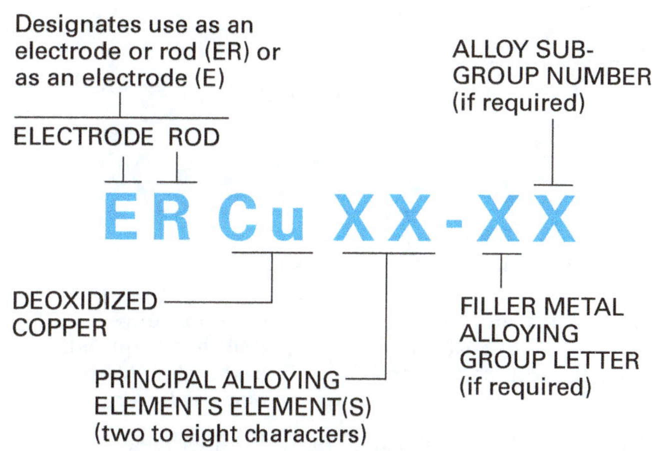

Figure 26 Typical AWS copper filler metal classification.

2.5.7 Titanium and Titanium Alloy Filler Metals

Titanium and titanium alloy filler metals (*Figure 29*) are covered by *AWS Specification A5.16*. These filler metals are generally used with GTAW, GMAW, SAW, and PAW processes. Titanium is sensitive to embrittlement by oxygen, nitrogen, and hydrogen at temperatures above 500°F (260°C). Like aluminum, titanium requires weld cleaning and high-purity gas shielding, and an especially adequate postflow shielding, to avoid embrittlement. Titanium can be successfully fusion-welded to zirconium, tantalum, niobium, and vanadium. Titanium should not be fusion-welded to copper, iron, nickel, or aluminum. Like magnesium, titanium fines are also flammable, so don't allow them to accumulate.

Rod and Electrode Designations

The letter R that follows the E in an AWS classification number indicates that the classification is used as a metal filler rod as well as an electrode. Electrode wire is used in GMAW. Since a tungsten-based electrode is used in GTAW, filler metal classifications with an ER must be used.

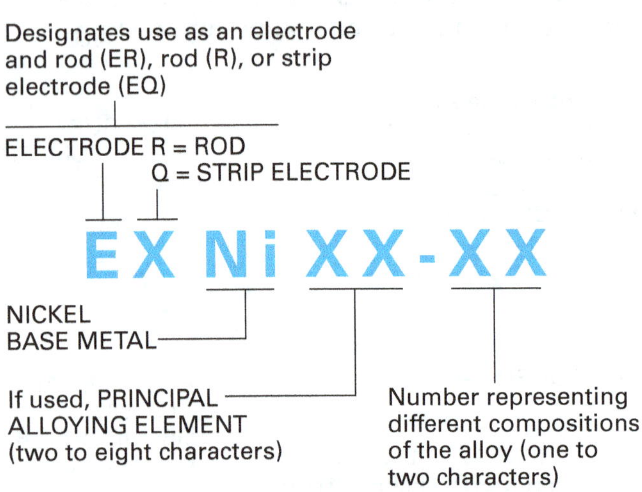

Figure 27 AWS classification for nickel filler metals.

2.5.8 Stainless Steel Flux-Cored Electrodes and Rods

Figure 30 shows the *AWS A5.22* specification format for stainless steel flux-cored electrodes and rods.

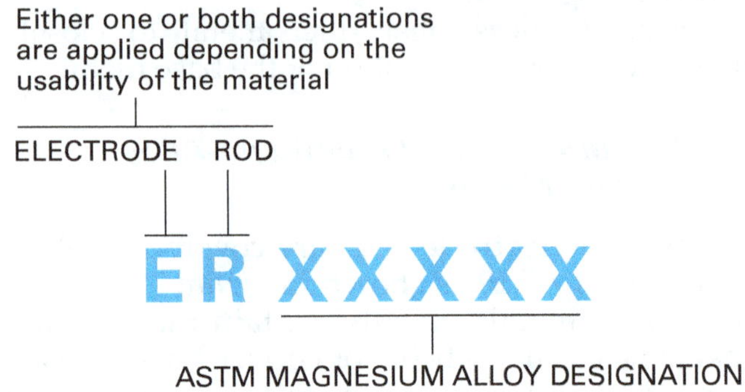

Figure 28 AWS classification for magnesium filler metals.

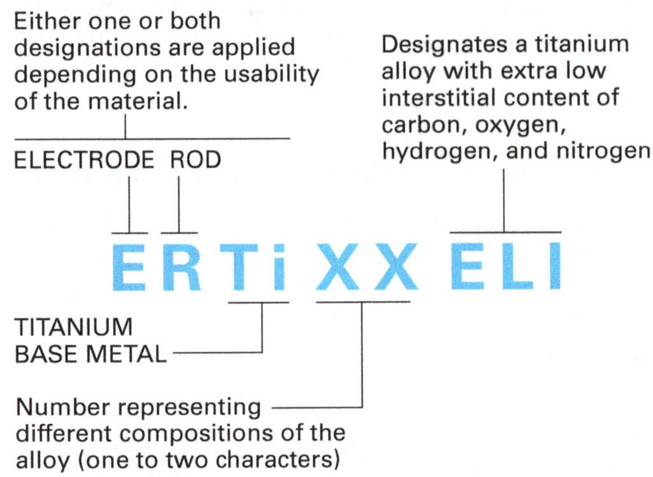

Figure 29 AWS classification for titanium filler metals.

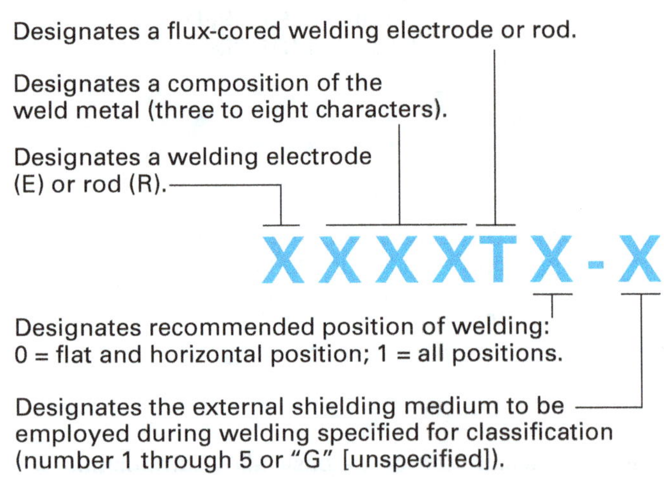

Figure 30 AWS classification for stainless steel flux-cored electrodes and rods.

Magnesium Fires

While ignition of magnesium is a very remote possibility when welding, the fire will cease when the heat source is removed. Ignition of the weld pool is prevented by the gas shielding used in GTAW, GMAW, and PAW processes. Most magnesium fires occur when fines from grinding or filing, or chips from machining, are allowed to accumulate. Take care to prevent the accumulation of fines on clothing. Graphite-based or salt-based powders recommended for extinguishing magnesium fires should be stored in the work area. If large amounts of fines are produced, collect them in a waterwash-type of dust collector designed for use with magnesium. Follow special precautions for the handling of wet magnesium fines.

Additional Resources

AWS B1.10, Guide for the Nondestructive Examination of Welds. Miami, FL: American Welding Society.

AWS B1.11, Guide for the Visual Examination of Welds. Miami, FL: American Welding Society.

AWS D3.5, Guide for Steel Hull Welding. Miami, FL: American Welding Society.

AWS D3.7, Guide for Aluminum Hull Welding. Miami, FL: American Welding Society.

AWS Safety and Health Fact Sheet 27. Miami, FL: American Welding Society.

Lincoln Electric website: **http://www.lincolnelectric.com**

Modern Welding Technology. 6th Edition. Howard B. Cary. Englewood Cliffs, NJ: Prentice Hall, Inc.

OSHA 1910.269, Appendix *C, Protection from Step and Touch Potentials.* Current Edition. Washington, DC: Occupational Safety and Health Administration (OSHA).

OSHA 19260.351, Arc Welding and Cutting. Current Edition. Washington, DC: Occupational Safety and Health Administration (OSHA).

Tungsten Electrodes: An **Arc-Zone.com**® Technical Focus Paper (pdf). Carlsbad, CA: **Arc-Zone.com**, Inc. **https://www.arc-zone.com/azc-scip/TungstenElectrodes.pdf**

Tungsten Guidebook (pdf). Newbury Park, CA: Diamond Ground Products, Inc. **http://www.diamond-ground.com/TungstenGuidebook2013.pdf**

Welding Handbook. Vol. 1. Welding Science & Technology. 9th Edition. Miami, FL: American Welding Society, 2001.

Welding Handbook. Vol. 2, Part 1: Welding Processes. Miami, FL: American Welding Society, 2004.

2.0.0 Section Review

1. A type of welding machine that is designed to produce either direct welding current only or both alternating and direct welding currents is a(n) _____.
   a. transformer welding machine
   b. transformer-rectifier welding machine
   c. alternator welding machine
   d. inverter power source

2. How many hoses does a water-cooled GTAW torch have?
   a. One
   b. Two
   c. Three
   d. Four

3. The device that contains an assembly of fine screens that eliminate turbulence by straightening the gas flow from a GTAW gas nozzle is called a _____.
   a. laminar flow gas lens
   b. shielding gas supply hose
   c. closed-loop gas rectifier
   d. gas lens collet body

4. In GTAW, small amounts of hydrogen are sometimes mixed with argon for deeper penetration when welding _____.
   a. titanium
   b. aluminum
   c. carbon steel
   a. stainless steel

5. To increase strength, deoxidize the weld metal, and match the base metal composition, most copper electrodes contain _____.
   a. other elements
   b. gas lens filters
   c. magnesium castings
   d. anodized quenchers

SECTION THREE

3.0.0 WELDING EQUIPMENT SETUP

Objective

Explain how to set up for GTAW welding.

a. Explain how to select and position the welding machine.
b. Explain how to connect and set up the shielding gas flow rate.
c. Explain how to select and prepare the tungsten electrode.
d. Explain how to select and install the nozzle along with the tungsten electrode.

Performance Tasks

3. Connect the shielding gas and set the flow rate.
4. Select and prepare the tungsten electrode.
5. Break down and reassemble a GTAW torch.

To weld safely and efficiently, all the necessary welding equipment must be properly set up. The following sections explain how to properly set up GTAW equipment.

3.1.0 The GTAW Welding Machine

Select the welding machine and current type for the base metal to be welded, considering the following factors:

- GTAW requires a constant-current power source. SMAW power sources and welding machines are commonly used for GTAW.
- Consider the welding current type required, AC or DC; generally, carbon steels, stainless steels, and alloy steels are welded with DC, while aluminum and magnesium are welded with AC at high frequency.
- Determine the maximum amperage required.
- Determine if there is an electrical power source and appropriate receptacles in which to plug a welding machine. Otherwise, an engine-driven welding machine is needed.

Because of the limited length of GTAW torch cables, the shielding gas supply must be located reasonably close to the welding site. The welding machine can be some distance away, but unless remote controls are installed, it will not be convenient to operate.

Select a site where the GTAW equipment will not be in the way but will be protected from welding, grinding, or cutting sparks. There should be good air circulation to keep the welding machine cool. The environment should be free from explosive or corrosive fumes and as free as possible from dust and dirt. Welding machines have internal cooling fans that will pull these particles into the machines if they are present. The site should also be free of standing water or water leaks. If an engine-driven generator is used, locate it so that it can be easily refueled and serviced. Also, make sure that the area around the generator is well ventilated.

There should be easy access to the site so that the equipment can be started, stopped, or adjusted as needed. If the machine will be plugged into an outlet, be sure that the outlet has been properly installed and properly grounded by a licensed electrician. Also, be sure to identify the location of the electrical power disconnect switch or circuit breaker before plugging the welding machine into the outlet.

3.2.0 The Shielding Gas

The hose from the shielding gas regulator and flowmeter connects to the welding machine gas solenoid if the welding machine is designed for GTAW. If a standard welding machine is used, the shielding gas hose connects to the torch cable. To connect the shielding gas, perform the following steps:

Step 1 Identify the shielding gas required by referring to the WPS or site quality standard.

Step 2 Locate a cylinder of the correct gas or mixture, and secure it nearby.

WARNING!

Be sure to secure the gas cylinder first so it cannot fall over. Do not remove the cylinder cap until the cylinder is secured.

Step 3 Remove the cylinder's protective cap, momentarily crack open the cylinder valve to blow out any dirt, and then close it again.

Step 4 Using a regulator or combination regulator/flowmeter with the correct CGA connection for the cylinder, mount the regulator on the cylinder.

Step 5 Connect the gas hose to the flowmeter and to the gas solenoid on the welding machine or to the end of the torch cable.

> **NOTE**
> If the gas hose is connected to the torch cable and the torch does not contain a gas shutoff valve, install a valve in the line between the torch cable and the regulator/flowmeter.

Step 6 Check to be sure the flowmeter adjusting valve is closed.

Step 7 Very slowly open the cylinder valve slightly. After the pressure gauge indicates the cylinder pressure and is stable, open the valve completely.

Figure 31 shows how to connect the shielding gas.

3.2.1 Setting the Shielding Gas Flow Rate

Gas flow rate is measured in cubic feet per hour (ft^3/hr or cfh) or liters per minute (L/min), which represents the volume of gas flowing from the torch nozzle. Flow rate is important because it affects the quality and the cost of a weld. A flow rate that is too low will not shield the weld zone adequately and will result in a poor-quality weld. An excessively high gas flow rate wastes expensive gas and can generate turbulence in the gas column above the weld. This turbulence can pull air into the weld zone and cause oxidation and weld contamination. The nozzle must be large enough to gently flood the weld pool with inert gas. As a general rule, larger nozzles require higher flow rates. Welding specifications contain nozzle sizes and shielding gas flow rates.

Factors that may affect the shielding gas flow rate include the following:

- *Drafts* – The flow rate must be increased in a drafty location to maintain the gas shield around the weld zone.
- *Specific gas used* – For example, helium usually requires a higher flow rate than argon because helium rises much faster because of its low density.
- *Welding current* – High welding currents require higher flow rates.
- *Nozzle size (exit opening)* – Larger nozzles require higher flow rates.
- *Weld joint type* – Welds on flat surfaces require higher flow rates than welds in deep grooves or fillets.
- *Welding speed* – Fast advance speeds require higher flow rates than slower advance speeds.
- *Weld position* – Vertical- and horizontal-position welds require higher flow rates than flat or overhead welds.

Adjust the flow for the type and thickness of the base metal being welded, or per applicable WPS. To set the shielding gas flow rate, perform the following steps.

Step 1 Start the gas flow by depressing the remote control foot pedal or hand-operated switch to open the gas solenoid on the welder, or manually open the hand valve on the torch.

> **CAUTION**
> Some flowmeters are preset for a specific gas. These flowmeters do not require any adjustment, but they can only be used with the specific shield gases indicated on the flowmeter. Due to density differences in the various shielding gases, using this type of flowmeter with the wrong gas will result in incorrect flow rate readings.

Step 2 Set the gas flow by turning the flowmetering adjustment knob out until the ball inside the glass tube rises to the desired flow rate (*Figure 32*). The flow rate is usually read from the top of the floating ball. However, always check the manufacturer's instructions to determine the correct adjustment method.

Step 3 When the proper gas flow has been achieved, release the remote foot pedal or hand switch, or manually close the valve on the torch to stop the gas flow.

> **NOTE**
> Some flowmeters are equipped with several scales of different calibrations around the same sight tube for monitoring the flows of different types (densities) of gases. Be sure to rotate the scales or read the correct side for the gas type being used.

Step 4 Check for leaks before starting to weld.

The following sections provide general recommendations for flow rates for various types of base metals. Note that these flow rates can vary; always check the WPS for the appropriate flow rate.

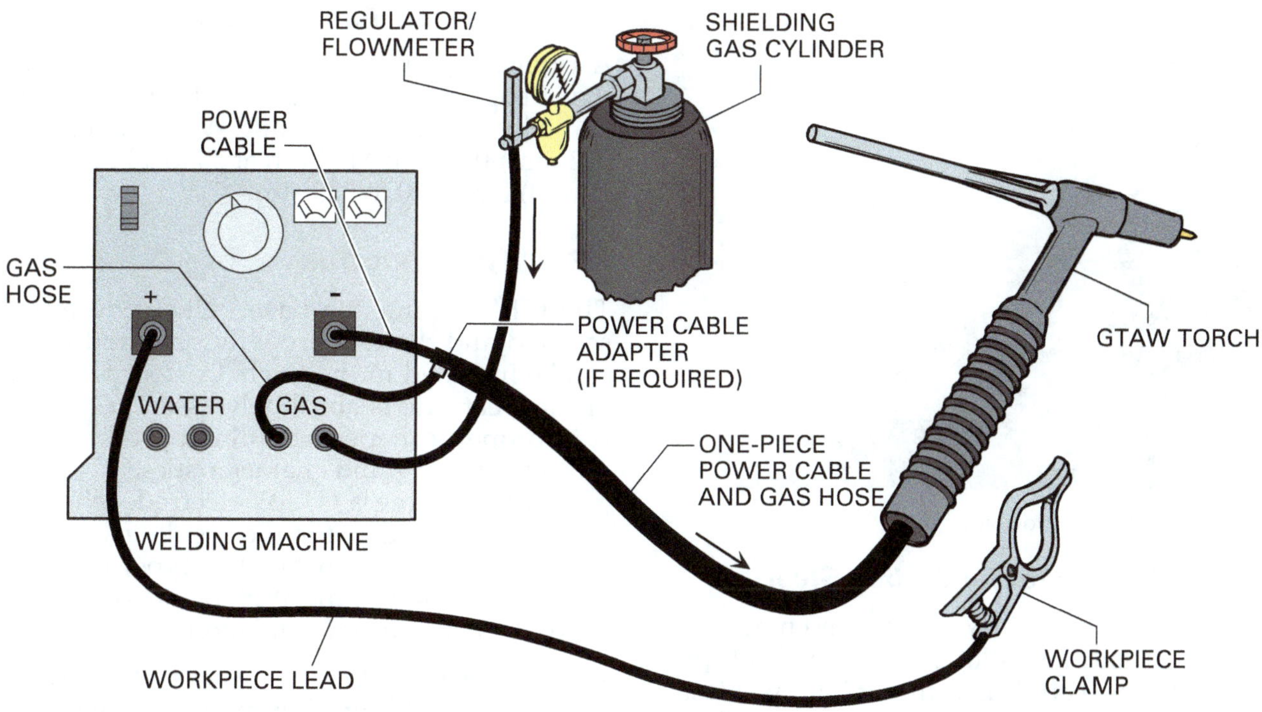

WARNING: IF A POWER CABLE ADAPTER IS USED IT MUST BE COVERED BY AN INSULATING RUBBER BOOT.

SINGLE-HOSE, GAS-COOLED TORCH SETUP

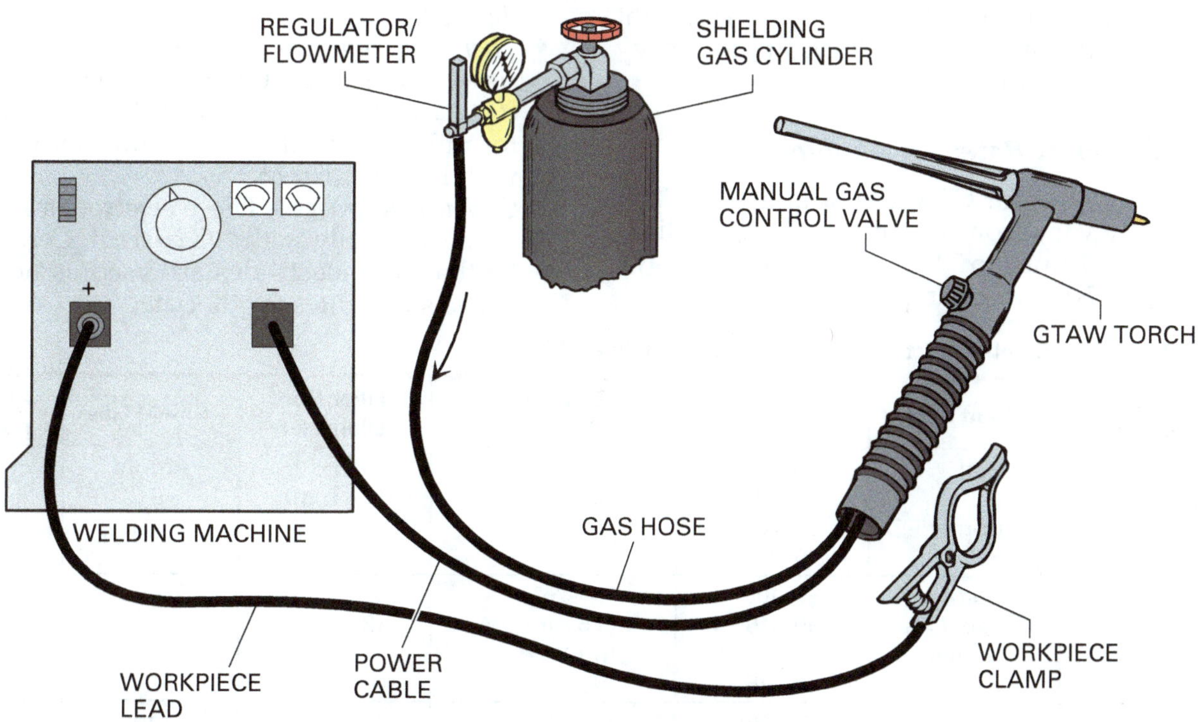

DOUBLE-HOSE, GAS-COOLED TORCH SETUP

Figure 31 Connecting the shielding gas.

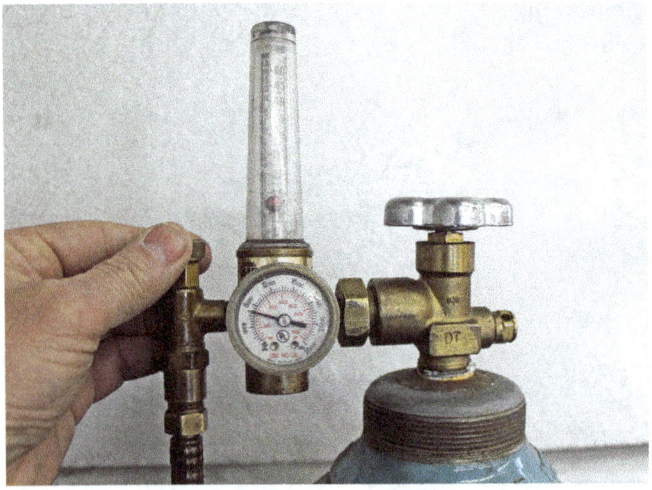

Figure 32 Adjusting the flowmeter.

3.2.2 Gas Flow Rates for Carbon Steel

Argon flow rates for GTAW on low- and medium-carbon steels and low-alloy steels (*Table 3*) typically vary from 15 cfh (7 L/min) for light welding at 60 amps, to 20 cfh (9 L/min) for heavier welding at 200 or more amps.

3.2.3 Gas Flow Rates for Stainless Steel

Argon flow rates for GTAW on stainless steel (*Table 4*) typically vary from 10 cfh (5 L/min) for light welding at 80 amps to 15 cfh (7 L/min) for heavy welding at 350 or more amps.

3.2.4 Gas Flow Rates for Aluminum

Argon flow rates for GTAW on aluminum (*Table 5*) typically vary from 15 cfh (7 L/min) for light welding at 60 amps to 30 cfh (14 L/min) for heavy welding at 350 or more amps.

3.2.5 Gas Flow Rates for Copper

A typical argon flow rate for GTAW on deoxidized copper up to 1/8" (3.2 mm) thick is 15 cfh (7 L/min) from 110 to 250 amps. With copper over 1/8" (3.2 mm) thick, use helium at 30 to 40 cfh (14 to 19 L/min) from 190 to 525 amps.

3.3.0 The Electrode

The GTAW tungsten-based electrode must be selected and the end properly prepared before it can be installed in the torch. Correct preparation of the electrode is absolutely essential. If the tip is improperly shaped, it will not produce the required arc shape and characteristics.

When selecting a GTAW electrode, choose the type recommended for the welding current required by the base metal. The type of electrode will be specified in the WPS or site quality standards. General recommendations are as follows:

- *Pure tungsten (EWP)* – These electrodes are used for welding aluminum and magnesium with AC and high frequency.
- *Zirconiated tungsten (EWZr)* – Used for welding aluminum and magnesium with AC and high frequency for welds where tungsten inclusions are not tolerated and higher current capacity is desired.
- *Thoriated tungsten (EWTh-1 or EWTh-2), ceriated tungsten (EWCe-2), or lanthanated tungsten (EWLa-1, -1.5, or -2)* – These electrodes are used for mild steels, alloy steels, and stainless steels with DCEN current.
- *Ceriated electrodes* – These electrodes also operate successfully with AC current. Ceriated and lanthanated electrodes are used as non-radioactive substitutes for thoriated electrodes.

Table 3 Carbon Steel—Direct Current Electrode Negative (DCEN)

Metal Thickness	Joint Type	Amperage	Electrode Diameter	Filler Rod Diameter	Shield Gas	Cubic Feet per Hour
1/16" (1.6 mm)	Butt Lap Corner Fillet	60–70 70–90 60–70 70–90	1/16" (1.6 mm)	1/16" (1.6 mm)	Argon	15
1/8" (3.2 mm)	Butt Lap Corner Fillet	80–100 90–115 80–100 90–115	1/16"–3/32" (1.6 mm–2.4 mm)	3/32" (2.4 mm)	Argon	15
3/16" (5 mm)	Butt Lap Corner Fillet	115–135 140–165 115–135 140–170	3/32" (2.4 mm)	1/8" (3.2 mm)	Argon	20
1/4" (6 mm)	Butt Lap Corner Fillet	160–175 170–200 160–175 175–210	1/8" (3.2 mm)	5/32" (4 mm)	Argon	20

Table 4 Stainless Steel—Direct Current Electrode Negative (DCEN)

Metal Thickness	Joint Type	Amperage (Flat Position)	Electrode Diameter	Filler Rod Diameter	Shield Gas	Cubic Feet per Hour
1⁄16" (1.6 mm)	Butt Lap Corner Fillet	80–100 100–120 80–100 90–110	1⁄16" (1.6 mm)	1⁄16" (1.6 mm)	Argon	10
1⁄8" (3.2 mm)	Butt Lap Corner Fillet	120–140 130–150 120–140 130–150	1⁄16" (1.6 mm)	3⁄32" (2.4 mm)	Argon	10
3⁄16" (5 mm)	Butt Lap Corner Fillet	200–250 225–275 200–250 225–275	3⁄32" or 1⁄8" (2.4 mm or 3.2 mm)	1⁄8" (3.2 mm)	Argon	15
1⁄4" (6 mm)	Butt Lap Corner Fillet	275–350 300–375 275–350 300–375	1⁄8" (3.2 mm)	5⁄32" (4 mm)	Argon	15
1⁄2" (12.5 mm)	Butt Lap Fillet	350–450 375–475 375–475	1⁄8" or 3⁄16" (3.2 mm or 4.8 mm)	3⁄16" (4.8 mm)	Argon	15

Table 5 Aluminum—Alternating Current with High Frequency Stabilization

Metal Thickness	Joint Type	Amperage (Flat Position)	Electrode Diameter	Filler Rod Diameter	Shield Gas	Cubic Feet per Hour
1⁄16" (1.6 mm)	Butt Lap Corner Fillet	60–85 70–90 60–85 75–100	1⁄16" (1.6 mm)	1⁄16" (1.6 mm)	Argon	15
1⁄8" (3.2 mm)	Butt Lap Corner Fillet	125–150 130–160 120–140 130–160	3⁄32" (2.4 mm)	3⁄32" or 1⁄8" (2.4 mm or 3.2 mm)	Argon	20
3⁄16" (5 mm)	Butt Lap Corner Fillet	180–225 190–240 180–225 190–240	1⁄8" (3.2 mm)	1⁄8" (3.2 mm)	Argon	20
1⁄4" (6 mm)	Butt Lap Corner Fillet	240–280 250–320 240–280 250–320	3⁄16" (4.8 mm)	1⁄8" or 3⁄16" (3.2 mm or 4 mm)	Argon	25
1⁄2" (12.5 mm)	Butt Lap Corner Fillet	400–450 400–450 400–450 420–470	3⁄16" or 1⁄4" (4.8 mm or 6.4 mm)	3⁄16" or 1⁄4" (4.8 mm or 6.4 mm)	Argon	30

- *Rare earth electrodes (EW)* – These electrodes are used as non-radioactive substitutes for 2-percent thoriated electrodes. They start and re-ignite well and have very good service life for applications in which welding cycles of at least 15 minutes are used.
- *Cryogenically treated thoriated tungsten electrodes* – These electrodes are used as non-radioactive substitutes for 2-percent thoriated electrodes. The cryogenically treated electrodes are easier to ignite and have longer tip life, longer overall electrode life, and higher amperage tolerance than non-treated 2-percent thoriated tungsten.

Select the size of electrode rated for the amperage to be used. *Table 6* lists examples of electrode current ranges by electrode size and type when used with argon shielding gas under various GTAW modes.

3.3.1 Preparing the Electrode

The different types of GTAW welding current require different-shaped electrode tips. For manual DCEN welding, a taper-ground and truncated point is the best shape. For manual DCEP welding, a rounded end is the best shape. For manual AC welding, a tapered and balled end is the best shape. *Figure 33* shows GTAW electrode end shapes.

Electrode tips must be formed on all new electrodes and on used electrodes that have damaged or contaminated tips. The contaminated section must be reshaped. Do not prepare the end of the electrode that contains the color-code band; otherwise you will not be able to identify the electrode's type in the future.

The degree of taper on an electrode tip affects weld penetration and bead width. A long, narrow taper (small angle, more pointed) produces a wide bead with shallow penetration. A short, wide taper (large angle, less pointed) produces a narrow bead and deep penetration. During any particular welding operation, it is important to maintain the same taper on the electrode during sharpening or replacement so that the bead characteristics remain the same.

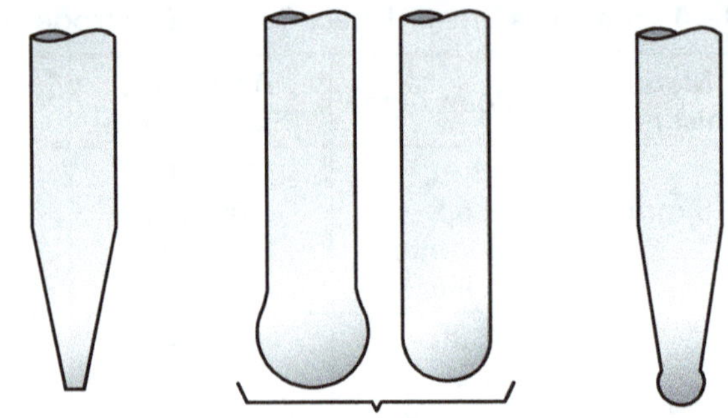

Figure 33 GTAW electrode end shapes for different current types.

3.3.2 Pointing the Electrode Tip

For DCEN welding, electrodes can be made pointed by dipping the red-hot end into a special chemical powder, or they can be tapered by grinding or sanding.

Perform the following steps to use a chemical powder to sharpen a tungsten electrode:

Step 1 Place the tungsten electrode in the torch so that it extends about 1 inch (≈25 mm) from the gas nozzle.

Table 6 Electrode Current Ranges

Electrode Diameter	GTAW Modes			
	AC*		Pure Tungsten, Rare Earth, or Thoriated Tungsten Electrodes	
	Pure Tungsten	Thoriated or Rare Earth Electrodes**	DCEN	DCEP
0.020" (0.5 mm)	5–15	8–20	8–20	–
0.040" (1.0 mm)	10–60	15–80	15–80	–
1⁄16" (1.6 mm)	50–100	70–150	70–150	10–20
3⁄32" (2.4 mm)	100–160	140–235	150–250	15–30
1⁄8" (3.2 mm)	150–210	225–325	250–400	25–40
5⁄32" (4.0 mm)	200–275	300–425	400–500	40–55
3⁄16" (4.8 mm)	250–350	400–525	500–800	55–80
1⁄4" (6.4 mm)	325–475	500–700	800–1,000	80–125

* Maximum values shown have been determined using an unbalanced AC power source. If a balanced AC power source is used, either reduce these values by about 30 percent or use the next size electrode. This is necessary because of the higher heat input to the electrode in a balanced AC setup.

** Balled electrode tip ends can best be sustained at these current levels.

Step 2 Heat the tungsten electrode by shorting it on a copper plate or by striking an arc. Be sure shielding gas is flowing to protect the tungsten electrode.

Step 3 When the tip of the tungsten electrode is cherry-red for a distance of about four times the electrode diameter, place the end of the electrode in the chemical powder. For example, if the electrode diameter is ⅛" (3.2 mm), about ½" (13 mm) of the length should be cherry-red.

Step 4 Hold the end of the tungsten electrode in the powder for the time recommended on the container for the diameter of the electrode being pointed. The chemical powder will dissolve the tungsten electrode and form a sharp point at the proper angle. For an un-tapered (new or broken off) tungsten, the procedure may have to be repeated a number of times to obtain the desired shape.

Reading Torch Flow Rate

Welders must maintain accurate and consistent torch flow rates to create consistent welds and adhere to a WPS. The device shown here can be used to quickly measure shielding gas flow directly at the torch tip. This simple peashooter style is durable and convenient. Digital models offer a bit more sophistication. They can also be mounted on the gas line, providing a constant digital display of flow rate.

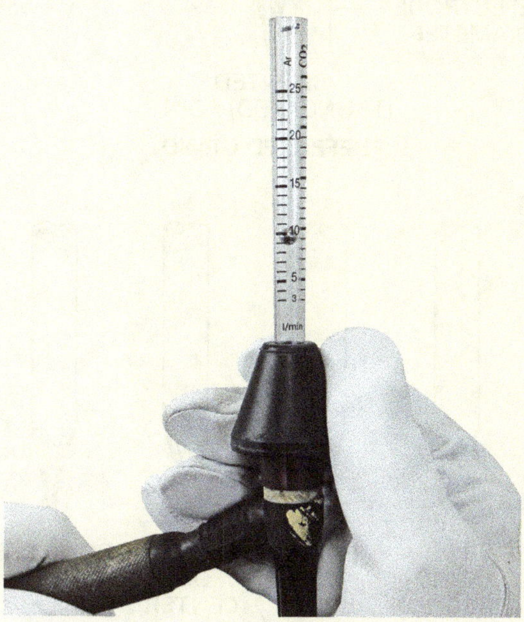

There Is No Universal Electrode Size

Contrary to some popular beliefs, the ³⁄₃₂" (2.4 mm) electrode is not a universal size that can be used for everything. Always use the electrode size recommended for the amperage required. Using an electrode that's too big will result in arc starting and stability problems. Using an electrode that is too small can result in tungsten spitting (tungsten melting and migration into the weld).

Figure 34 shows the chemical powder used for sharpening.

CAUTION

Always wear clean gloves when handling tungsten electrodes to avoid contaminating the electrode.

More often, tungsten electrodes are sharpened by grinding or sanding. Since tungsten electrodes are very hard, they are best ground with diamond-impregnated silicon carbide or alumina oxide grinding wheels or sanding belts. Use 80 grit for fast shaping and 120 grit for finishing.

WARNING!

Observe proper respiratory, hazard, and environmental contamination procedures when using a grinder. Dust from tungsten electrode grinding can be hazardous and may contain radioactive elements. Refer to the manufacturer's SDS/MSDS to determine any specific hazards. Safety glasses must be worn when using a grinder.

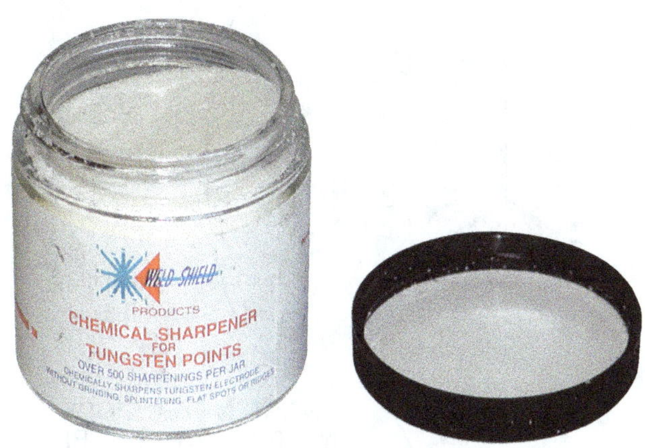

Figure 34 Tungsten sharpening powder.

Follow these steps to point the end of a tungsten electrode using a grinder:

> **CAUTION**
>
> Use a grinding wheel designated only for tungsten. Be careful to never use the grinding wheel for any other material, or contaminants will be deposited on the tungsten electrode and the grinding wheel will have to be discarded. Always wear safety glasses when using a grinder.

Step 1 With the color band end held farthest from the grinder, hold the tungsten electrode against the grinder so that the tungsten electrode is in the plane of the wheel and makes an angle of about 30 degrees with the wheel or belt surface (*Figure 35*).

Step 2 Slowly rotate the tungsten electrode as it is ground to keep the point on the center line of the axis. Orient the grind marks toward the point along the long dimension of the tungsten electrode, not around the point. For DCEN electrodes, the length of the taper should be between two and three times the electrode diameter. The point should be truncated to the approximate tip diameter shown in *Table 7*.

Step 3 Check the ground end to make sure the tapered length is correct and that the point is in the center of the tungsten electrode and not toward one side. Correct the shape if necessary.

Figure 36 shows preferred and incorrect grinds for prepared pointed electrode ends.

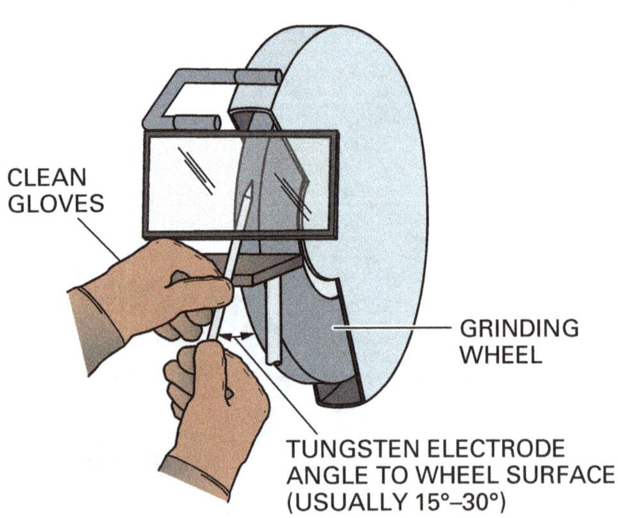

Figure 35 Proper grinding angle and tungsten electrode orientation.

Table 7 Tip Diameters

Electrode Diameter	Diameter at Tip
0.040" (1.0 mm)	0.005" (0.13 mm)
0.040" (1.0 mm)	0.010" (0.25 mm)
0.062" (1.6 mm)	0.020" (0.51 mm)
0.062" (1.6 mm)	0.030" (0.76 mm)
0.093" (2.4 mm)	0.030" (0.76 mm)
0.093" (2.4 mm)	0.045" (1.14 mm)
0.125" (3.2 mm)	0.045" (1.14 mm)
0.125" (3.2 mm)	0.060" (1.52 mm)

3.3.3 Balling the Electrode Tip

Only pure tungsten (EWP) or zirconiated tungsten (EWZr) can be balled, or spherical. The thorium in thoriated tungstens or other rare earth (except cerium) tungstens inhibits the formation of a ball. If welders attempt to ball most rare earth tungstens, the tungsten forms a number of small protrusions at the tip instead of a ball. These protrusions cause the arc to be unstable.

If the electrode is to be used for lower-amperage DCEP welding, the electrode end should be tapered about two diameters in length and be quite blunt. A ball will form on the blunt end when welding.

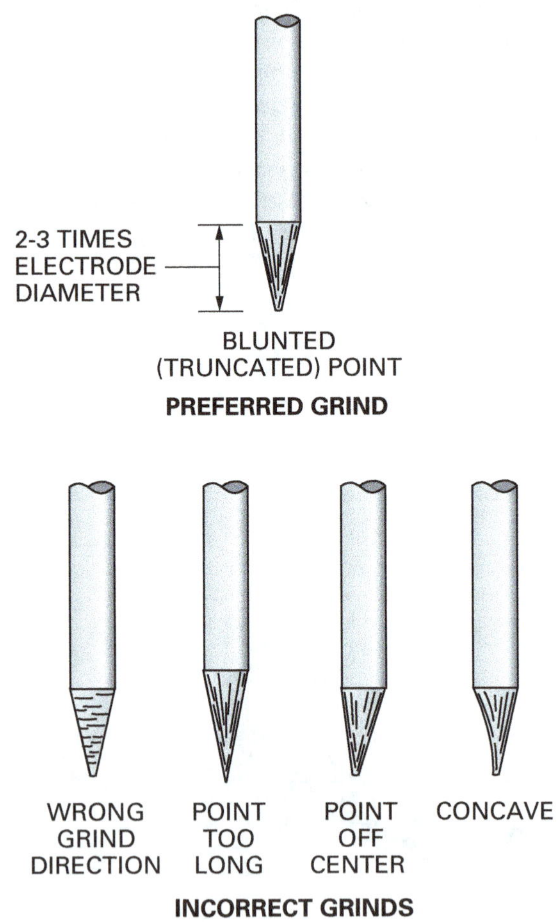

Figure 36 Properly and improperly prepared pointed electrode ends.

For AC welding, the end is balled with or without tapering. Balling is done by arcing the electrode over clean copper or another clean metal. Do not arc over carbon, because it will contaminate the electrode.

AC or DCEP can be used to ball an electrode. When balling an electrode, start off with a low current and gradually increase it until the end of the electrode starts to melt and form a hemispherical (half-round) end or slight ball. The ball should be no more than 1½ times the electrode diameter. Using DCEP will cause a ball to form at a much lower amperage setting than it will be formed using AC current. Using DCEP, only about 35 to 45 amps of welding current will be required to ball 3/32" (2.4 mm) to 1/8" (3.2 mm) diameter electrodes. *Figure 37* shows ideal, acceptable, and unacceptable prepared balled electrode ends.

CAUTION

Using excess amperage will cause the ball to melt and drop off the electrode. Be especially careful when using DCEP as it will generate much more heat on the electrode than the same AC setting.

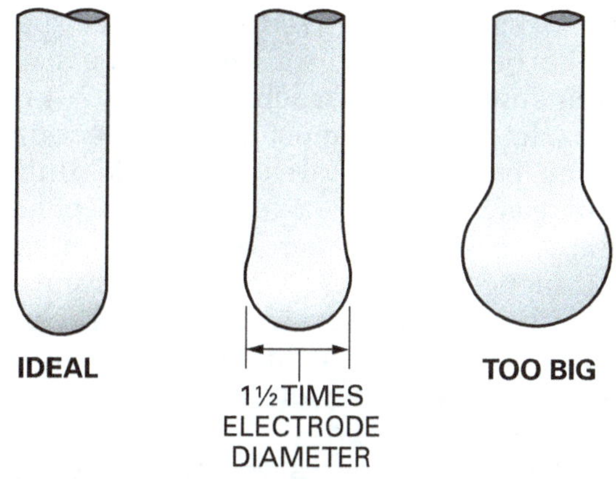

Figure 37 Properly and improperly prepared balled, nontapered electrode ends for AC welding.

3.4.0 Selecting and Installing the Nozzle

Typically, a range of nozzle sizes and styles is available for a given torch in order to accommodate different electrode sizes and welding applications. Nozzles are made for close clearances, for long reaches, for reaching deep into narrow places, and for covering wide beads. Discard nozzles with chips, cracks, or metal buildup on the end because they can affect the gas flow pattern and produce weld defects.

Diamond-Wheel Grinders

Diamond wheel grinders are used to prepare tungsten electrodes. Both stationary and handheld, portable grinders are available. A stationary water-bath diamond-wheel grinder (*A*) can be set to different electrode grind angles as necessary. A water bath captures the tungsten dust that is generated during grinding so the dust can be disposed of later. A handheld, portable diamond-wheel grinder (*B*), can also be set to different electrode grind angles as necessary. The dust generated during grinding is captured inside the grinder's fully enclosed grind head. The grinder also comes with an attachment that allows the grinder to be connected to a shop vacuum so accumulated dust can be removed from the grinding chamber.

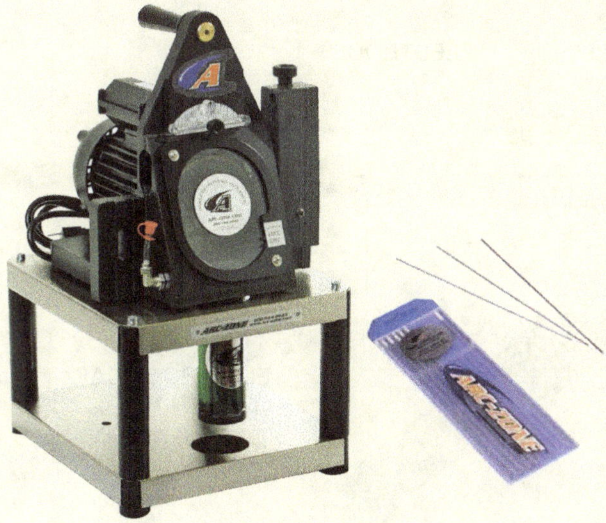

(A) STATIONARY GRINDER

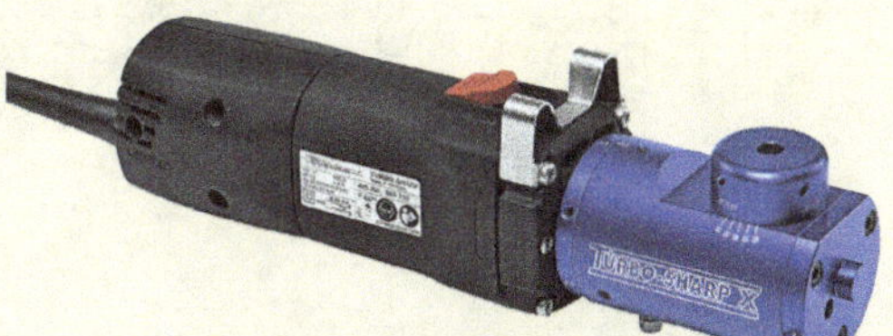

(B) HANDHELD GRINDER

The nozzle must be sized for the electrode diameter to be used and for any unique welding requirements, such as reaching into a V-groove. The nozzle diameter must be large enough to cover the entire weld area with shielding gas. A wide, easy flow of shielding gas is generally preferable to a narrow jet, which is apt to create turbulence and draw air into the weld zone. Small-diameter nozzles tend to overheat and break easily. However, a small-diameter nozzle is useful for steadying the torch and arc for the root pass in a V-groove weld.

To select the proper size and shape nozzle, always refer to the manufacturer's nozzle recommendations for the torch being used. In general, the nozzle's inside diameter should be about three times the diameter of the electrode being used.

3.4.1 Installing the Electrode

The electrode is usually installed after the nozzle because the electrode stickout (ESO) is measured from the end of the nozzle. The electrode is clamped in place in the torch body by the collet assembly. The collet is a cylindrical clamp that tightens as it is pulled or pushed (depending on the torch design) into a tapered holder as the threaded electrode cap is screwed on and tightened. To adjust the electrode stickout or to remove the electrode, loosen the cap assembly. Collet designs vary with torch manufacturers (*Figure 38*).

CAUTION

Handle tungsten electrodes with clean gloves to avoid contamination of the electrode.

Follow these steps to install the electrode:

Step 1 Remove the electrode cap from the rear of the torch.

Step 2 Check the collet size and collet body. If it is not the correct size for the electrode to be installed, remove and replace it with the correct size.

Step 3 Replace the electrode cap, but do not tighten it. Screw the cap only about one turn to leave the collet clamp loose.

Step 4 Insert the electrode into the collet from the nozzle end of the torch, and adjust it so that it extends from the nozzle about 1 to 1½ times the electrode diameter. Insert the color-coded end first to prevent destroying the electrode markings.

CAUTION

If the electrode cap is overtightened, the threads will be damaged as the torch heats with use. If the cap is too loose, the collet and electrode will overheat and damage the torch.

Step 5 Handtighten the electrode cap to lock the electrode in place.

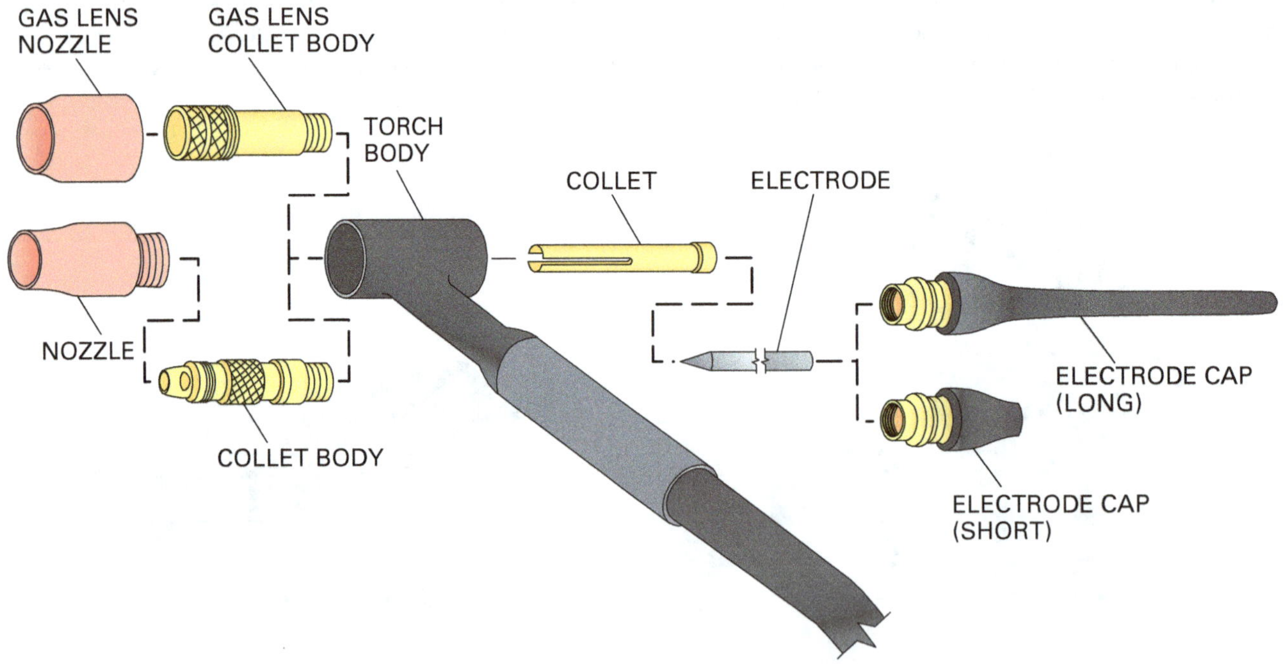

Figure 38 Collet assembly of a typical GTAW torch.

Electrode Stickout

Various manufacturers recommend stickout distances that differ from flush at the end of the cup to distances equal to the inside diameter of the cup, depending on the application and task. For instance, a root pass in a deep, narrow groove can be accomplished more easily with a narrow cup and a stickout that is about equal to the inside diameter of the cup. However, the use of a gas lens allows a longer tungsten stickout for welding under the same circumstances.

Additional Resources

AWS B1.10, Guide for the Nondestructive Examination of Welds. Miami, FL: American Welding Society.

AWS B1.11, Guide for the Visual Examination of Welds. Miami, FL: American Welding Society.

AWS D3.5, Guide for Steel Hull Welding. Miami, FL: American Welding Society.

AWS D3.7, Guide for Aluminum Hull Welding. Miami, FL: American Welding Society.

Lincoln Electric website: **http://www.lincolnelectric.com**

Modern Welding Technology. 6th Edition. Howard B. Cary. Englewood Cliffs, NJ: Prentice Hall, Inc.

OSHA 1910.269, Appendix *C, Protection from Step and Touch Potentials.* Current Edition. Washington, DC: Occupational Safety and Health Administration (OSHA).

OSHA 19260.351, Arc Welding and Cutting. Current Edition. Washington, DC: Occupational Safety and Health Administration (OSHA).

Welding Handbook. Vol. 1. Welding Science & Technology. 9th Edition. Miami, FL: American Welding Society, 2001.

Welding Handbook. Vol. 2, Part 1: Welding Processes. Miami, FL: American Welding Society, 2004.

3.0.0 Section Review

1. Before plugging a welding machine into an electrical outlet, be sure to identify the location of the _____.
   a. supplemental oxygen source
   b. backup generator
   c. alternator back cap
   d. electrical disconnect

2. When connecting the shielding gas for GTAW, if the gas hose is connected to the torch cable and the torch does not contain a gas shutoff valve, install a valve in the line between the torch cable and the _____.
   a. welding machine's gas solenoid
   b. gas cylinder's regulator/flowmeter
   c. gas cylinder's protective cap
   d. remote control foot pedal

3. Welders should select the size of GTAW electrode rated for the _____.
   a. amperage to be used
   b. voltage of the welding machine
   c. temperature at the work site
   d. required type of shielding gas

4. In general, the inside diameter of the nozzle should be about how many times the diameter of the GTAW electrode that is being used?
   a. One
   b. Two
   c. Three
   d. Four

SUMMARY

This module provided an overview of GTAW safety practices, equipment, shielding gas characteristics, electrode types and preparation, filler metal types and classifications, and equipment setup used for GTAW. A good understanding of these topics is essential for selecting and setting up GTAW equipment safely and efficiently, and it provides the foundation for making high-quality GTAW welds.

Review Questions

1. When making GTAW welds, always use safety glasses with a(n) _____.
   a. flame-retardant coating
   b. air-supplied respirator
   c. full face shield or a helmet
   d. OSHA-approved dust mask

2. When welding, the gas that should never be released in a large amount as compressed air around flammable materials is _____.
   a. oxygen
   b. argon
   c. carbon dioxide
   d. nitrogen

3. When a GTAW arc is struck across a gap, the resistance generates heat in the range of _____.
   a. 1,000°F to 2,000°F (550°C to 1,100°C)
   b. 2,000°F to 5,000°F (1,100°C to 2,750°C)
   c. 6,000°F to 10,000°F (3,300°C to 5,500°C)
   d. 15,000°F to 18,000°F (8,300°C to 10,000°C)

4. The duty cycle of a welding machine is based on a period of how many minutes?
   a. 5
   b. 10
   c. 15
   d. 20

5. What percent duty cycle do most heavy-duty industrial machines used for manual welding have?
   a. 50
   b. 100
   c. 50 to 60
   d. 60 to 100

6. An air-cooled GTAW torch that contains a manual gas control valve is typically used if the power source for the torch does not have a(n) _____.
   a. backup generator for the shielding gas supply
   b. electric solenoid valve to automatically control the shielding gas flow
   c. backflow prevention valve in the line leading to the electrode
   d. overload connection to a heat exchanger or cooling fan

7. Which of the following types of electrodes is radioactive?
   a. Lanthanated
   b. Thoriated
   c. Ceriated
   d. Zirconiated

8. How many times heavier is argon shielding gas than helium?
   a. Five
   b. Eight
   c. Ten
   d. Twenty

9. The filler metal used for manual GTAW is generally supplied in diameters of 1⁄16" (1.6 mm) to 1⁄4" (6.4 mm) and lengths of _____.
   a. 48" (1,219 mm)
   b. 36" (914 mm)
   c. 24" (610 mm)
   d. 12" (305 mm)

10. The aluminum alloy filler metal rod ER4043, widely used in GTAW, also contains _____.
    a. iron
    b. flux
    c. silicon
    d. tungsten

11. When setting up GTAW equipment, the shielding gas supply must be located reasonably close to the welding site because of the _____.
    a. limited length of GTAW torch cables
    b. risk of acetylene dispersion at the work site
    c. weight and immobility of the gas cylinders
    d. need to refuel the torch frequently

12. When a standard welding machine is used for a GTAW operation, the shielding gas hose connects to the _____.
    a. torch cable
    b. electrode back cap
    c. retaining cup
    d. nozzle gasket

13. The type of GTAW electrode to be used for a welding operation is specified in _____.

    a. OSHA regulations
    b. ASTM and ANSI appendices
    c. SDS/MSDS and NFPA requirements
    d. WPS or site quality standards

14. When using sanding to taper the point on a tungsten electrode, what size grit should a welder use for finishing?

    a. 60
    b. 80
    c. 120
    d. 200

15. A GTAW electrode is usually installed after the nozzle because the _____.

    a. nozzle grounds the entire torch assembly
    b. electrode stickout (ESO) is measured from the end of the nozzle
    c. electrode back cap threads onto the nozzle to form a seal
    d. nozzle determines the maximum outer diameter of the electrode

Trade Terms Introduced in This Module

Lift-arc technology: A type of GTAW technology in which the tungsten electrode may be touched to the work, but the arc will not strike until the tip is lifted away to an appropriate distance.

Potentiometer: A three-terminal resistor with an adjustable center connection; used for variable voltage control.

Truncated: Having the pointed top (vertex) of a conical shape cut off, leaving only a plane section parallel to the base of the cone.

Tungsten: A nonstandard common name for the tungsten electrode used in GTAW; also, the principal metal from which the electrodes are constructed.

Additional Resources

This module is intended as a thorough resource for task training. The following reference works are suggested for further study.

AWS B1.10, Guide for the Nondestructive Examination of Welds. Miami, FL: American Welding Society.

AWS B1.11, Guide for the Visual Examination of Welds. Miami, FL: American Welding Society.

AWS D3.5, Guide for Steel Hull Welding. Miami, FL: American Welding Society.

AWS D3.7, Guide for Aluminum Hull Welding. Miami, FL: American Welding Society.

AWS Safety and Health Fact Sheet 27. Miami, FL: American Welding Society.

Modern Welding Technology. 6th Edition. Howard B. Cary. Englewood Cliffs, NJ: Prentice Hall, Inc., 2004.

OSHA 1910.269, Appendix C, *Protection from Step and Touch Potentials*. Current Edition. Washington, DC: Occupational Safety and Health Administration (OSHA).

OSHA 1926.351, Arc Welding and Cutting. Current Edition. Washington, DC: Occupational Safety and Health Administration (OSHA).

Tungsten Electrodes: an **Arc-Zone.com** ® Technical Focus Paper (pdf). Carlsbad, CA: **Arc-Zone.com**, Inc. **https://www.arc-zone.com/azc-scip/TungstenElectrodes.pdf**

Tungsten Guidebook (pdf). Newbury Park, CA: Diamond Ground Products, Inc. **http://www.diamond-ground.com/TungstenGuidebook2013.pdf**

Welding Handbook. Vol. 1. Welding Science & Technology. 9th Edition. Miami, FL: American Welding Society, 2001.

Welding Handbook. Vol. 2, Part 1: Welding Processes. Miami, FL: American Welding Society, 2004.

Lincoln Electric website: **http://www.lincolnelectric.com** offers sources for products and training.

Figure Credits

The Lincoln Electric Company, Cleveland, OH, USA, Module opener, Figures 1, 5, 7–9, 12, 15–18, 19, 20, SA01

Topaz Publications, Inc., Figures 13 (photo), 34, SA02

AWS A3.0:2001, Figure 36, adapted with permission from the American Welding Society (AWS), Miami, FL, USA, Figure 14

Zachry Industrial, Inc., Figures 19A, 21

AWS A.18/A5.18M:2005, Figure A.1, Reproduced with permission from the American Welding Society (AWS), Miami, FL, USA, Figure 22

Terry Lowe, Figure 32, SA04

Arc-Zone.com®, Inc., SA05A

CK Worldwide, Inc., SA05B

Section Review Answer Key

Answer	Section Reference	Objective
Section One		
1. c	1.1.3	1a
2. a	1.2.2	1b
Section Two		
1. b	2.1.2	2a
2. c	2.2.0	2b
3. d	2.3.1	2c
4. d	2.4.1	2d
5. a	2.5.4	2e
Section Three		
1. d	3.1.0	3a
2. b	3.2.0	3b
3. a	3.3.0	3c
4. c	3.4.0	3d

NCCER CURRICULA — USER UPDATE

NCCER makes every effort to keep its textbooks up-to-date and free of technical errors. We appreciate your help in this process. If you find an error, a typographical mistake, or an inaccuracy in NCCER's curricula, please fill out this form (or a photocopy), or complete the online form at **www.nccer.org/olf**. Be sure to include the exact module ID number, page number, a detailed description, and your recommended correction. Your input will be brought to the attention of the Authoring Team. Thank you for your assistance.

Instructors – If you have an idea for improving this textbook, or have found that additional materials were necessary to teach this module effectively, please let us know so that we may present your suggestions to the Authoring Team.

NCCER Product Development and Revision
13614 Progress Blvd., Alachua, FL 32615
Email: curriculum@nccer.org
Online: www.nccer.org/olf

❑ Trainee Guide ❑ Lesson Plans ❑ Exam ❑ PowerPoints Other ____________

Craft / Level: Copyright Date:

Module ID Number / Title:

Section Number(s):

Description:

Recommended Correction:

Your Name:

Address:

Email: Phone:

GMAW – Aluminum Plate

OVERVIEW

Gas metal arc welding (GMAW) is one of the more common welding processes now used. As with most welding systems, GMAW equipment is available in many different sizes and varieties. However, the basic operating principles of GMAW apply to all makes and models of equipment. This module describes how GMAW equipment is used to weld aluminum plate and examines different techniques used to produce fillet and V-groove plate welds.

Module 29401

Trainees with successful module completions may be eligible for credentialing through the NCCER Registry. To learn more, go to **www.nccer.org** or contact us at 1.888.622.3720. Our website has information on the latest product releases and training, as well as online versions of our *Cornerstone* magazine and Pearson's product catalog.

Your feedback is welcome. You may email your comments to **curriculum@nccer.org**, send general comments and inquiries to **info@nccer.org**, or fill in the User Update form at the back of this module.

This information is general in nature and intended for training purposes only. Actual performance of activities described in this manual requires compliance with all applicable operating, service, maintenance, and safety procedures under the direction of qualified personnel. References in this manual to patented or proprietary devices do not constitute a recommendation of their use.

29401 V5

29401
GMAW – Aluminum Plate

Objectives

When you have completed this module, you will be able to do the following:

1. Identify aluminum-welding metallurgical considerations and the types of weld problems commonly encountered.
   a. Describe various aluminum alloys and their characteristics.
   b. Identify key characteristics of aluminum welding using GMAW.
   c. Describe the types of problems commonly encountered in aluminum welding.
2. Explain how to prepare for GMAW of aluminum plate and describe specific welding techniques.
   a. Identify common welding safety practices.
   b. Explain how to prepare the area and materials for GMAW practice.
   c. Describe how to prepare GMAW equipment for aluminum welding.
   d. Describe how to accomplish terminations and restarts and create basic GMAW weld beads.
   e. Describe how to make aluminum fillet welds in various positions.
   f. Describe how to make aluminum plate V-groove welds in various positions.

Performance Tasks

Under the supervision of the instructor, you should be able to do the following:

1. Complete the following GMAW aluminum-plate welding tasks:
   - Stringer beads
   - Weave beads
   - Weld terminations
   - Weld restarts
   - Overlapping beads
2. Complete GMAW fillet welds on aluminum plate in the following positions:
   - 1F
   - 2F
   - 3F
   - 4F
3. Complete GMAW V-groove welds on aluminum plate with backing in the following positions:
   - 1G
   - 2G
   - 3G
   - 4G

Trade Terms

Anodic coating
Artificial aging
Cast aluminum
Coalescence
Ingot
Liquation cracks
Natural aging
Solidification cracks
Solution heat treatment
Wrought aluminum

Industry Recognized Credentials

If you are training through an NCCER-accredited sponsor, you may be eligible for credentials from NCCER's Registry. The ID number for this module is 29401. Note that this module may have been used in other NCCER curricula and may apply to other level completions. Contact NCCER's Registry at 888.622.3720 or go to **www.nccer.org** for more information.

Contents

Figures and Tables

Figures and Tables (Continued)

SECTION ONE

1.0.0 ALUMINUM ALLOYS AND THE GMAW PROCESS

Objective

Identify aluminum-welding metallurgical considerations and the types of weld problems commonly encountered.

a. Describe various aluminum alloys and their characteristics.
b. Identify key characteristics of aluminum welding using GMAW.
c. Describe the types of problems commonly encountered in aluminum welding.

Trade Terms

Anodic coating: An artificial buildup of aluminum oxides on the surface of aluminum alloys that improves corrosion resistance. This coating is many times thicker than a coating of naturally occurring oxides.

Artificial aging: A process following solution heat treatment (see *solution heat treatment*) and other processes that allows the alloy to rest for a prescribed period of time at an elevated temperature.

Cast aluminum: Aluminum alloys that are molded into their final form by one of various methods such as sand-casting, die-casting, or pressure die-casting. Cast aluminum contains high levels of silicon to enhance the ability of the material to form complicated castings.

Coalescence: The growing together or growth into one body of the materials being joined.

Ingot: A mass of metal shaped for convenient storage and transport.

Liquation cracks: Liquation is the process of separation by melting, such as in an alloy when one element melts while the others remain solid. Liquation cracks result from localized melting in areas such as the grain boundaries, while adjacent grains remain solid. Both solidification and liquation cracking may be referred to as *hot cracking*.

Natural aging: A process that follows solution heat treatment (see *solution heat treatment*) and other processes that allows the alloy to simply rest at room temperature for a prescribed period of time, such as 96 hours.

Solidification cracks: Cracks that occur as the filler and base metal solidify, when the remaining liquid weld metal is unable to fill the spaces between areas of solidifying weld metal. Both solidification and liquation cracking may be referred to as *hot cracking*.

Solution heat treatment: A process applied to a heat-treatable aluminum alloy that raises the temperature of the alloy to roughly 1,000°F (538°C) for a set period of time, causing the alloy elements to disperse more consistently throughout the metal. The metal is then quenched with water to ensure the dispersed elements remain in place within the metal structure, rather than precipitate out.

Wrought aluminum: Aluminum alloys that are initially formed into ingots or similar shapes and are then worked into their desired shape through bending, rolling, extrusion, or other mechanical methods.

Gas metal arc welding (GMAW) is an arc welding process that uses a continuous, consumable, solid wire electrode for the filler metal, as well as, and a shielding gas to protect the weld zone (*Figure 1*). The GMAW process is commonly used to make welds on aluminum, carbon steel, low-alloy steel, and stainless steel. Some typical GMAW equipment is shown in *Figure 2*.

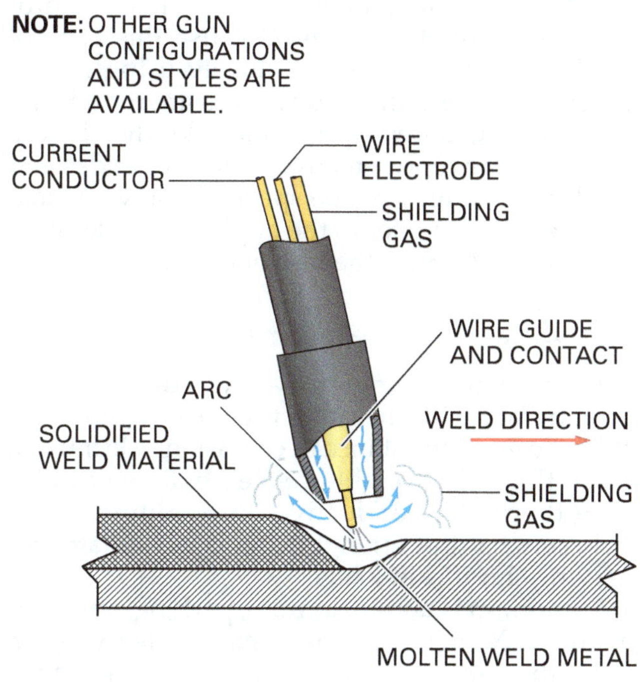

Figure 1 GMAW welding process.

(A) INVERTER WELDING MACHINE

(B) AIR-COOLED GUN

(C) WATER-COOLED GUN

Figure 2 Typical inverter-based GMAW equipment.

GMAW is a fast and effective method for producing high-quality welds. Because this type of welding can be done continuously using a wire feeder, discontinuities and restarts are reduced.

For the purposes of this module, the dimensions used are representative of various codes and standards and may not be specific to any particular code. Always refer to the applicable code, standard, or site WPS for the task at hand.

1.1.0 Aluminum Metallurgy

Aluminum and aluminum alloys are often used in manufacturing because they are lightweight and have good resistance to corrosion. Aluminum is one-third the weight of steel, but is six times more thermally conductive. Pure aluminum is fairly soft, has low strength, and is not suitable for carrying significant loads.

Aluminum is non-toxic, non-sparking, and non-magnetic. Molten aluminum has a silvery sheen under the glow of the welding arc. When heated, aluminum does not change color like steel does.

Aluminum is a highly reactive element. This explains why it is never found in nature in its free state, even though it is the most common metal on earth. Aluminum has a strong chemical attraction toward oxygen. This causes aluminum to oxidize, forming aluminum oxide (the aluminum version of rust) as soon as it is exposed to air. Aluminum oxide is extremely hard and has a high melting point. It is also porous, allowing it to hold moisture and contaminants, which can result in weld porosity. This tendency is especially true of aluminum alloys containing magnesium.

Aluminum can be strengthened by the addition of small amounts of alloy metals to the base metal. Alloys are also used to enhance various other properties of aluminum. Alloys with common alloying elements are grouped together into a series. The alloys in each series are further defined by their ability to be treated by heat.

Heat-treatable and non-heat-treatable aluminum alloys each react differently to welding. Strain hardening is a method of increasing the strength of a metal by cold-working it. Alloys that respond to cold-working are said to be strain-hardenable, separating them from heat-treatable alloys. *Table 1* shows aluminum alloy characteristics by series, classification, and major alloying elements.

Table 1 Aluminum Alloy Characteristics

Series	Classification	Major Alloying Elements	Characteristics
1xxx 1050A	Non-heat-treatable	High purity with small amounts of iron and silicon	Relatively soft, low-strength, nonload carrying; excellent corrosion resistance
2xxx 2014A	Heat-treatable	Copper	High strength; prone to liquation under welding heat
3xxx 3103	Non-heat-treatable	Manganese	Stronger and more ductile than 1xxx grades; excellent corrosion resistance
4xxx 4043A	Filler wires	Silicon	Excellent corrosion resistance
5xxx 5083	Non-heat-treatable	Magnesium	Good ductility; excellent corrosion resistance; good for situations involving impact and shock-loading; not recommended for service with prolonged exposure to high temperatures
6xxx 6082	Heat-treatable	Silicon and magnesium	Good as-welded strength; easily extruded; not recommended for exposure to electrolytes, such as seawater
7xxx 7020	Heat-treatable	Zinc	Good weldability and high as-welded strength
8xxx 8090	Non-heat-treatable and heat-treatable	Other	Not normally welded

Aluminum metallurgy can be divided into the following categories:

- Non-heat-treatable alloys
- Heat-treatable alloys
- Filler metal alloys
- Cast aluminum
- Wrought aluminum

1.1.1 Non-Heat-Treatable Alloys

The strength of non-heat-treatable aluminum alloys depends on elements such as silicon (Si), iron (Fe), manganese (Mn), and magnesium (Mg). These alloying elements are found in the 1xxx, 3xxx, and 5xxx series of alloys. When welded, these alloys may lose the effects of work hardening, which results in softening of the heat-affected zone (HAZ) next to the weld.

Additional strengthening comes from strain hardening in the forming process, such as by cold-rolling. Welds in non-heat-treatable alloys generally have excellent ductility and can withstand severe distortion before failing, and welds in heat-treatable alloys generally have lower ductility. The basic temper designations for non-heat-treatable and heat-treatable aluminum alloys, such as castings, are as follows:

- *O* – Annealed, recrystallized
- *F* – As fabricated
- *H1(x)* – Strain hardened only
  - where $x = 2, 4, 6, 8$, or 9, representing ¼, ½, ¾, full hardness, or extra hardness
- *H2(x)* – Strain hardened and then partially annealed
  - where $x = 2, 4, 6, 8$, or 9, representing ¼, ½, ¾, full hardness, or extra hardness
- *H3(x)* – Strain hardened and then stabilized
  - where $x = 2, 4, 6, 8$, or 9, representing ¼, ½, ¾, full hardness, or extra hardness
- *W(HR)* – Solution heat treated
  - where HR = fractional and full hours of aging at room temperature after heat treatment
- *T(N)* – Thermally treated to produce tempers other than O, F, or H
  - where N = 1 through 10, which indicates a specific treatment, sometimes followed by another digit representing a variation of the treatment

1.1.2 Heat-Treatable Alloys

Heat-treatable alloys contain copper, magnesium, zinc, or silicon. These elements become more soluble when heated. Heat-treatable aluminum alloys develop their properties through solution heat treatment and quenching, followed by either natural aging or artificial aging.

Welding heat applied over an extended period of time reduces the strength of heat-treatable aluminum alloys. As a result, factors such as welding speed and cooling rate have an influence on the final properties of an alloy. Because of the time factor, the HAZ in heat-treatable alloys is usually only partially annealed. Welding redistributes the hardening components in the HAZ, reducing strength locally (close to the weld). The 2xxx, 6xxx, 7xxx, and some of the 8xxx series contain the principal heat-treatable alloying elements. The heat-treatable temper designations include O, F, W, and T1 through T10.

Postweld heat treatment increases the strength of the HAZ, but it also reduces the amount of ductility. Strength improvement depends somewhat on the filler metal used. Preheating is seldom recommended for welding heat-treatable alloys.

Joint thickness can also have an effect on weld properties in heat-treatable alloys. Thinner metal can be welded with a minimum of heat input and with a high cooling rate. This results in higher as-welded strength. For thicker metal, a number of stringer beads are preferable to a few large beads.

1.1.3 Cast Aluminum

Cast aluminum has high silicon content for increased fluidity. It does not anodize as well as wrought aluminum. Cast aluminum is identified by three-digit grade numbers, such as 206, 319, and 356. *Table 2* lists the numbers representing the alloying elements used with cast aluminum. The first number indicates the alloying element. The second and third numbers represent the specific alloy by formula. The number after the decimal point shows whether the alloy composition is for the final casting (*.0*) or for the ingot (*.1* or *.2*, depending on purity). Modifications to cast aluminum are shown by a letter in front of the number. An example of a typical silicon-magnesium alloy for final casting is A356.0. The same alloy in ingot form would be A356.1 or A356.2.

Table 2 Alloy Numbers for Cast Aluminum

Alloy Series	Principal Alloying Elements
1xx.x	Essentially pure aluminum
2xx.x	Copper
3xx.x	Silicon plus copper and/or magnesium
4xx.x	Silicon
5xx.x	Magnesium
6xx.x	Unused series
7xx.x	Zinc
8xx.x	Tin
9xx.x	Other elements

1.1.4 Wrought Aluminum

Wrought-aluminum alloys are identified using four numbers. *Table 3* shows the alloy series numbers and the principal alloying elements used with the various wrought-aluminum alloys. The first number indicates the principal alloying element or elements. The second number signifies some change to the alloy (a zero indicates that there was no change). The third and fourth numbers identify the specific alloy by formula. For example, Alloy 5054 is an aluminum-magnesium alloy without modifications. It has the assigned alloy number of 54. Alloy 5154 is the same alloy after modification. The only exception is the 1xxx series, which is for pure aluminum. In that series, the last two numbers show the degree of purity above 99 percent. For example, Alloy 1060 has a purity of 99.6 percent.

Most of the wrought grades in the 1xxx, 3xxx, 5xxx, 6xxx, and the medium-strength 7xxx series (for example, 7020) can be fusion-welded using both the GTAW and GMAW processes. The 5xxx series alloys in particular have excellent weldability. Wrought aluminum is commonly formed into plates or sheets; rods, bars, and wire; and extruded products such as tubing.

1.2.0 Characteristics of Aluminum Welding

The preheating of aluminum is not usually required for GMAW. However, the high thermal conductivity of aluminum demands a high rate of heat input for welding. In some cases, very thick sections of aluminum may require preheating, typically at less than 250°F (121°C). Using a higher welding current, adding helium to the shielding gas mix, or more precisely controlling other welding variables are all ways to generate an appropriate amount of GMAW welding heat for aluminum.

Aluminum GMAW is done with direct current electrode positive (DCEP) constant-voltage units. When welding thick aluminum that requires deep penetration, a shielding gas of helium mixed with argon is sometimes used, but at a greater cost. Pulsed GMAW is used to reduce heat input and prevent burn-through in thin sections of aluminum.

Table 3 Alloy Numbers for Wrought Aluminum

Alloy Series	Principal Alloying Elements
1xxx	Aluminum (99% minimum and greater)
2xxx	Copper
3xxx	Manganese
4xxx	Silicon
5xxx	Magnesium
6xxx	Magnesium and Silicon
7xxx	Zinc
8xxx	Other elements
9xxx	Unused series

Aluminum welding current is mainly related to thickness because the current must be high enough to give the required penetration. If the weld puddle becomes too large, the welding speed may need to be increased beyond that which is comfortable for the welder. To prevent melt-through on thin materials and on root passes, backing strips or rings made of stainless steel or a nonmetallic material are sometimes used. The backing strips or rings are often made with a shallow groove in the surface. The groove ensures that penetrating welds have the required size and shape when cool. On advanced welding machines used for GMAW, pulsing control can also be used to help eliminate melt-through problems.

1.2.1 Surface Preparation

Aluminum picks up hydrogen gas from condensation, which can form either during a rapid rise in temperature or when the metal is cooled below the dew point of the environment. Because materials are often stored in unheated areas, special care should be taken with aluminum to raise its temperature slowly. Another recommendation is to store aluminum in the vertical position to allow any condensation to run off. To reduce the risk of condensation forming on the surface, allow aluminum to stabilize at room temperature for one hour before welding. Avoid bringing cold material into a warmer, more humid environment, expecting to weld it immediately.

Aluminum combines with oxygen from the atmosphere to form a surface film of very hard oxide. This process can even occur at room temperature. Pure aluminum melts at 1,220°F (660°C), but its oxide melts at 3,600°F to 3,900°F (1,980°C to 2,150°C). Aluminum oxide is harder than most of the substances it encounters. It can actually damage a milling cutter if the oxides are trapped in the welds. Because of its hardness, aluminum oxide is often used for the grit in grinding wheels. Aluminum oxide also tends to be hygroscopic (moisture-absorbing). Under the extreme heat of welding, the moisture breaks down to free hydrogen, which often causes porosity in the weld. As a safeguard against porosity, all oxidation must be thoroughly removed from the base metal before welding.

WARNING!

Grinding wheels used on aluminum must be made for aluminum only. They should not be used on any other type of metal or on surfaces contaminated with grease or oil. Using other types of grinding wheels on aluminum may cause the grinding wheels to shatter.

Proper cleaning of welding surfaces cannot be stressed enough. Oxides, greases, and oil films can act as contaminants and cause unsound welds. Aluminum should be welded as soon as possible after cleaning or within one-half hour of cleaning. Otherwise the weld zone may have to be cleaned again. If oxides are not removed to the fullest extent possible, the weld will be subject to poor fusion and inclusions. This is due to the reduced wetting action of the molten weld pool as it comes in contact with any remaining oxide-contaminated weld joint surfaces. Aluminum should be hand-brushed between passes with a stainless steel brush. Make sure that the brush has not been used on other materials or for any other purpose.

Water, oil, and grease represent the contaminants generally found on aluminum. Wire brushing before these contaminants are removed only pushes them into the surface or damages the aluminum. Oil and grease must be removed before cleaning aluminum with a wire brush.

Both water-based aluminum deoxidizers and solvent cleaners such as acetone can be used to remove oil and grease. When using water-based cleaners, all water must be removed before welding. Most hydrocarbon-based solvents are highly volatile and evaporate quickly. Water-based cleaners however, must be thoroughly wiped away or heat-dried before welding. Review the Safety Data Sheet (SDS) for any additional safety hazards or precautions associated with these cleaners.

Preweld cleaning solutions, also referred to as *chemical cleaners*, can remove both the aluminum oxide and any contaminants. Their disadvantage is that deoxidizers for aluminum are very aggressive chemicals; they usually contain dilute acids that can etch the surface. They can be dangerous to the people using them as a result. It is very important to follow the manufacturer's instructions and take the necessary safety precautions. Always review and follow the recommendations of the SDS when using cleaning solvents. When using deoxidizers, the metal must be completely rinsed and dried. This avoids creating a surface film of hydrated oxide.

WARNING!

Be very careful when using preweld chemical-cleaning solutions. They can harm not only the environment, but also the people using them. Make sure that all cleaning solutions are properly contained. Always be sure to wear the appropriate PPE. Refer to the applicable SDS and the product label for instructions on how to properly dispose of any used cleaning solutions and the materials used to apply them.

1.2.2 Filler Metal Alloys

Filler metal composition is determined in part by the weldability and properties of the base metal. Other factors to consider include corrosion resistance, exposure of the weldment to sustained high temperatures, and the requirements for anodic coating. For non-heat-treatable alloys, matching fillers are often used. For low-alloy materials and heat-treatable alloys, nonmatching fillers are used to prevent solidification cracking.

Filler metal alloys ER4043 and ER5356 are used in roughly 85 percent of aluminum welds. They are also the most readily available filler metal products for aluminum. Your instructor will likely determine the use of a specific filler metal that is most appropriate for your practice coupons. When there is a WPS or site quality standard, the filler rod type will be specified. It will be given as an AWS classification and/or manufacturer's standard. It will sometimes include a manufacturer's name. When a WPS for the weld exists, the filler metal that is specified must be used.

> **CAUTION**
>
> Filler metal specifications in the WPS or site quality standards must be followed. Failure to use the specified filler metal will result in a weld quality issue and the potential for rejection.

There are many different types of aluminum base metals and filler metals. *AWS Standard A5.10/A5.10M: 2012 (ISO 18273:2004 MOD), Welding Consumables—Wire Electrodes, Wires and Rods for Welding of Aluminum and Aluminum Alloys—Classification* provides a guide for the selection of bare aluminum filler metals for specific aluminum base metals and alloys. *Table 4A* and *Table 4B* have been taken directly from the standard. Charts and tables provided by filler metal manufacturers also assist in selecting the correct filler metal from their product line.

High-quality aluminum welds can only be achieved if the filler metal that is used is clean and of high quality. Contaminated filler metal transfers the contamination to the weld pool.

Table 4A AWS Filler Metal Guide for General Purpose Welding of Aluminum (Part 1)

Table A.1
Guide to the Choice of Filler Metal for General Purpose Welding

Base Metal	201.0 206.0 224.0	319.0, 333.0, 354.0, 355.0, C355.0	356.0, A356.0 357.0, A357.0 413.0, 443.0, A444.0	511.0, 512.0, 513.0, 514.0, 535.0	7004, 7005, 7039, 710.0, 712.0	6009 6010 6070	6005, 6061, 6063, 6101, 6151, 6201, 6351, 6951	5456	5454
1060, 1070, 1080, 1350	ER4145	ER4145	ER4043[a,b]	ER5356[c,d]	ER5356[c,d]	ER4043[a,b]	ER4043[b]	ER5356[d]	ER4043[b,d]
1100, 3003, Al[c], 3003	ER4145	ER4145	ER4043[a,b]	ER5356[c,d]	ER5356[c,d]	ER4043[a,b]	ER4043[b]	ER5356[d]	ER4043[b,d]
2014, 2036	ER4145[e]	ER4145[e]	ER4145	—	—	ER4145	ER4145	—	—
2219	ER2319[a]	ER4145[e]	ER4145[b,c]	ER4043	ER4043	ER4043[a,b]	ER4043[a,b]	—	ER4043[b]
3004, Al[c], 3004	—	ER4043[b]	ER4043[b]	ER5356[f]	ER5356[f]	ER4043[b]	ER4043[b,f]	ER5356[d]	ER5356[f]
5005, 5050	—	ER4043[b]	ER4043[b]	ER5356[f]	ER5356[f]	ER4043[b]	ER4043[b,f]	ER5356[d]	ER5356[f]
5052, 5652[i]	—	ER4043[b]	ER4043[f]	ER5356[f]	ER5356[f]	ER4043[b]	ER5356[c,f]	ER5356[f]	ER5356[f]
5083	—	—	ER5356[c,d]	ER5356[d]	ER5356[d]	—	ER5356[d]	ER5356[d]	ER5356[d]
5086	—	—	ER5356[c,d]	ER5356[d]	ER5356[d]	—	ER5356[d]	ER5356[d]	ER5356[d]
5154, 5254[i]	—	—	ER4043[f]	ER5356[f]	ER5356[f]	—	ER5356[f]	ER5356[f]	ER5356[f]
5454	—	ER4043[b]	ER4043[f]	ER4043[f]	ER4043[f]	ER4043[b]	ER5356[c,f]	ER5356[f]	ER5554[c,f]
5456	—	—	ER5356[c,d]	ER5356[d]	ER5356[d]	—	ER5356[d]	ER5356[d]	
6005, 6061, 6063 6101, 6151, 6201 6351, 6951	ER4145	ER4145[b,c]	ER4043[b,f,g]	ER5356[f]	ER5356[c,f]	ER4043[a,b,g]	ER4043b[a,f,g]		
6009, 6010, 6070	ER4145	ER4145[b,c]	ER4043[a,b,g]	ER4043	ER4043	ER4043[a,b,g]			
7004, 7005, 7039 710.0, 712.0	—	ER4043[b]	ER4043[b,f]	ER5356[f]	ER5356[f]				
511.0, 512.0, 513.0 514.0, 535.0	—	—	ER4043[f]	ER5356[f]					
356.0, A356.0, 357.0 A357.0, 413.0 443.0, A444.0	ER4145	ER4145[b,c]	ER4043[b,h]						
319.0, 333.0 354.0, 355.0 C355.0	ER4145[c]	ER4145[b,c,h]							
201.0, 206.0, 224.0	ER2319[a,h]								

Note: See *Table 4B* for referenced footnotes.

Filler metal must be used as soon as possible after removal from the shipping or storage packaging. When this is not possible, the filler metal must be stored in airtight storage containers to prevent further oxidation. Also remember that heating aluminum and other metals accelerates the oxidation process dramatically. This is especially true when the humidity is high.

1.3.0 Weld Problems

Before fusion welding can take place, all aluminum oxide must be removed. Remember that the melting point of aluminum oxide is about three times that of aluminum. Because of this, the welding heat needed to break down the oxide causes the aluminum to melt long before the oxide does, and coalescence cannot take place. Complete removal of the oxide prior to welding is essential to avoid this problem.

Silicon

One of the advantages to having silicon in aluminum alloys is that it expands as it cools. This characteristic helps prevent cracking in aluminum-alloy welds, as the silicon expands to help fill spaces that might otherwise be subject to solidification or liquation cracking.

The following are the most common problems in aluminum welding:

- Porosity
- Cracking
- Poor weld bead profile

Table 5 shows the causes, contributing factors, and corrective measures related to porosity and cracking in GMAW. These issues are explored in greater depth in the sections that follow.

Table 4B AWS Filler Metal Guide for General Purpose Welding of Aluminum (Part 2)

Table A.1 (Continued)
Guide to the Choice of Filler Metal for General Purpose Welding

Base Metal	5154 5254[i]	5086	5083	5052 5652[i]	5005 5050	3004 Alc. 3004	2219	2014 2036	1100 3003 Alc. 3003	1060 1070 1080 1350
1060, 1070, 1080, 1350	ER5356[c,d]	ER5356[d]	ER5356[d]	ER4043[b,d]	ER1100[b,c]	ER4043[b,d]	ER4145[b,c]	ER4145	ER1100[b,c]	ER1188[b,c,h,j]
1100, 3003, Al[c], 3003	ER5356[c,d]	ER5356[d]	ER5356[d]	ER4043[b,d]	ER1100[b,c]	ER4043[b,d]	ER4145[b,c]	ER4145	ER1100[b,c]	
2014, 2036	—	—	—	—	ER4145	ER4145	ER4145[e]	ER4145[e]		
2219	ER4043	—	—	ER4043[b]	ER4043[a,b]	ER4043[a,b]	ER2319[a]			
3004, Al[c], 3004	ER5356[f]	ER5356[d]	ER5356[d]	ER5356[c,f]	ER5356[c,f]	ER5356[c,f]				
5005, 5050	ER5356[f]	ER5356[d]	ER5356[d]	ER5356[c,d]	ER5356[c,f]					
5052, 5652[i]	ER5356[f]	ER5356[d]	ER5356[d]	ER5654[c,f,i]						
5083	ER5356[d]	ER5356[d]	ER5183[d]							
5086	ER5356[d]	ER5356[d]								
5154, 5254[i]	ER5356[f,i]									

Note: ISO Classifications different from AWS Classifications have not been added to this table.

[a] *ER4145 may be used for some applications.*
[b] *ER4047 may be used for some applications.*
[c] *ER4043 may be used for some applications.*
[d] *ER5183, ER5356, or ER5556 may be used.*
[e] *ER2319 may be used for some applications. It can supply high strength when the weldment is postweld solution heat-treated and aged.*
[f] *ER5183, ER5356, ER5554, ER5556, and ER5654 may be used. In some cases, they provide: (1) improved color match after anodizing treatment, (2) highest weld ductility, and (3) higher weld strength. ER5554 is suitable for sustained elevated temperature service.*
[g] *ER4643 and ER4943 will provide higher strength than ER4043 in 1/2 in [12 mm] and thicker groove welds in 6XXX base alloys when postweld solution heat-treated and aged. They will also provide higher strength fillet welds than ER4043 in the as welded, postweld aged, or postweld heat-treated and aged conditions.*
[h] *Filler metal with the same analysis as the base metal is sometimes used. The following wrought filler metals possess the same chemical composition limits as cast filler alloys: ER4009 and R4009 as R-C355.0; ER4010 and R4010 as R-A356.0; and R4011 as R-A357.0.*
[i] *Base metal alloys 5254 and 5652 are used for hydrogen peroxide service. ER5654 filler metal is used for welding both alloys for service temperatures below 150°F [66°C].*
[j] *ER1100 may be used for some applications.*

Notes:

1. *ER4047, ER4643, or ER4943 may be used in some applications when alloy ER4043 is specified.*
2. *Service conditions such as immersion in fresh or salt water, exposure to specific chemicals, or a sustained high temperature (over 150°F [66°C]) may limit the choice of filler metals. Filler metals ER5183, ER5356, and ER5556 are not recommended for sustained elevated temperature service.*
3. *Recommendations in this table apply to gas shielded arc welding processes. For oxyfuel gas welding, only ER1188, ER1100, ER4043, ER4047, and ER4145 filler metals are ordinarily used.*
4. *Where no filler metal is listed, the base metal combination is not recommended for welding.*

Table 5 Causes and Cures for Porosity and Cracking in GMAW

Cause	Contributing Factors	Corrective Measures
Hydrogen	Hydrogen is the result of dirt-containing oils or other hydrocarbons; moisture in atmosphere or on metal; a hydrated oxide film on metal; or moisture in gas or gas lines. Base metal may be the source of entrapped hydrogen (the thicker the metal, the greater the possibility of hydrogen). Spatter contributes to hydrogen formation.	Degrease and mechanically or chemically remove oxide from weld area. Avoid humidity; use dry metal or wipe dry. Reduce moisture content of gas. Check gas and water lines for leaks. Increase gas flow to compensate for increased hydrogen in thicker sections. To minimize spatter, adjust welding conditions.
Impurities	Cleaning or other compounds, especially those containing calcium.	Use recommended cleaning compounds; keep work free of contaminants.
Incomplete root penetration	Insufficient welding current or preheat.	Use a higher welding current, or redesign the joint geometry. Preheat.
Temperature	Running too cool tends to increase porosity that is caused by premature solidification of molten metal.	Maintain the proper current, arc-length, and torch-travel speed relationship.
Welding speed	Too great a welding speed may increase porosity.	Decrease the welding speed and establish and maintain proper arc-length and current relationship.
Solidification time	Quick freezing of the weld pool entraps any gases present, causing porosity.	Establish the correct welding current and speed. If the work is appreciably below room temperature, use supplemental heating.
Chemical composition of weld material	Pure aluminum weld metal is more susceptible to porosity than is an aluminum alloy.	If porosity is excessive, try an alloy filler material.
Cracking	Such causes of porosity as temperature, welding time, and solidification may also be contributing causes of cracking. Other causes may be discontinuous welds, welds that intersect, repair welds, cold-working either before or after welding, and weld-metal composition. In general, crack-sensitive alloys include those containing 0.4 to 0.6% silicon,1.5 to 3.0% magnesium, or 1.0% copper.	Lower current and faster speeds often prevent cracking. However, a change to a filler alloy that brings the weld-metal composition out of the cracking range is recommended where possible.

Manufacturers' Data for Filler Metals

Most manufacturers of filler metal publish paperback welding guides that include information about their filler materials. These booklets can be obtained through your distributor free or for a small charge. An increasing number of manufacturers also offer apps for filler metal selection that can be used on smartphones, tablets, and similar devices.

Cast Aluminum

Cast aluminum alloys are given their final form by one of several methods. Sand-casting is a popular process, where a mold is made of sand that can be reused numerous times. Molten aluminum is then poured into the mold. Pressure die casting processes use molds made of steel, and the molten aluminum is injected into the mold under pressure. Casting is best for complex shapes, such as the one shown below.

1.3.1 Porosity

Aluminum welds are highly prone to porosity. The main cause of porosity is the absorption of hydrogen into the weld pool. The hydrogen forms separate pores in the weld metal as the weld pool solidifies. Some common sources of hydrogen are hydrocarbons and the atmosphere. Another is the moisture from contaminants on the parent material and on the surface of the filler wire. In some cases, a torch gas flow that is set too high can cause porosity in the weld. It does this by creating turbulence that draws air into the gas, causing porosity. *Figure 3* shows porosity caused by contamination in a casted aluminum alloy 356 weld.

Thoroughly clean the surface to reduce the risk of porosity. Proper cleaning techniques include solvent degreasing and chemical etching. Surface cleansing should be followed by mechanical cleaning by grinder buffing or wire brushing.

1.3.2 Cracking

Cracking occurs in aluminum alloys because of high stress. The stresses are generated across the weld by the high degree of thermal expansion, which is twice that of steel. Thermal expansion is the fractional increase in size due to a rise in temperature. Cracking can also be traced to the amount of contraction that occurs when the weld cools and solidifies. Contraction is usually 5 percent more in aluminum than in similar welds on steel. High-strength aluminum alloys (such as 7010 and 7050) and most of the 2xxx series of alloys are not recommended for fusion welding. This is because welding these alloys commonly results in both **solidification cracks** and **liquation cracks**. However, these types of cracks can also occur in other alloys.

Solidification cracks (*Figure 4*), also referred to as *hot cracks*, form in the center of the weld and run along its centerline. Solidification cracks also occur in the weld crater at the final termination. There are two main causes of solidification cracks. One is the incorrect combination of filler wire and base metal. The other is welding under high-restraint conditions.

The use of a non-matching, crack-resistant filler metal (usually from the 4xxx and 5xxx series alloys) can reduce the risk of solidification cracking. However, the use of these fillers can also result in a weld of lower strength, and the weld may not respond to heat treatment. The weld bead must be thick enough to withstand the stresses caused by contraction as the weld solidifies.

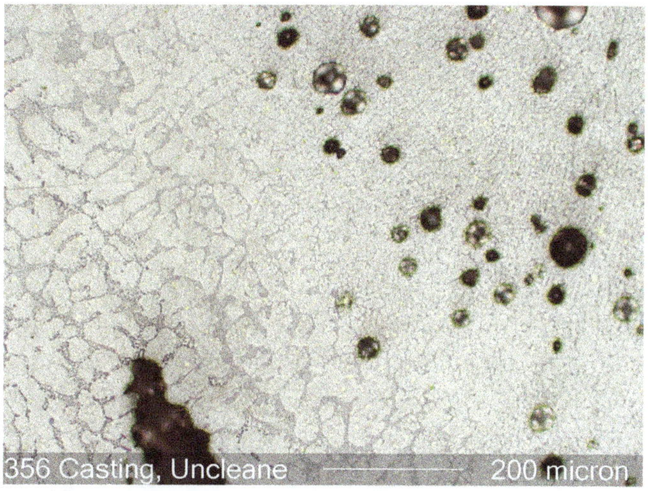

Figure 3 Aluminum porosity example.

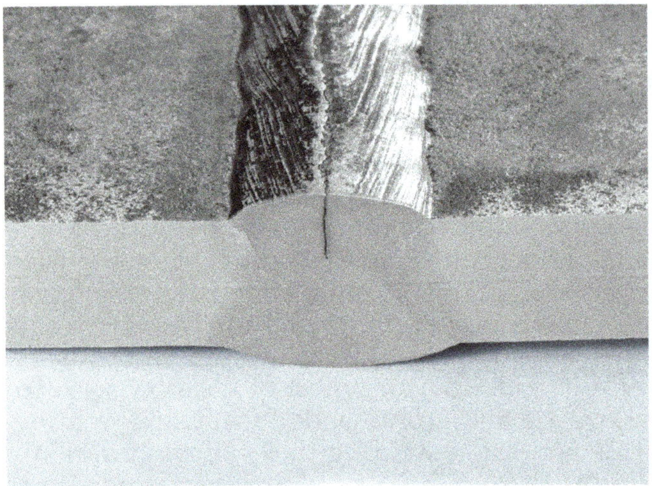

Figure 4 Solidification crack in an aluminum weld.

There are several ways to reduce the degree of restraint on the weld. Methods include correct edge preparation, accurate joint setup, more compatible filler alloy, and the correct weld sequence. Higher weld speeds that reduce the size of the HAZ can also reduce solidification cracking.

Preheating reduces stress on the solidifying weld metal. It does this by reducing the temperature change across the HAZ and by permitting faster joining. However, to reduce the risk of stress cracking, preheating should only be used when the joint is unrestrained. Preheating should be limited to temperatures under 250°F (121°C) to preserve the mechanical properties of the base metal.

Liquation cracks occur in the HAZ at the grain boundaries when low-melting-point films are formed. These films are not strong enough to withstand the contraction stresses created when the weld metal solidifies and cools. Heat-treatable alloys, such as the 6xxx, 7xxx, and 8xxx series, are more likely than other alloys to crack in this way.

Use of a filler metal that has a lower melting temperature than the base metal can reduce the risk of liquation cracking. An example would be to use a 4xxx filler metal to weld a 6xxx-series-alloy base metal. However, 4xxx filler metal should not be used to weld high-magnesium alloys, such as 5083. This is because excessive magnesium silicide may form at the fusion boundary, causing a loss of ductility and an increased risk of cracking.

1.3.3 Poor Weld Bead Profile

A poor weld bead profile can result from setting incorrect welding parameters. It can also be traced to poor welding techniques. Some common weld profile defects are lack of fusion, lack of penetration, and undercut. Aluminum alloys are especially prone to profile imperfections. This tendency is due to their high thermal conductivity and the rapidly solidifying weld pool.

Wrought Aluminum

Wrought alloys start out as cast ingots or billets (think bars, blocks, or bricks of metal), which are then formed into the desired shape mechanically by rolling, extruding, or forging. The material may be formed hot or cold, depending on the requirements. The fencing material shown here is constructed of wrought aluminum.

Additional Resources

Aluminum GMAW Welding Guide (PDF). 2014. Cleveland, OH: The Lincoln Electric Company. **www.lincolnelectric.com.** Last accessed May 26, 2016.

Guide for Aluminum Welding. 2013. Troy, OH: Hobart Brothers Corporation.

AWS C5.6-89R Recommended Practices for Gas Metal Arc Welding. Miami, FL: American Welding Society.

GMAW Welding Guide (PDF). Current Edition. Cleveland, OH: The Lincoln Electric Company. **www.lincolnelectric.com.** Last accessed May 26, 2016.

OSHA 1926.351, Arc Welding and Cutting. Current Edition. Washington, DC: Occupational Safety and Health Administration (OSHA).

The Procedure Handbook of Arc Welding. 14th Edition. Cleveland, OH: The James F. Lincoln Arc Welding Foundation, 2000.

Welding Handbook, Volumes 1 and *2.* 2001. Miami, FL: American Welding Society.

Welding Aluminum: Theory and Practice. 2002. New York, NY: The Aluminum Association.

1.0.0 Section Review

1. How much more thermally conductive is aluminum when compared to steel?
   a. Twice as conductive
   b. Three times as conductive
   c. Five times as conductive
   d. Six times as conductive

2. Which of the following is a correct statement about aluminum oxide?
   a. Aluminum oxide melts at a much lower temperature than aluminum.
   b. Aluminum oxide is hard enough to use as the grit in grinding wheels.
   c. Aluminum oxide only forms while the aluminum is very hot.
   d. Once aluminum is properly cleaned, aluminum oxide will no longer form.

3. Cracks in aluminum welds that form along the centerline due to expansion and contraction are called _____.
   a. solidification cracks
   b. liquation cracks
   c. oxide cracks
   d. fusion cracks

Section Two

2.0.0 GMAW of Aluminum Plate

Objective

Explain how to prepare for GMAW of aluminum plate and describe specific welding techniques.

a. Identify common welding safety practices.
b. Explain how to prepare the area and materials for GMAW practice.
c. Describe how to prepare GMAW equipment for aluminum welding.
d. Describe how to accomplish terminations and restarts and create basic GMAW weld beads.
e. Describe how to make aluminum fillet welds in various positions.
f. Describe how to make aluminum plate V-groove welds in various positions.

Performance Tasks

1. Complete the following GMAW aluminum-plate welding tasks:
   - Stringer beads
   - Weave beads
   - Weld terminations
   - Weld restarts
   - Overlapping beads
2. Complete GMAW fillet welds on aluminum plate in the following positions:
   - 1F
   - 2F
   - 3F
   - 4F
3. Complete GMAW V-groove welds on aluminum plate with backing in the following positions:
   - 1G
   - 2G
   - 3G
   - 4G

This section reviews the safety practices related to welding and explains how to prepare the aluminum plate practice coupons. Guidance in completing aluminum welds using the GMAW process in various positions is also provided.

2.1.0 Safety Practices

Before proceeding with welding coupon preparation and practice, it is important to review common safety practices associated with the craft. Safety considerations are a critical part of preparing for any welding operation.

The following is a summary of safety procedures and practices that must be observed while cutting or welding. Be advised that this is only a summary. Complete safety coverage is provided in the *Welding Level One* module, "Welding Safety." That module should be studied, if not already completed, before continuing. Above all, be sure to wear appropriate protective clothing and equipment at all times when welding or cutting.

2.1.1 Protective Clothing and Equipment

Protective clothing and equipment are a necessary part of every welding task. Follow the guidelines below as a general rule, and ensure that you are familiar with any unique requirements your employer may have regarding personal protective equipment (PPE).

- A welding hood equipped with a properly shaded lens is best for all forms of welding. The welding hood lens must be an appropriate shade for the application.
- Wear proper leather or flame-retardant clothing along with welding gloves for protection from flying sparks, molten metal, and heat.
- Wear high-top safety shoes or boots. Make sure that the tongue and lace area of the footwear is covered by a pant leg. If the tongue and lace area is exposed or if the footwear must be protected from burn marks, wear spats under the pant leg or leather chaps over the front and top of the footwear.
- Wear a 100-percent cotton cap without any mesh material. If a hard hat is required for the environment, use one that allows the attachment of rear deflector material and a face shield. A hard hat with a rear deflector is generally preferred when working overhead, and may be required by some employers and job sites.
- Wear earmuffs, or at least earplugs, to prevent sparks from entering the ear canals.

2.1.2 Housekeeping and Fall Protection

Welding is often performed outside in the weather and in the construction environment. Production work completed inside large, open shops exposes a welder to equipment moving across the floor and overhead. Welders often have to crawl under or reach over objects to complete their work.

At times, welders must work at height. Fall protection equipment is required for all work at height (*Figure 5*). Always keep fall protection devices untangled and properly attached. Combined with the other equipment a welder must carry, there is a great deal to manage, but fall protection equipment must always be the priority. Fall protection equipment saves lives. Never attach any kind of welding gear to the harness that could interfere with its ability to perform its intended function.

Regardless of where you are working, always keep the work area as neat as possible. This makes the area safer and helps to reduce distractions. Mindful actions, such as making sure that welding cables are not tripping hazards, contribute to a safer environment. Electrode stubs and unused electrodes must not be left on the floor. They can cause someone to lose their footing and fall. Parts waiting to be welded or left over from previous work are a tripping hazard and should be kept out of the walking area. If the parts have sharp edges, they could cut welding cable insulation or torch gas hoses.

Figure 5 Working at height.

Lightweight Welding Gloves

Soft, flexible, leather gloves are used for light-duty work and for operations involving GMAW, GTAW, brazing, and soldering where more flexibility and dexterity is required. These welding processes produce less heat and spatter than others. More dexterity is required for the welding techniques and for operating torch controls. A wide variety of styles, using various materials, is available.

2.1.3 Fire and Explosion Prevention

Flammable gases are commonly found around welding and cutting operations, as are sparks and other significant sources of ignition. For that reason, and because many fire-related incidents over the years have been traced back to welding operations, it is crucial that fire and explosion prevention be given consistent attention. Follow these guidelines to maintain a safe environment:

- Never carry matches or gas-filled lighters. Sparks can cause the matches to ignite or the lighter to explode in your pocket, causing serious injury.
- Never use oxygen to blow dust or dirt from clothing. Traces of the oxygen can remain trapped in the fabric for some time. If a spark hits the fabric during this time, the clothing can burn rapidly and violently out of control.
- Make sure that any flammable material in the work area is removed or shielded by a fire-resistant covering. Approved fire extinguishers must be available before attempting any heating, welding, or cutting operations. If a hot-work permit and a fire watch are required, be sure these are in place before beginning.
- Ensure that all site safety requirements are being observed.
- Never release a large amount of oxygen or use oxygen in place of compressed air. It can cause rapid and uncontrolled combustion around flammable materials or sparks. Keep oxygen away from oil, grease, and other petroleum products.

- Never release a large amount of fuel gas (especially acetylene). Because propane is heavier than air, it tends to concentrate in and along low areas, and can ignite at a considerable distance from the release point. Acetylene is lighter than air, but is even more dangerous than propane. When mixed with air or oxygen, acetylene will ignite at much lower concentrations than any other fuel gas.
- Although shielding gases are not toxic, any gas released in a large quantity can displace the oxygen in a small room or enclosed work area, creating an atmosphere with little or no oxygen. Avoid large releases of any type of gas, regardless of its characteristics.
- To prevent fires, maintain a neat and clean work area, and make sure that any metal scrap or slag is cold before disposal.
- Before cutting containers such as tanks or barrels, check to see if they have contained any explosive, hazardous, or flammable materials, including petroleum products, citrus products, or chemicals that decompose into toxic fumes when heated. As a standard practice, always clean and then fill any tanks or barrels with water, or purge them with a flow of inert gas such as nitrogen to displace any oxygen. Containers must be cleaned by steam cleaning, flushing with water, or washing with detergent until all traces of the material have been removed. Proper procedures for cutting or welding hazardous containers are described in the *American Welding Society (AWS) F4.1, Safe Practices for the Preparation of Containers and Piping for Welding and Cutting.*

GOING GREEN

Conserve Aluminum Practice Coupons

Aluminum is expensive and difficult to obtain for practice purposes. To conserve resources and reduce waste, completely use all weld coupons until all surfaces have been welded upon. Cutting coupons apart and reusing the pieces also conserves materials. Use flat material that cannot be cut into weld coupons to practice running beads. Note that stringer beads can be practiced while leaving enough room between them to practice weave beads as well. Any aluminum material that is no longer suitable for welding practice should be recycled.

WARNING!

Welding or cutting must never be performed on drums, barrels, tanks, vessels, or other containers until they have been emptied and cleaned thoroughly to eliminate all flammable materials and all substances (such as detergents, solvents, greases, tars, or acids) that might produce flammable, toxic, or explosive vapors when heated. Clean containers only in well-ventilated areas. Vapors can accumulate during cleaning, causing explosions and serious personal injury.

2.1.4 Work Area Ventilation

Many gases are introduced to the atmosphere through welding processes. Some of these gases are harmful, and all can displace or combine with atmospheric oxygen in a confined space. Follow these guidelines, as well as those of your employer, for work area ventilation:

- Make sure to follow all confined space procedures before conducting any welding or cutting in a confined space. Use powered extraction systems as needed.
- Never use oxygen to provide ventilation in confined spaces.
- Always perform cutting or welding operations in a well-ventilated area.

Cutting or welding operations involving zinc or cadmium materials or coatings result in toxic fumes. Another significant concern is Chromium 6, commonly referred to as *hexavalent chromium*, which is used in stainless steel. It releases toxic fumes when heated. Manganese is another hazardous metal used in stainless steel. It is also used in some welding rods, increasing the risk of exposure.

For long-term cutting or welding of materials containing these components, always wear an approved full-face, supplied-air respirator that uses breathing air supplied from outside the work area. For occasional short-term exposure, a HEPA-rated or metal-fume filter may be used on a standard respirator.

Powered fume-extraction equipment must be used when welding materials that contain zinc, cadmium, hexavalent chromium, manganese, and similar materials that produce hazardous fumes if the area is not naturally well-ventilated. *Figure 6* shows a powered air-purification respirator used with fume extraction equipment.

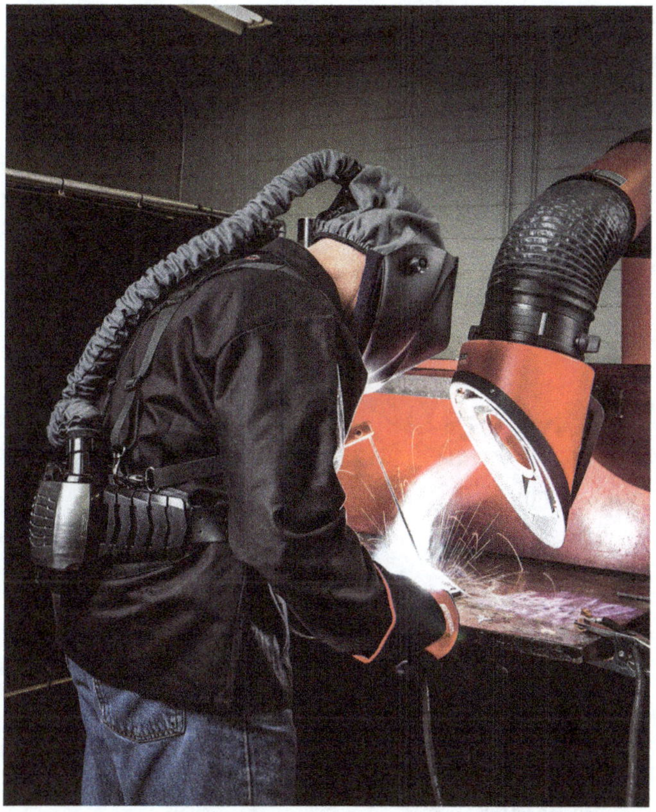

Figure 6 Powered air-purification respirator (PAPR) and localized fume extraction equipment.

2.2.0 Preparations for Welding

Before welding can begin, the area must be readied, the welding equipment must be set up, and the metal to be welded must be prepared. The following sections explain how to prepare for welding.

To practice welding, you will need a welding table, bench, or stand. The welding surface can be steel, but an aluminum surface is preferred. Provisions must be available for placing weld coupons out of position.

To set up the area for welding, follow these steps:

Step 1 Make sure that the area is properly ventilated. Make use of doors, windows, and fans.

Step 2 Check the area for fire hazards. Remove any flammable materials before proceeding.

Step 3 Locate the nearest fire extinguisher. Do not proceed unless the extinguisher is charged and you know how to use it.

Step 4 Set up welding screens around the welding area.

2.2.1 Preparing Aluminum Coupons

Because aluminum is easily contaminated, proper coupon preparation is a priority. Aluminum coupons must be cut to the desired lengths, cleaned, and beveled to the proper angles before they are positioned for welding. An aluminum backing must also be cut and cleaned. The basic shapes can be cut by shearing or sawing.

The importance of cleaning aluminum before welding cannot be overstated. There are two related facts about aluminum that help explain why. First, aluminum does not give the welder a visual indication that it is nearing its melting point through color change, as many other common metals do. With no indication that the melting point is approaching, it is easy to apply too much heat. Second, aluminum oxide—the "skin" that forms on the outside—has a melting point roughly three times higher than the aluminum itself. Excessive weld contamination and poor weld quality result when the aluminum base metal is not thoroughly cleaned.

The preferred technique for beveling and/or adding a root face is mechanical machining. Grinding is also acceptable, although machining the grooves and root faces of aluminum is the cleanest and most uniform way to prepare the material for GMAW. Thermal cutting techniques are not considered acceptable. If thermal cutting techniques must be used to cut aluminum alloys, at least ⅛" (3 mm) of the cut surface must be removed by machining or grinding before welding.

WARNING!

Grinding wheels used on aluminum must be made for aluminum only. They should not be used on any other type of metal or on surfaces contaminated with grease or oil. Using other types of grinding wheels on aluminum may cause the grinding wheels to shatter.

CAUTION

Grinding wheels, files, and similar tools used on aluminum should not be used on any other type of metal or on surfaces contaminated with grease or oil.

Practice coupons are best cut from aluminum plate that is ¼" thick (6 mm plate). If this size is not readily available, aluminum plate up to ⅜" thick (10 mm plate) can be used. Material up to ½" thick (12 mm plate) can be used with the joint design described herein, if welding equipment of sufficient current capacity is available. Material thicknesses greater than ½" (>12 mm plate) can be used, but a different joint design is required, and welding equipment of extreme current capacity is required when argon gas is used alone. Although it is expensive, helium can be added to the mixture to significantly reduce the welding current required.

NOTE

For welding thin aluminum materials (less than ¼" or 6 mm), there are a number of other acceptable joint designs from which to choose that do not require a backing. The joint design used in this module, however, is better suited for practice on the ¼" and thicker material specified above.

Clean oil and grease from the coupons and the metal backing with an approved chemical such as acetone. After oil and grease have been removed, use a stainless-steel wire brush to remove all other contaminants and oxidation. Use only light pressure on the brush.

WARNING!

To prevent injury when using cleaning solvents, always comply with the safety guidelines found on the SDS for the solvents being used.

The coupons must be prepared to allow the following welds:

- *Stringer beads, weave beads,* and *overlapping beads* – The plate coupons (simple pads) can be any size or shape that can be easily handled.
- *Fillet welds* – Cut the metal into 4" × 7" (102 × 178 mm) rectangles for the base and 3" × 7" (76 × 178 mm) rectangles for the web (*Figure 7*).
- *V-groove welds with backing* – A properly prepared aluminum V-groove weld joint with backing is illustrated in *Figure 8*. Note that this joint design is not recommended for material thicker than ½" (12 mm plate). As shown in *Figure 9*, cut two pieces of metal into 3" × 7" (76 × 178 mm) rectangles with one (or both) of the 7" lengths beveled at 30 degrees. Grind a 1/32" to ⅛" (0.8 to 3.2 mm) root face on the bevel as directed by your instructor. Note that a thicker root face makes it more difficult to achieve proper penetration and consistent fusion. Cut backing strips 1" × 8" (25 × 203 mm) long from the same alloy as the beveled pieces (*Figure 10*). The backing should be sufficiently thick to prevent melt-through.

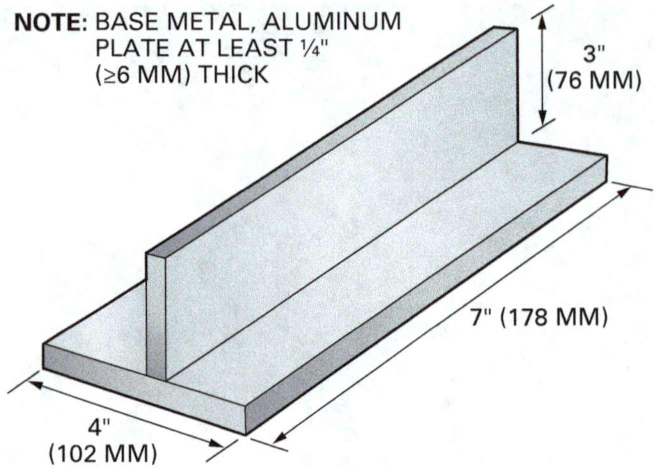

Figure 7 Fillet weld coupons.

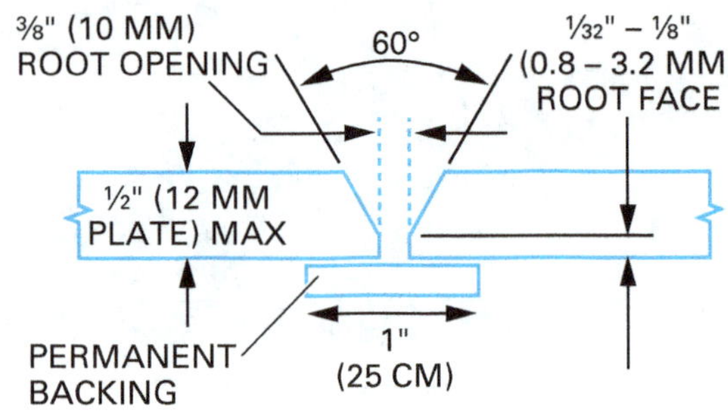

FIT TOLERANCES	
ROOT OPENING	+¼"; –1/16" (+6.4 MM; –1.6 MM)
GROOVE ANGLE	+10 DEGREES; –5 DEGREES

Figure 8 GMAW aluminum weld-joint preparation.

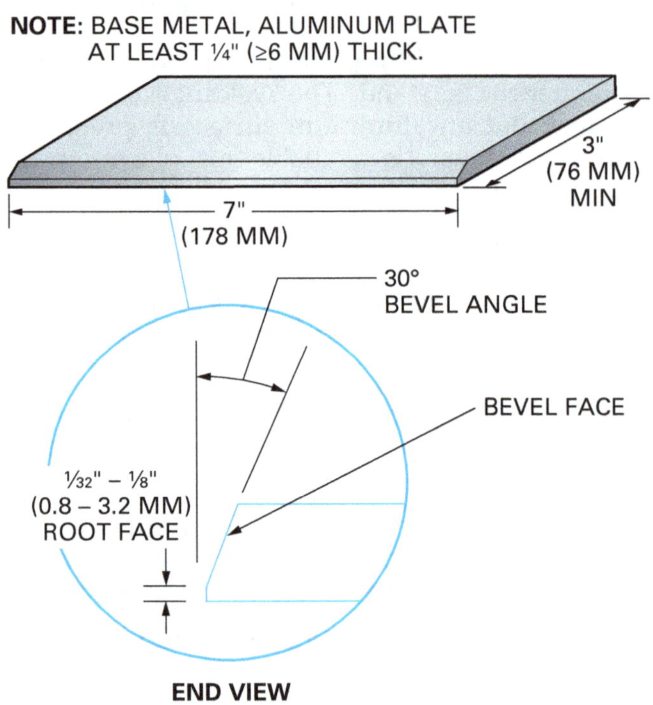

Figure 9 Preparing the V-groove weld coupon.

Observe the following steps to prepare each V-groove weld coupon, with backing, for welding. Note that coupons and backing strips should be welded within 30 minutes of their preparation to prevent the excessive formation of aluminum oxide:

Step 1 Cut and bevel the coupon components as shown using any of the acceptable methods discussed in this section. If the bevel has been cut thermally, remember to machine-cut the bevel back roughly ⅛" (3 mm) before machining the root face to its proper dimension.

Step 2 Make sure that the surface of the backing strip under the root opening has been thoroughly cleaned of all oxide coating in the same manner as the coupon. Center the beveled strips on the backing strip with a ⅜" (10 mm) root opening as shown in *Figure 10*, and tack weld them in place. If desired, use a piece of metal or rod with the required thickness as a temporary spacer to set the root opening. Place the tack welds on the reverse side of the joint in the lap formed by the backing strip and the beveled plate. Use three to four ½" (13 mm) long tack welds on each beveled plate. Be certain that the backing strip is tight against the beveled plates.

2.2.2 Alternate Joint Preparation

An alternate joint preparation can be used when welding in the horizontal position. This variation has one plate beveled at about 45 degrees and the other at about 15 degrees. The plate beveled at 45 degrees is placed above the 15-degree plate with a ⅜" root opening (*Figure 11*). Note that the degrees shown are approximate.

> **NOTE**
>
> Check with your instructor about whether to use the standard V-groove preparation or the alternate preparation for horizontal weld coupons.

Backing Strip Length

When the backing strip is ½" to 1" longer than the plates at each end, it allows you to start and stop the bead outside of the weld groove.

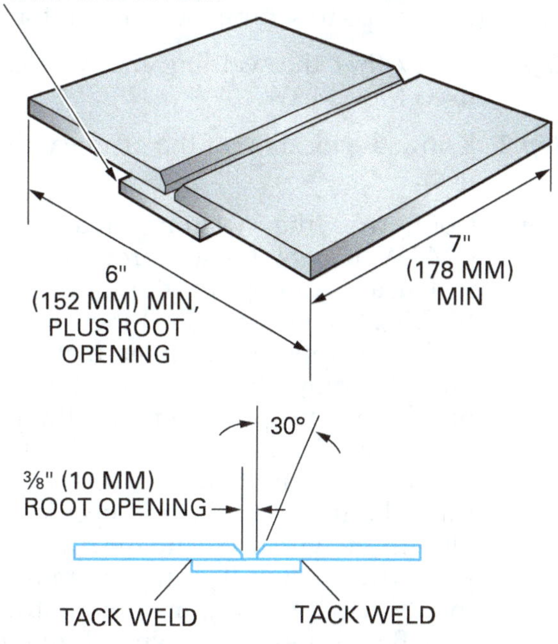

Figure 10 V-groove weld coupon with metal backing.

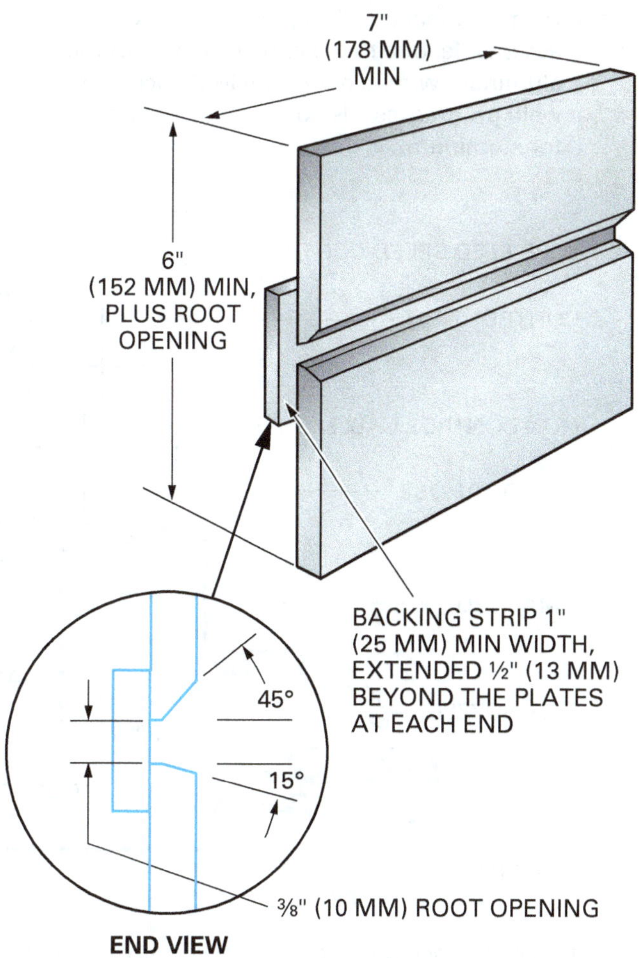

Figure 11 Alternate horizontal aluminum weld coupons.

2.3.0 The Welding Machine

Identify a proper welding machine for GMAW. Follow these steps to set the machine up for use:

Step 1 Verify that the welding machine can be used for GMAW.

Step 2 Know the location of the primary disconnect.

Step 3 Configure the welding machine for GMAW (*Figure 12*) as directed by your instructor. Configure the gun polarity (DCEP). Connect and load the wire feeder with the appropriate filler metal as required. Equip the gun with the correct nozzle for the application and the proper liner material and contact tube for the diameter of wire being used. The contact tube should be designed specifically for aluminum wire to ensure the size matches properly. Push-pull guns are most commonly used. Spool guns are another option. Both types are shown in *Figure 13*.

CAUTION

Using contact tips designed for steel wire is not acceptable, due to the expansion rate of the aluminum wire; binding will likely occur as the weld progresses. Use only contact tips designed for aluminum.

NOTE

The wire feeder drive rollers should be of the U-groove style for aluminum wire, unlike the V-groove rollers used for carbon steel wire.

Step 4 In accordance with the manufacturer's instructions, configure and load the wire feeder and gun with the proper type and diameter of electrode wire.

CAUTION

To prevent contaminants on the electrode wire from contaminating the welds, always keep the wire clean and dry when it is in storage or in the machine.

Step 5 Connect the proper shielding gas (argon or argon/helium) for the application as specified by the wire electrode manufacturer, WPS, site quality standards, or your instructor.

Step 6 Connect the clamp of the workpiece lead to the workpiece.

Step 7 Turn on the welding machine. Purge the gun as directed by the gun manufacturer's instructions.

Step 7 Set the initial welding voltage and wire feed speed as recommended by the manufacturer for the type and size of electrode wire being used.

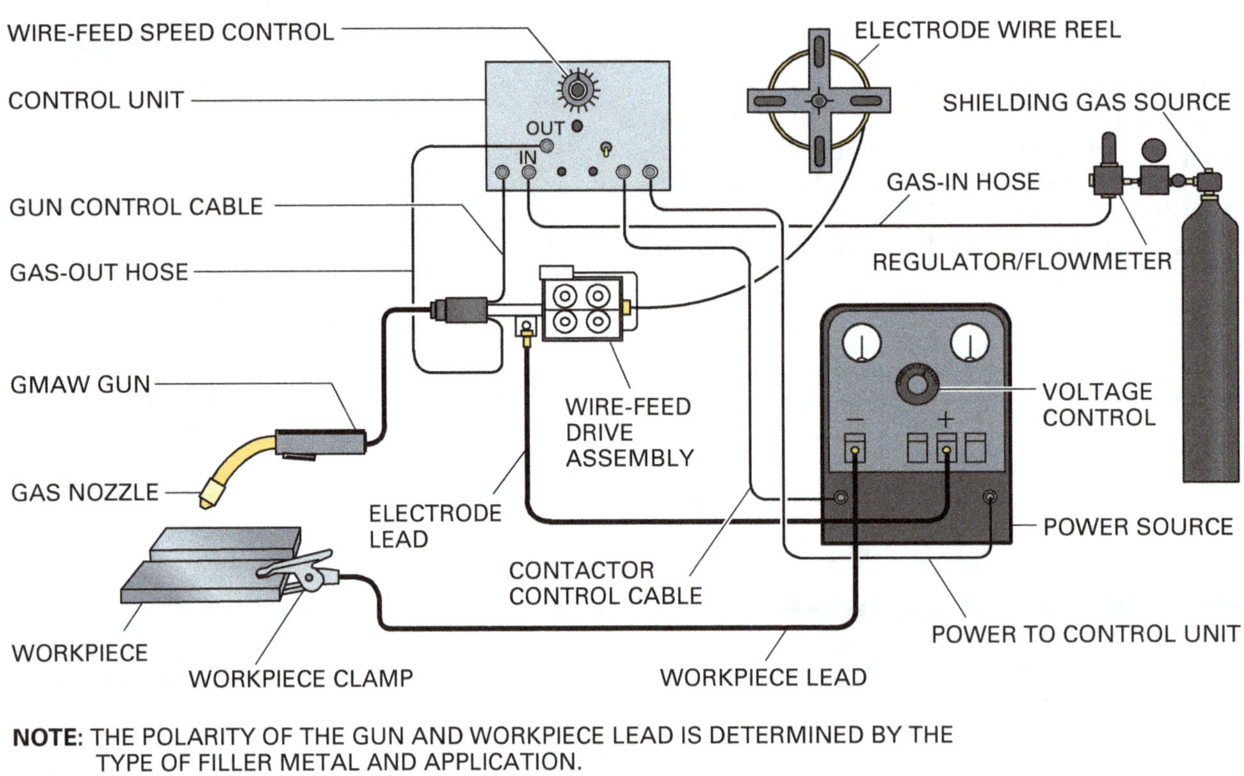

Figure 12 Configuration diagram of a typical GMAW machine.

(A) PUSH-PULL GUN

(B) SPOOL GUN

Figure 13 Push-pull and spool guns for GMAW.

2.3.1 Electrode Extension, Stickout, and Standoff Distance

Electrode extension is the length of the wire that extends beyond the tip of the welding gun's contact tip. For high-conductivity metal wires, the preheating effect of wire resistance is minimal. Wire speed and voltage settings have a more direct effect on weld penetration and deposition rate.

Either spray transfer or pulsed spray transfer can be used for aluminum welding. GMAW electrode extension for spray transfer varies from ½" to 1" (13 to 25 mm). *Figure 14* shows the typical electrode extension for the spray-transfer GMAW gun configuration. It also shows various gun terms and components. Stickout is the distance from the gas nozzle or insulating nozzle to the end of the electrode. Standoff distance is the distance from the gas nozzle or insulating nozzle to the workpiece.

Wire Electrode Manufacturer's Recommendations

Always obtain and follow the manufacturer's recommendations for shielding gas and for the initial setting of the welding voltage and wire feed speed parameters. These balanced parameters are critical and are based on the welding position, size, and composition of the solid wire.

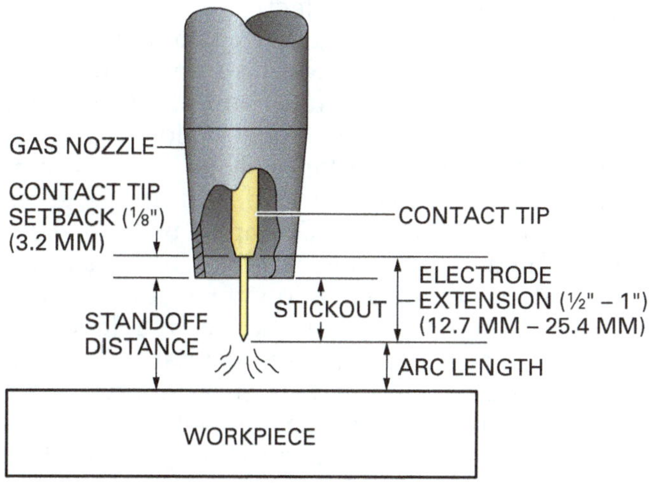

Figure 14 GMAW gun configured for spray transfer.

Contact-tip setback is important in the spray or pulse transfer mode of GMAW. The setback protects the contact tip from the additional heat created during the spray or pulse transfer mode. Reducing the heat helps to keep the wire from melting in the contact tip.

2.3.2 Welding Voltage

Arc length (also shown in *Figure 14*) is determined by the voltage, which is set at the welding machine. Arc length is the distance from the wire electrode tip to the base metal or to the molten pool at the base metal. If voltage is set too high, the arc will be too long. This can cause the wire to melt and fuse to the contact tube. Voltage that is too high can also cause porosity and excessive spatter. Voltage must be increased or decreased as the wire feed speed is increased or decreased. Set the voltage to maintain consistent spray transfer; a minimum arc voltage is needed to maintain spray transfer. However, penetration is not directly related to voltage. Penetration will increase as voltage is increased for a time, but penetration will actually decrease if the voltage is increased above its optimum setting.

2.3.3 Welding Amperage

When using a standard constant-voltage welding machine, the electrode feed speed controls the welding amperage, after the initial recommended setting. The welding machine provides the amperage needed to melt the wire electrode while maintaining the selected welding voltage. Within limits, when the wire electrode feed speed is increased, the welding amperage and deposition rate are also increased. This results in higher welding heat, deeper penetration, and higher beads. When the wire electrode feed speed is decreased, the welding amperage automatically decreases. With lower welding amperage and less heat, the deposition rate drops, and the weld beads are smaller with less penetration.

Note that some constant-voltage welding machine used for GMAW provide varying degrees of current modification. Examples include slope and induction adjustments. Welding machines with a slope adjustment allow the welder to vary the amount of amperage change in relation to the voltage range of the unit. Standard constant-voltage GMAW units have a current slope fixed by the manufacturer for general welding applications and conditions. Pulse-transfer units allow the peak current for the pulse and the background current between pulses to be adjusted to match specific welding applications and/or wire electrode requirements.

2.3.4 Weld Travel Speed

The travel speed must be considered before beginning a weld. Weld travel speed is the speed at which the electrode tip passes across the base metal in the direction of the weld. It is measured in inches per minute (ipm). The metric values for travel speed are millimeters or centimeters per minute. Travel speed has a great effect on penetration and bead size. Slower travel speeds build higher beads with deeper penetration. Faster travel speeds build smaller beads with less penetration. Ideally, the welding parameters should be adjusted so that the electrode tip is positioned at the leading edge of the weld puddle during travel.

2.3.5 Gun Position

The gun position in relation to the direction of the weld is defined for aluminum as shown in *Figure 15*. A push angle of 5 to 10 degrees is used to enhance base metal cleaning ahead of the weld puddle.

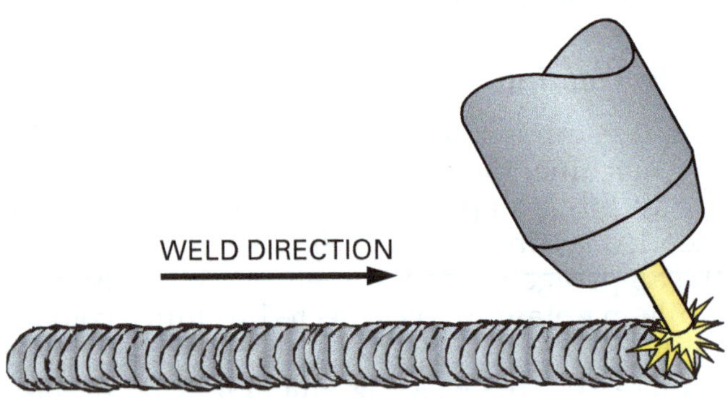

Figure 15 Gun work and travel angles.

2.3.6 Gas Nozzle Cleaning

As the welding machine is being used, weld spatter builds up on the gas nozzle, diffuser, and contact tube. The gas nozzle must be cleaned from time to time with a reamer, round file, or the tang of a standard file. If it is not properly cleaned, it will restrict the shielding gas flow and cause porosity in the weld.

2.4.0 Welding Beads

The two basic bead types are stringer beads and weave beads. Aluminum welds are easily contaminated. Because of this, extreme care must be taken in cleaning the base metals just before starting either type of bead. Always refer to the WPS for cleaning instructions and for the proper chemicals to use to remove oil and grease.

2.4.1 Stringer Beads

Stringer beads (*Figure 16*) are made with little or no side-to-side motion of the gun. Practice running stringer beads in the flat position. Experiment with different push angles and electrode stickouts.

Observe the following steps to make stringer beads:

Step 1 Clean the base metal thoroughly before welding.

WARNING

To prevent personal injury when using cleaning solvents, always comply with the SDS guidelines for the cleaning solvents being used.

Step 2 Hold the gun at the desired angle with the electrode tip directly over the point where the weld will begin. Pull the gun trigger.

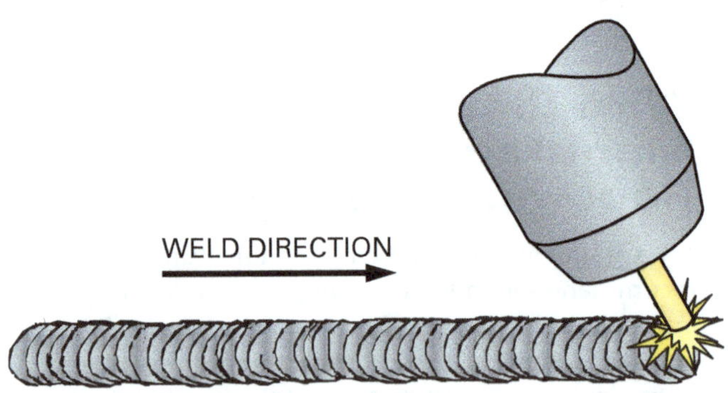

Figure 16 Stringer bead.

Travel Speed and Wire Feed Speed

New welders are tempted to turn down the wire feed speed if they have difficulty controlling the weld puddle. Wire electrodes must be run at certain balanced parameters that cannot be changed individually. Voltage, current (if variable), wire feed speed, and travel speed are adjusted and balanced together to control the weld puddle.

Step 3 Hold the arc in place until the weld puddle begins to form.

Step 4 Slowly advance the arc while maintaining the proper work and travel angles. Stay in the leading edge of the puddle to maintain cleaning action and to ensure fusion.

Step 5 Stop forward movement and back up about ⅛" (3 mm).

Step 6 Release the trigger once the crater has been filled.

Step 7 Inspect the bead to ensure it exhibits the following characteristics:

- Straightness
- Uniform appearance
- Smooth, flat transition with complete fusion at the toes of the weld
- Crater filled
- No porosity
- No undercut
- No underfill
- No cracks

Step 8 Continue practicing stringer beads until you can make acceptable welds every time.

2.4.2 Weave Beads

The side-to-side motions of the electrode used to make weave beads are called *oscillations* (*Figure 17*). The width of a weave bead is determined by the amount of side-to-side motion.

When making a weave bead, take special care at the toes to ensure proper fusion to the base metal. Do this by slowing down or pausing slightly at the edges. Pausing at the edges will also reduce undercut and give the weld the proper profile.

Practice running weave beads in the flat position. Experiment with different weave motions, push angles, and stickouts.

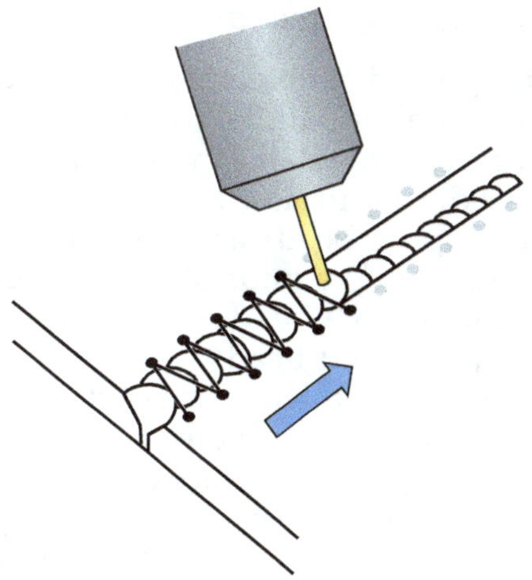

Figure 17 Weave bead motions.

CAUTION

Do not exceed the recommended weave bead width, which is often specified in the welding code or WPS used at your site.

Observe to following steps to make weave beads:

Step 1 Clean the base metal before welding.

Step 2 Hold the gun at the desired angle with the electrode tip directly over the point where the weld will begin. Pull the gun trigger.

Step 3 Hold the arc in place until the weld puddle begins to form.

Step 4 Slowly advance the arc in a weaving motion while maintaining the proper work and travel angles.

Step 5 Finish the bead by pausing until the crater has been filled, and then release the trigger.

Step 6 Inspect the bead to ensure it exhibits the following characteristics:

- Straightness
- Uniform appearance
- Smooth, flat transition with complete fusion at the toes of the weld
- Crater filled
- No porosity
- No overlap
- No underfill
- No undercut
- No cracks

Step 7 Continue practicing weave beads until you can make acceptable welds every time. Clean the beads between each pass.

2.4.3 Overlapping Beads

Overlapping beads are made by depositing connective weld beads parallel to one another. Each successive bead overlaps the previous bead to form a raised, flat surface. This is also called *padding*. Overlapping beads are used to build up a surface and to make multiple-pass welds. Both stringer and weave beads can be overlapped. Properly overlapped beads, when viewed from the end, form a relatively flat surface. Properly overlapped beads, when viewed from their ends, form a nearly flat, level surface (*Figure 18*).

Follow these steps to weld overlapping stringer or weave beads using aluminum wire electrodes and the appropriate shielding gas:

Step 1 Mark out a square on a piece of aluminum.

Step 2 Clean and deoxidize the base metal.

Step 3 Weld a stringer or weave bead along one edge.

Step 4 Clean the bead with a stainless steel brush.

Step 5 Position the gun at a work angle of 10 to 15 degrees toward the side of the previous bead to get proper fusion. Pull the trigger.

Step 6 Keep running overlapping stringer or weave beads until the square has been covered.

Step 7 Continue building layers of stringer or weave beads, one on top of the other, until your technique has been perfected.

2.4.4 Weld Terminations

A weld termination normally leaves a crater. Most welding codes require the crater be filled to the full cross section of the weld when making a termination. This can be difficult because most terminations are at the edge of a plate where welding heat tends to build up. This makes it harder to fill the crater.

To terminate a weld, observe the following steps (the first three steps are further illustrated in *Figure 19*):

Step 1 Start to bring the gun up to a 0-degree travel angle.

Step 2 Slow the forward travel as you approach the end of the weld.

Step 3 Stop the forward movement and back up about ⅛" (3 mm).

> **CAUTION**
>
> Do not immediately remove the gun from the weld until the puddle has solidified. The shielding gas post-flow that continues after welding has stopped protects the molten aluminum. If the gun and shielding gas are withdrawn before the crater has solidified, contamination and defects crater can occur.

Step 4 Release the trigger when the crater has been filled.

Step 5 Inspect the termination. The crater should be filled to the full cross section of the weld.

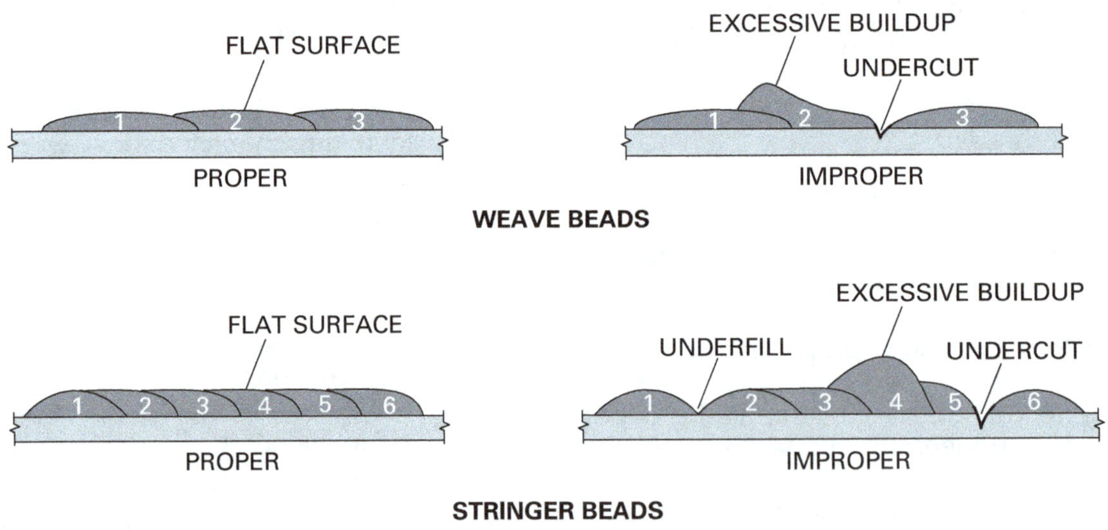

Figure 18 Proper and improper overlapping beads.

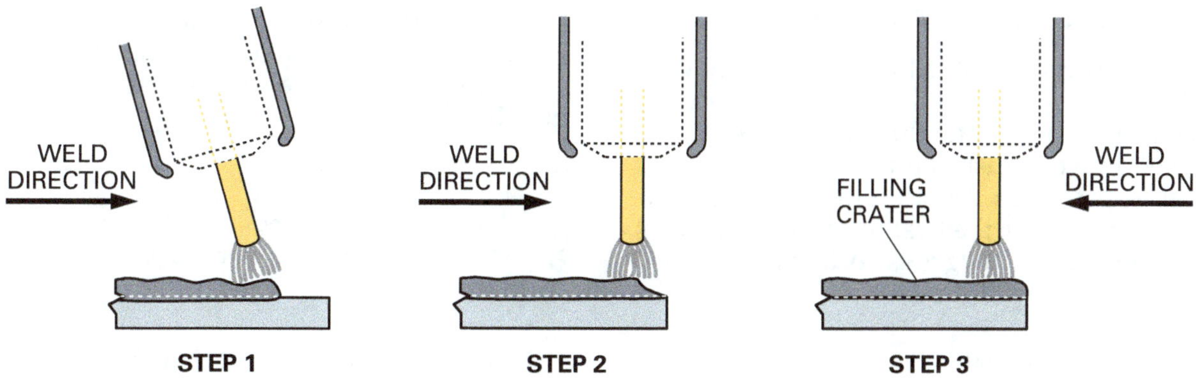

Figure 19 Terminating a weld.

2.4.5 Weld Restarts

A restart is the junction where a new weld connects to and continues the bead of a previous weld. A restart must blend smoothly with the previous weld and must not stand out or create a weld defect. The technique for making a GMAW restart is the same for both stringer and weave beads.

Follow these steps to make a restart:

Step 1 Clean the base metal before welding.

Step 2 Hold the gun at the proper angle while restarting the arc directly over the center of the crater. Remember that arc strikes outside of the weld area are not acceptable.

Step 3 Move the electrode tip in a small circular motion over the crater to fill it with a molten puddle.

Step 4 Move to the leading edge of the puddle. As soon as the puddle fills the crater, continue the stringer or weave bead pattern.

Step 5 Inspect the restart.

A properly made restart blends into the bead and is hard to detect. If the restart has undercut, not enough time was spent in the crater to fill it. If the undercut is on one side or the other, use more of a side-to-side motion as you move back into the crater. If the restart has a lump, it was overfilled. It is likely that too much time was spent in the crater before the forward motion was resumed. On restarts, start the weld just beyond the crater, establish the arc, and move back toward the crater. This will prevent the development of a lump in this area.

2.5.0 Fillet Welds

Fillet welds require little base metal preparation, since there are no bevels, root faces, or root openings involved. However, the weld area must still be thoroughly cleaned and any rough edges must be removed. Any dross from thermal cutting will cause porosity in the weld. For this reason, codes require that this material be removed entirely prior to welding. As noted previously, thermal cutting is not recommended for aluminum coupons.

The most common fillet welds are made in lap and T-joints. The weld position for plate is determined by the axis of the weld and the orientation of the workpiece. The positions for fillet welding on plate (*Figure 20*) are flat (1F), horizontal (2F), vertical (3F), and overhead (4F). In the 1F and 2F positions, the weld axis can be inclined up to 15 degrees. Any weld axis inclination for the other positions varies with the rotational position of the weld face as specified in the AWS standards.

Fillet welds can be concave or convex, depending on the WPS or site quality standards. Welding codes require a fillet weld to have a uniform concave or convex face, although a slightly nonuniform face is acceptable. The convexity of a fillet weld or individual surface bead must not exceed that permitted by the applicable code or standard. A fillet weld with profile defects is unacceptable and must be repaired (*Figure 21*).

WARNING!

Wear gloves when handling practice coupons and be sure to use pliers when the coupons are hot.

2.5.1 Practicing Flat (1F) Position Fillet Welds

Practice 1F-position (flat) fillet welds by making multiple-pass (six-pass) convex fillet welds in a T-joint. Use an appropriate aluminum filler metal as directed by your instructor. When making flat fillet welds, pay close attention to the work angle and travel speed. For the first bead, the gun work angle is vertical (45 degrees to both plate surfaces). Adjust the work angle for all subsequent beads (*Figure 22*). Clean each completed bead with a stainless steel wire brush before starting the next bead.

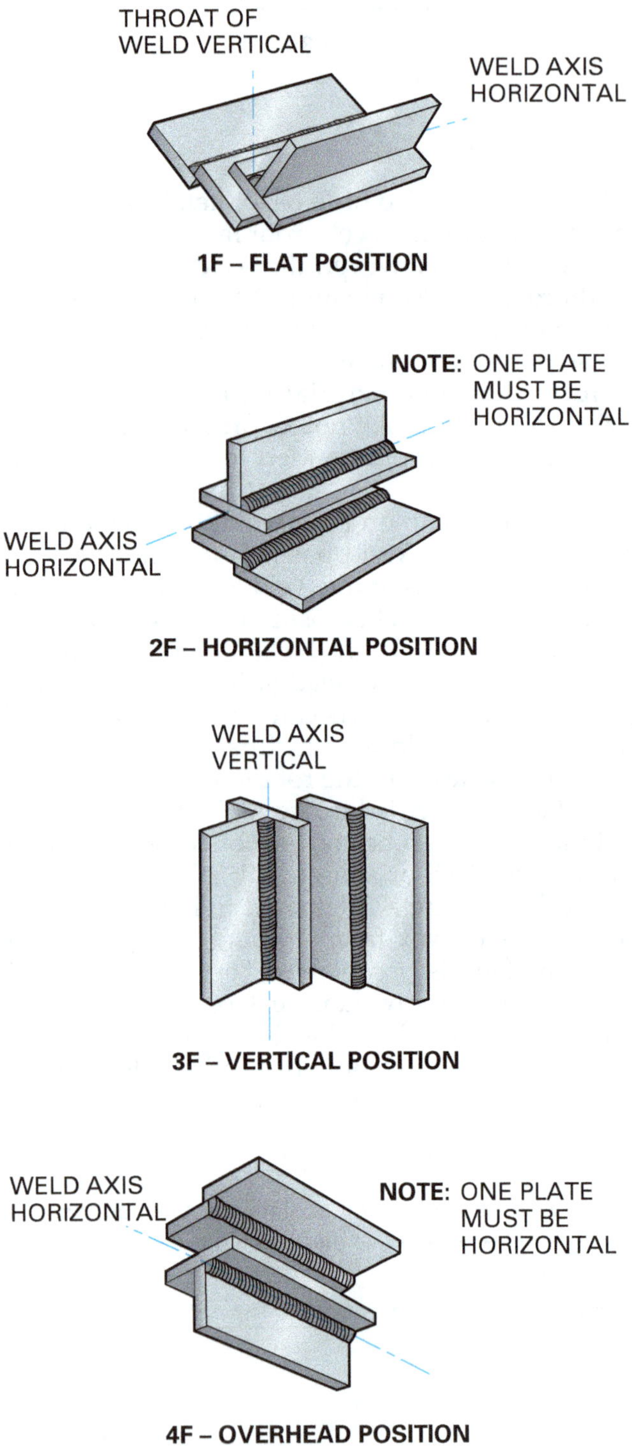

Figure 20 Fillet welding positions.

> **NOTE**
> In the following steps, clean all weld beads before beginning the next bead.

Follow these steps to make a flat fillet weld:

Step 1 Tack two plates together to form a T-joint for a fillet weld coupon (*Figure 23*). Clean the tack welds.

Step 2 Clamp or tack weld the coupon in the 1F position (flat).

Step 3 Run the first bead along the root of the joint using a 45-degree work angle and a 5- to 10-degree push angle.

Step 4 Run the second bead along a toe of the first weld, overlapping about 75 percent of the first bead. Alter the work angle as shown in *Figure 22*. Use a 5- to 10-degree push angle with a slight oscillation.

Step 5 Run the third bead along the other toe of the first weld. Be sure to fill the groove created when the second bead was applied. Use the work angle shown in *Figure 22* and a 5- to 10-degree push angle with a slight oscillation.

Step 6 Run the fourth bead along the outside toe of the second weld, overlapping about half of the second bead. Use the work angle shown in *Figure 22* and a 5- to 10-degree push angle with a slight oscillation.

Step 7 Run the fifth bead along the inside toe of the fourth weld, overlapping about half of the fourth bead. Use the work angle shown in *Figure 22* and a 5- to 10-degree push angle with a slight oscillation.

Step 8 Run the sixth bead along the toe of the fifth weld. Be sure to fill the groove created when the fifth bead was run. Use the work angle shown in *Figure 22* and a 5- to 10-degree push angle with a slight oscillation.

Step 9 Have your instructor inspect the weld. The weld is acceptable if it exhibits the following characteristics:

- Uniform appearance on the bead face
- Craters and restarts filled to the full cross section of the weld

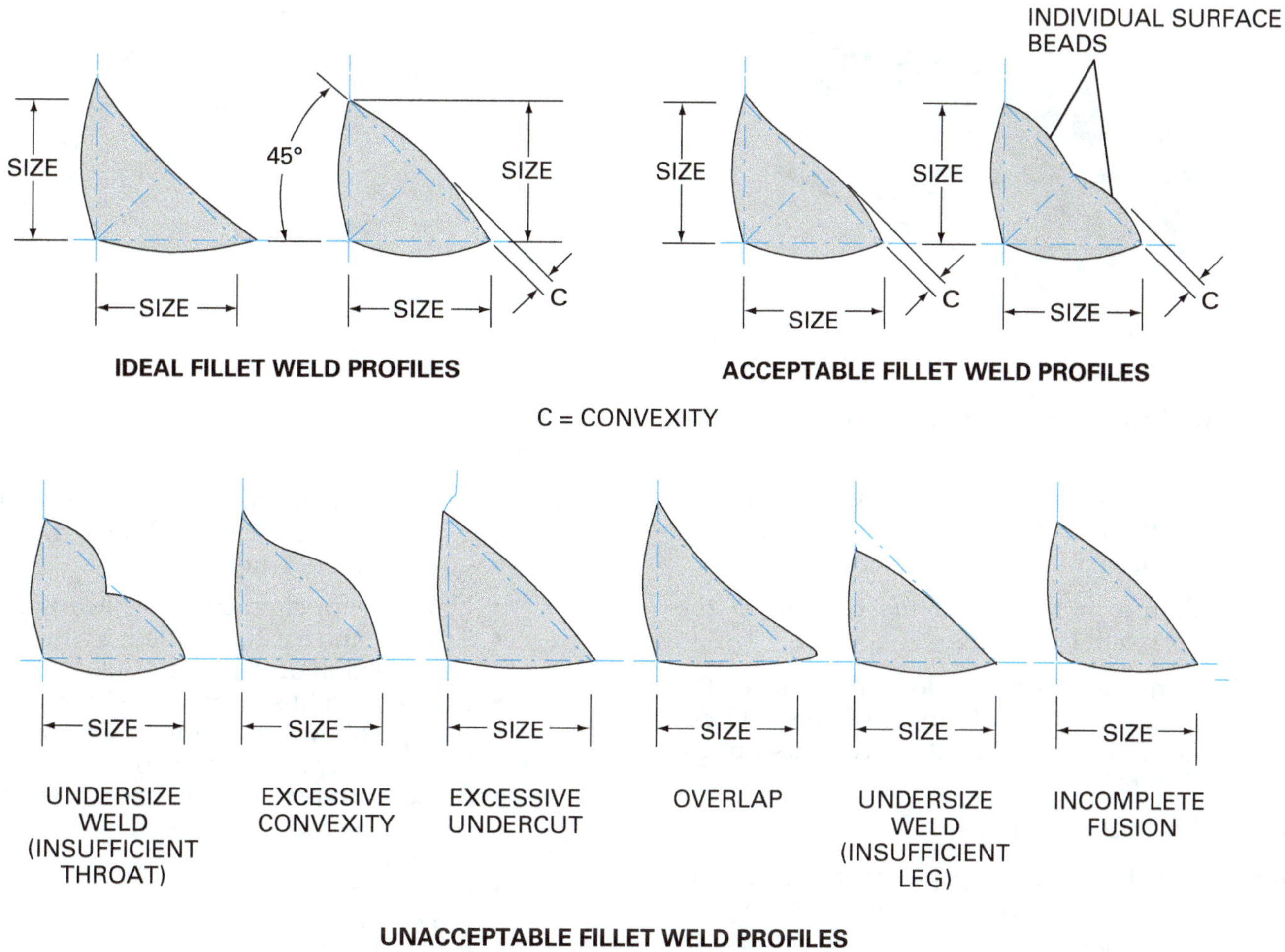

Figure 21 Acceptable and unacceptable fillet weld profiles.

- Uniform weld size of ±1⁄16" (1.6 mm)
- Acceptable weld profile in accordance with the applicable code or standard
- Smooth transition with complete fusion at the toes of the weld
- No porosity
- No undercut
- No overlap
- No underfill
- No inclusions
- No cracks

2.5.2 Practicing Horizontal (2F) Position Fillet Welds

Practice horizontal (2F) fillet welding by placing multiple-pass fillet welds in a T-joint. Use an aluminum filler metal as directed by your instructor. When making horizontal fillet welds, pay close attention to the work angle. For the first bead, the electrode work angle is 45 degrees. The work angle is adjusted for all other welds.

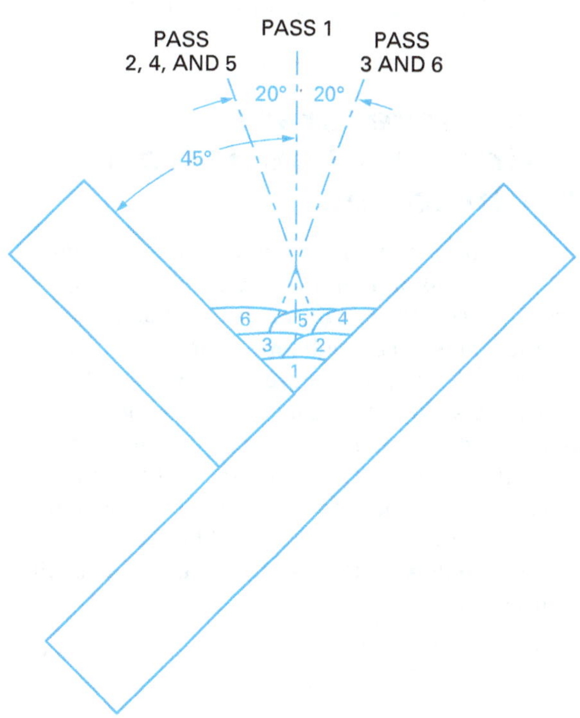

Figure 22 Multiple-pass 1F bead sequence and work angles.

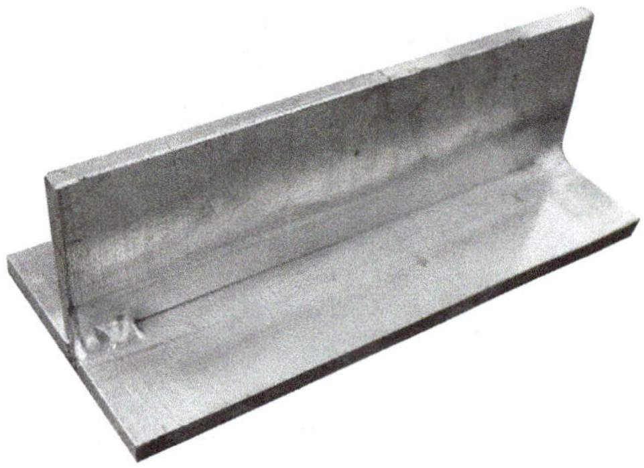

Figure 23 Fillet weld coupon.

Follow these steps to make a horizontal fillet weld:

Step 1 Tack two plates together to form a T-joint for the fillet weld coupon. Clean the tack welds.

Step 2 Clamp or tack weld the coupon in the horizontal position.

Step 3 Run the first bead along the root of the joint using a work angle of about 45 degrees and a 5- to 10-degree push angle (*Figure 24*).

Step 4 Clean the weld.

Step 5 Run the remaining passes at the proper work angles using a 5- to 10-degree push angle and a slight oscillation, as shown in *Figure 24*. Overlap each previous pass. Clean the weld after each pass.

Alternate Crater-Filling or Elimination Techniques

One way to eliminate the crater when terminating a GMAW weld on aluminum is to use run-off tabs. In cases where tabs cannot be used or are impractical, such as with pipe welds, you can reduce the weld pool before extinguishing the arc by increasing the travel speed just before releasing the gun trigger. This results in a small crater that may be free of cracks. Making several quick arc starts into the crater as it cools is another technique for filling the crater without adding heat to increase its size. Discuss these techniques with your instructor.

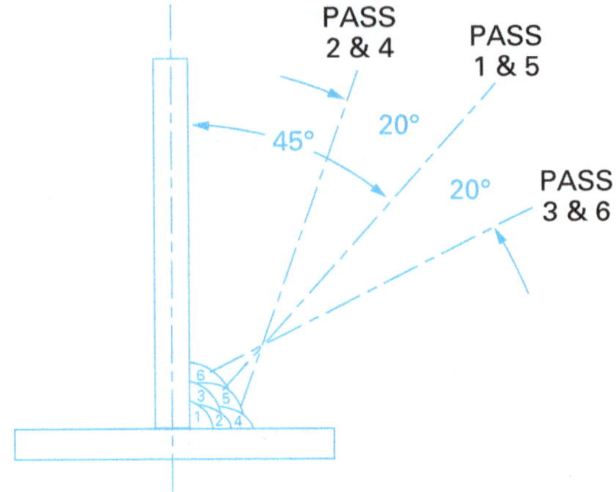

Figure 24 Multiple-pass 2F bead sequence and work angles.

Step 6 Have your instructor inspect the weld. The weld is acceptable exhibits the following characteristics:

- Uniform appearance on the bead face
- Craters and restarts filled to the full cross section of the weld
- Uniform weld size of ±1⁄16" (1.6 mm)
- Acceptable weld profile in accordance with the applicable code or standard
- Smooth transition with complete fusion at the toes of the weld
- No porosity
- No undercut
- No overlap
- No underfill
- No inclusions
- No cracks

2.5.3 Practicing Vertical (3F) Position Fillet Welds

Practice vertical (3F) fillet welding by placing multiple-pass fillet welds in a T-joint. Use an aluminum filler metal as directed by your instructor. Vertical welds are usually made by welding uphill from the bottom to the top using a push angle of 10 to 15 degrees. Note that this push angle is slightly greater than the usual 5 to 10 degrees. Because of the uphill welding and push angle, this type of welding is sometimes called vertical-up fillet welding. Either stringer or weave beads can be used for vertical welding. The site WPS or site quality standard will specify which technique to use on the job. Check with your instructor to see if you should practice stringer beads, weave beads, or both.

Tacking and Aligning Workpieces

When tacking workpieces together, position them to minimize distortion when the final welds are made. To achieve this, both sides of the workpieces are usually tacked with welds that are about ½" (13 mm) long. After the first tack weld, use a hammer or other tool to align the workpieces side-to-side and end-to-end. Then tack the opposite side. Tack the far ends of the workpieces in the same way. Intermediate tack welds can be made every 5" to 6" (13 to 15 cm) as needed to minimize longitudinal distortion.

Observe the following steps to make an uphill fillet weld using weave beads (refer to *Figure 25*):

Step 1 Tack two plates together to form a T-joint for the fillet weld coupon.

Step 2 Clamp or tack weld the coupon in the vertical position.

Step 3 Starting at the bottom, use a slightly oscillating motion to run the first bead along the root of the joint. Use a work angle of about 45 degrees and a 10- to 15-degree push angle. Pause in the weld puddle to fill the crater.

Step 4 Clean the weld.

Step 5 Run the remaining passes using a 10- to 15-degree push angle and a side-to-side weave technique with a 45-degree work angle. Use a slow motion across the face of the weld. Pause at each toe to penetrate and fill the crater. Clean the weld after each pass.

Step 6 Have your instructor inspect the weld. The weld is acceptable if it exhibits the following characteristics:

- Uniform appearance on the bead face
- Craters and restarts filled to the full cross section of the weld
- Uniform weld size of ±1⁄16" (1.6 mm)
- Acceptable weld profile in accordance with the applicable code or standard
- Smooth transition with complete fusion at the toes of the weld
- No porosity
- No undercut
- No overlap
- No underfill
- No inclusions
- No cracks

Vertical Weave Bead

To help control undercut on a root pass, an alternate pattern, such as a small triangular weave. By pausing at the edges, the previous undercut can be filled. This action also creates undercut at the existing weld pool, but that will be filled in by the next weave. This technique should only be applied during the root pass.

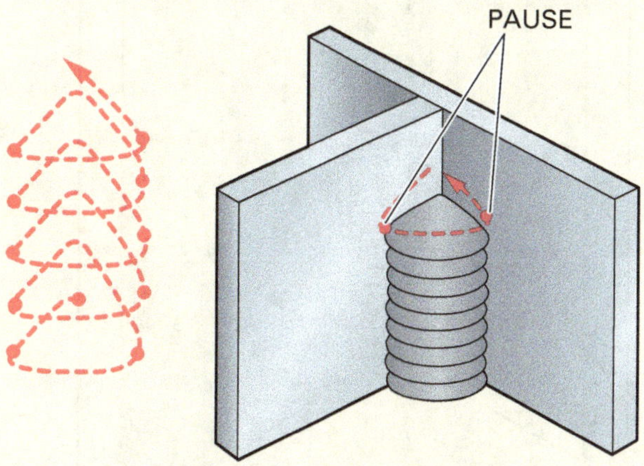

TRIANGULAR WEAVE PATTERN

Repeat vertical fillet (3F) welding using stringer beads. Use a slight oscillating motion, and pause slightly at each toe to prevent undercut. For stringer beads, use a 10- to 15-degree push angle and the required work angles. Refer to *Figure 25* for the bead sequence and work angles. Remember to maintain the 10- to 15-degree push angle as you weld uphill.

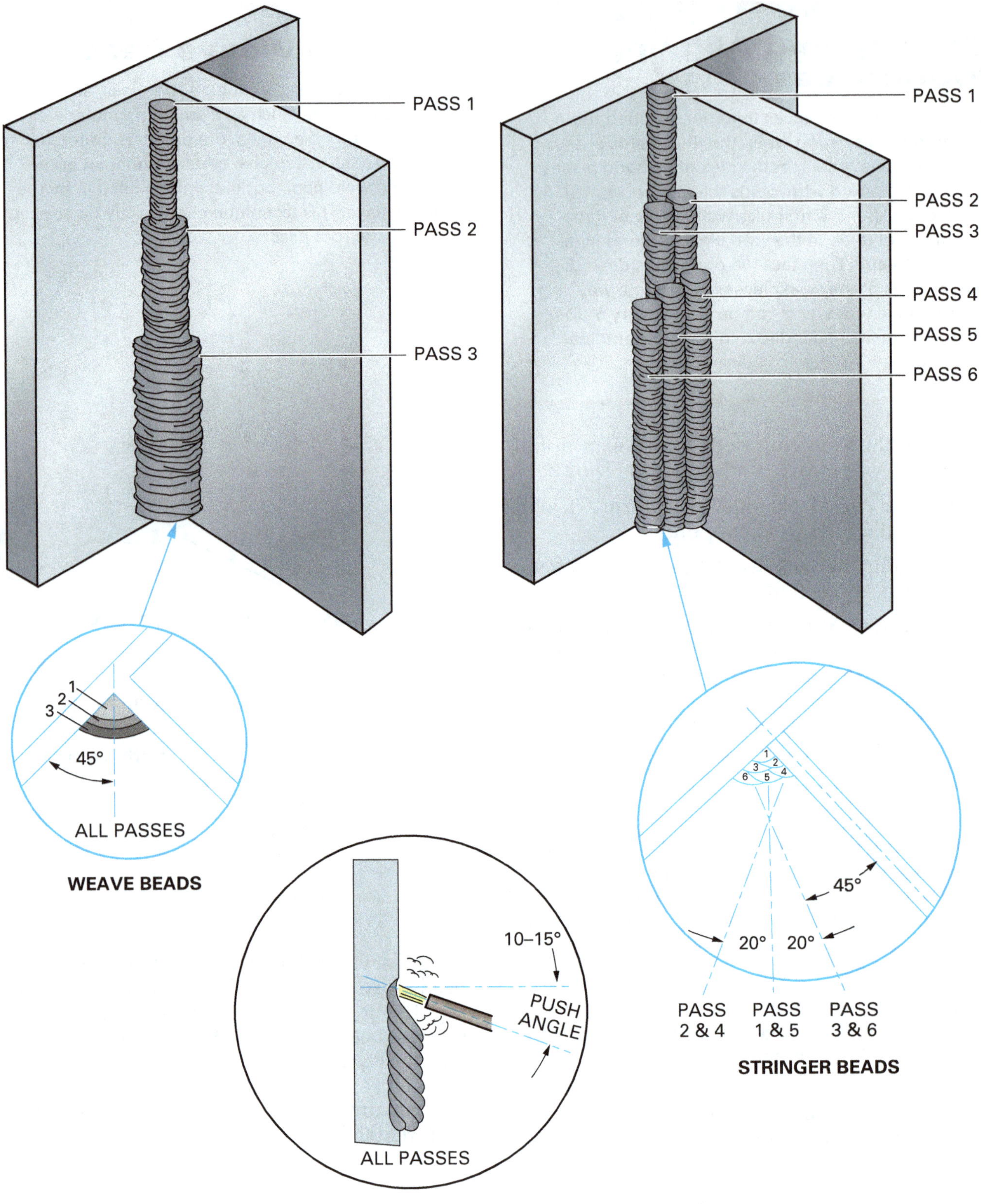

Figure 25 Multiple-pass 3F bead sequences and work angles for weave and stringer beads.

2.5.4 Practicing Overhead (4F) Position Fillet Welds

Practice overhead (4F) fillet welding by welding multiple-pass fillet welds in a T-joint. Use an aluminum filler metal as directed by your instructor. When making overhead fillet welds, pay close attention to the work angle. The work angle is about 45 degrees for the first bead, and then it is adjusted for all other welds.

Follow these steps to make an overhead fillet weld:

Step 1 Tack two plates together to form a T-joint for the fillet weld coupon.

Step 2 Clamp or tack weld the coupon so it is in the overhead position.

Step 3 Run the first bead along the root of the joint. Use a work angle of about 45 degrees and a 5- to 10-degree push angle.

Step 4 Clean the weld.

Step 5 Run the remaining passes using a 5- to 10-degree push angle and the work angles shown in *Figure 26*. Overlap each previous pass. Clean the weld after each pass.

Step 6 Have your instructor inspect the weld. The weld is acceptable if it exhibits the following characteristics:

- Uniform appearance on the bead face
- Craters and restarts filled to the full cross section of the weld
- Uniform weld size of ±1⁄16" (1.6 mm)
- Acceptable weld profile in accordance with the applicable code or standard
- Smooth transition with complete fusion at the toes of the weld
- No porosity
- No undercut
- No overlap
- No underfill
- No inclusions
- No cracks

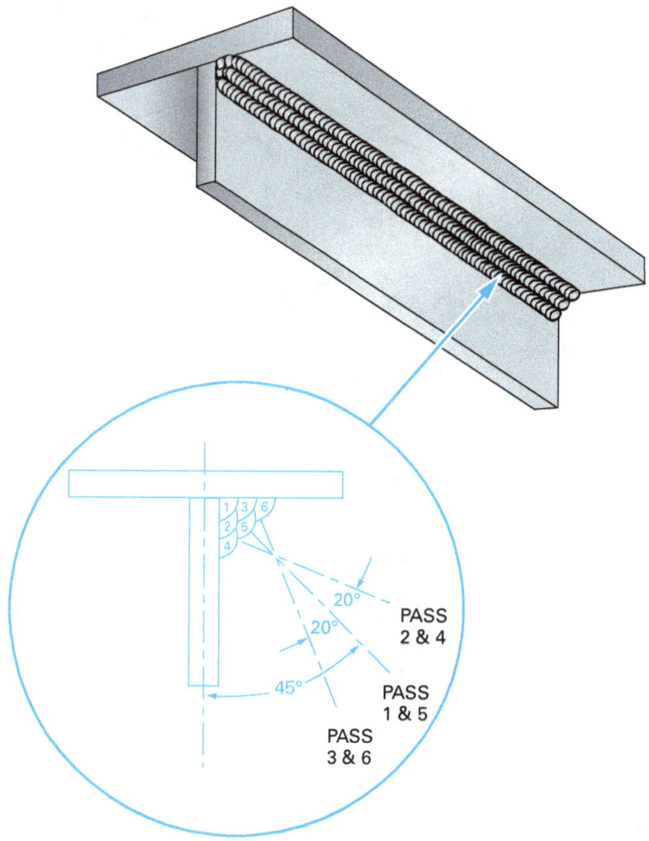

Figure 26 Multiple-pass 4F bead sequence and work angles.

2.6.0 V-Groove Plate Welds

The V-groove weld is a common groove weld made on both plate and pipe. The backing method of welding an aluminum V-groove joint using GMAW is presented in this module. Practicing V-groove welds on plate is good preparation for making the more difficult pipe welds later, which will be covered in another module.

V-groove welds with backing can be made in all positions. The weld position is determined by the axis of the weld and the orientation of the workpiece. The groove-weld positions are flat (1G), horizontal (2G), vertical (3G), and overhead (4G), as shown in *Figure 27*.

After completing the root pass, it should be cleaned and inspected. Use the proper cleaning method to remove the oxidation that tends to form along the joint. Inspect the root pass to ensure that it exhibits the following characteristics:

- Uniformly smooth face
- Complete fusion
- No excessive buildup
- No undercut
- No porosity

Practice making GMAW V-groove welds with backing in the 1G, 2G, 3G, and 4G positions. Use the proper size of aluminum filler wire for the passes. Pay special attention to filling the crater at the termination of the weld. Clean each completed bead with a stainless steel wire brush before starting the next bead.

WARNING!

Wear gloves when handling practice coupons and be sure to use pliers when the coupons are hot.

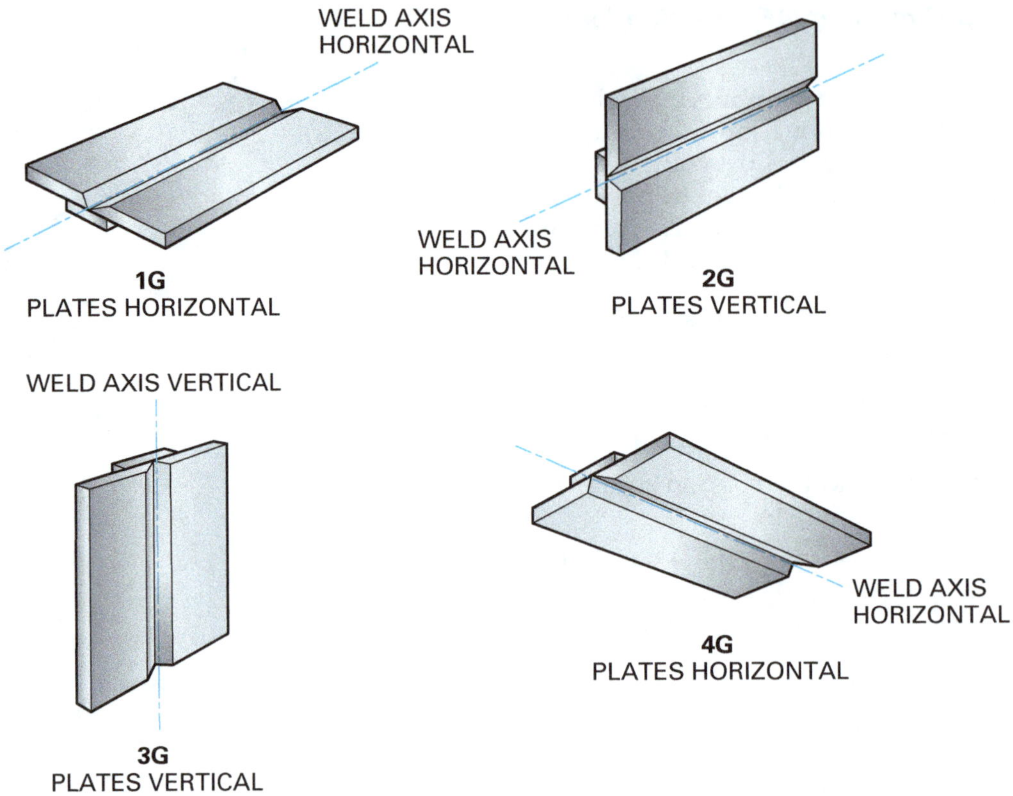

Figure 27 V-groove weld positions.

V-groove welds with backing should be made with slight face reinforcement, not exceeding 3⁄32" (2.4 mm) for materials up to 3⁄8" thick. There should be a gradual transition to the base metal at each toe. Groove welds must not have excessive face reinforcement, any underfill, undercut, or overlap (*Figure 28*). If a groove weld has any of these defects, it must be repaired.

2.6.1 Practicing Flat (1G) Position V-Groove Welds

Follow these steps to practice making V-groove welds with backing in the flat (1G) position; refer to *Figure 29* for work angles and the bead sequence:

Step 1 Tack weld the practice coupon and backing together.

Step 2 Clamp or tack weld the weld coupon in the flat position above the surface of the welding table.

Step 3 Use a 5- to 10-degree push angle and a 0-degree work angle to run the root pass with an appropriate aluminum filler metal.

Step 4 Clean the root pass.

Step 5 Run the remaining passes using a 5- to 10-degree push angle and the bead sequence and work angles shown in *Figure 29*. Clean the weld between each pass.

2.6.2 Practicing Horizontal (2G) Position V-Groove Welds

Follow these steps to practice making V-groove welds with backing in the horizontal (2G) position; refer to *Figure 30* for work angles and the bead sequence:

Step 1 Tack weld the practice coupon and backing together.

Step 2 Clamp or tack weld the coupon in the horizontal position.

Step 3 Use a 5- to 10-degree push angle to run the root pass with an appropriate aluminum filler metal.

Step 4 Clean the weld.

Step 5 Run the remaining passes using a 5- to 10-degree push angle and the bead sequences and work angles shown in *Figure 30*. Clean the weld between each pass.

Cooling Practice Coupons

Cooling coupons with water can affect the mechanical properties of the base metal. Only practice coupons can be cooled with water. Never cool test coupons with water.

2.6.3 Practicing Vertical (3G) Position V-Groove Welds

Follow these steps to practice making V-groove welds with backing in the vertical (3G) position; refer to *Figure 31* for work angles and the bead sequence. As is true when welding uphill in the 3F position, the push angle is slightly increased in the 3G position. Note that stringer or weave beads may be used for these welds, as specified by your instructor, WPS, or site quality standard:

Step 1 Tack weld the practice coupon and backing together.

Step 2 Clamp or tack weld the coupon in the vertical position.

Step 3 Run the root pass uphill using an appropriate aluminum filler metal and a stringer bead. Use a push angle of 10 to 15 degrees and a 0-degree work angle for the root pass.

Step 4 Clean the weld.

Step 5 Run the remaining passes uphill to complete the weld. Use a 5- to 10-degree push angle and the bead sequences and work angles shown in *Figure 31* for stringer or weave beads. Clean the weld between each pass.

2.6.4 Practicing Overhead (4G) Position V-Groove Welds

Follow these steps to practice making V-groove welds with backing in the overhead (4G) position; refer to *Figure 32* for work angles and the bead sequence:

Step 1 Tack weld the practice coupon and backing together.

Step 2 Clamp or tack weld the coupon in the overhead position.

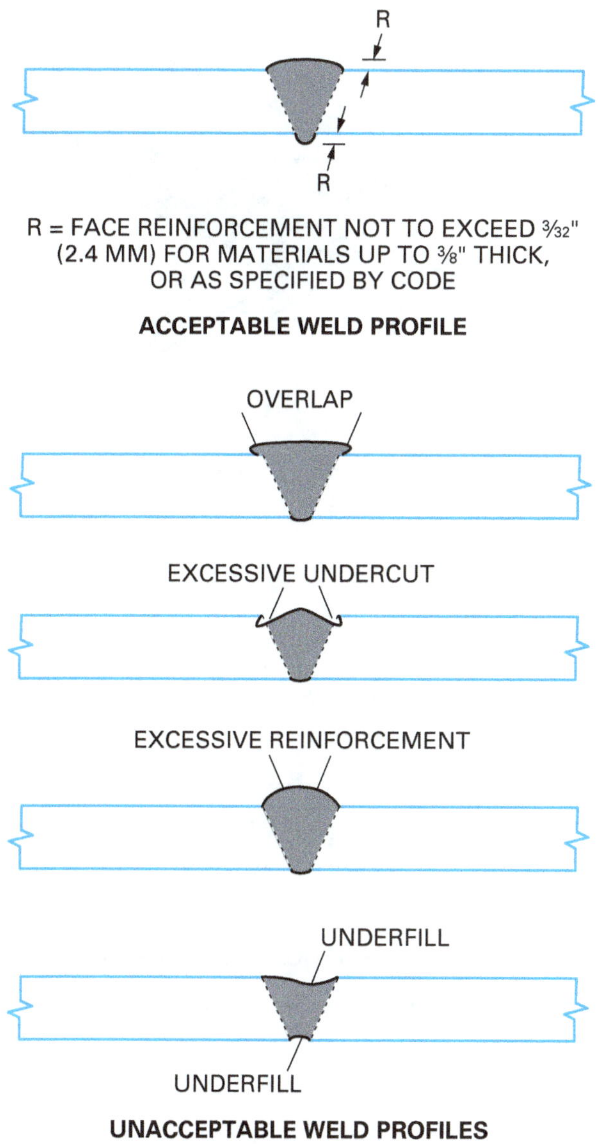

Figure 28 Acceptable and unacceptable V-groove weld profiles.

Step 3 Run the root pass using an appropriate aluminum filler metal. Use a 5- to 10-degree push angle and a 0-degree work angle for the root pass.

Step 4 Clean the weld.

Step 5 Run the remaining passes using a 5- to 10-degree push angle and the bead sequence and work angles shown in *Figure 32*. Clean the weld between each pass.

Specific V-Groove Weld Requirements

Refer to your site's WPS for specific requirements on groove welds. The information in this module is provided only as a general guideline. The WPS or site quality standards must be followed for all welds. Check with your supervisor if you are unsure of the specifications for your application.

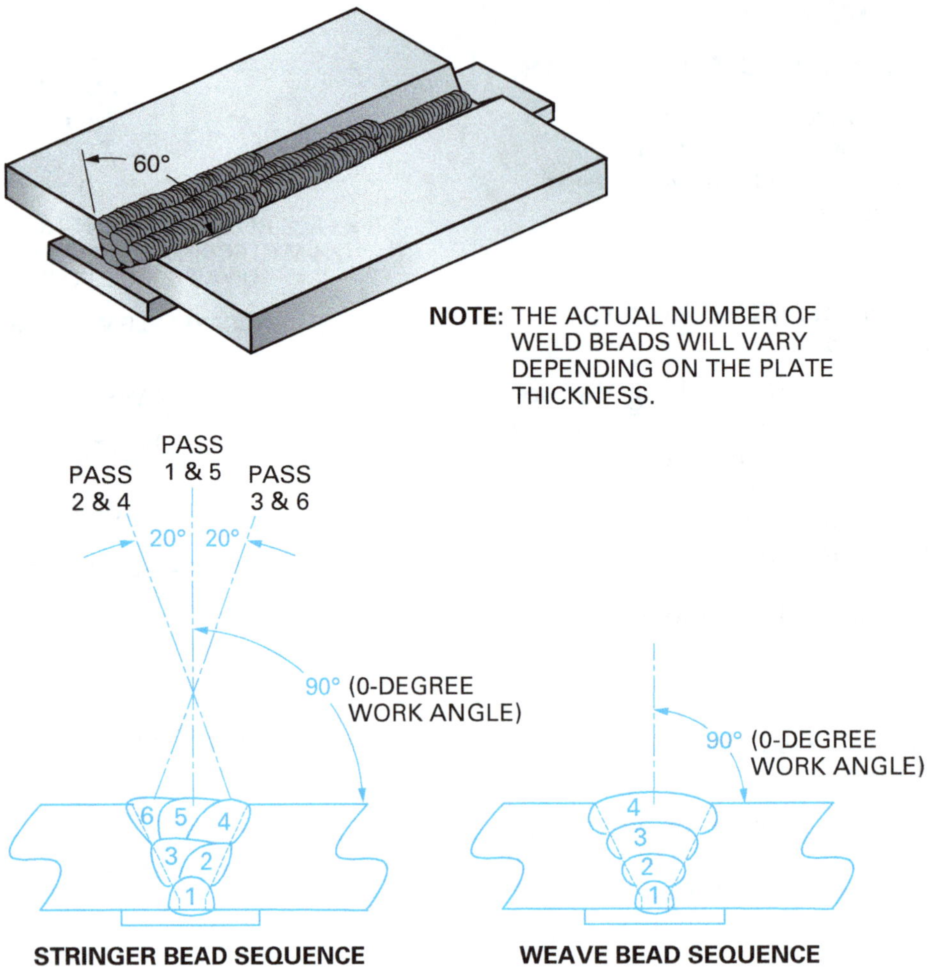

Figure 29 Multi-pass 1G bead sequence and work angles.

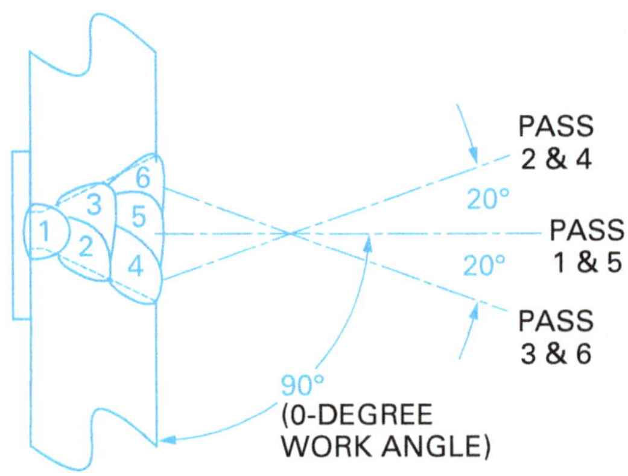

Figure 30 Multiple-pass 2G bead sequence and work angles.

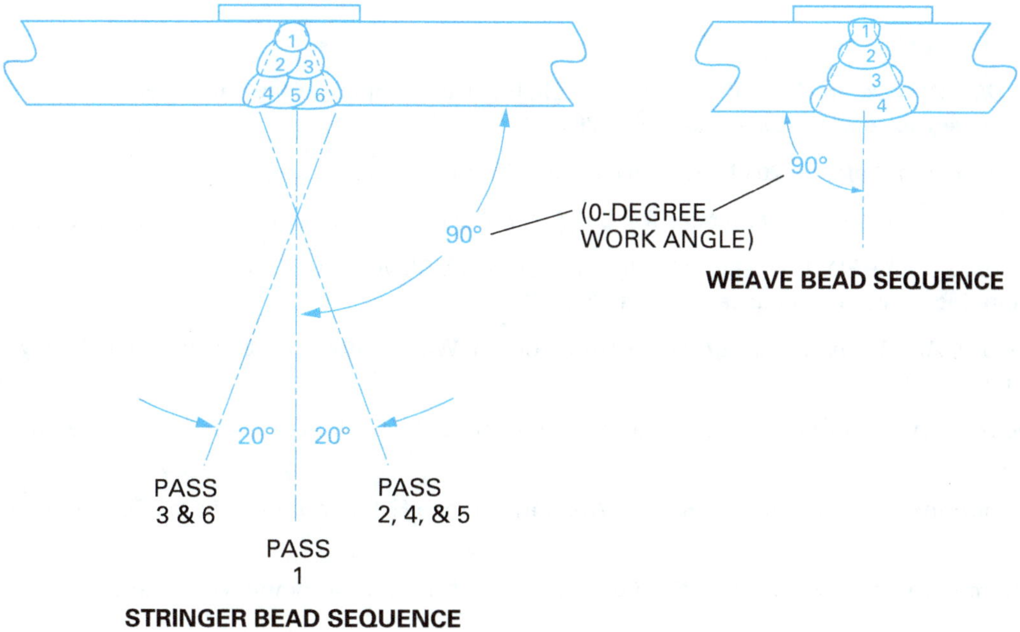

Figure 31 Multiple-pass 3G bead sequences and work angles.

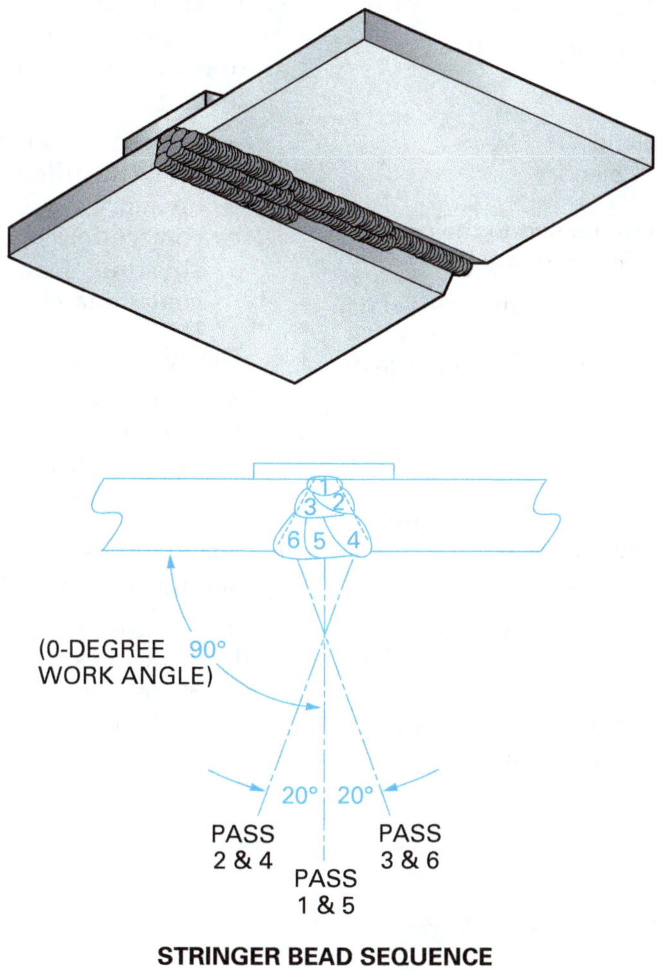

Figure 32 Multiple-pass 4G bead sequence and work angles.

Additional Resources

Aluminum GMAW Welding Guide (PDF). The Lincoln Electric Company, Cleveland, OH, USA. **www.lincolnelectric.com**. Last accessed May 26, 2016.

Guide for Aluminum Welding. 2013. Troy, OH: Hobart Brothers Corporation.

AWS C5.6-89R, Recommended Practices for Gas Metal Arc Welding. Miami, FL: American Welding Society.

GMAW Welding Guide (PDF). The Lincoln Electric Company, Cleveland, OH, USA. **www.lincolnelectric.com**. Last accessed May 26, 2016.

OSHA 1926.351, Arc Welding and Cutting. Current Edition. Washington, DC: Occupational Safety and Health Administration (OSHA).

The Procedure Handbook of Arc Welding. 2000. Cleveland, OH: The James F. Lincoln Arc Welding Foundation.

Welding Handbook, Volume 1 and *Volume 2.* Welding Science & Technology. Miami, FL: American Welding Society.

Welding Aluminum: Theory and Practice. 2002. New York, NY: The Aluminum Association.

2.0.0 Section Review

1. Caps worn during welding are best constructed of _____.
   a. 100-percent polyester
   b. 100-percent cotton
   c. a nylon and cotton blend
   d. a nylon and polyester blend

2. What is the negative factor related to the use of helium in the shielding gas for aluminum?
   a. It does not prevent oxide formation very well.
   b. It significantly increases the required welding current.
   c. It is more expensive than argon.
   d. Some aluminum alloys are damaged by it.

3. The proper practice for aluminum is to use a contact tube on the GMAW gun that is _____.
   a. one size larger than the wire diameter
   b. one size smaller than the wire diameter
   c. designed for stainless steel but still fits the aluminum wire
   d. designed specifically for the aluminum filler wire to ensure proper sizing

4. When terminating a weld bead, the welder should allow the weld to solidify before withdrawing the gun from the weld area because the _____.
   a. arc may need to be restarted quickly
   b. tip of the filler wire can drip off and cause an injury
   c. contact tip cools off too quickly
   d. shielding gas post-flow is need to prevent contamination

5. The 3F position refers to a(n) _____.
   a. horizontal V-groove weld
   b. horizontal fillet weld
   c. vertical fillet weld
   d. overhead V-groove weld

6. The typical work angle used for a GMAW aluminum weld in the 1G position is _____.
   a. 0 degrees
   b. 15 degrees
   c. 30 degrees
   d. 45 degrees

SUMMARY

A welder must possess a number of essential skills in order to perform basic welding jobs and to progress to more difficult welding procedures. These skills include setting up GMAW equipment, preparing the welding work area, running stringer and weave beads, and making acceptable fillet welds and open V-groove welds on aluminum plate in all positions. The ability to consistently produce acceptable welds requires an understanding of the necessary welding techniques as well as faithful, dedicated practice.

Review Questions

1. The GMAW process for aluminum uses which of the following for the filler metal?
   a. Short lengths of electrode
   b. Tin-coated aluminum rods
   c. A continuous wire electrode
   d. Nickel-coated aluminum rods

2. Aluminum is more thermally conductive than steel by a factor of _____.
   a. three
   b. four
   c. five
   d. six

3. Because of its strong chemical attraction to oxygen, aluminum begins to oxidize as soon as it is exposed to _____.
   a. water
   b. air
   c. carbon
   d. hydrogen

4. Cast aluminum has a high silicon content for increased _____.
   e. oxidation resistance
   f. corrosion resistance
   g. fluidity
   h. strength

5. Special care should be taken to raise the temperature of cold aluminum slowly because rapid rises in temperature cause _____.
   a. distortion
   b. loss of strength
   c. loss of ductility
   d. condensation

6. The main cause of porosity in aluminum welds is the _____.
   a. absorption of hydrogen into the weld pool
   b. absorption of argon into the weld pool
   c. use of helium in the shielding gas supply
   d. use of an incorrect filler metal

7. Although it is expensive, helium can be added to the shielding gas to _____.
   a. eliminate any chance of porosity
   b. double the welding speed
   c. weld aluminum plate without cleaning it
   d. significantly reduce the welding current

8. In GMAW, the length of wire extending beyond the tip of the welding gun's contact tip is called _____.
   a. setback
   b. stickout
   c. standoff distance
   d. electrode extension

9. When making a weave bead, take care at the toes to _____.
   a. avoid pauses on the edges of the base metal
   b. touch the electrode to the base metal
   c. avoid oscillations on the base metal
   d. ensure proper fusion to the base metal

10. The push angle of an aluminum GMAW weld is increased for which of the following weld positions?
    a. 1F
    b. 2F
    c. 3F
    d. 4G

Trade Terms Introduced in This Module

Anodic coating: An artificial buildup of aluminum oxides on the surface of aluminum alloys that improves corrosion resistance. This coating is many times thicker than a coating of naturally occurring oxides.

Artificial aging: A process following solution heat treatment (see *solution heat treatment*) and other processes that allows the alloy to rest for a prescribed period of time at an elevated temperature.

Cast aluminum: Aluminum alloys that are molded into their final form by one of various methods such as sand-casting, die-casting, or pressure die-casting. Cast aluminum contains high levels of silicon to enhance the ability of the material to form complicated castings.

Coalescence: The growing together or growth into one body of the materials being joined.

Ingot: A mass of metal shaped for convenient storage and transport.

Liquation cracks: Liquation is the process of separation by melting, such as in an alloy when one element melts while the others remain solid. Liquation cracks result from localized melting in areas such as the grain boundaries, while adjacent grains remain solid. Both solidification and liquation cracking may be referred to as hot cracking.

Natural aging: A process that follows solution heat treatment (see *solution heat treatment*) and other processes that allows the alloy to simply rest at room temperature for a prescribed period of time, such as 96 hours.

Solidification cracks: Cracks that occur as the filler and base metal solidify, when the remaining liquid weld metal is unable to fill the spaces between areas of solidifying weld metal. Both solidification and liquation cracking may be referred to as *hot cracking*.

Solution heat treatment: A process applied to a heat-treatable aluminum alloy that raises the temperature of the alloy to roughly 1,000°F (538°C) for a set period of time, causing the alloy elements to disperse more consistently throughout the metal. The metal is then quenched with water to ensure the dispersed elements remain in place within the metal structure, rather than precipitate out.

Wrought aluminum: Aluminum alloys that are initially formed into ingots or similar shapes and are then worked into their desired shape through bending, rolling, extrusion, or other mechanical methods.

Appendix

Performance Accreditation Tasks

The Performance Accreditation Tasks (PATs) correspond to and support learning objectives in *AWS EG2.0, Curriculum Guide for the Training of Welding Personnel; Level I—Entry Welder.*

PATs provide specific acceptable criteria for performance and help to ensure a true competency-based welding program for students.

The following tasks are designed to evaluate your ability to run stringer beads, weave beads, and overlapping beads. In addition, fillet and V-groove welds must be completed with GMAW equipment in four standard test positions using aluminum filler wire of the appropriate diameter and shielding gas. Perform each task when you are instructed to do so by your instructor. As you complete each task, show it to your instructor for evaluation. Do not proceed to the next task until told to do so. For AWS 2G and 5G certifications, refer to *AWS EG3.0, Guide for the Training and Qualification of Welding Personnel; Level II—Advanced Welder* for bend test requirements. For AWS 6G certifications, refer to *AWS EG4.0, Guide for the Training and Qualification of Welding Personnel; Level III—Expert Welder* for bend test requirements.

PAT 1 has no AWS correlation. PATs 2 through 5 correspond to *AWS EG3.0, Section 3.3.6.2, Unit #2 GMAW, Learning Objective #6*. PATs 6 through 9 correspond to *AWS EG3.0, Section 3.3.6.2, Unit #2 GMAW, Learning Objective #7.*

WELD A PAD ON ALUMINUM PLATE IN THE FLAT (1G) POSITION USING GMAW STRINGER BEADS

As directed by the instructor, use the GMAW process with the appropriate aluminum filler wire to make the following welds on aluminum plate: stringer beads, weave beads, weld restarts, weld terminations, and overlapping beads.

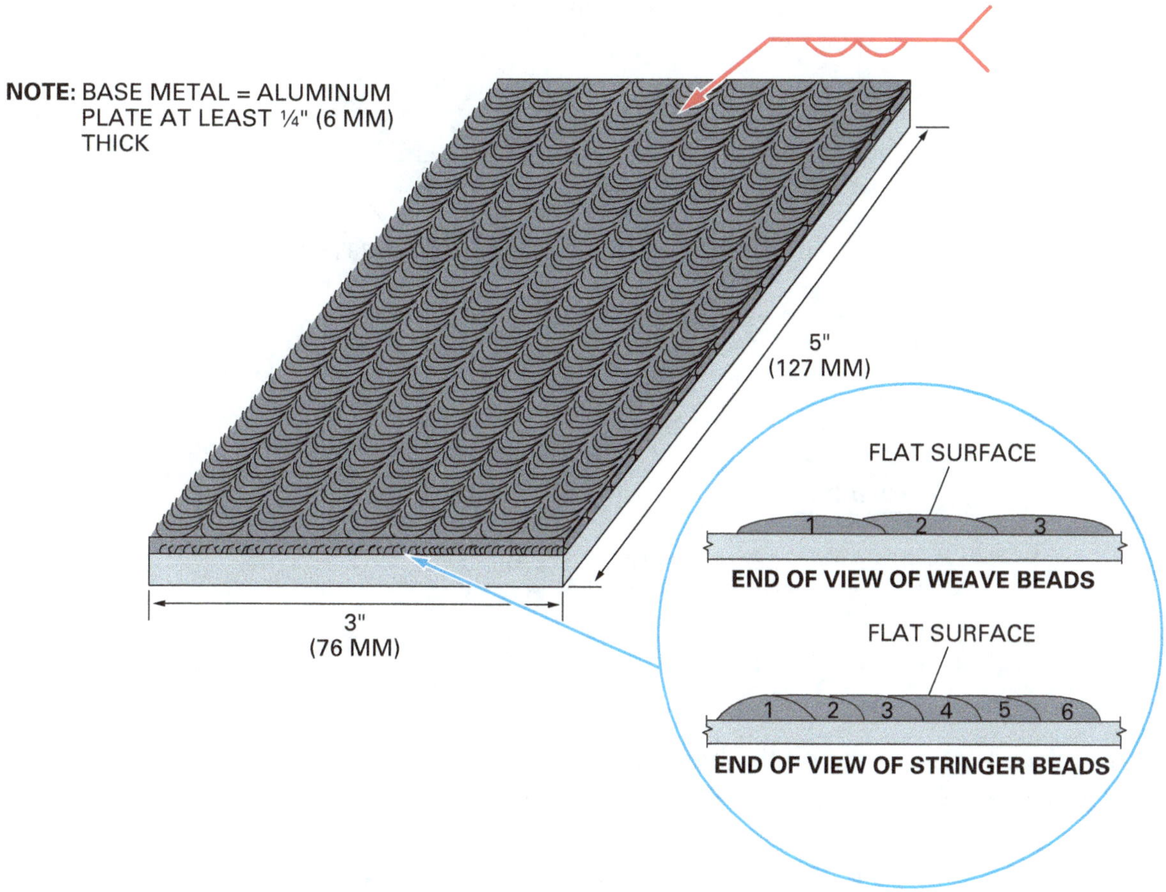

Criteria for Acceptance:

- Uniform rippled appearance on the bead face ____________
- Craters and restarts filled to the full cross section of the weld ____________
- Uniform weld width ±1⁄16" (1.6 mm) ____________
- Acceptable weld profile in accordance with the applicable code or standard ____________
- Smooth flat transition with complete fusion at the toes of the weld ____________
- No porosity ____________
- No undercut ____________
- No inclusions ____________
- No cracks ____________

MAKE MULTIPLE-PASS FILLET WELDS ON ALUMINUM PLATE IN THE FLAT (1F) POSITION

As directed by the instructor, use the GMAW process with the appropriate aluminum wire to make a six-pass fillet weld using stringer beads on aluminum plate, as shown.

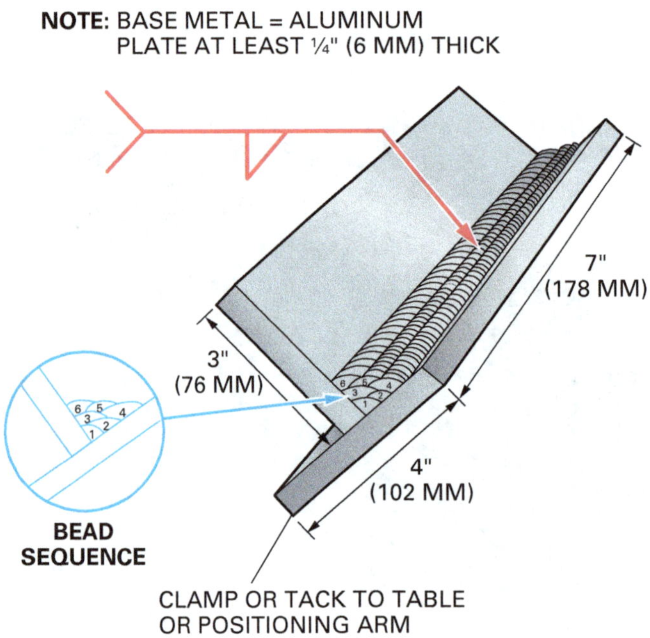

Criteria for Acceptance:

- Uniform rippled appearance on the bead face ____________
- Craters and restarts filled to the full cross section of the weld ____________
- Uniform weld width ±1⁄16" (1.6 mm) ____________
- Smooth flat transition with complete fusion at the toes of the weld ____________
- Acceptable weld profile in accordance with the applicable code or standard ____________
- No porosity ____________
- No undercut ____________
- No overlap ____________
- No inclusions ____________

MAKE MULTIPLE-PASS FILLET WELDS ON ALUMINUM PLATE IN THE HORIZONTAL (2F) POSITION

As directed by the instructor, use the GMAW process with the appropriate aluminum wire to make a six-pass fillet weld using stringer beads on aluminum plate, as shown.

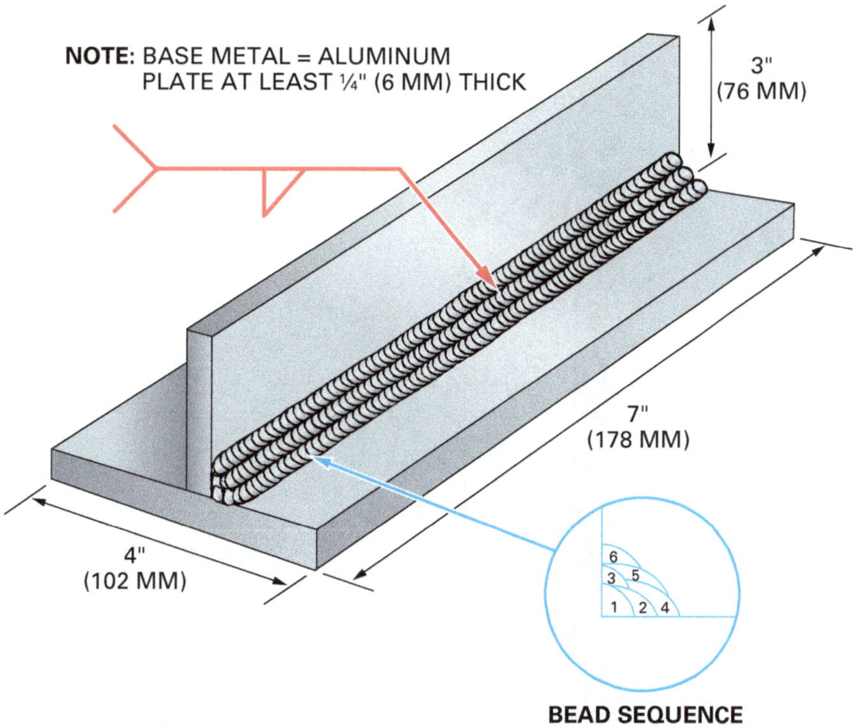

Criteria for Acceptance:

- Uniform rippled appearance on the bead face ____________
- Craters and restarts filled to the full cross section of the weld ____________
- Uniform weld width ±1⁄16" (1.6 mm) ____________
- Smooth flat transition with complete fusion at the toes of the weld ____________
- Acceptable weld profile in accordance with the applicable code or standard ____________
- No porosity ____________
- No undercut ____________
- No overlap ____________
- No inclusions ____________

MAKE MULTIPLE-PASS FILLET WELDS ON ALUMINUM PLATE IN THE VERTICAL (3F) POSITION

As directed by the instructor, use the GMAW process with the appropriate aluminum wire to make a six-pass fillet weld using stringer beads on aluminum plate, as shown.

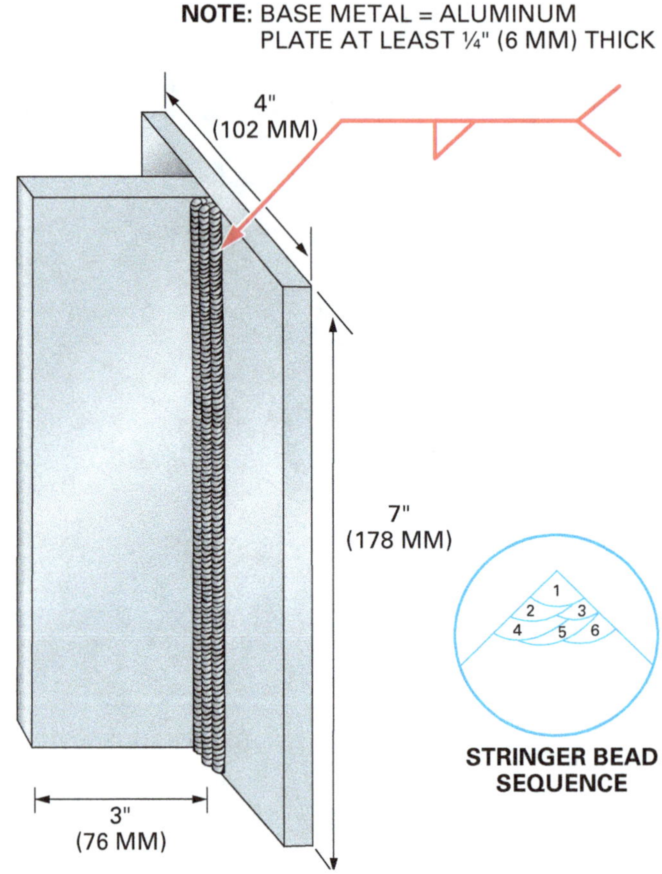

Criteria for Acceptance:

- Uniform rippled appearance on the bead face ________
- Craters and restarts filled to the full cross section of the weld ________
- Uniform weld width ±1⁄16" (1.6 mm) ________
- Smooth flat transition with complete fusion at the toes of the weld ________
- Acceptable weld profile in accordance with the applicable code or standard ________
- No porosity ________
- No undercut ________
- No inclusions ________
- No cracks ________

MAKE MULTIPLE-PASS FILLET WELDS ON ALUMINUM PLATE IN THE OVERHEAD (4F) POSITION

As directed by the instructor, use the GMAW process with the appropriate aluminum wire to make a six-pass fillet weld using stringer beads on aluminum plate, as shown.

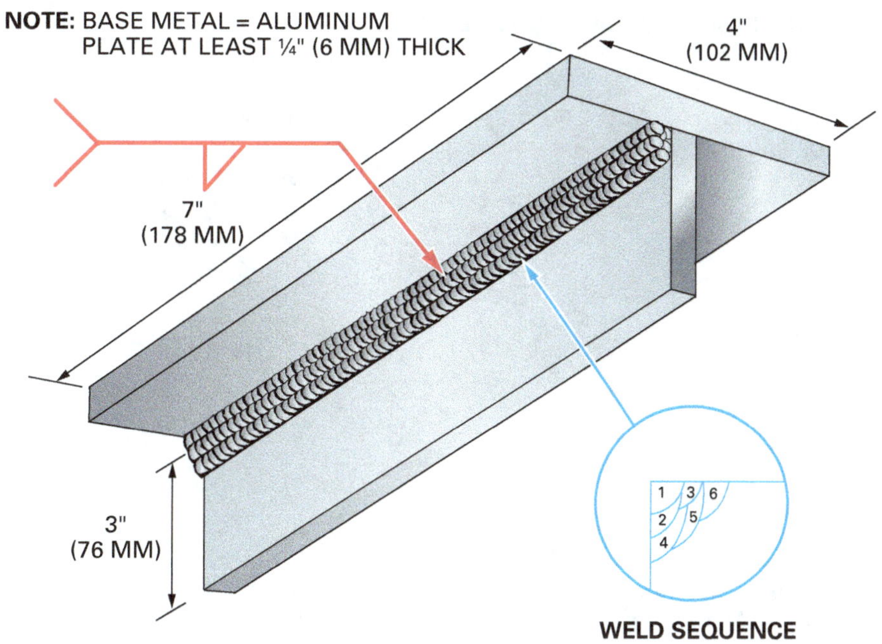

Criteria for Acceptance:

- Uniform rippled appearance on the bead face ________
- Craters and restarts filled to the full cross section of the weld ________
- Uniform weld width ±1⁄16" (1.6 mm) ________
- Smooth flat transition with complete fusion at the toes of the weld ________
- Acceptable weld profile in accordance with the applicable code or standard ________
- No porosity ________
- No undercut ________
- No inclusions ________
- No cracks ________

MAKE MULTIPLE-PASS V-GROOVE WELDS ON ALUMINUM PLATE WITH BACKING IN THE FLAT (1G) POSITION

As directed by the instructor, use the GMAW process with the appropriate aluminum wire to make a multiple-pass groove weld using stringer beads on aluminum plate with backing, as shown.

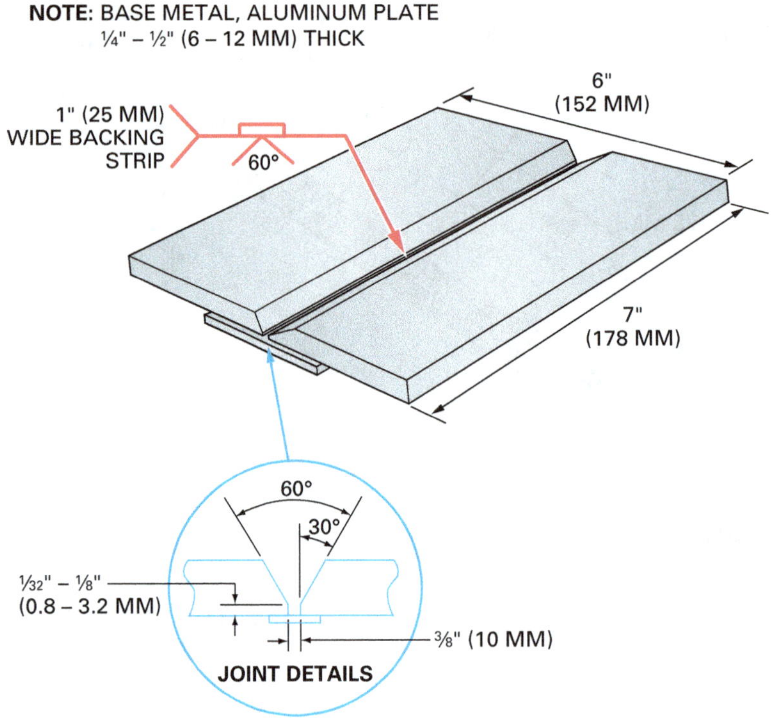

Criteria for Acceptance:

- Uniform rippled appearance on the bead face ______
- Craters and restarts filled to the full cross section of the weld ______
- Acceptable weld profile in accordance with the applicable code or standard ______
- Smooth flat transition with complete fusion at the toes of the weld ______
- No porosity ______
- No undercut ______
- No overlap ______
- No inclusions ______
- No cracks ______

MAKE MULTIPLE-PASS V-GROOVE WELDS ON ALUMINUM PLATE WITH BACKING IN THE HORIZONTAL (2G) POSITION

As directed by the instructor, use the GMAW process with the appropriate aluminum wire to make a multiple-pass groove weld on aluminum plate with a 1" (25 mm) backing, as shown. Either joint design may be used for the 2G position, at the discretion of the instructor.

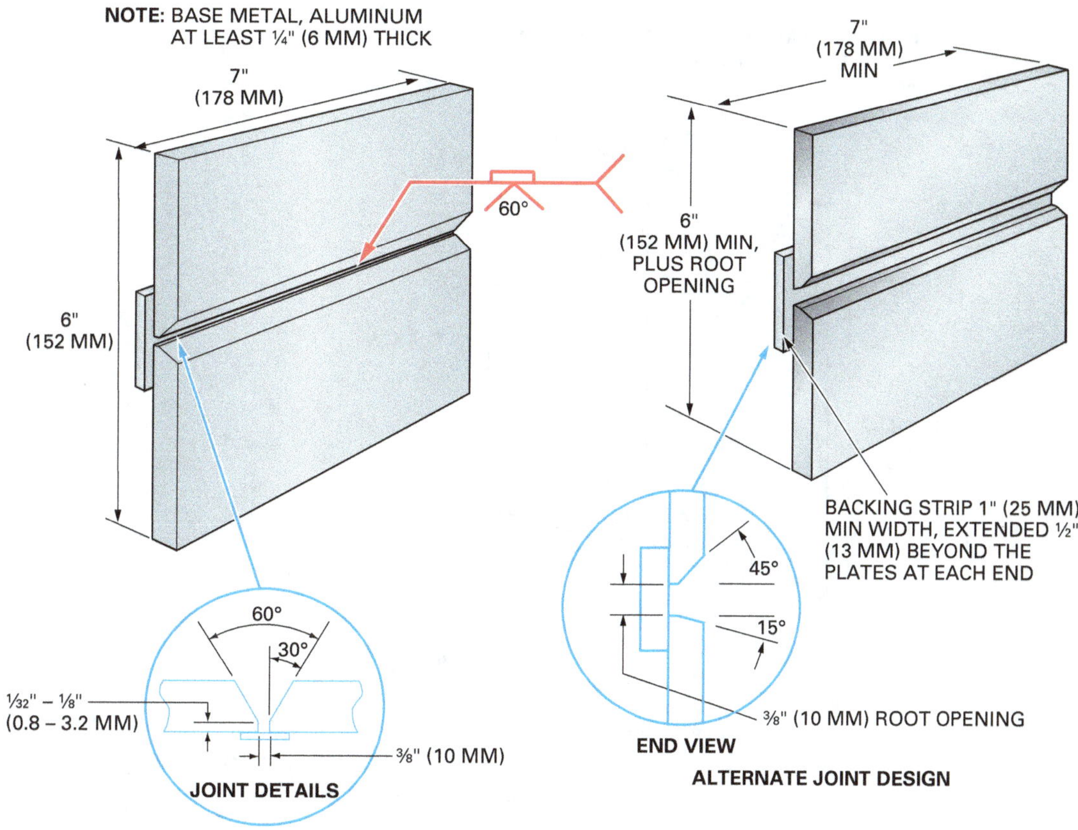

Criteria for Acceptance:

- Uniform rippled appearance on the bead face ____________
- Craters and restarts filled to the full cross section of the weld ____________
- Acceptable weld profile in accordance with the applicable code or standard ____________
- Smooth flat transition with complete fusion at the toes of the weld ____________
- No porosity ____________
- No undercut ____________
- No overlap ____________
- No inclusions ____________
- No cracks ____________

MAKE MULTIPLE-PASS V-GROOVE WELDS ON ALUMINUM PLATE WITH BACKING IN THE VERTICAL (3G) POSITION

As directed by the instructor, use the GMAW process with the appropriate aluminum wire to make a multiple-pass groove weld on aluminum plate with a 1" (25 mm) backing, as shown.

Note: Run the root vertical up.

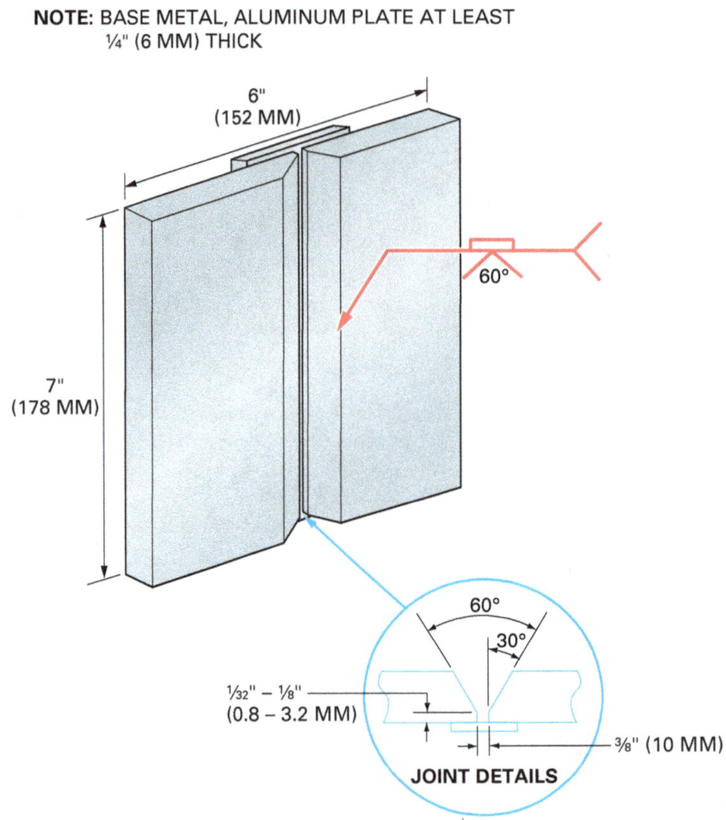

Criteria for Acceptance:

- Uniform rippled appearance on the bead face ________
- Craters and restarts filled to the full cross section of the weld ________
- Acceptable weld profile in accordance with the applicable code or standard ________
- Smooth flat transition with complete fusion at the toes of the weld ________
- No porosity ________
- No undercut ________
- No overlap ________
- No inclusions ________
- No cracks ________

MAKE MULTIPLE-PASS V-GROOVE WELDS ON ALUMINUM PLATE WITH BACKING IN THE OVERHEAD (4G) POSITION

As directed by the instructor, use the GMAW process with the appropriate aluminum wire to make a multiple-pass groove weld on aluminum plate with a 1" (25 mm) backing, as shown.

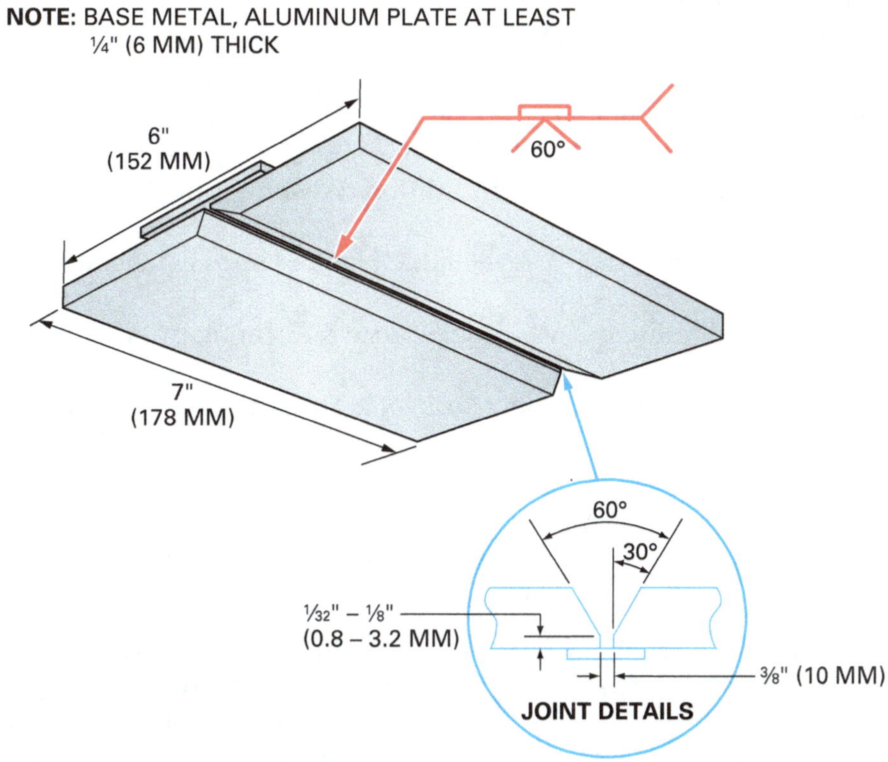

Criteria for Acceptance:

- Uniform rippled appearance on the bead face ______
- Craters and restarts filled to the full cross section of the weld ______
- Acceptable weld profile in accordance with the applicable code or standard ______
- Smooth flat transition with complete fusion at the toes of the weld ______
- No porosity ______
- No undercut ______
- No overlap ______
- No inclusions ______
- No cracks ______

Additional Resources

This module presents thorough resources for task training. The following reference material is recommended for further study.

Aluminum GMAW Welding Guide (PDF). The Lincoln Electric Company, Cleveland, OH, USA. **www.lincolnelectric.com**. Last accessed May 26, 2016.

Guide for Aluminum Welding. 2013. Troy, OH: Hobart Brothers Corporation.

AWS C5.6-89R, Recommended Practices for Gas Metal Arc Welding. Miami, FL: American Welding Society.

GMAW Welding Guide (PDF). The Lincoln Electric Company, Cleveland, OH, USA. **www.lincolnelectric.com**. Last accessed May 26, 2016.

OSHA 1926.351, Arc Welding and Cutting. Current Edition. Washington, DC: Occupational Safety and Health Administration (OSHA).

The Procedure Handbook of Arc Welding. 2000. Cleveland, OH: The James F. Lincoln Arc Welding Foundation.

Welding Handbook, Volume 1 and Volume 2. Welding Science & Technology. Miami, FL: American Welding Society.

Welding Aluminum: Theory and Practice. 2002. New York, NY: The Aluminum Association.

Figure Credits

The Lincoln Electric Company, Cleveland, OH, USA, Module Opener, Figures 2, 5, 6, 13

Microlabgallery.com, Figure 3

Courtesy of TWI Ltd. **www.twi-global.com**, Figure 4

© iStockphoto.com/loonger, SA01

© iStockphoto.com/IPGGutenbergUKLtd, SA02

John Tillman Company, SA03

AWS A5.10/A5.10M: 2012 (ISO 18273:2004 MOD), Table A1, reproduced with permission of the American Welding Society (AWS), Miami, Florida, Table 4

Terry Lowe, Figure 23

Section Review Answer Key

Answer	Section Reference	Objective
Section One		
1. d	1.1.0	1a
2. b	1.2.1	1b
3. a	1.3.2	1c
Section Two		
1. b	2.1.1	2a
2. c	2.2.1	2b
3. d	2.3.0	2c
4. d	2.4.4	2d
5. c	2.5.0	2e
6. a	2.6.1	2f

NCCER CURRICULA — USER UPDATE

NCCER makes every effort to keep its textbooks up-to-date and free of technical errors. We appreciate your help in this process. If you find an error, a typographical mistake, or an inaccuracy in NCCER's curricula, please fill out this form (or a photocopy), or complete the online form at **www.nccer.org/olf**. Be sure to include the exact module ID number, page number, a detailed description, and your recommended correction. Your input will be brought to the attention of the Authoring Team. Thank you for your assistance.

Instructors – If you have an idea for improving this textbook, or have found that additional materials were necessary to teach this module effectively, please let us know so that we may present your suggestions to the Authoring Team.

NCCER Product Development and Revision

13614 Progress Blvd., Alachua, FL 32615

Email: curriculum@nccer.org
Online: www.nccer.org/olf

❑ Trainee Guide ❑ Lesson Plans ❑ Exam ❑ PowerPoints Other ______________

Craft / Level: Copyright Date:

Module ID Number / Title:

Section Number(s):

Description:

Recommended Correction:

Your Name:

Address:

Email: Phone:

GTAW – Aluminum Plate

Overview

The GTAW process is unique in that the welder must add the filler metal using the opposite hand. The process is well suited for welding aluminum, using the freehand or walking-the-cup technique and both stringer and weave beads. This module provides insight into welding aluminum using the GTAW process and an opportunity to hone the necessary skills.

Module 29402

Trainees with successful module completions may be eligible for credentialing through the NCCER Registry. To learn more, go to **www.nccer.org** or contact us at 1.888.622.3720. Our website has information on the latest product releases and training, as well as online versions of our *Cornerstone* magazine and Pearson's product catalog.

Your feedback is welcome. You may email your comments to **curriculum@nccer.org**, send general comments and inquiries to **info@nccer.org**, or fill in the User Update form at the back of this module.

This information is general in nature and intended for training purposes only. Actual performance of activities described in this manual requires compliance with all applicable operating, service, maintenance, and safety procedures under the direction of qualified personnel. References in this manual to patented or proprietary devices do not constitute a recommendation of their use.

29402 V5

29402
GTAW – Aluminum Plate

Objectives

When you have completed this module, you will be able to do the following:

1. Explain how to prepare for GTAW aluminum welding and describe specific fillet and groove welding techniques.
   a. Describe common GTAW welding techniques.
   b. Identify common welding safety practices.
   c. Explain how to prepare the area and materials for aluminum GTAW practice.
   d. Explain how to prepare GTAW equipment for welding.
   e. Describe how to accomplish terminations and restarts and create basic GTAW weld beads.
   f. Describe how to make aluminum fillet welds in various positions.
   g. Describe how to make aluminum plate V-groove welds in various positions.

Performance Tasks

Under the supervision of your instructor, you should be able to do the following:

1. Weld a pad on aluminum plate in the flat position using GTAW stringer and weave beads.
2. Make multiple-pass GTAW fillet welds on aluminum plate in the following positions:
   - 1F
   - 2F
   - 3F
   - 4F
3. Make multiple-pass GTAW V-groove welds with backing on aluminum plate in the following positions:
   - 1G
   - 2G
   - 3G
   - 4G

Industry Recognized Credentials

If you are training through an NCCER-accredited sponsor, you may be eligible for credentials from NCCER's Registry. The ID number for this module is 29402. Note that this module may have been used in other NCCER curricula and may apply to other level completions. Contact NCCER's Registry at 888.622.3720 or go to **www.nccer.org** for more information.

Contents

Figures and Tables

SECTION ONE

1.0.0 ALUMINUM PLATE GTAW

Objective

Explain how to prepare for GTAW aluminum welding and describe specific fillet and groove welding techniques.

a. Describe common GTAW welding techniques.
b. Identify common welding safety practices.
c. Explain how to prepare the area and materials for aluminum GTAW practice.
d. Explain how to prepare GTAW equipment for welding.
e. Describe how to accomplish terminations and restarts and create basic GTAW weld beads.
f. Describe how to make aluminum fillet welds in various positions.
g. Describe how to make aluminum plate V-groove welds in various positions.

Performance Tasks

1. Weld a pad on aluminum plate in the flat position using GTAW stringer and weave beads.
2. Make multiple-pass GTAW fillet welds on aluminum plate in the following positions:
   - 1F
   - 2F
   - 3F
   - 4F
3. Make multiple-pass GTAW V-groove welds with backing on aluminum plate in the following positions:
   - 1G
   - 2G
   - 3G
   - 4G

Gas tungsten arc welding (GTAW) is an arc-welding process used for making high-quality welds on aluminum. This method of welding has also been referred to as *tungsten inert gas welding* (TIG welding). The term *tungsten* refers to the tungsten electrode that delivers the arc. *Figure 1* shows the GTAW process and equipment.

This module reviews GTAW equipment setup and explains how to prepare aluminum plate for both fillet welds and V-groove welds in the common positions.

1.1.0 GTAW Techniques

GTAW weld bead characteristics and quality are affected by several factors, each of which is influenced by the way the welder handles the torch. These factors include the following:

- Torch travel speed and arc length
- Torch angles
- Torch and filler metal handling techniques
- Consistent feeding of the filler metal

1.1.1 Torch Travel Speed and Arc Length

Torch travel speed and arc length affect the GTAW weld puddle and the penetration depth of the weld. Slower travel speeds allow more heat to concentrate and form larger, more deeply penetrating puddles. Faster travel speeds prevent heat buildup and form smaller, shallower puddles. Arc length is the major control for bead width. As the torch is raised, the arc length increases, the voltage increases, and the bead width increases as a result.

1.1.2 Torch Angles

The two basic torch angles that must be controlled when performing GTAW are the work angle and the travel angle. The definition of these angles is the same as for other methods of welding.

The torch work angle (*Figure 2*) is an angle of 90 degrees or less between a line perpendicular to the workpiece surface at the point of electrode contact and a plane determined by the electrode axis and the weld axis. For a T-joint or corner

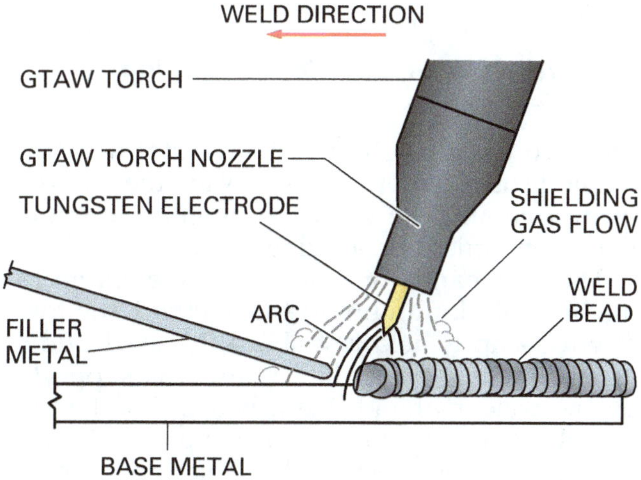

Figure 1 GTAW welding process.

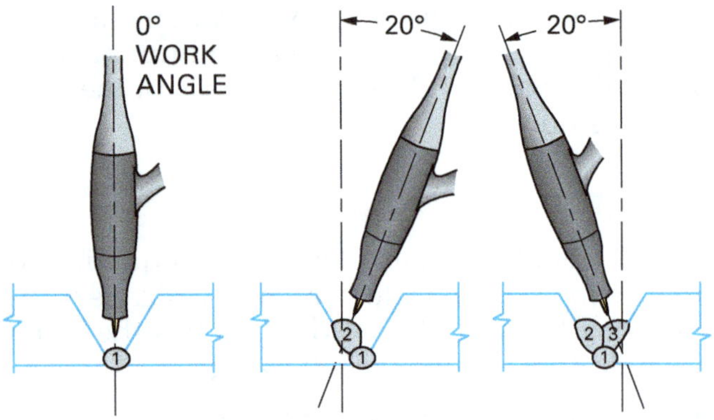

Figure 2 Typical torch work angles.

joint, the line is perpendicular to the non-butting member. As shown in *Figure 2*, when the torch is held perpendicular, it is referred to as a *0-degree work angle*. To properly position beads in a V-groove joint, the work angle should be changed to 20 degrees in either direction.

The torch travel angle is an angle of 90 degrees or less between the electrode axis and a line perpendicular to the weld axis at the point of electrode contact in a plane determined by the electrode axis and the weld axis. The travel angle used for GTAW is a push angle (*Figure 3*). A push angle is created when the torch is tilted back so that the electrode tip precedes the torch in the direction of the weld. In this position, the electrode tip and shielding gas are being directed ahead of the weld bead being formed. This provides a cleaning action during the positive portion of the AC voltage cycle. Push angles of 15 to 20 degrees are commonly used for aluminum GTAW.

CAUTION

Excessive push angles tend to cause air to be drawn into the weld from under the back edge of the torch nozzle. The air mixes with the shielding gas stream and contaminates the weld.

1.1.3 Torch and Filler Metal Handling Techniques

The two basic torch and filler metal handling methods used to perform GTAW are the freehand technique and the walking-the-cup technique. Both techniques are described here.

The freehand technique is the preferred method for aluminum. In this method, the torch electrode tip is held just above the weld puddle or base metal (*Figure 4*). The torch is supported by the welder's hand like a pencil. The hand is usually steadied by resting some part of it on or against the base metal. This helps to keep the proper arc length (distance from the electrode tip to the weld puddle). Due to the high thermal-transfer properties of aluminum, it may help to insert a piece of non-flammable insulation between the gloved hand and the workpiece. As needed, the welder can move the torch tip in a small circular motion within the molten puddle. This maintains the puddle size and advances the puddle. Filler metal is added as needed.

The GTAW filler metal rod is held in the hand that is not supporting the torch. Hold the rod at an angle of 15 to 20 degrees above the base metal surface and in line with the weld. When a 15- to 20-degree push angle is also used, the rod remains roughly perpendicular to the electrode axis. Always keep the tip of the rod within the shielding gas envelope to protect it from atmospheric contamination and to keep it preheated.

Insert the rod into the leading edge of the weld puddle, using extreme care not to touch the tungsten electrode to the end of the rod. If the tungsten electrode touches the rod or weld puddle, it will become contaminated with filler metal. The electrode must then be removed, the contaminated portion broken off, and the end reformed before welding can continue. To prevent contamination of the electrode by the filler metal, move the electrode to the back edge of the weld puddle as the rod is inserted into the weld puddle's leading edge. Timing those two actions helps prevent tungsten contamination.

In the walking-the-cup technique, the welder rests the edge of the torch nozzle (cup) against the base metal or groove edges to steady the torch and maintain a constant arc length. The torch is rocked from side to side on the edge of the cup as it is advanced. This action maintains the puddle size and heats both sides of the groove. The filler metal is added in the same manner used in the freehand technique, taking care not to contaminate the electrode.

The walking-the-cup technique can be very difficult to use when welding aluminum. Since the cup is in contact with surface, its diameter determines the distance of the tungsten from the workpiece. If the wrong cup size is chosen, the welder must adjust the travel angle to achieve the desired distance, possibly to an angle that causes a poor result. To achieve success with this method, it is important to choose the correct cup size. The cup also tends to stick to the freshly welded metal, creating an added challenge.

1.1.4 Consistent Feeding of the Filler Metal

Feeding the filler metal consistently as the weld progresses is important to ensure the weld puddle develops evenly and produces an even bead. Like all welding techniques, the technique described here requires practice to master but is well worth the effort.

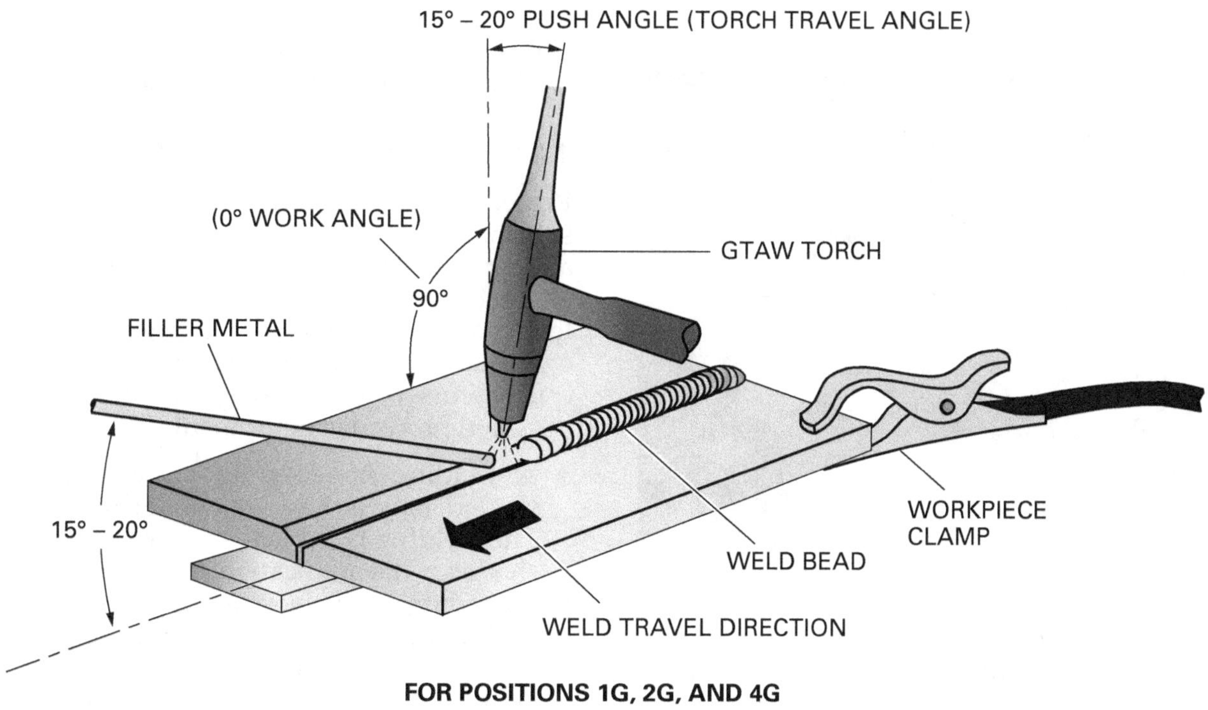

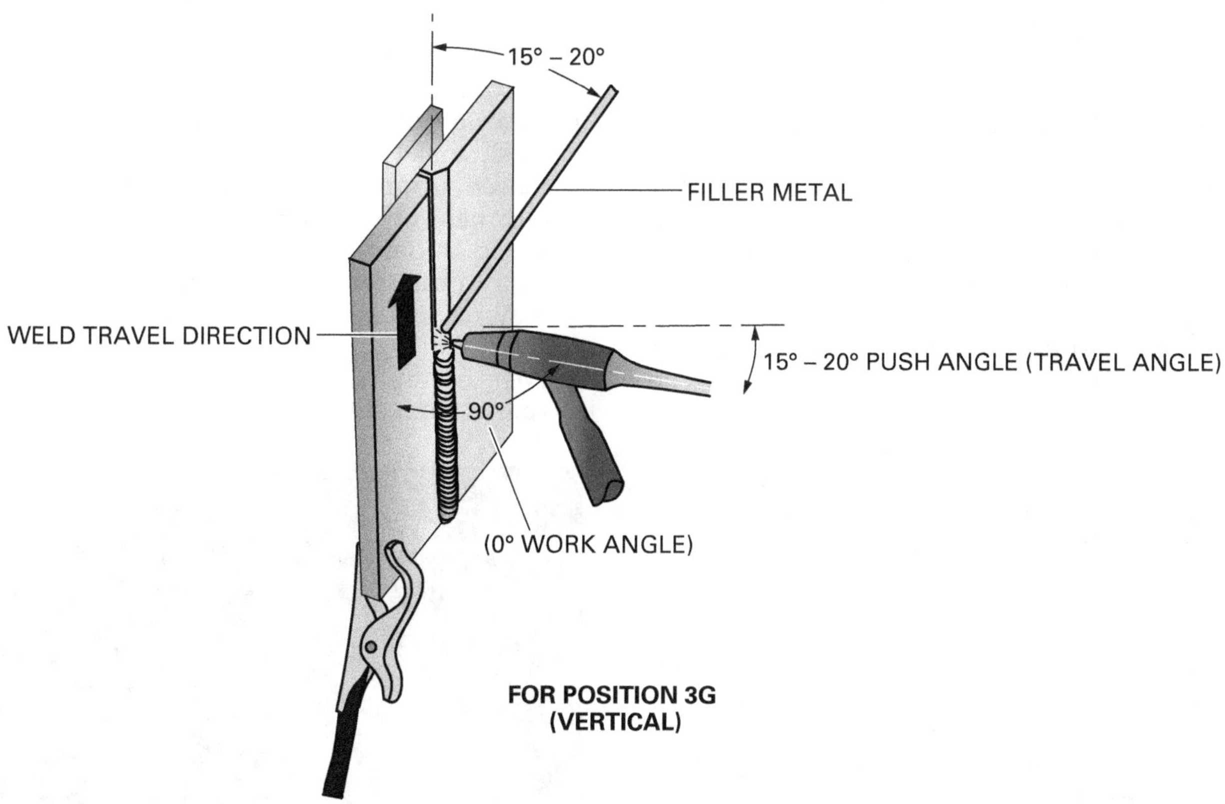

Figure 3 Torch travel angles.

Figure 4 Freehand technique.

Cradle the rod between the pointer finger and the middle finger, similar to the way you might hold a pencil. Then use the thumb to consistently push the rod forward in small motions. Once the thumb is extended pushing the wire forward, quickly bend and reposition the thumb to begin a new forward push. Consistent motion is important, so the thumb must be repositioned quickly to maintain the pace. The thumb can also be used to slowly roll the rod as well during the pushing motion. This technique is easily practiced without actively welding; simply practice with a rod at home or during free time to become comfortable with the technique before trying it on a weld.

Another technique to consider is referred to as the *finger-pull technique*. The rod is secured between the pointer and middle fingers, which are used to pull the electrode forward. At the end of the forward stroke, the fingers are slipped backwards along the rod to reposition, without pulling it backwards and out of the weld puddle in the process. Once the two fingers are repositioned, the pulling motion begins again.

1.2.0 Safety Practices

Before proceeding with welding coupon preparation and practice, it is important to review common safety practices associated with the craft. Safety considerations are a critical part of preparing for any welding operation.

The following is a summary of safety procedures and practices that must be observed while cutting or welding. Be advised that this is only a summary. Complete safety coverage is provided in the *Welding Level One* module, "Welding Safety." That module should be studied, if not already completed, before continuing. Above all, be sure to wear appropriate protective clothing and equipment at all times when welding or cutting.

1.2.1 Protective Clothing and Equipment

Protective clothing and equipment are a necessary part of every welding task. Be sure to wear appropriate protective clothing and equipment at all times when welding or cutting (*Figure 5*). Follow the guidelines below as a general rule, and ensure that you are familiar with any unique requirements your employer may have regarding personal protective equipment (PPE):

- A welding hood equipped with a properly shaded lens is best for all forms of welding. The welding hood lens must be an appropriate shade for the application.
- Wear proper leather or flame-retardant clothing along with welding gloves for protection from flying sparks, molten metal, and heat.
- Wear high-top safety shoes or boots. Make sure that the tongue and lace area of the footwear is covered by a pant leg. If the tongue and lace area is exposed or if the footwear must be protected from burn marks, wear spats under the pant leg or leather chaps over the front and top of the footwear.
- Wear a 100-percent cotton cap without any mesh material. If a hard hat is required for the environment, use one that allows the attachment of rear deflector material and a face shield. A hard hat with a rear deflector is generally preferred when working overhead, and may be required by some employers and job sites.
- Wear earmuffs, or at least earplugs, to prevent sparks from entering the ear canals.

Figure 5 GTAW welder with the proper PPE.

1.2.2 Housekeeping and Fall Protection

Welding is often performed outside in the weather and in the construction environment. Even production work completed inside large, open shops exposes a welder to equipment moving across the floor and overhead. Welders often have to crawl under or reach over objects to complete their work.

At times, welders must work at height. Fall protection equipment is required for all work at height (*Figure 6*). Always keep fall protection devices untangled and properly attached. Combined with the other equipment a welder must carry, there is a great deal to manage, but fall protection equipment must always be the priority. Fall protection equipment saves lives. Never attach any kind of welding gear to the harness that could interfere with its ability to perform its intended function.

Regardless of where you are working, always keep the work area as neat as possible. This makes the area safer and helps to reduce distractions. Mindful actions, such as making sure that welding cables are not tripping hazards, contribute to a safer environment. Electrode stubs and unused electrodes must not be left on the floor. They can cause someone to lose their footing and fall. Parts waiting to be welded or left over from previous work are a tripping hazard and should be kept out of the walking area. If the parts have sharp edges, they could cut welding cable insulation or torch gas hoses.

Figure 6 Working at height.

1.2.3 Fire and Explosion Prevention

Flammable gases are commonly found around welding and cutting operations, as are sparks and other significant sources of ignition. For that reason, and because many fire-related incidents over the years have been traced back to welding operations, it is crucial that fire and explosion prevention be given consistent attention. Follow these guidelines to maintain a safe environment:

- Never carry matches or gas-filled lighters. Sparks can cause the matches to ignite or the lighter to explode in your pocket, causing serious injury.

Lightweight Welding Gloves

Soft, flexible, leather gloves are used for light-duty work and for operations involving GMAW, GTAW, brazing, and soldering where more flexibility and dexterity is required. These welding processes produce less spatter than others. More dexterity is required for the welding techniques and for operating torch controls. A wide variety of styles, using various materials, is available.

- Never use oxygen to blow dust or dirt from clothing. Traces of the oxygen can remain trapped in the fabric for some time. If a spark hits the fabric during this time, the clothing can burn rapidly and violently out of control.
- Make sure that any flammable material in the work area is removed or shielded by a fire-resistant covering. Approved fire extinguishers must be available before attempting any heating, welding, or cutting operations. If a hot-work permit and a fire watch are required, be sure these are in place before beginning.
- Ensure that all site safety requirements are being observed.
- Never release a large amount of oxygen or use oxygen in place of compressed air. It can cause rapid and uncontrolled combustion around flammable materials or sparks. Keep oxygen away from oil, grease, and other petroleum products.
- Never release a large amount of fuel gas (especially acetylene). Because propane is heavier than air, it tends to concentrate in and along low areas, and can ignite at a considerable distance from the release point. Acetylene is lighter than air, but is even more dangerous than propane. When mixed with air or oxygen, acetylene will ignite at much lower concentrations than any other fuel gas.
- Although shielding gases are not toxic, any gas released in a large quantity can displace the oxygen in a small room or enclosed work area, creating an atmosphere with little or no oxygen. Avoid large releases of any type of gas, regardless of its characteristics.
- To prevent fires, maintain a neat and clean work area, and make sure that any metal scrap or slag is cold before disposal.
- Before cutting containers such as tanks or barrels, check to see if they have contained any explosive, hazardous, or flammable materials, including petroleum products, citrus products, or chemicals that decompose into toxic fumes when heated. As a standard practice, always clean and then fill any tanks or barrels with water, or purge them with a flow of inert gas such as nitrogen to displace any oxygen. Containers must be cleaned by steam cleaning, flushing with water, or washing with detergent until all traces of the material have been removed. Proper procedures for cutting or welding hazardous containers are described in the *American Welding Society (AWS) F4.1, Safe Practices for the Preparation of Containers and Piping for Welding and Cutting.*

WARNING!

Welding or cutting must never be performed on drums, barrels, tanks, vessels, or other containers until they have been emptied and cleaned thoroughly to eliminate all flammable materials and all substances (such as detergents, solvents, greases, tars, or acids) that might produce flammable, toxic, or explosive vapors when heated. Clean containers only in well-ventilated areas. Vapors can accumulate during cleaning, causing explosions and serious personal injury.

1.2.4 Work Area Ventilation

Many gases are introduced to the atmosphere through welding processes. Some of these gases are harmful, and all can displace or combine with atmospheric oxygen in a confined space. Follow these guidelines, as well as those of your employer, for work area ventilation:

- Make sure to follow all confined space procedures before conducting any welding or cutting in a confined space. Use powered extraction systems as needed.
- Never use oxygen to provide ventilation in confined spaces.
- Always perform cutting or welding operations in a well-ventilated area.

Cutting or welding operations involving zinc or cadmium materials or coatings result in toxic fumes. Another significant concern is Chromium 6, commonly referred to as *hexavalent chromium*, which is used in stainless steel. It releases toxic fumes when heated. Manganese is another hazardous metal used in stainless steel. It is also used in some welding rods, increasing the risk of exposure.

For long-term cutting or welding of materials containing these components, always wear an approved full-face, supplied-air respirator that uses breathing air supplied from outside the work area. For occasional short-term exposure, a HEPA-rated or metal-fume filter may be used on a standard respirator.

Powered fume-extraction equipment must be used when welding materials that contain zinc, cadmium, hexavalent chromium, manganese, and similar materials that produce hazardous fumes if the area is not naturally well-ventilated. *Figure 7* shows a powered air-purification respirator used with fume extraction equipment.

Figure 7 Powered air-purification respirator (PAPR) and localized fume extraction equipment.

1.3.0 Preparations for Welding

Before welding practice can take place, the work area and welding equipment prepared, and the coupons to be welded must be cut and cleaned.

The following sections explain how to prepare for GTAW of aluminum plate.

To practice welding, you will need a welding table, bench, or stand. The preferred welding surface is aluminum. Provisions must be available for mounting practice welding coupons out of position.

To set up the area for welding, follow these steps:

Step 1 Make sure that the area is properly ventilated. Use doors, windows, fans and localized fume extraction equipment as needed.

Step 2 Check the area for fire hazards. Remove any flammable materials before proceeding.

Step 3 Locate the nearest fire extinguisher. Do not proceed unless the extinguisher is charged and you know how to use it.

Step 4 Set up welding screens around the welding area.

1.3.1 Selecting Aluminum Filler Metals

The aluminum filler metal typically used for the GTAW process is an uncoated, solid metal filler rod. The filler metal used in aluminum GTAW is chosen based on its metallurgical compatibility with the base metal being welded. Filler metals are available for almost all types of aluminum.

Unlike steel, the aluminum filler metal will not develop a tensile strength equal to that of the base metal. As a result, aluminum weld joints are the weakest link in a welded assembly. This important fact is another reason why proper filler metal selection for aluminum is critical.

Aluminum filler metals are typically sold in cut lengths, most often 36" (91 cm) lengths. Some are available in 18" (46 cm) lengths, but rods can be cut to a shorter length if necessary. Common diameters are 1⁄16", 3⁄32", 1⁄8", 5⁄32", 3⁄16", and 1⁄4" (1.6, 2.4, 3.2, 4.0, 4.8, and 6.4 mm, respectively). The rods are commonly packaged in 5-, 10-, 25-, or 50-pound (2.3, 4.5, 9.1, 13.6, and 22.7 kg, respectively) packages.

All filler rods made in accordance with AWS specifications must have the AWS classification number printed on the shipping container. In addition, filler rod manufacturers frequently print their own unique classification name or number on the shipping container. On some jobs, welders must keep proof of the filler rods used on certain welds. One way to do this is to cut the filler rods in half and burn the rods from their cut ends. The rod ends showing the manufacturers' ID numbers can then be retained for proof that a welding procedure specification (WPS) or other specification has been followed.

Because there are so many types of filler metals available, the welder must be able to distinguish between them and select the correct filler metal for the job. There are two basic factors to consider when selecting aluminum filler metals: the WPS and the base metal composition.

If there is a WPS or site quality standard, the filler rod type will be specified. It will be given as an AWS classification and/or manufacturer's standard. It will sometimes include a manufacturer's name. When a WPS for the weld exists, the filler metal that is specified must be used.

CAUTION

Filler metal specifications in the WPS or site quality standards must be followed. Failure to use the specified filler metal will result in a weld quality issue and the potential for rejection.

GTAW in a Bubble

Some metals, such as titanium, molybdenum, and aluminum are more reactive to atmospheric contaminants than others. Problems in the dispersal of the purge gas from the torch, regardless of the reason, represent a larger problem for these metals. The best approach is to complete the weld with the entire workpiece located in an atmosphere of inert gas, preventing any possible contamination from the surrounding air. The CK purge chamber shown here provides that environment while also offering portability. The clear, collapsible tent is first evacuated (drawn into a vacuum) to remove the air, and then is filled with a purge gas, such as argon, from a perforated hose surrounding the base of the tent. The welder uses the openings to reach in and get the job done. Although rigid welding chambers are available, they do not offer the portability of a collapsible product. Complete kits are available as shown, which include a vacuum pump, a flowmeter/regulator, two collapsible tents, and other needed hardware.

Aluminum filler rods for GTAW are usually selected for their compatibility with the chemical composition of the base metal. There are many different types of aluminum base metals and filler metals. AWS Standard A5.10/A5.10M: 2012 (ISO 18273:2004 MOD), *Welding Consumables—Wire Electrodes, Wires and Rods for Welding of Aluminum and Aluminum-Alloys—Classification* provides a guide for the selection of bare aluminum filler metals for specific aluminum base metals and alloys. *Table 1A* and *Table 1B* have been taken directly from the standard. Charts and tables provided by filler metal manufacturers also assist in selecting the correct filler metal from their product line.

In many cases, only a few filler alloys are used to weld the majority of aluminum alloys. Filler alloys, such as ER5356 and ER4043, can be used for most aluminum welding applications. Always check the WPS or job site standards to determine the correct filler metal for the job.

High-quality aluminum welds cannot be achieved unless the filler metal used is clean and high-quality. Contaminated filler metal transfers the contamination to the weld puddle. Filler metal must be used as soon as possible after removal from the shipping or storage packaging. When this is not possible, the filler metal must be stored in airtight storage containers to prevent further oxidation of the rod. Remember that heating aluminum and other metals accelerates the oxidation process dramatically. Once the weld is in progress, the end of the rod becomes hot, resulting in contamination of the rod. For this reason, the tip of the rod should be kept in the shielding gas envelope of the weld to protect it.

For all welding exercises associated with this module, use 3⁄32", 1⁄8", 5⁄32", or 3⁄16" (2.4, 3.2, 4.0, or 4.8 mm) rods, per the aluminum filler material classifications shown in *Table 1A* and *Table 1B* (or their equivalents). Remove only a small number of filler rods at a time. Store the rest in the package to keep them clean. Before using a filler metal, inspect it for any contamination, such as oxidation, dirt, oil, grease, or burned ends. Contamination will cause defects in the weld. Use emery cloth or stainless-steel wool to remove oxides. Use acetone and a clean, oil-free rag to thoroughly clean the filler metal. Snip off any previously burned ends after the filler metal has been cleaned. If the filler metal cannot be cleaned, do not use it. Note that

Table 1A AWS Filler Metal Guide for General-Purpose Welding of Aluminum (Part 1)

Table A.1
Guide to the Choice of Filler Metal for General Purpose Welding

Base Metal	201.0 206.0 224.0	319.0, 333.0, 354.0, 355.0, C355.0	356.0, A356.0 357.0, A357.0 413.0, 443.0, A444.0	511.0, 512.0, 513.0, 514.0, 535.0	7004, 7005, 7039, 710.0, 712.0	6009 6010 6070	6005, 6061, 6063, 6101, 6151, 6201, 6351, 6951	5456	5454
1060, 1070, 1080, 1350	ER4145	ER4145	ER4043[a,b]	ER5356[c,d]	ER5356[c,d]	ER4043[a,b]	ER4043[b]	ER5356[d]	ER4043[b,d]
1100, 3003, Al[c], 3003	ER4145	ER4145	ER4043[a,b]	ER5356[c,d]	ER5356[c,d]	ER4043[a,b]	ER4043[b]	ER5356[d]	ER4043[b,d]
2014, 2036	ER4145[e]	ER4145[e]	ER4145	—	—	ER4145	ER4145	—	—
2219	ER2319[a]	ER4145[e]	ER4145[b,c]	ER4043	ER4043	ER4043[a,b]	ER4043[a,b]	—	ER4043[b]
3004, Al[c], 3004	—	ER4043[b]	ER4043[b]	ER5356[f]	ER5356[f]	ER4043[b]	ER4043[b,f]	ER5356[d]	ER5356[f]
5005, 5050	—	ER4043[b]	ER4043[b]	ER5356[f]	ER5356[f]	ER4043[b]	ER4043[b,f]	ER5356[d]	ER5356[f]
5052, 5652[i]	—	ER4043[b]	ER4043[f]	ER5356[f]	ER5356[f]	ER4043[b]	ER5356[c,f]	ER5356[f]	ER5356[f]
5083	—	—	ER5356[c,d]	ER5356[d]	ER5356[d]	—	ER5356[d]	ER5356[d]	ER5356[d]
5086	—	—	ER5356[c,d]	ER5356[d]	ER5356[d]	—	ER5356[d]	ER5356[d]	ER5356[d]
5154, 5254[i]	—	—	ER4043[f]	ER5356[f]	ER5356[f]	—	ER5356[f]	ER5356[f]	ER5356[f]
5454	—	ER4043[b]	ER4043[f]	ER4043[f]	ER4043[f]	ER4043[b]	ER5356[c,f]	ER5356[f]	ER5554[c,f]
5456	—	—	ER5356[c,d]	ER5356[d]	ER5356[d]	—	ER5356[d]	ER5356[d]	
6005, 6061, 6063 6101, 6151, 6201 6351, 6951	ER4145	ER4145[b,c]	ER4043[b,f,g]	ER5356[f]	ER5356[c,f]	ER4043[a,b,g]	ER4043b[a,f,g]		
6009, 6010, 6070	ER4145	ER4145[b,c]	ER4043[a,b,g]	ER4043	ER4043	ER4043[a,b,g]			
7004, 7005, 7039 710.0, 712.0	—	ER4043[b]	ER4043[b,f]	ER5356[f]	ER5356[f]				
511.0, 512.0, 513.0 514.0, 535.0	—	—	ER4043[f]	ER5356[f]					
356.0, A356.0, 357.0 A357.0, 413.0 443.0, A444.0	ER4145	ER4145[b,c]	ER4043[b,h]						
319.0, 333.0 354.0, 355.0 C355.0	ER4145[c]	ER4145[b,c,h]							
201.0, 206.0, 224.0	ER2319[a,h]								

Note: See *Table 1B* for referenced footnotes.

the filler metal and base metal should both be cleaned using the same procedures and cleaning products.

1.3.2 Preparing Aluminum Coupons

Because aluminum is easily contaminated, proper coupon preparation is a priority. Aluminum plate coupons must be cut to the desired lengths, cleaned, and beveled to the proper angles before they are positioned for welding. An aluminum backing must also be cut and cleaned. The basic shapes can be cut by shearing or sawing.

The importance of cleaning aluminum before welding cannot be overstated. There are two related facts about aluminum that help explain why. First, aluminum does not give the welder a visual indication that it is nearing its melting point through color change, as many other common metals do. With no indication that the melting point is approaching, it is easy to apply too much heat. Second, aluminum oxide—the "skin" that forms on the outside—has a melting point roughly three times higher than the aluminum itself. Excessive weld contamination and poor weld quality result when the aluminum base metal is not thoroughly cleaned.

The preferred technique for beveling and/or adding a root face is mechanical machining. Grinding is also acceptable, although machining the grooves and root faces of aluminum joints is the cleanest and most uniform way to prepare aluminum for GTAW. Thermal cutting techniques are not considered acceptable. If thermal cutting techniques must be used to cut aluminum alloys, at least ⅛" (3 mm) of the cut surface must be removed by machining or grinding before welding.

WARNING!

Grinding wheels used on aluminum must be made for aluminum only. They should not be used on any other type of metal or on surfaces contaminated with grease or oil. Using other types of grinding wheels on aluminum may cause the grinding wheels to shatter.

Table 1B AWS Filler Metal Guide for General-Purpose Welding of Aluminum (Part 2)

Table A.1 (Continued)
Guide to the Choice of Filler Metal for General Purpose Welding

Base Metal	5154 5254[i]	5086	5083	5052 5652[i]	5005 5050	3004 Alc. 3004	2219	2014 2036	1100 3003 Alc. 3003	1060 1070 1080 1350
1060, 1070, 1080, 1350	ER5356[c,d]	ER5356[d]	ER5356[d]	ER4043[b,d]	ER1100[b,c]	ER4043[b,d]	ER4145[b,c]	ER4145	ER1100[b,c]	ER1188[b,c,h,j]
1100, 3003, Al[c], 3003	ER5356[c,d]	ER5356[d]	ER5356[d]	ER4043[b,d]	ER1100[b,c]	ER4043[b,d]	ER4145[b,c]	ER4145	ER1100[b,c]	
2014, 2036	—	—	—	—	ER4145	ER4145	ER4145[e]	ER4145[e]		
2219	ER4043	—	—	ER4043[b]	ER4043[a,b]	ER4043[a,b]	ER2319[a]			
3004, Al[c], 3004	ER5356[f]	ER5356[d]	ER5356[d]	ER5356[c,f]	ER5356[c,f]	ER5356[c,f]				
5005, 5050	ER5356[f]	ER5356[d]	ER5356[d]	ER5356[c,d]	ER5356[c,f]					
5052, 5652[i]	ER5356[f]	ER5356[d]	ER5356[d]	ER5654[c,f,i]						
5083	ER5356[d]	ER5356[d]	ER5183[d]							
5086	ER5356[d]	ER5356[d]								
5154, 5254[i]	ER5356[f,i]									

Note: ISO Classifications different from AWS Classifications have not been added to this table.

a *ER4145 may be used for some applications.*
b *ER4047 may be used for some applications.*
c *ER4043 may be used for some applications.*
d *ER5183, ER5356, or ER5556 may be used.*
e *ER2319 may be used for some applications. It can supply high strength when the weldment is postweld solution heat-treated and aged.*
f *ER5183, ER5356, ER5554, ER5556, and ER5654 may be used. In some cases, they provide: (1) improved color match after anodizing treatment, (2) highest weld ductility, and (3) higher weld strength. ER5554 is suitable for sustained elevated temperature service.*
g *ER4643 and ER4943 will provide higher strength than ER4043 in 1/2 in [12 mm] and thicker groove welds in 6XXX base alloys when postweld solution heat-treated and aged. They will also provide higher strength fillet welds than ER4043 in the as welded, postweld aged, or postweld heat-treated and aged conditions.*
h *Filler metal with the same analysis as the base metal is sometimes used. The following wrought filler metals possess the same chemical composition limits as cast filler alloys: ER4009 and R4009 as R-C355.0; ER4010 and R4010 as R-A356.0; and R4011 as R-A357.0.*
i *Base metal alloys 5254 and 5652 are used for hydrogen peroxide service. ER5654 filler metal is used for welding both alloys for service temperatures below 150°F [66°C].*
j *ER1100 may be used for some applications.*

Notes:

1. *ER4047, ER4643, or ER4943 may be used in some applications when alloy ER4043 is specified.*
2. *Service conditions such as immersion in fresh or salt water, exposure to specific chemicals, or a sustained high temperature (over 150°F [66°C]) may limit the choice of filler metals. Filler metals ER5183, ER5356, and ER5556 are not recommended for sustained elevated temperature service.*
3. *Recommendations in this table apply to gas shielded arc welding processes. For oxyfuel gas welding, only ER1188, ER1100, ER4043, ER4047, and ER4145 filler metals are ordinarily used.*
4. *Where no filler metal is listed, the base metal combination is not recommended for welding.*

CAUTION

Grinding wheels, files, and similar tools used on aluminum should not be used on any other type of metal or on surfaces contaminated with grease or oil.

Practice coupons are best cut from aluminum plate that is ¼" thick (6 mm plate). If this size is not readily available, aluminum plate up to ⅜" thick (10 mm plate) can be used. Material up to ½" thick (12 mm plate) can be used with the joint design described herein, if welding equipment of sufficient current capacity is available. Material thicknesses greater than ½" (>12 mm plate) can be used, but a different joint design is required, and welding equipment of extreme current capacity is required when argon gas is used alone. Although it is expensive, helium can be added to the mixture to significantly reduce the welding current required.

NOTE

For welding thin aluminum materials (less than ¼" or 6 mm), there are a number of other acceptable joint designs from which to choose that do not require a backing. The joint design used in this module, however, is better suited for practice on the ¼" and thicker material specified above.

Clean oil and grease from the coupons and the metal backing with an approved chemical such as acetone. After oil and grease have been removed, use a stainless-steel wire brush to remove all other contaminants and oxidation. Use only light pressure on the brush.

WARNING!

To prevent injury when using cleaning solvents, always comply with the safety guidelines found on the SDS for the solvents being used.

Manufacturers' Data for Filler Metals

Most filler metal manufacturers publish paperback welding guides that include information about their filler materials. These booklets can be obtained through your distributor for free or for a small charge. An increasing number of manufacturers also offer apps for filler metal selection that can be used on smartphones, tablets, and similar devices.

The coupons must be shaped to allow the following welds:

- *Stringer beads, weave beads,* and *overlapping beads* – The plate coupons can be any size or shape that can be easily handled.
- *Fillet welds* – Cut the metal into 4" × 5" (102 × 127 mm) rectangles for the base and 3" × 5" (76 × 127 mm) rectangles for the web (*Figure 8*).
- *V-groove welds with backing* – A properly prepared aluminum V-groove weld joint with backing is illustrated in *Figure 9*. Note that this joint design is not recommended for material thicker than ½" (12 mm plate). As shown in *Figure 10,* cut the metal into 3" × 5" (76 × 127 mm) rectangles with one (or both) of the 5" lengths beveled at 30 degrees. Grind a 1⁄32" to ⅛" (0.8 to 3.2 mm) root face on the bevel as directed by your instructor. Note that a thicker root face makes it more difficult to achieve proper penetration and consistent fusion. Cut backing strips 1" × 6" (25 × 152 mm) long from the same alloy as the beveled pieces (*Figure 11*). The backing strip should be sufficiently thick to prevent melt-through.

Observe the following steps to prepare each V-groove weld coupon, with backing, for welding.

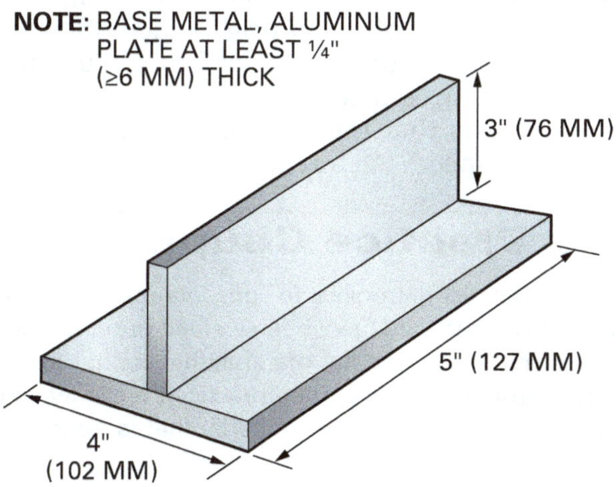

Figure 8 Fillet weld coupons.

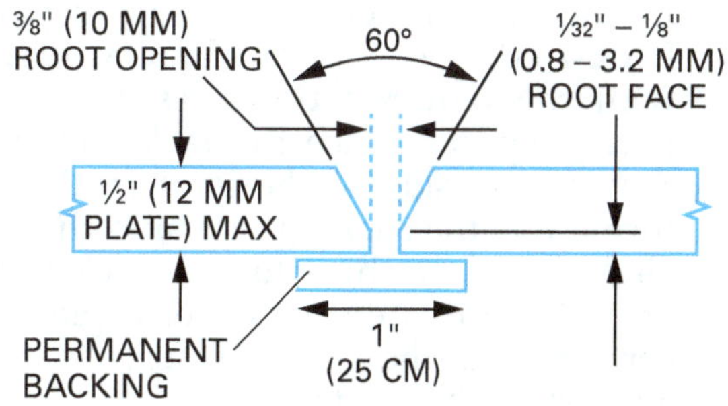

FIT TOLERANCES	
ROOT OPENING	+¼"; –1⁄16" (+6.4 MM; –1.6 MM)
GROOVE ANGLE	+10 DEGREES; –5 DEGREES

Figure 9 GTAW aluminum weld-joint preparation.

Note that coupons and backing strips should be welded within 30 minutes of their preparation to prevent the excessive formation of aluminum oxide:

Step 1 Cut and bevel the coupon components (if required) as shown using any of the acceptable methods discussed in this section. If the bevel has been cut thermally, remember to machine-cut the bevel back roughly ⅛" (3 mm) before machining the root face to its proper dimension.

Step 2 Make sure that the surface of the backing strip under the root opening has been cleaned of all oxide coating in the same manner as the coupon. Center the beveled strips on the backing strip with a ⅜" (9.5 mm) root opening as shown in

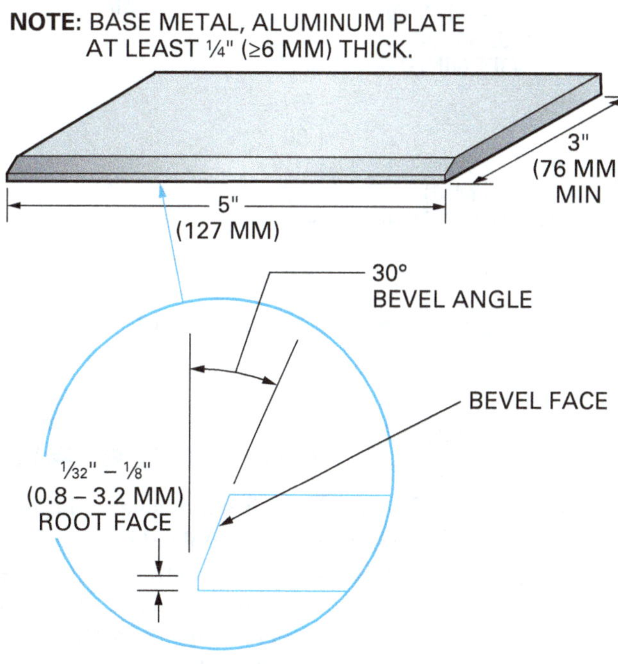

Figure 10 Preparing the V-groove weld coupon.

Figure 11, and tack weld them in place. If desired, use a piece of metal or rod with the required thickness as a temporary spacer. Place the tack welds on the reverse side of the joint in the lap formed by the backing strip and the beveled plate. Use three to four tack welds of ½" (13 mm) on each beveled plate. Be certain that the backing strip is tight against the beveled plates.

When welding in the horizontal position, an alternate joint preparation (*Figure 12*) can be used. The alternate joint has one plate beveled at about 45 degrees and the other plate at about 15 degrees. As a result, the total groove angle remains at 60 degrees. The standard bevel angles are two 30-degree angles, which reflect the same groove angle of 60 degrees. The plates are positioned with the 45-degree beveled plate above the 15-degree plate. The other dimensions of the joint remain the same.

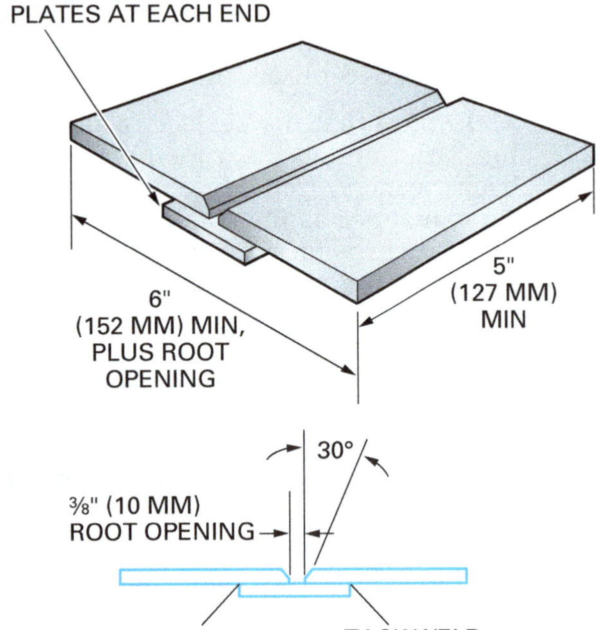

Figure 11 V-groove weld coupon with metal backing.

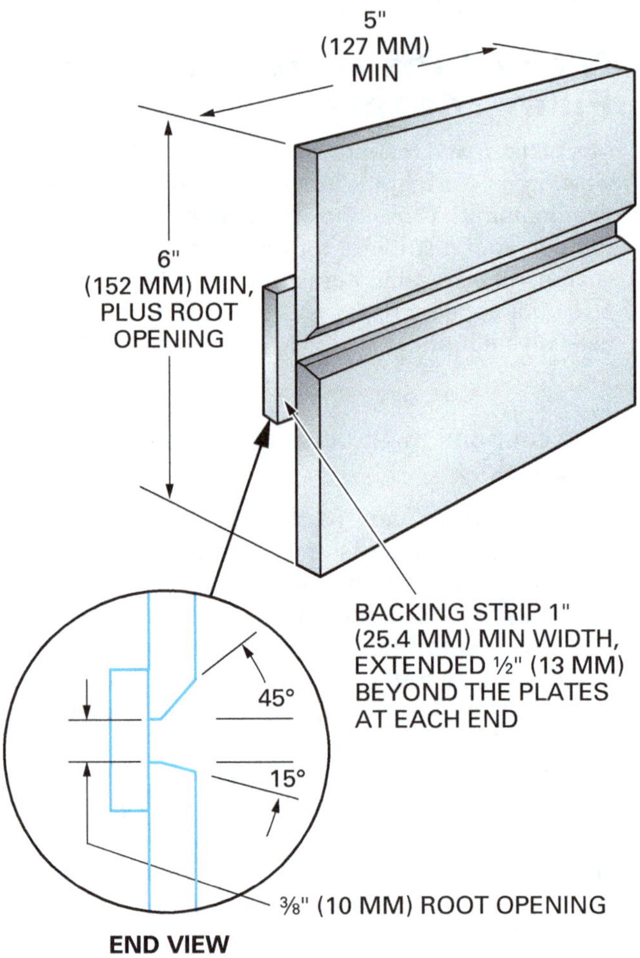

Figure 12 Alternate horizontal aluminum weld coupons.

NOTE

Check with your instructor about whether to use the standard V-groove preparation or the alternate preparation for horizontal weld coupons.

1.4.0 Welding Equipment

GTAW of aluminum requires a significant amount of welding current. Ensure that the equipment selected is capable of providing sufficient current at a reasonable duty cycle. Welding aluminum at welding-current levels less than those suggested means that the weld cannot progress at the proper speed. As a result, the workpiece will

GOING GREEN

Conserve Aluminum Practice Coupons

Aluminum is expensive and difficult to obtain for practice purposes. To conserve resources and reduce waste, completely use all weld coupons until all surfaces have been welded upon. Cutting coupons apart and reusing the pieces also conserves materials. Use flat material that cannot be cut into weld coupons to practice running beads. Note that stringer beads can be practiced while leaving enough room between them to practice weave beads as well. Any aluminum material that is no longer suitable for welding practice should be recycled.

overheat and create an inferior weld. Because of the higher currents associated with aluminum welding, water-cooled torches are recommended. Of course, this will not reduce the temperature of the workpiece itself.

Conventional AC machines used for GTAW are typically based on sine-wave or square-wave technology. Newer inverter-technology units have additional features, such as high-frequency start only and current sequencing (start/stop ramping) that help control puddle formation at the beginning of the weld and also help in filling the termination crater. These machines provide additional control over important variables that affect the quality of aluminum welds. The ability to vary both the AC wave balance and frequency is preferred for aluminum welding. Control over the balance between the electrode negative (EN) and electrode positive (EP) portions of the AC sine wave allow the operator to control two important functions:

- As the wave is unbalanced to the electrode-negative side, the current-carrying capacity of the electrode is increased and the amount of cleaning action is decreased.
- When welding on new material that is properly cleaned, the wave can be set at 80 percent EN or more. This allows the use of a smaller electrode at higher welding currents.

When welding aluminum with GTAW, a narrow band of the workpiece cleaned by the action of the arc should appear at the toes on each side of the weld bead. *Figure 13* shows examples of stringer beads applied with various EN settings. Note that the plate was not cleaned at all prior to the test to provide for a better view of the cleaning action. The lower bead reflects a 95-percent EN setting that provided insufficient cleaning. The center bead was applied at an 85-percent EN setting that provided sufficient cleaning. Finally, the upper bead was the result of a 60-percent EN setting that produced an excessive amount of cleaning action. Again, properly cleaned aluminum does not need to rely heavily on the arc for cleaning action, allowing for a higher EN setting that provides other advantages.

Conventional welding machines in the United States use 60-cycle electrical power as supplied by the line current. (Many other countries use 50 cycles.) Newer inverter machines have the ability to vary the frequency, potentially increasing it to 300 cycles per second or even more. This eliminates the need to use balled pure- or zirconiated-tungsten electrodes and permits the use of a truncated EWTh2 or other tungsten electrode. It also provides better arc control and allows for smaller weld sizes. Frequencies of 120 to 140 cycles are typically used to weld aluminum.

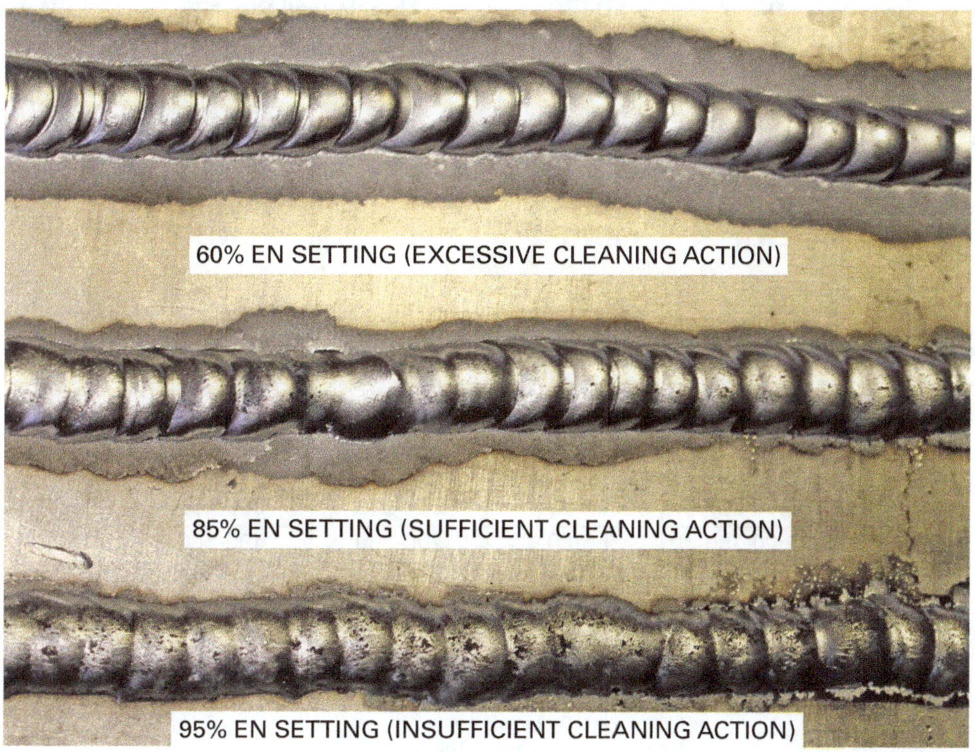

Figure 13 Cleaning action resulting from various EN settings.

Although pure tungsten (EWP) electrodes are used for welding aluminum with AC and high-frequency machines, zirconiated-tungsten electrodes (EWZr) are often preferred due to their higher current-carrying capacity and lower susceptibility to contamination. They also hold a balled tip better when welding aluminum. *Table 2* shows the current ranges for various electrode types and sizes.

Once an appropriate welding machine has been selected, follow these steps to prepare it:

Step 1 Verify that the welding machine can be used for GTAW. If a water-cooled torch will be used, select an appropriate cooling unit.

Step 2 Select an air-cooled or water-cooled GTAW torch. Make sure that it is compatible with the welding machine and any optional cooling unit. Water-cooled torches are preferred. Equip the torch with an appropriate cup size for the technique to be used (freehand or walking-the-cup).

Step 3 Know the location of the primary power disconnect switch.

Step 4 Configure the welding machine for aluminum GTAW welding (*Figure 14*) as directed by your instructor. Configure the torch polarity to AC. Equip the torch with a properly prepared 3/32" or 1/8" (2.4 or 3.2 mm) EWP or EWZr tungsten electrode.

Step 5 Connect the proper argon or argon/helium shielding gas for the application as directed by your instructor, filler metal manufacturer, WPS, or site quality standards.

Step 6 Connect the clamp of the workpiece lead to the workpiece.

Step 7 Turn on the welding machine. Adjust the flowmeter and purge the torch according to the manufacturer's instructions.

Step 8 Set the initial welding current for the application.

1.4.1 Selecting Shielding Gas

Shielding gas is selected according to the material being welded. Argon is the most commonly used shielding gas. Argon can be used for welding a wide range of materials, including aluminum, titanium, steel, and stainless steel.

Table 2 Typical Current Ranges for Tungsten Electrodes

Electrode Diameter		Direct Current Amperes		Alternating Current Amperes			
		DCEN	DCEP	Unbalanced Wave		Balanced Wave	
in.	mm	EWCe-2 EWLa-1 EWLa-1.5 EWLa-2 EWP EWTh-1 EWTh-2	EWCe-2 EWLa-1 EWLa-1.5 EWLa-2 EWP EWTh-1 EWTh-2	EWP	EWCe-2 EWLa-1 EWLa-1.5 EWLa-2 EWTh-1 EWTh-2 EWZr	EWP	EWCe-2 EWLa-1 EWLa-1.5 EWLa-2 EWTh-1 EWTh-2 EWZr
0.010	0.30	up to 15	(2)	up to 15	up to 15	up to 15	up to 15
0.020	0.50	5-20	(2)	5-15	5-20	10-20	5-20
0.040	1.00	15-80	(2)	10-60	15-80	20-30	20-60
1/16	1.60	70-150	10-20	50-100	70-150	30-80	60-120
3/32	2.40	150-250	15-30	100-160	140-235	60-130	100-180
1/8	3.20	250-400	25-40	150-210	225-325	100-180	160-250
5/32	4.00	400-500	40-55	200-275	300-400	160-240	200-320
3/16	4.80	500-750	55-80	250-350	400-500	190-300	290-390
1/4	6.40	750-1000	80-125	325-450	500-630	250-400	340-525

(1) All values are based on the use of argon as the shielding gas.Other current values may be employed depending on the shielding gas, type of equipment, and application.

(2) These combinations are not commonly used.

EWCe-2 = Tungsten + 2% Ceria
EWLa-1 = Tungsten + 1% Lanthana
EWLa-1.5 = Tungsten + 1.5% Lanthana
EWLa-2 = Tungsten + 2% Lanthana

EWP = Pure Tungsten
EWTh1 = Tungsten +1% Thoria
EWTh2 = Tungsten +2% Thoria
EWZr = Tungsten + Zirconia

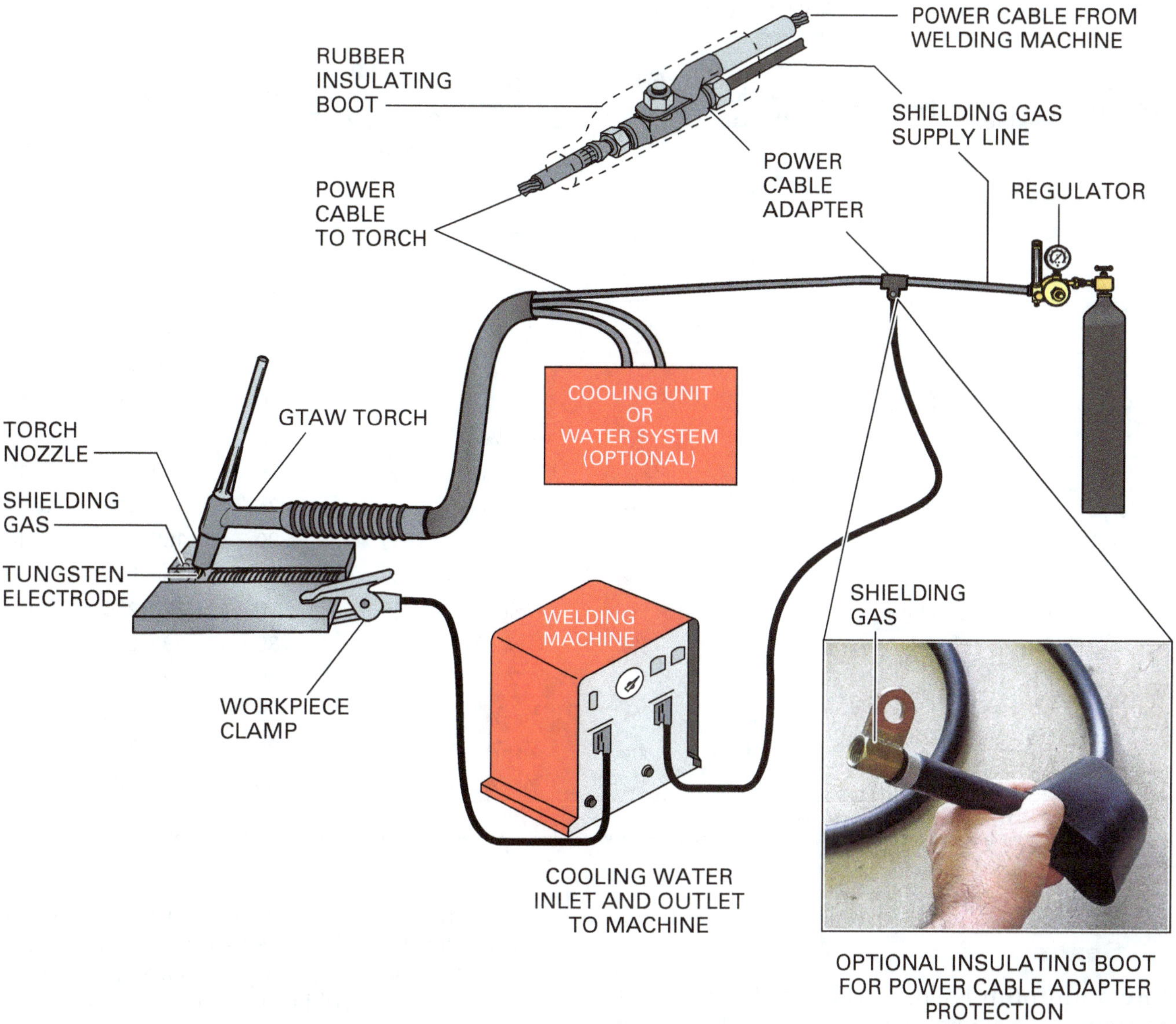

Figure 14 Configuration diagram of typical GTAW welding equipment.

Helium and helium/argon mixtures raise the temperature of the arc. This rise in temperature promotes faster welding speeds and deeper weld penetration. However, helium is expensive and can cause difficulty in starting the arc.

Impurities in the shielding gas or mixture with other gases, such as carbon dioxide (CO_2), will contaminate the weld and change the color of the aluminum (usually to black).

CAUTION

Always use the correct gas when performing GTAW. Accidental use of a gas mix intended for GMAW can cause rapid destruction of the tungsten and contamination of the weld.

1.4.2 GTAW Welding Parameters for Aluminum Plate

Different welding techniques and settings are used for different welding situations. The tables in this section provide the common parameters for aluminum welding in each respective position. The specifications for each weld are determined by a number of factors. *Table 3A* and *Table 3B* show the parameters for aluminum GTAW fillet and lap welds. *Table 3B* has been derived from *Table 3A*, with the units changed to the appropriate metric values. Note that the parameters for aluminum groove welds with backing are considered the same as the parameters for fillet welds. Thus a separate table for groove welds has not been developed.

Table 3A Parameters for Making GTAW Fillet and Lap Welds on Aluminum Plate

Aluminum Thickness (inch)	Weld Position[1]	Preheat[2] (F)	Weld Passes	Filler Diameter (inch)	Pure Tungsten Electrode Diameter (inch)	Gas Cup Inside Diameter (inch)	Argon Flow (cfh)	AC (amps)	Arc Travel Speed (ipm)
1⁄16	F, V, H	None	1	3⁄32	1⁄16–3⁄32	3⁄8	16	70–100	8–10
	O		1	3⁄32	1⁄16–3⁄32	3⁄8	20	65–90	8–10
3⁄32	F	None	1	3⁄32–1⁄8	1⁄8–5⁄32	3⁄8	18	110–145	8–10
	V, H		1	3⁄32	3⁄32–1⁄8	3⁄8	18	90–125	8–10
	O		1	3⁄32	3⁄32–1⁄8	3⁄8	20	110–135	8–10
1⁄8	F	None	1	1⁄8	1⁄8–5⁄32	7⁄16	20	135–175	10–12
	V, H		1	1⁄8	3⁄32–1⁄8	3⁄8	20	115–145	8–10
	O		1	1⁄8	3⁄32–1⁄8	7⁄16	25	125–155	8–10
3⁄16	F	None	1	5⁄32	5⁄32–3⁄16	1⁄2	25	190–245	8–10
	V, H		1	5⁄32	5⁄32–3⁄16	1⁄2	25	175–210	8–10
	O		1	5⁄32	5⁄32–3⁄16	1⁄2	30	185–225	8–10
1⁄4	F	None	1	3⁄16	3⁄16–1⁄4	1⁄2	30	240–295	8–10
	V, H		1	3⁄16	3⁄16	1⁄2	30	220–265	8–10
	O		1	3⁄16	3⁄16	1⁄2	35	230–275	8–10
3⁄8	F	Optional up to 250°F maximum	2	3⁄16	1⁄4	5⁄8	35	325–375	8–10
	V		2	3⁄16	3⁄16–1⁄4	5⁄8	35	280–315	8–10
	H		3	3⁄16	3⁄16–1⁄4	5⁄8	35	270–300	8–10
	O		3	3⁄16	3⁄16–1⁄4	5⁄8	40	290–335	8–10

[1] F = flat; V = vertical; H = horizontal; O = overhead

[2] Preheating at excessive temperatures or for extended periods of time will reduce weld strength. This is particularly true for alloys in heat-treated tempers.

1.5.0 GTAW Terminations, Restarts, and Beads

Before aluminum welding practice begins, it is important to understand how to properly handle restarts and terminations, as well as how to create weave beads and overlapping beads to build a pad.

1.5.1 Weld Terminations

A weld termination is made at the end of a weld and usually when the filler metal rod has been consumed. A termination normally results in a crater that must be filled. Most welding codes require the crater to be filled to the full cross-section of the weld to prevent crater cracking.

Filling it to this level can be difficult. Most terminations are at the edge of a plate where welding heat tends to build up. This makes it harder to fill the crater. Filling the crater is easier when remote control equipment with a potentiometer is used. The potentiometer allows the welding current to be reduced manually as the end of the weld is approached. The control can be foot-operated or located on the torch handle.

In addition to using a manually operated potentiometer to reduce weld current, some welding machines are equipped with a downslope control. This control is set to automatically reduce weld current over a specific time period. The timing of the downslope setting is initiated through a switch on the torch. The welder simply activates the downslope control timing cycle as the termination begins, and the machine begins to reduce the welding current as programmed.

Follow these steps to make a termination using equipment with a potentiometer for current control:

Step 1 As you approach the end of the weld, slowly reduce the welding current. The amount of reduction in welding current is determined by the width of the weld puddle. If the puddle begins to widen, there is too much heat. If the puddle starts to become narrower, there is not enough heat.

Step 2 Continue to reduce the welding current while adding filler metal until the crater has been filled.

Table 3B Parameters for Making GTAW Fillet and Lap Welds on Aluminum Plate (Metric Values)

Aluminum Thickness (inch)	Weld Position[1]	Preheat[2] (F)	Weld Passes	Filler Diameter (mm)	Pure Tungsten Electrode Diameter (mm)	Gas Cup Inside Diameter (mm)	Argon Flow (lpm)	AC (amps)	Arc Travel Speed (mm/m)
1/16	F, V, H	None	1	2.4	1.6–2.4	9.5	7.6	70–100	200–250
	O		1	2.4	1.6–2.4	9.5	9.4	65–90	200–250
3/32	F	None	1	2.4–3.2	3.2–4.0	9.5	8.5	110–145	200–250
	V, H		1	2.4	2.4–3.2	9.5	8.5	90–125	200–250
	O		1	2.4	2.4–3.2	9.5	9.4	110–135	200–250
1/8	F	None	1	3.2	3.2–4.0	11.0	9.4	135–175	250–300
	V, H		1	3.2	2.4–3.2	9.5	9.4	115–145	200–250
	O		1	3.2	2.4–3.2	11.0	11.8	125–155	200–250
3/16	F	None	1	4.0	4.0–4.8	12.5	11.8	190–245	200–250
	V, H		1	4.0	4.0–4.8	12.5	11.8	175–210	200–250
	O		1	4.0	4.0–4.8	12.5	14.2	185–225	200–250
1/4	F	None	1	4.8	4.8–6.4	12.5	14.2	240–295	200–250
	V, H		1	4.8	4.8	12.5	14.2	220–265	200–250
	O		1	4.8	4.8	12.5	16.5	230–275	200–250
3/8	F	Optional up to 120°C maximum	2	4.8	6.4	16	16.5	325–375	200–250
	V		2	4.8	4.8–6.4	16	16.5	280–315	200–250
	H		3	4.8	4.8–6.4	16	16.5	270–300	200–250
	O		3	4.8	4.8–6.4	16	18.9	290–335	200–250

[1] F = flat; V = vertical; H = horizontal; O = overhead

[2] Preheating at excessive temperatures or for extended periods of time will reduce weld strength. This is particularly true for alloys in heat-treated tempers.

Step 3 Stop the arc and hold the torch in place, allowing the post-flow of gas to protect both the weld metal and electrode as they cool. The end of the filler metal rod should also remain in the gas flow.

Follow these steps to make a weld termination without remote control equipment:

Step 1 As you approach the end of the weld, add filler metal at a faster rate. The filler metal will absorb excess heat.

Step 2 Continue to add filler metal until the crater has been filled, and then stop the arc.

Step 3 Hold the torch in place, allowing the post-flow of gas to protect both the weld metal and electrode as they cool. The end of the filler metal rod should also remain in the gas flow.

CAUTION

Do not remove the torch or filler metal from the torch shielding-gas flow until the puddle has solidified and cooled enough to be unaffected by the atmosphere. The gas post-flow that continues after the welding has stopped protects the molten metal and the electrode. If the torch and its shielding-gas flow are removed from the weld area before the weld has solidified, crater porosity or cracks can occur. Also, the end of the filler metal will be contaminated, and the contaminated end will have to be discarded. Finally, the tungsten electrode can oxidize (turn a bluish-black color), forcing redressing of the electrode.

Step 4 Inspect the weld termination to ensure it has the following characteristics:

- Crater filled to the full cross section of the weld
- No crater cracks
- No porosity
- No undercut

1.5.2 Weld Restarts

A restart is the junction where a new weld connects to the bead of a previous weld to continue the weld. A restart must be made so that it blends smoothly with the previous weld and does not stand out. An improperly made restart will create a weld discontinuity. Whenever possible, avoid restarts by running a bead the full length of the weld joint.

A properly made restart blends into the bead and is hard to detect. If the restart has undercut, not enough time was spent in the crater to fill it. If the undercut is on one side or the other, use more of a side-to-side motion as you move back into the crater. If the restart has a lump, it was overfilled; too much time was spent in the crater before the forward motion was resumed.

Follow these steps to make a restart:

Step 1 Clean the area of the restart using an approved cleaning method.

Step 2 Hold the torch at the proper angle and arc distance. Restart the arc directly over the center of the crater. Remember that welding codes do not allow arc strikes outside of the area that is to be welded.

Step 3 Move the electrode tip in a small circular motion over the crater, and add filler metal when the molten puddle is the same size as the crater.

Step 4 As soon as the puddle fills the crater, advance the puddle slightly. Continue to add filler metal as needed.

Step 5 Inspect the restart.

Step 6 Continue to practice restarts until they are correct.

1.5.3 Weave Beads

The two basic bead types used in GTAW are weave beads and stringer beads. Stringer and weave beads are used to make welds, while overlapping stringer and weave beads are used to build pads. Always check the WPS or site quality standards to determine whether stringer or weave beads should be used.

Weave beads (*Figure 15*) are made by working the weld puddle back and forth across the axis of the weld. This produces a weld bead much wider than the puddle width. The width of a weave bead is determined by the amount of oscillation or cross motion. Weave beads put more heat into the base metal.

When making a weave bead, take special care at the toes to ensure proper fusion to the base metal. Do this by slowing down or pausing slightly at the edges. Pausing at the edges will also flatten out the weld and give it the proper profile.

Practice weave beads in the flat position first. Experiment with different weave motions and push angles.

> **CAUTION**
>
> Do not exceed the recommended weave bead width. The width of weave beads is often specified in the welding code or WPS used at the site.

Follow these steps to create weave beads:

Step 1 Hold the torch at the desired angle with the electrode tip directly over the point where the weld will begin.

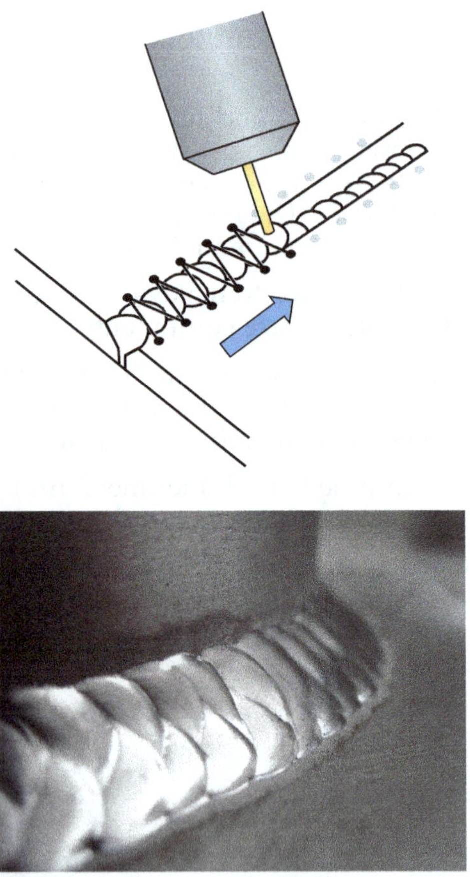

Figure 15 Weave bead.

Step 2 Lower the torch and filler rod to the weld.

Step 3 Hold the arc in place until the weld puddle begins to form. Add filler metal.

Step 4 Slowly advance the arc in a weaving motion while maintaining the torch angle and adding filler metal.

Step 5 Continue to weld until a bead about 2" to 3" (5 to 8 cm) long is formed.

Step 6 Stop and then fill the crater while slowly reducing amperage with a current control to allow the weld puddle to solidify. Turn off the current. Hold the filler rod and torch in place until both the tungsten electrode and weld cool under the protection of the gas envelope.

Step 7 Inspect the bead to ensure it has the following characteristics:

- Straightness of the bead
- Uniform appearance of the bead face; face and root reinforcement depth in accordance with applicable codes
- Smooth, flat transition with complete fusion at the toes of the weld
- Termination crater filled
- No porosity
- No undercut
- No overlap
- No overfill
- No cracks

Step 8 Continue practicing weave beads until you can make acceptable beads every time.

WARNING! Use pliers and gloves to handle hot practice coupons. Gloves should remain on when placing hot practice coupons in water. Steam will rise off the coupons and can burn or scald unprotected hands.

1.5.4 Overlapping Beads

Overlapping beads are made by depositing connective weld beads parallel to one another. Each successive bead overlaps the toe of the previous bead to form a raised, flat surface. This is also called *padding*. Overlapping beads are used to build up surfaces and to make multiple-pass welds. Both stringer and weave beads can be overlapped. Properly overlapped beads, when viewed from their ends, form a nearly flat, level surface (*Figure 16*).

Follow these steps to make GTAW overlapping stringer beads using ⅛" aluminum filler metal:

Step 1 Mark out a square on a piece of aluminum.

Step 2 Deoxidize the base metal.

Step 3 Weld a stringer bead along one edge.

Step 4 Clean the bead with a stainless-steel brush. Make sure that the base metal is cleaned with the brush between each pass.

Thin Materials

When welding thin materials, stop about 1" or 2" from the edge without concern for the crater. Then back weld from the edge back to the crater and fill it as described.

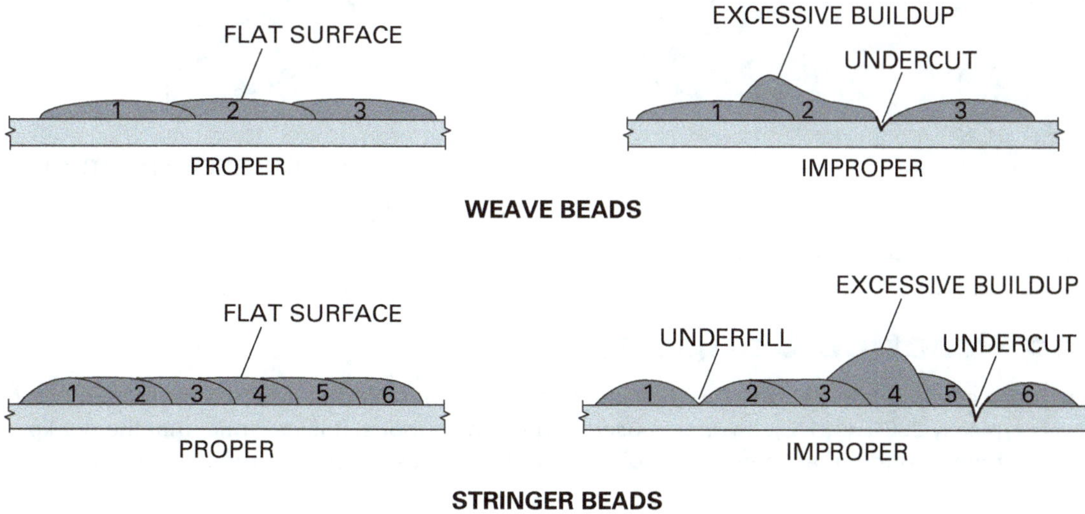

Figure 16 Proper and improper overlapping beads.

Step 5 Run a second bead along the previous bead. Be sure to overlap the previous bead to obtain a good fusion and produce a flat profile.

Step 6 Continue running overlapping stringer beads until the face of the square piece of metal has been covered.

Step 7 Continue building layers of stringer beads (pads), one on top of the other, until your technique has been perfected.

Continue to weld overlapping beads. Build a pad using stringer and weave beads as shown in *Figure 17.*

1.6.0 Fillet Welds

Fillet welds generally require less base metal preparation since bevels are not needed. However, the cleaning process must be thorough.

The most common fillet welds are made in lap and T-joints. The weld position for plate is determined by the axis of the weld and the orientation of the workpiece. The positions for fillet welding, shown in *Figure 18,* are flat (1F), horizontal (2F), vertical (3F), and overhead (4F). In the 1F and 2F positions, the weld axis can be inclined up to 15 degrees. Any weld axis inclination for the other positions varies with the rotational position of the weld face as specified in AWS standards.

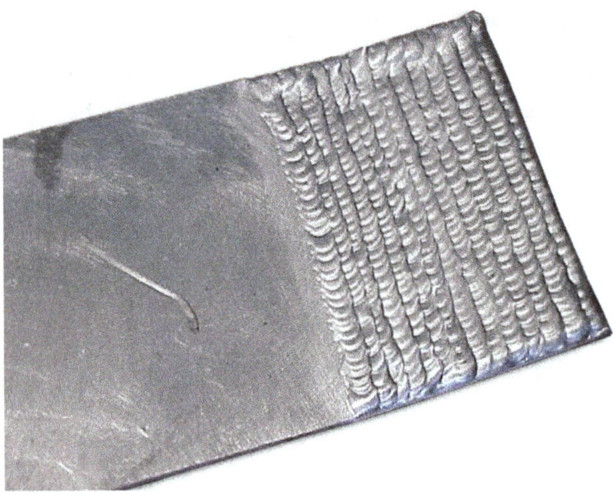

Figure 17 Pad built with stringer and weave beads.

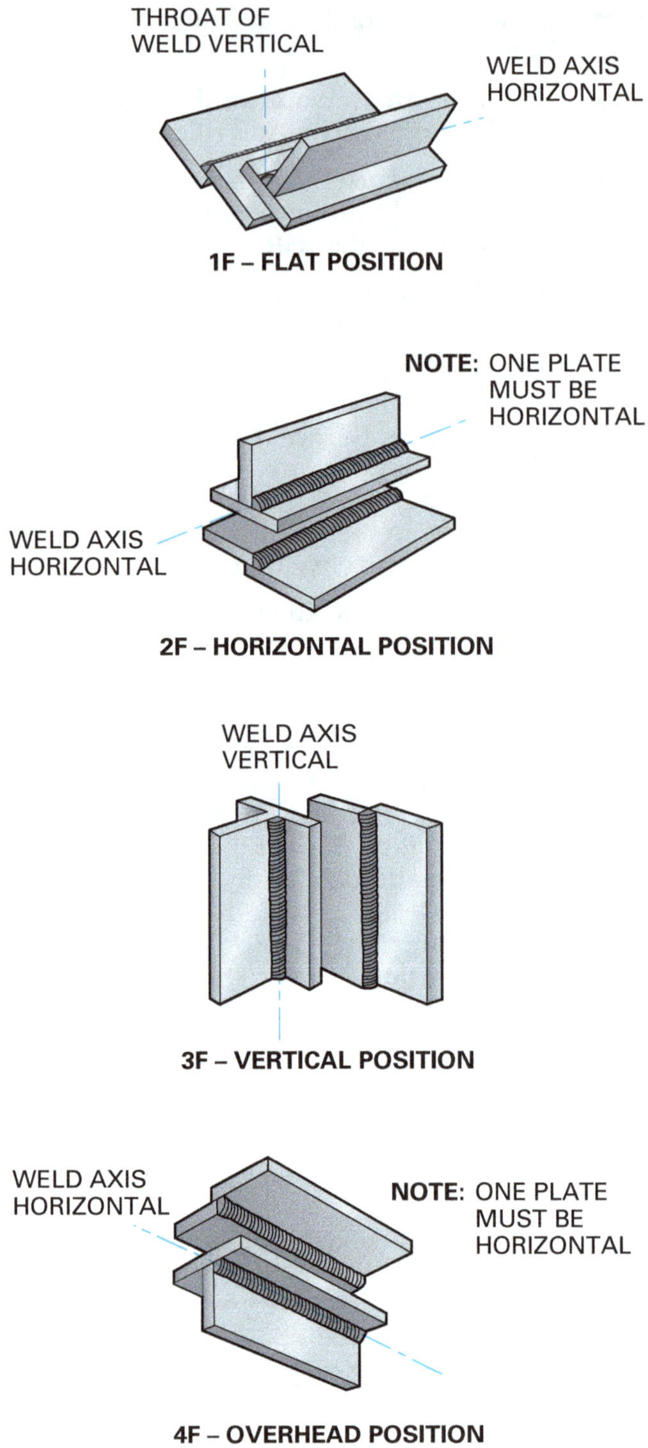

Figure 18 Fillet welding positions.

Cooling Practice Coupons

Aluminum practice coupons can be cooled with water. Remember that the maximum interpass temperature of aluminum is typically 250°F (121°C). If water is used to quench coupons, it is essential that the workpiece be thoroughly dried before welding resumes. Any moisture on the surface causes hydrogen contamination of the weld.

Fillet welds can be concave or convex, depending on the WPS or site quality standards. The welding codes require a fillet weld to have a uniform concave or convex face. However, a slightly non-uniform face is acceptable. The convexity of a fillet weld or individual surface bead must not exceed that permitted by the applicable code or standard. A fillet weld with profile defects is unacceptable and must be repaired. Ideal, acceptable, and unacceptable profiles are shown in *Figure 19.*

1.6.1 Practicing Flat (1F) Position Fillet Welds

Practice flat (1F) position fillet welds. Make multiple-pass (six-pass) convex fillet welds in a T-joint using an appropriate aluminum filler metal. When making flat fillet welds, pay close attention to the torch work angle and travel speed. For the first bead, the work angle should be approximately 45 degrees to both plate surfaces, slightly favoring the non-butting member to accommodate its greater heat loss. The work angle is adjusted for all subsequent beads for proper bead placement. *Figure 20* shows the torch angles for fillet welds in the 1F position. Clean each completed bead with a stainless-steel wire brush before starting the next.

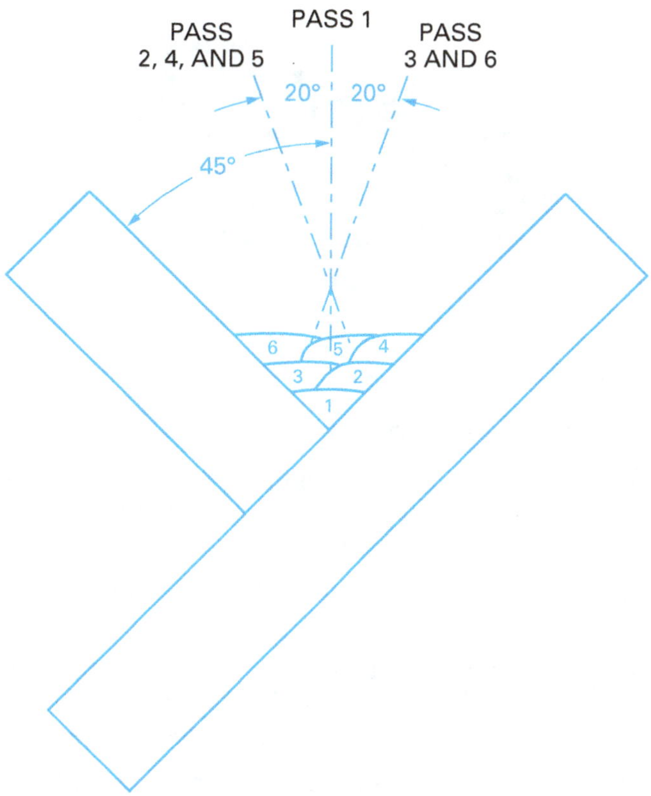

Figure 20 Multiple-pass 1F bead sequence and work angles.

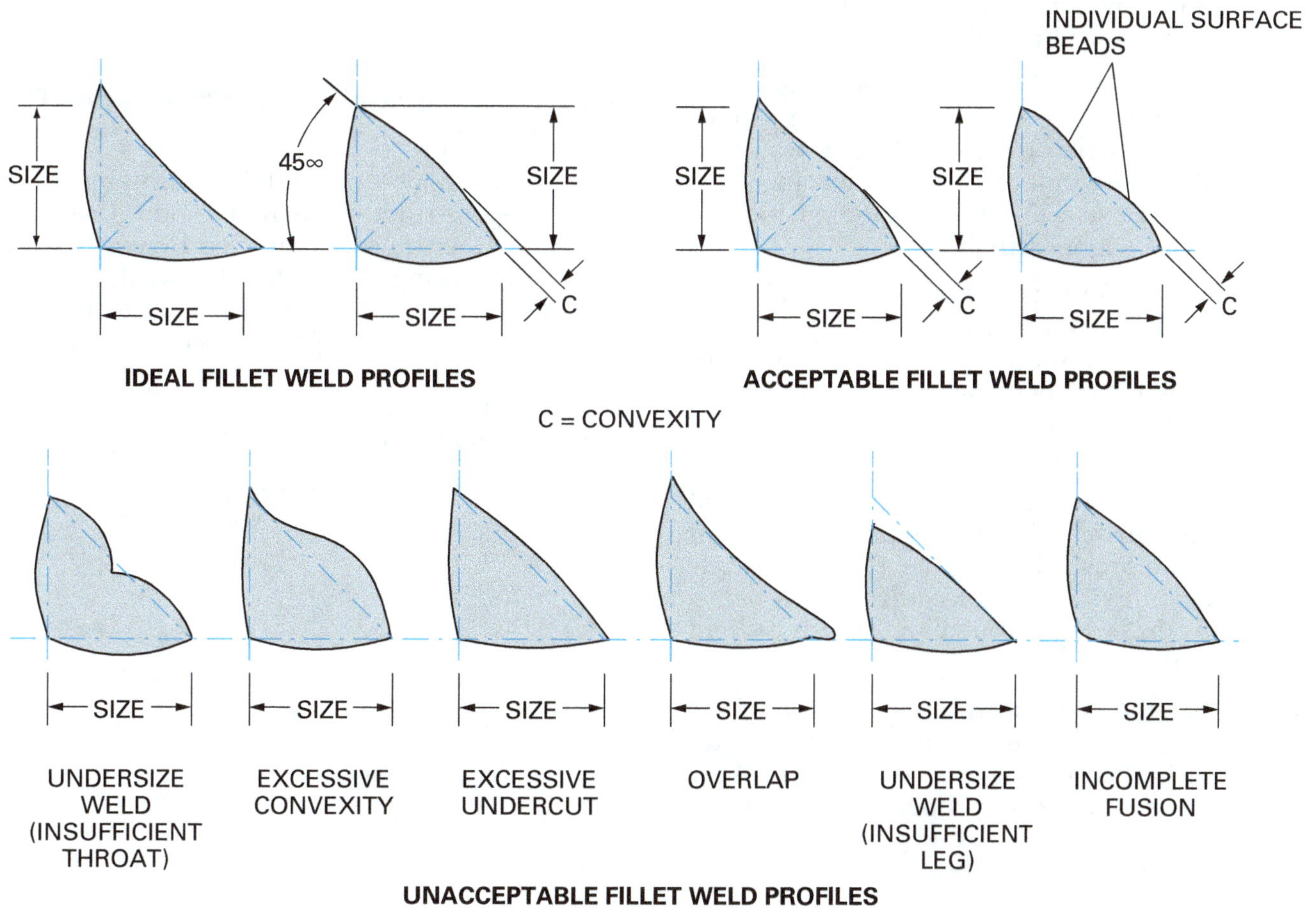

Figure 19 Ideal, acceptable, and unacceptable fillet weld profiles.

Figure 21 Fillet weld coupon.

Follow these steps to make a flat fillet weld:

Step 1 Tack two plates together to form a T-joint for the fillet weld coupon (*Figure 21*). Clean the tack welds.

Step 2 Clamp or tack weld the coupon in the 1F position (flat).

Step 3 Run the first bead along the root of the joint. Use a work angle of approximately 45 degrees and a 15- to 20-degree push angle.

Step 4 Run the second bead along a toe of the first weld, overlapping about 75 percent of the first bead. Alter the work angle as shown in *Figure 20*. Use a 15- to 20-degree push angle with a slight oscillation.

Step 5 Run the third bead along the other toe of the first weld. Be sure to fill the groove created when the second bead was run. Alter the work angle as shown in *Figure 20* and use a 15- to 20-degree push angle with a slight oscillation.

Step 6 Run the fourth bead along the outside toe of the second weld, overlapping about half of the second bead. Alter the work angle for proper bead placement and use a 15- to 20-degree push angle with a slight oscillation.

Step 7 Run the fifth bead along the inside toe of the fourth weld, overlapping about half of the fourth bead. Alter the work angle for proper bead placement and use a 15- to 20-degree push angle with a slight oscillation.

Step 8 Run the sixth bead along the toe of the fifth weld. Be sure to fill the groove created when the fifth bead was run. Alter the work angle for proper bead placement and use a 15- to 20-degree push angle with a slight oscillation.

T-Joint Heat Dissipation

In T-joints, the welding heat dissipates more rapidly in the thicker or non-butting member. On the root pass, the arc may have to be slightly more concentrated on the thicker or the non-butting member to compensate for heat loss.

Step 9 Have your instructor inspect the weld. The weld is acceptable if it exhibits the following characteristics:

- Uniform appearance of the bead face
- Craters and restarts filled to the full cross section of the weld
- Uniform weld size of ±1⁄16" (±1.6 mm)
- Acceptable weld profile in accordance with the applicable code or standard
- Smooth transition with complete fusion at the toes of the weld
- No porosity
- No undercut
- No overlap
- No inclusions
- No cracks

1.6.2 Practicing Horizontal (2F) Position Fillet Welds

Practice horizontal (2F) fillet welding by placing multiple-pass fillet welds in a T-joint. Use an appropriate filler metal as directed by your instructor. When making horizontal fillet welds, pay close attention to the torch angles. For the first bead, the work angle should be approximately 45 degrees to both plate surfaces, slightly favoring the non-butting member to accommodate its greater heat loss. The work angle is adjusted for all other passes.

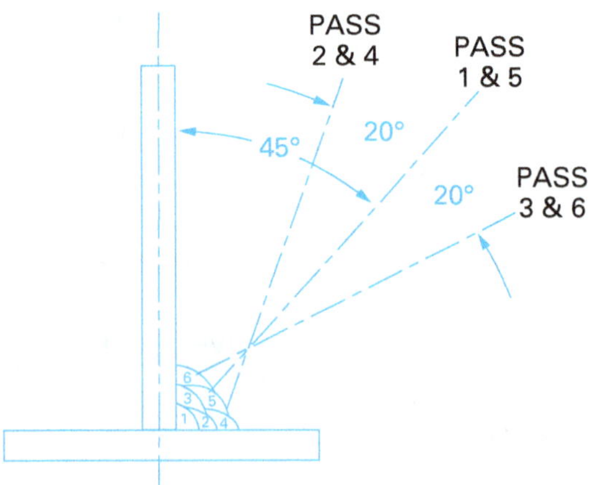

Figure 22 Multiple-pass 2F bead sequence and work angles.

Follow these steps to make a horizontal fillet weld:

Step 1 Tack two plates together to form a T-joint for the fillet weld coupon. Clean the tack welds.

Step 2 Clamp or tack weld the coupon in the horizontal position.

Step 3 Run the first bead along the root of the joint. Use a work angle of about 45 degrees and a 15- to 20-degree push angle (*Figure 22*).

Step 4 Clean the weld.

Step 5 Run the remaining passes at the proper work angles using a 15- to 20-degree push angle and a slight oscillation. Overlap each previous pass. Clean the weld after each pass.

Step 6 Have your instructor inspect the weld. The weld is acceptable if it exhibits the following characteristics:

- Uniform appearance on the bead face
- Craters and restarts filled to the full cross section of the weld
- Uniform weld size of ±1⁄16" (±1.6 mm)
- Acceptable weld profile in accordance with the applicable code or standard
- Smooth transition with complete fusion at the toes of the weld
- No porosity
- No undercut
- No overlap
- No inclusions
- No cracks

1.6.3 Practicing Vertical (3F) Position Fillet Welds

Practice vertical (3F) fillet welding by placing multiple-pass fillet welds in a T-joint. Vertical welds are usually made by welding uphill from the bottom to the top using a push angle. For the first bead, the work angle should be approximately 45 degrees to both plate surfaces, slightly favoring the non-butting member to accommodate its greater heat loss. Because of the uphill welding and push angle, this type of weld is sometimes called *vertical-up fillet welding*.

Either stringer or weave beads can be used for vertical welding. On the job, the site WPS or site quality standard will specify which technique to use.

NOTE

Check with your instructor to see if you should run stringer beads or weave beads, or practice both techniques.

Follow these steps to make an uphill fillet weld:

Step 1 Tack two plates together to form a T-joint for the fillet weld coupon.

Step 2 Clamp or tack weld the coupon in the vertical position.

Step 3 Run the first bead along the root of the joint, starting at the bottom. Use a work angle of about 45 degrees and a 15- to 20-degree push angle (*Figure 23*). Pause in the weld puddle to fill the crater.

Step 4 Clean the weld.

Step 5 Run the remaining passes using a 45-degree work angle and a 15- to 20-degree push angle with a side-to-side weave technique. Use a slow motion across the face of the weld. Pause at each toe for penetration and to fill the crater. Clean the weld after each pass.

Step 6 Have your instructor inspect the weld. The weld is acceptable if it exhibits the following characteristics:

- Uniform appearance on the bead face
- Craters and restarts filled to the full cross section of the weld
- Uniform weld size of ±1⁄16" (±1.6 mm)
- Acceptable weld profile in accordance with the applicable code or standard
- Smooth transition with complete fusion at the toes of the weld
- No porosity
- No undercut
- No overlap
- No inclusions
- No cracks

Raising the Working Height of a GTAW Torch

In some cases, slightly raising the torch height (increasing the arc length) can be used to widen the weld beads, instead of relying on the technique of oscillating the torch.

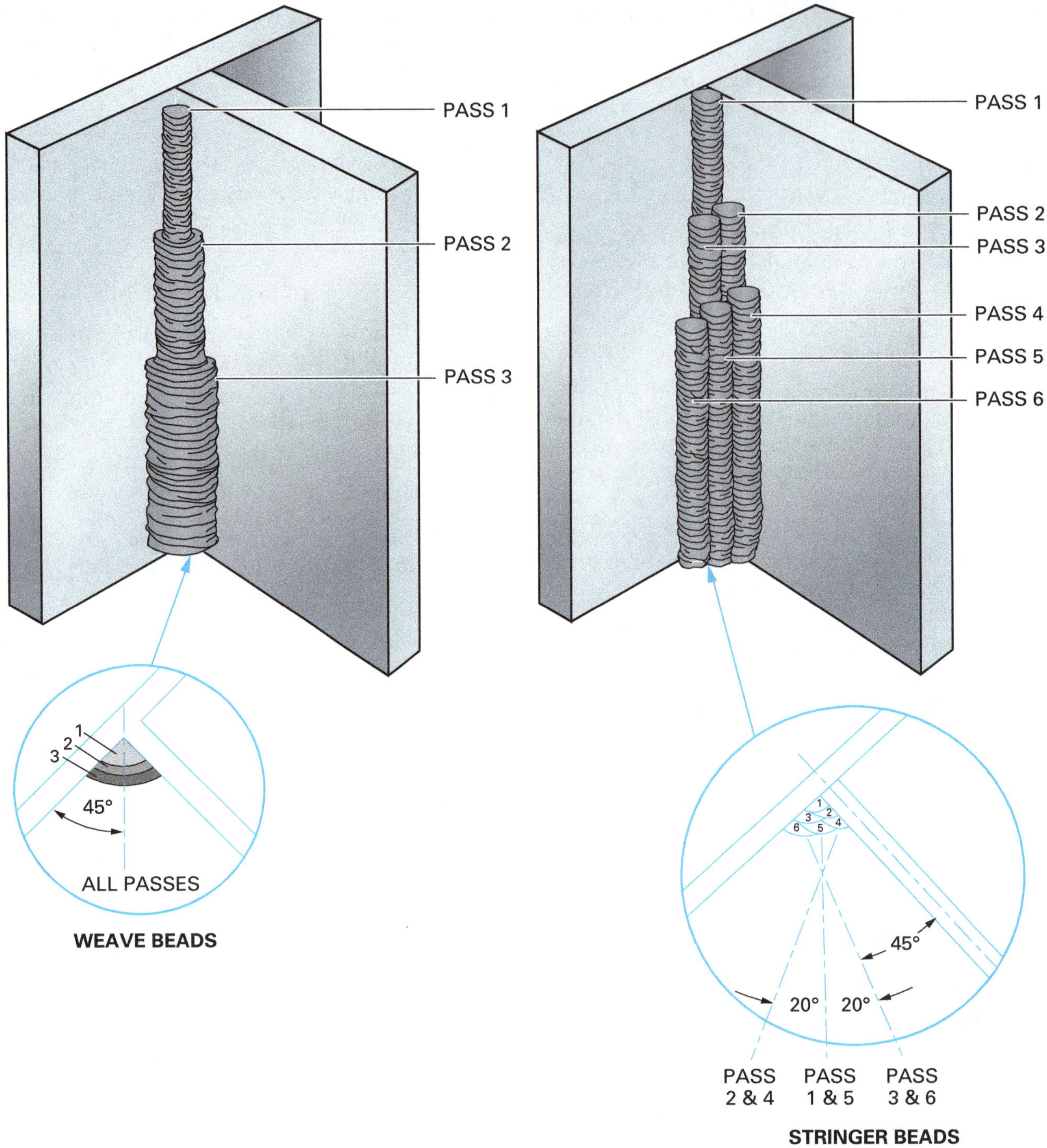

Figure 23 Multiple-pass 3F bead sequences and work angles for stringer and weave beads.

Repeat vertical fillet (3F) welding using stringer beads and a slight oscillating (side-to-side) motion. Pause briefly at each toe to prevent undercut. For stringer beads, use a 15- to 20-degree push angle and adjust the work angle to properly position each bead.

1.6.4 Practicing Overhead (4F) Position Fillet Welds

Practice overhead (4F) position fillet welds by making multiple-pass fillet welds in a T-joint. When making overhead fillet welds, pay close attention to the work angle. For the first bead, the work angle should be approximately 45 degrees to both plate surfaces, slightly favoring the non-butting member to accommodate its greater heat loss. The work angle is adjusted for all other passes.

Follow these steps to make an overhead fillet weld:

Step 1 Tack two plates together to form a T-joint for the fillet weld coupon.

Step 2 Clamp or tack weld the coupon so that it is in the overhead position.

Step 3 Run the first bead along the root of the joint. Use a work angle of about 45 degrees and a 15- to 20-degree push angle.

Step 4 Clean the weld.

Step 5 Run the remaining passes at the proper work angles using a 15- to 20-degree push angle and a slight oscillation (*Figure 24*). Overlap each previous pass. Clean the weld after each pass.

Step 6 Have your instructor inspect the weld. The weld is acceptable if it exhibits the following characteristics:

- Uniform appearance on the bead face
- Craters and restarts filled to the full cross section of the weld
- Uniform weld size of ±1⁄16" (±1.6 mm)
- Acceptable weld profile in accordance with the applicable code or standard
- Smooth transition with complete fusion at the toes of the weld
- No porosity
- No undercut
- No overlap
- No inclusions
- No cracks

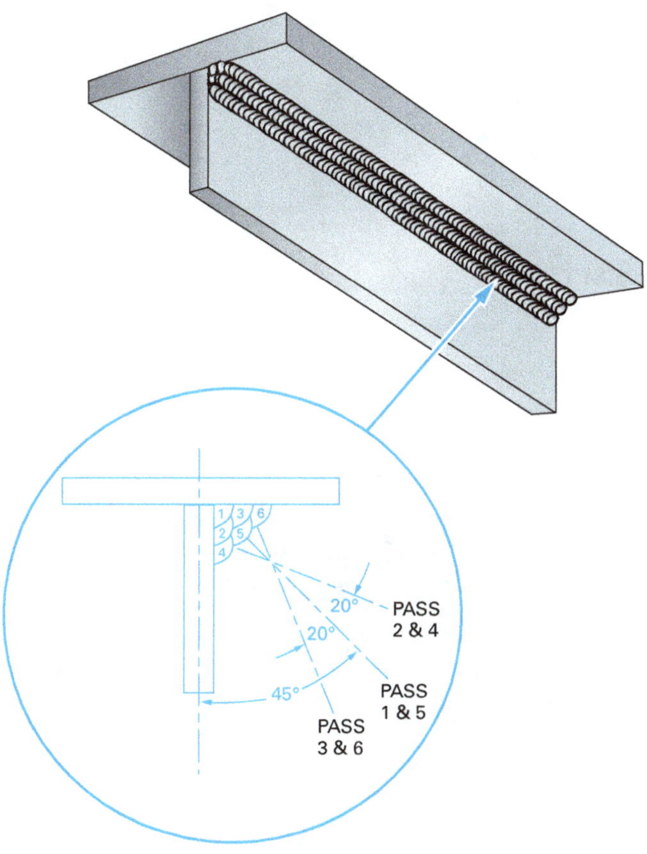

Figure 24 Multiple-pass 4F bead sequence and work angles.

1.7.0 V-Groove Plate Welds

The V-groove weld is a common groove weld used on both plate and pipe. The backing method of welding an aluminum V-groove joint with GTAW is presented in this module. Practicing V-groove welds on plate is good preparation for making the more difficult pipe welds.

For the root pass, use a 1⁄8" (3.2 mm) aluminum filler metal rod. After the root pass has been completed, clean and inspect it. Use the appropriate cleaning method to remove the oxidation that tends to form along the joint. Inspect the root pass for the following criteria:

- Uniform, smooth face
- No excessive buildup
- No undercut
- No overlap
- No porosity
- Proper fusion

V-groove welds with backing should be made with slight face reinforcement, not exceeding 3⁄32" (2.4 mm) for materials up to 3/8" thick. There should be a gradual transition to the base metal at each toe. Groove welds must not have excessive face reinforcement, any underfill, undercut, or overlap (*Figure 25*). If a groove weld has any of these defects, it must be repaired.

V-groove welds with backing can be made in all positions (*Figure 26*). The weld position for plate is determined by the axis of the weld and the orientation of the workpiece. Groove-weld positions are flat (1G), horizontal (2G), vertical (3G), and overhead (4G).

During the practice sessions that follow, use 1⁄8" aluminum filler metal for the root as well as the other passes unless your instructor directs otherwise. Pay special attention to filling the crater at the termination of the weld. Clean each completed bead with a stainless-steel wire brush before starting the next bead.

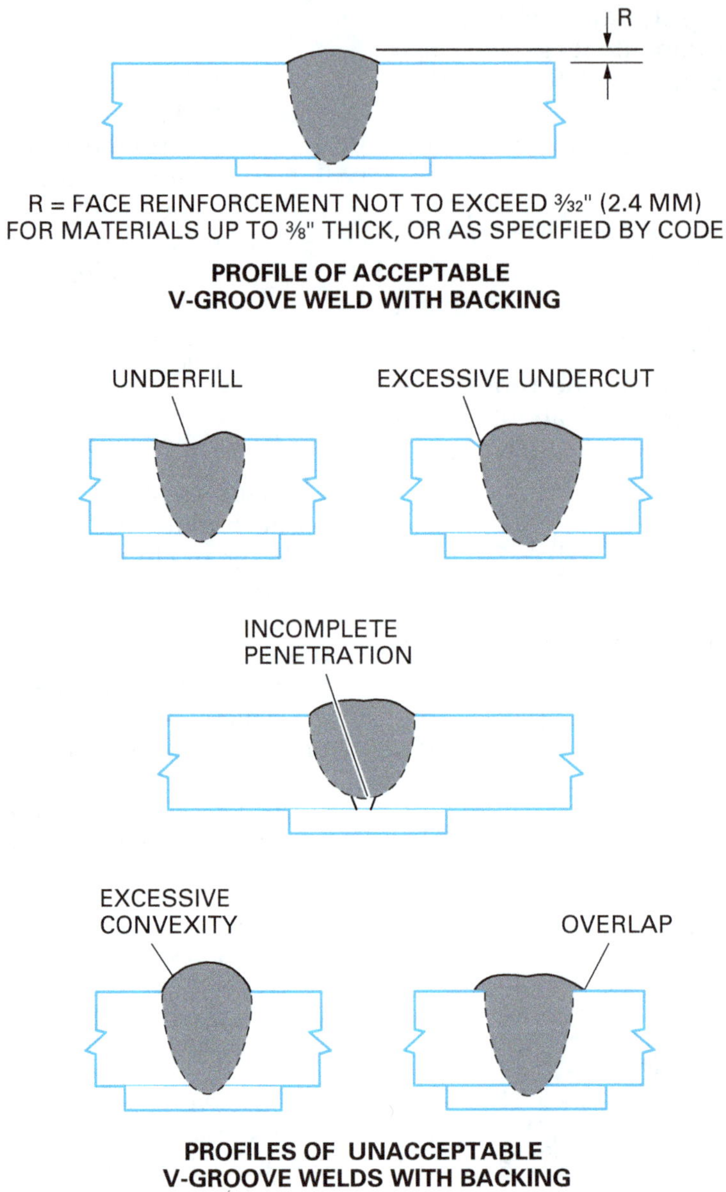

Figure 25 Acceptable and unacceptable V-groove weld profiles.

Modified U-Groove Joint

Instead of a backing strip, a modified U-groove joint can be used when welding aluminum plate. The modified U-groove joint shown is the same as the one that is used for GTAW of aluminum pipe. Use a degreaser on the back of the joint, but do not brush the metal; leaving a bit of oxide in place on the back actually helps support the weld.

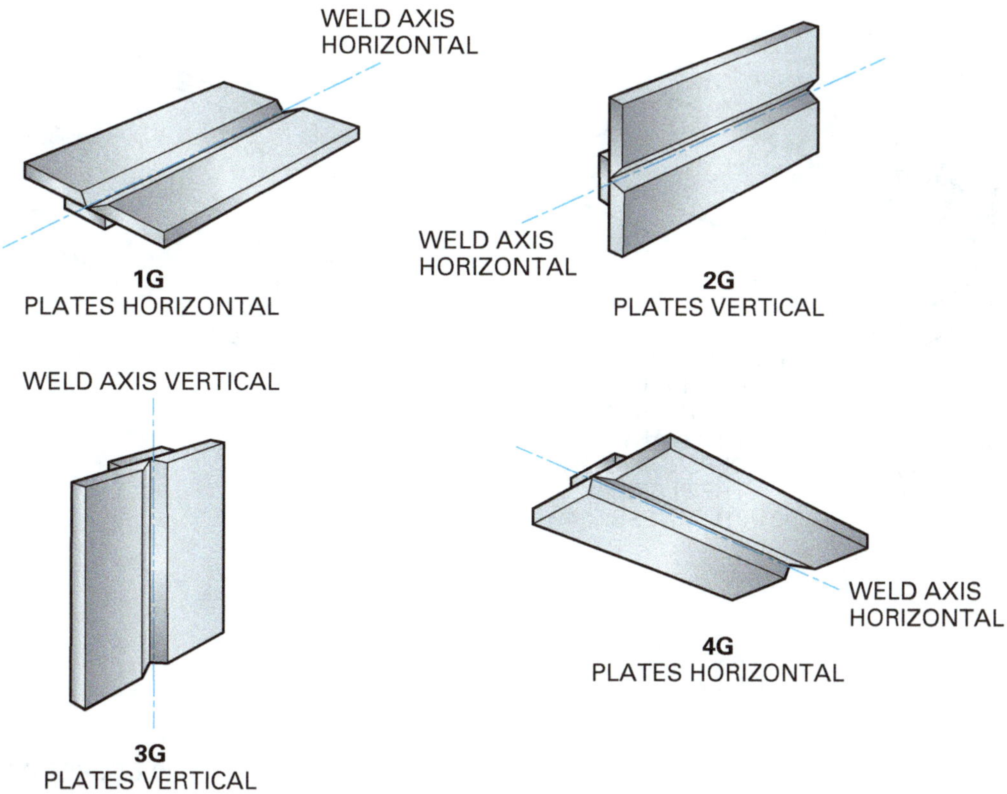

Figure 26 V-groove weld positions.

1.7.1 Practicing Flat (1G) Position V-Groove Welds with Backing

Follow these steps to make V-groove welds with backing in the flat (1G) position (*Figure 27*):

Step 1 Tack weld the practice coupon together as explained earlier.

Step 2 Clamp or tack weld the assembled coupon in the flat position above the welding table surface.

Step 3 Use a 15- to 20-degree push angle and a 0-degree work angle to weld the root pass with an appropriate aluminum filler metal. Feed the filler metal at a 15- to 20-degree angle to the plate.

Step 4 Clean the root pass.

Step 5 Run the remaining passes using a 10-degree to 15-degree push angle and the bead sequence and work angles shown in *Figure 27.* Hold the filler metal about 15 to 20 degrees above the plate face. Clean the weld between each pass.

1.7.2 Practicing Horizontal (2G) Position V-Groove Welds with Backing

Follow these steps to practice V-groove welds with backing in the horizontal (2G) position (*Figure 28*):

Step 1 Tack weld the practice coupon together as described earlier.

Step 2 Clamp or tack weld the assembled coupon in the horizontal position.

Step 3 Run the root pass with an appropriate aluminum filler metal. Use a 15- to 20-degree push angle and a 15- to 20-degree filler metal angle with the plate surface. Use the appropriate work angle for Pass 1 shown in *Figure 28.*

Step 4 Clean the weld.

Step 5 Run the remaining passes using a 15- to 20-degree push angle and the bead sequence and work angles shown in *Figure 28.* Clean the weld between each pass.

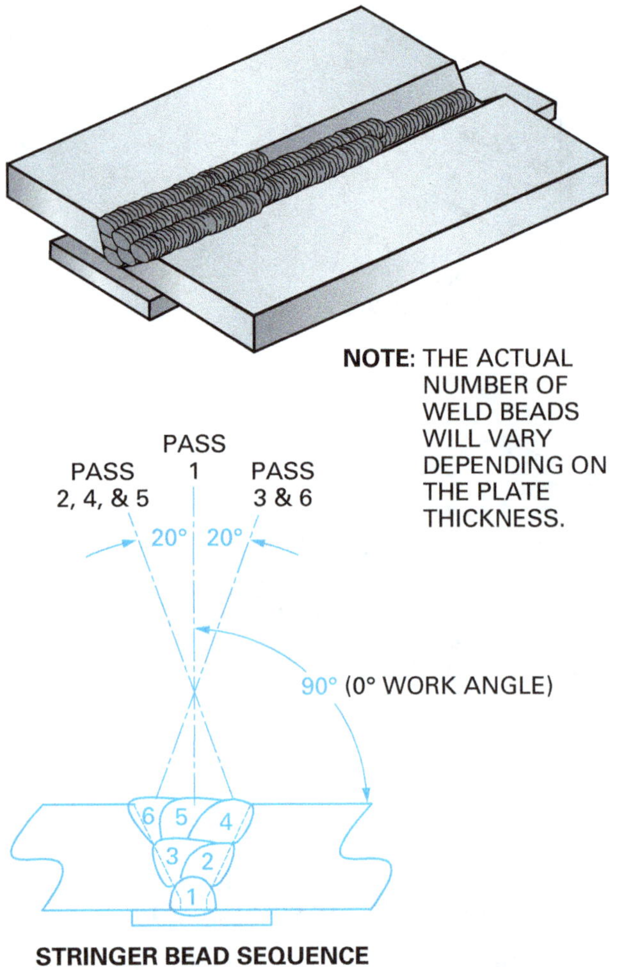

Figure 27 Multiple-pass 1G bead sequence and work angles.

1.7.3 Practicing Vertical (3G) Position V-Groove Welds with Backing

Follow these steps to practice V-groove welds with backing in the vertical (3G) position (*Figure 29*):

Step 1 Tack weld the practice coupon together as described earlier.

Step 2 Clamp or tack weld the assembled coupon in the vertical position.

Step 3 Run the root pass uphill using an appropriate aluminum filler metal. Use a push angle of 15 to 20 degrees and a filler metal angle of 15 to 20 degrees to the plate surface. Use a 0-degree work angle for Pass 1 as shown in *Figure 29*.

Step 4 Clean the weld.

Step 5 Run the remaining passes uphill to complete the weld. Use a 15- to 20-degree push angle and the bead sequence and work angles shown in *Figure 29*. Clean the weld between each pass.

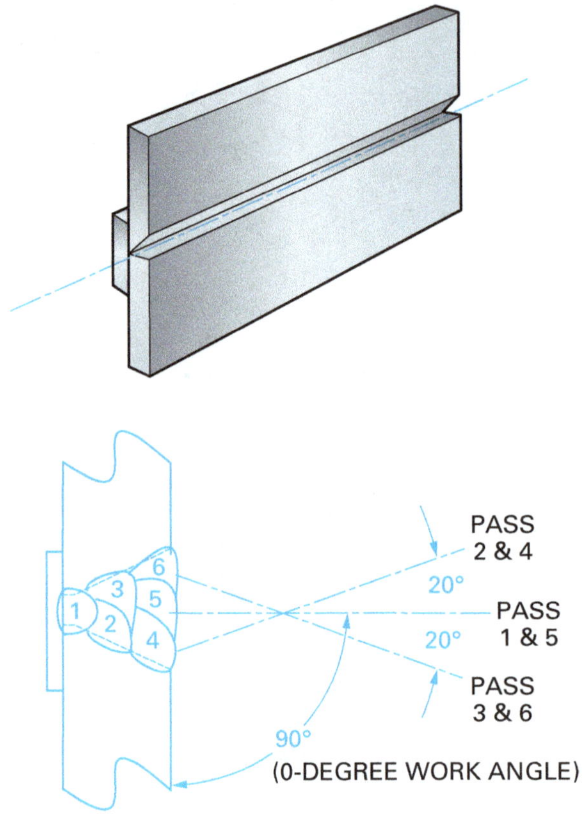

Figure 28 Multiple-pass 2G bead sequence and work angles.

1.7.4 Practicing Overhead (4G) Position V-Groove Welds with Backing

Follow these steps to practice V-groove welds with backing in the overhead (4G) position (*Figure 30*):

Step 1 Tack weld the practice coupon together as explained earlier.

Step 2 Clamp or tack weld the coupon in the overhead position.

Step 3 Run the root pass using an appropriate aluminum filler metal. Use a 15- to 20-degree push angle and a 15- to 20-degree filler metal angle with the plate surface. Use a 0-degree work angle for Pass 1, as shown in *Figure 30*.

Step 4 Clean the weld.

Step 5 Run the remaining passes using a 15- to 20-degree torch push angle and the bead sequence and work angles shown in *Figure 30*. Clean the weld between each pass.

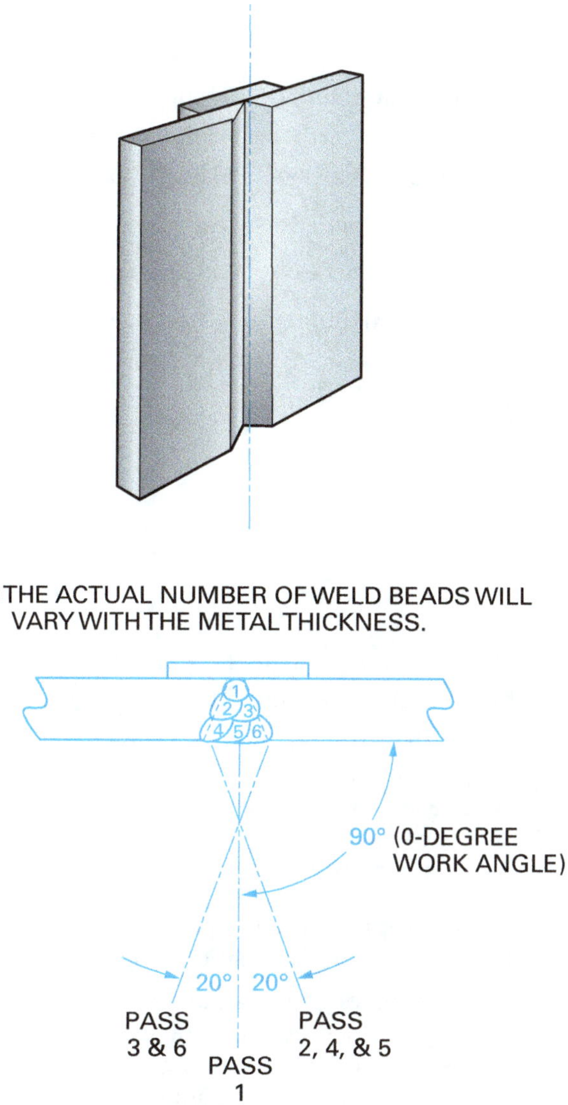

Figure 29 Multiple-pass 3G bead sequence and work angles.

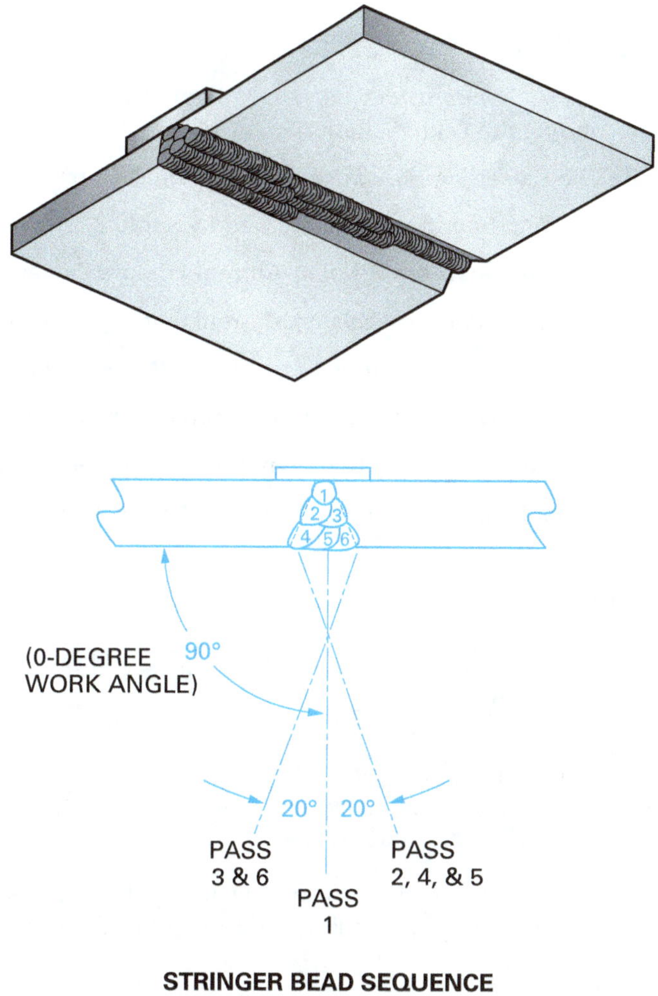

Figure 30 Multiple-pass 4G bead sequence and work angles.

Additional Resources

OSHA Standard 1926.351, Arc Welding and Cutting. Current Edition. Washington, DC: Occupational Safety and Health Administration.

The Procedure Handbook of Arc Welding. Current Edition. Cleveland, OH: The Lincoln Electric Company.

Welding Handbook, Volume 1 and Volume 2. Miami, FL: The American Welding Society.

Gas Tungsten Arc Welding. Current Edition. Cleveland, OH: The James F. Lincoln Arc Welding Foundation.

Gas Tungsten Arc Welding. Current Edition. Appleton, WI: Miller Electric Manufacturing Company.

AWS D1.2/D1.2M, *Structural Welding Code—Aluminum.* Miami, FL: American Welding Society.

The following websites offer resources for products and training:

The Lincoln Electric Company, **www.lincolnelectric.com**.

Miller Electric Manufacturing, **www.millerwelds.com**.

1.0.0 Section Review

1. A slow travel speed in GTAW forms weld puddles that are _____.
   a. small and shallow
   b. large and penetrate deeply
   c. contaminated by the argon
   d. oxidized and gel-like

2. For occasional, short-term welding of materials containing manganese, a welder should wear a _____.
   a. damp bandana over the mouth and nose
   b. tight-fitting disposable dust mask
   c. full-face, supplied-air respirator
   d. HEPA-rated filter on a respirator

3. The two basic factors to be considered when selecting aluminum filler metals are the base metal composition and the _____.
   a. WPS
   b. equipment manufacturer
   c. base metal manufacturer
   d. Aluminum Association

4. The typical range of frequencies used for aluminum GTAW with an inverter welding machine is _____.
   a. 50 to 60 cycles
   b. 90 to 100 cycles
   c. 120 to 140 cycles
   d. 280 to 300 cycles

5. Filling the crater of an aluminum weld termination is easier when the GTAW torch is equipped with a(n) _____.
   a. potentiometer
   b. phase inverter
   c. wire feeder
   d. oversized tungsten

6. The appropriate travel angle for an aluminum fillet weld in the 4F position is a _____.
   a. 40- to 50-degree drag angle
   b. 15- to 20-degree push angle
   c. 40- to 50-degree push angle
   d. 15- to 20-degree drag angle

7. Aluminum V-groove welds on ¼" plate should have slight face reinforcement, not exceeding _____.
   a. 1⁄32" (0.8 mm)
   b. 1⁄16" (1.6 mm)
   c. 3⁄32" (2.4 mm)
   d. ⅛" (3.2 mm)

SUMMARY

Welders must understand aluminum metallurgy and be familiar with the characteristics of the base metal before they can successfully perform aluminum GTAW. Welding aluminum plate using the GTAW process requires practice and a different approach to current control than when working with metals with different thermal properties.

All forms of welding require that the equipment and workspace be properly prepared. The correct filler metal must be chosen, and the area made as safe as possible.

Current control is important during terminations and restarts on aluminum. When manually-operated current control using a potentiometer is available to the welder, terminations in aluminum become easier.

Continued practice with aluminum plate and fillet welds will help welders master the necessary techniques and produce excellent weld beads.

Review Questions

1. To maintain the desired puddle size and advance the puddle when using the freehand technique, the welder may _____.
   a. leave the end of the filler rod in the puddle
   b. place the electrode in contact with the puddle
   c. preheat the filler metal rod
   d. move the torch tip in a small circular motion

2. The aluminum filler metal typically used for the GTAW process is a(n) _____.
   a. dual-shielded continuous wire
   b. uncoated, solid metal rod
   c. carbon-coated rod
   d. composite-core rod

3. To clean contamination such as oil, dirt, and grease from aluminum filler metal rods, you should use a clean rag and _____.
   a. a citrus-based cleaner
   b. silicone spray
   c. hydrochloric acid
   d. acetone

4. When preparing beveled aluminum plate to be welded with a backing strip, how wide must the root opening be?
   a. 1⁄32" (1 mm)
   b. 1⁄16" (1.6 mm)
   c. 1⁄8" (3 mm)
   d. 3⁄8" (9.5 mm)

5. Helium and helium/argon shielding-gas mixtures enable faster welding speeds and deeper penetration because helium _____.
   a. raises the temperature of the arc
   b. lowers the melting point of aluminum
   c. protects a larger area than argon alone
   d. combines with oxygen to eliminate oxidation

6. Filling the crater of a weld termination is made easier by _____.
   a. slowly increasing the welding current
   b. slowly decreasing the welding current
   c. rapidly increasing the arc length
   d. rapidly decreasing the arc length

7. When welding weave beads, to ensure proper fusion to the base metal at the toes of the weld, the welder should _____.
   a. decrease the arc length
   b. change to a drag angle
   c. stay well away from the toes
   d. slow down or pause briefly

8. The weld position in a fillet weld is determined by the orientation of the workpiece and the _____.
   a. size of the electrode
   b. axis of the weld
   c. type of joint
   d. welder's preference

9. In the 2F position, the first bead of the joint should be made with a work angle of approximately _____.
   a. 20 degrees
   b. 45 degrees
   c. 60 degrees
   d. 70 degrees

10. Aluminum plate V-groove welds in the 2G position should be made using a filler metal rod angle of _____.
    a. 90 degrees
    b. 45 degrees
    c. 15 to 20 degrees
    d. 5 to 10 degrees

Appendix

Performance Accreditation Tasks

The Performance Accreditation Tasks (PATs) correspond to and support learning objectives in the *AWS EG2.0, Curriculum Guide for the Training of Welding Personnel; Level I—Entry Welder.*

PATs provide specific acceptable criteria for performance and help to ensure a true competency-based welding program for students.

The following tasks are designed to evaluate your ability to run stringer beads, weave beads, overlapping beads, and to make fillet and multiple-pass V-groove plate welds with GTAW equipment in four standard test positions using an aluminum filler wire of the appropriate diameter and shielding gas. Perform each task when you are instructed to do so by your instructor.

As you complete each task, show it to your instructor for evaluation. Do not proceed to the next task until told to do so. For AWS 3F and 4F certifications, as well as 2G, 3G, and 4G certifications, refer to *AWS EG3.0, Guide for the Training and Qualification of Welding Personnel; Level II—Advanced Welder* for bend test requirements.

PAT 1 has no direct AWS correlation. PATs 2 and 3 correspond to *AWS EG2.0, 3.3.1, Module 7, Key Indicator 15 (1F and 2F)*. PATs 4 and 5 correspond to *AWS 3.0, 3.3.6.4, Unit #4, GTAW, Learning Objective #6 (3F and 4F)*. PAT 6 corresponds to *AWS EG2.0, 3.3.1, Module 7, Key Indicator 16 (1G)*. PATs 7 through 9 correspond to *AWS EG3.0, 3.3.6.4, Unit #4, GTAW, Learning Objective #7 (2G, 3G, and 4G).*

WELD A PAD ON ALUMINUM PLATE IN THE FLAT (1G) POSITION USING GTAW STRINGER BEADS

As directed by the instructor, use the GTAW process with the appropriate aluminum filler wire to make the following welds on aluminum plate: stringer beads, weave beads, weld restarts, weld terminations, and overlapping beads.

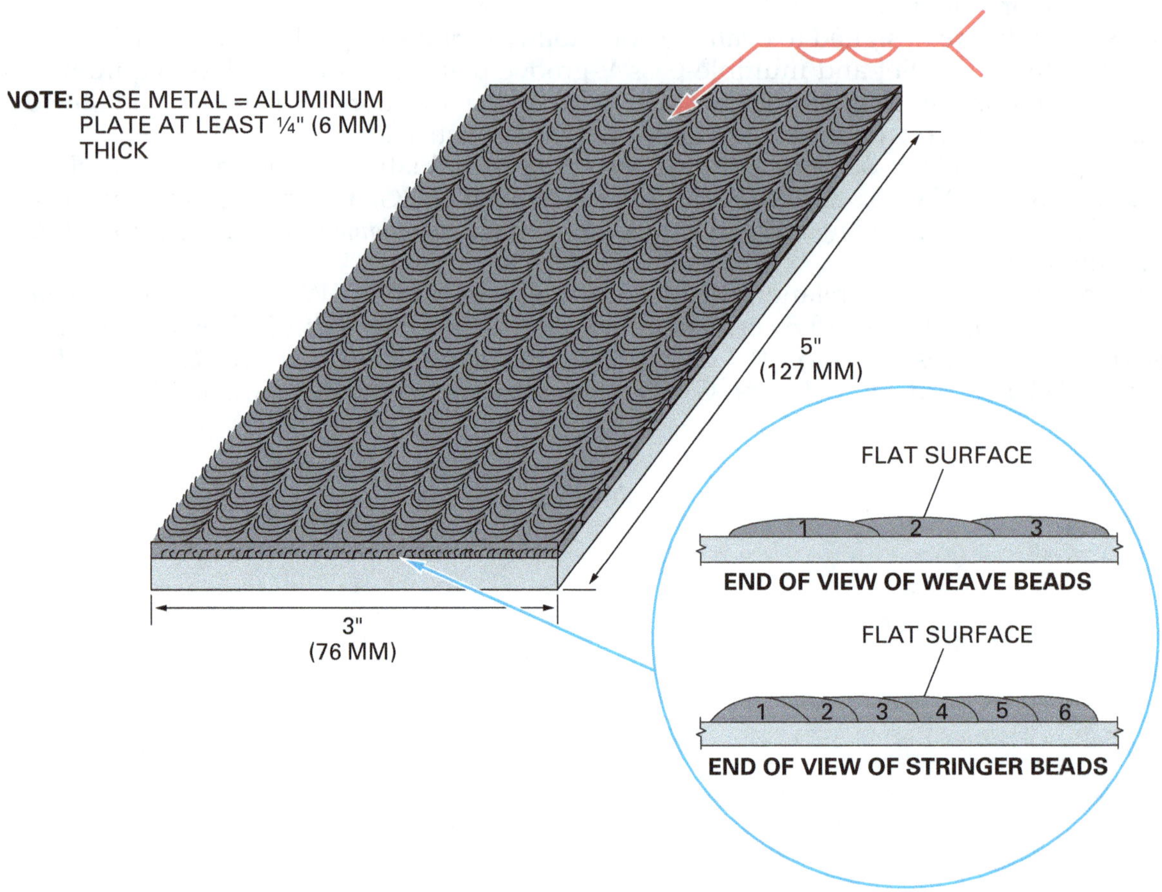

Criteria for Acceptance:

- Uniform rippled appearance on the bead face ________
- Craters and restarts filled to the full cross section of the weld ________
- Uniform weld width ±1⁄16" (1.6 mm) ________
- Acceptable weld profile in accordance with the applicable code or standard ________
- Smooth flat transition with complete fusion at the toes of the weld ________
- No porosity ________
- No undercut ________
- No inclusions ________
- No cracks ________

MAKE MULTIPLE-PASS FILLET WELDS ON ALUMINUM PLATE IN THE FLAT (1F) POSITION

As directed by the instructor, use the GTAW process with the appropriate aluminum wire to make a six-pass fillet weld using stringer beads on aluminum plate, as shown.

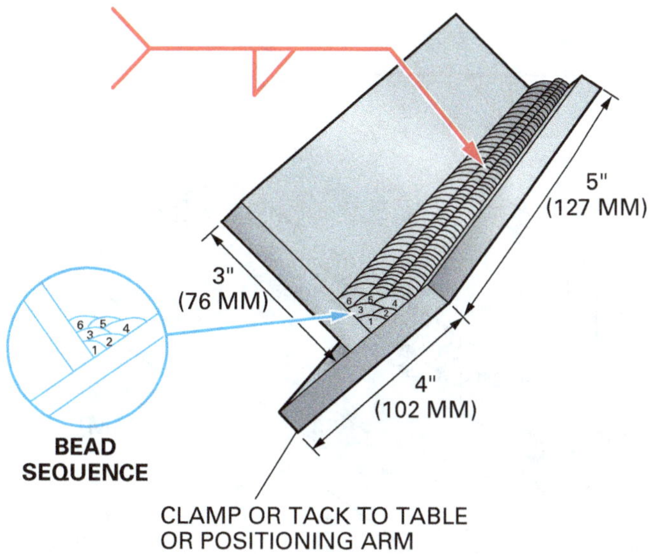

Criteria for Acceptance:

- Uniform rippled appearance on the bead face ________
- Craters and restarts filled to the full cross section of the weld ________
- Uniform weld width ±1⁄16" (1.6 mm) ________
- Smooth flat transition with complete fusion at the toes of the weld ________
- Acceptable weld profile in accordance with the applicable code or standard ________
- No porosity ________
- No undercut ________
- No overlap ________
- No inclusions ________
- No cracks ________

MAKE MULTIPLE-PASS FILLET WELDS ON ALUMINUM PLATE IN THE HORIZONTAL (2F) POSITION

As directed by the instructor, use the GTAW process with the appropriate aluminum wire to make a six-pass fillet weld using stringer beads on aluminum plate, as shown.

NOTE: BASE METAL = ALUMINUM
PLATE AT LEAST ¼" (6 MM) THICK

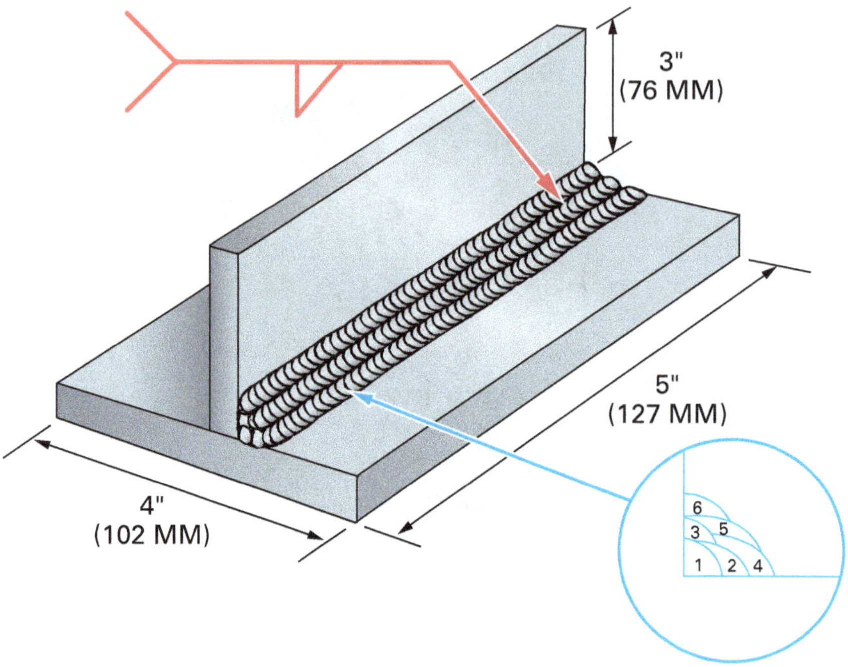

BEAD SEQUENCE

Criteria for Acceptance:

- Uniform rippled appearance on the bead face ______
- Craters and restarts filled to the full cross section of the weld ______
- Uniform weld width ±1⁄16" (1.6 mm) ______
- Smooth flat transition with complete fusion at the toes of the weld ______
- Acceptable weld profile in accordance with the applicable code or standard ______
- No porosity ______
- No undercut ______
- No overlap ______
- No inclusions ______
- No cracks ______

MAKE MULTIPLE-PASS FILLET WELDS ON ALUMINUM PLATE IN THE VERTICAL (3F) POSITION

As directed by the instructor, use the GTAW process with the appropriate aluminum wire to make a six-pass fillet weld using stringer beads on aluminum plate, as shown.

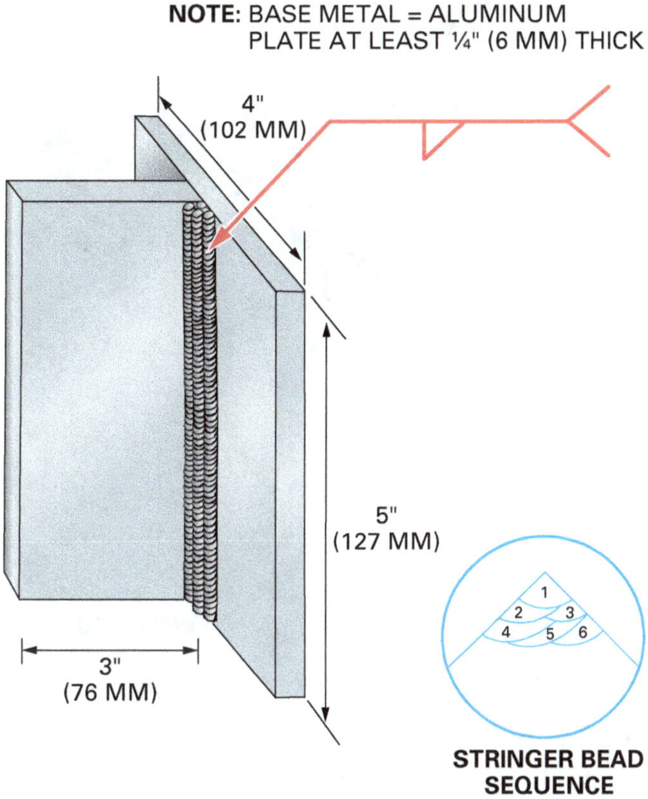

Criteria for Acceptance:

- Uniform rippled appearance on the bead face ______
- Craters and restarts filled to the full cross section of the weld ______
- Uniform weld width ±1⁄16" (1.6 mm) ______
- Smooth flat transition with complete fusion at the toes of the weld ______
- Acceptable weld profile in accordance with the applicable code or standard ______
- No porosity ______
- No undercut ______
- No overlap ______
- No inclusions ______
- No cracks ______

MAKE MULTIPLE-PASS FILLET WELDS ON ALUMINUM PLATE IN THE OVERHEAD (4F) POSITION

As directed by the instructor, use the GTAW process with the appropriate aluminum wire to make a six-pass fillet weld using stringer beads on aluminum plate, as shown.

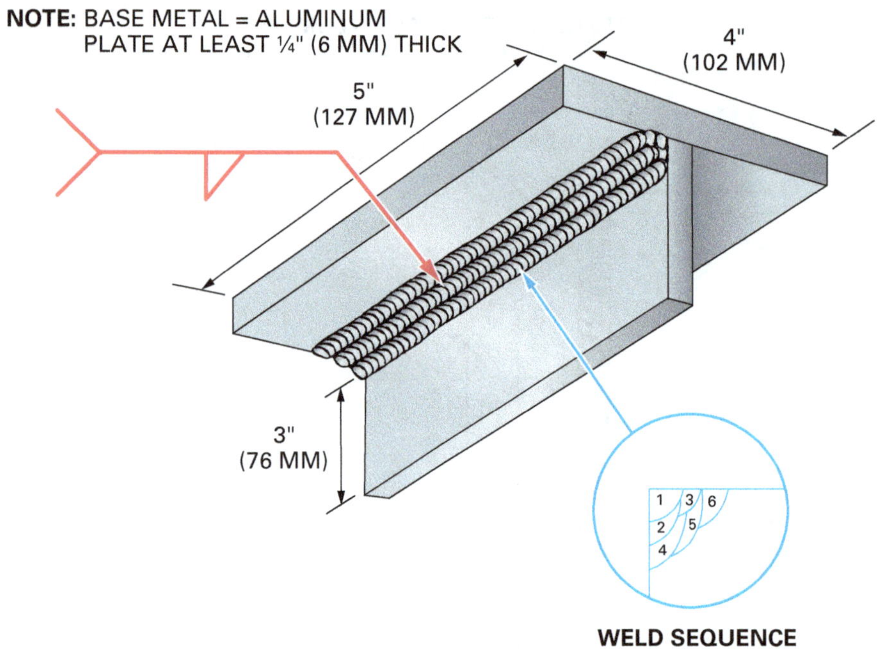

Criteria for Acceptance:

- Uniform rippled appearance on the bead face ____________
- Craters and restarts filled to the full cross section of the weld ____________
- Uniform weld width ±1⁄16" (1.6 mm) ____________
- Smooth flat transition with complete fusion at the toes of the weld ____________
- Acceptable weld profile in accordance with the applicable code or standard ____________
- No porosity ____________
- No undercut ____________
- No overlap ____________
- No inclusions ____________
- No cracks ____________

MAKE MULTIPLE-PASS V-GROOVE WELDS ON ALUMINUM PLATE WITH BACKING IN THE FLAT (1G) POSITION

As directed by the instructor, use the GTAW process with the appropriate aluminum wire to make a multiple-pass groove weld using stringer beads on aluminum plate with backing, as shown.

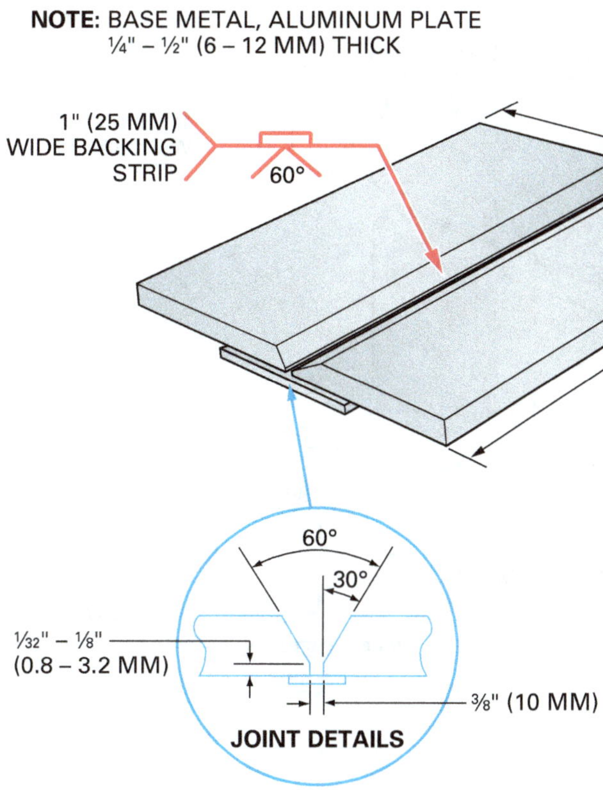

Criteria for Acceptance:

- Uniform rippled appearance on the bead face ____________
- Craters and restarts filled to the full cross section of the weld ____________
- Acceptable weld profile in accordance with the applicable code or standard ____________
- Smooth flat transition with complete fusion at the toes of the weld ____________
- No porosity ____________
- No undercut ____________
- No overlap ____________
- No inclusions ____________
- No cracks ____________

MAKE MULTIPLE-PASS V-GROOVE WELDS ON ALUMINUM PLATE WITH BACKING IN THE HORIZONTAL (2G) POSITION

As directed by the instructor, use the GTAW process with the appropriate aluminum wire to make a multiple-pass groove weld on aluminum plate with backing, as shown.

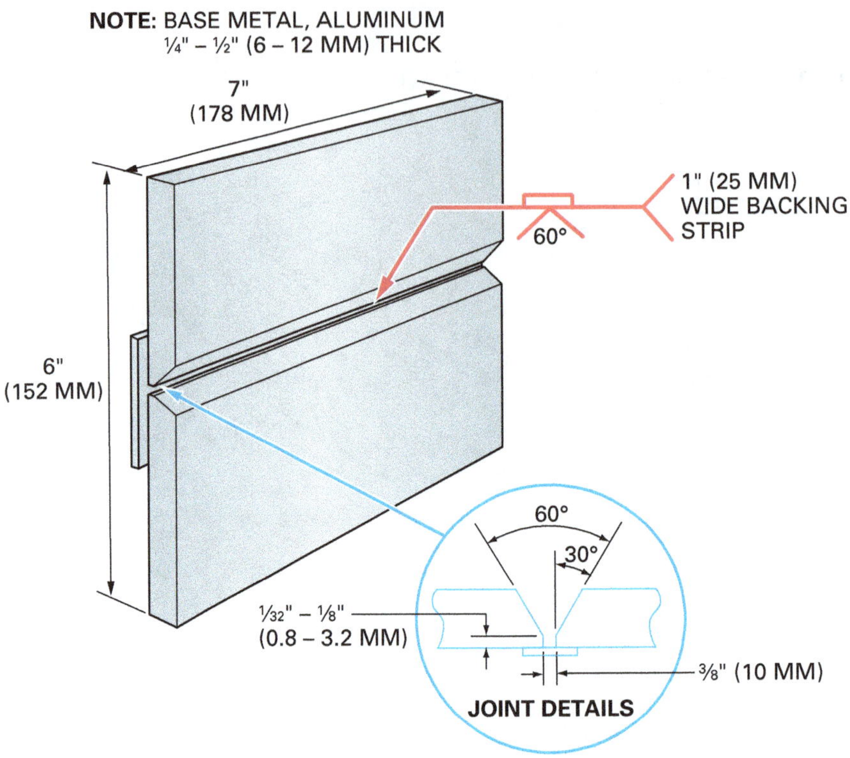

Criteria for Acceptance:

- Uniform rippled appearance on the bead face __________
- Craters and restarts filled to the full cross section of the weld __________
- Acceptable weld profile in accordance with the applicable code or standard __________
- Smooth flat transition with complete fusion at the toes of the weld __________
- No porosity __________
- No undercut __________
- No overlap __________
- No inclusions __________
- No cracks __________

MAKE MULTIPLE-PASS V-GROOVE WELDS ON ALUMINUM PLATE WITH BACKING IN THE VERTICAL (3G) POSITION

As directed by the instructor, use the GTAW process with the appropriate aluminum wire to make a multiple-pass groove weld on aluminum plate with backing, as shown.

Note: Run the root vertical up.

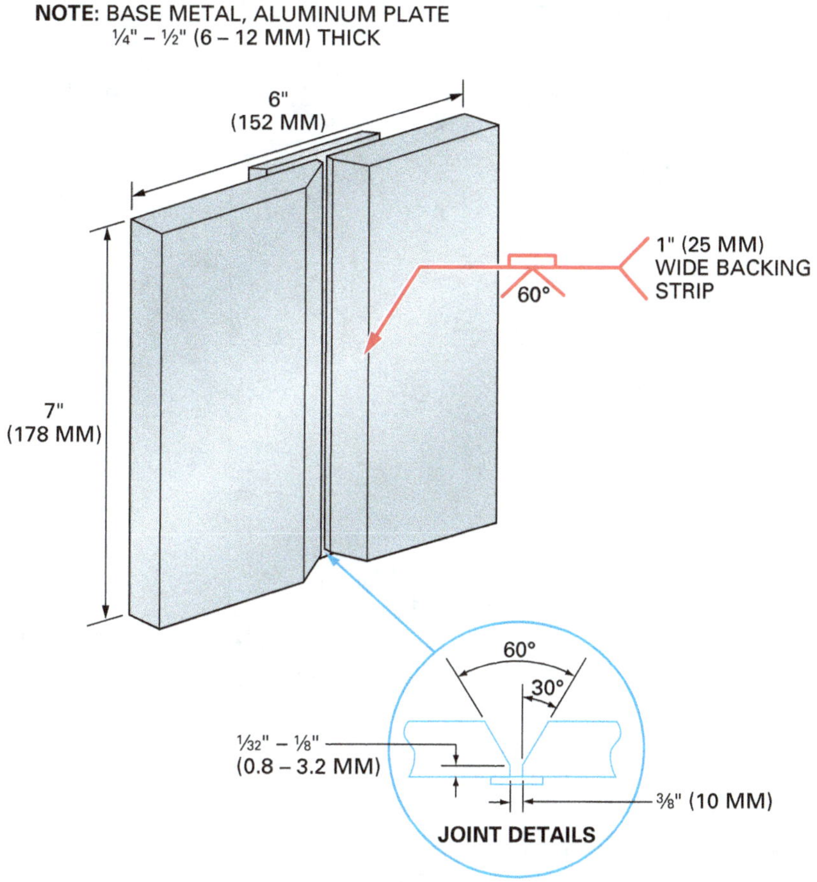

Criteria for Acceptance:

- Uniform rippled appearance on the bead face ________
- Craters and restarts filled to the full cross section of the weld ________
- Acceptable weld profile in accordance with the applicable code or standard ________
- Smooth flat transition with complete fusion at the toes of the weld ________
- No porosity ________
- No undercut ________
- No overlap ________
- No inclusions ________
- No cracks ________

MAKE MULTIPLE-PASS V-GROOVE WELDS ON ALUMINUM PLATE WITH BACKING IN THE OVERHEAD (4G) POSITION

As directed by the instructor, use the GTAW process with the appropriate aluminum wire to make a multiple-pass groove weld on aluminum plate with backing, as shown.

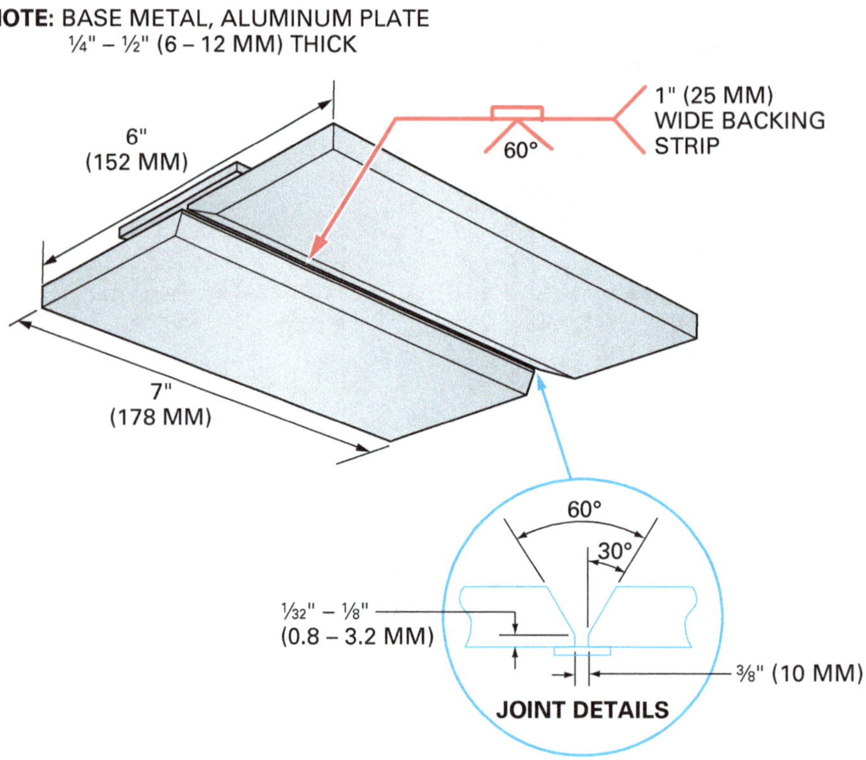

Criteria for Acceptance:

- Uniform rippled appearance on the bead face ______
- Craters and restarts filled to the full cross section of the weld ______
- Acceptable weld profile in accordance with the applicable code or standard ______
- Smooth flat transition with complete fusion at the toes of the weld ______
- No porosity ______
- No undercut ______
- No overlap ______
- No inclusions ______
- No cracks ______

Additional Resources

This module presents thorough resources for task training. The following reference material is recommended for further study.

OSHA Standard 1926.351, Arc Welding and Cutting. Current Edition. Washington, DC: Occupational Safety and Health Administration.

The Procedure Handbook of Arc Welding. Current Edition. Cleveland, OH: The Lincoln Electric Company.

Welding Handbook, Volume 1 and Volume 2. Miami, FL: The American Welding Society.

Gas Tungsten Arc Welding. Current Edition. Cleveland, OH: The James F. Lincoln Arc Welding Foundation.

Gas Tungsten Arc Welding. Current Edition. Appleton, WI: Miller Electric Manufacturing Company.

AWS D1.2/D1.2M, *Structural Welding Code—Aluminum.* Miami, FL: American Welding Society.

The following websites offer resources for products and training:

The Lincoln Electric Company, **www.lincolnelectric.com**.

Miller Electric Manufacturing, **www.millerwelds.com**.

Figure Credits

The Lincoln Electric Company, Cleveland, OH, USA, Module Opener, Figures 4, 6, 7, 15 (photo), Table 2

Miller Electric Manufacturing Co., Figure 5

John Tillman Company, SA01

CK Worldwide, Inc., SA02

AWS A5.10/A5.10M: 2012 (ISO 18273:2004 MOD), Table A1, reproduced with permission of the American Welding Society (AWS), Miami, Florida, Table 1

Terry Lowe, Figures 13, 17, 21, SA03

Topaz Publications, Inc., Figure 14 (photo)

The Aluminum Association, Inc., Table 3

Section Review Answer Key

Answer	Section Reference	Objective
Section One		
1. b	1.1.1	1a
2. d	1.2.4	1b
3. a	1.3.1	1c
4. c	1.4.0	1d
5. a	1.5.1	1e
6. b	1.6.4	1f
7. c	1.7.0	1g

NCCER CURRICULA — USER UPDATE

NCCER makes every effort to keep its textbooks up-to-date and free of technical errors. We appreciate your help in this process. If you find an error, a typographical mistake, or an inaccuracy in NCCER's curricula, please fill out this form (or a photocopy), or complete the online form at **www.nccer.org/olf**. Be sure to include the exact module ID number, page number, a detailed description, and your recommended correction. Your input will be brought to the attention of the Authoring Team. Thank you for your assistance.

Instructors – If you have an idea for improving this textbook, or have found that additional materials were necessary to teach this module effectively, please let us know so that we may present your suggestions to the Authoring Team.

NCCER Product Development and Revision

13614 Progress Blvd., Alachua, FL 32615

Email: curriculum@nccer.org
Online: www.nccer.org/olf

❑ Trainee Guide ❑ Lesson Plans ❑ Exam ❑ PowerPoints Other ____________________

Craft / Level: Copyright Date:

Module ID Number / Title:

Section Number(s):

Description:

Recommended Correction:

Your Name:

Address:

Email: Phone:

GTAW – Aluminum Pipe

Overview

The GTAW process is unique in that the welder must add the filler metal using the opposite hand. The process is well suited for welding aluminum, using the freehand or walking-the-cup technique. Welders must understand that a significant amount of welding current is required for welding aluminum, especially when working with thick materials. This module provides insight into welding aluminum pipe using the GTAW process and the modified U-groove joint, which does not require a metal backing or a backing gas.

Module 29403

Trainees with successful module completions may be eligible for credentialing through the NCCER Registry. To learn more, go to **www.nccer.org** or contact us at 1.888.622.3720. Our website has information on the latest product releases and training, as well as online versions of our *Cornerstone* magazine and Pearson's product catalog.

Your feedback is welcome. You may email your comments to **curriculum@nccer.org**, send general comments and inquiries to **info@nccer.org**, or fill in the User Update form at the back of this module.

This information is general in nature and intended for training purposes only. Actual performance of activities described in this manual requires compliance with all applicable operating, service, maintenance, and safety procedures under the direction of qualified personnel. References in this manual to patented or proprietary devices do not constitute a recommendation of their use.

29403 V5

29403
GTAW – Aluminum Pipe

Objectives

When you have completed this module, you will be able to do the following:

1. Explain how to prepare for GTAW aluminum pipe welding and describe specific pipe-welding techniques.
   a. Describe common GTAW pipe-welding techniques.
   b. Identify common welding safety practices.
   c. Explain how to prepare the area and materials for aluminum GTAW practice.
   d. Explain how to prepare GTAW equipment for welding.
   e. Describe techniques and positions related to aluminum pipe GTAW.
   f. Describe how to make modified U-groove GTAW welds on aluminum pipe.

Performance Tasks

Under the supervision of your instructor, you should be able to do the following:

1. Make GTAW modified U-groove welds on aluminum pipe in the following positions:
   - 2G
   - 5G
   - 6G

Industry Recognized Credentials

If you are training through an NCCER-accredited sponsor, you may be eligible for credentials from NCCER's Registry. The ID number for this module is 29403. Note that this module may have been used in other NCCER curricula and may apply to other level completions. Contact NCCER's Registry at 888.622.3720 or go to **www.nccer.org** for more information.

Contents

Figures and Tables

Section One

1.0.0 Aluminum Pipe GTAW

Objective

Explain how to prepare for GTAW aluminum pipe welding and describe specific pipe-welding techniques.

a. Describe common GTAW pipe-welding techniques.
b. Identify common welding safety practices.
c. Explain how to prepare the area and materials for aluminum GTAW practice.
d. Explain how to prepare GTAW equipment for welding.
e. Describe techniques and positions related to aluminum pipe GTAW.
f. Describe how to make modified U-groove GTAW welds on aluminum pipe.

Performance Tasks

1. Make GTAW modified U-groove welds on aluminum pipe in the following positions:
   - 2G
   - 5G
   - 6G

Gas tungsten arc welding (GTAW) uses an arc between a tungsten electrode and the base metal to melt the base and filler metals. The electrode, the arc, and the molten base metal are shielded from atmospheric contamination by a flow of inert gas from the torch nozzle (*Figure 1*). The filler metal is a handheld rod compatible with the base metal. The filler metal is manually fed into the leading edge of the weld puddle. GTAW produces high-quality welds without slag or oxidation. Because there is no flux, there is no corrosion due to slag entrapment. As a result, minimal postweld cleaning is needed.

A major use of GTAW is to make high-quality welds on aluminum piping. The GTAW process allows greater control of root penetration and fill than almost any other process. For this reason, GTAW is often used to make the root pass on aluminum pipe, even when the remaining passes are made with GMAW, which has a higher filler-metal deposition rate.

The V-groove and modified U-groove welds are typically used for joining medium- and thick-

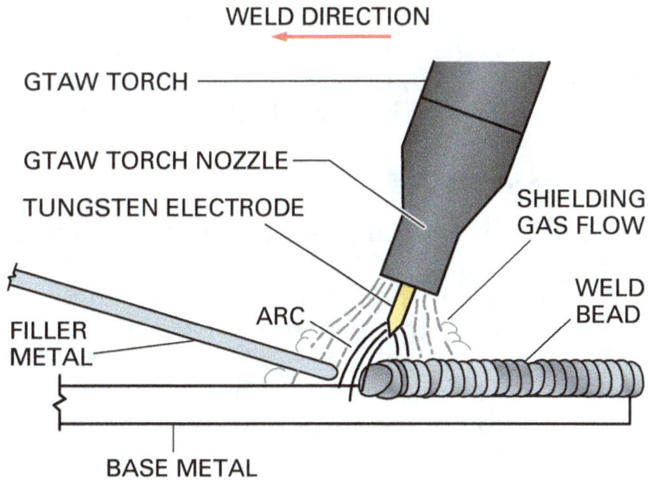

Figure 1 GTAW process.

walled pipe used in various piping systems. Piping containing or carrying material that can cause injury to people or damage the environment, or those that operate at elevated pressures, may be categorized as critical piping. Injury or damage occurs when the piping system or other system components fail to contain the material as designed. Welds in critical piping may be required to meet more stringent requirements. Piping that is used for hydronic heating and air conditioning applications, simple water systems, and similar applications may be assessed using less stringent requirements.

This module provides guidance in the application of modified U-groove welds on aluminum pipe in the 2G, 5G, and 6G welding positions. The dimensions and specifications in this module are representative of codes in general and may not be specific to any particular code. Always follow the proper codes for your site.

1.1.0 GTAW Techniques

GTAW weld bead characteristics and quality are affected by several factors, each of which is influenced by the way the welder handles the torch. These factors include the following:

- Torch travel speed and arc length
- Torch angles
- Torch and filler metal handling techniques
- Consistent feeding of the filler metal

1.1.1 Torch Travel Speed and Arc Length

Torch travel speed and arc length affect the GTAW weld puddle and the penetration depth of the weld. Slower travel speeds allow more heat to concentrate and form larger, more deeply penetrating puddles. Faster travel speeds prevent heat buildup and form smaller, shallower puddles. Arc length is the major control for bead width. As the torch is raised, the arc length and voltage increase, causing the bead width to increase as well.

1.1.2 Torch Angles

The two basic torch angles that must be controlled when performing GTAW are the work angle and the travel angle. The definition of these angles is the same as for other methods of welding.

The torch work angle (*Figure 2*) is an angle of 90 degrees or less between a line perpendicular to the workpiece surface at the point of electrode contact and a plane determined by the electrode axis and the weld axis. For a T-joint or corner joint, the line is perpendicular to the non-butting member. For pipe, the plane is determined by the electrode axis and a line tangent to the pipe surface at the same point. As shown in *Figure 2*, when the torch is held perpendicular, it is referred to as a *0-degree work angle*. To properly position beads in a groove joint, the work angle may be changed 20 degrees in either direction.

The torch travel angle is an angle of 90 degrees or less between the electrode axis and a line perpendicular to the weld axis at the point of electrode contact in a plane determined by the electrode axis and the weld axis. For pipe, the plane is determined by the electrode axis and a line tangent to the pipe's surface at the same point. The travel angle used for GTAW is a push angle (*Figure 3*). A push angle is created when the torch is tilted back so that the electrode tip precedes the torch in the direction of the weld. In this position, the electrode tip and shielding gas are being directed ahead of the weld bead being formed. This provides a cleaning action on the positive portion of the AC voltage cycle. Push angles of 15 to 20 degrees are commonly used for aluminum GTAW.

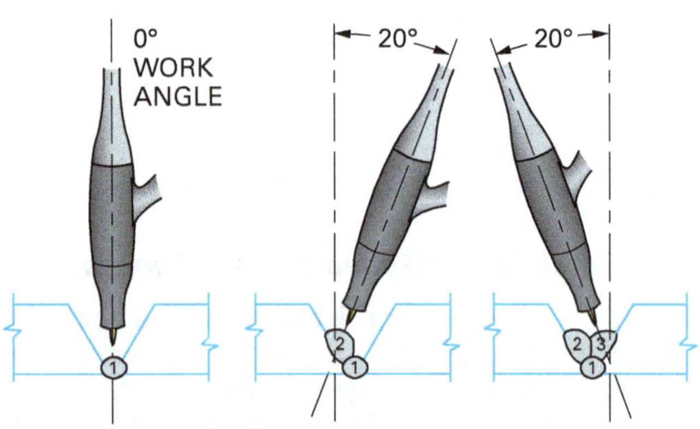

Figure 2 Typical GTAW torch work angles.

Weld Puddle Size

When performing GTAW in the horizontal, vertical, and overhead positions, you must keep the weld puddle smaller than it is in the flat position. The weld puddle is generally held in the joint by surface tension. When the puddle is too large, gravitational force pulling down on the puddle overcomes surface tension and the liquid metal runs out of the joint.

1.1.3 Torch and Filler Metal Handling Techniques

The two basic torch and filler metal handling methods used to perform GTAW are the freehand technique and the walking-the-cup technique. Both techniques are described here.

The freehand technique is the preferred method for aluminum. In this method, the torch electrode tip is held just above the weld puddle or base metal (*Figure 4*). The torch is supported by the welder's hand like a pencil. The hand is usually steadied by resting some part of it on or against the base metal. This helps to keep the proper arc length (distance from the electrode tip to the weld puddle). Due to the high thermal-transfer properties of aluminum, it may help to insert a piece of non-flammable insulation between the gloved hand and the workpiece. As needed, the welder can move the torch tip in a small circular motion within the molten puddle. This maintains the puddle size and advances the puddle. Filler metal is added as needed.

The GTAW filler metal rod is held in the hand that is not supporting the torch. Hold the rod at an angle of 15 to 20 degrees above the base metal surface and in line with the weld. When a 15- to 20-degree push angle is also used, the rod remains roughly perpendicular to the electrode axis. Always keep the tip of the rod within the shielding gas envelope to protect it from atmospheric contamination and to keep it preheated.

Insert the rod into the leading edge of the weld puddle, using extreme care not to touch the tungsten electrode to the end of the rod. If the tungsten electrode touches the rod or weld puddle, it will become contaminated with filler metal. The electrode must then be removed, the contaminated portion broken off, and the end reformed before welding can continue. To prevent contamination of the electrode by the filler metal, move the electrode to the back edge of the weld puddle as the rod is inserted into the weld puddle's lead-

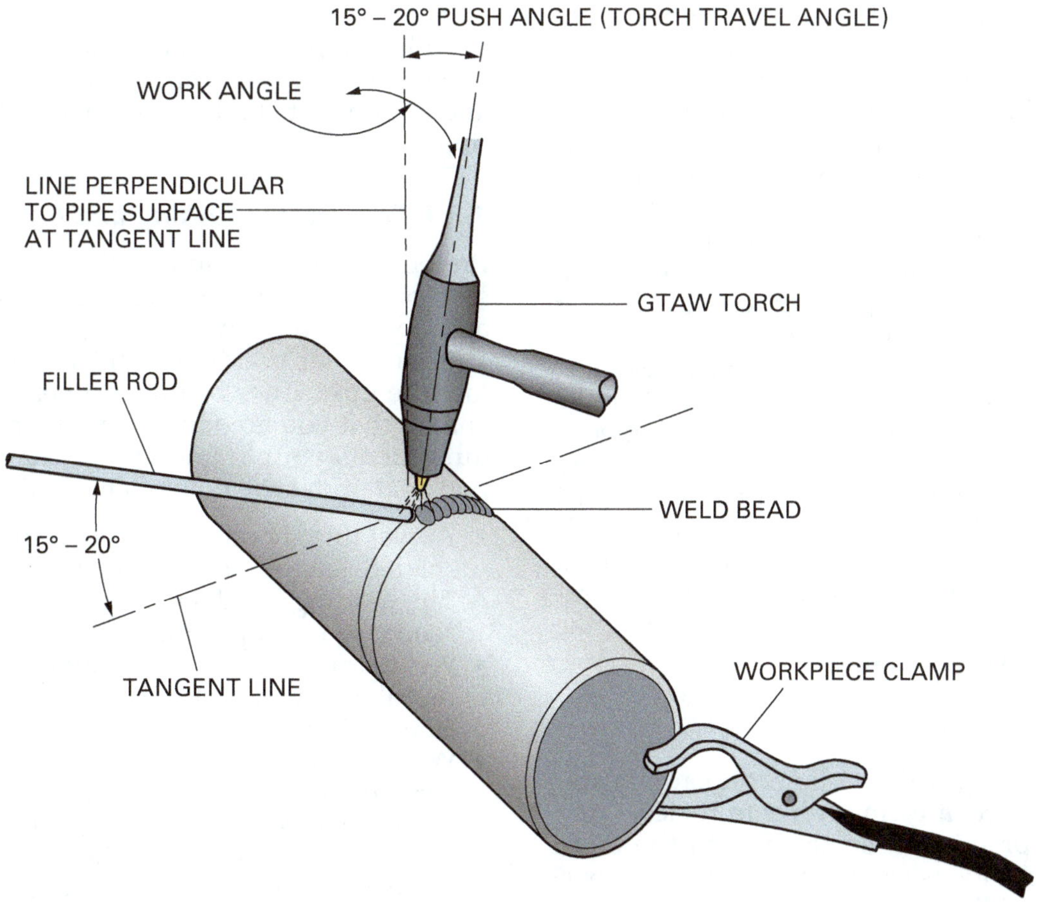

Figure 3 GTAW torch travel angles.

Figure 4 Freehand technique.

ing edge. Timing those two actions helps prevent tungsten contamination.

In the walking-the-cup technique, the welder rests the edge of the torch nozzle (cup) against the base metal or groove edges to steady the torch and maintain a constant arc length. The torch is rocked from side to side on the edge of the cup as it is advanced. This action maintains the puddle size and heats both sides of the groove. The filler metal is added in the same manner used in the freehand technique, taking care not to contaminate the electrode.

The walking-the-cup technique can be very difficult to use when welding aluminum. Since the cup is in contact with surface, its diameter determines the distance of the tungsten from the workpiece. If the wrong cup size is chosen, the welder must adjust the travel angle to achieve the desired distance, possibly to an angle that causes a poor result. To achieve success with this method, it is therefore important to choose the correct cup size. The cup also tends to stick to the freshly welded metal, creating an added challenge.

1.1.4 Consistent Feeding of the Filler Metal

Feeding the filler metal consistently as the weld progresses is important to ensure the weld puddle develops evenly and produces an even bead. Like all welding techniques, the technique described here requires practice to master but is well worth the effort.

Cradle the rod between the pointer finger and the middle finger, similar to the way you might hold a pencil. Then use the thumb to consistently push the rod forward in small motions. Once the thumb is extended pushing the wire forward, quickly bend and reposition the thumb to begin a new forward push. Consistent motion is important, so the thumb must be repositioned quickly to maintain the pace. The thumb can also be used to slowly roll the rod as well during the pushing motion. This technique is easily practiced without actively welding; simply practice with a rod at home or during free time to become comfortable with the technique before trying it on a weld.

Another technique to consider is called the *finger-pull technique*. The rod is secured between the pointer and middle fingers, which are used to pull the electrode forward. At the end of the forward stroke, the fingers are slipped backwards along the rod to reposition, without pulling it backwards and out of the weld puddle in the process. Once the two fingers are repositioned, the pulling motion begins again.

1.2.0 Safety Practices

Before proceeding with welding coupon preparation and practice, it is important to review common safety practices associated with the craft. Safety considerations are a critical part of preparing for any welding operation.

The following is a summary of safety procedures and practices that must be observed while cutting or welding. Be advised that this is only a summary. Complete safety coverage is provided in the *Welding Level One* module, "Welding Safety." That module should be studied, if not already completed, before continuing. Above all, be sure to wear appropriate protective clothing and equipment at all times when welding or cutting.

1.2.1 Protective Clothing and Equipment

Protective clothing and equipment are a necessary part of every welding task. Be sure to wear appropriate protective clothing and equipment at all times when welding or cutting (*Figure 5*). Follow the guidelines below as a general rule, and ensure that you are familiar with any unique requirements your employer may have regarding personal protective equipment (PPE).

- A welding hood equipped with a properly shaded lens is best for all forms of welding. The welding hood lens must be an appropriate shade for the application.
- Wear proper leather or flame-retardant clothing along with welding gloves for protection from flying sparks, molten metal, and heat.
- Wear high-top safety shoes or boots. Make sure that the tongue and lace area of the footwear is covered by a pant leg. If the tongue and lace area is exposed or if the footwear must be protected from burn marks, wear spats under the pant leg or leather chaps over the front and top of the footwear.
- Wear a 100-percent cotton cap without any mesh material. If a hard hat is required for the environment, use one that allows the attachment of rear deflector material and a face

Heat Protection

GTAW welding often places the welder's hand in direct contact with the workpiece to steady the torch and maintain a consistent arc length. However, especially when welding aluminum, the workpiece can become extremely hot. Even the best gloves are not enough to effectively protect the welder's hand, leading to the possibility of injury and more weld-bead terminations due to discomfort.

There are several products available that can provide extra protection against heat as well a surface that slides across the workpiece easily, without being too clumsy and interfering with the welder's dexterity. Some are designed to fit over a portion of the hand or multiple fingers, while other products slide over a single finger. One thing is certain—many GTAW welders are thankful for them and can weld very long beads without burns or discomfort.

Figure 5 GTAW welder with the proper PPE.

shield. A hard hat with a rear deflector is generally preferred when working overhead, and may be required by some employers and job sites.

- Wear earmuffs, or at least earplugs, to prevent sparks from entering the ear canals.

1.2.2 Housekeeping and Fall Protection

Welding is often performed outside in the weather and in the construction environment. Even production work completed inside large, open shops exposes a welder to equipment moving across the floor and overhead. Welders often have to crawl under or reach over objects to complete their work.

At times, welders must work at height. Fall protection equipment is required for all work at height (*Figure 6*). Always keep fall protection devices untangled and properly attached. Combined with the other equipment a welder must carry, there is a great deal to manage, but fall protection equipment must always be the priority. Fall protection equipment saves lives. Never attach any kind of welding gear to the harness that could interfere with its ability to perform its intended function.

Regardless of where you are working, always keep the work area as neat as possible. This makes the area safer and helps to reduce distractions. Mindful actions, such as making sure that welding cables are not tripping hazards, contrib-

Friction Stir Welding

A marriage between steel and aluminum has long been a problem in the welding world, due to their extreme differences in thermomechanical properties. Fusion welding of these materials tends to form compounds that are too brittle for reliability. Friction stir welding (FSW) is changing that, and the process is equally effective with aluminum alone.

Friction stir welding works by applying a tool to the surface of the two metals along the desired joint. The process is autogenous (requires no filler metal). Through friction alone, enough heat is generated to join the two metals, although they never fully melt. The process has already been used in a few specialized applications, such as the front sub-frame of the 2013 Honda Accord and the external fuel tank on the space shuttle. Boeing presently uses the process in aircraft structures and automotive wheel manufacturers are making use of it. The process creates less distortion; doesn't produce a significant heat-affected zone; and no loss of strength associated with the use of a filler metal is suffered. A number of videos of the process can be found through an internet search.

Figure 6 Working at height.

ute to a safer environment. Electrode stubs and unused electrodes must not be left on the floor. They can cause someone to lose their footing and fall. Parts waiting to be welded or left over from previous work are a tripping hazard and should be kept out of the walking area. If the parts have sharp edges, they could cut welding cable insulation or torch gas hoses.

1.2.3 Fire and Explosion Prevention

Flammable gases are commonly found around welding and cutting operations, as are sparks and other significant sources of ignition. For that reason, and because many fire-related incidents over the years have been traced back to welding operations, it is crucial that fire and explosion prevention be given consistent attention. Follow these guidelines to maintain a safe environment:

- Never carry matches or gas-filled lighters. Sparks can cause the matches to ignite or the lighter to explode in your pocket, causing serious injury.
- Never use oxygen to blow dust or dirt from clothing. Traces of the oxygen can remain trapped in the fabric for some time. If a spark hits the fabric during this time, the clothing can burn rapidly and violently out of control.
- Make sure that any flammable material in the work area is removed or shielded by a fire-resistant covering. Approved fire extinguishers must be available before attempting any heating, welding, or cutting operations. If a hot-work permit and a fire watch are required, be sure these are in place before beginning.
- Ensure that all site safety requirements are being observed.
- Never release a large amount of oxygen or use oxygen in place of compressed air. It can cause rapid and uncontrolled combustion around flammable materials or sparks. Keep oxygen away from oil, grease, and other petroleum products.
- Never release a large amount of fuel gas (especially acetylene). Because propane is heavier than air, it tends to concentrate in and along low areas, and can ignite at a considerable distance from the release point. Acetylene is lighter than air, but is even more dangerous than propane. When mixed with air or oxygen, acetylene will ignite at much lower concentrations than any other fuel gas.
- Although shielding gases are not toxic, any gas released in a large quantity can displace the oxygen in a small room or enclosed work area, creating an atmosphere with little or no oxygen. Avoid large releases of any type of gas, regardless of its characteristics.
- To prevent fires, maintain a neat and clean work area, and make sure that any metal scrap or slag is cold before disposal.
- Before cutting containers such as tanks or barrels, check to see if they have contained any explosive, hazardous, or flammable materials, including petroleum products, citrus products, or chemicals that decompose into toxic fumes when heated. As a standard practice, always clean and then fill any tanks or barrels with water, or purge them with a flow of inert gas such as nitrogen to displace any oxygen. Containers must be cleaned by steam cleaning, flushing with water, or washing with detergent until all traces of the material have been removed. Proper procedures for cutting or welding hazardous containers are described in the *American Welding Society (AWS) F4.1, Safe Practices for the Preparation of Containers and Piping for Welding and Cutting.*

WARNING!

Welding or cutting must never be performed on drums, barrels, tanks, vessels, or other containers until they have been emptied and cleaned thoroughly, eliminating all flammable materials and all substances (such as detergents, solvents, greases, tars, or acids) that might produce flammable, toxic, or explosive vapors when heated. Clean containers only in well-ventilated areas. Vapors can accumulate during cleaning, causing explosions and serious personal injury.

1.2.4 Work Area Ventilation

Many gases are introduced to the atmosphere through welding processes. Some of these gases are harmful, and all can displace or combine with atmospheric oxygen in a confined space. Follow these guidelines, as well as those of your employer, for work area ventilation:

- Make sure to follow all confined space procedures before conducting any welding or cutting in a confined space. Use powered extraction systems as needed.
- Never use oxygen to provide ventilation in confined spaces.
- Always perform cutting or welding operations in a well-ventilated area.

Cutting or welding operations involving zinc or cadmium materials or coatings result in toxic fumes. Another significant concern is Chromium 6, commonly referred to as *hexavalent chromium*, which is used in stainless steel. It releases toxic fumes when heated. Manganese is another hazardous metal used in stainless steel. It is also used in some welding rods, increasing the risk of exposure.

For long-term cutting or welding of materials containing these components, always wear an approved full-face, supplied-air respirator that uses breathing air supplied from outside the work area. For occasional short-term exposure, a HEPA-rated or metal-fume filter may be used on a standard respirator.

Powered fume-extraction equipment must be used when welding materials that contain zinc, cadmium, hexavalent chromium, manganese, and similar materials that produce hazardous fumes if the area is not naturally well-ventilated. *Figure 7* shows a powered air-purification respirator used with fume extraction equipment.

1.3.0 Preparations for Welding

Before welding practice can take place, the work area and welding equipment must be prepared, and the coupons to be welded must be cut and cleaned. The following sections explain how to prepare for GTAW of aluminum pipe.

To practice welding, you will need a welding table, bench, or stand. The preferred welding surface is aluminum. Provisions must be available for mounting practice welding coupons out of position.

To set up the area for welding, follow these steps:

Step 1 Make sure that the area is properly ventilated. Use doors, windows, fans and localized fume extraction equipment as needed.

Step 2 Check the area for fire hazards. Remove any flammable materials before proceeding.

Step 3 Locate the nearest fire extinguisher. Do not proceed unless the extinguisher is charged and you know how to use it.

Step 4 Set up welding screens around the welding area.

1.3.1 Aluminum Filler Metals

Filler metals are selected to be compatible with the base metal to be welded. The WPS or site standards will specify the proper filler metal type and size to use.

CAUTION

Always use filler metal as specified by the WPS or site quality standards, if applicable, to prevent weld discontinuities and satisfy the requirements of the weld.

Unlike steel, the aluminum filler metal will not develop a tensile strength equal to that of the base metal. As a result, aluminum weld joints are the weakest link in a welded assembly. This important fact is another reason why proper filler metal selection for aluminum pipe is critical.

Figure 7 Powered air-purification respirator (PAPR) and localized fume extraction equipment.

For all welding exercises associated with this module, weld the aluminum pipe coupons using 3/32" and/or 1/8" aluminum filler metal rods as specified by your instructor. Remove only a small number of filler rods at a time. Keep the remainder in the package to keep them clean. Before using the filler metal, check it for burned ends or contamination, such as oxidation, dirt, oil, or grease, all of which can cause weld discontinuities. Clean the filler metal with a clean rag and acetone. Use emery cloth or stainless steel wool to remove oxides if necessary, and snip any burned ends. If the filler metal cannot be cleaned effectively, do not use it.

1.3.2 Practice Pipe Weld Coupons

Pipe weld coupons should be cut from Schedule 40 aluminum pipe that is 3" to 12" (DN80 to DN300) in nominal diameter. Each welded joint requires two coupons of the same size. *Figure 8* shows typical AWS aluminum-pipe specifications for V-groove and modified U-groove joint preparation. The modified U-groove joint is preferred when there is no backing.

Welding practice for this module should be done using the modified U-groove joint shown in *Figure 8*. The V-groove joint shown can be used, although a backing gas is needed. A backing gas is not required for the modified U-groove joint.

Aluminum is easily contaminated. Proper coupon preparation is very important. Aluminum pipe coupons must be cut to the desired lengths, cleaned, and beveled to the proper angles before they are positioned for welding. The preferred technique for beveling and/or adding a root face is mechanical machining. Grinding is also acceptable, although machining the grooves and root faces of aluminum pipe is the cleanest and most uniform way to prepare pipe for GTAW. Thermal cutting techniques are not considered acceptable. If thermal cutting techniques must be used to cut aluminum alloys, at least 1/8" (3 mm) of the cut surface must be removed by machining or grinding before welding.

WARNING!

Grinding wheels used on aluminum must be made for aluminum only. Using other types of grinding wheels on aluminum may cause the grinding wheels to shatter.

CAUTION

Grinding wheels, files, and similar tools used on aluminum should not be used on any other type of metal or on surfaces contaminated with grease or oil.

Clean oil and grease from the coupons with an approved substance such as acetone. After oil and grease have been removed, use a stainless-steel wire brush to remove all other contaminants and oxidation. Use only light pressure on the brush.

NOTE

When using the modified U-groove joint, it is best to clean the backside of the weld (the inside of the pipe) with acetone or other approved solvent to remove surface contaminants, but to avoid cleaning with wire brushes or grinders. The high melting point of the thin film of oxidation left behind can help provide a form of backside protection to the joint.

WARNING!

To prevent injury when using cleaning solvents, always comply with the SDS guidelines for the solvents being used.

To prepare pipe coupons for groove weld joints, follow these steps:

Step 1 Bevel or groove the end of the pipe, as shown in *Figure 8*, using any acceptable method discussed in this section.

GOING GREEN

Conserve Aluminum Materials

Aluminum is expensive and difficult to obtain for practice purposes. To conserve resources and reduce waste, completely use all weld coupons until all surfaces have been welded upon. Cutting coupons apart and reusing the pieces conserves materials. Use flat material that cannot be cut into weld coupons to practice running beads. Note that stringer beads can be practiced while leaving enough room between them to practice weave beads as well. Any aluminum material that is no longer suitable for welding practice should be recycled.

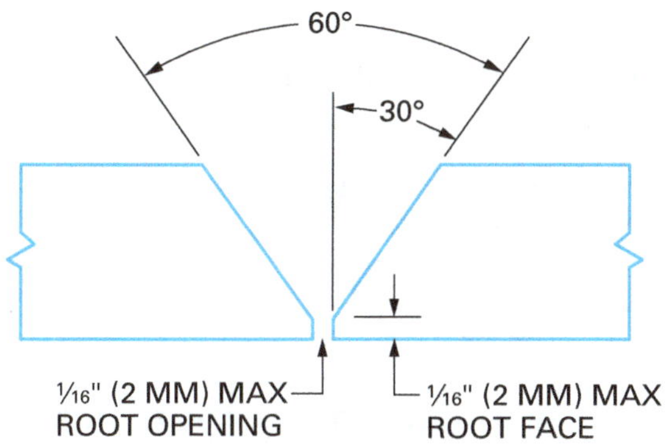

STANDARD V-GROOVE (WALL THICKNESS 3⁄4" OR LESS)

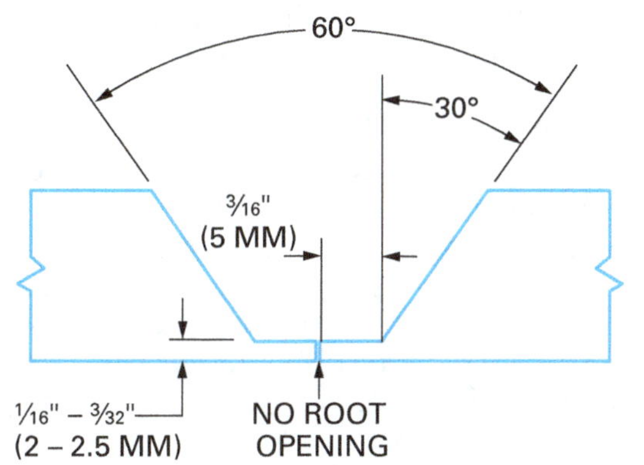

MODIFIED U-GROOVE (PREFERRED) FOR AC GTAW

Figure 8 AWS aluminum pipe groove specifications for use without backing.

Step 2 Check the groove angle and root face for accuracy. There should be no dross. The angle and dimensions should be as shown in *Figure 8*, with no notches deeper than 1⁄16" (1.6 mm).

Step 3 Cut off the section with the beveled pipe end. The piece should be a minimum of 4" (10 cm) long.

Step 4 Prepare a second, matching section of pipe to complete the coupon assembly.

Step 5 Use an approved solvent to remove any grease or oil within 1" (2.5 cm) of the weld joint. Clean the outside of the aluminum pipe with a grinder or stainless steel brush as needed. As noted above, do not use a grinder or wire brush on the inside of the pipe when using the modified U-groove joint; the layer of oxidation inside can actually help provide some backing protection.

CAUTION

To prevent contamination, do not use a stainless steel wire brush on anything else after it has been used on aluminum.

Step 6 Align the two pipe coupon sections so that their inside surfaces are even all around. The coupons must also be aligned so that the resulting welded pipe assembly is straight. Align small-diameter pipe by clamping both pieces to a piece of angle iron. Larger pipe can be aligned with the aid of a pipe alignment jig (*Figure 9*) or by holding a straightedge across the joint, parallel to the pipe axis. The straightedge must be used on all four quadrants of the pipe in case one or both sections are distorted.

Step 7 When the pipe ends are properly aligned with no root opening, make the first of four tack welds.

CAUTION

When welding pipe, avoid making arc strikes outside of the weld groove on the surface of the pipe. An arc strike can cause a hardened spot that can crack as the pipe expands and contracts. An arc strike on the pipe surface is considered a defect, which will require repair or reworking.

Step 8 Make the second tack weld on the opposite side from the first, making sure there is no root opening.

Step 9 Check the root again, and weld the third tack midway between the first two tacks.

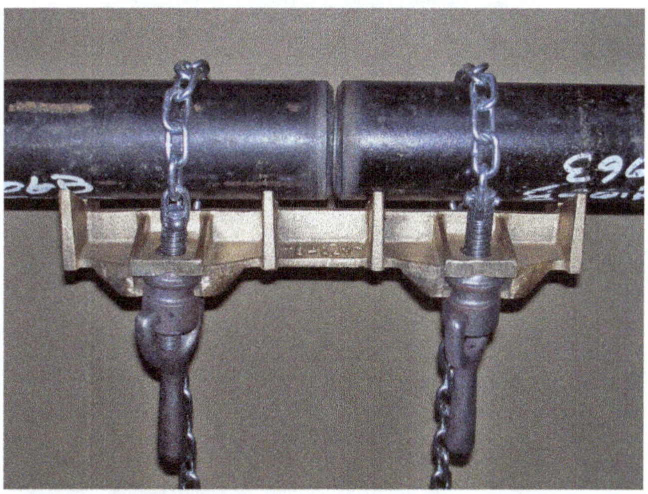

Figure 9 Pipe alignment jig (carbon steel pipe shown).

Step 10 Weld the fourth tack opposite the third tack and midway between the first and second tacks. There should now be four tack welds evenly spaced at 90 degrees around the pipe coupon (*Figure 10*).

Step 11 Feather the tack welds with the edge of a grinding wheel. Feathering the ends of the tack welds with a grinder helps to fuse the root pass into the tack weld.

1.4.0 Welding Equipment

Welding aluminum with GTAW typically requires equipment that provides alternating current (AC) with an argon shielding gas. Aluminum GTAW requires a significant amount of welding current due to its thermal conductivity; heat rapidly flows away from the weld joint. Ensure that the equipment selected is capable of providing sufficient current at a reasonable duty cycle. Welding aluminum at welding-current levels less than those required means that the weld cannot progress at the proper speed. As a result, the workpiece will overheat and create an inferior weld. Because of the higher currents associated with aluminum welding, water-cooled torches are recommended.

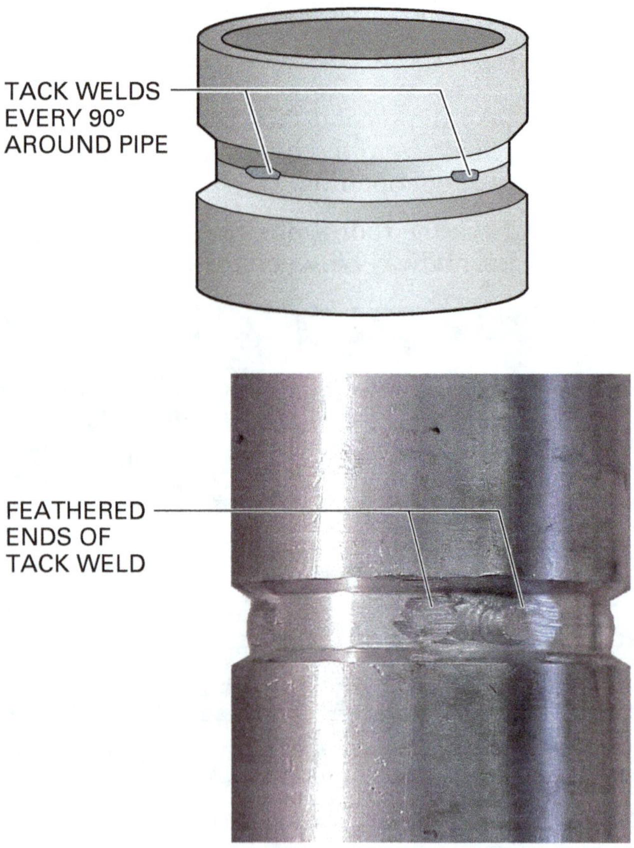

Figure 10 Modified U-groove weld coupons tacked together.

> **Welding Large-Diameter Pipe**
>
> The practice coupons on large-diameter pipe should be greater than 6" in length. They need more metal to help absorb the increased heat used to make these welds.

Of course, this will not reduce the temperature of the workpiece itself.

Conventional AC machines used for GTAW are typically based on sine-wave or square-wave technology. Newer inverter-technology units have additional features, such as high-frequency start only and current sequencing (start/stop ramping) that help control puddle formation at the beginning of the weld and also help in filling the termination crater. These machines provide additional control over important variables that affect the quality of aluminum welds. The ability to vary both the AC wave balance and frequency is preferred for aluminum welding. Control over the balance between the electrode negative (EN) and electrode positive (EP) portions of the AC sine wave allow the operator to control two important functions:

- As the wave is unbalanced to the electrode-negative side, the current-carrying capacity of the electrode is increased and the amount of cleaning action is decreased.
- When welding on new material that is properly cleaned, the wave can be set at 80 percent EN or more. This allows the use of a smaller electrode at higher welding currents.

Conventional welding machines in the United States use 60-cycle electrical power as supplied by the line current. (Many other countries use 50 cycles.) Newer inverter machines have the ability to vary the frequency, potentially increasing it to 300 cycles per second or even more. This eliminates the need to use a balled pure- or zirconiated-tungsten electrodes and permits the use of a truncated EWTh2 or other tungsten electrode. It also provides better arc control and permits smaller weld sizes. Frequencies of 120 to 140 cycles are typically used to weld aluminum.

Although pure tungsten (EWP) electrodes are used for welding aluminum with AC and high-frequency machines, zirconiated-tungsten electrodes (EWZr) are often preferred due to their higher current-carrying capacity and lower susceptibility to contamination. They also hold a balled tip better when welding aluminum. *Table 1* shows the current ranges for various electrode types and sizes.

Table 1 Typical Current Ranges for Tungsten Electrodes

Electrode Diameter		Direct Current Amperes		Alternating Current Amperes			
		DCEN	DCEP	Unbalanced Wave		Balanced Wave	
		EWCe-2 EWLa-1 EWLa-1.5 EWLa-2 EWP EWTh-1 EWTh-2	EWCe-2 EWLa-1 EWLa-1.5 EWLa-2 EWP EWTh-1 EWTh-2	EWP	EWCe-2 EWLa-1 EWLa-1.5 EWLa-2 EWTh-1 EWTh-2 EWZr	EWP	EWCe-2 EWLa-1 EWLa-1.5 EWLa-2 EWTh-1 EWTh-2 EWZr
in.	mm						
0.010	0.30	up to 15	(2)	up to 15	up to 15	up to 15	up to 15
0.020	0.50	5-20	(2)	5-15	5-20	10-20	5-20
0.040	1.00	15-80	(2)	10-60	15-80	20-30	20-60
1/16	1.60	70-150	10-20	50-100	70-150	30-80	60-120
3/32	2.40	150-250	15-30	100-160	140-235	60-130	100-180
1/8	3.20	250-400	25-40	150-210	225-325	100-180	160-250
5/32	4.00	400-500	40-55	200-275	300-400	160-240	200-320
3/16	4.80	500-750	55-80	250-350	400-500	190-300	290-390
1/4	6.40	750-1000	80-125	325-450	500-630	250-400	340-525

(1) All values are based on the use of argon as the shielding gas.Other current values may be employed depending on the shielding gas, type of equipment, and application.

(2) These combinations are not commonly used.

EWCe-2 = Tungsten + 2% Ceria
EWLa-1 = Tungsten + 1% Lanthana
EWLa-1.5 = Tungsten + 1.5% Lanthana
EWLa-2 = Tungsten + 2% Lanthana
EWP = Pure Tungsten
EWTh1 = Tungsten +1% Thoria
EWTh2 = Tungsten +2% Thoria
EWZr = Tungsten + Zirconia

Once an appropriate welding machine has been selected, follow these steps to prepare it:

Step 1 Verify that the welding machine can be used for GTAW. If a water-cooled torch will be used, select an appropriate cooling unit.

Step 2 Select an air-cooled or water-cooled GTAW torch. Make sure that it is compatible with the welding machine and any optional cooling unit. Water-cooled torches are preferred. Equip the torch with an appropriate cup size for the technique to be used (freehand versus walking-the-cup).

Step 3 Know the location of the primary power disconnect switch.

Step 4 Configure the welding machine for aluminum GTAW welding (*Figure 11*) as directed by your instructor. Configure the torch polarity to AC. Equip the torch with a properly prepared $\frac{3}{32}$" or $\frac{1}{8}$" (2.4 or 3.2 mm) EWP or EWZr tungsten electrode.

Step 5 Connect the proper argon or argon/helium shielding gas for the application as directed by your instructor, filler metal manufacturer, WPS, or site quality standards.

Aluminum Pipe Weld Backing

Open-root and consumable-insert welding do not generally work well with aluminum piping. When you need to make an open-root groove weld, the welder must back it with a temporary or permanent backing ring; otherwise, the root opening must be eliminated from the standard V-groove joint configuration. No gas backing is required when the modified U-groove joint is used, or when a temporary or permanent backing ring is used with a V-groove joint. The modified U-groove joint is recommended for manual AC GTAW of pipe.

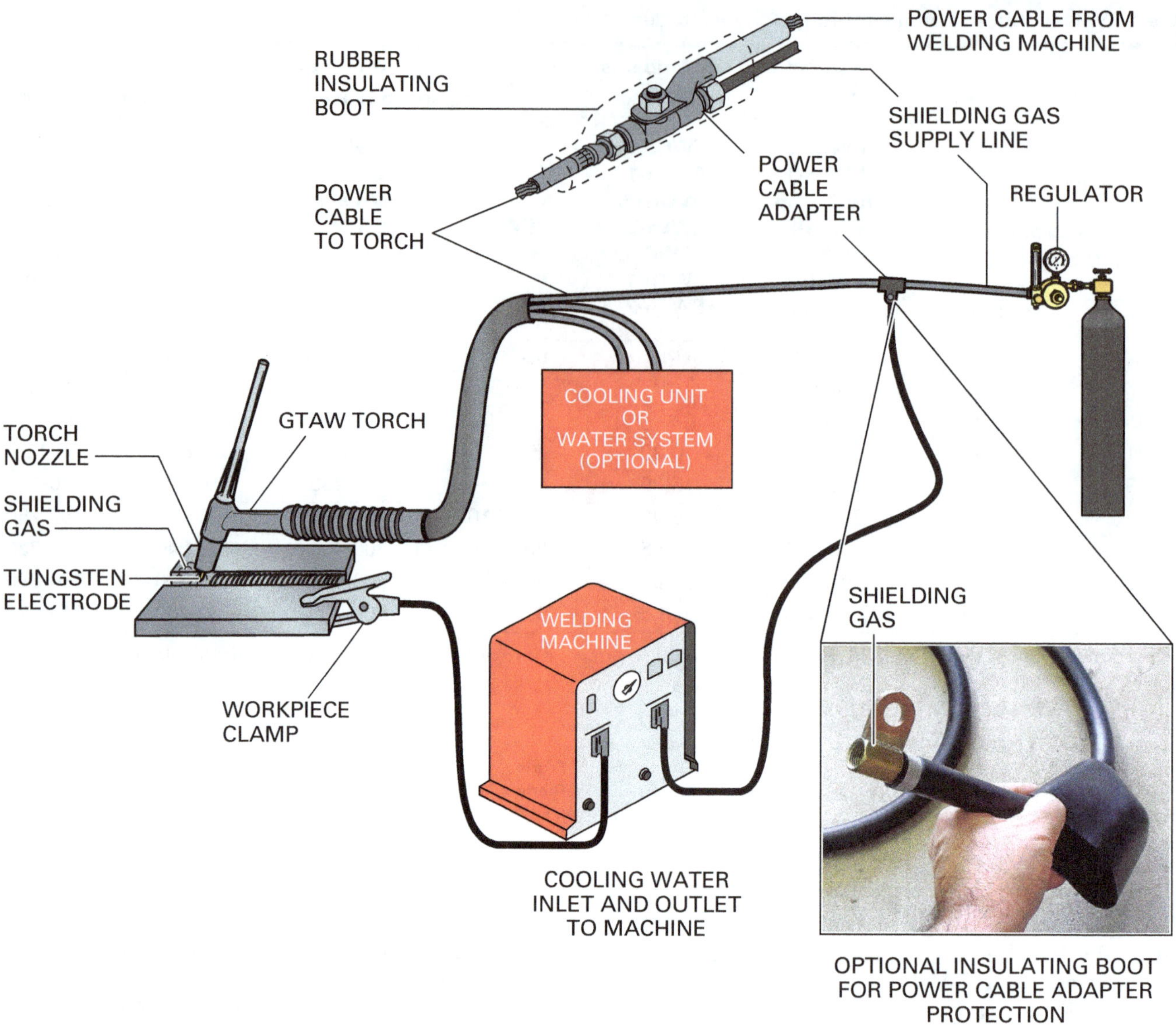

Figure 11 Configuration diagram of typical GTAW welding equipment.

Step 6 Connect the clamp of the workpiece lead to the workpiece.

Step 7 Turn on the welding machine. Adjust the flowmeter and purge the torch according to the manufacturer's instructions.

Step 8 Set the initial welding current for the application.

1.4.1 Selecting a Shielding Gas

A shielding gas is selected according to the material being welded. Argon is the most commonly used shielding gas. Argon can be used for welding a wide range of materials, including aluminum, titanium, steel, and stainless steel.

Helium and helium/argon mixtures raise the temperature of the arc. This rise in temperature promotes faster welding speeds and deeper weld penetration. However, some disadvantages include that helium is expensive and can cause difficulty in starting the arc.

Impurities in the shielding gas or mixture with other gases, such as carbon dioxide (CO_2), will contaminate the weld and change the color of the aluminum (usually to black).

CAUTION

Always use the correct gas when performing GTAW. Accidental use of a gas mix intended for GMAW can cause rapid destruction of the tungsten and contamination of the weld.

1.5.0 GTAW Techniques and Positions for Aluminum Pipe

Aluminum pipe is commonly welded with alternating current (AC) using argon as a shielding gas. In some cases, an argon/helium mix is used for increased penetration, although adding helium significantly increases material costs. Higher currents are used when welding aluminum because of high thermal conductivity in the base metal. The modified U-groove joint is preferred for aluminum pipe joints without backing because it reduces heat transfer away from the weld joint. The joint design also allows complete root penetration and fusion with a smaller weld puddle.

A smaller weld puddle is more easily controlled in the vertical and overhead positions by surface tension and molecular attraction than a large puddle. The challenge is to establish a weld puddle that penetrates and fuses with the pipe without falling through to the inside. The most difficult part of welding aluminum is getting the proper penetration and fusion for the root pass when a backing material is not there to prevent the puddle from falling through.

1.5.1 Pipe Groove-Weld Test Positions

Groove welds may be made in all positions on pipe. The weld position is determined by the axis of the pipe. Four standard weld-test positions are used with pipe (*Figure 12*).

To determine the quality of a pipe weld, specific sections of pipe are removed for destructive testing. For positions 1G-ROTATED and 2G, two specimens are removed. They must be taken from opposite sides of the pipe, as shown in *Figure 13*. To destructively test 5G and 6G welds, the test specimens are cut from the four regions shown. Once removed, the specimens are put through different bend tests to determine the quality of the welds.

1.5.2 Acceptable and Unacceptable Pipe Weld Profiles

Pipe groove welds should be made with slight reinforcement and a gradual transition to the base metal at each toe. The root pass should have complete penetration. The face reinforcement can range from being flush to a maximum of 3/32" (2.4 mm) for materials up to 3/8" (9.6 mm) thick. Excessively large face reinforcement reduces the pipe's strength.

Versatile GTAW Machine

Miller's Dynasty 350TIGRunner is a versatile machine designed to provide superior arc performance with complete control over the arc shape. Auto-Line™ allows the unit to be connected to any voltage worldwide from 208 to 575 volts. The unit also operates on power supplies using a 50 or 60Hz frequency. Extended AC balance controls the amount of oxide cleaning, which is essential for high-quality welds on aluminum.

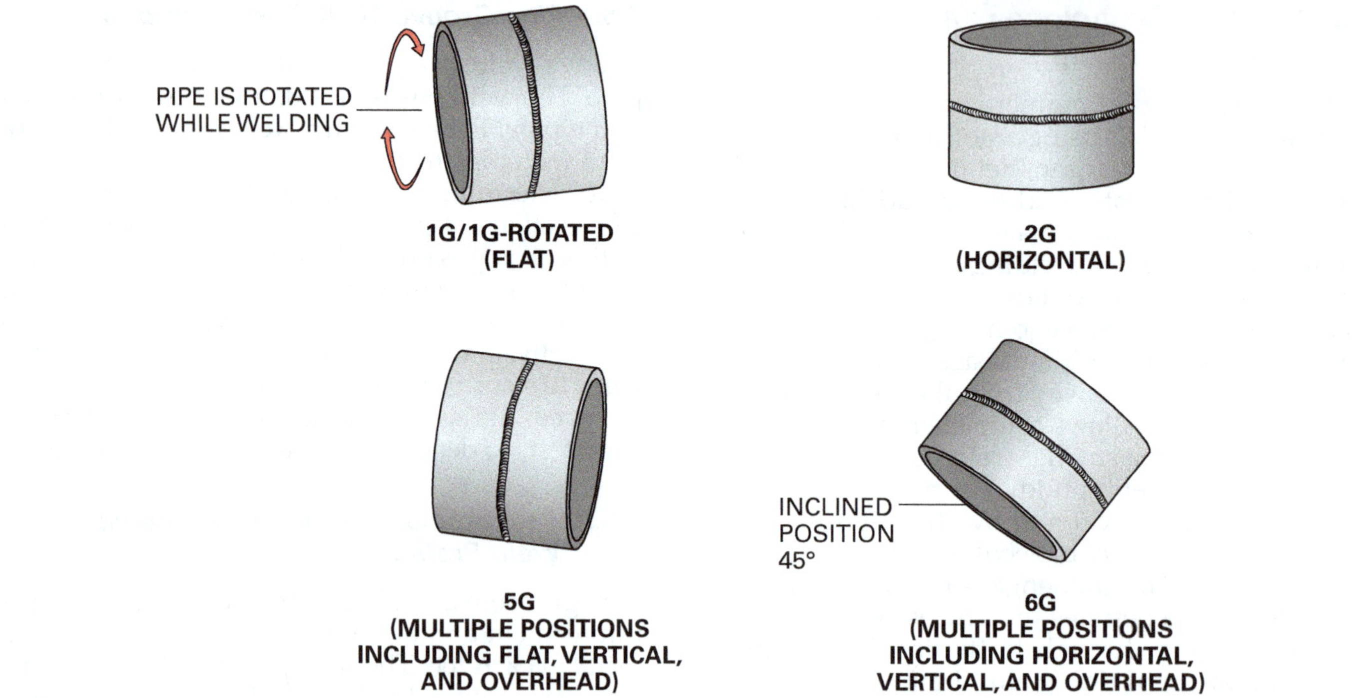

Figure 12 Four basic pipe groove-weld test positions.

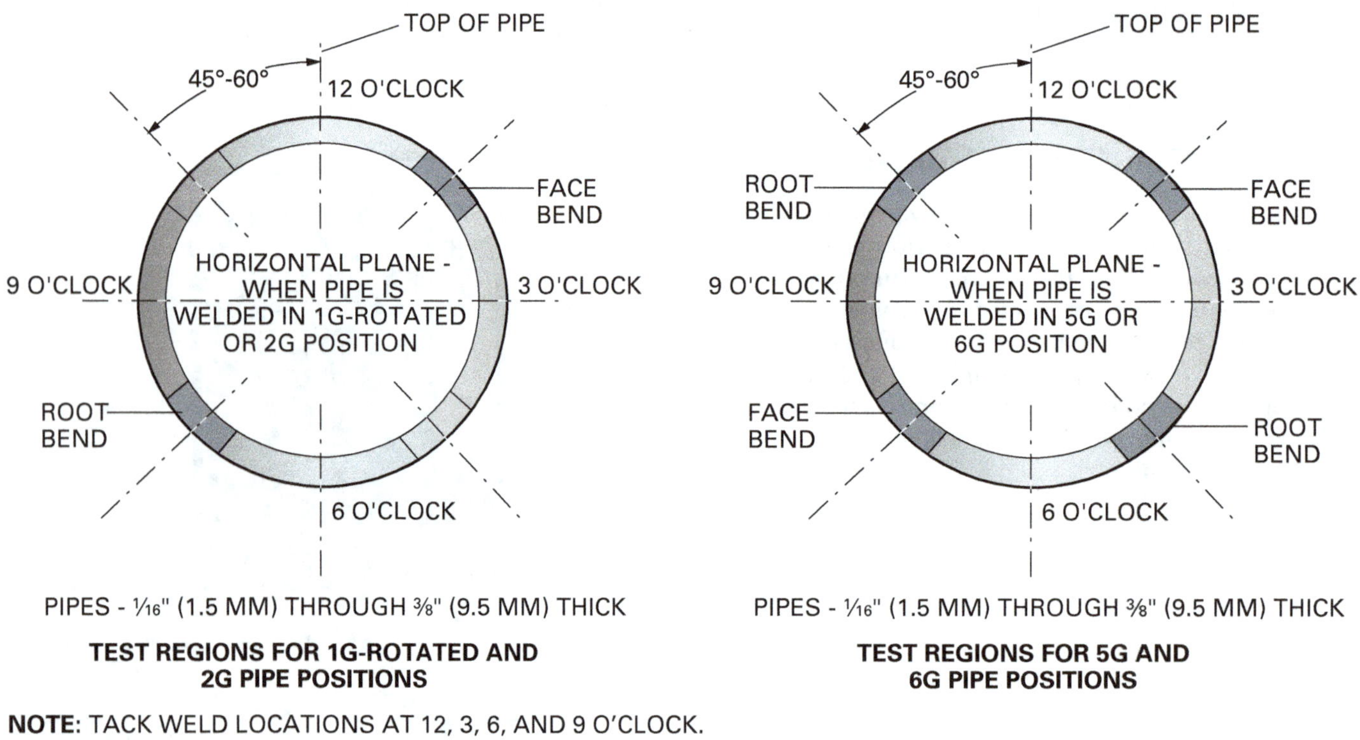

Figure 13 Test specimen regions for pipe positions 1G-ROTATED, 2G, 5G, and 6G.

Completed pipe welds should be inspected to ensure they exhibit the following characteristics; if a weld does not meet these criteria, or of it has any of the defects shown in *Figure 14*, it must be repaired:

- Straightness of the bead
- Uniform appearance of the bead face; face and root reinforcement depth in accordance with applicable codes
- Smooth, flat transition with complete fusion at the toes of the weld
- Termination crater filled
- No porosity
- No undercut
- No overlap
- No overfill
- No cracks

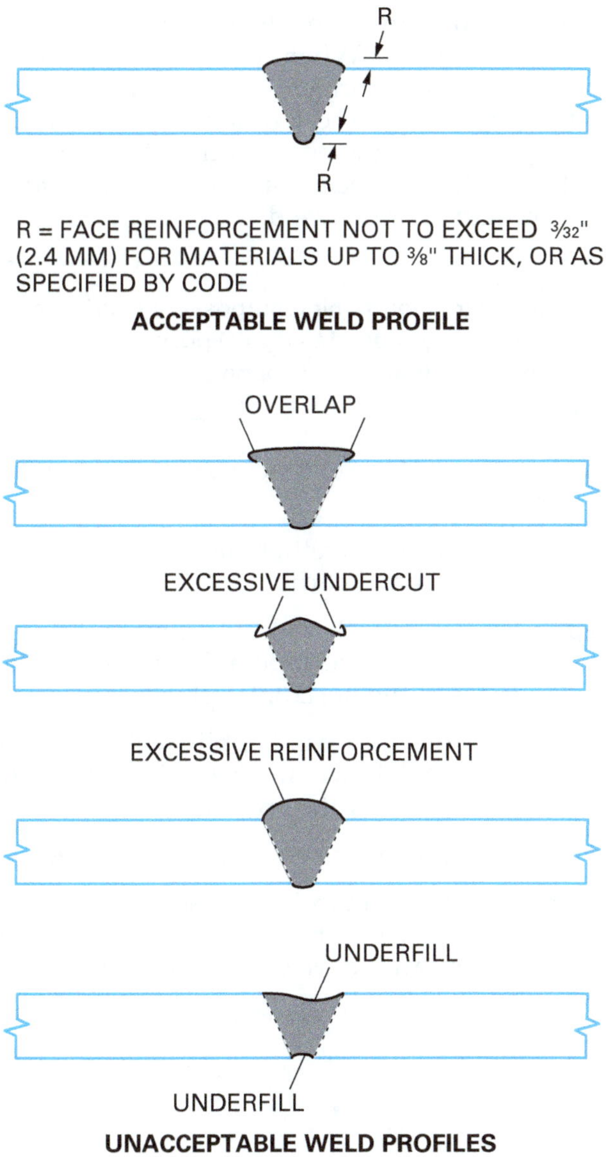

Figure 14 Acceptable and unacceptable pipe groove weld profiles.

1.6.0 Practicing Modified U-Groove Pipe Welds

The following sections explain how to perform modified U-groove welds in these positions (note that the 1G-ROTATED position is not practiced in this module):

- Horizontal (2G)
- Multiple (5G)
- Multiple inclined (6G)

1.6.1 Practicing Horizontal (2G) Position Pipe Welds

Practice 2G position pipe welds using the bead sequences and work angles shown in *Figure 15*. Use argon shielding gas and a filler rod of a diameter specified by your instructor. The torch should be held at a 15- to 20-degree push angle. To better control the thickness of the root reinforcement, the torch work angle can be dropped slightly from the angles shown in *Figure 15*, but not more than 10 degrees. Note that the work angle changes for Passes 2 through 8 to properly place the weld bead. For the

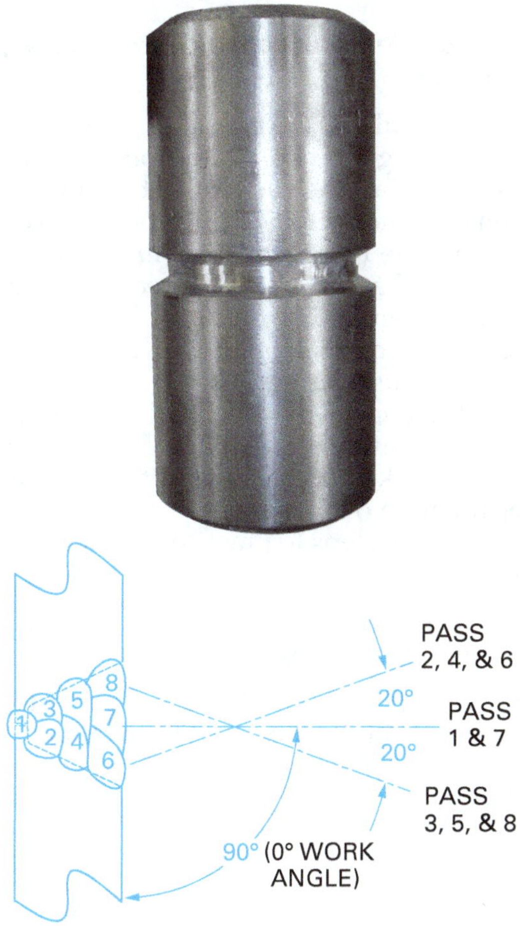

Figure 15 Multiple-pass 2G bead sequences and work angles for the modified U-groove joint.

root pass, use a slight circular motion to form the weld puddle. Be sure to fuse into the tack welds.

When running the remaining passes, use stringer beads and either the walking-the-cup or the freehand torch-handling technique. Ensure proper fusion at the toes of the weld bead. Take special care to fill the crater at the termination of the weld.

Follow these steps to practice pipe welds in the 2G position:

Step 1 Tack weld the practice coupons together as previously explained.

Step 2 Clamp or tack weld the coupon into position with the axis of the pipe vertical.

Step 3 Run the root pass (Pass 1) as shown in *Figure 15*, starting on a tack weld. Continue this procedure until the root pass has been completed.

Step 4 Brush the root pass to clean the weld.

Step 5 Run the remaining passes at the appropriate work angles. Overlap the passes, and clean each weld pass before beginning the next one.

1.6.2 Practicing Multiple (5G) Position Pipe Welds

Practice the multiple (5G) position pipe welds. Use argon shielding gas and a filler rod of a diameter specified by your instructor. Start on a tack weld positioned at the bottom of the pipe, and weld uphill toward the top. Refer to *Figure 16*. The torch should be held at a 15- to 20-degree push angle for the root and remaining passes.

When running the remaining passes as shown in *Figure 17*, use stringer beads and either the walking-the-cup or the freehand torch-handling technique. Ensure proper fusion at the toes of the weld bead. Take special care to fill the crater at the termination of the weld.

Follow these steps to practice pipe welds in the 5G position:

Step 1 Tack weld the practice coupons together as explained earlier.

Step 2 Clamp or tack weld the coupon into position with the pipe axis horizontal.

Step 3 Run the root pass uphill with a stringer bead, as shown in *Figure 16*. Repeat the procedure for the opposite side of the pipe.

Step 4 Brush the root pass to clean the weld.

Step 5 Run the remaining passes uphill at the proper angles using the bead sequence shown in *Figure 17*. Overlap the passes, and clean the weld after each pass.

1.6.3 Practicing Multiple Inclined (6G) Position Pipe Welds

Practice the multiple (6G) position pipe welds. Use argon shielding gas and a filler rod of a diameter specified by your instructor. Start on a tack weld positioned at the bottom of the pipe, and weld uphill toward the top.

The torch should be at approximately a 15- to 20-degree push angle for the root and remaining passes (*Figure 18*). When running the remaining passes, use stringer beads and either the walking-the-cup or the freehand torch-handling technique. Ensure proper fusion at the toes of the weld bead. Take special care to fill the crater at the termination of the weld.

NOTE

If required for training purposes, a restricting ring may be added to the 6G position coupon to form a 6GR position coupon.

Follow these steps to practice pipe welds in the 6G position:

Step 1 Tack weld together the practice coupon as explained earlier.

Step 2 Clamp or tack weld the coupon into position with the pipe axis inclined 45 degrees to the horizontal plane.

Step 3 Run the root pass uphill with a stringer bead, as shown in *Figure 16*. Repeat the procedure for the opposite side of the pipe.

Step 4 Brush the root pass to clean the weld.

Step 5 Run the remaining passes uphill at the proper work angles. Overlap the passes, and clean the weld after each pass.

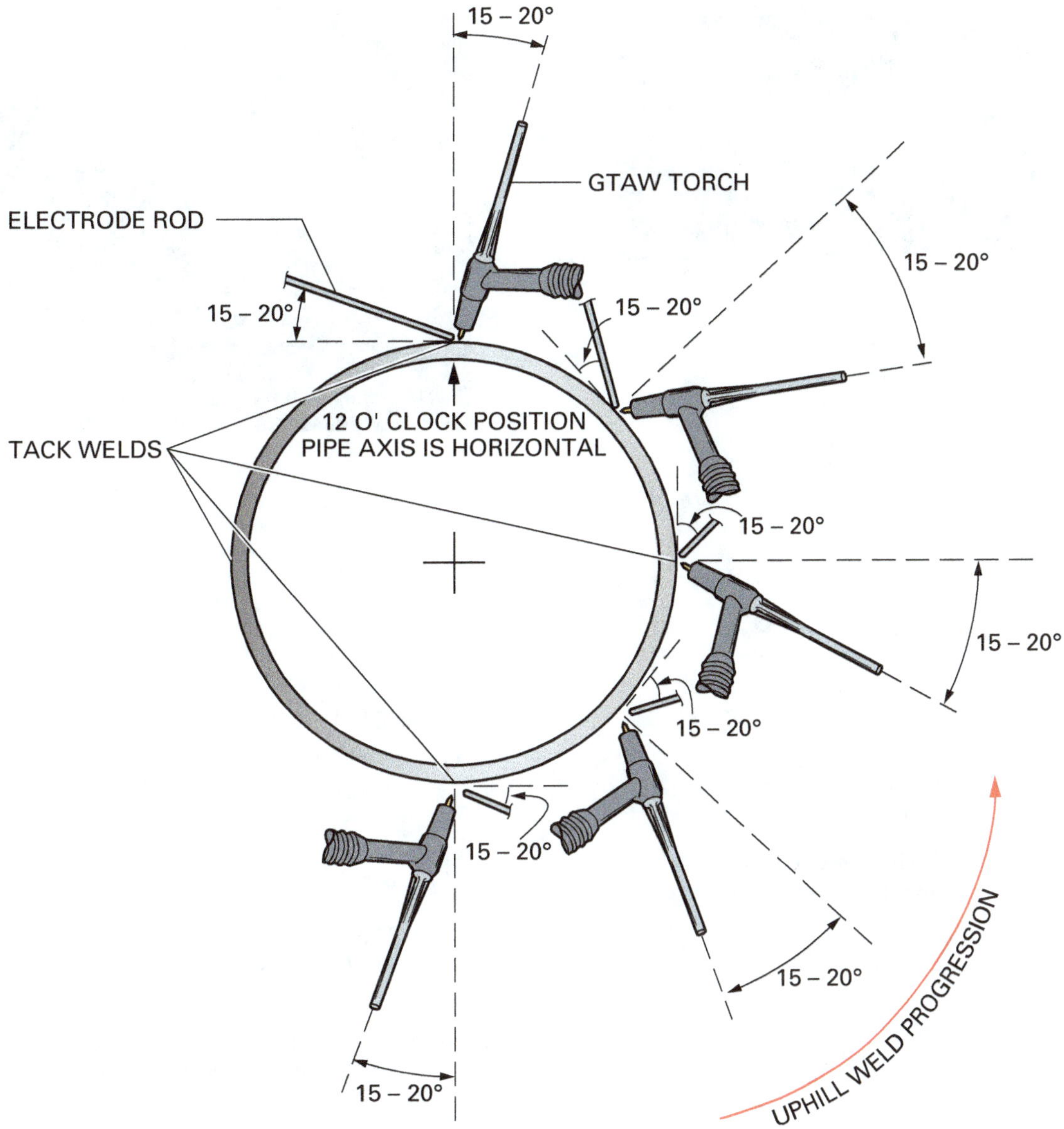

Figure 16 GTAW torch and electrode rod angles for the 5G position.

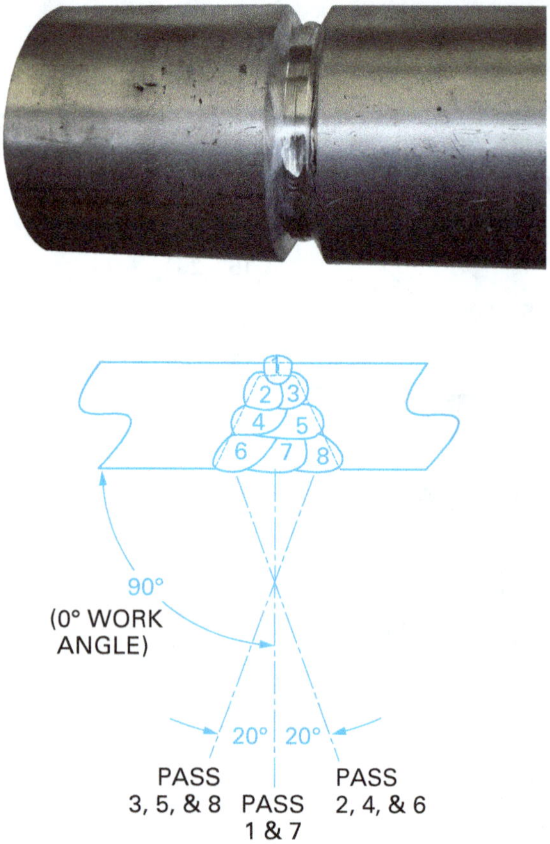

Figure 17 Multiple-pass 5G bead sequences and work angles for the modified U-groove joint.

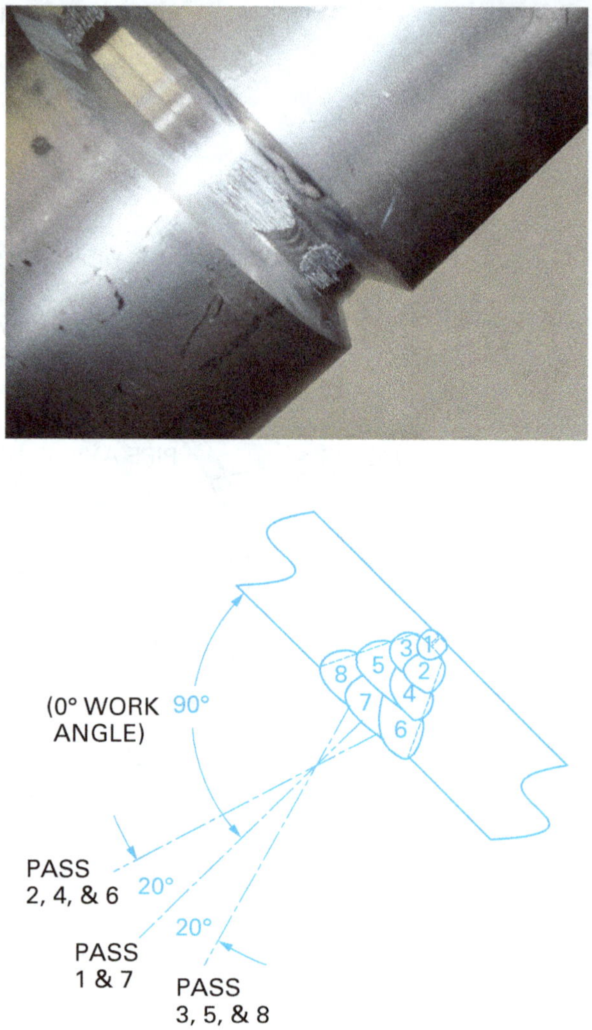

Figure 18 Multiple-pass 6G bead sequences and work angles for the modified U-groove joint.

Additional Resources

OSHA Standard 1926.351, Arc Welding and Cutting. Current Edition. Washington, DC: Occupational Safety and Health Administration.

The Procedure Handbook of Arc Welding. Current Edition. The Lincoln Electric Company, Cleveland, OH, USA.

Welding Handbook, Volume 1 and Volume 2. Miami, FL: The American Welding Society.

Gas Tungsten Arc Welding. Current Edition. Cleveland, OH: The James F. Lincoln Arc Welding Foundation.

Gas Tungsten Arc Welding. Current Edition. Appleton, WI: Miller Electric Manufacturing Company.

AWS D1.2/D1.2M Structural Welding Code—Aluminum. Miami, FL: American Welding Society.

The following websites offer resources for products and training:

The Lincoln Electric Company, **www.lincolnelectric.com**

Miller Electric Manufacturing, **www.millerwelds.com**

1.0.0 Section Review

1. The two basic torch-handling techniques used to perform GTAW are _____.
   a. freehand and on-the-wire
   b. freehand and walking-the-cup
   c. side-to-side and freehand
   d. on-the-wire and side-to-side

2. Which of the following statements about welding safety is true?
   a. Oxygen is a good and safe choice for blowing dust from clothing.
   b. Wearing ear plugs at any time while welding creates an unacceptable safety issue.
   c. A neat, clean work area reduces hazards and distractions.
   d. GTAW can be done using the lowest level of shading available in the welding hood lens.

3. The preferred GTAW joint for aluminum pipe without a backing is the _____.
   a. double U-groove joint
   b. double V-groove joint
   c. open-root V-groove joint
   d. modified U-groove joint

4. When setting an inverter-type welding machine for use on clean aluminum, a wave balance setting of 80 percent EN or higher allows the welder to use a(n) _____.
   a. smaller electrode at a higher current
   b. larger electrode at a lower current
   c. blunt electrode
   d. EWP electrode

5. Pipe groove welds should have slight reinforcement, not exceeding _____.
   a. 1⁄32" (0.8 mm)
   b. 1⁄16" (1.6 mm)
   c. 3⁄32" (2.4 mm)
   d. 1⁄8" (3.2 mm)

6. When welding in the 5G position, clamp or tack weld the weld coupon into position with the pipe axis _____.
   a. vertical
   b. horizontal
   c. inclined at 5 degrees to the horizontal plane
   d. inclined at 45 degrees to the horizontal plane

SUMMARY

The ability to make groove welds on aluminum pipe in all positions is one of the more challenging skills a welder will develop. Aluminum base metal requires significant welding-current levels, and the techniques associated with it require practice to master. Proper cleaning of the base metal and filler rod is essential for aluminum welding to be successful.

U- and V-groove joints, with the root opening eliminated, are commonly used on aluminum pipe. Open-root joints typically require a backing. Groove welds on pipe can be made in the 1G-ROTATED, 2G, 5G, and 6G positions. Practice welding in the appropriate positions until you can consistently produce acceptable welds.

Review Questions

1. As the GTAW torch is raised, arc length and voltage _____.
   a. increase, and the bead width also increases
   b. decrease, and the bead width also decreases
   c. increase, but the bead width decreases
   d. decrease, but the bead width increases

2. The torch travel angle that is normally used for GTAW of aluminum is a _____.
   a. push angle of 10 to 15 degrees
   b. push angle of 15 to 20 degrees
   c. drag angle of 10 to 15 degrees
   d. drag angle of 15 to 20 degrees

3. When you are making groove welds on aluminum pipe, the weld puddle size can be increased using a small _____.
   a. uphill motion
   b. downhill motion
   c. circular motion
   d. up and down motion

4. When welding aluminum pipe, the preferred welding table surface is _____.
   a. low-alloy steel
   b. high-carbon steel
   c. aluminum
   d. stainless steel

5. When welding aluminum, the high thermal conductivity of the base metal requires the use of _____.
   a. higher welding currents
   b. lower welding currents
   c. higher welding voltage
   d. lower welding voltage

6. One disadvantage of using helium as a shielding gas, alone or as part of a mixture, is that the _____.
   a. arc may be more difficult to start
   b. melting point of the aluminum rises
   c. aluminum workpiece is blackened
   d. travel speed is reduced

7. The shielding gas to use to increase welding speed and achieve deeper penetration is a mix of _____.
   a. helium and nitrogen
   b. argon and carbon dioxide
   c. helium and carbon dioxide
   d. argon and helium

8. The modified U-groove weld is preferred for joints without backing because it _____.
   a. allows for a larger weld puddle
   b. prevents complete root penetration
   c. allows for a larger gap in the root opening
   d. reduces heat transfer away from the weld joint

9. Pipe groove weld positions are determined by the _____.
   a. axis of the pipe
   b. length of the pipe
   c. weight of the pipe
   d. thickness of the pipe

10. When welding in the 6G position, clamp or tack weld the pipe weld coupon into position with the pipe axis inclined _____.
    a. 10 degrees to the horizontal plane
    b. 15 degrees to the horizontal plane
    c. 30 degrees to the horizontal plane
    d. 45 degrees to the horizontal plane

Appendix

PERFORMANCE ACCREDITATION TASKS

The Performance Accreditation Tasks (PATs) correspond to and support learning objectives in the *AWS EG3.0, Guide for the Training and Qualification of Welding Personnel; Level II—Advanced Welder*.

PATs provide specific acceptable criteria for performance and help to ensure a true competency-based welding program for students.

The following tasks are designed to evaluate your ability to run groove welds on aluminum pipe with GTAW equipment in three standard test positions, using an aluminum filler rod of the appropriate diameter and argon shielding gas. Perform each task when you are instructed to do so. As you complete each task, show it to your instructor for evaluation. Do not proceed to the next task until told to do so. For AWS 2G and 5G certifications, refer to *AWS EG3.0, Guide for the Training and Qualification of Welding Personnel; Level II—Advanced Welder* for bend test requirements. For AWS 6G certifications, refer to *AWS EG4.0, Guide for the Training and Qualification of Welding Personnel; Level III—Expert Welder* for bend test requirements.

PATs 1 and 2 correspond to *AWS EG3.0, 3.3.6.4, Unit #4, GTAW, Learning Objective #14 (2G and 5G)*. PAT 3 corresponds to *AWS EG3.0, 3.3.6.5, Unit #5, GTAW, Learning Objective #12*.

MODIFIED U-GROOVE WELDS ON ALUMINUM PIPE IN THE 2G POSITION

As directed by the instructor, use the GTAW process with the appropriate aluminum filler wire, argon shielding gas, and stringer beads to make a multiple-pass groove weld on aluminum pipe in the 2G position.

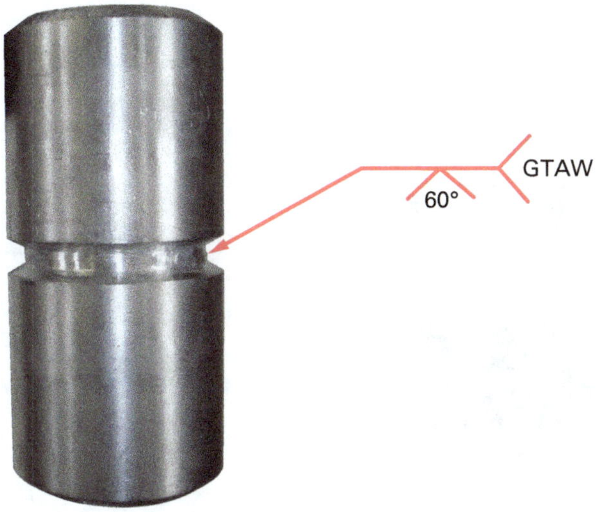

MODIFIED U-GROOVE

Criteria for Acceptance:

- Uniform appearance on the bead face ________
- Craters and restarts filled to the full cross section of the weld ________
- Acceptable weld profile in accordance with the applicable codes and standards ________
- Smooth transition with complete fusion at the toes of the weld ________
- No porosity ________
- No undercut ________
- No cracks ________
- No overlap ________
- No incomplete fusion ________

MODIFIED U-GROOVE WELDS ON ALUMINUM PIPE IN THE 5G POSITION

As directed by the instructor, use the GTAW process with the appropriate aluminum filler wire, argon shielding gas, and stringer beads to make a multiple-pass groove weld on aluminum pipe in the 5G position.

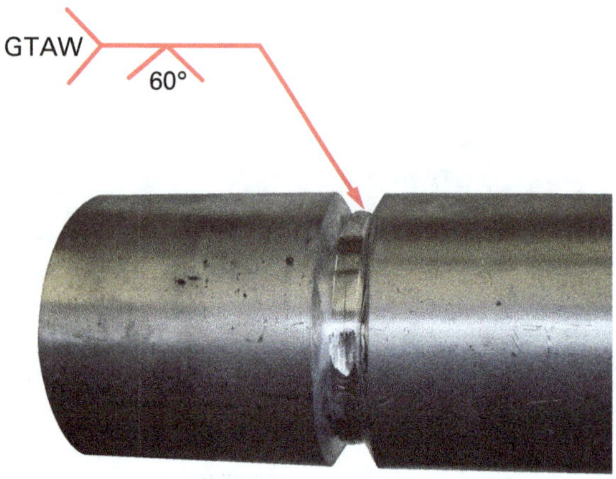

MODIFIED U-GROOVE

Criteria for Acceptance:

- Uniform appearance on the bead face ________
- Craters and restarts filled to the full cross section of the weld ________
- Acceptable weld profile in accordance with the applicable codes and standards ________
- Smooth transition with complete fusion at the toes of the weld ________
- No porosity ________
- No undercut ________
- No cracks ________
- No overlap ________
- No incomplete fusion ________

MODIFIED U-GROOVE WELDS ON ALUMINUM PIPE IN THE 6G (OR 6GR) POSITION

As directed by your instructor, use the GTAW process with the appropriate aluminum filler wire, argon shielding gas, and stringer beads to make a multiple-pass groove weld on aluminum pipe in the 6G (or 6GR) position.

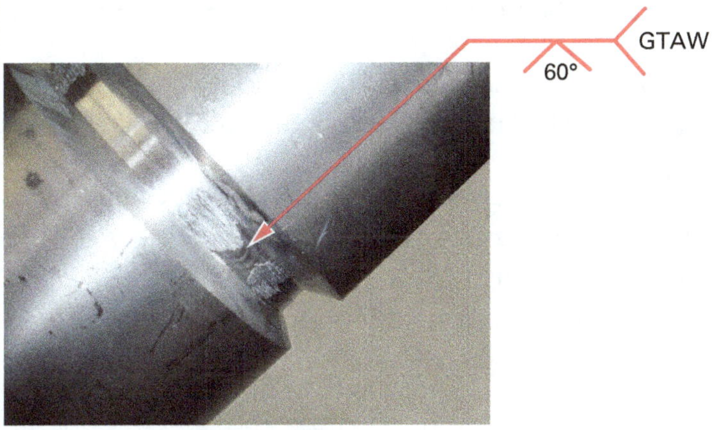

NOTE: IF REQUIRED FOR QUALIFICATION PURPOSES, A RESTRICTING RING MAY BE ADDED TO THE 6G POSITION COUPON TO FORM A 6GR POSITION COUPON.

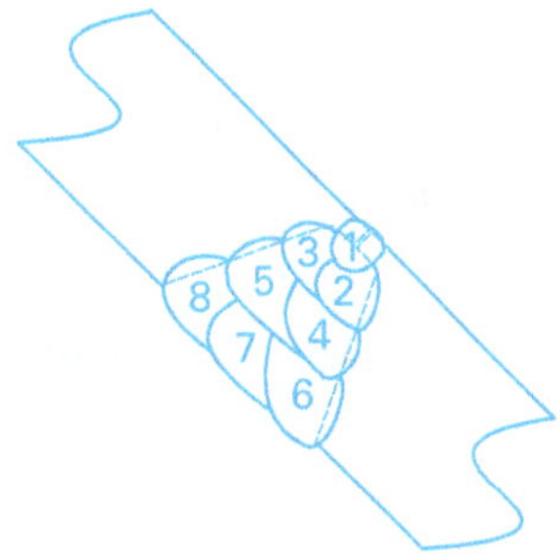

MODIFIED U-GROOVE

Criteria for Acceptance:

- Uniform appearance on the bead face __________
- Craters and restarts filled to the full cross section of the weld __________
- Acceptable weld profile in accordance with the applicable codes and standards __________
- Smooth transition with complete fusion at the toes of the weld __________
- No porosity __________
- No undercut __________
- No cracks __________
- No overlap __________
- No incomplete fusion __________

Additional Resources

This module presents thorough resources for task training. The following reference materials are recommended for further study.

OSHA Standard 1926.351, Arc Welding and Cutting. Current Edition. Washington, DC: Occupational Safety and Health Administration.

The Procedure Handbook of Arc Welding. Current Edition. The Lincoln Electric Company, Cleveland, OH, USA.

Welding Handbook, Volume 1 and Volume 2. Miami, FL: The American Welding Society.

Gas Tungsten Arc Welding. Current Edition. Cleveland, OH: The James F. Lincoln Arc Welding Foundation.

Gas Tungsten Arc Welding. Current Edition. Appleton, WI: Miller Electric Manufacturing Company.

AWS D1.2/D1.2M Structural Welding Code—Aluminum. Miami, FL: American Welding Society.

The following websites offer resources for products and training:

The Lincoln Electric Company, **www.lincolnelectric.com**

Miller Electric Manufacturing, **www.millerwelds.com**

Figure Credits

The Lincoln Electric Company, Cleveland, OH, USA, Module Opener, Figures 4, 6, 7, Table 1

Griffin Heat Shield Co., **www.griffinheatshield.com**, SA01

Courtesy of TWI Ltd. **www.twi-global.com**, SA02

Miller Electric Manufacturing Co., Figure 5, SA03

Courtesy of H & M Pipe Beveling Machine Company, Inc., Figure 9

Terry Lowe, Figures 10 (photo), 15 (photo), 17 (photo), 18 (photo), Performance Accreditation Tasks 1 through 3 (photos)

Topaz Publications, Inc., Figure 11 (photo)

Section Review Answer Key

Answer	Section Reference	Objective
Section One		
1. b	1.1.3	1a
2. c	1.2.2	1b
3. d	1.3.2	1c
4. a	1.4.0	1d
5. c	1.5.2	1e
6. b	1.6.2	1f

NCCER CURRICULA — USER UPDATE

NCCER makes every effort to keep its textbooks up-to-date and free of technical errors. We appreciate your help in this process. If you find an error, a typographical mistake, or an inaccuracy in NCCER's curricula, please fill out this form (or a photocopy), or complete the online form at **www.nccer.org/olf**. Be sure to include the exact module ID number, page number, a detailed description, and your recommended correction. Your input will be brought to the attention of the Authoring Team. Thank you for your assistance.

Instructors – If you have an idea for improving this textbook, or have found that additional materials were necessary to teach this module effectively, please let us know so that we may present your suggestions to the Authoring Team.

NCCER Product Development and Revision
13614 Progress Blvd., Alachua, FL 32615

Email: curriculum@nccer.org
Online: www.nccer.org/olf

❑ Trainee Guide ❑ Lesson Plans ❑ Exam ❑ PowerPoints Other ________________

Craft / Level: Copyright Date:

Module ID Number / Title:

Section Number(s):

Description:

Recommended Correction:

Your Name:

Address:

Email: Phone:

GMAW – Aluminum Pipe

Overview

Gas metal arc welding (GMAW) is one of the more common welding processes. GMAW equipment is available in many different sizes and types, however, the basic operating principles of GMAW apply to all makes and models of equipment. This module describes how GMAW equipment is used to weld aluminum pipe, and presents the techniques used to produce V-groove pipe welds with backing.

Module 29404

Trainees with successful module completions may be eligible for credentialing through the NCCER Registry. To learn more, go to **www.nccer.org** or contact us at 1.888.622.3720. Our website has information on the latest product releases and training, as well as online versions of our *Cornerstone* magazine and Pearson's product catalog.

Your feedback is welcome. You may email your comments to **curriculum@nccer.org**, send general comments and inquiries to **info@nccer.org**, or fill in the User Update form at the back of this module.

This information is general in nature and intended for training purposes only. Actual performance of activities described in this manual requires compliance with all applicable operating, service, maintenance, and safety procedures under the direction of qualified personnel. References in this manual to patented or proprietary devices do not constitute a recommendation of their use.

29404 V5

29404
GMAW – Aluminum Pipe

Objectives

When you have completed this module, you will be able to do the following:

1. Explain how to prepare for GMAW of aluminum pipe and describe specific welding techniques.
   a. Identify common welding safety practices.
   b. Describe how to prepare GMAW equipment for aluminum pipe welding.
   c. Explain how to prepare the area and materials for aluminum pipe GMAW.
   d. Describe how to make aluminum V-groove pipe welds using GMAW.

Performance Tasks

Under the supervision of the instructor, you should be able to do the following:

1. Make GMAW V-groove welds on aluminum pipe with backing in the following positions:
   - 2G
   - 5G
   - 6G

Industry Recognized Credentials

If you are training through an NCCER-accredited sponsor, you may be eligible for credentials from NCCER's Registry. The ID number for this module is 29404. Note that this module may have been used in other NCCER curricula and may apply to other level completions. Contact NCCER's Registry at 888.622.3720 or go to **www.nccer.org** for more information.

Contents

Figures

Section One

1.0.0 Aluminum Pipe GMAW

Objective

Explain how to prepare for GMAW of aluminum pipe and describe specific welding techniques.

a. Identify common welding safety practices.
b. Describe how to prepare GMAW equipment for aluminum pipe welding.
c. Explain how to prepare the area and materials for aluminum pipe GMAW.
d. Describe how to make aluminum V-groove pipe welds using GMAW.

Performance Tasks

1. Make GMAW V-groove welds on aluminum pipe with backing in the following positions:
   - 2G
   - 5G
   - 6G

Gas metal arc welding (GMAW) is an arc welding process that uses a continuous, consumable wire electrode as filler metal and a shielding gas to protect the weld zone (*Figure 1*). The GMAW process is commonly used to make welds on aluminum, as well as carbon steel, low-alloy steel, and stainless steel. *Figure 2* shows typical GMAW equipment. When welding aluminum at current settings of 200 amps or more, a water-cooled gun is preferred to minimize buildup in the gun and contact tip. Note that the contact tips used for aluminum are different than those used with steel wire. They are slightly larger to accommodate the expansion of the aluminum wire as it heats up.

CAUTION

Using contact tips designed for steel wire is not acceptable. Due to the expansion rate of the aluminum wire, binding will likely occur as the weld progresses. Use only contact tips designed for aluminum.

GMAW is a fast and effective method for producing high-quality welds. Because this type of welding allows the process to continue for the length of most weld beads, discontinuities and restarts are significantly reduced. Note that it is quite common to use GTAW for the root pass and GMAW to complete the remaining passes. However, GMAW can be used for root passes as well.

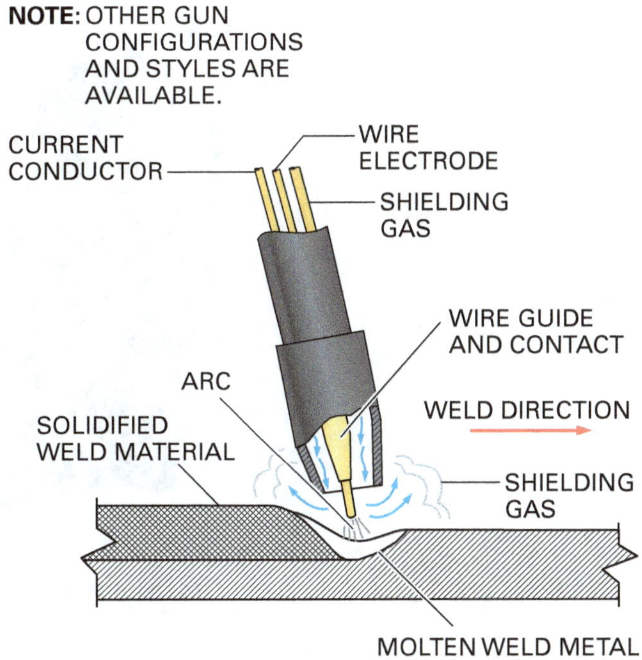

Figure 1 GMAW process.

For the purposes of this module, the dimensions used are representative of various codes and standards and may not be specific to any particular code. Always refer to the WPS, applicable code, and/or site standard to ensure welds are completed as specified.

1.1.0 Safety Practices

Before proceeding with welding coupon preparation and practice, it is important to review common safety practices associated with the craft. Safety considerations are a critical part of preparing for any welding operation.

The following is a summary of safety procedures and practices that must be observed while cutting or welding. Be advised that this is only a summary. Complete safety coverage is provided in the *Welding Level One* module, "Welding Safety." That module should be studied, if not already completed, before continuing. Above all, be sure to wear appropriate protective clothing and equipment at all times when welding or cutting.

1.1.1 Protective Clothing and Equipment

Protective clothing and equipment are a necessary part of every welding task. Follow the guide-

Figure 2 Inverter-type welding machine with GMAW guns.

lines below as a general rule, and ensure that you are familiar with any unique requirements your employer may have regarding personal protective equipment (PPE).

- A welding hood equipped with a properly shaded lens is best for all forms of welding. The welding hood lens must be an appropriate shade for the application.
- Wear proper leather or flame-retardant clothing along with welding gloves for protection from flying sparks, molten metal, and heat.
- Wear high-top safety shoes or boots. Make sure that the tongue and lace area of the footwear is covered by a pant leg. If the tongue and lace area is exposed or if the footwear must be protected from burn marks, wear spats under the pant leg or leather chaps over the front and top of the footwear.
- Wear a 100-percent cotton cap without any mesh material. If a hard hat is required for the environment, use one that allows the attachment of rear deflector material and a face shield. A hard hat with a rear deflector is generally preferred when working overhead, and may be required by some employers and job sites.
- Wear earmuffs, or at least earplugs, to prevent sparks from entering the ear canals.

1.1.2 Housekeeping and Fall Protection

Welding is often performed outside in the weather and in the construction environment. Even production work completed inside large, open shops exposes a welder to equipment moving across the floor and overhead. Welders often have to crawl under or reach over objects to complete their work.

At times, welders must work at height. Fall protection equipment is required for all work at height (*Figure 3*). Always keep fall protection devices untangled and properly attached. Combined with the other equipment a welder must carry, there is a great deal to manage, but fall protection equipment must always be the prior-

Figure 3 Working at height.

ity. Fall protection equipment saves lives. Never attach any kind of welding gear to the harness that could interfere with its ability to perform its intended function.

Regardless of where you are working, always keep the work area as neat as possible. This makes the area safer and helps to reduce distractions. Mindful actions, such as making sure that welding cables are not tripping hazards, contribute to a safer environment. Electrode stubs and unused electrodes must not be left on the floor. They can cause someone to lose their footing and fall. Parts waiting to be welded or left over from previous work are a tripping hazard and should be kept out of the walking area. If the parts have sharp edges, they could cut welding cable insulation or torch gas hoses.

1.1.3 Fire and Explosion Prevention

Flammable gases are commonly found around welding and cutting operations, as are sparks and other significant sources of ignition. For that reason, and because many fire-related incidents over the years have been traced back to welding operations, it is crucial that fire and explosion prevention be given consistent attention. Follow these guidelines to maintain a safe environment:

- Never carry matches or gas-filled lighters. Sparks can cause the matches to ignite or the lighter to explode in your pocket, causing serious injury.
- Never use oxygen to blow dust or dirt from clothing. Traces of the oxygen can remain trapped in the fabric for some time. If a spark hits the fabric during this time, the clothing can burn rapidly and violently out of control.
- Make sure that any flammable material in the work area is removed or shielded by a fire-resistant covering. Approved fire extinguishers must be available before attempting any heating, welding, or cutting operations. If a hot-work permit and a fire watch are required, be sure these are in place before beginning.
- Ensure that all site safety requirements are being observed.
- Never release a large amount of oxygen or use oxygen in place of compressed air. It can cause rapid and uncontrolled combustion around flammable materials or sparks. Keep oxygen away from oil, grease, and other petroleum products.
- Never release a large amount of fuel gas (especially acetylene). Because propane is heavier than air, it tends to concentrate in and along low areas, and can ignite at a considerable distance from the release point. Acetylene is lighter than air, but is even more dangerous than propane. When mixed with air or oxygen, acetylene will ignite at much lower concentrations than any other fuel gas.
- Although shielding gases are not toxic, any gas released in a large quantity can displace the oxygen in a small room or enclosed work area, creating an atmosphere with little or no oxygen. Avoid large releases of any type of gas, regardless of its characteristics.
- To prevent fires, maintain a neat and clean work area, and make sure that any metal scrap or slag is cold before disposal.
- Before cutting containers such as tanks or barrels, check to see if they have contained any explosive, hazardous, or flammable materials, including petroleum products, citrus products, or chemicals that decompose into toxic fumes when heated. As a standard practice, always clean and then fill any tanks or barrels with water, or purge them with a flow of inert gas such as nitrogen to displace any oxygen. Containers must be cleaned by steam cleaning, flushing with water, or washing with detergent until all traces of the material have been removed. Proper procedures for cutting or welding hazardous containers are described in the *American Welding Society (AWS) F4.1, Safe Practices for the Preparation of Containers and Piping for Welding and Cutting.*

> **WARNING!**
>
> Welding or cutting must never be performed on drums, barrels, tanks, vessels, or other containers until they have been emptied and cleaned thoroughly to eliminate all flammable materials and all substances (such as detergents, solvents, greases, tars, or acids) that might produce flammable, toxic, or explosive vapors when heated. Clean containers only in well-ventilated areas. Vapors can accumulate during cleaning, causing explosions and serious personal injury.

1.1.4 Work Area Ventilation

Many gases are introduced to the atmosphere through welding processes. Some of these gases are harmful, and all can displace or combine with atmospheric oxygen in a confined space. Follow these guidelines, as well as those of your employer, for work area ventilation:

- Make sure to follow all confined space procedures before conducting any welding or cutting in a confined space. Use powered extraction systems as needed.
- Never use oxygen to provide ventilation in confined spaces.
- Always perform cutting or welding operations in a well-ventilated area.

Cutting or welding operations involving zinc or cadmium materials or coatings result in toxic fumes. Another significant concern is Chromium 6, commonly referred to as *hexavalent chromium,* which is used in stainless steel. It releases toxic fumes when heated. Manganese is another hazardous metal used in stainless steel. It is also used in some welding rods, increasing the risk of exposure.

For long-term cutting or welding of materials containing these components, always wear an approved full-face, supplied-air respirator that uses breathing air supplied from outside the work area. For occasional short-term exposure, a HEPA-rated or metal-fume filter may be used on a standard respirator.

Powered fume-extraction equipment must be used when welding materials that contain zinc, cadmium, hexavalent chromium, manganese, and similar materials that produce hazardous fumes if the area is not naturally well-ventilated. *Figure 4* shows a powered air-purification respirator used with fume extraction equipment.

Figure 4 Powered air-purification respirator (PAPR) and localized fume extraction equipment.

Remember What Grounding Really Means

In the welding trade, the term *ground* is often used to identify the workpiece lead from the welding machine. Welding leads are connected to the positive and negative terminals of the welding machine. The polarity settings of the machine determine the direction of current flow. Neither of the two leads from the welding machine provides any level of actual ground service for the equipment, nor any electrical protection for the welder. True grounding is achieved through a wire lead connected to the machine frame that is then connected to the facility power grounding system.

It is best to only use the term *ground* as it relates to the safe connection of the welding machine to the power system, providing a path for power to flow to ground should the machine structure become energized. *ANSI Z49.1, Safety in Welding, Cutting, and Allied Processes* requires the work table to be grounded as well.

1.2.0 The Welding Machine

Identify a proper welding machine for GMAW, and follow these steps to set it up for use:

Step 1 Verify that the welding machine can be used for GMAW.

Step 2 Know the location of the primary disconnect.

Step 3 Configure the welding machine for GMAW as directed by your instructor (*Figure 5*), ensuring the proper DCEP polarity is established. Equip the gun with the correct nozzle for the application. Also equip the gun with the correct liner material and contact tube, designed for aluminum wire, for the diameter of the wire in use.

> **CAUTION**
>
> Using contact tips designed for steel wire is not acceptable. Due to the expansion rate of the aluminum wire, binding will likely occur as the weld progresses. Use only contact tips designed for aluminum.

Step 4 In accordance with the manufacturer's instructions, configure and load the wire feeder and gun with solid electrode wire of the proper diameter as directed by your instructor.

> **NOTE**
>
> The wire feeder drive rollers should be of the U-groove style for aluminum wire, unlike the V-groove rollers used for carbon steel wire.

> **CAUTION**
>
> To prevent the electrode wire from contaminating the welds, always keep the wire clean and dry when it is in storage or in the machine.

Step 5 Connect the proper shielding gas (argon or argon/helium) for the application as specified by the wire electrode manufacturer, WPS, site quality standards, and/or your instructor.

Step 6 Connect the clamp of the workpiece lead to the workpiece.

Step 7 Turn on the welding machine, and purge the gun as directed by the gun manufacturer's instructions.

Step 8 Set the initial welding voltage and wire feed speed as recommended by the manufacturer or the WPS for the type and size electrode wire being used.

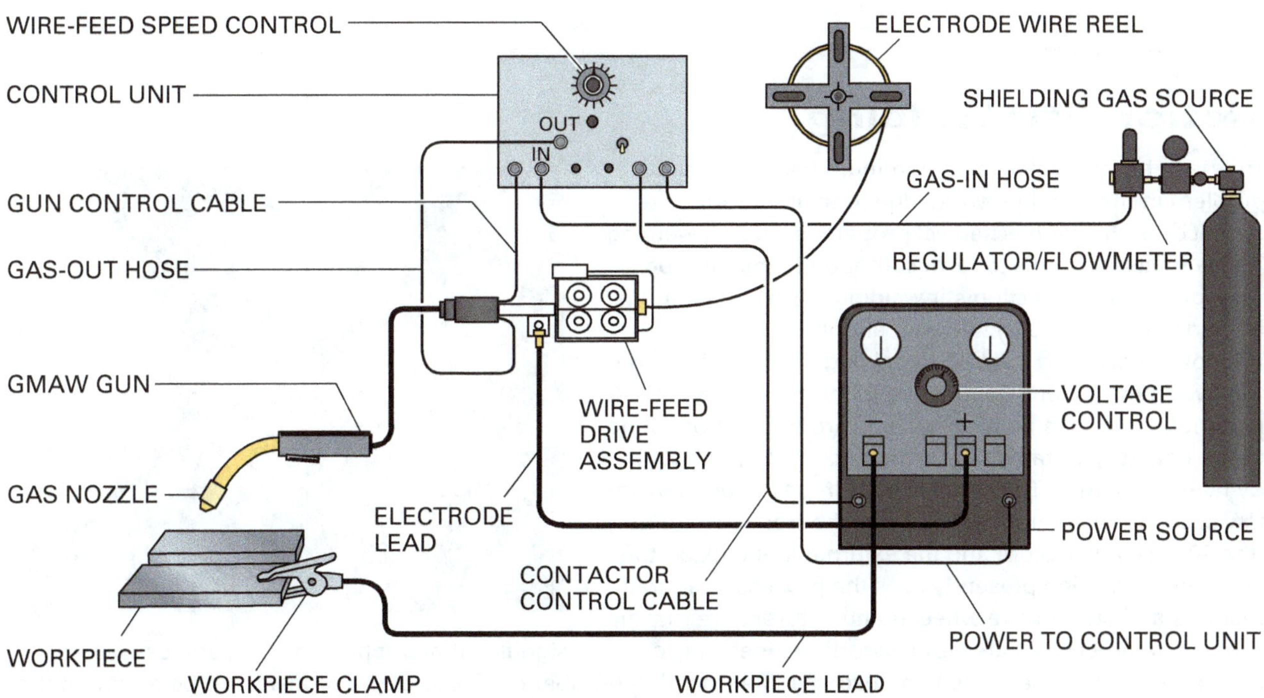

NOTE: THE POLARITY OF THE GUN AND WORKPIECE LEAD IS DETERMINED BY THE TYPE OF FILLER METAL AND APPLICATION.

Figure 5 Configuration diagram of a typical GMAW machine.

1.2.1 Welding Voltage

Arc length is determined by the voltage, which is set at the power source. Arc length is the distance from the wire electrode tip to the base metal or to the molten pool at the base metal (*Figure 6*). If voltage is set too high, the arc will be too long. This can cause the wire to melt and fuse to the contact tube. Voltage that is set too high can also cause porosity and excessive spatter. Voltage must be increased or decreased as the wire feed speed is increased or decreased. Be sure to set the voltage to maintain consistent spray transfer.

1.2.2 Welding Amperage

When using a standard constant-voltage welding machine, the electrode feed speed controls the welding amperage after the initial recommended setting. The welding machine provides the amperage needed to melt the wire electrode while maintaining the selected welding voltage. Within limits, when the wire electrode feed speed is increased, the welding amperage and deposition rate are also increased. This results in higher welding heat, deeper penetration, and higher beads. When the wire electrode feed speed is decreased, the welding amperage automatically decreases. With lower welding amperage and less heat, the deposition rate drops, and the weld beads are smaller with less penetration.

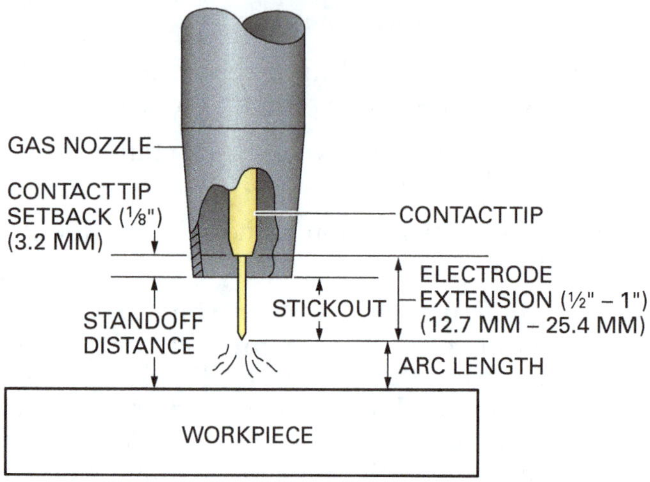

Figure 6 GMAW gun configured for spray transfer.

Note that some constant-voltage welding machines used for GMAW provide varying degrees of current modification. Examples include slope and induction adjustments. Welding machines with a slope adjustment allow the welder to vary the amount of amperage change in relation to voltage range of the unit. Standard constant-voltage GMAW units have a current slope fixed by the manufacturer for general welding applications and conditions. Pulse transfer units allow the peak current for the pulse and the background current between pulses to be adjusted to match specific welding applications and/or wire electrode requirements.

Friction Stir Welding

A marriage between steel and aluminum has long been a problem in the welding world, due to their extreme differences in thermomechanical properties. Fusion welding of these materials tends to form compounds that are too brittle for reliability. Friction stir welding (FSW) is changing that, and it is equally effective with aluminum alone.

Friction stir welding works by applying a tool to the surface of the two metals along the desired joint. The process is autogenous (requires no filler metal). Through friction alone, enough heat is generated to join the two metals, although they never fully melt. The process has already been used in a few specialized applications, such as the front sub-frame of the 2013 Honda Accord and the external fuel tank on the space shuttle. Boeing presently uses the process in aircraft structures and automotive wheel manufacturers are making use of it. The process creates less distortion, doesn't produce a significant heat-affected zone, and causes no loss of strength associated with the use of a filler metal. A number of videos of the process can be found on the Internet.

1.2.3 *Electrode Extension, Stickout, and Standoff Distance*

Electrode extension is the length of the wire that extends beyond the tip of the welding gun's contact tip. For high-conductivity metal wires, the preheating effect of wire resistance is minimal. Wire speed and voltage settings have a more direct effect on the amount of weld penetration and deopsition rate.

Either spray transfer or pulsed spray transfer can be used for aluminum welding. GMAW electrode extension for spray transfer varies from ½" to 1" (13–25 mm). *Figure 6* shows the typical electrode extension for the spray-transfer GMAW gun configuration. It also shows various gun terms and components. Stickout is the distance from the gas nozzle or insulating nozzle to the end of the electrode. Standoff distance is the distance from the gas nozzle or insulating nozzle to the workpiece. Contact tip extension or setback is usually dependent on the chosen transfer mode (spray transfer or pulsed spray transfer).

1.2.4 *Gas Nozzle Cleaning*

As the welding machine is used, weld spatter builds up on the gas nozzle, diffuser, and contact tip. The gas nozzle must be cleaned from time to time with a reamer, round file, or the tang of a standard file. If it is not properly cleaned, it will restrict the shielding gas flow and cause porosity in the weld.

1.3.0 Preparations for Welding

Before welding can begin, the area must be readied, the welding equipment must be set up, and the metal to be welded must be prepared. The following sections explain how to prepare for welding.

To practice welding, you will need a welding table, bench, or stand. The welding surface can be steel, but an aluminum surface is preferred. Provisions must be available for placing weld coupons out of position.

To set up the area for welding, follow these steps:

Step 1 Make sure that the area is properly ventilated. Make use of doors, windows, and fans.

Step 2 Check the area for fire hazards. Remove any flammable materials before proceeding.

Wire Electrode Manufacturer's Recommendations

Always follow the manufacturer's recommendations for using shielding gas and for the initial setting of the welding voltage and wire feed speed parameters. These balanced parameters are critical and are based on the welding position, size, and composition of the solid wire.

Step 3 Locate the nearest fire extinguisher. Do not proceed unless the extinguisher is charged and you know how to use it.

Step 4 Set up welding screens around the welding area.

1.3.1 *Practice Welding Coupons*

Because aluminum is easily contaminated, proper coupon preparation is very important. Aluminum pipe coupons must be cut to the desired lengths, cleaned, and beveled to the proper angles before they are positioned for welding. A backing ring must also be cut and cleaned. The pipe can be cut by machining or sawing.

The importance of cleaning aluminum before welding cannot be overstated. There are two related facts about aluminum that help explain why. First, aluminum does not give the welder a visual indication that it is nearing its melting point through color change, as many other common metals do. With no indication that the melting point is approaching, it is easy to apply too much heat. Second, aluminum oxide—the "skin" that forms on the outside—has a melting point roughly three times higher than the aluminum itself. Excessive weld contamination and poor weld quality result when the aluminum base metal is not thoroughly cleaned.

The preferred technique for beveling and/or adding a root face is mechanical machining. Grinding is also acceptable, although machining the grooves and root faces of aluminum pipe is the cleanest and most uniform way to prepare pipe for GMAW. Thermal cutting techniques are not considered acceptable. If thermal cutting techniques must be used to cut aluminum alloys, at least ⅛" (3 mm) of the cut surface must be removed by machining or grinding before welding.

> **WARNING!**
> Grinding wheels used on aluminum must be made for aluminum only. They should not be used on any other type of metal or on surfaces contaminated with grease or oil. Using other types of grinding wheels on aluminum may cause the grinding wheels to shatter.

> **CAUTION**
> Grinding wheels, files, and similar tools used on aluminum should not be used on any other type of metal or on surfaces contaminated with grease or oil.

Pipe weld coupons should be cut from Schedule 40 aluminum pipe that is from 6" to 12" (DN150 to DN300) in diameter. Each welded joint requires two coupons of the same size. *Figure 7* shows typical AWS aluminum pipe joint specifications. Depending on the joint configuration and the WPS for a given weld, note that a consumable insert may be required.

To prepare pipe coupons for V-groove weld joints with backing, follow these steps:

Step 1 Bevel the end of the pipe and apply a root face as shown in *Figure 7* using any acceptable method discussed in this section.

Step 2 Check the groove and root face. The groove angle should be as shown in *Figure 7*.

Step 3 Cut off a section of the beveled pipe end (4" minimum length). For 1G-ROTATED welding, you may need to cut longer coupons to fit the rollers, if used.

> **NOTE**
> For this module, the practice pipe coupons will be tacked with a root opening and a permanent backing ring, without the use of any backing gas.

Step 4 Use an approved solvent and a clean rag to remove any grease or oil within a minimum of 1" from the weld joint. Clean the inside and outside of the coupons with a grinder or stainless steel brush. Also thoroughly clean the backing ring.

> **WARNING!**
> To prevent injury when using cleaning solvents, always comply with the SDS guidelines for the solvents in use.

> **CAUTION**
> If you use a grinding wheel, make sure it is a special high-speed grinding wheel approved for use on aluminum. Make sure that it, and any other cleaning tools such as wire brushes, have not been used on any other metal or used to remove hydrocarbon contaminants.

Step 5 Use a permanent backing ring compatible with the size and alloy of the aluminum pipe. Insert the ring inside one of the beveled edges of a pipe coupon section. Make an allowance for the root opening (up to ¼" or 6.4 mm) if there are no pins or pegs on the backing ring, and center the backing ring at the joint edge. Clamp the ring to the pipe at several places. Ensure that the backing ring is appropriately sized for the pipe and fits snug.

Step 6 Tack weld the ring to the pipe coupon at four equally spaced positions around the pipe. If the weld ring has three pins instead of four, as shown in *Figure 7*, three tack welds are normally made instead. The tack welds should be narrow and should not exceed 1" in length. Clean the tack welds.

Step 7 Place the beveled edge of the other pipe coupon section over the backing ring. Adjust for the proper root opening. The coupons must also be aligned so that the resulting welded pipe assembly is straight. Align small-diameter pipe by clamping both pieces to a piece of angle iron. Larger pipe can be aligned with the aid of a pipe alignment jig (*Figure 8*) or by holding a straightedge across the joint, parallel to the pipe axis. The straightedge must be used on all four quadrants of the pipe in case one or both sections are distorted. Then make the first of four tacks parallel to one of the existing tack welds on the ring.

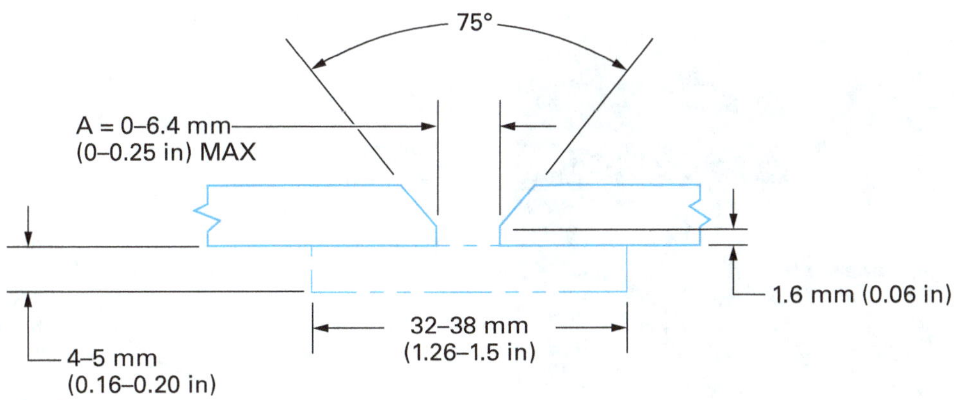

EDGE PREPARATION

PIPING DIMENSIONS							Electrode Diameter		Current DCEP	Argon Flow		Number of Passes[a]
Nominal Pipe Diameter Size Number			Outside Diameter (OD)		Wall Thickness							
DN	NPS	Sch.	mm	(in)	mm	(in)	mm	(in)	amp	L/min	(ft³/h)	A=0
100	4	40	114.3	(4.50)	6.0	(0.24)	1.2	(3⁄64)	190–210	20	(42)	2
125	5	40	141.3	(5.56)	6.6	(0.26)	1.2	(3⁄64)	205–225	20	(42)	2
150	6	40	168.3	(6.63)	7.1	(0.28)	1.2	(3⁄64)	215–235	20	(42)	2
200	8	40	219.1	(8.63)	8.2	(0.32)	1.6	(1⁄16)	215–235	25	(53)	3
250	10	40	273.1	(10.75)	9.3	(0.37)	1.6	(1⁄16)	215–235	25	(53)	3
300	12	40	323.9	(12.75)	10.3	(0.41)	1.6	(1⁄16)	240–260	25	(53)	3

[a] More passes required when A = 6 mm (0.24).
Note: Root opening = 0 for no backing ring or removable backing ring and 6 mm (0.24 in) for permanent backing ring.

(A) AWS ALUMINUM PIPE JOINT SPECIFICATIONS

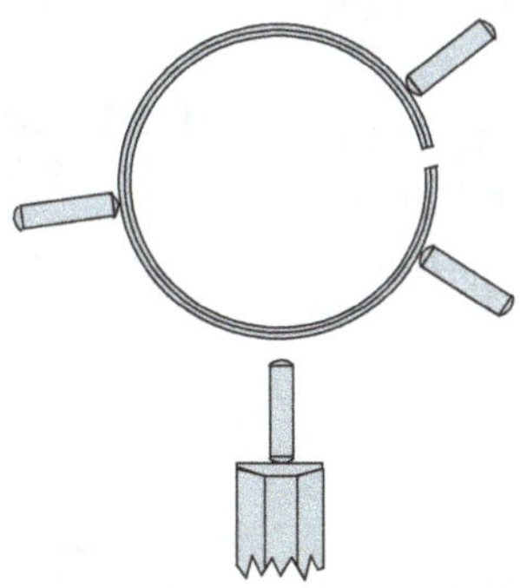

(B) TYPICAL SPLIT WELD RING WITH PEGS

Figure 7 AWS aluminum pipe joint specifications.

Figure 8 Pipe alignment jig (carbon steel pipe shown).

Step 7 After the first tack weld, check the root opening on the opposite side. Adjust the root opening as necessary, and make the second tack weld on the opposite side from the first tack.

Step 8 Check the root opening again, and weld the third tack midway between the first two tacks.

Step 9 Apply the fourth tack opposite the third tack and midway between the first and second tacks. There should now be four sets of tack welds evenly spaced (90 degrees) around the pipe coupon as shown in *Figure 9*.

Step 10 Clean and feather the tack welds with the edge of a grinding wheel. Feathering the ends of the tack welds with a grinder helps ensure that the tack welds become fused into the root pass.

1.4.0 V-Groove Pipe Welds

The V-groove weld with backing is the most common groove weld made on aluminum pipe. These welds are typically used for joining medium- and thick-walled pipe used in various piping systems. Piping containing or carrying material that can cause injury to people or damage the environment, or those that operate at elevated pressures, may be categorized as critical piping. Injury or damage occurs when the piping system or other system components fail to contain the material as designed. Welds in critical piping may be required to meet more stringent requirements. Piping that is used for hydronic heating and air conditioning applications, simple water systems, and similar applications may be assessed using less stringent requirements.

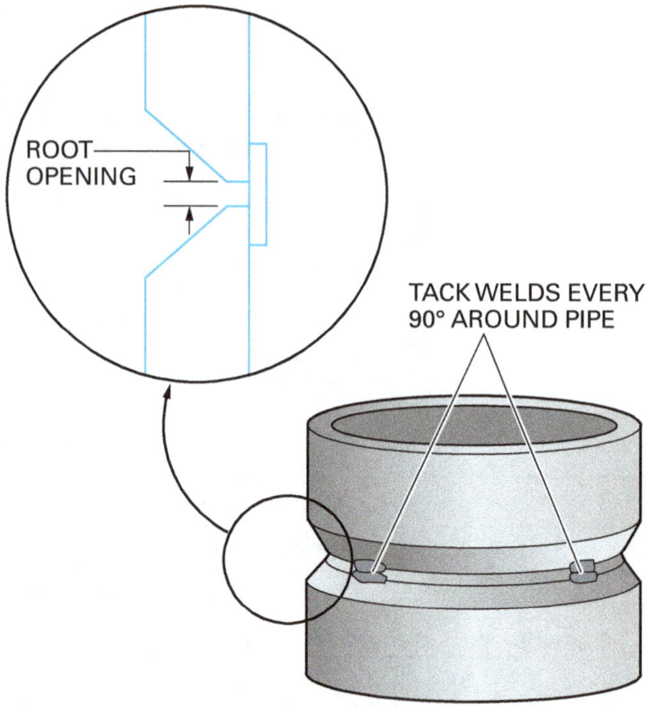

Figure 9 Tacked V-groove weld coupon with backing.

Groove welds may be made in all positions on pipe. The weld position is determined by the axis of the pipe. Four standard weld test positions are used with pipe, as shown in *Figure 10*. The sections that follow explain how to perform these V-groove pipe welds in the four standard positions.

1.4.1 *Acceptable and Unacceptable Pipe Weld Profiles*

Pipe groove welds should be made with slight reinforcement and a gradual transition to the base metal at each toe. The root pass should have complete penetration. The face reinforcement can range from being flush to a maximum of $\frac{3}{32}$" (2.4 mm) for materials up to $\frac{3}{8}$" (9.6 mm) thick. Excessively large face reinforcement reduces the pipe's strength.

Travel Speed and Wire Feed Speed

New welders are tempted to turn down the wire feed speed if they have difficulty controlling the weld puddle. Wire electrodes must be run at certain balanced parameters that cannot be changed individually. Voltage, current (if variable), wire feed speed, and travel speed are adjusted and balanced together to control the weld puddle.

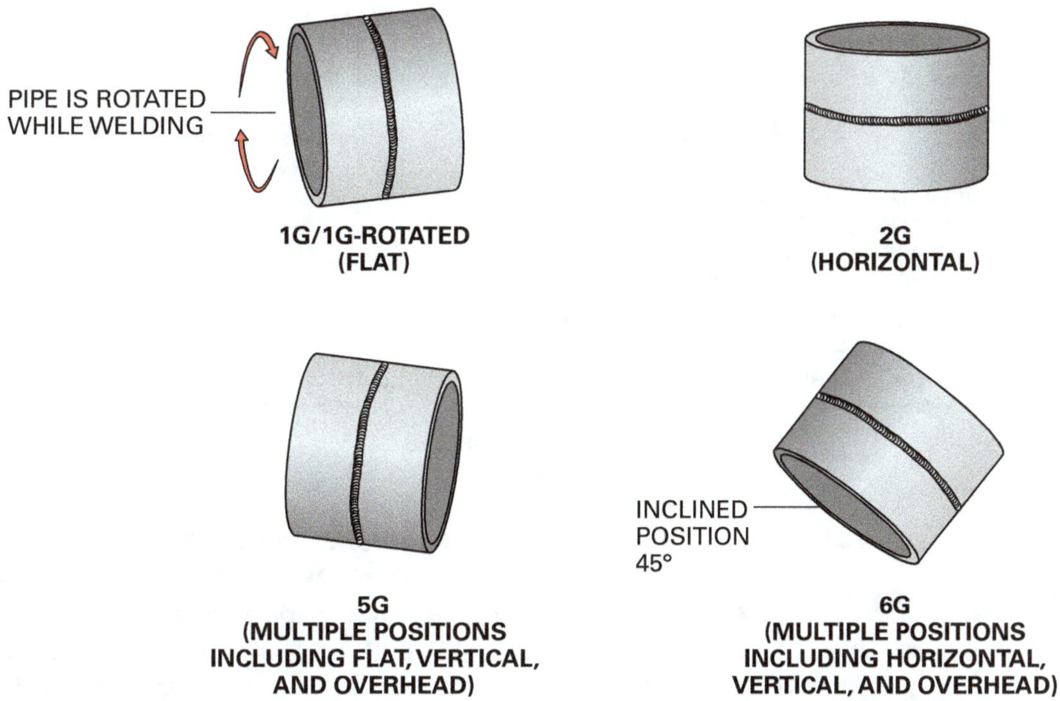

Figure 10 Four basic pipe groove weld test positions.

Completed pipe welds should be inspected to ensure they exhibit the following characteristics:

- Straightness of the bead
- Uniform appearance of the bead face; face and root reinforcement depth in accordance with applicable codes
- Smooth, flat transition with complete fusion at the toes of the weld
- Termination crater filled
- No porosity
- No undercut
- No overlap
- No overfill
- No cracks

If a weld has any of the defects described above and shown in *Figure 11*, it must be repaired.

1.4.2 Weld Travel Speed and Gun Positions

The travel speed must be considered before beginning a weld. Weld travel speed is the speed at which the electrode tip passes across the base metal in the direction of the weld. It is measured in inches per minute (ipm). The metric value for travel speed may be millimeters or centimeters per minute. Travel speed has a great effect on penetration and bead size. Slower travel speeds build higher beads with deeper penetration. Faster travel speeds build smaller beads with less penetration. Ideally, the welding parameters should be adjusted so that the electrode tip is positioned at the leading edge of the weld puddle during travel.

The gun travel angle in relation to the direction of the weld is defined as shown in *Figure 12*. A push angle is used to allow the shielding gas to cover and protect the base metal ahead as the weld advances.

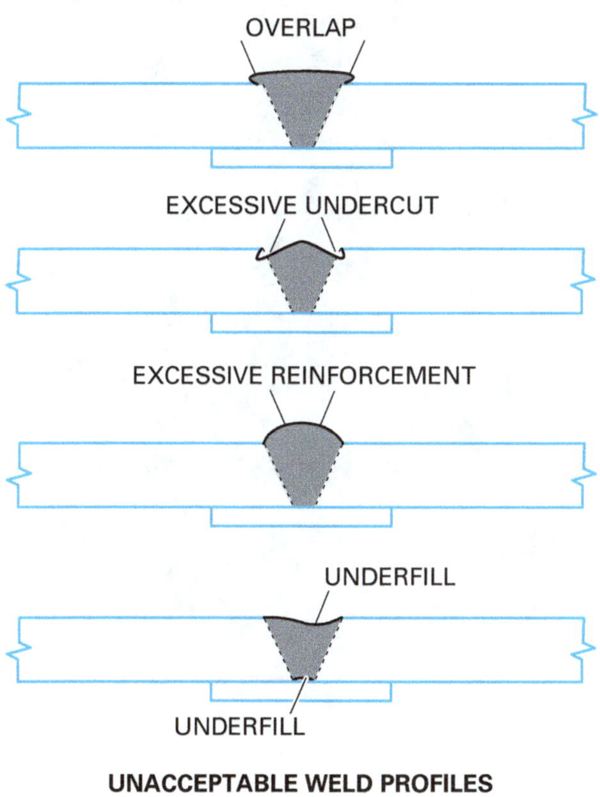

Figure 11 Acceptable and unacceptable pipe groove-weld profiles.

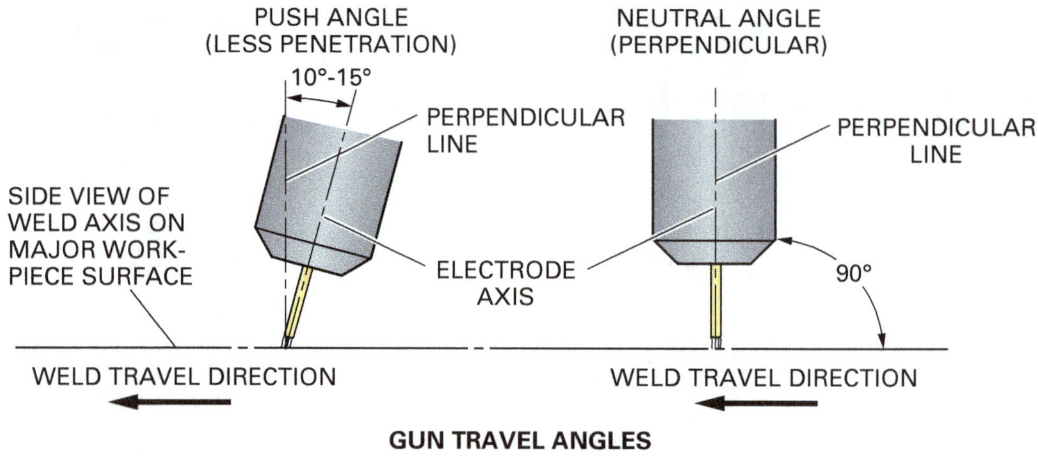

Figure 12 Gun travel and work angles.

Preparing Large Pipe

Regardless of the pipe size, there is always a way to cut and bevel it with precision. Clamshell lathes, for example, come in two pieces that are clamped together on the pipe and adjusted to ensure an accurate cut and/or bevel. Clamshell lathes like this one can accommodate pipe sizes up to an astonishing 15 feet (DN4500). Clamshell lathes are generally hydraulically or pneumatically powered.

1.4.3 Practicing Flat (1G-ROTATED) Position V-Groove Welds

Practice the 1G-ROTATED position V-groove pipe welds with backing. Use a shielding gas and aluminum filler wire of a diameter specified by your instructor. Ensure that the contact tip used is the proper size for the aluminum filler wire.

For the root pass, keep the gun work angle at 90 degrees (a 0-degree work angle) to the pipe axis with a 10- to 15-degree push angle. Be sure to fuse into the tack welds. Clean the face of the root pass with a wire brush.

When running the remaining passes, use stringer or weave beads with a slight side-to-side motion (oscillation) to ensure fusion at the toes of the weld bead. Try to keep the wire at the leading edge of the weld puddle to ensure proper penetration. Take special care to fill the crater at the termination of the weld. Run all passes at or near the top of the pipe as it is rotated.

Follow these steps to practice V-groove pipe welds with backing in the 1G-ROTATED position:

Step 1 Prepare and tack weld the practice pipe weld coupon together as described previously.

Step 2 Position the pipe weld coupon horizontally on two sets of rollers at a comfortable welding height.

Step 3 Make sure that the workpiece clamp is attached directly to the pipe coupon. This will prevent the welding current from passing through the roller bearings or arcing between the rollers and the pipe coupon.

Welding Large-Diameter Pipe

The practice coupons on large-diameter pipe must be greater than 6" in length. They need more metal to help absorb the increased heat used to make these welds.

CAUTION

Failure to attach the workpiece clamp to the pipe coupon can result in variations in welding current, damage to the roller bearings, and arcing on the rollers and pipe coupon.

Step 4 Run the root pass using a slight side-to-side oscillation. Place a tack weld at the 11 o'clock position. Start the weld bead on the tack weld, and advance toward the 12 o'clock position.

Step 5 Roll the pipe as necessary to keep the weld in the flat position.

Step 6 Clean the weld.

Step 7 Use the same rolling procedure to make the remaining passes. Use stringer or weave beads as applicable. Pay particular attention to ensure proper fusion and prevent excess buildup. Overlap the passes and, if using stringer beads, follow the sequence as shown in *Figure 13*. Clean the weld after each pass.

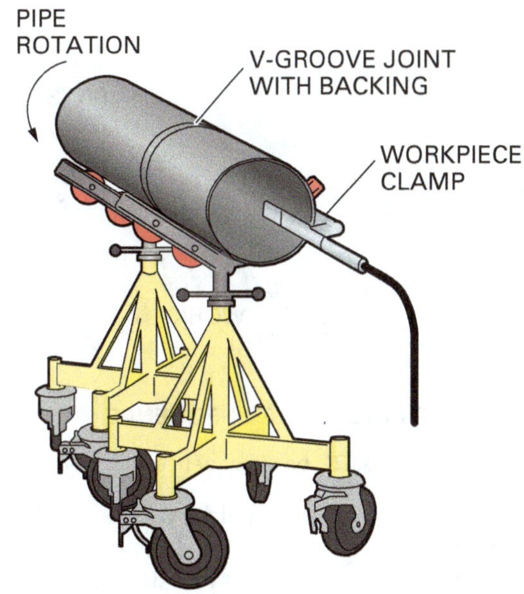

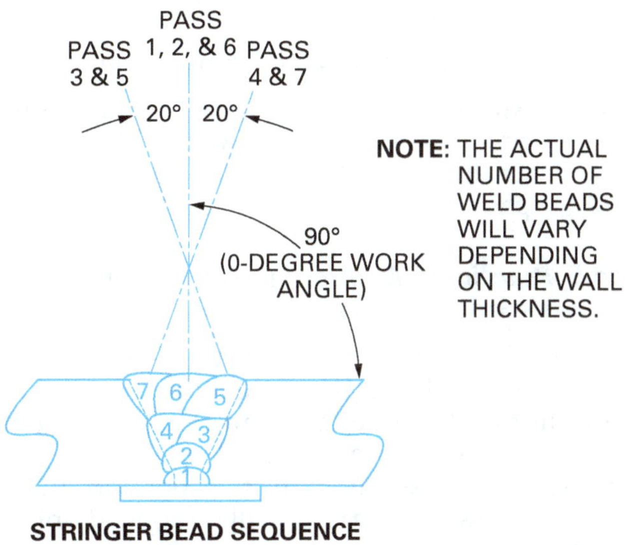

Figure 13 Multiple-pass 1G-ROTATED bead sequence and work angles.

1.4.4 Practicing Horizontal (2G) Position V-Groove Welds

Practice making V-groove pipe welds with backing in the 2G position. Use a shielding gas and an aluminum filler wire of a diameter specified by your instructor.

For the root pass, the gun should be held at a 10- to 15-degree push angle and at 90 degrees (0-degree work angle) to the surface of the pipe. To prevent the weld puddle from sagging, the gun work angle can be dropped slightly, but not more than 10 degrees. Be sure to fuse into the tack welds.

When running the remaining passes, use stringer beads and a very slight side-to-side motion to ensure proper fusion at the toes of the weld bead. Try to keep the wire electrode at the leading edge of the weld puddle. Take special care to fill the crater at the termination of the weld. Refer to *Figure 14* for the bead sequence and proper work angles.

Follow these steps to practice V-groove pipe welds with backing in the 2G position:

Step 1 Prepare and tack weld the practice pipe weld coupon together as described previously.

Step 2 Clamp or tack weld the pipe coupon with the axis of the pipe vertical.

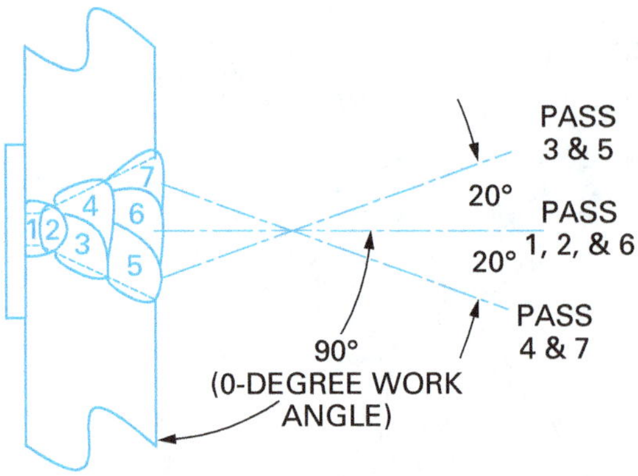

Figure 14 Multiple-pass 2G bead sequence and work angles.

Step 3 Starting on a tack weld, run the root pass as shown in *Figure 14*. Clean the crater between any necessary restarts. Continue this procedure until the root pass has been completed.

Step 4 Clean the weld with a wire brush.

Step 5 Run the remaining passes at the appropriate work angles. Overlap the passes as shown, and clean the weld after each pass.

1.4.5 Practicing Multiple (5G) Position V-Groove Welds

Practice the multiple (5G) position V-groove pipe welds with backing. Use a shielding gas and an aluminum filler wire of a diameter specified by your instructor. The gun should be at about a 10- to 15-degree push angle for the root and all remaining passes. All passes will be applied uphill. *Figure 15* shows the travel positions needed for the passes that are made on both sides of the pipe.

When running the remaining passes as shown in *Figure 16*, use weave beads or stringer beads with a slight side-to-side motion to ensure proper fusion at the toes of the weld. The instructor may direct practice using both techniques. Keep the wire electrode at the leading edge of the weld puddle. Be sure to fill the crater at the termination of the weld.

Follow these steps to practice V-groove pipe welds with backing in the 5G position:

Step 1 Prepare and tack weld the practice pipe weld coupon together as described previously.

6GR Test Position

Welding pipe in the 6G position is a common welding test. A ring may be added to test for restricted accessibility (6GR).

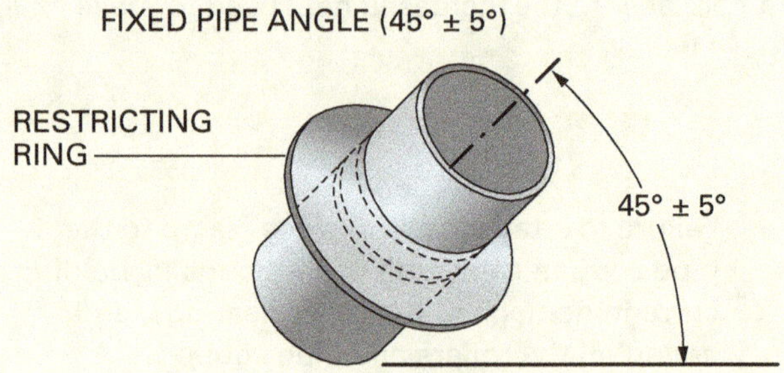

Step 2 Clamp or tack weld the pipe weld coupon into position with the pipe axis horizontal.

Step 3 Run the root pass with a stringer bead, modifying the travel angle as necessary.

Step 4 Clean the weld with a wire brush.

Step 5 Run the remaining passes as stringer or weave beads in an uphill direction at the appropriate work angles. Overlap the passes as shown in *Figure 16*, and clean the weld after each pass.

1.4.6 Practicing Multiple-Inclined (6G) Position V-Groove Welds

Practice the 6G multiple-inclined (45-degree) position V-groove pipe welds with backing. Use a shielding gas and an aluminum filler wire that has a diameter specified by your instructor. The gun should be at about a 10- to 15-degree push angle for the root and all remaining passes. All passes will be applied uphill. *Figure 15* shows the travel positions needed for the passes that are made on both sides of the pipe.

When running the remaining passes, use weave beads or stringer beads with a slight side-to-side motion to ensure fusion at the toes of the weld bead. Keep the wire electrode at the leading edge of the weld puddle. Be sure to fill the crater

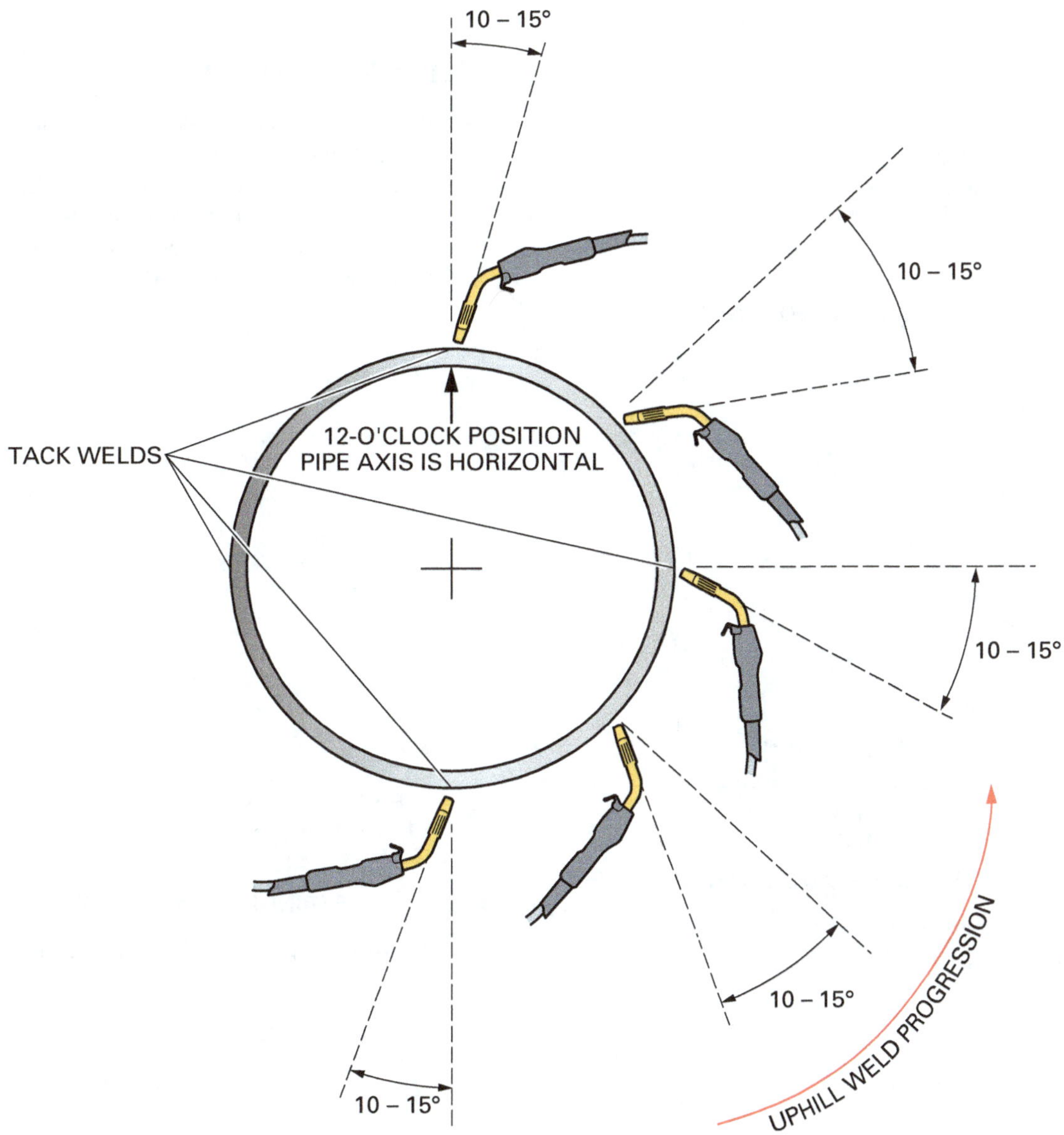

Figure 15 Multiple-pass weld progression for the 5G position.

at the termination of the weld. *Figure 17* shows the stringer bead sequence and work angles for the 6G position.

Follow these steps to practice V-groove pipe welds with backing in the 6G position:

> **NOTE**
> If desired or required for training purposes, a restricting ring may be added to the 6G position coupon to form a 6GR position coupon.

Step 1 Prepare and tack weld the practice pipe weld coupon together as described previously.

Step 2 Clamp or tack weld the pipe weld coupon into position with the pipe axis inclined 45 degrees to the horizontal plane.

Step 3 Run the root pass with a stringer bead, modifying the travel angle as necessary.

Step 4 Clean the weld with a wire brush.

Step 5 Run the remaining passes in an uphill direction at the appropriate work angles. Overlap the passes, and clean the weld after each pass.

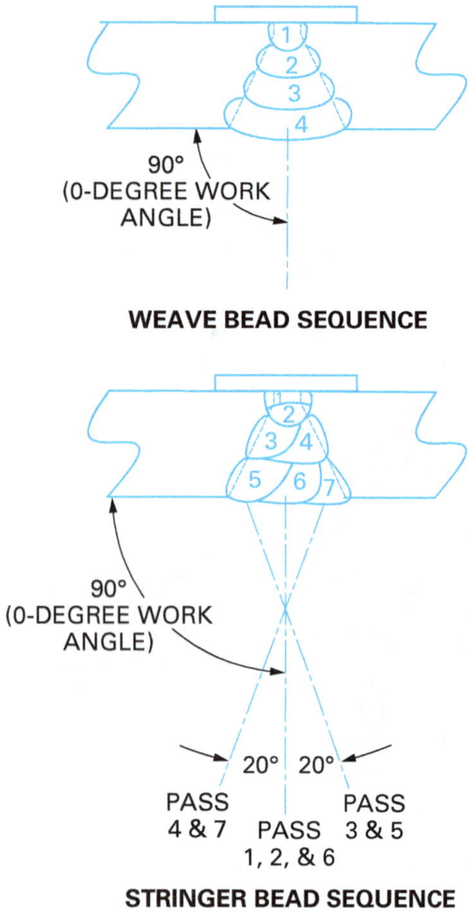

Figure 16 Multiple-pass 5G bead sequences and work angles.

Arc Voltage

A minimum arc voltage is needed to maintain the spray-transfer welding process. However, penetration is not directly related to voltage. Penetration will increase with voltage over a limited range, but it will actually decrease if the voltage is increased above its optimum.

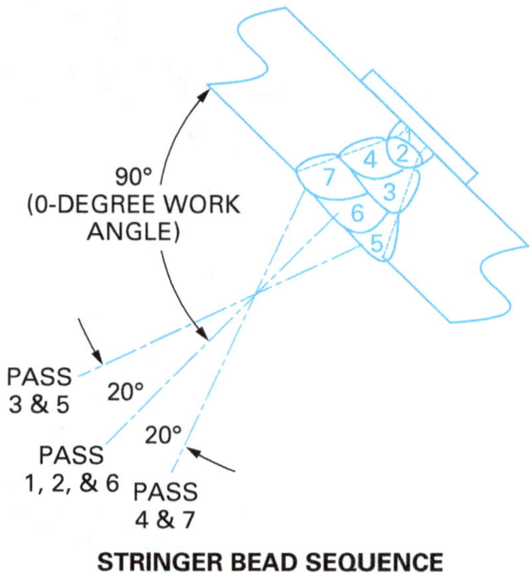

Figure 17 Multiple-pass 6G bead sequence and work angles.

Additional Resources

Aluminum GMAW Welding Guide (PDF). The Lincoln Electric Company, Cleveland, OH, USA. **www.lincolnelectric.com.** Last accessed May 26, 2016.

Guide for Aluminum Welding. 2013. Troy, OH: Hobart Brothers Corporation.

AWS C5.6-89R, Recommended Practices for Gas Metal Arc Welding. Miami, FL: American Welding Society.

GMAW Welding Guide (PDF). The Lincoln Electric Company, Cleveland, OH, USA. **www.lincolnelectric.com.** Last accessed May 26, 2016.

OSHA 1926.351, Arc Welding and Cutting. Current Edition. Washington, DC: Occupational Safety and Health Administration (OSHA).

The Procedure Handbook of Arc Welding. 2000. Cleveland, OH: The James F. Lincoln Arc Welding Foundation.

Welding Handbook, Volume 1 and Volume 2. Current Editions. Welding Science & Technology. Miami, FL: American Welding Society.

Welding Aluminum: Theory and Practice. 2002. New York, NY: The Aluminum Association.

1.0.0 Section Review

1. Caps worn during welding are best constructed of _____.
   a. 100-percent polyester
   b. 100-percent cotton
   c. a nylon and cotton blend
   d. a nylon and polyester blend

2. When welding amperage is increased, the result is _____.
   a. deeper penetration
   b. less penetration
   c. shorter beads
   d. reduced heat

3. The melting point of aluminum oxide is roughly _____.
   a. one-third that of the aluminum base metal
   b. the same as the aluminum base metal
   c. twice that of the aluminum base metal
   d. three times that of the aluminum base metal

4. When running the remaining passes in the 6G position, you can ensure fusion at the toes of the weld by using a _____.
   a. slight front-to-back motion
   b. slight side-to-side motion
   c. downhill motion
   d. circular motion

SUMMARY

This module covers the basic information needed for an experienced welder to prepare for and complete aluminum pipe welds using the GMAW process. As with all types of welding, a significant amount of practice is required to consistently produce acceptable welds on aluminum pipe. Safety must always be the priority when welding, regardless of the process or technique used.

Review Questions

1. For the filler metal, the GMAW process for aluminum uses _____.
   a. continuous, consumable wire electrode
   b. nickel-coated aluminum rods
   c. short lengths of electrode
   d. tin-coated aluminum rods

2. Chromium 6 is also referred to as _____.
   a. cadmium
   b. Chromalloy
   c. hexavalent chromium
   d. magnetic chromium

3. The GMAW applications normally used for aluminum are spray transfer and _____.
   a. globular transfer
   b. deposit transfer
   c. short-circuiting transfer
   d. pulsed spray transfer

4. Contact tip extension or setback is usually dependent on the _____.
   a. chosen transfer mode
   b. the arc length
   c. amperage
   d. voltage

5. If the weld spatter that accumulates in the gas nozzle is not removed, it will restrict the shielding gas flow to the weld and cause _____.
   a. cold fusion
   b. porosity
   c. cracking
   d. worm holes

6. Pipe weld coupons should be cut from Schedule 40 aluminum pipe that has a diameter of _____.
   a. 1" to 3" (DN25 to DN80)
   b. 6" to 12" (DN150 to DN300)
   c. 3" to 18" (DN80 to DN450)
   d. 12" to 16" (DN300 to DN400)

7. When preparing pipe coupons for V-groove weld joints using a backing ring with four pins or pegs, tack weld the ring to the first section of pipe at _____.
   a. two equally spaced positions
   b. four equally spaced positions
   c. six equally spaced positions
   d. eight equally spaced positions

8. Faster travel speeds during GMAW welding of aluminum _____.
   a. prevent termination craters
   b. overheat the contact tip
   c. build taller beads with deeper penetration
   d. build smaller beads with less penetration

9. To ensure proper penetration when welding aluminum pipe, keep the filler metal wire _____.
   a. just above the weld puddle
   b. at the trailing edge of the weld puddle
   c. at the leading edge of the weld puddle
   d. at the center of the weld puddle

10. For GMAW of aluminum pipe, the appropriate push angle for a V-groove weld in the 1G-ROTATED position is _____.
    a. 15 to 20 degrees
    b. 10 to 15 degrees
    c. 0 to 5 degrees
    d. 0 degrees

Appendix

Performance Accreditation Tasks

The Performance Accreditation Tasks (PATs) correspond to and support learning objectives in *AWS EG3.0, Guide for the Training and Qualification of Welding Personnel: Level II—Advanced Welders.*

PATs provide specific acceptable criteria for performance and help to ensure a true competency-based welding program for students.

The following tasks are designed to evaluate your ability to run V-groove welds with GMAW equipment in three standard test positions using aluminum filler wire of the appropriate diameter and shielding gas. Perform each task when you are instructed to do so. As you complete each task, show it to your instructor for evaluation. Do not proceed to the next task until told to do so. For AWS 2G and 5G certifications, refer to *AWS EG3.0, Guide for the Training and Qualification of Welding Personnel; Level II—Advanced Welder* for bend test requirements. For AWS 6G certifications, refer to *AWS EG4.0, Guide for the Training and Qualification of Welding Personnel; Level III—Expert Welder* for bend test requirements.

PATs 1 and 2 have no AWS correlation. PAT 3 corresponds to *AWS EG4.0, Section 3.3.6.3, Unit #3 GMAW, Learning Objective #8 (6G).*

V-GROOVE WELDS ON ALUMINUM PIPE WITH BACKING IN THE 2G POSITION

Using aluminum filler wire of the appropriate diameter, proper shielding gas, and stringer beads make V-groove welds on aluminum pipe in the 2G position.

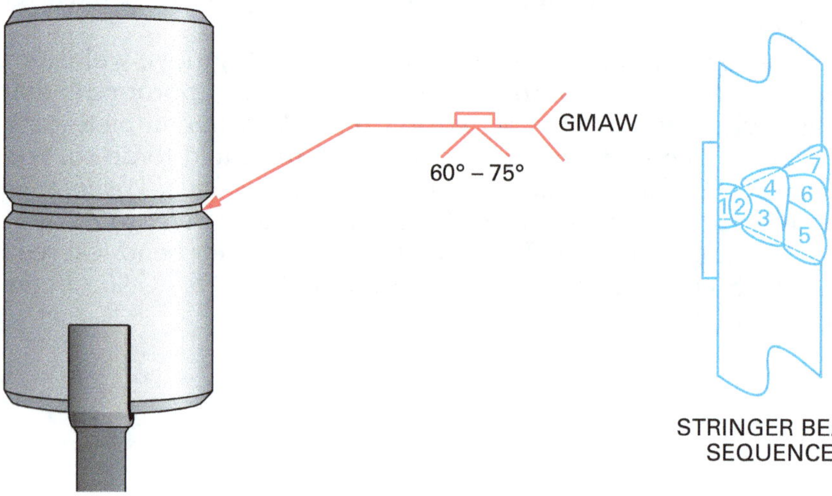

STRINGER BEAD SEQUENCE

Criteria for Acceptance:

- Uniform appearance on the bead face ________
- Craters and restarts filled to the full cross section of the weld ________
- Acceptable weld profile in accordance with the applicable codes and standards ________
- Smooth transition with complete fusion at the toes of the weld ________
- No porosity ________
- No undercut ________
- No underfill ________
- No cracks ________
- No overlap ________
- No incomplete fusion ________

V-GROOVE WELDS ON ALUMINUM PIPE WITH BACKING IN THE 5G POSITION

Using aluminum filler wire of the appropriate diameter, proper shielding gas, and stringer or weave beads, make V-groove welds on aluminum pipe in the 5G position.

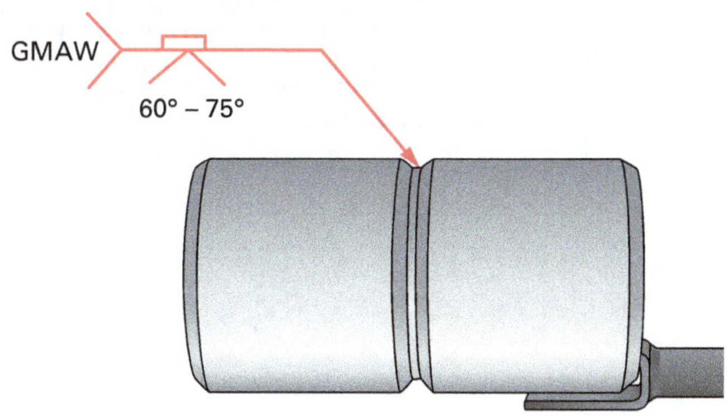

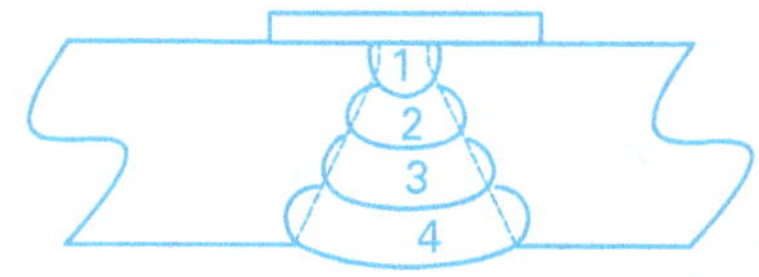

WEAVE BEAD SEQUENCE

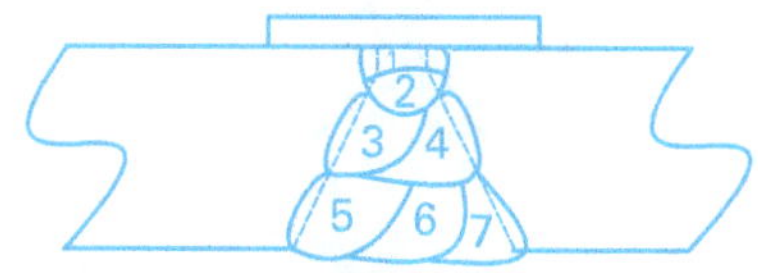

STRINGER BEAD SEQUENCE

Criteria for Acceptance:

- Uniform appearance on the bead face ________
- Craters and restarts filled to the full cross section of the weld ________
- Acceptable weld profile in accordance with the applicable codes and standards ________
- Smooth transition with complete fusion at the toes of the weld ________
- No porosity ________
- No undercut ________
- No underfill ________
- No cracks ________
- No overlap ________
- No incomplete fusion ________

V-GROOVE WELDS ON ALUMINUM PIPE WITH BACKING IN THE 6G (OR 6GR) POSITION

Using aluminum filler wire of the appropriate diameter, proper shielding gas, and stringer beads, make V-groove welds on aluminum pipe in the 6G (or 6GR) position.

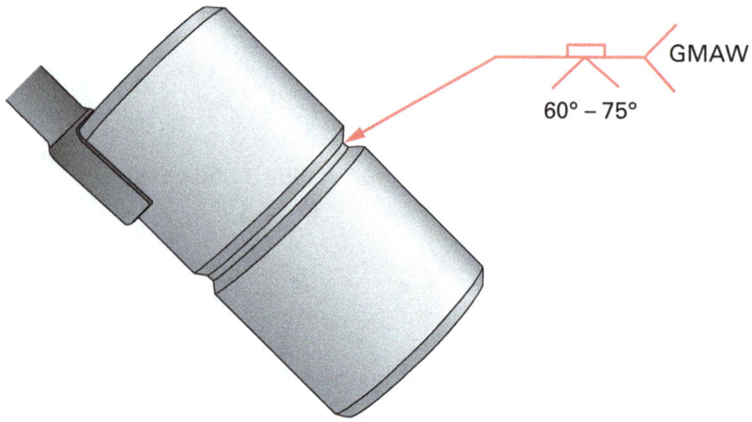

NOTE: IF REQUIRED FOR QUALIFICATION PURPOSES, A RESTRICTING RING MAY BE ADDED TO THE 6G POSITION COUPON TO FORM A 6GR POSITION COUPON.

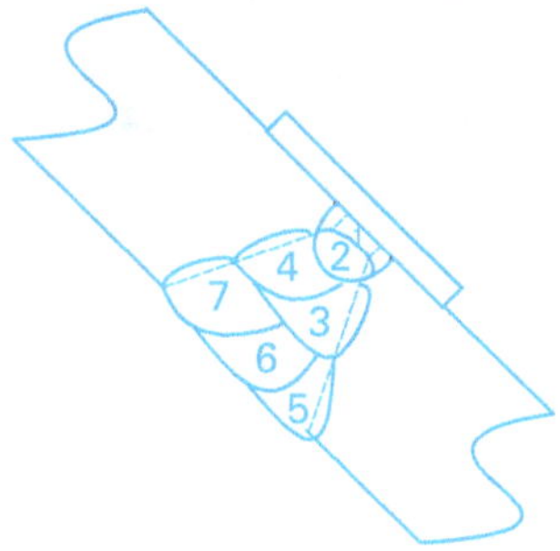

Criteria for Acceptance:

- Uniform appearance on the bead face ____________
- Craters and restarts filled to the full cross section of the weld ____________
- Acceptable weld profile in accordance with the applicable codes and standards ____________
- Smooth transition with complete fusion at the toes of the weld ____________
- No porosity ____________
- No undercut ____________
- No underfill ____________
- No cracks ____________
- No overlap ____________
- No incomplete fusion ____________

Additional Resources

This module presents thorough resources for task training. The following reference material is recommended for further study.

Aluminum GMAW Welding Guide (PDF). The Lincoln Electric Company, Cleveland, OH, USA. **www.lincolnelectric.com**. Last accessed May 26, 2016.

Guide for Aluminum Welding. 2013. Troy, OH: Hobart Brothers Corporation.

AWS C5.6-89R, Recommended Practices for Gas Metal Arc Welding. Miami, FL: American Welding Society.

GMAW Welding Guide (PDF). The Lincoln Electric Company, Cleveland, OH, USA. **www.lincolnelectric.com**. Last accessed May 26, 2016.

OSHA 1926.351, Arc Welding and Cutting. Current Edition. Washington, DC: Occupational Safety and Health Administration (OSHA).

The Procedure Handbook of Arc Welding. 2000. Cleveland, OH: The James F. Lincoln Arc Welding Foundation.

Welding Handbook, Volume 1 and Volume 2. Current Editions. Welding Science & Technology. Miami, FL: American Welding Society.

Welding Aluminum: Theory and Practice. 2002. New York, NY: The Aluminum Association.

Figure Credits

Miller Electric Manufacturing Co., Module Opener

The Lincoln Electric Company, Cleveland, OH, USA, Figures 2–4

Courtesy of TWI Ltd. **www.twi-global.com**, SA01

AWS D10.7M/D10.7:2008, Table 12, reproduced with permission of the American Welding Society (AWS), Miami, Florida, Figure 7A

Courtesy of Tri Tool Inc., SA02

Courtesy of H & M Pipe Beveling Machine Company, Inc., Figure 8

Section Review Answer Key

Answer	Section Reference	Objective
Section One		
1. b	1.1.1	1a
2. a	1.2.2	1b
3. d	1.3.1	1c
4. b	1.4.6	1d

NCCER CURRICULA — USER UPDATE

NCCER makes every effort to keep its textbooks up-to-date and free of technical errors. We appreciate your help in this process. If you find an error, a typographical mistake, or an inaccuracy in NCCER's curricula, please fill out this form (or a photocopy), or complete the online form at **www.nccer.org/olf**. Be sure to include the exact module ID number, page number, a detailed description, and your recommended correction. Your input will be brought to the attention of the Authoring Team. Thank you for your assistance.

Instructors – If you have an idea for improving this textbook, or have found that additional materials were necessary to teach this module effectively, please let us know so that we may present your suggestions to the Authoring Team.

NCCER Product Development and Revision

13614 Progress Blvd., Alachua, FL 32615

Email: curriculum@nccer.org
Online: www.nccer.org/olf

❑ Trainee Guide ❑ Lesson Plans ❑ Exam ❑ PowerPoints Other ____________________

Craft / Level: Copyright Date:

Module ID Number / Title:

Section Number(s):

Description:

Recommended Correction:

Your Name:

Address:

Email: Phone:

Glossary

Acetone: A colorless organic solvent that is volatile and extremely flammable. In reference to the soldering and welding process, it is used as a carrier for acetylene gas in cylinders.

Addendum: Supplementary information typically used to describe corrections or revisions to documents.

Alloy: A metal that has had other elements added to it that substantially change its mechanical properties; Any substance made up of two or more metals.

Ambient temperature: The room temperature of the atmosphere completely surrounding an object on all sides.

American Iron and Steel Institute (AISI): An industry organization responsible for preparing standards for steel and steel alloys that are based upon the Society of Automotive Engineers (SAE) code system.

American Society for Testing and Materials (ASTM International): An organization that developed a code system for identifying and labeling steels.

Amperage: The unit of measure used for the intensity of an electric current; often abbreviated to *amp* or *amps*.

Annealed: To free from internal stress by heating and gradually cooling.

Annealing: A postweld heat treatment for stress relieving weldments that is done at temperatures approximately 100°F (38°C) above the critical temperature with a prolonged holding period.

Anodic coating: An artificial buildup of aluminum oxides on the surface of aluminum alloys that improves corrosion resistance. This coating is many times thicker than a coating of naturally occurring oxides.

Arc blow: The deflection of an arc from its intended path by magnetic forces.

Arc burn: Burn to the skin produced by brief exposure to intense radiant heat and ultraviolet light.

Arrow line: The line drawn at an angle from the reference line (either end or both ends) to an arrowhead at the location of the weld.

Artificial aging: A process following solution heat treatment (see *solution heat treatment*) and other processes that allows the alloy to rest for a prescribed period of time at an elevated temperature.

Austenitic: Containing austenite (a solid solution of carbon or of carbon and other elements in a ferrous alloy), added through heating.

Austenitizing: Heating a steel up to or above the transformation temperature range to achieve partial or complete transformation to austenite grain structure.

Axially: In a straight line along an axis.

Backing: A weldable or nonweldable material used behind a root opening to allow defect-free welding at the open root of a joint

Base metal: Metal to be welded, cut, or brazed.

Blind hole: A drilled hole that does not go all the way through the object.

Bonded: The permanent joining of metallic parts to form an electrically conductive path that will assure electrical continuity and the capacity to safely conduct any current likely to be imposed on it.

Brazing: A method of joining metal using heat and a filler metal with a melting point above 842°F (450°C). Unlike welding, the base metal is not melted during the brazing process.

Brazing: A method of joining metals with a nonferrous filler metal using heat above 842°F but below the melting point of the base metals being joined; also known as hard soldering.

Capillary action: The movement of a liquid along the surface of a solid in a kind of spreading action.

Carbon steel: An alloy of iron combining iron and usually less than 1 percent carbon to provide hardness.

Cast aluminum: Aluminum alloys that are molded into their final form by one of various methods such as sand-casting, die-casting, or pressure die-casting. Cast aluminum contains high levels of silicon to enhance the ability of the material to form complicated castings.

Casting: The metal object produced by pouring molten metal into a mold.

Castings: Something cast; any article that has been cast in a mold.

Certified mill test report (CMTR): A quality assurance report provided to a customer by a metal manufacturer that shows the chemical content and testing results of the metal being purchased; also called a mill test report (MTR).

Coalescence: The growing together or growth into one body of the materials being joined.

Code: A document that establishes the minimum requirements for a product or process. Codes can be, and often are, adopted as laws.

Coefficient: A numerical measure of a physical or chemical property that is constant for a system under specified conditions, such as the coefficient of friction.

Coefficient of thermal expansion (linear): The change in length per unit length of material for a 1°C change in temperature.

Composite metal-core wire electrode: A filler wire electrode with a hollow core containing powdered materials, primarily metals.

Constituents: The elements and compounds, such as metal oxides, that make up a mixture or alloy.

Consumable insert: Preplaced filler metal that is completely fused into the root of the joint during welding, becoming part of the weld.

Counterbore: To enlarge a portion of a drilled hole so that a socket head fastener will seat flush or below the surface.

Countersink: A hole with tapered sides and a wider opening that allows a flat-head fastener to seat flush to the surface of the material in which the hole is made.

Critical temperature: The temperature at which iron crystals in a ferrous-based metal transform from being face-centered to body-centered. This dramatically changes the strength, hardness, and ductility of the metal.

Crystalline: Like a crystal; having a uniform atomic structure throughout the entire material.

Cup depth: The distance that a tube inserts into a fitting, usually determined by a stop inside the fitting.

Defect: A discontinuity or imperfection that renders a part of the product or the entire product unable to meet minimum acceptable standards or specifications.

Discontinuity: A change or break in the shape or structure of a part that may or may not be considered a defect, depending on the code; The deformation, or interruption, of the crystal structure inside a metal.

Distortion: The expansion and contraction of welded parts caused by the heating and subsequent cooling of the weld joint.

Dross: A waste byproduct of molten metal.

Ductile: The term used to describe a material that can be bent or shaped without breaking.

Ductility: The characteristic of metal that allows it to be stretched, drawn, or hammered without breaking.

Duty cycle:: The percentage of time a plasma arc cutting machine can cut without overheating within a ten-minute period.

Electric arc: The flow of an electrical current across an air gap or gaseous space.

Electrically grounded: Connected to Earth or to some conducting body that serves in place of the earth.

Electrode: The point from which a welding arc is produced.

Elevation: A height measurement above sea level or other identified point; also, the vertical sides of an object in an orthographic view.

Embrittled: Metal that has been made brittle and that will tend to crack with little bending.

Essential variable: Items in a welding procedure specification (WPS) that cannot be changed without requalifying the WPS.

Ferritic: Steel containing less than 0.10 percent carbon and is magnetic. This steel can't be hardened via heat treatment.

Ferrous: Relating to iron or an alloy that is mostly iron; a metal containing iron.

Fillet: A rounded internal corner or shoulder of filler metal often appearing at the meeting point of a piece of tubing and a fitting when the joint is soldered or brazed.

Flash burn: Burns to the eyes sometimes called welder's flash; caused by exposure to intense radiant heat and ultraviolet light.

Flashback arrestors: A valve that prevents a flame from traveling back from the torch tip and into the hoses.

Flux: A chemical substance that prevents oxides from forming on the surface of metals as they are heated for soldering, brazing, or welding; A material used to dissolve or prevent the formation of oxides and other undesirable substances on a weld joint.

Fume plume: The fumes, gases, and particles from the consumables, base metal, and base metal coating during the welding process.

Galvanized steel: Carbon steel dipped in zinc to inhibit corrosion.

Globular transfer: A GMAW process in which the electrode wire does not touch the base metal, and the filler metal is transferred randomly in the arc as large, irregular globules.

Hardenable materials: Metals that have the ability to be made harder by heating and then cooling.

Hardenability: A characteristic of a metal that enables it to become hard, usually through heat treatment.

Heat-affected zone (HAZ): The portion of the base metal workpiece that is not melted but has had its properties affected by welding or other heat-intensive operation such as oxy-fuel cutting.

High-low: The discrepancy in the internal alignment of two sections of pipe.

Homogeneity: The quality or state of having a uniform structure or composition throughout.

Hot shortness: The condition of metal when it proves to be very brittle and unbendable at red heat but can be bent without showing signs of brittleness when cold or at white heat. This condition is often a result of high sulfur and phosphorus content.

Immediate danger to life and health (IDLH): As defined by OSHA, this term refers to an atmosphere that poses an immediate threat to life, would cause irreversible adverse health effects, or would impair an individual's ability to escape from a dangerous atmosphere.

Inclusion: Foreign matter introduced into and remaining in a weld.

Inductance: An electrical circuit property, adjustable on some FCAW power supplies, that influences the pinch effect.

Ingot: A mass of metal shaped for convenient storage and transport.

Interpass temperature: In a multiple-pass weld, the temperature of the base metal at the time the next weld pass is started; it is usually the same temperature that is used for preheating.

Interpass temperature control: The process of maintaining the temperature of the weld zone within a given range during welding.

Isometric view: A drawing depicting a three-dimensional view of an object.

Laminations: Cracks in the base metal formed when layers separate.

Level: A line on a horizontal axis parallel with the earth's surface (horizon). Also refers to the tool used to check if an object is level.

Lift-arc technology: A type of GTAW technology in which the tungsten electrode may be touched to the work, but the arc will not strike until the tip is lifted away to an appropriate distance.

Liquation cracks: Liquation is the process of separation by melting, such as in an alloy when one element melts while the others remain solid. Liquation cracks result from localized melting in areas such as the grain boundaries, while adjacent grains remain solid. Both solidification and liquation cracking may be referred to as hot cracking.

Load : The amount of force applied to a material or a structure.

Malleable: Able to be hammered or pressed into another shape without breaking; Capable of being extended or shaped by hammering or by pressure from rollers.

Mark: A system of letters and numbers used with structural steel to identify the type of member and where it is placed.

Martensite: A solid solution of carbon in alpha-iron that is formed when steel is cooled so rapidly that the change from austenite to pearlite is suppressed; responsible for the hardness of quenched steel.

Martensitic: Steel that shares some characteristics with ferritic, but has a higher levels of carbon, up to a full 1 percent. It can be tempered and hardened and is used where strength is more important than a resistance to oxidation.

Mechanical properties: The characteristics or traits that indicate how flexible and strong a metal will be under stress.

Melt-through: Complete joint penetration.

Metallic: Having characteristics of metal, such as ductility, malleability, luster, and heat and electrical conductivity.

Mill test report (MTR): A quality assurance report provided to a customer by a metal manufacturer that shows the chemical content and testing results of the metal being purchased; also called a certified mill test report (CMTR).

Natural aging: A process that follows solution heat treatment (see *solution heat treatment*) and other processes that allows the alloy to simply rest at room temperature for a prescribed period of time, such as 96 hours.

Non-essential variable: Items in a welding procedure specification (WPS) that can be changed without requalifying the WPS.

Nonferrous: A group of metals and metal alloys that contain no iron.

Nonferrous metal: A metal, such as aluminum, copper, or brass, that lacks sufficient quantities of iron to be affected by its properties.

Notch toughness: The impact load required to break a test piece of weld metal measured by a notch toughness test (Charpy V-notch test). This test may be performed on metal below room temperature, at which point it may be more brittle; The ability of a material to absorb energy in the presence of a flaw such as a notch

Orthographic projection: A method of developing multiple flat views of an object by projecting lines out from the object at right angles from the face.

Oxidation: The process by which the oxygen in the air combines with metal, producing tarnish and rust.

Oxide: The scale that forms on metal surfaces when they are exposed to oxygen or air containing oxygen.

Peening: The mechanical working of metal using repeated hammer blows, typically using a ball-shaped hammer head such as that found on a ball peen hammer. Peening may be done for both metallurgical and appearance reasons.

Phase: In a three-phase power supply system, the sine waves of the voltage on each of the three separate conductors is displaced from each of the others by 120 degrees, although all conductors are carrying alternating current at the same frequency and are synchronized. Each of the three sine waves is referred to as a phase of the power source.

Pinch effect: The tendency of the end of the wire electrode to melt into a droplet during short-circuit GMAW welding.

Piping porosity: A form of porosity having a length greater than its width and that is approximately perpendicular to the weld face.

Pitch: The center-to-center distance between welds.

Plasma: A fourth state of matter (not solid, liquid, or gas) created by heating a gas to such a high temperature that it boils electrons off the gas molecules (ionization).

Plumb: A line on a vertical axis perpendicular to the earth's surface (horizon).

Porosity: Gas pockets, or voids in the weld metal.

Postweld heat treatment (PWHT): The process of heat treating the base metal and weld after the weld has been made.

Potential: The relative electrical voltage difference between two points of reference; the electromotive force or difference in potential between two points, expressed in volts.

Potentiometer: A three-terminal resistor with an adjustable center connection; used for variable voltage control.

Preheat weld treatment (PHWT): The process of heating the base metal before making a weld.

Procedure qualification record (PQR): The document containing the results of the nondestructive and destructive testing required to qualify a WPS.

Pulsed transfer: A GMAW welding process in which the welding power is cycled from a low level (background level) to a higher level (welding level), resulting in lower average voltage and current; this is a variation of the spray transfer process.

Purge gas: An inert gas, such as nitrogen, used to drive oxygen away from a weld site.

Quench: To rapidly cool a hot metal using air, water, or oil.

Radiographic: Describes images made by passing X-rays or gamma rays through an object and recording the variations in density on photographic film.

Reference line: The horizontal line in the center of the welding symbol from which all elements of the welding symbol are referenced. The reference line is one of the most important elements of the welding symbol.

Residual stress: The strain that occurs in a welded joint after the welding has been completed, as a result of mechanical action, thermal action, or both; Stress remaining in a weldment as a result of heat.

Root face: A small flattened area on the end of a bevel for a groove weld.

Root opening: The space between the base metal pieces at the bottom or root of the joint.

Scale: A method of drawing an object larger or smaller than its actual size while keeping the proportions the same.

Shielding gas: A gas such as argon, helium, or carbon dioxide used to protect the welding electrode wire from contamination in GMAW and FCAW welding.

Short-circuiting transfer: A GMAW welding process in which the electrode wire shorts against the base metal. The welding current repeatedly and rapidly melts the wire and establishes an arc, which is extinguished when the wire again touches the base metal.

Sintered: A process of making parts from a powdered metal in which the metal is molded and then heated in a furnace without melting to fuse it into a solid metal.

Society of Automotive Engineers (SAE): An automotive society responsible for creating an early system for classifying carbon steels.

Solder: A fusible alloy used to join metals, with melting points below 800°F.

Soldering: A method of joining metals with a nonferrous filler metal using heat below 842°F and below the melting point of the base metals being joined; also known as soft soldering or sweating.

Solenoid valve: A valve used to control the flow of gases or liquids that is opened or closed by the action of an energized electromagnet.

Solidification cracks: Cracks that occur as the filler and base metal solidify, when the remaining liquid weld metal is unable to fill the spaces between areas of solidifying weld metal. Both solidification and liquation cracking may be referred to as hot cracking.

Solution heat treatment: A process applied to a heat-treatable aluminum alloy that raises the temperature of the alloy to roughly 1,000°F (538°C) for a set period of time, causing the alloy elements to disperse more consistently throughout the metal. The metal is then quenched with water to ensure the dispersed elements remain in place within the metal structure, rather than precipitate out.

Specific heat per unit volume: The quantity of heat required to raise one unit mass of a substance by one unit degree.

Specification: A document that defines in detail the work to be performed or the materials to be used in a product or process.

Spray transfer: A GMAW welding process in which the electrode wire does not touch the base metal, and the filler metal is transferred axially (in a straight line) in the arc as fine droplets.

Stainless steel: An iron-based alloy usually containing at least 11 percent chromium.

Standard: A document that defines how a code is to be implemented.

Stickout: The length of electrode that projects from the contact tube, measured from the arc tip to the contact tube.

Strain hardening: A method of increasing the strength of a metal by cold-working it.

Stress-relief heat treatment: A technique used to reduce welding stresses and to increase the ductility and toughness of the weld metal and the heat-affected zone (HAZ).

Supplemental essential variable: The variables that must be considered when notch toughness requirements are invoked.

Surfacing: The application by welding, brazing, or thermal spraying of a layer of material to a surface to obtain desired properties or dimensions.

Tack weld: A weld made to hold parts of a weldment in proper alignment until the final weld is made.

Tempering: The process of reheating quench hardened or normalized steel to a temperature below the transformation range and then cooling it at any rate desired.

Tempering: To impart strength or toughness to (steel or cast iron) by heating and cooling.

Tensile strength: The maximum stress or force that a material can withstand without breaking; The measure of the ability of a material to withstand a longitudinal stress without breaking.

Through hole: A drilled hole that goes completely through an object.

Tolerance: The allowance over or under the specified dimension, value, or other parameter.

Transducer: A device that converts one form of energy into another.

Truncated: Having the pointed top (vertex) of a conical shape cut off, leaving only a plane section parallel to the base of the cone.

Tungsten: A nonstandard common name for the tungsten electrode used in GTAW; also, the principal metal from which the electrodes are constructed.

Ultraviolet (UV) radiation: Invisible rays capable of causing burns. UV rays from the sun are the causes of sunburn.

Underbead cracking: Subsurface cracking in the base metal under or near the weld.

Unified numbering system (UNS): An industry-accepted numbering system for identifying ferrous metals and alloys.

Ventricular fibrillation: Irregular heart rhythm characterized by a rapid fluttering heartbeat.

Weathering steel: Steel alloy that, under specific conditions, is designed to form a very dense oxide layer on its outer surfaces, which retards further oxidation.

Web: The plate joining the flanges of a girder, joist, or rail.

Weld symbol: A graphic character connected to the reference line of a welding symbol specifying the weld type.

Welding procedure qualification: The demonstration that welds made following a specific process can meet prescribed standards.

Welding procedure specification (WPS): The document containing all the detailed methods and practices required to produce a sound weld.

Welding symbol: A graphical representation of the specifications for producing a welded joint; includes a reference line and arrow line and many also include a weld symbol.

Weldment: An assembly that is fastened together by welded joints.

Wetting: A process that reduces the surface tension so that molten (liquid) solder flows evenly throughout the joint.

Wide flange (WF) beam: A metal I-beam that has an extra wide flange at the top and bottom.

Work hardening: The increase in strength that sometimes accompanies plastic deformation of a metal.

Wrought: Formed or shaped by hammering or rolling.

Wrought aluminum: Aluminum alloys that are initially formed into ingots or similar shapes and are then worked into their desired shape through bending, rolling, extrusion, or other mechanical methods.

www.ingramcontent.com/pod-product-compliance
Lightning Source LLC
Jackson TN
JSHW061054170926
108221JS00012BA/36